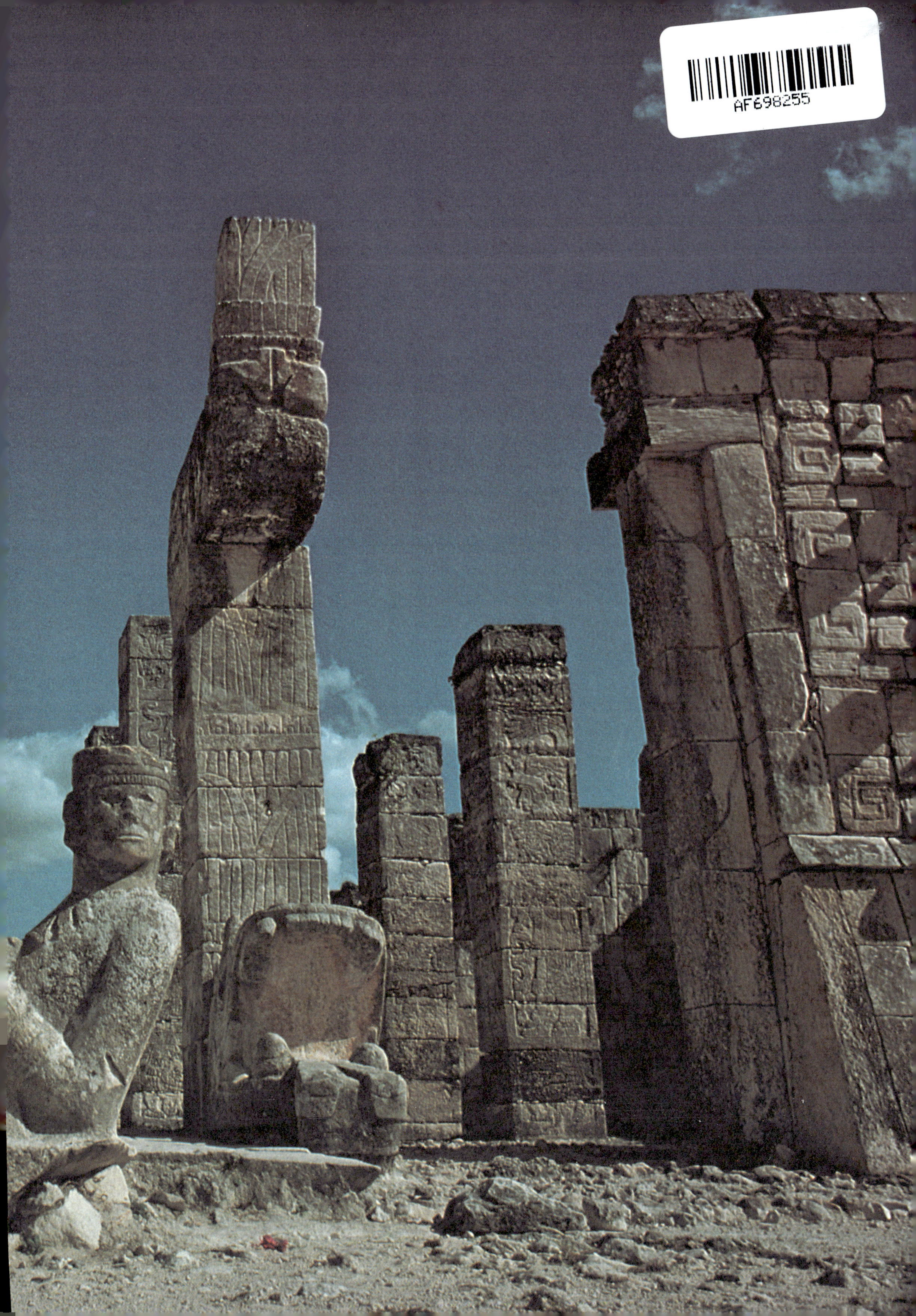

(FRONT COVER) *Detail of Stela B at Copan*
(BACK COVER) *Maya Lord Smoking, Island of Jaina (private collection)*

The Smoking Gods

(INSIDE COVER) *Upper platform, Temple of the Warriors, Chichén Itzá*

THE SMOKING GODS

Tobacco in Maya Art, History, and Religion

by

Francis Robicsek

Foreword by

Michael D. Coe

and by

Barbara A. Goodnight

Photographs by the Author

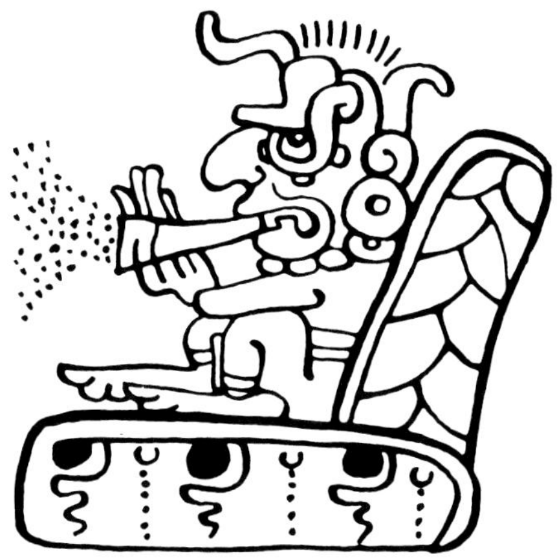

University of Oklahoma Press : Norman

By Francis Robicsek

Copán: Home of the Mayan Gods (New York, 1972)
A Study in Maya Art and History: The Mat Symbol (New York, 1975)
(editor) *Biología Centrali-Americana,* by Alfred P. Maudslay (New York, 1976)
The Smoking Gods: Tobacco in Maya Art, History, and Religion (Norman, 1978)

Library of Congress Catalog Card Number: 78-64904

Publishing Division of the University. Manufactured in the U.S.A. First edition.

To the people of Guatemala

Foreword

Michael D. Coe

The use of tobacco among the Mayas and their cousins in the rest of the New World not only was a social custom, as it is in our society nowadays, but had far-reaching implications in religion and medicine and a lasting impact on art. It is most surprising, therefore, that this important subject has received so little attention in the past. On the role that tobacco played in the life of the ancient Mayas, only two short studies have been written; the unpublished manuscript of Ariel Baynes, a work of very little, if any, scientific value, and Sir J. Eric S. Thompson's four-page report, *Some Uses of Tobacco Among the Maya,* which appeared in the 1946 volume of the *Notes of Middle American Archaeology and Ethnology of the Carnegie Institution.* Twenty-four years later Thompson reiterated his thoughts on tobacco in a somewhat extended form in his book *Maya History and Religion.* While his work was certainly distinguished by the expertise of a keen ethnographer, unfortunately it also suffered from scarcity of ethnohistoric data and complete absence of archaeological material on this important subject.

Francis Robicsek's book is certainly a definitive work on the use of tobacco among the Mayas and their cousins in the rest of the New World. As in his previous books, *Copán: Home of the Maya Gods* and *A Study in Maya Art and History: The Mat Symbol,* Dr. Robicsek has called upon historical, visual, linguistic, and ethnological evidence to throw light upon his chosen subject. Our ethnologist colleagues, such as Johannes Wilbert and Peter T. Furst, have been telling us for a long time that tobacco had a deep religious and even ecstatic meaning for the American Indian. This book shows that only by acknowledging factors such as tobacco in native mental life can one understand certain scenes depicted in Classic Maya art. By ignoring the mental preoccupations of American Indians, archaeologists are in grave danger of misinterpreting and even overlooking those parts of ancient culture that were most meaningful to the natives themselves.

In this and in several other aspects, Dr. Robicsek's book is a lasting contribution to Americanist studies.

Foreword

Barbara A. Goodnight

In this book Francis Robicsek undertakes the study of a significant but heretofore neglected aspect of Maya studies: the place of tobacco in the culture of the ancient Mayas. In *The Smoking Gods* he makes a significant contribution toward filling that void. The book is a visible expression of his continuing interest in and deepening knowledge of Maya culture. It is based on a thorough examination of historical, ethnological, and archaeological material, including documentation of the use of tobacco among the ancients, as well as the meanings tobacco seems to have had for them. It is a scholarly work, at once informative, entertaining, and rich in visual effect. As such it has appeal for professional archaeologists as well as for members of the general reading public.

In these pages some intriguing questions are raised that can be expected to stimulate further research. It contributes significantly to the understanding of ancient Maya culture, as well as to the literature on the incidence and importance of tobacco use throughout the history of mankind. In addition, the author offers insightful interpretations of the symbolism depicted on various Maya artifacts that will challenge scholars interested in iconography.

The Department of Sociology and Anthropology at the University of North Carolina at Charlotte is pleased to recognize Dr. Robicsek's accomplishments as a student of archaeology and Maya culture. It is our wish that he will continue his studies—and that his readers will share our appreciation of them.

Acknowledgments

At the completion of this book I wish to express my most sincere appreciation to many friends and scholars: first of all to General Carlos Arana Osorio, former President of Guatemala; to Jorge Castillo V.; to Karl-Heinz Nottebohm; and to more than forty other scholars and collectors who opened the doors of their institutions and homes and allowed me access to their valuable collections of artifacts and manuscripts.

I will remain especially indebted to Linda Schele for her most valuable participation in this book, which included both the interpretation of the glyphic texts and critiques of my notes on the iconography of the archaeological material. Without her valuable contribution, this book would not have become a reality.

I am very grateful to Michael D. Coe, Thomas Barthel, Peter T. Furst, and Barbara A. Goodnight for reviewing my manuscript and to David H. Kelley for commenting on my chapter on God K.

For their gracious cooperation with my research efforts I am thankful to the staff of the American Tobacco Institute, Washington, D. C.; E. Wyllys Andrews V, Director, Middle American Research Institute, Tulane University; Elizabeth P. Benson, Director, Center for Pre-Columbian Studies, Dumbarton Oaks; Paul A. Clifford, Curator, Pre-Columbian Collection, Duke University; William R. Coe and David Sedat, Department of Anthropology, University of Pennsylvania; Gordon F. Ekholm, Curator Emeritus, American Museum of Natural History; Eric von Euw, Peabody Museum, Harvard University; Dora Gonzales, Director, Museo de Arqueología e Etnología de Guatemala; Gillette G. Griffin, Curator, Pre-Columbian Art, Princeton Art Museum; Juan Gnilluf, Tegucigalpa, Honduras; Donald M. Hales, University of California at Los Angeles; Dr. Henry Hawley, Curator of Post-Renaissance and Decorative Arts, Cleveland Art Museum; John M. Keshishian, George Washington University; Floyd G. Lounsbury, Professor of Anthropology, Yale University; John Lunsford, Curator, Dallas Museum of Fine Arts; Luis Lujan Muñoz, Director, Instituto de Antropología y Historia, Republic of Guatemala; and Douglas Dillon, Chairman of the Board of Trustees, Metropolitan Museum of Art.

I wish to thank Merle Greene Robertson, who guided me with her vast experience of Maya symbolism, and Trudy Duby Blom, whose knowledge of the Lacandones is excelled only by her love for this disappearing tribe and who provided me with valuable material on their life and customs. I will remain grateful to the staff of the Arents Collection, New York City Public Library, especially to Jeffrey Kaimowitz; to Thomas K. Niehaus, Director, and Martha Robertson, of the Latin American Library, Tulane University, for their valuable help; Cynthia Tyson; Nell Franklin; William Poveromo; my daughter and son, Suzanne and John Christopher, for their editorial assistance; María Zsigmond, Sandor Darvas, and James Scancarelli for their high-quality illustrations.

Last, but certainly not least, I wish also to express my thanks to Nicholas M. Hellmuth and his organization, the Foundation for Latin American Research, for his most unselfish and collegial assistance, which made it possible for me to obtain essential photographs and research materials.

Francis Robicsek

Charlotte, North Carolina

Contents

Illustrations

Color Plates

Illustrations

Black and White

Figure Page

Figure Page

The Maya Realm

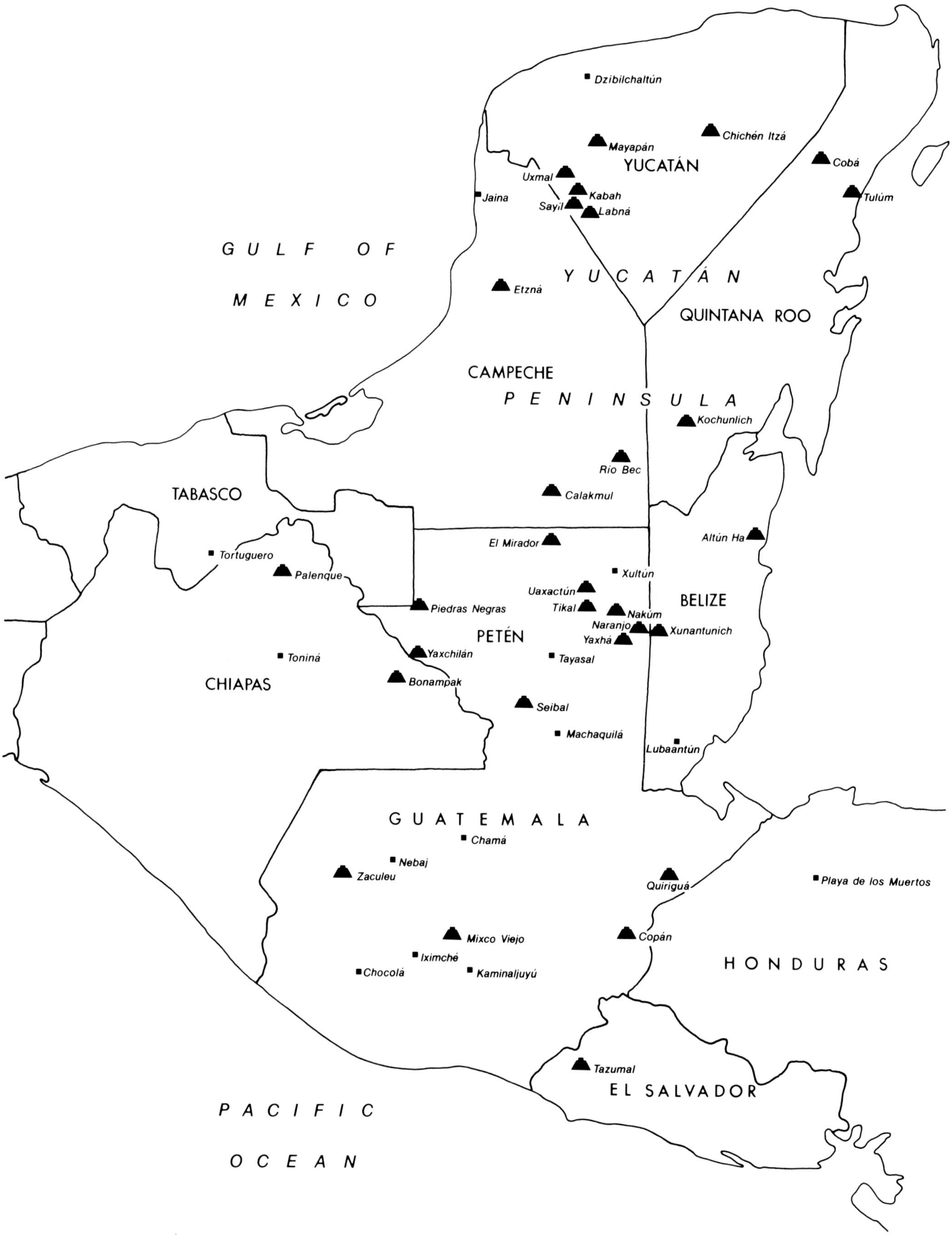

PART ONE

Historical and Ethnological Background

Modern science was born when substitution of limited questions for general questions began—when man stopped asking how the universe was created, what matter consists of, and what the essence of life is—and began wondering how a body falls, how water runs through a pipe, and how blood circulates through the blood vessels. It was astounding how limited questions led increasingly to more and more general answers.

François Jacobs

Tobacco was pleasing to the gods and had magical powers.

Sir J. Eric S. Thompson

The views the Maya had must be pieced together from a number of Mesoamerican sources.

Michael D. Coe

Of the cultural traits of aboriginal . . . America, that of smoking tobacco was, perhaps the most universal.

J. C. H. Kings

It is hoped that the following pages will demonstrate something of the essential interplay between nature and culture . . . .

Peter T. Furst

1. Tobacco in the New World

The area occupied by the Mayas before the Conquest included most of present Guatemala, all of Belize, a portion of El Salvador, and Honduras, as well as the Mexican states of Yucatán, Campeche, and Chiapas and the territory of Quintana Roo. The Maya race, which in ancient times developed the most advanced aboriginal civilization in the New World, is still very much alive, and today Central America is inhabited by almost two million Indians of the Maya linguistic stock.

Maya arts and science reached their astonishing heights during the span of the Classic epoch, which began during the third century A.D. and ended at the turn of the ninth century. This period was followed by a cultural decline that was sudden in some areas and gradual in others. When five hundred years later the Spaniards landed in Yucatán, they found a people that, though still very much aware of many of its ancient traditions, was as far from the cultural achievements and science of its ancestors as the Egyptian fellahs of the fifteenth century from the pyramid builders of the pharaohs.

The arriving Spaniards found the Mayas ardent tobacco smokers, using an herb and engaging in a habit theretofore unknown to the Western world. According to ethnological and archaeological evidence, this custom of the Mayas not only was a social pastime, as it is today, but also had significant religious and mythological implications; it deeply penetrated folklore and heavily influenced art.

That is the subject of this book.

The genus *Nicotiana* (tobacco), together with seventy-four other genera, belongs to the nightshade family (Solanaceae), which encompasses not only "innocent" and familiar vegetables like the Irish potato and the eggplant but also several herbs of narcotic properties such as the henbane. It consists of seventy-three species, all but two native to the American continent. The latter two *(N. fragrans* and *N. suaveolens)* grow wild in Australia but were not used for smoking there before the arrival of the white man.[1] The genus was named after Jean Nicot, of Nîmes, a sixteenth-century French ambassador to Lisbon, who is said to have received seeds of tobacco from the Portugese scientist Damião de Goes and sent them as a present to Catherine de' Medici, Queen of France.

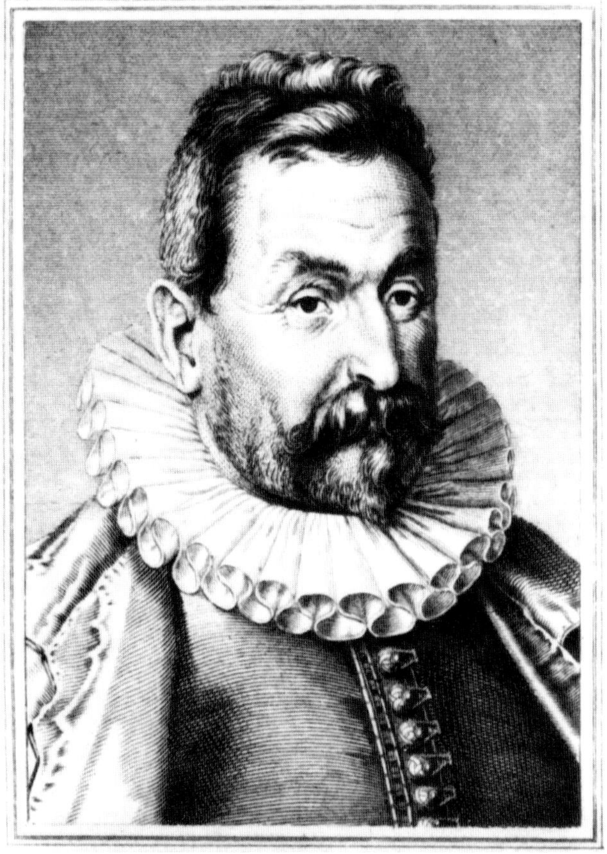

Figure 1. Jean Nicot, French ambassador to the Royal Court of Portugal (after Arents)

Several of the tobacco species were known and cultivated by the natives of the New World long before the Conquest.[2] Besides *N. tabacum,* to which today most varieties of commercial tobacco belong, they also grew *N. rustica,* or weed tobacco, a plant native to Mesoamerica and western South America, as well as *N. attenuata,* a tobacco used north of the Sonora Desert in prehistoric times. *N. bigelovi* var. *quadrivalvis* was the tobacco of the Mandan, Hidatsa, Arikara, and Dakota Indians. *N. bigelovii* var. *multivalvis* was the sacred tobacco of the Crows. All these species are now largely replaced by *tabacum,* which was probably first cultivated in the South American tropical lowlands and then spread from there to the West Indies, Central America, and southern Mexico.[3] *N. rustica,* on the other hand, probably originated in Mesoamerica and gradually migrated from there to the North American continent.[4] This was the tobacco known in the Mississippi Valley as early as the first millennium B.C.[5] The *tabacum* plant grows two to nine feet high, with wide-spreading ovate, oblong, or lanceolate leaves. The leaves are alternately attached to the stalk spirally so that the ninth leaf overhangs the first and the tenth leaf the second. The distance between the leaves on the stalk is about two inches in ordinary varieties. The flowers are in large clusters with corollas of rose color or white tinged with pink, about two inches long, and funnel-shaped. The pods have two valves.[6] The *rustica* plant differs from the *tabacum* in that, when full grown, it is shorter and has broader suckers and axillary shoots. The corolla tube is short with distinctly rounded lobes, the flowers are yellow, and the seeds are about three times larger than those of the *tabacum.*

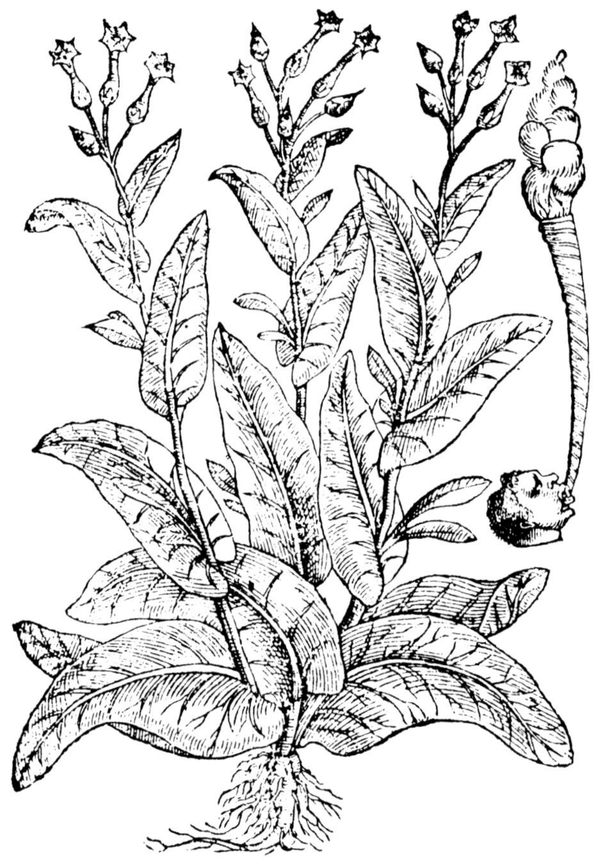

Figure 2. The first published picture of Nicotiana tabacum *and of tobacco smoking (after Peña and l'Obel)*

All scientific evidence indicates that the customs of smoking, chewing, and drinking tobacco were first practiced on the American continents. In Australia, the only other area where the genus *Nicotiana* is thought to be native, the aborigines have chewed the leaves of *N. suaveolens* probably since ancient times.[7] Witness to the great popularity and wide dissemination of the *Nicotianae* among the native tribes of the New World is the fact that no less than six hundred words in as many American aboriginal languages and dialects have been used to designate tobacco, including three *(zig, kutz,* and *mai)* in different Maya tongues.[8]

The First Reports

The first written record of tobacco in a European language is probably that of Columbus, who on October 12, 1492, made a note in his journal that, when he reached the beaches of San Salvador, "the natives brought fruit, wooden spears, and certain dried leaves [tobacco] which gave off a distinct fragrance." The Spaniards ate the fruit but threw the leaves away. Three days later, after leaving San Salvador, the admiral made the following note in his diary:

> While I was traveling between two islands, namely, Santa María and the larger one I named Fernandina [Cuba], I saw a lonely man in a small canoe. He was sailing from Santa María to Fernandina. He gave me a small piece of bread, about the size of a fist, a bottle of water, powdered red earth, and some dry leaves they must have kept in special esteem, because they gave me the same in San Salvador.[9]

The remarkable custom of inhaling the smoke of aromatic herbs (other than tobacco) rolled within an outer leaf was popular in ancient India but did not become widespread on the European continent. Therefore, when the Spaniards first encountered this strange activity, they had difficulty even

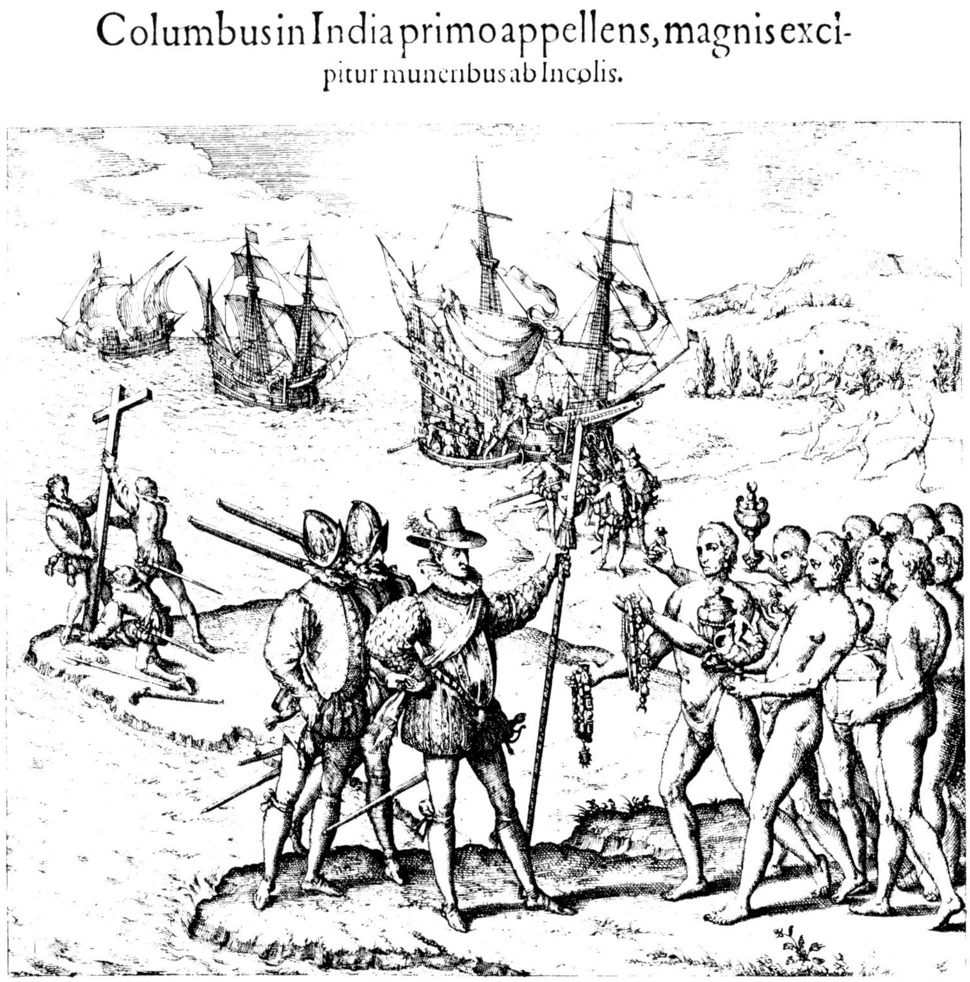

Figure 3. Columbus' first landing in the New World (after Cortí)

describing it. The Europeans who witnessed the act of tobacco smoking for the first time were probably two Spaniards, Rodríguez de Jérez and Luis Torres, who, according to the entry for November 6 in Columbus' logbook, "had been put ashore to survey the country."[10] The historian Bartolomé de las Casas gave the following account of the episode:

> The two Christians met many men and women who were carrying glowing coal in their hands, as well as good-smelling herbs. They were dried plants, rolled in a large dry leaf. They looked like small muskets made of paper that children play with during the Easter festivities. They set one end on fire and inhaled and drank the smoke on the other. It is said that this way they became sleepy and drunk, but also they got rid of their tiredness. The people called these small muskets *tabacos.*[11]

In this account the emissaries are identified merely as *"dos Cristianos."* According to other sources Torres was a learned Jew who spoke not only Chaldean but also Hebrew and Arabic and who, Columbus felt sure, would therefore be able to communicate with any delegate of the Great Khan of Cathay (China) whom he might meet. Jérez later became a confirmed tobacco smoker, probably the first man in Europe to do so. When he returned to his home town, Ayamonte, the smoke emanating from his mouth and nose terrified his neighbors to such a degree that they notified the holy inquisitors that the devil had got hold of their countryman. The unfortunate Spaniard spent seven years in prison before he could prove his innocence.[12]

As the Spaniards cut their way with sword and cunning across their newly conquered New World territory, they encountered the custom of tobacco smoking wherever they went. Gonzalo Fernández de Oviedo y Valdés, in his *Historia general y natural de las Indias Occidentales,* described the strange habit as follows:

> . . . when they begin to drink, the *cacique* takes a small pack of tobacco *(un manojo de tabaco),* about six inches long and as thick as a man's finger. These leaves are rolled together and tied with a thread. They exert great care in making these rolls. They light up one of the ends, and they place the other end in their mouth; they inhale the smoke, hold it for a while in themselves and they blow it out through their mouth and nostrils. These rolls burn slowly all day. They call these *yapoquete.* In Haiti or in Hispaniola they call those *tabaco.*[13]

Tobacco, an item of luxury, played an important part in Mesoamerican social activity and probably was also used extensively in everyday life. According to Francisco Javier Clavijero, the expelled Jesuit historian, the custom of the "after-dinner smoke" was well known even in pre-Columbian times, and after meals the Aztec lords "composed themselves to sleep with the smoke of tobacco."[14]

Bernal Díaz del Castillo, in his *Historia verdadera de la conquista de la Nueva España,* described a feast held by Montezuma II:

> . . . very handsome women served Montezuma when he was at the table with maize cakes kneaded with eggs and other nourishing ingredients. These maize cakes were pure white and were brought in on plates covered with clean napkins. They brought him a different kind of bread also in a long ball kneaded with other kinds of nourishing food and *pachol* cake, as they call it in that country, which is a kind of wafer. They also placed on the table three tubes, much painted and gilded, in which they put liquid amber mixed with some herbs which are called tobacco. When Montezuma finished his dinner, the singing and dancing were over and the cloths had been removed, he would inhale the smoke from one of the tubes. He took very little of it and then fell asleep.[15]

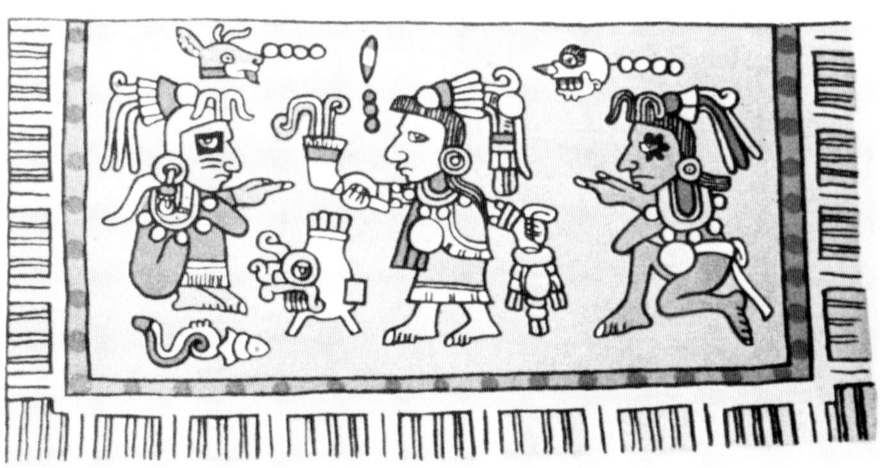

Figure 4. Scene in the Mixtec Codex Nuttall *showing three deities, the one in the center holding a smoking pipe*

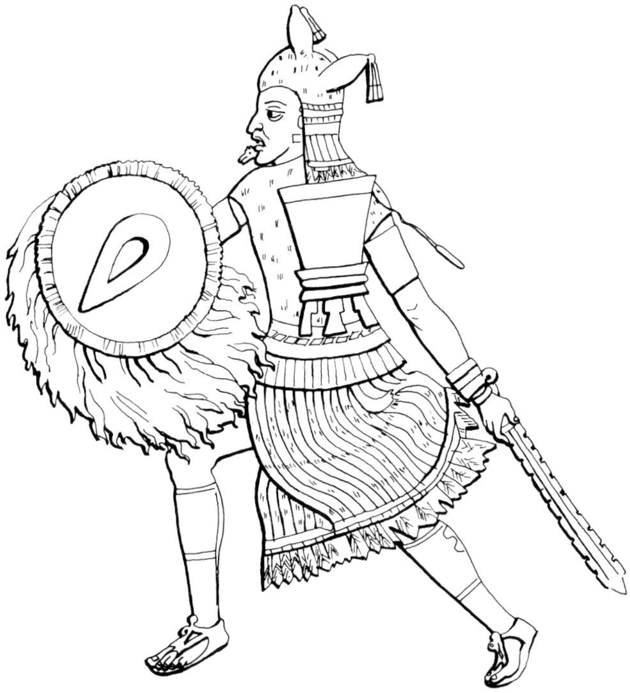

Figure 5. Nezahualcoyotl, king of Tetzcoco (Codex Tixlixochitl)

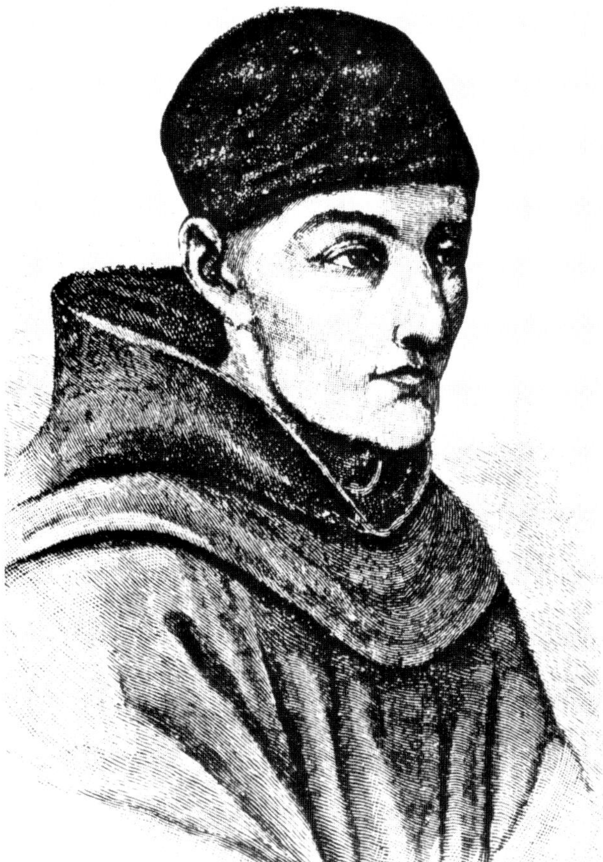

Figure 6. Fray Bernardino Sahagún (after Cortí)

Cigar smoking was customary in other royal houses of Mesoamerica. Juan de Torquemada, describing a feast of Netzahualcoyotl, the Aztec king of Tabasco, wrote of a youth from Tulacingo who served the guests cane cigarettes called *molcajetes.*[16] Another early historian, Fray Bernardino Sahagún, mentioned the use of tobacco at a banquet given by a wealthy Aztec merchant:

> When he had thus spoken, thereupon he spread out his possessions, his goods, in order to display all which was to be used when needed. First he set down for this purpose the cacao beans, the divine ear spice, the tubes of tobacco, the turkeys, the sauce dishes, the carrying baskets, the earthen cups, the wood or the combustible cane to burn, on which the tamales would cook. All this he planed in the house. And when indeed all [and] everything was at hand which would be needed, when he was dissatisfied with nothing more, thereupon he had the people forget [their preoccupations]; he notified the old man, the old woman, and all who were to assist him, the kinsmen and the singers; seven at a time he informed them.
>
> The tobacco server, to perform his task, bore the tobacco [tube] in his right hand; thus he held it: there where it was sealed off; not by the tube [end]. And he went bearing the bowl for the tobacco tubes in his left hand. First he offered one of the tobacco tubes. He said: "My beloved noble, here is thy cane of tobacco." Then the guest took it up; he placed it between his fingers to smoke it. This denoted the spear thrower or the spear; war equipment: valor. And the bowl for tobacco tubes stood for the shield, wherefore [he] bore it in his left hand. He went holding it only by its rim, to carry it. He laid it before perhaps the commanding general; perhaps the *atempanecatl;* all the lords, and the eagle warrior guides, or the noblemen: indeed all who were invited.[17]

Like the Aztecs, the Indians of the northern plains also regarded some of the tobacco utensils as symbolic weapons, such as arrows and spear throwers.[18] Among the tribes east of the Rocky Mountains tobacco was the favorite offering to the spirits, and no ceremony could take place without it. Smoking was a part of all solemn occasions, such as tribal councils, and was a necessary part of most rituals.

Tobacco smoking had another important aspect: it became an essential area of contact between the natives and the conquistadors, who soon grew accustomed to the use of tobacco and became ardent smokers themselves. The practice was first frowned upon by both religious and worldly leaders. Las Casas wrote: "I knew Spaniards who were accustomed to take [tobacco], and, being reprimanded for it by telling them it was vice, they replied that they were unable to cease using it. I do not know what pleasure or benefit they found in it."[19] That even the Spanish clergy was not immune to this "vice" is readily evident in the dictum of the Council

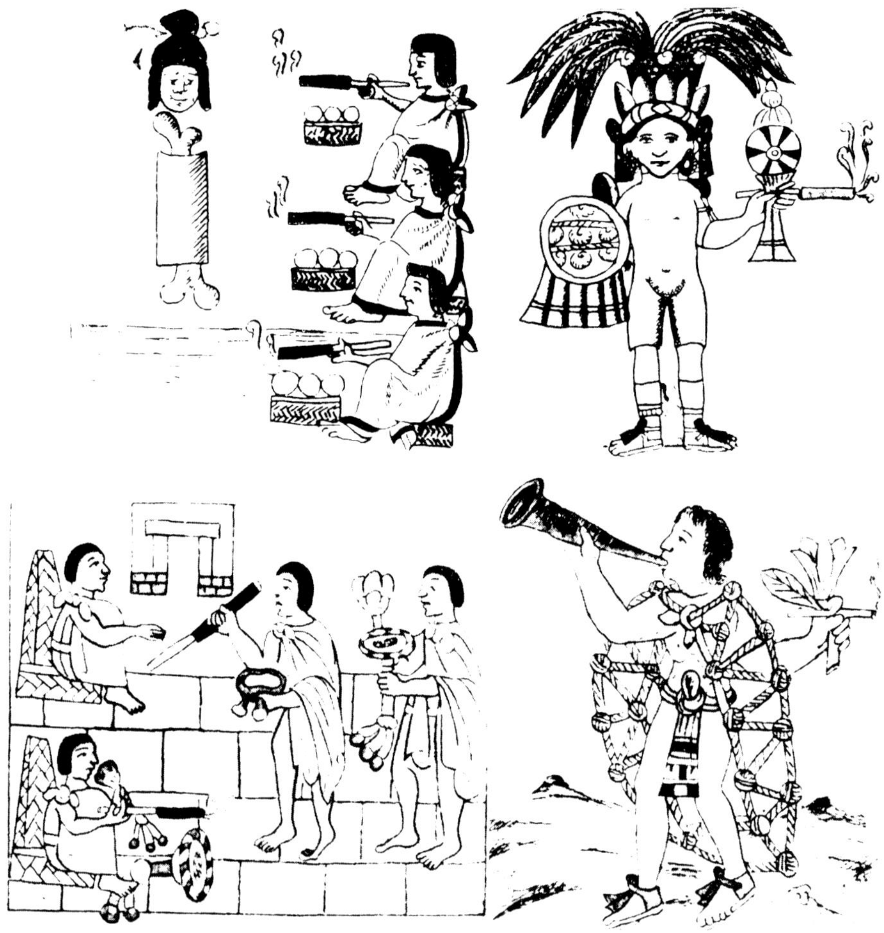

Figure 7. Tobacco-smoking scenes in the Florentine Codex

of Lima, held in 1583 under Gregory XIII, the supreme pontiff, approved by the authority of Pope Sixtus V. At the meeting the high-ranking prelates felt it necessary to forbid the priests to use tobacco before administering the Sacrament: "Under pain of mortal sin no priest, before celebrating and administering the Communion should take tobacco in smoke or powder not even for medicinal reasons."[20]

Just as today, there was in the sixteenth century a determined antismoking lobby. One of its members, Girolamo Benzoni, an Italian traveler who visited the New World in 1541, raised his indignant voice:

> See what a pestiferous and wicked poison from the devil this must be! It happened several times to me that, going through the provinces of Guatemala and Nicaragua, I have entered the house of an Indian who had taken this herb, which in the Mexican language is called *tabaco*, and, immediately perceiving the sharp, fetid smell, I was obliged to go away in haste and seek some other place.[21]

In spite of the displeasure of the leaders of the Spanish clergy, the habit of tobacco smoking spread rapidly among the newcomers to the New World and was soon exported to Europe by Portuguese, Spanish, Dutch, and later English sailors. No one can say with certainty just when the tobacco plant and the custom of smoking first reached Europe, but it is likely that introductions of tobacco (in seeds or whole plants) occurred in Belgium, France, Portugal, and Spain before 1560.[22] It is also said that tobacco as an ornamental plant was imported to Spain in 1518 by Hernando de Toledo and five years later to Lisbon by Diego Colón.[23] It is also probable that a variety of *N. rustica* was cultivated as early as 1554 in the physic garden of the Dutch botanist Rembert Dodoens, who in his book of herbs *(Cruydeboek)* called it *Hyoschiamus luteus*, or yellow henbane.[24] The Dutch, trying to break Spain's hegemony over tobacco sales, began smuggling tobacco from the West Indies to Europe, mainly to England, beginning about 1575, and they soon became such enthusiastic smokers themselves that when in 1580 the port city Enkhuizen issued a decree for the observance of a special day of prayer, the officials not only forbade indulgence in alcoholic

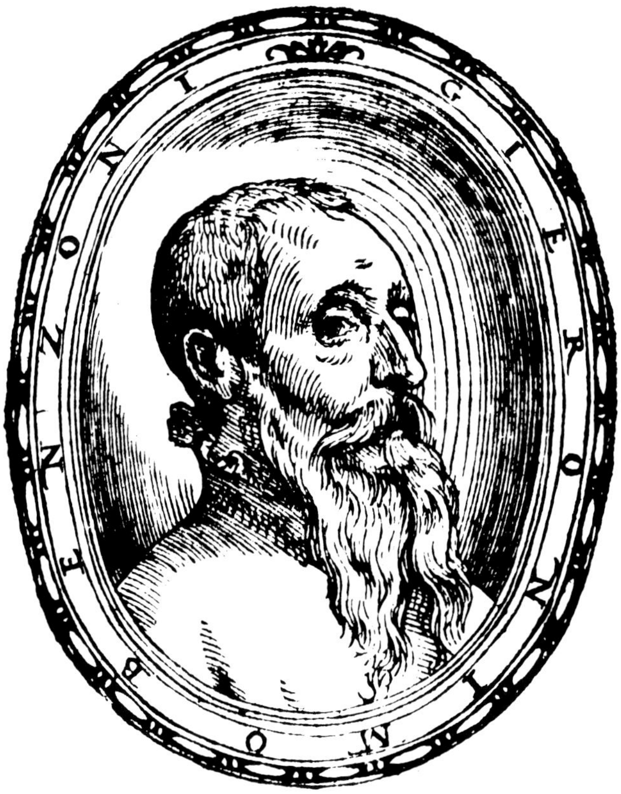

Figure 8. Girolamo Benzoni (after Arents)

beverages but also ordered the citizens to abstain from tobacco: "Besides, no alehouse keepers, topesters or others shall be allowed to serve their customers . . . with beer, wine or French brandy of any kind whatsoever, or supply or sell tobacco."[25]

In 1594 the historian Emanuel van Meteren gave an excellent description of how tobacco was used in Holland:

> As a result of these various Voyages of divers Nations from the West Indies, from Brazil and from the Shores of Peru there was brought a dried herb which we call Nicotiana, and in the Indies it is called Tobacco, smoke-dried, which only now has become known in Europe within the last 20 or 25 years, where it arrives in many shapes, bound and rolled. This they use, dried to powder, with a silver or earthenware Pipe, specially made or baked therefore, one finger long, and in front a little bowl about the size of a Hazelnut, which they fill with the Powder and then light it with a live coal or candle. And at

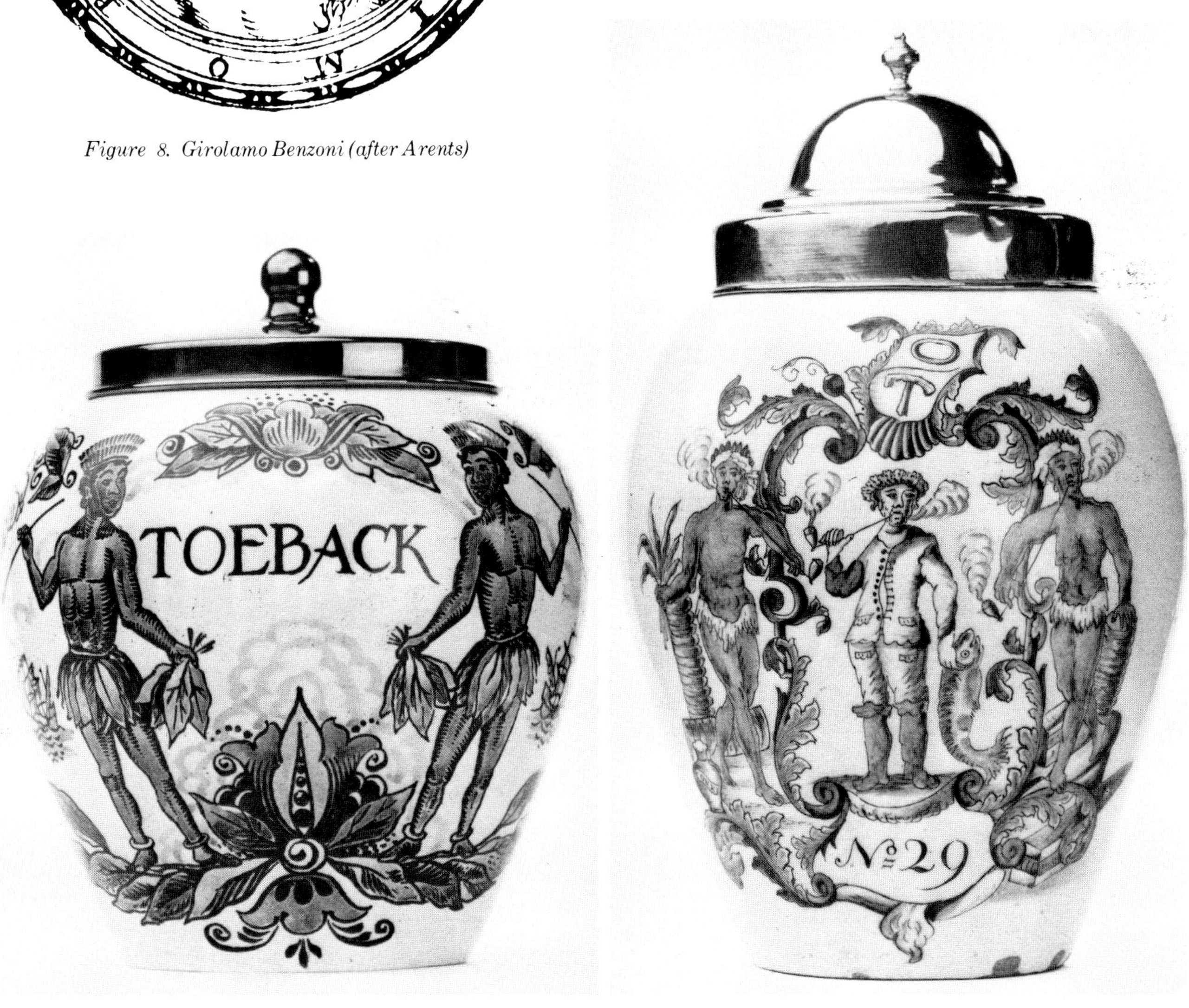

Figure 9. Old Dutch tobacco jars decorated with tobacco plants, smoking Indians, and colonists (Douwe Egberts Museum, Utrecht)

the other end they suck through the hollow Pipe, the Smoke of it into the body and the mouth and, having these full of smoke, puff it out again through the nostrils. And they do this until it is all burned out. And they do this as often as they like. And they call this: drinking a person's health with a Pipe of Tobacco. And everywhere they have, especially in England, Houses and People who live by it, only having public houses for Tobacco, which they take at any hour of the day, after dinner, before dinner, early in the morning, and late at night.[26]

Some tobacco undoubtedly also reached England directly through the raiders of Sir John Hawkins and Sir Francis Drake, who plundered the Spanish Main, where they got a "great store of tobacco."[27] The first report in English about the use of tobacco among the natives of the New World came from the pen of Thomas Hariot, surveyor-historian of the group of English colonists who settled Roanoke Island in 1584:

There is an herb called *uppowoc* which sows itself. In the West Indies it has several names, according to different places where it grows and is used, but the Spaniards generally call it *tobacco*. Its leaves are dried, made into powder, and then smoked by being sucked through clay pipes into the stomach and head.[28]

The strange new custom was not accepted among the English without resistance. King James I himself expressed his royal displeasure and indignation about smoking in his famous *Counterblaste to Tobacco*, a pamphlet he wrote in 1604:

And now good Country man, let us (I pray you) consider what honour or policie can move us to imitate the slevish Indians, especially in so vile and stinking custome.... I say without blushing, [why do we] abase ourselves so farre, as to imitate these beastly Indians, slaves to the Spaniards, refuse to the world, and as yet aliens to the holy Covenant of God? Why do we not as well imitate them in walking naked as they doe?... Yea, why do we not denie God and adore the Devill, as they doe?[29]

That the good king was not very successful in his campaign against tobacco is demonstrated by a report, published eight years later, that there existed the "trade of Selling Tobacco in London, and neare about London; if a man may believe what is confidently reported, there are found to be upward 7,000 houses that doth live by that trade."[30] To gain an idea of the great popularity tobacco achieved in Europe within an unparalleled short time, one may read the eulogy on tobacco by the French playwright Molière in his comedy *Don Juan*, written in 1665:

Sganarele (with a tobacco box in his hand): Whatever Aristotle and the whole body of philosophers may say, there's nothing comparable to tobacco; 'tis the reigning passion of your better sort of people, and he who lives

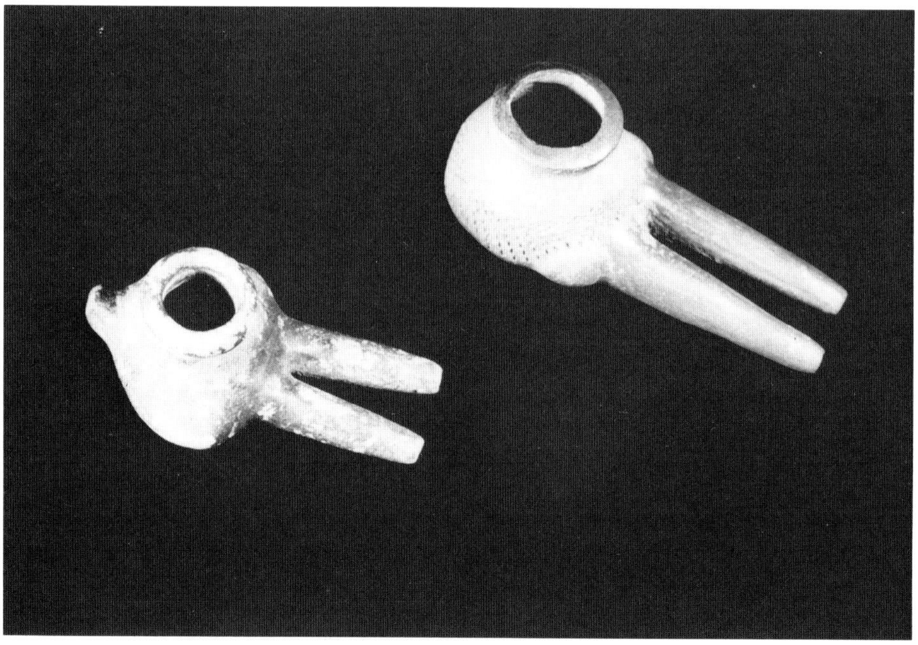

Figure 10. Nose pipes from Guanacaste, Costa Rica (unless otherwise noted photography by the author)

without tobacco deserves not to live; it not only exhilarates and purges human brains; it also trains the mind to virtue, and by this one learns to become well bred. Don't you see plainly, from the time one takes it, in what an obliging manner one uses it with all the world, and how one is delighted to give it right and left wherever one comes? One doesn't even wait to be asked for it, but anticipates people's wishes; so true it is that tobacco inspires all who take it with sentiments of honor and virtue.[31]

It soon became evident that the custom of tobacco smoking, for good or bad, was to stay in both the New World and the Old.

Tobacco Smoking in America

In America the habit of tobacco smoking probably developed from the custom of incense burning during religious and secular ceremonial rites. In feeding the fire, the shamans used dry twigs, resins, and leaves of aromatic plants including tobacco, which in its wild state often attained a remarkable luxuriance of growth. While they were thus engaged, the acolytes would naturally blow on the embers, inhale the smoke, and so become aware of the pleasing effects of the herb.[32] It may have been from this primitive form of "smoke sniffing" that the three principal ways of smoking tobacco—in pipes, cigars, and cigarettes—evolved. All North American Indians except the Pueblos and some of the California tribes were exclusively pipe smokers.[33] Their pipes were made mainly of clay and stone, but early records indicate that wood, horn, and bone were also used. These "classic" pipes with bowls set angularly upon the stem were found in abundance in the mounds of the northern and southwestern plains. The Indians of Mexico were primarily cigar and cigarette smokers, but they used pipes too, a habit probably imported from North America,[34] and they also enjoyed tobacco in the forms of snuff and chew.[35] South of the Mexican influence cigar and cigarette smoking was very strongly in favor.[36] The only known true pipes in Central America are the elbow types from Nicaragua and Costa Rica.[37] Some of these pipes have two stems, indicating that they may have been used for snuffing.[38] A few pipes have also been found in Colombia, Venezuela, Chile, Argentina, Paraguay, and Brazil, but not in Ecuador, Peru, or Bolivia.[39]

When the Spaniards made their first contact with the Mayas, they found them smoking a primitive cigar made of long, thick tobacco leaves, similar to the cigar smoked today by the Lacandón Indians. The cigar that was and still is preferred in Central America has three parts, filler, binder, and

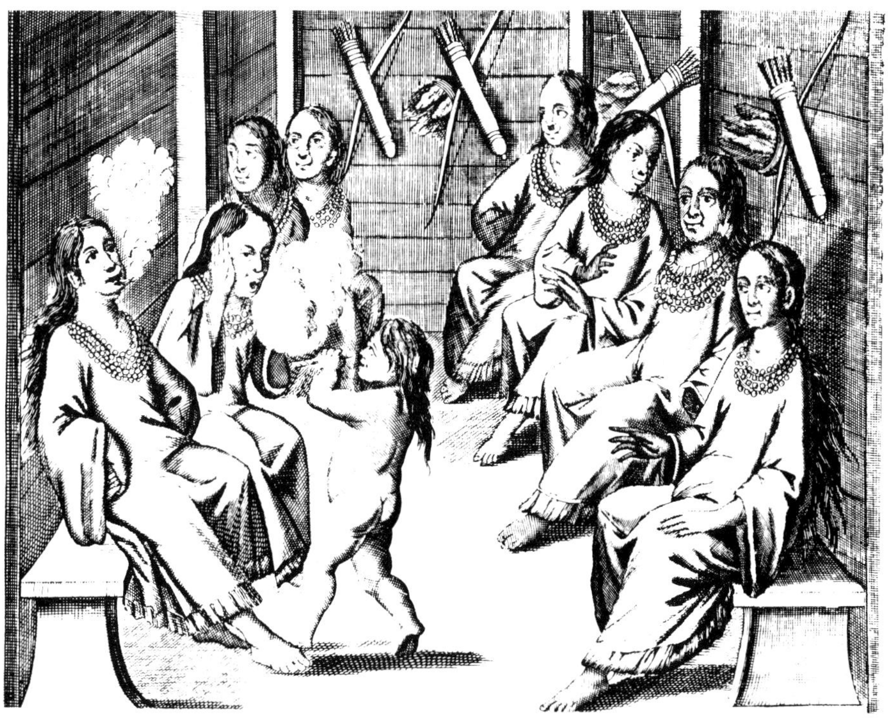

Figure 11. A peculiar way to smoke (after Wafer)

wrapper. The filler is the inner core that forms the body and shape of the cigar. The binder is the leaf in which the filler is wrapped to form what is known as the "cigar bunch." The wrapper or outer covering consists of a ribbon leaf wrapped spirally around the cigar bunch.

The term *cigarette* denotes a small cigar made of fine-cut tobacco rolled in a wrapper or other material. Cigarettes are generally not only smaller but also milder than cigars. The Mesoamerican Indians usually smoked (and some still smoke) corn-husk cigarettes or tobacco-stuffed tubes of reed or cane. Paper-wrapped cigarettes were not invented in America but were improvised in the sixteenth century by the beggars of Seville, who picked up discarded cigar butts, shredded them, and rolled them in scraps of paper. These "poor-man smokes" became known as *papeletes* or *cigarillos*. The word "cigarette" appeared first in Giovanni Casanova's memoirs, published in 1767.[40]

In the Andean highlands and along the west coast of South America tobacco is replaced by another herb, coca *(Erythroxylum coca)*, which is usually chewed, not smoked.

The various ways the American natives smoked tobacco strongly influenced European smoking habits. The English, who made their first contacts with the northeastern tribes, are predominantly pipe smokers even today. The Mediterranean peoples, who learned the use of tobacco from the natives of Mesoamerica and South America, still prefer cigars and cigarettes. Besides these "conventional" ways of smoking, there were also some very unusual ways. For example, it is said that the Dakotas and the Crees in early times did not use pipes but inhaled smoke from burning tobacco that had been placed in a hole in the ground.[41] Perhaps the most curious method of smoking was recorded by John Wafer, an Englishman who, passing through the Isthmus of Panama in 1681 and trying to avoid contact with the hostile Spaniards, took refuge among the natives. He saw the Indians fabricate long rolls of tobacco leaves and smoke them in a most unusual manner:

> Their way of Smoaking when they are in Company together is thus: A Boy lights one end of a Roll and burns it to a Coal, wetting the part next to it to keep it from wasting too fast. The End so lighted he puts into his Mouth, and blows the Smoak through the whole length of the Roll into the Face of every one of the Company or Council, tho' there be 2 or 300 of them. Then they, sitting in their usual Posture upon Forms, make, with their Hands held hollow together, a kind of Funnel around their Mouth and Noses.[42]

Son, bring me a firefly of the night. Its odor shall pass to the north and to the west.
The Book of Chilam Balam of Chumayel

Early Spanish accounts of smoking among the Maya are not very informative.
Sir J. Eric S. Thompson

The historian composes a meaning from tradition.
George Kubler

2. Tobacco and the Mayas

Early Historical Records

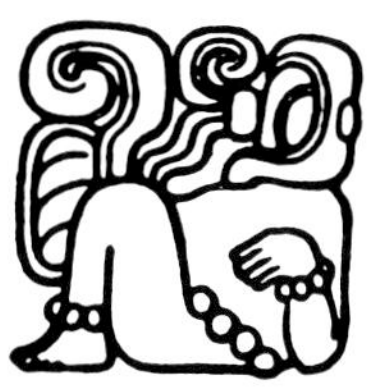

Early records relating to the use of tobacco among the Mayas at the time of the Conquest are, to say the least, highly questionable and certainly confusing. The first Western observer of Maya smoking habits was probably Juan Díaz, chaplain of Juan de Grijalva's expedition, which landed on the island Cozumel, off the coast of Yucatán, in 1518.[1] Díaz reported that the natives presented to Captain Grijalva reeds that "gave off delicate odor on being burned."[2] This episode was also mentioned by Oviedo y Valdés in *La Historia general de las Indias* (published in Seville in 1535), who probably relied on Díaz' account:

> Then he [the old chief] gave to the captain and to each one of the Christians who were seated a small tube burning at one end. These were so constructed that after being lit they slowly consumed themselves without giving forth a flame, until they had ceased burning, just as do the incense sticks of Valencia. And they were of fragrant odor, as is the smoke which comes from them. The Indians made signs to the Christians that they may not allow that smoke to be lost or pass away as one does who is taking *tabaco*.[3]

While some authors doubted that these tubes contained tobacco, others found it reasonable to accept the opinion that they were a kind of reed cigarette filled with tobacco[4]. An account of the Grijalva expedition has been rendered by another participant named Díaz, the historian Bernal Díaz de Castillo. He spent some time in the West Indies before he sailed with Grijalva, and he was familiar not only with the tobacco herb but also, according to his own records, with the custom of smoking tobacco.[5] Giving a detailed and probably a more accurate account of the expedition than chaplain Díaz, he mentioned several times that the Spaniards were "fumigated with incense by the Maya priests" as an act of courtesy and submission, but on no occasion did he imply that the herb may have been tobacco. He described the first such episode, which occurred at the mouth of the Tabasco River:

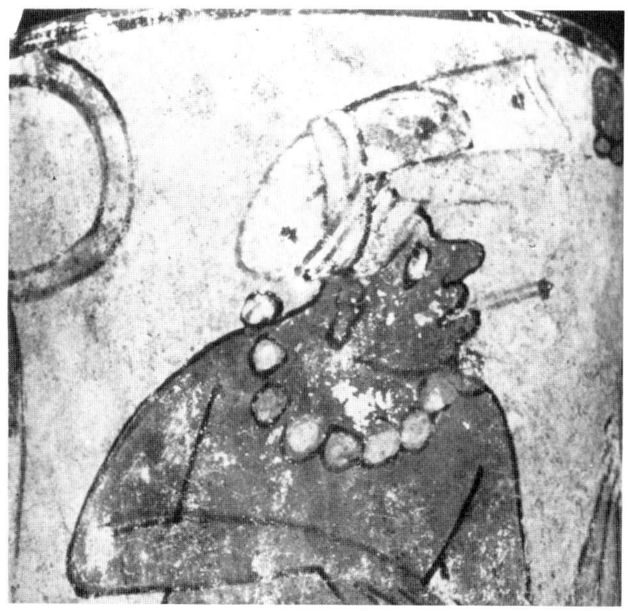

Figure 12. Smoking Maya lord, detail on a Classic vase from Tikal (Tikal Museum)

> On the following day more than thirty Indians came to the headland of the palm trees where we encamped, bringing roasted fish and fowls, *sapota* fruit and maize cakes, also braziers with live coals, and incense, with which they perfumed us all.[6]

The Spaniards were again "fumigated" a few days later on the banks of the Jamapai River. There the Indians brought clay braziers with burning coals and "incensed [them] with a sort of resin."[7] Incense, rather than tobacco, burning appears to have been the ritual on the third occasion too:

> That day they [the Mayas] sacrificed two boys, cutting open their chests and offering their blood and hearts to that accursed idol. And these priests came to incense us with the same incense as they used on Tezcatlipoca, for at the time of our arrival they were burning something that smelled like incense before him. But we would not allow them to burn it over us.[8]

Another account that Thompson believed referred to tobacco smoking is that of Diego de Landa, the first bishop of Yucatán.[9] Landa, who was an implacable enemy of the ancient Maya religion and a destroyer of anything even remotely related to it, was also an assiduous chronicler of Mayan affairs. In his *Relación de las cosas de Yucatán* he described the puberty rite of Maya children: "Behind walked the rest of the assistants of the priest, holding a bunch of flowers and a pipe, which the Indians are accustomed to suck, with each one of these they threatened each child nine times, and afterwards gave each in turn the flowers to smell and the smoke to inhale."[10] Thompson concurred with the translator's interpretation that Landa was referring to tobacco smoking and commented, "One may conjecture that Landa wished to say that each boy was given a few puffs at a cigar."[11] This explanation is possible but not necessarily correct. Landa's original Spanish text never says "pipe" *(tubo, cano)*, but rather "dense smoke" *(humazo)*, which may refer again to incense rather than tobacco. Indeed, incense is more likely, because in the previous chapter Landa clearly stated that the priest who was performing the ritual was seated by a brazier and that each of the children came up to it:

> Afterwards or before they placed in the middle another little stool on which the priest sat down with a brazier and with a little ground maize and their incense. The boys and girls came in order and the priest put into their hands a little ground maize and incense and they threw it into the brazier.[12]

Another no less confusing early reference is that of the soldier-historian Juan de Villagutierre de Soto-Mayor, who described an early field campaign of Spanish soldiers against the Lacandón Indians.

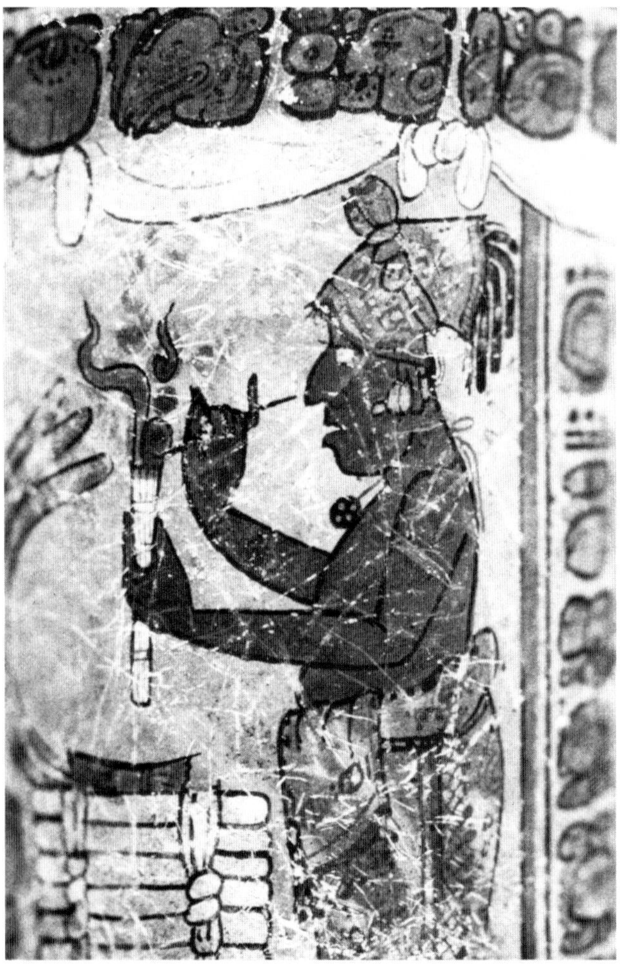

Figure 13. Attendant aerating a large cigar, detail on a vase from the late Classic period, probably from northern Petén

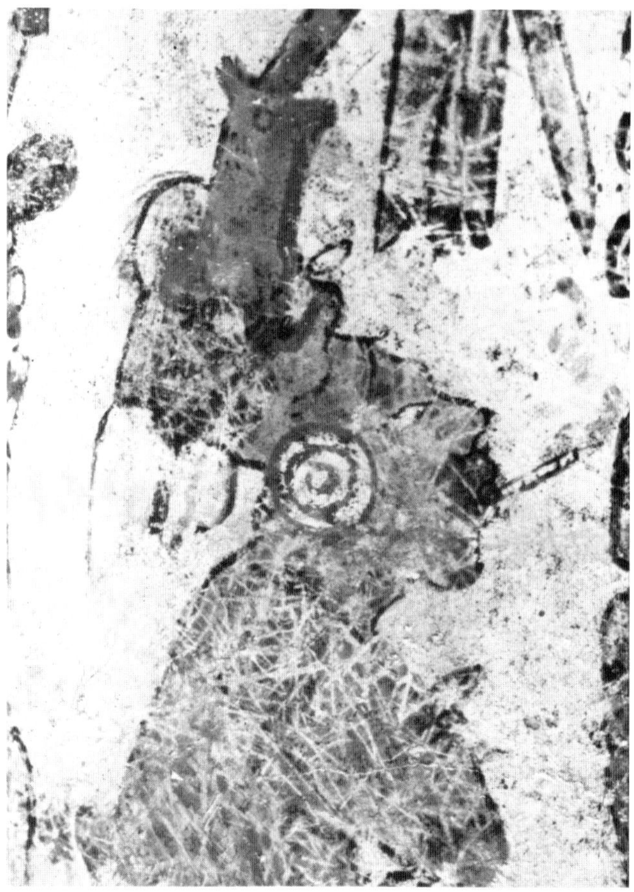

Figure 14. A Maya courtier smoking, detail on a vase from the late Classic period, probably from northern Petén

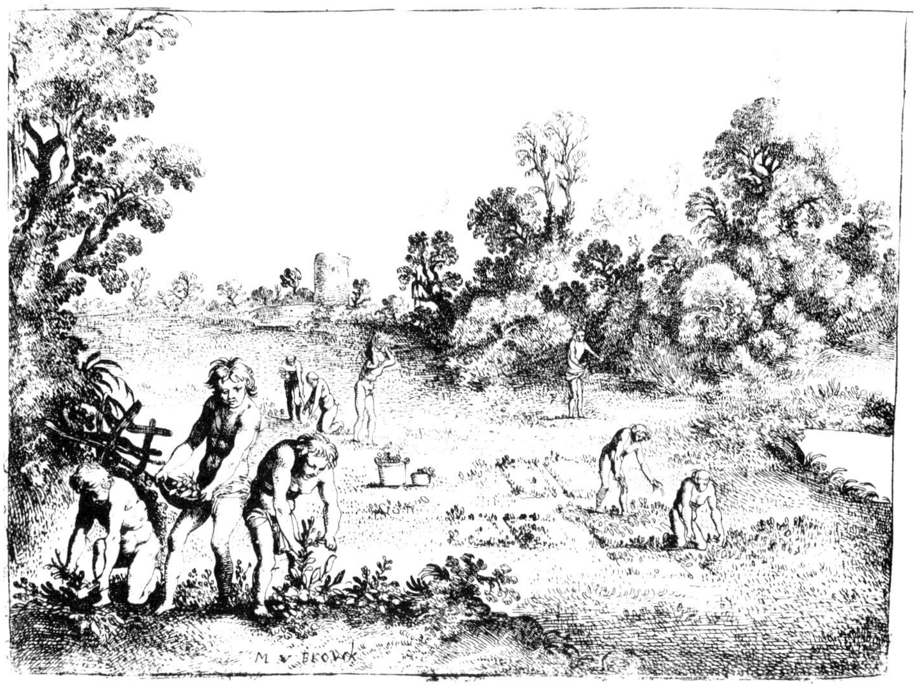

Figure 15. Tobacco cultivation in America at the time of the Conquest (after Neander)

When a force of infantry led by Captain Melchor Rodríguez approached the village of Dolores, the natives fled in disarray, leaving behind most of their belongings, among them "quantities of cigars made of clay, which were being fabricated by the natives, curiously painted."[13] What exactly he meant by these "clay cigars" is hard to determine. While one may speculate that they were not cigars at all but cigar holders, molds, or simply cigar-shaped ceramic objects, it is more likely that there may have been an error in either observation or description. Such a mistake is probable, because the source on which the noted Guatemalan historian Francisco de Paula García de Peláez drew in describing the same incident in 1851 simply mentioned these objects as ordinary cigars: "[The Lacandones left behind] utensils made out of stone, knitting needles, cooking and eating ware, cigars twisted into the leaves of the nance tree (*Malphigia glabra*) which they called *puquiet*." He even describes their "cigarette lighter": "They lighted the cigars by rubbing a cork adjusted to a small mill."[14]

Nicholas M. Hellmuth, also much interested in the subject, believes that the original source used by both Villagutierre de Soto-Mayor and García de Peláez is a sixteenth-century manuscript by Nicolas de Valenzuela.[15] It describes these *puquietes* as follows:

> [They were] occupied with *puquietes*, which are some kind of a cigar, more than 3 *quartas* [one *quarta* or *cuarta* is one-fourth of the Spanish yard] of length, and about a thumb in thickness, fabricated using the leaves of the nance tree, and varnished with clay that looks like *ochre*, and on it they painted in different colors. They stuffed it with tobacco and completed it with a tip. They made it in a way that they neither lit nor wasted the tip of another, not even the ash, because with good skill and with quickness, the end was returned to their open mouth, and tapping their teeth lightly, they swallowed it [ashes?]. This was generally admired [by the Spaniards].[16]

Apocryphal Maya Scripts

The Spanish missionaries regarded as one of their important tasks that of teaching the Mayas to write in their native tongue using the Western alphabet. While their converts were supposed to use this newly acquired knowledge only for the purpose of their new religion, during the two centuries after the Conquest the Mayas recorded a rich collection of their historical events, myths, legends, and folklore. Among these apocryphal chronicles cigar smoking is mentioned in both the *Popol Vuh* and *The Book of Chilam Balam*.

In the *Popol Vuh* (Book of the Council), the sacred book of the Quiché Mayas, cigar smoking plays an important part in one of the adventures of the Hero Twins Hunahpú and Xbalanqué:

First they went to the House of Darkness,
And so they came to be given a torch.
It was burning
When it arrived
With a cigar for each of them
(R uq huhub ki zik')
From the messengers of I Death
"These are your torches,
Said the Lord,
They must come and give back the torches
in the morning.
Together with the cigars.
(R uq ri zik')
They must come back whole.
The Lord says,"
The messenger said when they came.
"Very well," they said then.
But they did not light the torch.
Just a red substitute was put up.
It was a parrot's tail,
And it looked like a torch to the guards.
And then for their cigars,
(Are chi q'u ri zik')
They just put a fire bug on the ends of
their cigars
(Sa q'aq a chikop x ki koho ch u vi zik')
One whole night they keep them lit
"We have beaten them," said the guards.
But the torches weren't finished at all;
It just looked like it.
They went back and gave them to the lords.[17]

The legend of the "fake cigars" with which the Hero Twins fooled the guards has been passed down into the life of the modern Mayas. It is imitated in the Chorti ceremonies of today, and in Quiché folklore Christ faced the same ordeal and solved it in the same manner.[18]

Reference to cigar smoking is also made in *The Book of Chilam Balam of Chumayel* (Book of the Jaguar Priest of Chumayel), a book as revered by the Yucatecs as the *Popol Vuh* is by the Quichés. In the Chumayel version of the chronicle the chieftain is probing the ability and witness of his son by asking him riddles:

"Son bring me the firefly of the night. Its odor shall pass to the north and to the west. Bring with it the beckoning tongue of the jaguar." "It is well, father." What he asks for is a smoking tube filled with tobacco [*hex lic u katice, chamal*]. The beckoning tongue of the jaguar for which he asks is fire.[19]

In another chapter of the book (the prophecy for Katún 5 Ahau) reference is made again to (tobacco?) smoking: "Then came . . . the end of the cigar [*chamal*], after the lord of the world was created . . . he heard the dance . . . ."[20] Roys noted on this prophecy:

Chamal, a roll or tube of tobacco for smoking (Motul). The modern Maya believe in four supernatural protectors, the Balams, who move abroad at night. The shooting stars are believed to be glowing stumps of the cigars smoked by the Balams, which they throw away.[21]

Tobacco Culture in the Maya Area

The little information available on aboriginal tobacco culture in the New World is related primarily to the natives of North America.[22] There is no written record of tobacco cultivation among the Mayas at the time of the Conquest, but it is certain that cultivation derived from the earlier custom of gathering several species of wild tobacco, such as *sylvestris*, *otophera*, and *paniculata*.[23]

Today tobacco is one of the most important plants cultivated by the Mayas. In the Maya area it grows about five feet tall and is harvested once a year. It is cultivated principally in the highlands, but it will also grow in the lowlands if properly irrigated. If the methods used today in isolated smaller villages are analogous to those of pre-Conquest times, then it is probable that the ancient Mayas sowed the tobacco seeds in large *ollas* or wooden *cañoes* or in special small plots, just as the Chortis do today.[24] The young sprouts are later transplanted to the fields, usually in October or November. In ancient times the Mayas probably also had some ritual customs related to tobacco planting, as the North American Indians still had at the turn of the nineteenth century. We know, for example, that the Winnebagos planted sacred gourds in the small protected patches where they grew their tobacco for ritual and sacrificial purposes. In the eastern part of North America it was also customary among the Indians to mark or fence off their tobacco fields to prevent the intrusion of women or other influences that might destroy the ritual powers of the tobacco crop.[25]

The Maya tobacco grower cuts the leaves in April. They are spread on hides or blankets to dry or are hung on strings, as is customary among the Lacandón and Yucatec tribes.[26] The highland Maya cures his tobacco by a method that has apparently changed little over the centuries. Before the leaves are cut, he sets up drying poles six to seven feet tall, fifteen feet apart, with transverse rods bridging the gaps between them. On the rods he suspends the green tobacco leaves on strings. The tobacco dries in the sun for about two weeks, and is then pressed between large, flat stones covered with leaves of other plants. After five days the stalks are cut, and the tobacco leaves are again placed in the stone

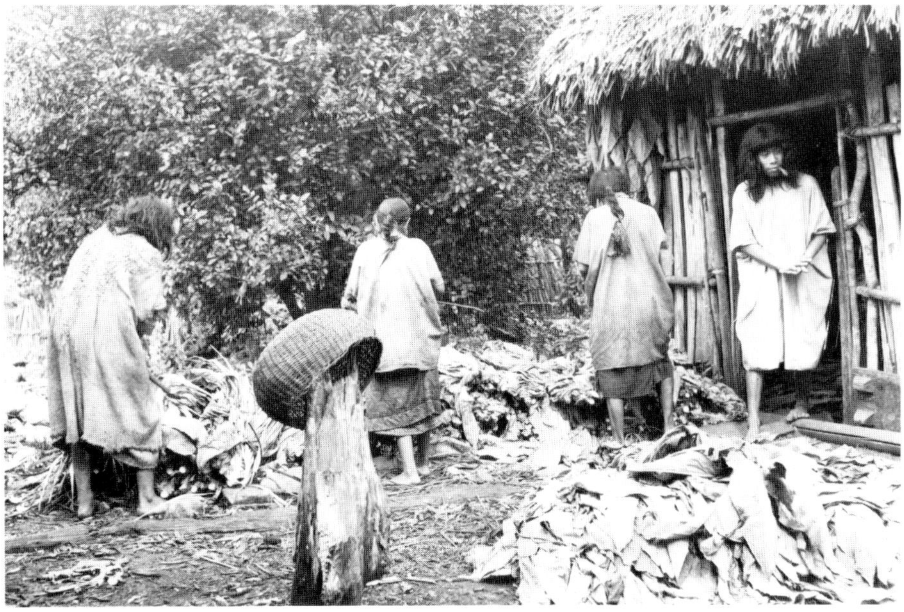

Figure 16. Tobacco culture among the Lacandones. 1. The selection of the leaves (photography by Gertrude Duby Blom)

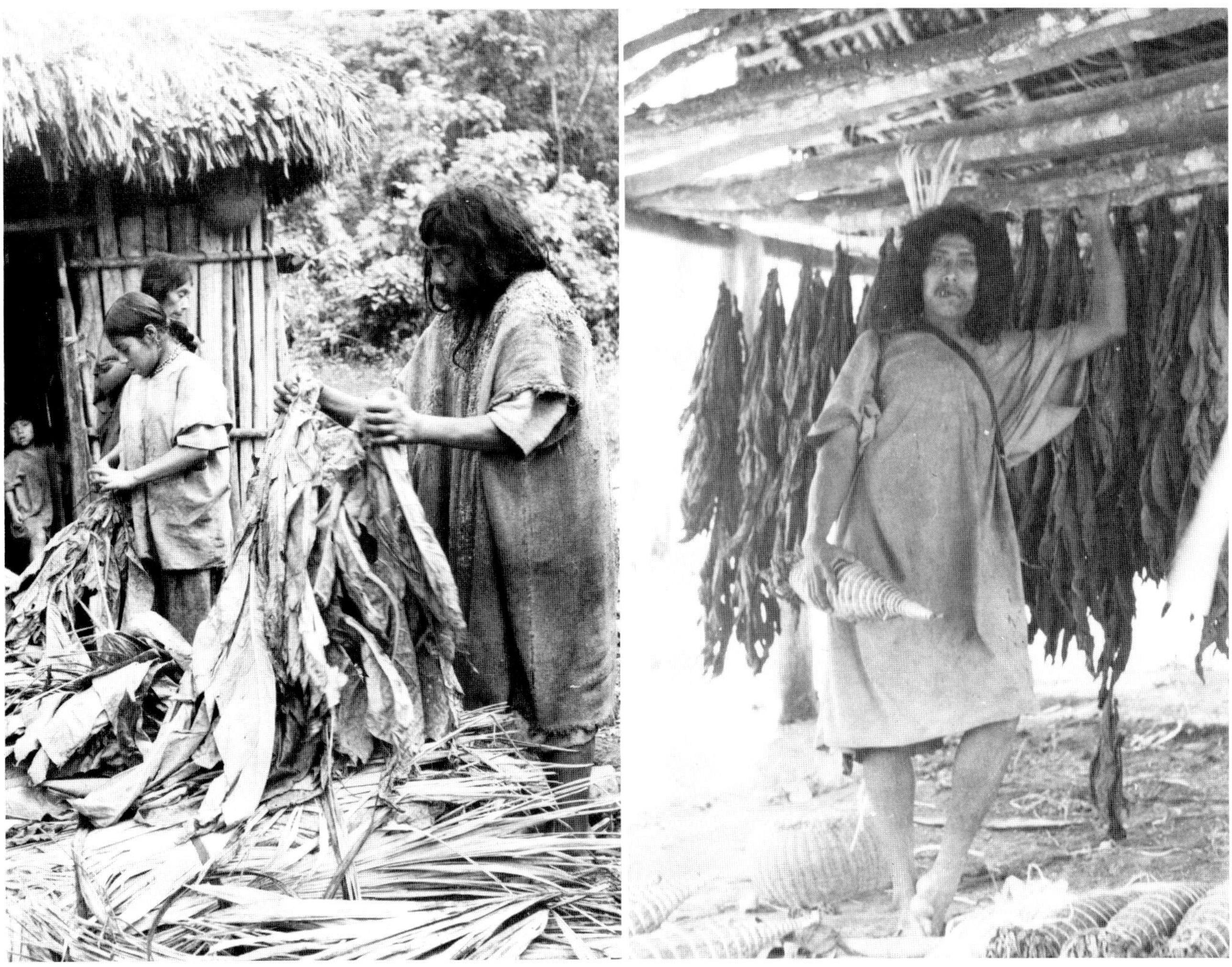

Figure 17. Tobacco culture among the Lacandones. 2. Preparation for drying and the tobacco shed (photography by Gertrude Duby Blom)

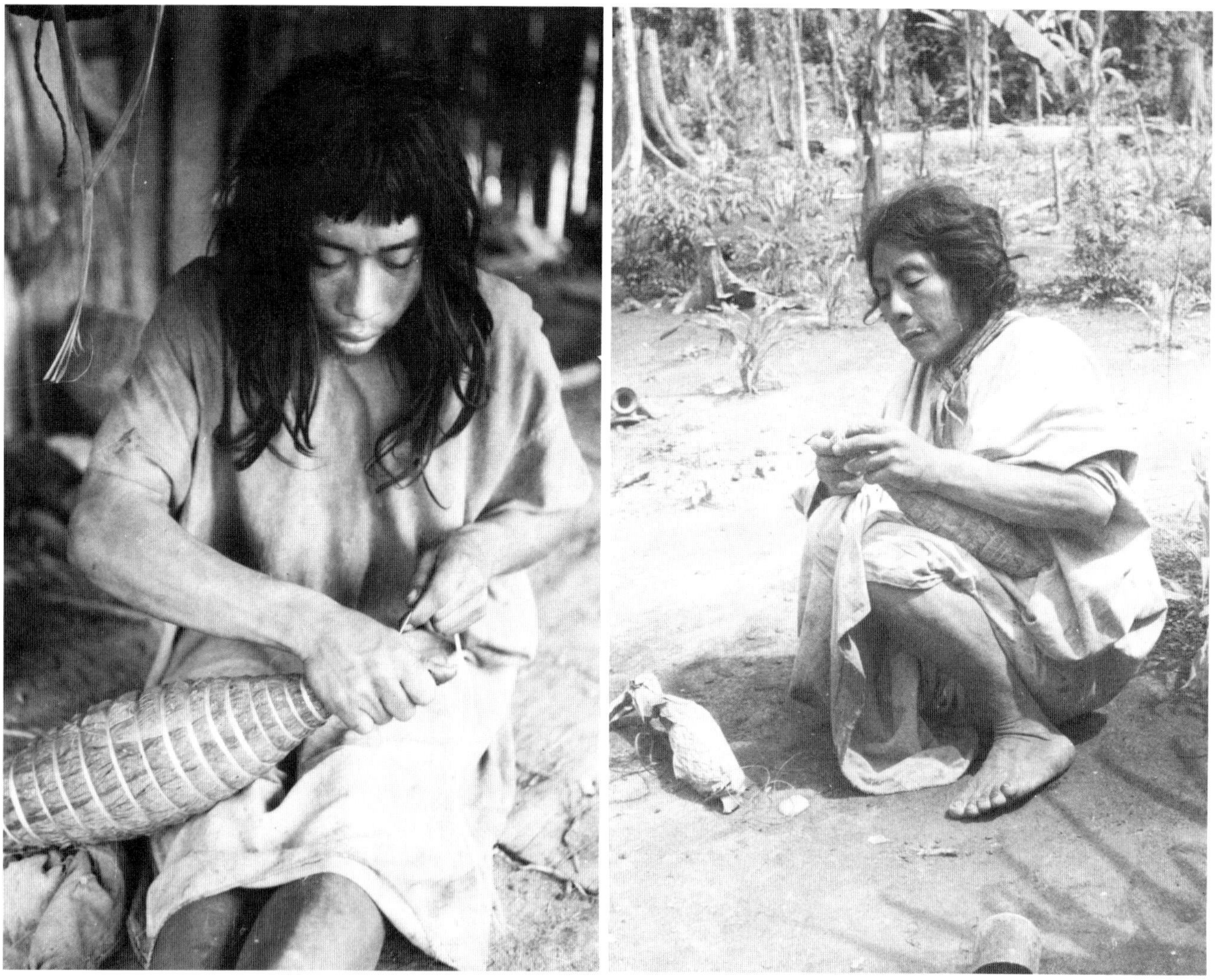

Figure 18. Tobacco culture among the Lacandones. 3. Preparation of the tobacco bundle (photography by Gertrude Duby Blom)

press. The tobacco is removed from the press every two days to allow it to cool and is then pressed again. This process continues for about a month, until the leaves no longer become heated. Then they are completely cured and ready to be made into cigars or cigarettes.[27]

Besides growing it for their own use, the Mayas have long used tobacco as an important trading commodity.[28] While no contemporary record of the ancient Maya tobacco dealer exists, Sahagún gave a detailed description of the Aztec colleague of the Maya tobacco merchant:

> The seller of fine tobacco sells fine tobacco . . . small tobacco. He rubs it between his hands. He who finds fine tobacco chews it. And some prefer wormwood to tobacco. Fine tobacco affects one; it makes one drunk, it aids one's digestion, it dispels one's fatigue.[29]

That the tobacco remained an important economic product of the Mayas even after the Conquest is demonstrated in the proclamations of Antonio Oliver and Joseph Merino y Caballos, former governors of Yucatán, each of whom in his *edición*, written in the Maya language, one in 1774 and the other in 1783, referred to himself as "Brigadier of the soldiers of our great King, Governor and Captain General here in the entire Province, Commander of the assemblage of soldiers, regulars, and rural militia in charge of the Santa Cruzada, deputy for the tobacco payments [*u tubul kutz*] for our great King."[30]

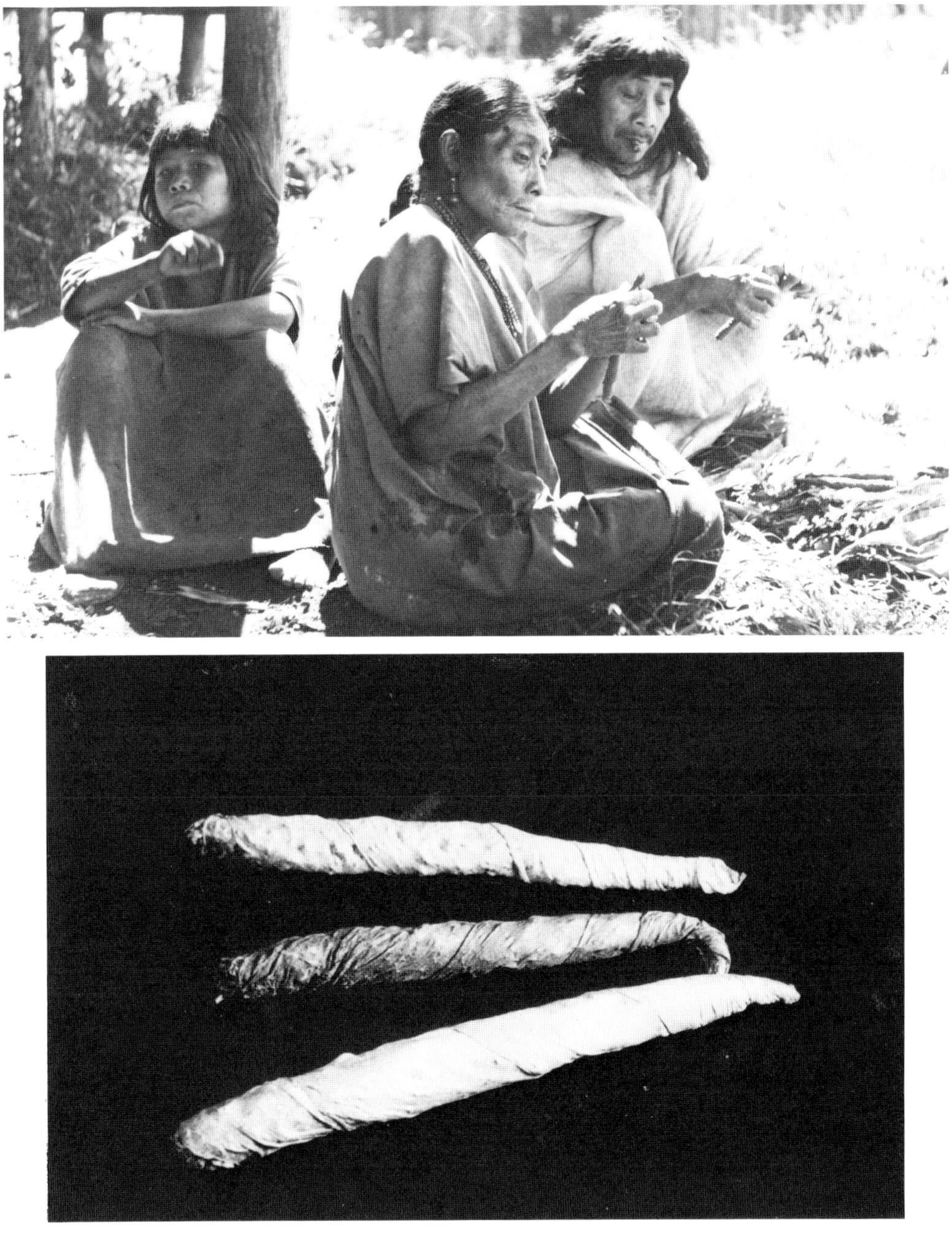

Figure 19. Tobacco culture among the Lacandones. 4. Cigar making and cigars (upper photo by Gertrude Duby Blom)

The Maya as a Tobacco Smoker

As stated before, the smoking habits of the ancient Mayas very probably derived from the incense ceremonies of the medicine men and priests. Apparently they were exclusively cigar and cigarette smokers. John Cockburn, who traveled the Maya realm extensively in 1735, wrote: "These gentlemen gave us seegars to smoke. These are leaves of tobacco rolled up in such a manner that they serve both for pipe and tobacco itself; they know no other way, here for there is no such a thing as a tobacco pipe throughout New Spain."[31]

The absence of the pipe from the Maya area is also noteworthy in both artistic representations and archaeological finds. There are only five possible exceptions, all of them highly doubtful. The first "Maya pipes" were found by G. Boggs, who, working with the Peabody Museum expedition in 1944, excavated a few hollow ceramic tubes at Tazumal, San Salvador, and identified them as "pipes." In this labeling, however, he was not certain, adding that

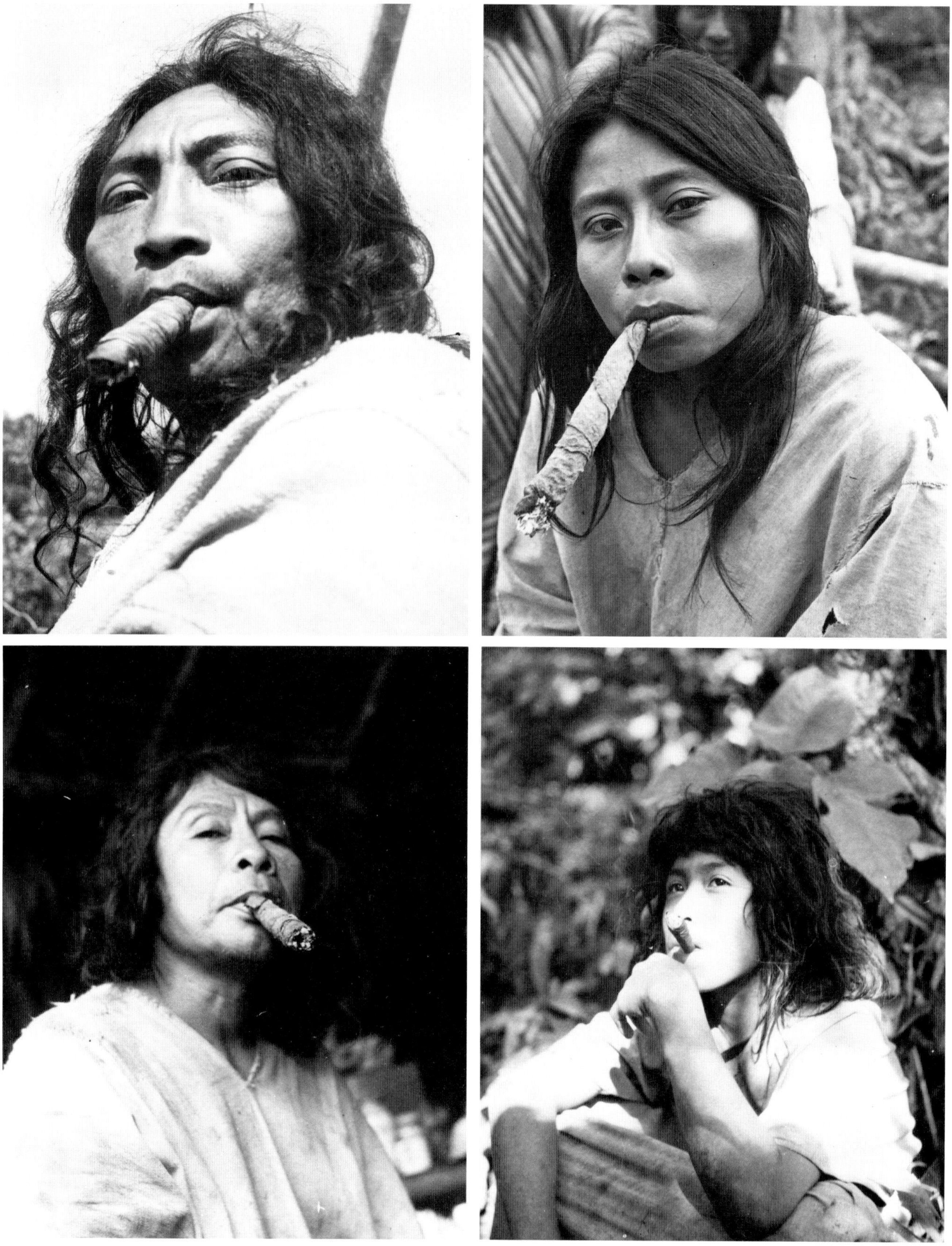

Figure 20. Smoking Lacandones (photography by Gertrude Duby Blom)

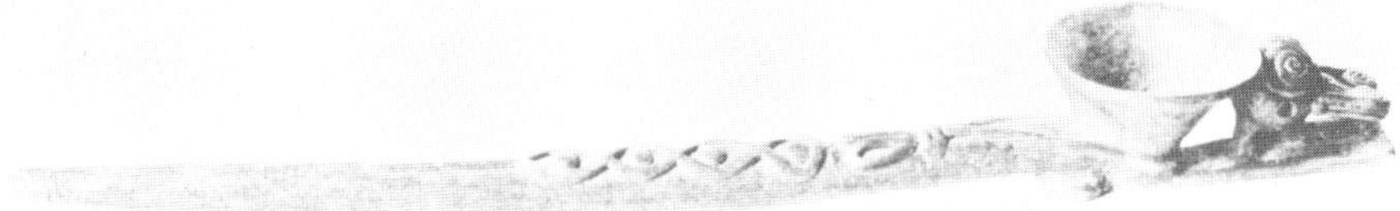

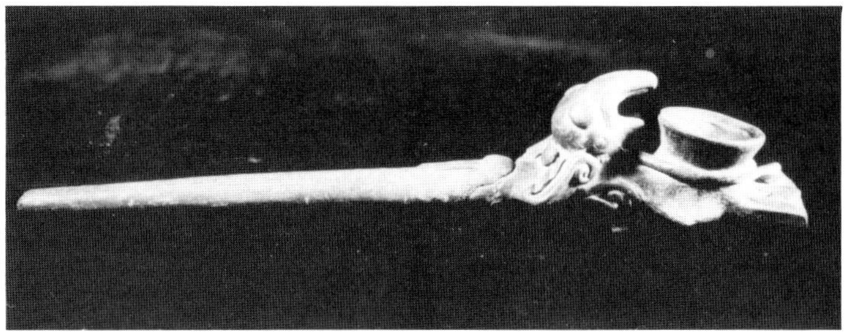

Figure 21. Long-stemmed pipes from the post-Classic period. The upper one was excavated by Morris at Chichén Itzá; the lower was found at Michoacán. Note the similarity in both shape and artistic style

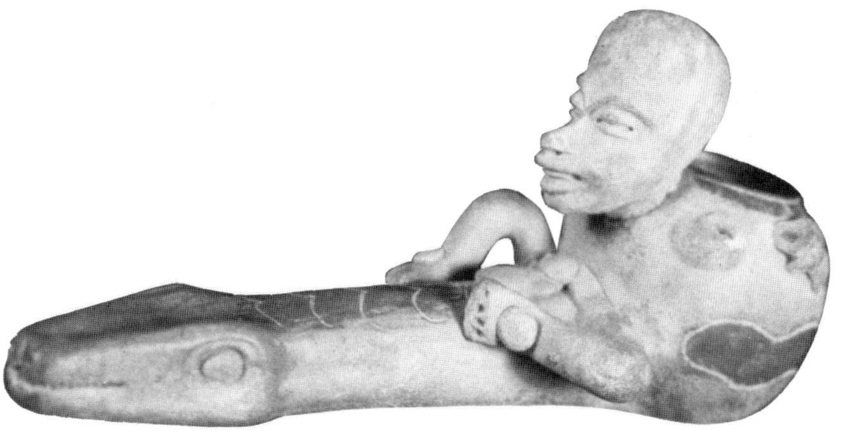

Figure 22. Pipe said to be found in the Quiché area. From the style and pigmentation of the object it appears to be of Mexican origin. Thermoluminescence studies indicate that it was made before the third century A.D.

"whether they were for smoking is not demonstrable because of incompleteness of the specimen."[32] While these tubes may indeed have been used as pipestems and may thus have served as a transition between the elbow pipe and the reed cigarette, it is more probable that they were used for some purpose other than smoking.

A "Maya pipe" was also found in San Salvador. It is said to have been excavated by Silvanus Miller, a mining engineer at the Flamengo gold mine in 1882.[33] It was described as an elbow pipe made of slate and carved with three faces. The unusualness and completely "out-of-context" nature of this find makes its Maya provenance highly doubtful.[34]

An important bona fide pipe found in the Maya realm is a long-stemmed ceramic object discovered at Chichén Itzá.[35] While George W. Brainerd identified it as a pipe, Thompson believed that it may be either a pipe or an incense burner.[36] The object clearly is of a recognizable type manufactured in Michoacán, western Mexico, an area known for its long-stemmed pipes, and was probably imported to Yucatán by Toltec warriors, thus representing a "record for long-distance trade in ancient pottery"—fifteen hundred kilometers.[37]

A very interesting find relating to the smoking habits of the Mayas is a group of three pipes found in the Guatemalan highlands near Quiché. Each of the pipes, which are now scattered in different private collections, is painted brown on a white base. The bowl is formed by the crouching figure of a man, the stem by the straight body of a serpent. Both the style and the painting technique are so definitely "un-Maya" that their origin is still uncertain.

A Maya archaeological object that may relate to pipe smoking is a polychrome plate, probably from Petén, portraying an anthropomorphic jaguar holding a long tubular object (pipe? blowgun?) to its mouth and blowing two curlicues, perhaps of smoke (see Plate 17). The identification of the tube as a pipe is tempting but not certain.

Though the Spaniards found no pipes, several different types of cigars are mentioned in the early manuscripts, some of them made entirely of tobacco, others rolled in leaves of other plants.[38] Perhaps the earliest description of how the Mayas made such cigars is included in the Valenzuela Manuscript:

> There are in all [the houses] forms for fabricating cigars or *puquietes* that are some little sticks of wood, strong [and] the thickness of a finger, diminishing in dimension from the head below, or from one to the other extremity. And in these they fabricate said *puquietes* with leaves of nance and covered with different colors and from said forms they pass them to another portion of clay like sugar mold. And in it they introduce a like hundred in which they place them, and put the cigars in order to dry them. And there were so many there that we found that also I understand [that] they are not only for their use but also for selling.[39]

An early Yucatec almanac describes another method of Maya cigar making:

> These tubers [of the *cocom plant*] were cut in narrow strips, which were carbonized on a hot griddle. The material was pulverized and placed in a gourd or vessel. *Atole de comote* [sweet potato and corn gruel] and a little honey were added so that the whole formed a thick paste. The cigar maker then took leaves of zapote [*Achras zapota*] or, were he a resident of Chichancama, allspice [*Pimenta officinalis*] and split them lengthwise. Each half-leaf was then rolled on a small spindle-shaped stick about the thickness of a turkey quill, and tied with henequen thread. The surface was then coated with *cocom* paste, except an inch at the thinner end which was left uncoated to serve as a mouthpiece. After the whole was dried in the sun, the stick was pulled out and regular cigar tobacco was inserted in its place. The end was doubled over to prevent the tobacco from spilling, and the cigar was ready for enjoyment.[40]

Today the Chorti Maya cigar maker, instead of using *cocom* and spices, sprinkles each leaf with *aguardiente*, a strong alcoholic drink, or dips it in molasses mixed with essence of anise. The veins of the leaves are then cut out, and the leaves are halved. The fine parts of the leaves are used as wrappers, and the other parts as fillers. The whole is rolled up and then wrapped in a single fine leaf to make a finished cigar (*puro*). The ends of the *puro* are glued

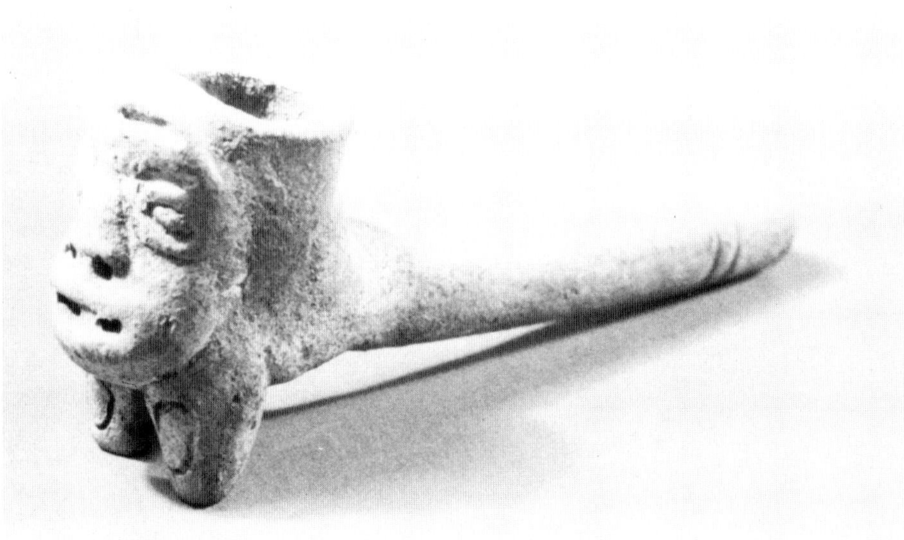

Figure 23. Modern Maya clay pipe from the Guatemalan highlands (Collection of Bertoldo Nasthusius)

together with the gummy milk of the *yupay* fruit. The cigars are tied in bundles and usually taken to one of the large weekly open markets for sale.[41] The Quiché, like his Chorti brother, is also an ardent cigar smoker. The Quichés may have originated the "smoke-filled rooms" of politics, for their councils "were illuminated by fat-pine torches and accompanied by fat cigars."[42]

As far as the cigarette is concerned, it would be a tempting evolutionary scheme to suggest that in ancient Central America, just as in Europe, the earlier method of tobacco use was in the form of the cigar and that from it later evolved the cigarette. Archaeological evidence, however, indicates that the Mayas may have smoked cigarettes even during the Classic times. While God L on the Classic relief at Palenque is indeed puffing on what appears to be a large funnel-shaped cigar and cigars often appear also on ceramic vessels, most of the lords portrayed on painted ceramics from the early Classic epoch are holding or smoking slender white cigarettes. It is probable that this aboriginal cigarette was made with a corn-husk wrapper (*cigarillo de tusa*), which is still used throughout Mesoamerica. It usually contains much less tobacco than the modern commercial variety. The Chorti Maya makes these cigarettes by fine cutting the tobacco leaves with a sharp machete or with a pair of scissors and then wrapping them in small sections of corn shucks. They seldom smoke the cigarettes themselves but usually sell them in bundles to the Ladinos (Indians or mixed bloods who adopted the Spanish way of life).[43] Another variety of aboriginal cigarette is the so-called cane cigarette that was popular not only in the Maya realm but also in the Aztec area.[44] It is made by stuffing a short piece of reed or cane with powdered tobacco. A band of fabric or string is tied to it, with a flap left free by which it is held.[45] The cane cigarette was regarded by some commentators as an intermediate form between the crude cigar and the pipe, but it is probably an individual development of one particular area.[46]

Nowadays the Mayas' smoking habits are changing rapidly. More and more of them, especially those living in and around cities, smoke commercial cigarettes, and some members of the highland nations also smoke pipes. Some of the village-dwelling Chortis, for example, now use pipes that are made to fit a bit of cigar, for the tobacco is generally bought in that form. The pipes are of the common elbow variety, the bowl and cane being made of one piece. They are either carried in a shoulder bag or suspended on a twine around the neck. New pipes are usually buried in damp soil for a week or so to give them a slightly brownish color or are rubbed with the root of *camotillo* and lime to stain them red.[47]

Consuming Tobacco in Ways Other Than Smoking

At the time of the Conquest the Mayas were found to enjoy tobacco not only by smoking but also by chewing. Some believe that this may have been a relatively recent habit, an element like the introduction of the habit of sleeping in hammocks and the growing of bitter manioc and the pineapple, which appears to have begun in the Maya realm shortly before the arrival of the Spaniards.[48] According to legends still told in the Huehuetenango area, however, tobacco chewing was a custom among the ancient Maya shamans.[49] They usually mixed lime with their chewing tobacco, a custom that may be a transmutation of the Andean custom of mixing lime with coca.[50] Coca chewing on the Mesoamerican subcontinent extended as high as the lakes of Nicaragua. The lime, which chemically reacts with both tobacco and coca by releasing their alkaloid content, was usually obtained by burning seashells and was carried in pear-shaped gourds.[51] This peculiarly American Indian custom of chewing tobacco or coca mixed with lime could be compared with similar practices in India and neighboring countries where betel nut is chewed.[52]

The first record of tobacco chewing in America comes from the man for whom the continents of the Western Hemisphere are named, the explorer Amerigo Vespucci. In 1499, on Margarita Island, off the coast of Venezuela, he saw the Indians chewing a "green herb like cattle to such an extent that they scarcely talk."[53] The habit was soon widely adopted by European sailors, chiefly because it enabled them to use tobacco without the fire hazard of smoking at sea.[54]

Chewing, licking, and even drinking tobacco were sometimes practiced in conjunction with, and sometimes instead of, smoking among the Tzeltal, Tzotzil, and Jacalteca Mayas.[55] The Tzeltal carries his chew, made of powdered green tobacco mixed with lime and chili, which he calls *mai*, in a little gourd. He dips it with a small stick and licks or chews it to "strengthen his teeth" and also as a stimulant to help him withstand the fatigue of the road.[56] An interesting account of tobacco chewing among the ancient Mayas is recorded in the early Spanish manuscript the *Descripción de San Bartolomé*:

> [They] dry it [tobacco], grind it, mix it with a little lime, ... bring it to their mouth because apparently it gives them strength and cures their thirst. They use it in their homes, and also when they are walking or working in their cornfield. They also have it on their person and in their mouth. It is also frequently used to defend themselves against attacks of a snake or other poisonous animal. They also make with it *cañutillos* [cigarettes?] that they light up

and smoke. They found these very healthy against any kind of disease.[57]

Eating or drinking tobacco in the form of a syrupy juice may be another of the early methods of consuming the herb. This brew was produced by steeping or boiling the leaves and was taken either by mouth or through the nostrils. By this ingestion the active ingredients were absorbed more quickly.

Drinking tobacco juice was popular not only in America but also in medieval Europe. On the Noordermarkt in Amsterdam the following rhyme was found on the signboard of one of the early tobacco shops:

You exalted tobacco drinkers
Who drink tobacco till the end of your lives
And who want to buy tobacco and pay for it:
On Sundays, you will get the pipe as a reward.[58]

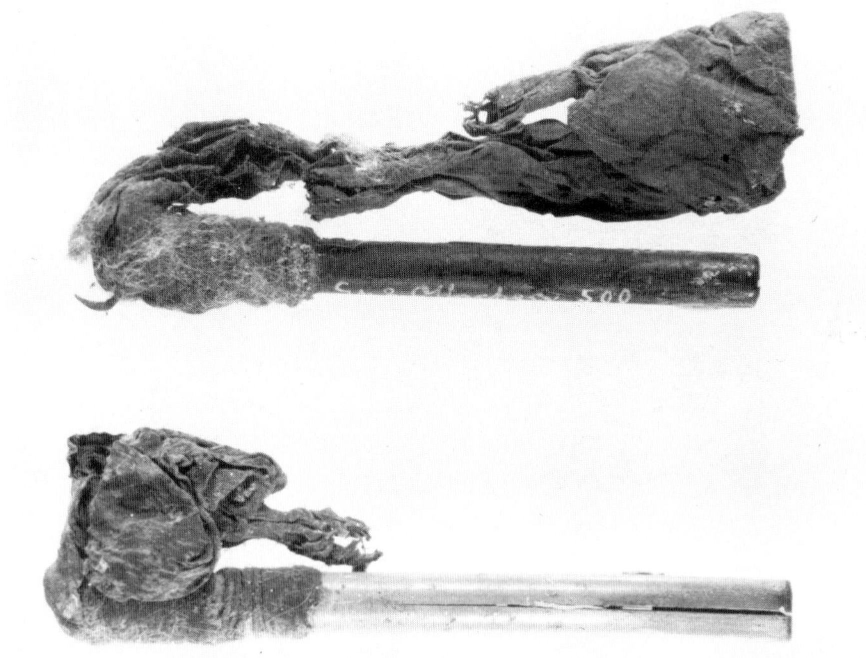

Figure 24. Pre-hispanic enema syringes found at Olleache, Peru (Ethnographic Museum of Stockholm; after Wassen)

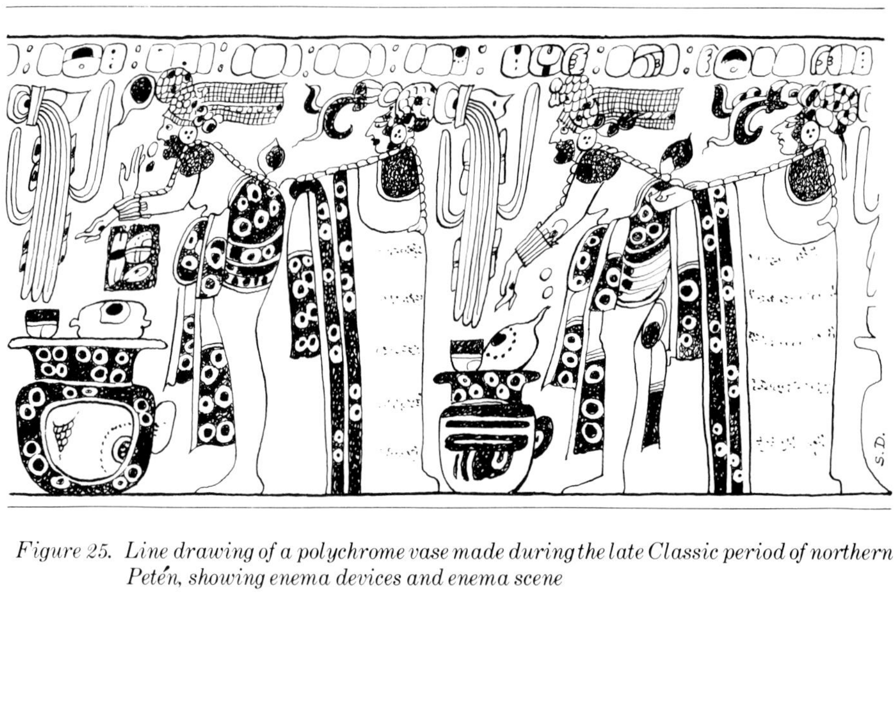

Figure 25. Line drawing of a polychrome vase made during the late Classic period of northern Petén, showing enema devices and enema scene

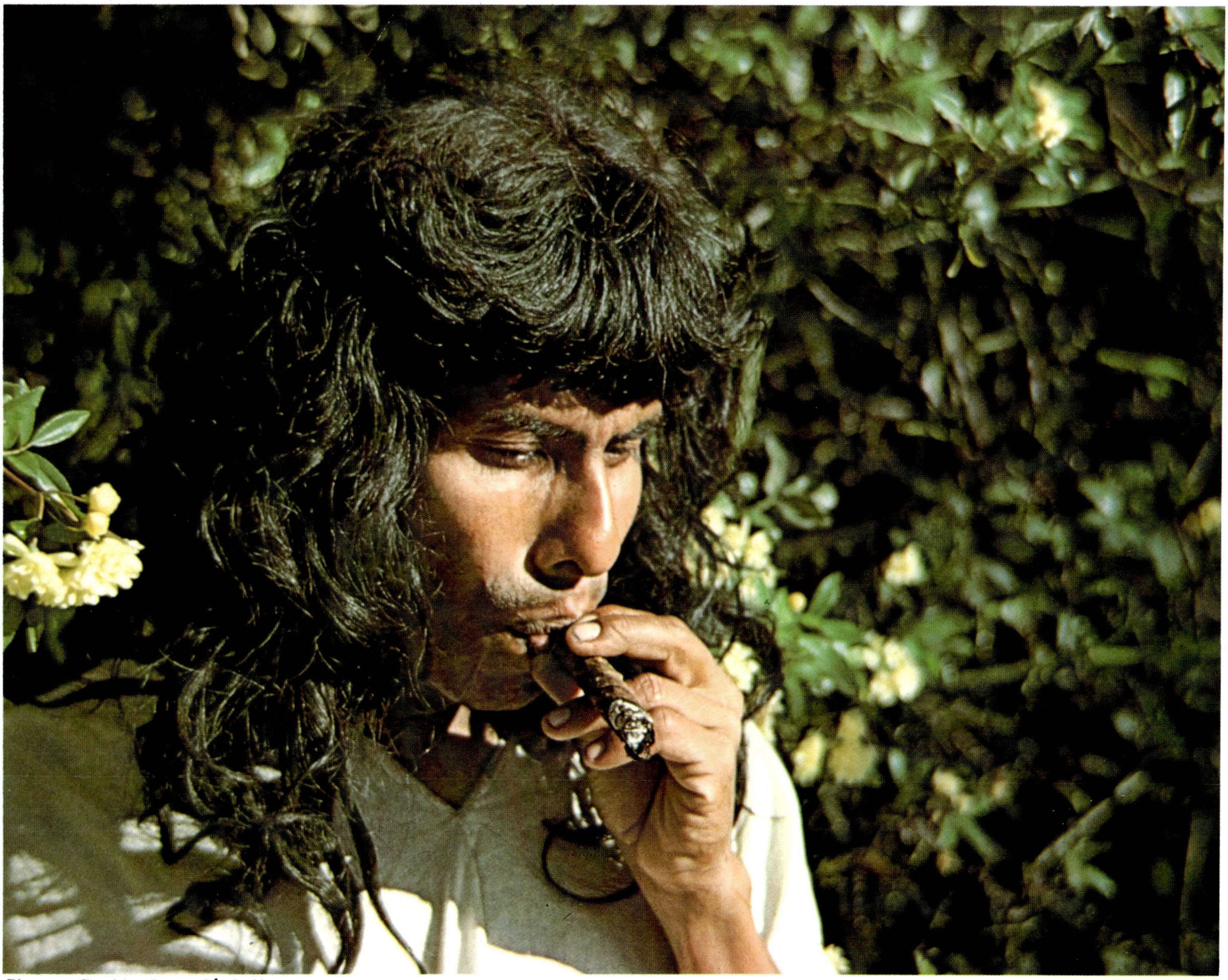

Plate 1. Smoking Lacandón Indian

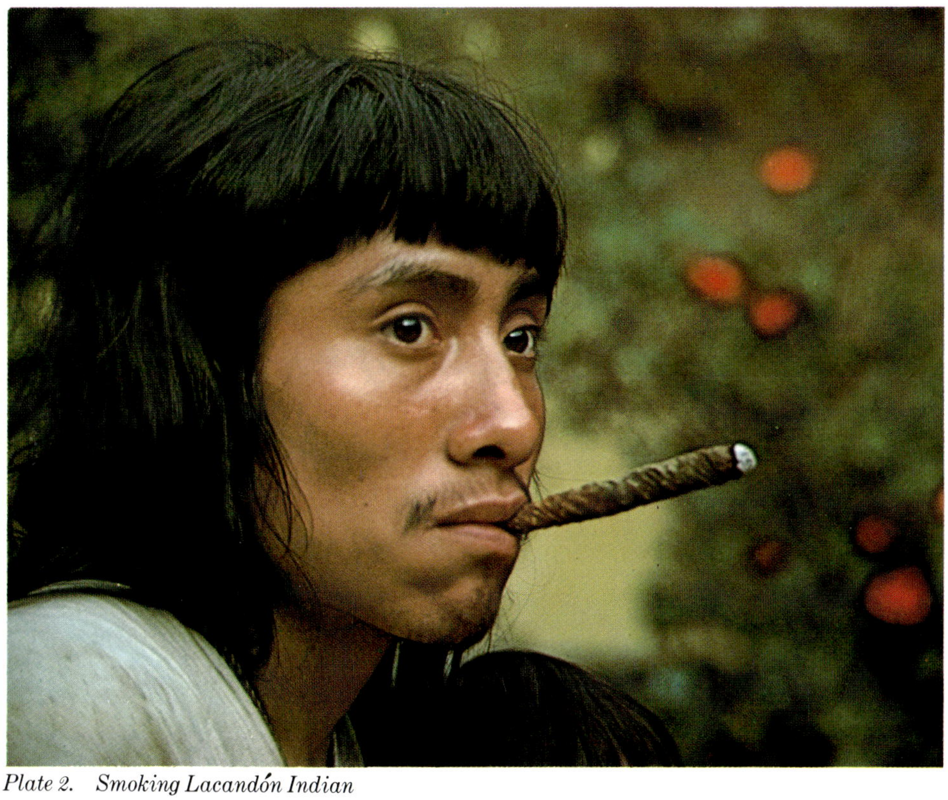

Plate 2. Smoking Lacandón Indian

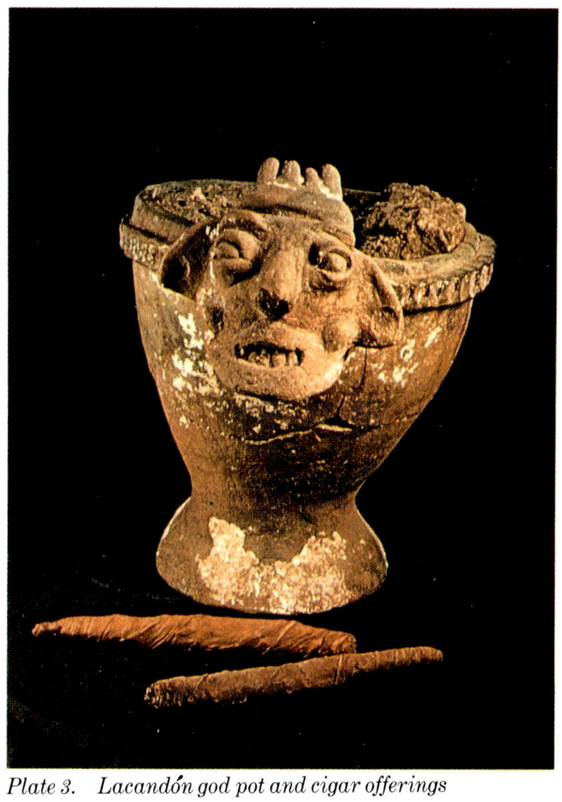

Plate 3. Lacandón god pot and cigar offerings

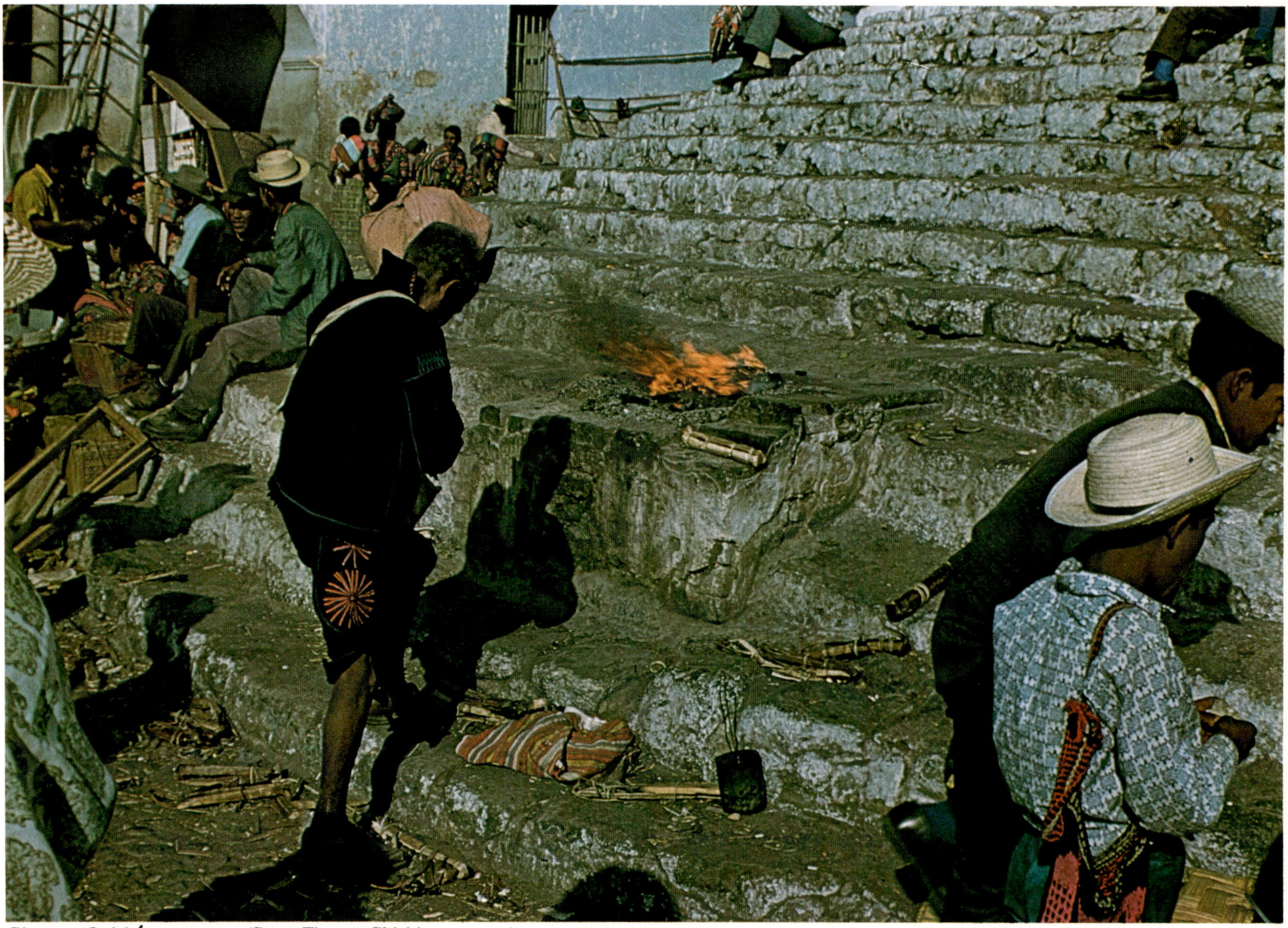

Plate 4. Quiché prayer-man (Santo Thomas Chichicastenango)

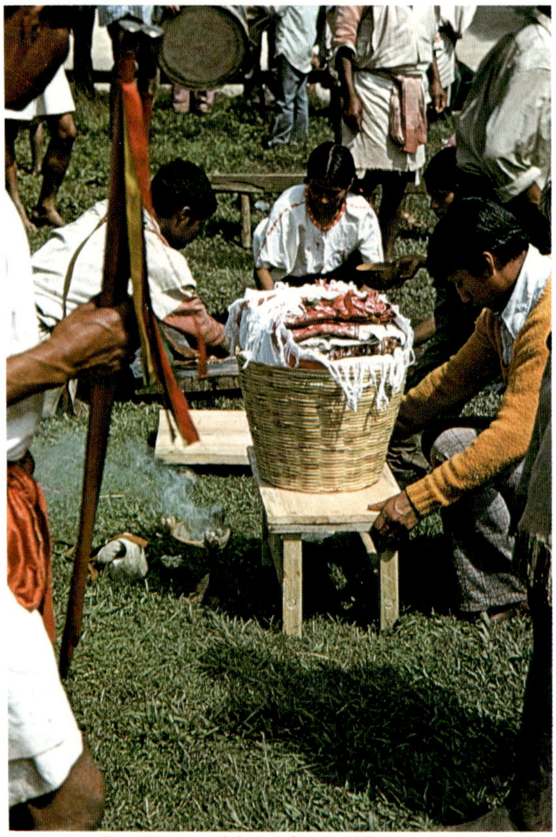

Plate 5. Bundle worship at Santa Marta, Chiapas

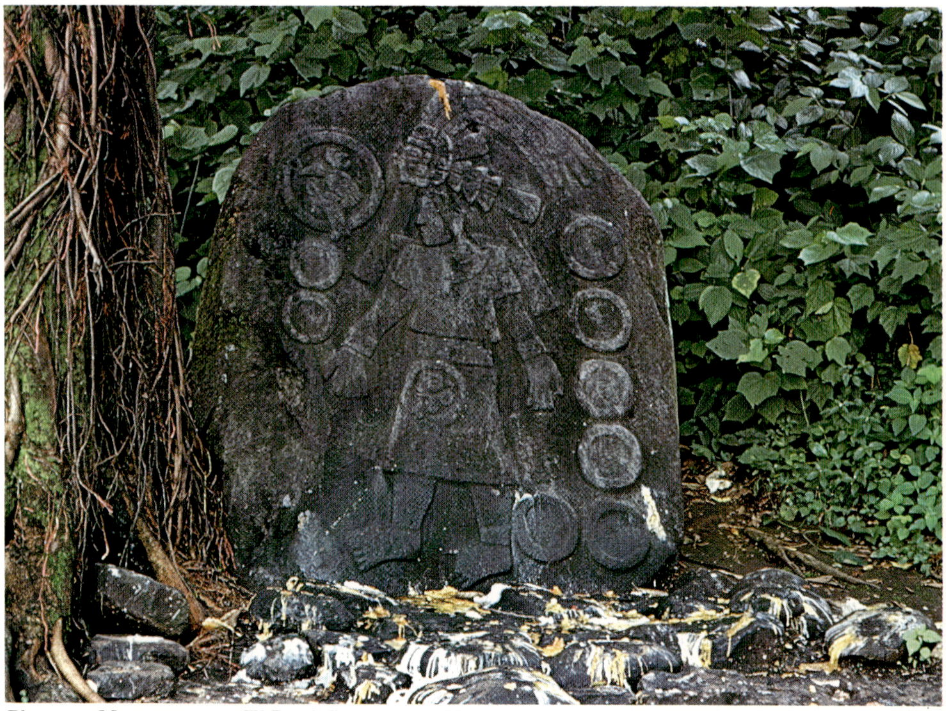

Plate 6. Monument 2 at El Baul, Guatemala. The stela, which is still an object of worship among the local Indian population, stands on a ground covered with candle droppings, offerings of corn leaves and cigars.

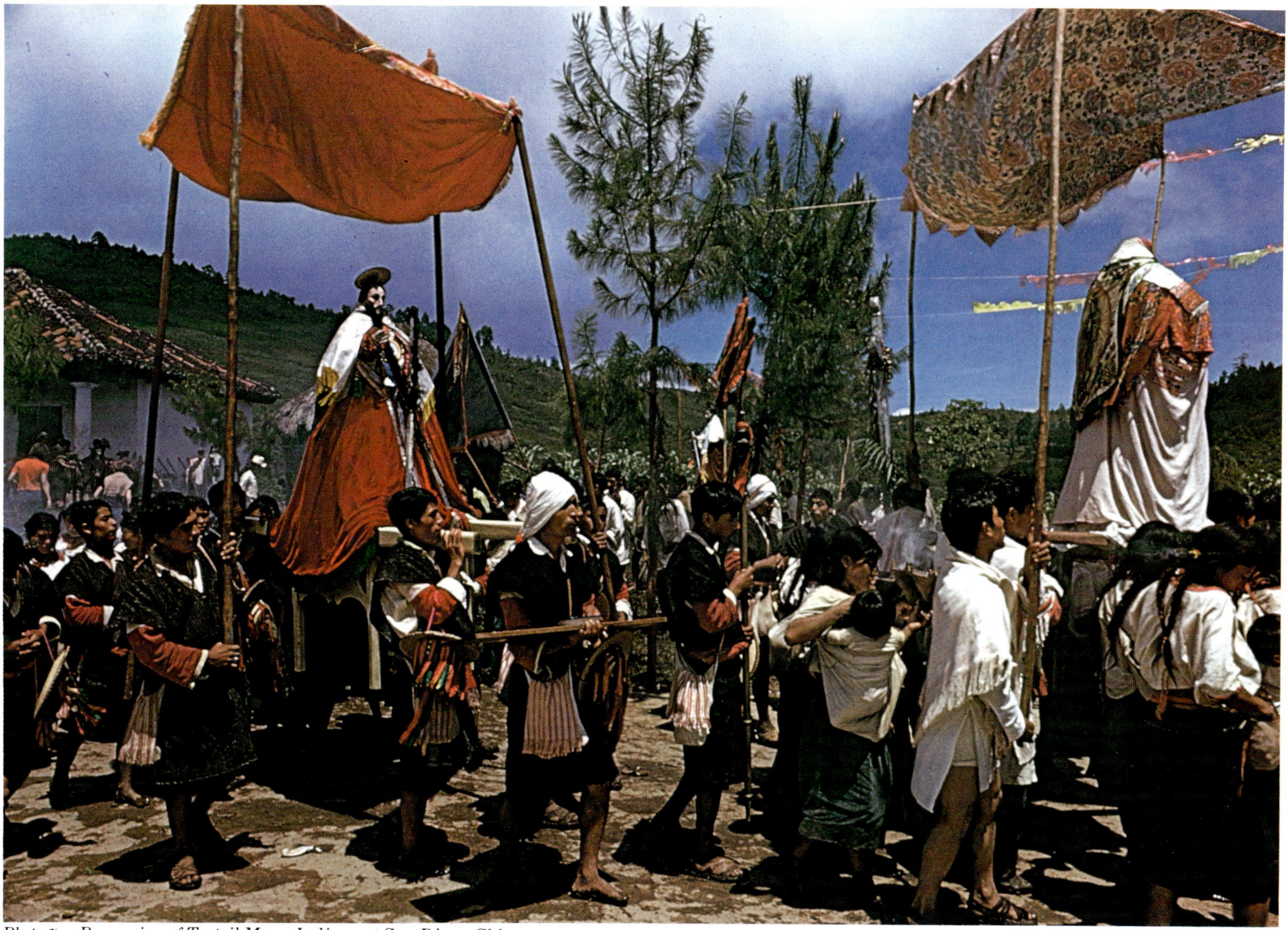

Plate 7. Procession of Tzotzil Maya Indians at San Diego, Chiapas

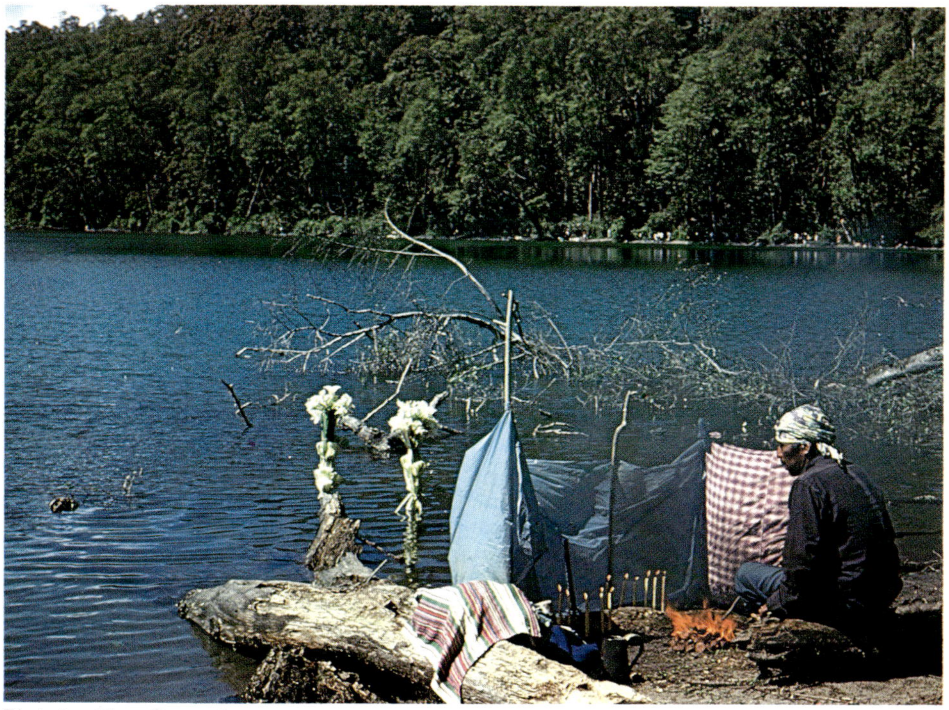

Plate 8. Mam Curandero

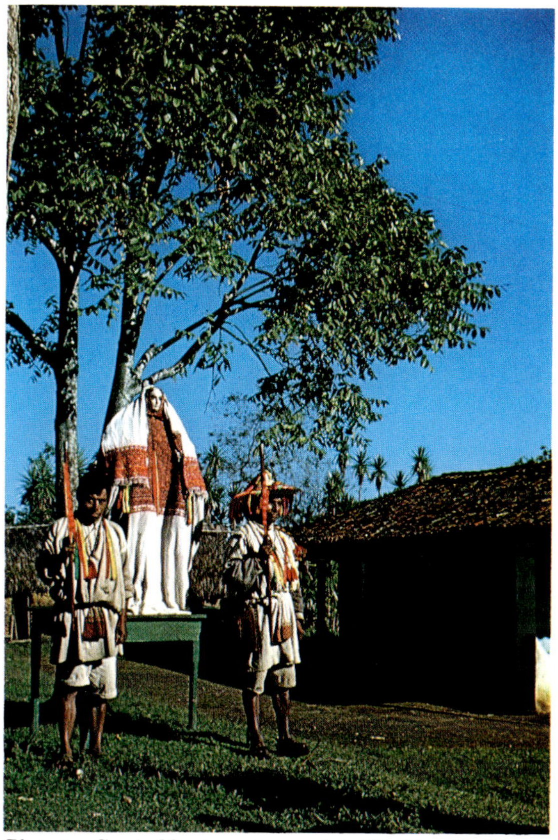

Plate 9. Santa Marta, patron saint of the Tzotzil village

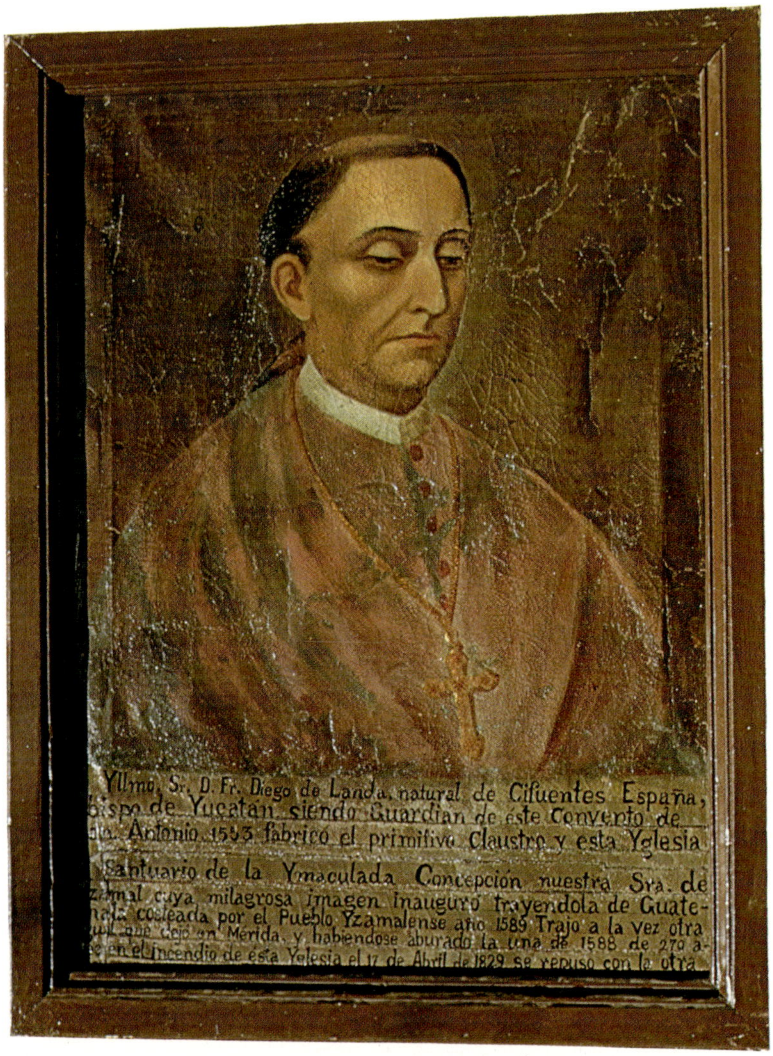

Plate 10. Don Diego de Landa, first bishop of Yucatán (Painting in the sacristy of the Cathedral of Izabal)

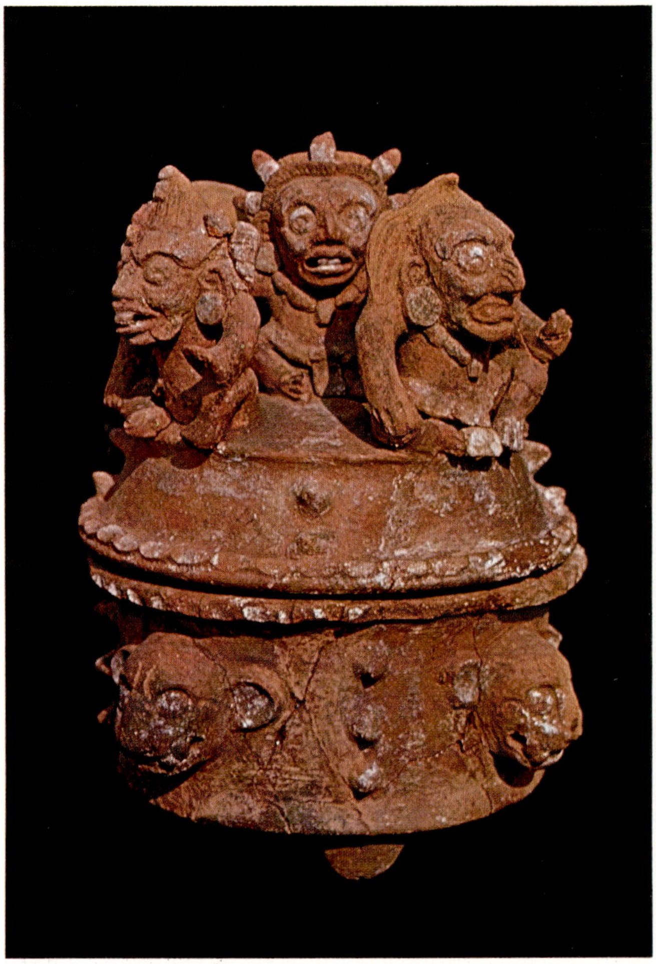

Plate 11. Lidded container from the Guatemalan Highlands with deities facing the four directions

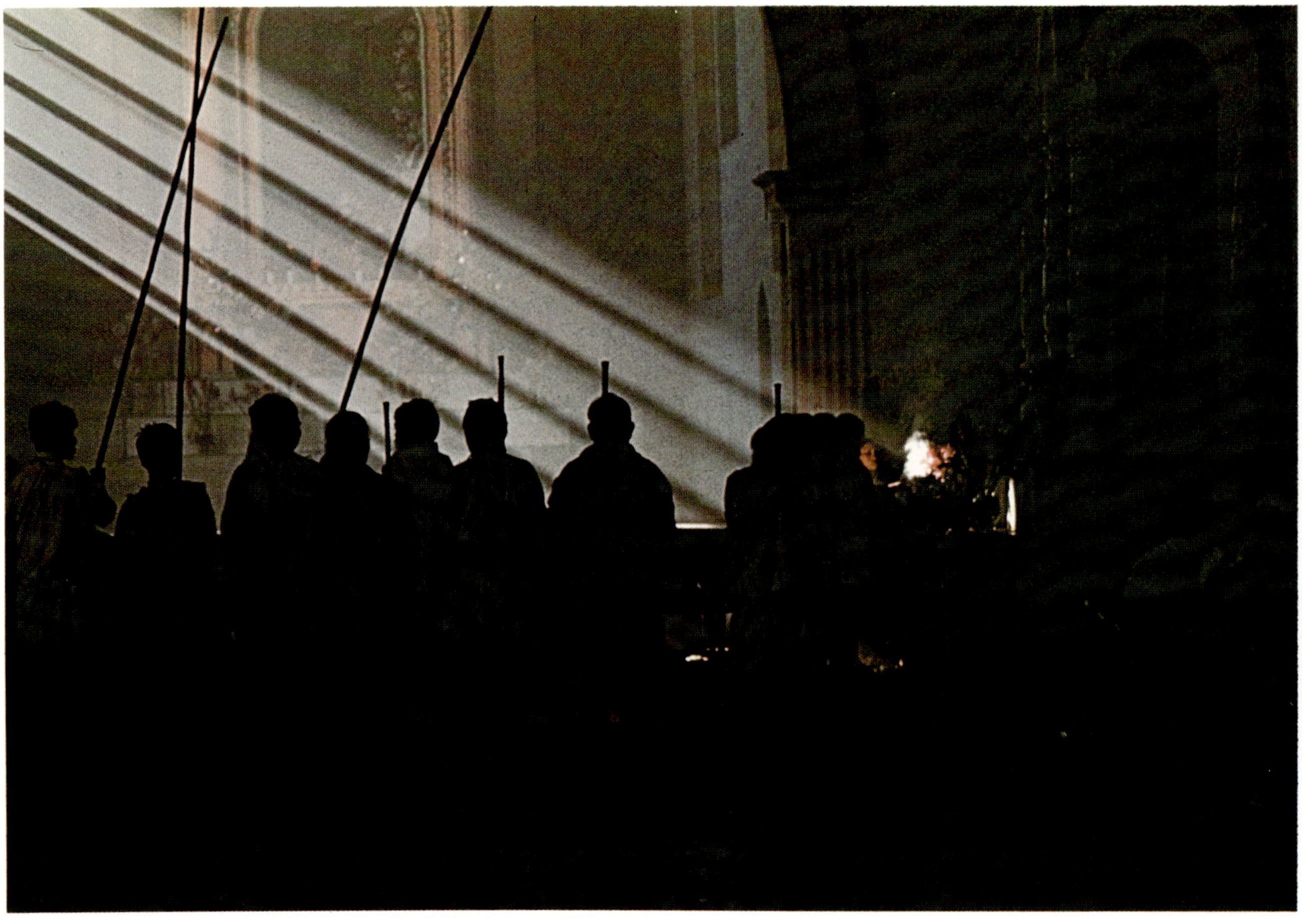

Plate 12. Morning worship in the church of Magdalena, Chiapas

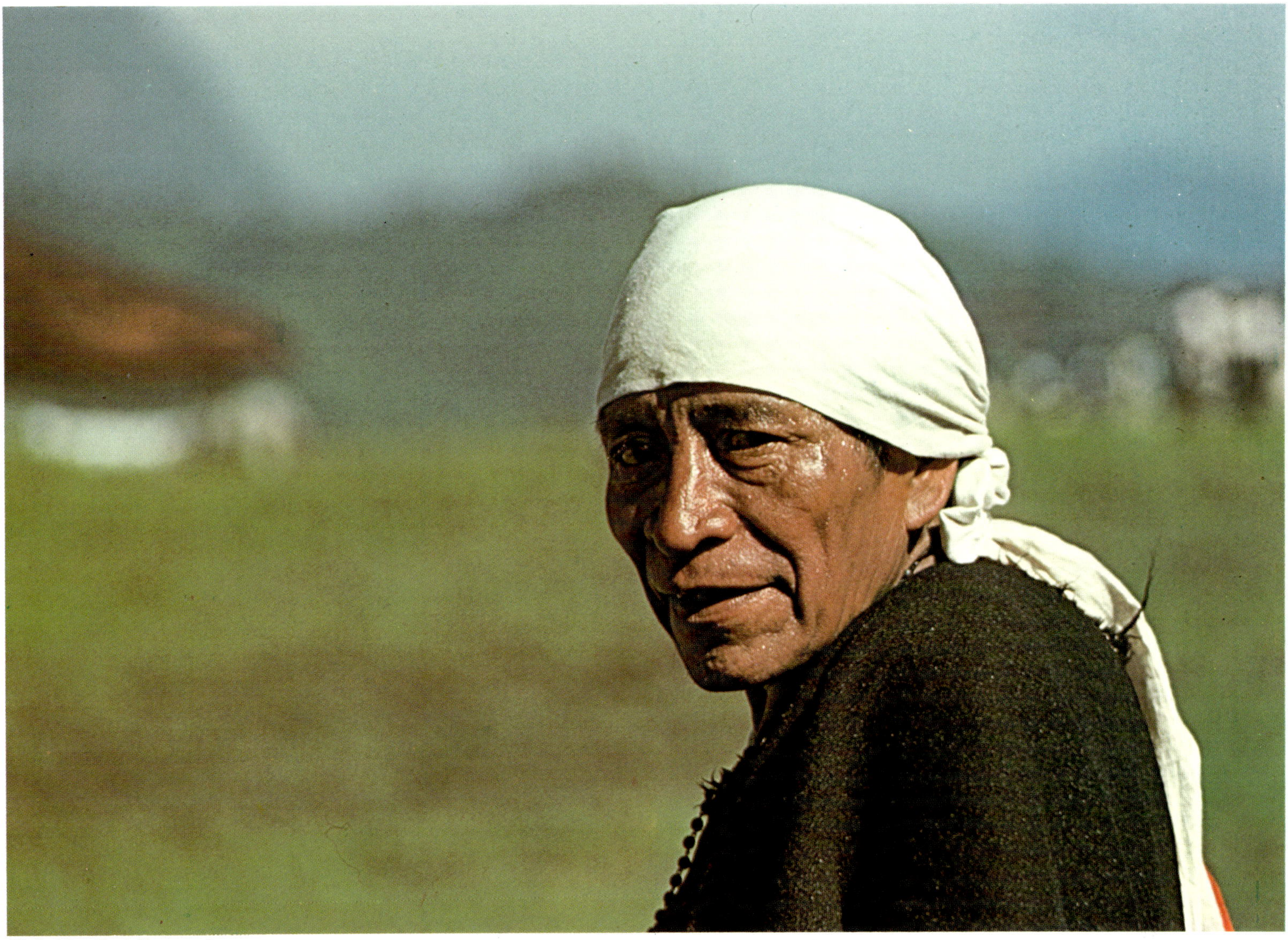
Plate 13. Tzotzil curandero

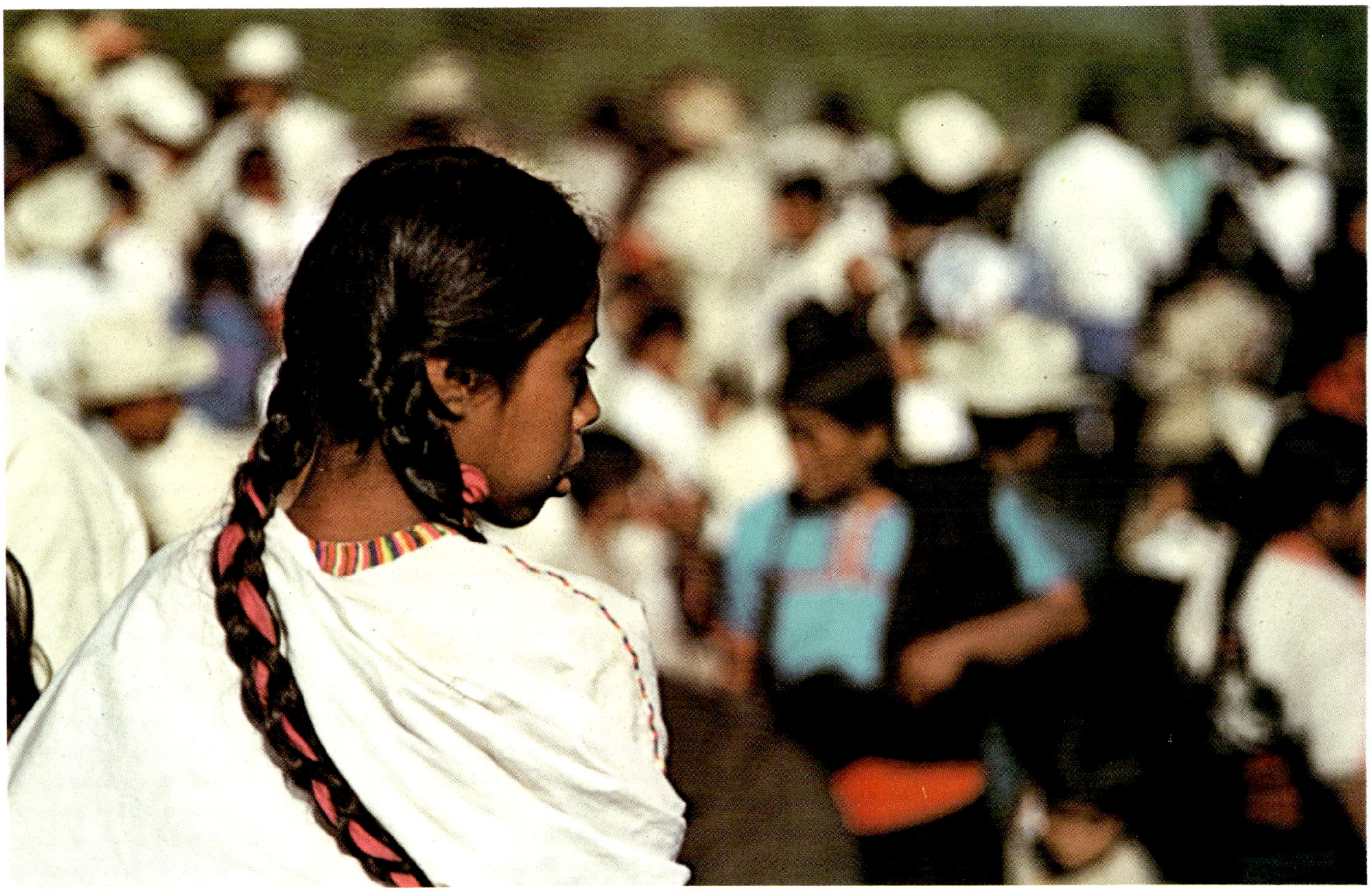
Plate 14. Tzotzil girl

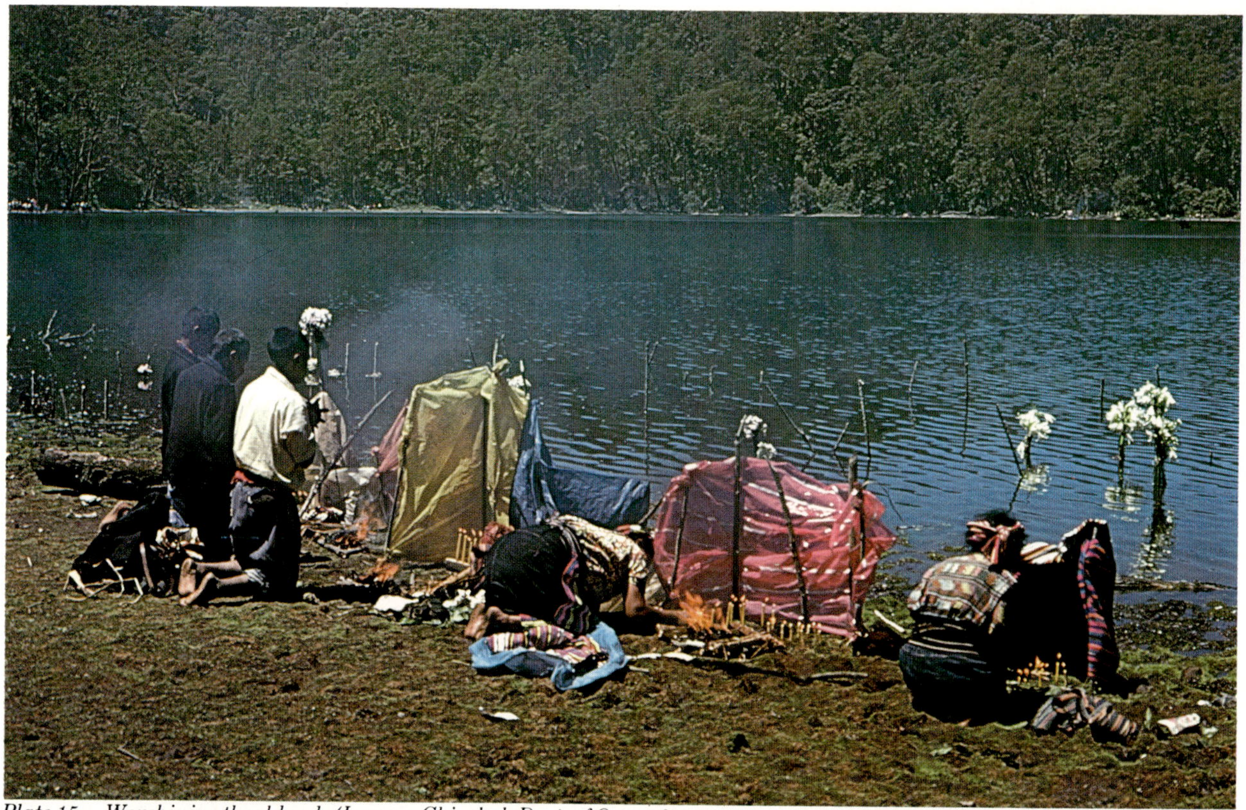

Plate 15. Worshiping the old gods (Laguna Chicabal, Dept. of Quetzaltenango)

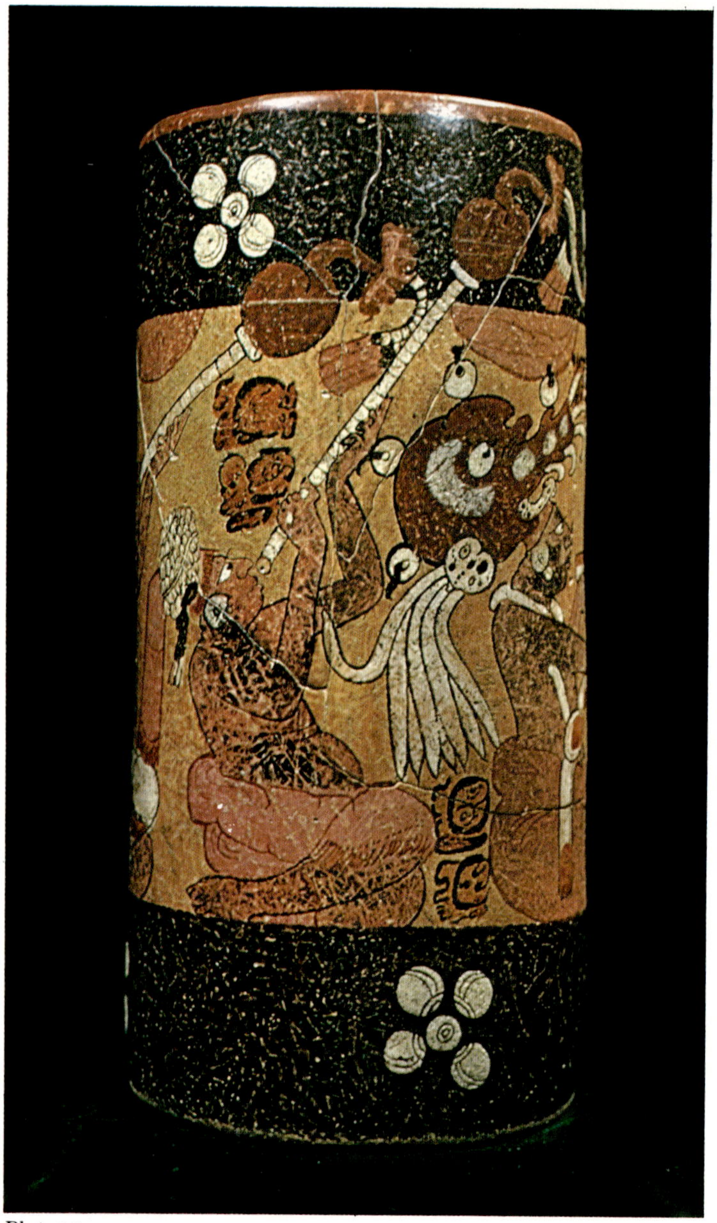

Plate 16.

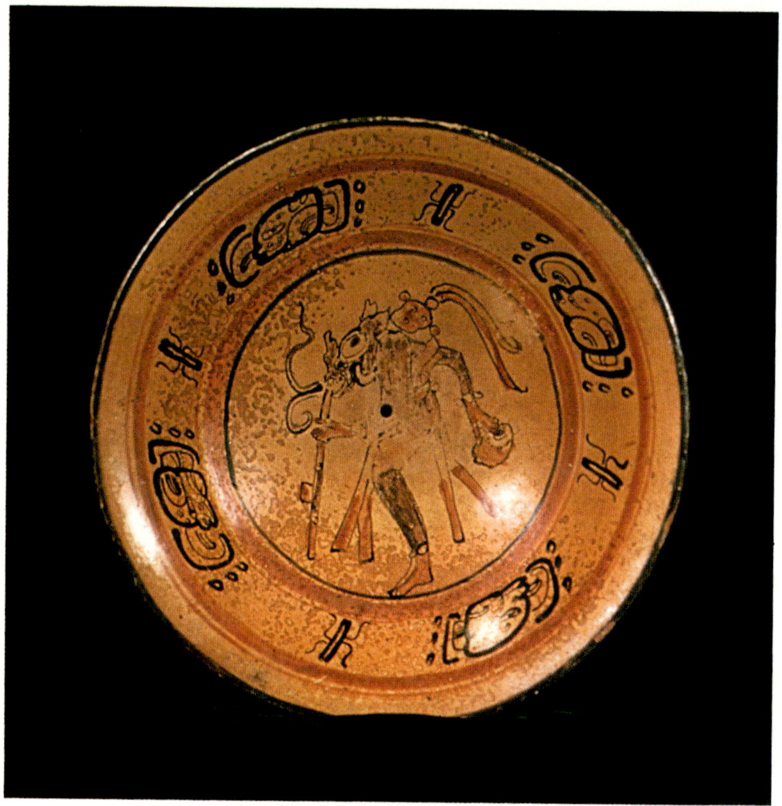

Plate 17.

Plate 16. Polychrome vase painted with seated personages holding long tubular objects, probably trumpets (Late Classic Period, Petén, private collection)

Plate 17. Polychrome plate with an anthropomorphic jaguar holding a long tube, probably a blowgun or a pipe (Classic Period, Petén, private collection.)

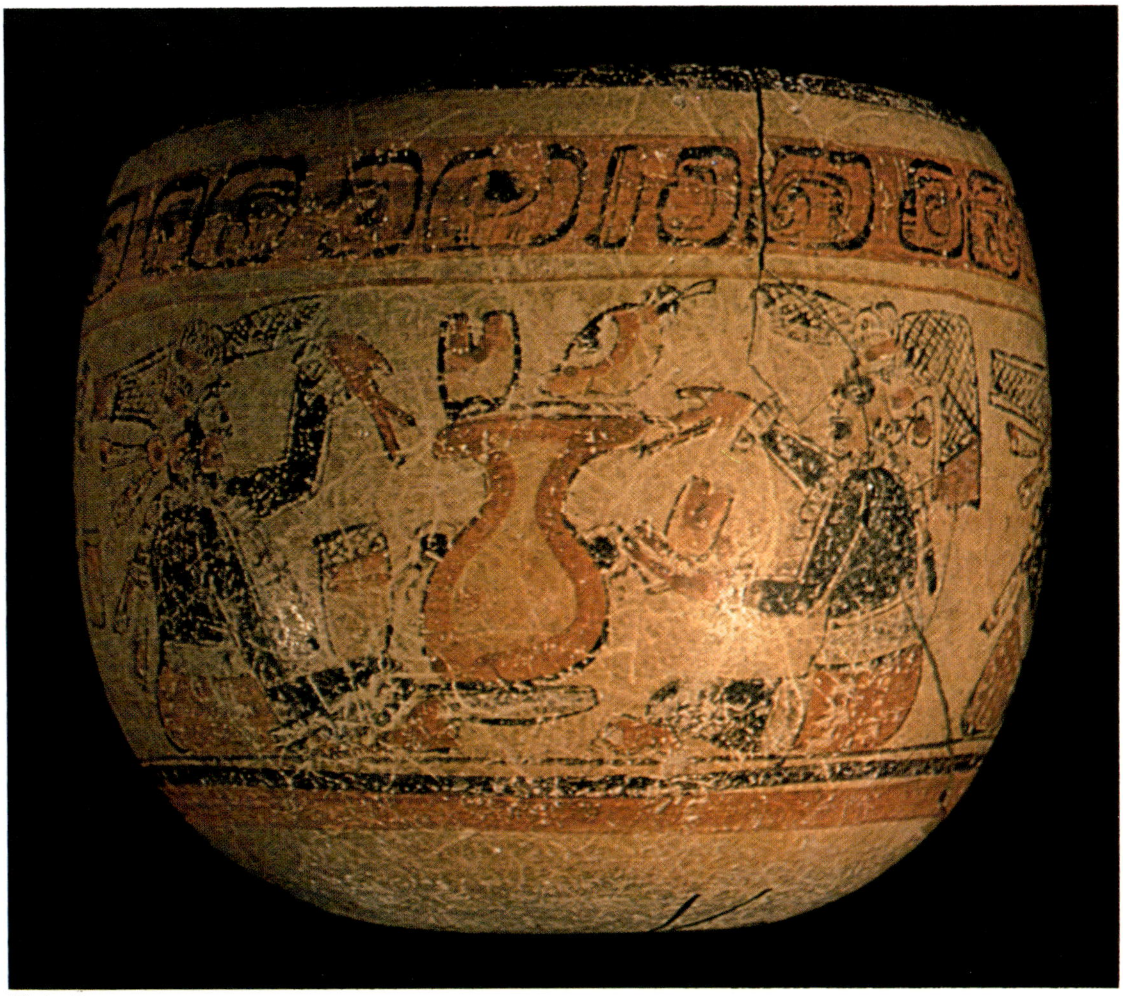

Plate 18.

Plate 18. *Polychrome bowl from the Late Classic Period of Petén with seated figures and enema paraphernalia (Museo de Popol Vuh, Ela and Jorge Castillo collection.)*

Plate 19. *Polychrome vase with an (hallucinogenic?) enema scene (Late Classic Period, Petén, private collection)*

Plate 20. *Man receiving (hallucinogenic?) clyster (Classic Period, Guatemalan Highlands, November Collection)*

Plate 21. *Polychrome bowl depicting self-administered clyster (Guatemalan Highlands, Late Classic Period, private collection)*

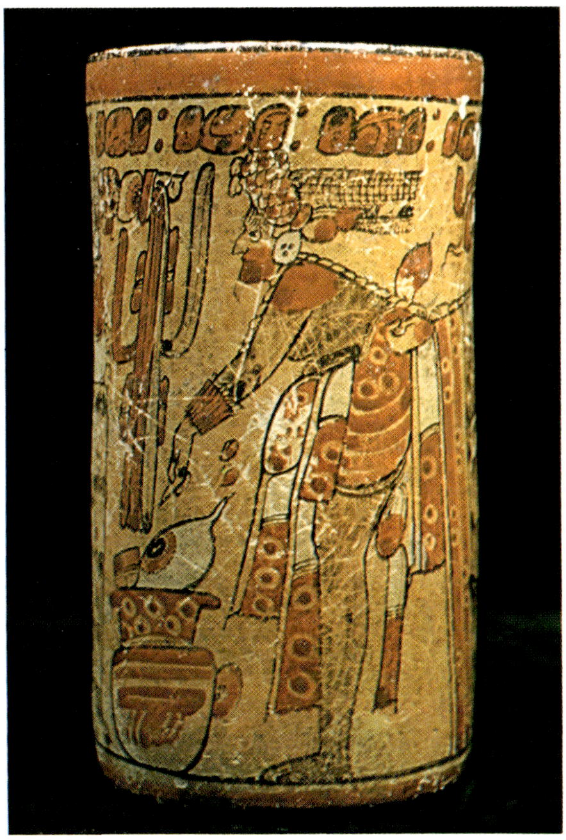

Plate 19.

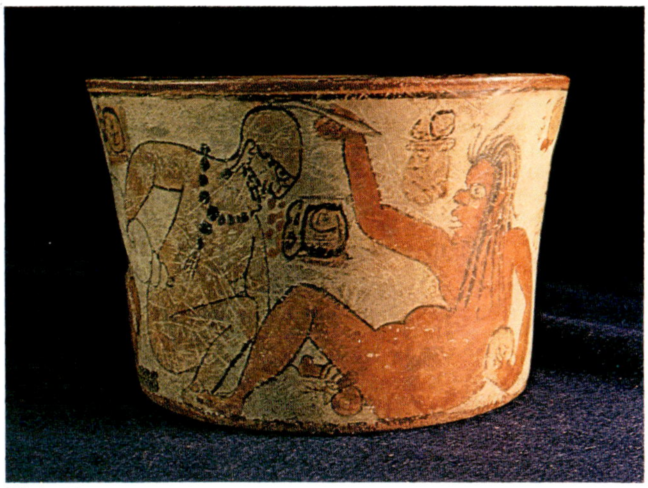

Plate 20.

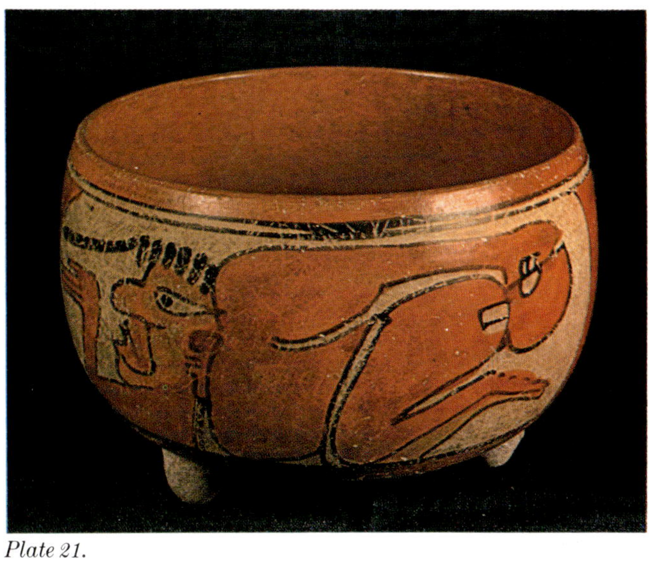

Plate 21.

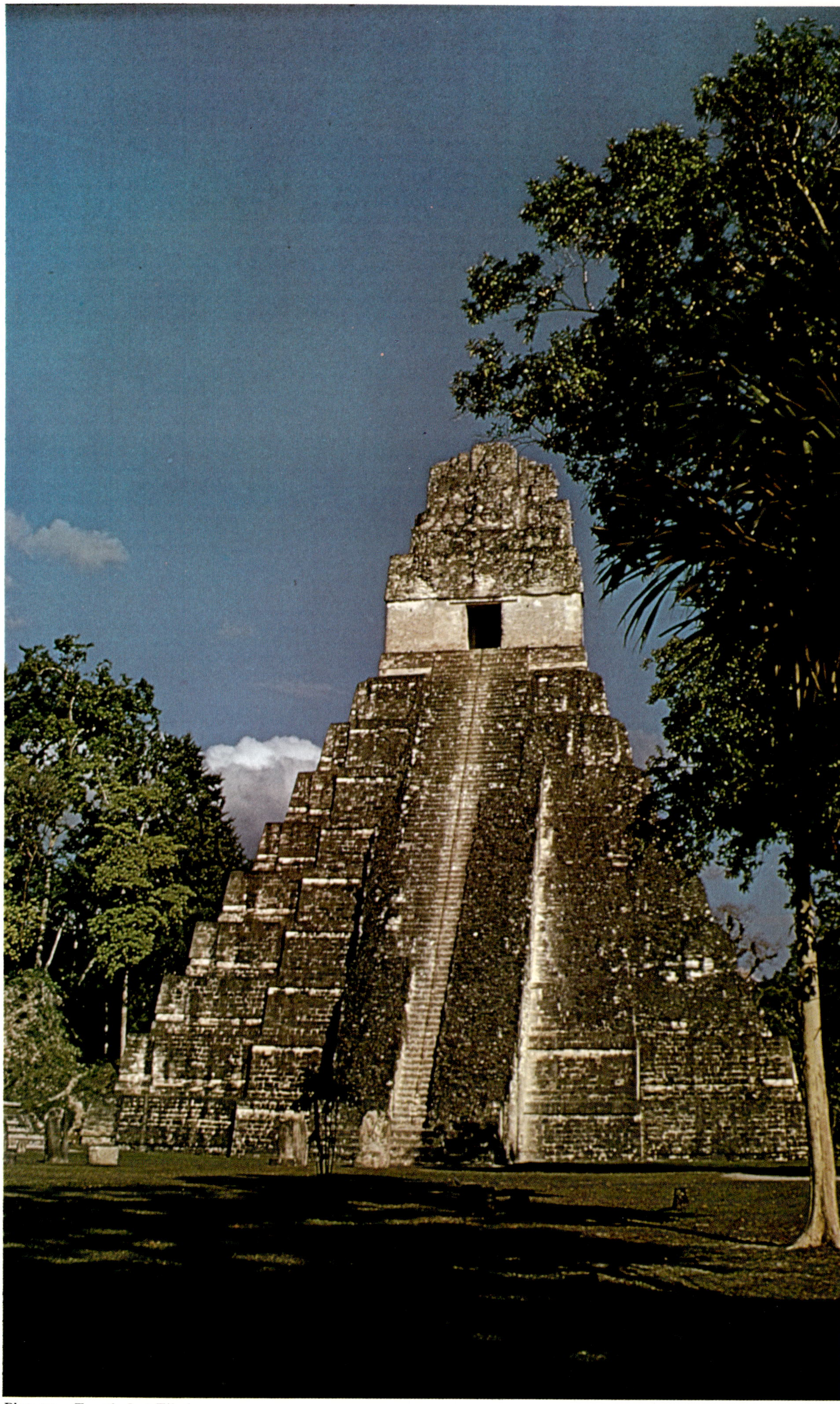

Plate 22. Temple I at Tikal

An even more unusual use of tobacco among the aboriginal peoples of America is mentioned by Peter T. Furst in *Hallucinogens and Culture.* He and Michael D. Coe quoted sources claiming that several aboriginal American cultures, such as the Tihuanaco, the Inca, and the Aztec, used rectal injections of different herbal extracts to induce intoxication.[59] Karsten describes the administration of herbal enemas among the Jivaros: ". . . this strong medicine was injected into the rectum of the child by means of an enema syringe consisting of a hen's [*sic*] bladder attached to a piece of the stem of the *sanahoria* plant [*Daucus*], which is hollow. Syringes of this kind are commonly used by the Indians."[60] Furst also commented on herbal clysters:

> It has only recently come to light that the ancient Maya, too, employed enemas. Enema syringes or narcotic clysters, and even enema rituals, were discovered to be represented in Maya art, an outstanding example being a large painted vase dating A.D. 600–800, on which a man is depicted carrying an enema syringe, applying an enema to himself, and having a woman applying it to him. As a result of this newly discovered scene, archaeologist M. D. Coe was able to identify a curious object held by a jaguar deity on another painted Maya vessel as an enema syringe. If the enemas of the ancient Maya were like those of Peruvian Indians, intoxicating or hallucinogenic, they might have been compounds of fermented *balché,* (honey mead), itself a very sacred beverage, fortified with tobacco or with morning glory seed infusions. Of course they could have been a tobacco infusion alone.[61]

Rectal administration of tobacco is also mentioned in the *Relación de Texcoco* and in the Cruz and Badianus Manuscript and has recently been discussed by Coe and Furst.[62]

The custom of hallucinogenic enemas was also followed in Mesoamerica. It is described in the records of one of Cortes' companions, who wrote of the Huaztecs, a Mesoamerican tribe of Maya linguistic stock:

> [The Huaztecs] worship figures showing different forms of pleasure between a man and a woman and figures of human beings with their legs lifted different ways. . . . the men are great sodomites, cowards, and—bored drinking wine with their mouths—lie down and, extending their legs, have the wine poured into their anus through a tube until the body is full.[63]

Those who find the Maya tobacco enemas bizarre or unbelievable should read the treatise *Historiarum anatomicarum et medicarum rariorum,* published in Copenhagen in 1661, in which the noted anatomist Thomas Bartholin recommeded not only tobacco-juice but also tobacco-smoke enemas to his unfortunate Danish patients. The beginning of the chapter "The Tobacco Clyster and a New Instrument for It" reads:

> Those who have swallowed tobacco by accident can testify to its purgative effect. This property is employed in the tobacco clyster used as an enema. My beloved brother, Erasmus, has shown me the method. Smoke from two pipes (filled with tobacco) is blown into the intestines. A suitable instrument for this was devised by the ingenious English.[64]

The tobacco enema as a panacea for various illnesses showed up again and again in medical literature over the next two hundred years. On February 25, 1769, Jacob Bicker Raye, of Amsterdam, wrote in his diary how a woman who fell into one of the canals at Flueweelenburgwal was resuscitated with a tobacco-smoke enema: "They used a new way of administering enemas from behind with a tobacco-smoke syringe, warming her as much as possible, and as a result she recovered completely within two hours." Still another unusual application was recorded by a French physician named Buc'hoz, who used intravaginal insufflation of tobacco smoke to cure hysteria.[65]

Tobacco Smoking in Glyphic Texts

While tobacco and tobacco smoking are mentioned very often in post-Conquest Maya manuscripts, we know far less about ancient Maya ideograms that may refer to tobacco, the act of smoking, or associated subjects such as fire or smoke. Writing of the hieroglyph for "fire," David Kelley, in "Maya Fire Glyphs," ably summarized the data accumulated during the past century.[66] In 1902, Eduard Seler identified the glyphic compound T122:563a as the sign for "fire."[67] Thirty-five years later Beyer identified glyph T563a as the "firewood bundle" and the T122 affix as the flames arising from the burning wood.[68] Thompson concurred in this identification in his later works.[69] Kelley does not accept the first part of the compound T122 as "flames" but considers it the ideograph for "smoke." He also identifies glyph T1035 as the head variant for the compound T122:563a.[70] In spite of these efforts, however, there is still no completely satisfactory reading of the fire glyph, because T563a appears in a number of compounds that apparently do not refer to flames at all. Because of this, most epigraphers accept the reading of T563 as "fire" only if it is accompanied by the T122 "smoke" prefix—producing a variant on an old saying: "If there is no smoke, there is no fire."

We are somewhat more fortunate with the grapheme for "smoking." In the *Madrid Codex* (*Codex Tro-Cortesianus*), an illustrated Maya manuscript from post-Classic times, three pages show different deities smoking large cigars. All the accompanying texts begin with the compound

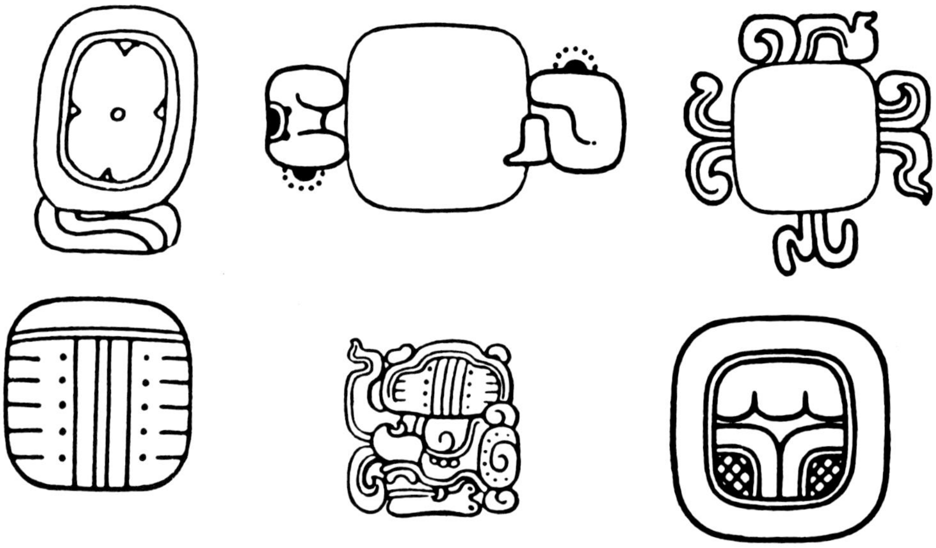

Figure 26. Hieroglyphs associated with fire, smoke, and smoking. Left to right: kin, *T227, T122, T563a, T1035b, and* akbal *(after J. E. S. Thompson)*

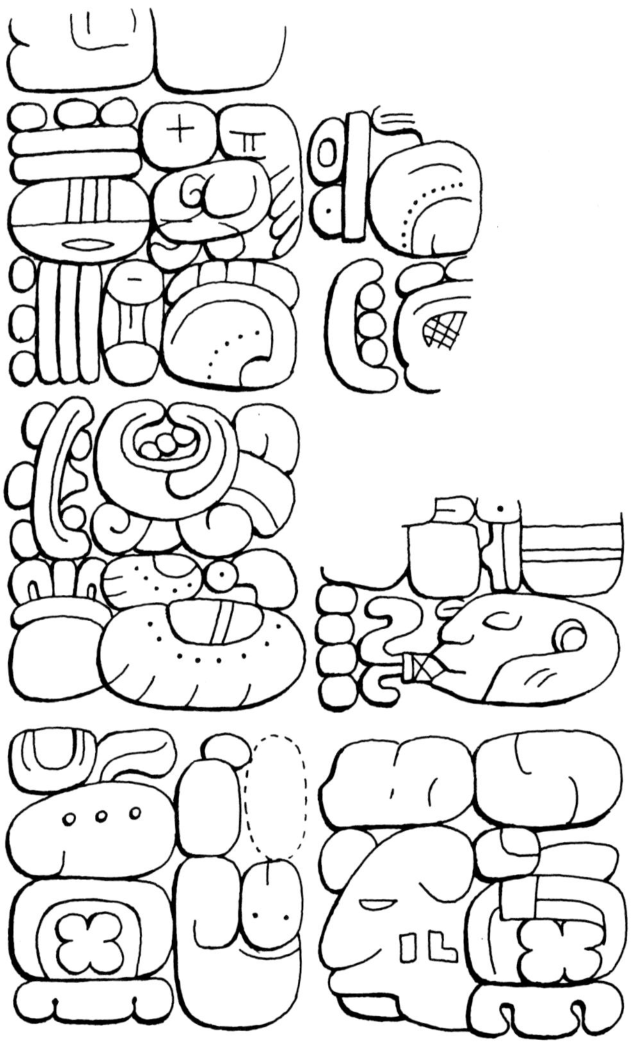

Figure 27. Detail, hieroglyphic text on a recently discovered stela at Cobá. Note the glyph block with the smoking head (drawn after Eric von Euw)

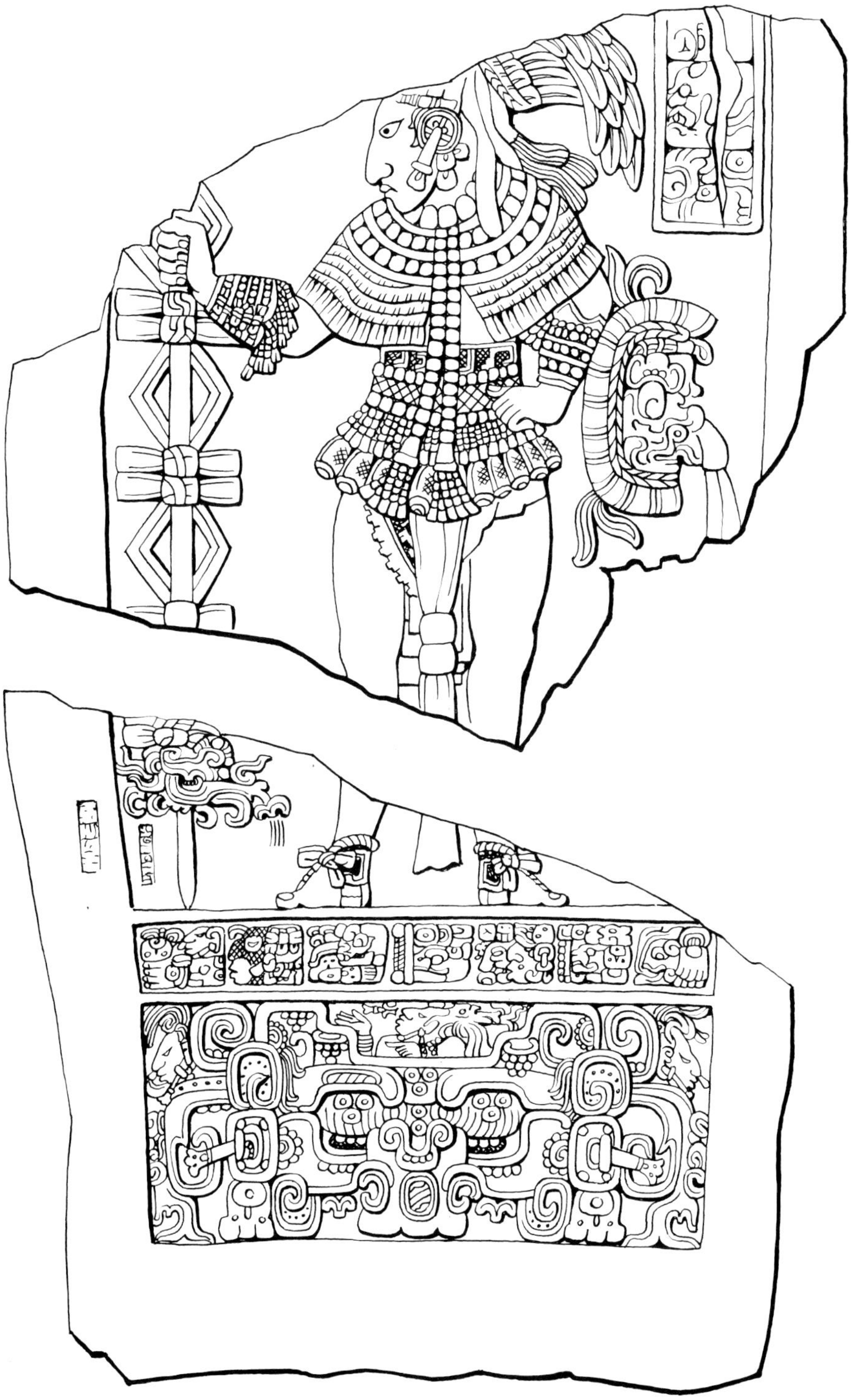

Figure 28. Line rendering of Stela 1 at Bonampak

T227:504. At the turn of the century Eduard Seler suggested that this glyphic compound designates "smoking,"[71] a theory that was elaborated by David Kelley:

T227 is a seated man; *nak* in Tzotzil is "to sit;" the second glyph is the well known glyph for "night," Yucatec *nak* "belly." For the combination T227.561:23 Knorozov reads *nak chaan*, citing Motul *nak caan*, "capa del cielo" (layer of heaven), a plausible reading. T504, the *akbal* glyph is read by Knorozov simply *ak*. Cordan follows this reading completely, giving *nakak* "to smoke." This seems to be entirely derived meaning for which he cites the comparative evidence of Yucatec *nakba* "to burn" and Chontal *nuk'tum* "he smokes."[72]

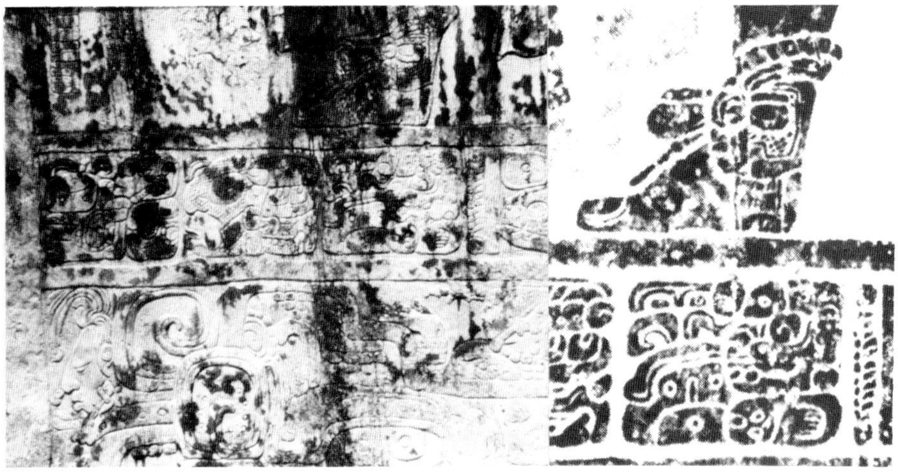

Figure 29. Detail of glyph panel of Stela 1 at Bonampak (the rubbing by Merle Greene Robertson)

A very important development in the search for the glyph compound designating smoking was the discovery of the ruins of a stela in one of the structures of Cobá. The monument had probably been recarved,[73] and only the lower portion of it now remains. It is decorated with a human figure on the front and five rows and four columns of glyphs on the back. One of the glyphs is without doubt a human head with a smoking cigar in his mouth. To our knowledge this hieroglyph, which has been recorded by Eric von Euw, is the only known example of a pictorial Maya symbol for smoking.[74]

In the search for a possible symbol for tobacco Thompson in 1939 suggested that the *kin* glyph that stands for "day," "sun," or "Sun God" also represents a four-petaled flower, or a five-petaled flower converted to a four-petaled one—four being the number of the Sun God. Although originally he thought that it was the five-petaled tobacco flower, later he changed his opinion and said that it was the five-petaled Plumiera, known in Maya as *nicte*, a flower of great erotic and religious importance in Mesoamerica.[75] This leaves us, for the time being at any rate, without a known glyph for "tobacco."

We come only to sleep
We come only to dream
It is not true, it is not true
That we come to this earth to live
We become as grass in the springtime
With the flourishing
With the opening of petals
In the heart
But our body is more like the rosevine
It blossoms and then grows dry
Anonymous Nahua poet

They use it in their sacrificial offerings to their imaginary gods....
Juan Ruis de Alarcón y Mendoza

The American Indian religion is based on direct psychodynamic and pharmacological experience.
Weston La Barre

3. Tobacco and Religion

Plants whose properties place the user in an unusual state have always been looked upon by the natives of the New World as endowed with supernatural powers. Such plants, including tobacco, played an important part in ceremonies and folklore. Tobacco was eminently fitted for religious rituals: "It was aromatic and therefore suitable as an incense; it was beautiful in bloom; and it was consumed by fire—the cleanser; it mysteriously disappeared into the great void, the abode of gods and departed spirits, to whom the breath of its smoke was sweet."[1]

It is probable that the use of tobacco was at first only esoteric and was confined to shamans, priests, and medicine men. Only later, through migrations, the interchange of tribal customs, the abundance of

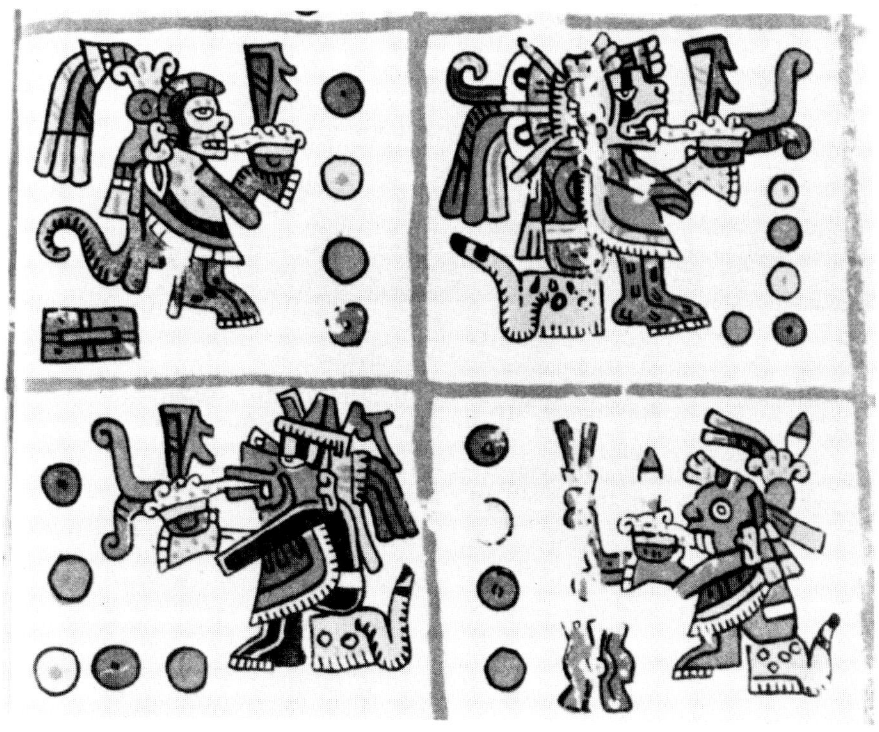

Figure 30. Smoking deities in the Mexican Codex Vaticanus B

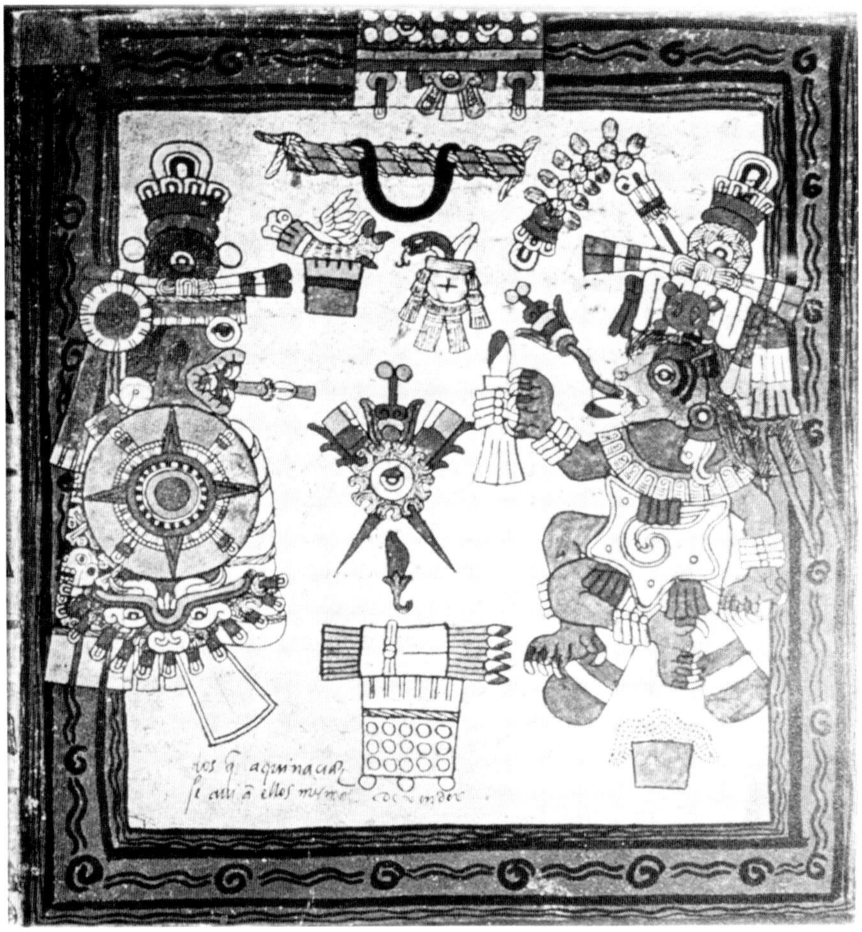

Figure 31. Illustration in the Codex Borbonicus *showing the sun god Tlachitonatiuh and Quetzalcoatl as a coyote. The sun god is smoking a cigar*

tobacco in some areas, the ease by which the practice of smoking was acquired, and European influence, the plant passed from its sacerdotal exclusiveness to general use. Even after it became widespread among the Mesoamerican Indians, its essential sacrosanct character remained unchanged. The role of tobacco among the American Indians has been eloquently described by Robert Hall:

> The main evidence of antiquity is pervading holiness of tobacco. It was a sacrifice, a ritual fumigant, a good-will offering, and a sacrament. It was used to seal treaties, friendships, and solemn, binding agreements, to begin war, conclude peace, and legitimize covenants of every description between man and man, between man and the supernatural. Tobacco was used in rites of curing and in rites of human sacrifice.[2]

Tobacco was an important ceremonial object among the Aztecs, who regarded it as the incarnation of the goddess Cihuacoahuatl.[3] They believed that the body of the goddess was composed of tobacco.[4] They also used tobacco in the rites honoring their war god, Huitzilopochtli.[5] In the *Tratado de las supersticiones de los naturales de esta Nueva España*, written in 1629, a drink called *ololiuhqui* is mentioned of which tobacco was an ingredient and which was used as an offering to the gods:

> . . . they used it in their sacrificial offerings to their imaginary gods, which is either voluntary for them or they are ordered. For their personal well being, they offer the best they can, what is theirs or valuable. However, who does not know what else to offer, he gives a drink they call *ololiuhqui* or *pezote* or tobacco as it is known in this area.[6]

The custom of offering tobacco to the Aztec gods is also evident from the laments of Fray Diego Durán, one of the contemporary Spanish missionaries:

> Every day [the goddess Xiuhtecutli] was presented with offerings of bread, pulque, viands of meat, and porridge. At some times these were offered from devotion, [at] others because of a vow, at others for one's health, and [at] others in search of prosperity or for the sake of one's

children—for a thousand superstitions regarding the crackling of fire, the creaking of the embers, the leaping of sparks, the smoking [of the fire]. Thus it was sprinkled with pulque and was fed incense or tobacco (which is a type of plant with which the Indians deaden the flesh so as not to feel bodily fatigue). These beliefs, superstitious inventions, and lies had been made up by the deceivers and were held as trustworthy and true. Indeed, it has been an effort to uproot [these beliefs]. May it please the Almighty Lord that they now be torn up, [at least] in part![7]

Tobacco, tobacco gourds, and pouches have been described as symbols of divinity among several Mesoamerican nations. Both in the *Codex Mendoza* and in the *Codex Florentino* several of the figures participating in sacrificial rites are pictured carrying tobacco gourds (*ye tecomotl*) and pouches (*ye guachtli*).[8] The gourds were made of the hard-shelled fruits of *Lagenaria* vines and *Crescentia* trees and, together with the sleeveless jacket (*tlemaitl*), were the insignia of the Aztec priesthoods.[9]

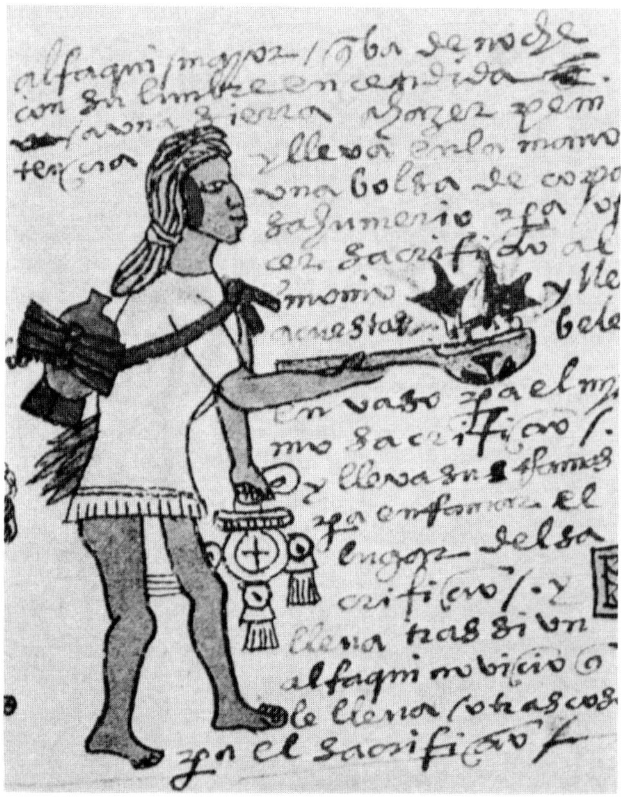

Figure 32. Aztec priest with tobacco gourd (Codex Mendoza)

Durán described the attire of the priest serving the god Tezcatlipoca:

Over this surplice he wore a sleeveless jacket of red leather similar to a penitent's garment. Instead of sleeves he wore what looked like wings with other long ornaments hanging in similar fashion. From these wings hung wide ribbons from which was suspended, in the middle of the back, a medium-sized gourd. The gourd was perforated with holes, and in each one of these had been placed a flower. It was also adorned with little balls of tobacco and soot. This gourd was called Iyetecon [His Little Gourd].[10]

The Aztec priests who prepared the victim to be sacrificed to the Goddess Toci were also distinguished by their tobacco gourds, and they were described as "old men who wore white tunics reaching to their feet, their backs adorned with small gourds filled with tobacco and other magic objects."[11] Tobacco gourds completely covered with gold leaf and probably worn by the priests were discovered in the tombs of Monte Albán.[12]

Smoking, probably of tobacco, is also mentioned in the description of the beautiful Aztec ceremony the "Dance of the Flowers":

The dance they most enjoyed was one in which they crowned and adorned themselves with flowers. A house of flowers was erected for the dance on the main pyramid at the temple of the great divinity Huitzilopochtli. They also erected artificial trees covered with fragrant flowers where they seated the goddess Xochiquetzalli. During the dance some boys dressed up as birds, and others [dressed] as butterflies descended [from the trees]. They were richly decked with fine green, blue, red, and yellow feathers. These youths ascended the trees, climbing from limb to limb, sucking the dew of the flowers. Then the "gods" appeared, each wearing robes such as the idols wore on the altars—a man or a woman dressed in the guise of each. With their blowguns in their hands they went around shooting at the "birds" who were in the trees. Then the Goddess of the Flowers—Xochiquetzal—came out to receive [the gods], took them by the hand, making them sit next to her, treating them with great honor and respect, as such divinities deserved. There she presented them with flowers and gave them to smoke, and then she made her representatives come to amuse them.[13]

This pre-Columbian ballet certainly must have been a stunning spectacle, and even the otherwise phlegmatic Durán could not resist expressing his admiration: "This was the most solemn dance in the land, and I rarely see another one danced today....[14]

Other Mexican tribes, among them the Cuicatecs, used wild tobacco in their religious rituals. The ceremonies were usually performed in caves or on hilltops and were accompanied by prayers and supplications to the gods.[15]

The Tlaxcalans also included tobacco (*picietl*) as a valuable offering to their god of war, Camaxtli:

He [the chief of the Tlaxcalans] gathered all of his soldiers, the volunteers and all other people friendly to his cause, and together they went to the mountain Tepeticpac to pray in the temple to their idol Camaxtli. They offered [the idol] bamboo spears, darts made of wood and arrows, together with other things of which they make their weapons. After the offering was made, they prayed to the gods to help them in their task. They prayed with cries, with fasting, and with sacrifices. After the God gave them favorable omens, he requested further offerings.

Then they bring to the temple a fair maiden whose breasts are unequal in size, and she is given a potion which causes her to secrete milk. They take the milk and place it in the vase of the God. This chalice is made either of expensive wood or from a stone of ebony color, found only in that area, which is called *teotetl*, which means "stone of God." They take the milk of the maiden, together with the offerings of bamboo, spearheads, receptacles, nerve of the deer [bowstrings?] receptacles, and they put it all on the altar in the temple of Camaxtli. They cover it all with branches of the laurel. They make all these offerings as part of their evil superstition. Then they offer other things, such as cut paper, thorns, thistle and *picietl*, a herb similar to henbane. During this period the Chichimecs do not shed their own blood, neither do they offer bloody [human?] sacrifice. They only offer cut white paper, perfumed things, snakes, rabbits, which are killed in front of their idol Camaxtli. They also offer at the same time the leftovers, thistle, and *picietl*.[16]

Tobacco also played an important role in the ceremony of declaration of war in Michoacán. The *Relación de Mechuacán* describes the elaborate ritual held after the cacique announced the state of war. At midnight the high priest, impersonator of the fire god, burned a number of bowls filled with the herb *andamuqua*, or tobacco. "After the ceremony their scouts took two bowls of tobacco, eagle feathers and two blood-stained arrows to the enemy camp as the sign of beginning of hostilities."[17]

Among the Mazatecs, another Mesoamerican Indian nation, the *curandero* uses "paste of powdered tobacco and lime to render pregnant women invulnerable to witchcraft."[18] A similar custom was recorded among the Aztecs by Sahagún: "The pregnant woman, when she walked in the night, . . . and also the men, carried on their breast *picietl* to prevent danger to the child, while he was inside of his mother, and they said that if they did not do this, then the child will be born with an illness called *ayomama* or with another called *cuetz-palicuiztl*."[19]

Besides using tobacco as a protection against witchcraft, the Aztecs also believed that it could be used to cast spells and to protect them from wild animals.[20] It was thought to be especially effective against the poisonous ones, such as scorpions and snakes. Early Spanish historians recorded that the natives thought that "if you camped out in the open and you had (tobacco) in your hands and in your mouth, you were safe from poisonous animals."[21] The Totonacs of Papantla also believed that tobacco protects one from snakes and the spirits of the dead and offered it to the supernatural rulers of the forests.[22]

Maybe the tallest snake-tobacco tale was recorded by Sahagún. He described a snake hunt in which the magic powers of tobacco were used:

[The Indian] wraps up fine tobacco in paper; he throws it at (the serpent) or else he fills small jars; to throw small jars at it so that the small jars suddenly burst and the fine tobacco suddenly scatters. Thus the (serpent) stretches out enfebled. Then he puts the end of a stick in a paper, a rag dipped in fine tobacco, he inserts it into the mouth of the serpent, so that it can't do anything. He then captures and kills it.[23]

While no one would doubt the accuracy of the friar's reportage, his story certainly resembles the folk tale that claims that the best method for hunting birds is to sprinkle salt on their tails, after which they cannot fly away and can be easily killed.

The Huichol Indians of Mexico also revere tobacco, regarding it as a prized possession of "Grandfather Fire." During their pilgrimages to Real Catorce (San Luis Potosí) it is still their custom to make small tobacco balls, touch them with plumes, and wrap them in corn husks. They carry these ball cigarettes in special tobacco gourds tied to their quivers. The custom symbolizes the birth of tobacco. At the end of the pilgrimage they burn the tobacco in honor of Grandfather Fire.[24] The Tarahumaras believed that the magical power of tobacco smoke wards off evil.[25]

Tobacco and tobacco smoke were used extensively in divination among the North and South American Indians and those of the Caribbean area.[26] Tobacco was given as an offering to the gods by the Seminoles, the Iroquois, some of the tribes east of the Rocky Mountains, and the nations of northwestern Venezuela.[27] The Yibaros of Río Pastesa and the Tucunas believed that tobacco augmented the magical properties of their bodies.[28] The mountain tribes of Peru, the Jivaros and the Taparos, drank tobacco juice to prepare themselves for initiation into shamanhood.[29] The Menominees buried their dead with a supply of tobacco to placate their gods.[30] The Tewas of San Ildefonso and the Omahas used tobacco to ensure the success of their hunt.[31] The Tewas also used tobacco in their rain rituals to appease the gods of cloudbursts.[32] Tobacco smoke was used in shamanistic divination by the Venezuelan tribes, the Florida Indians, the Guajiros of Colombia, several Caribbean groups, the Cumanos of the Orinoco River, the Warroas, the Uaupes (Puinavis), the Arawaks, the Caribs, and several other tribes.[33]

The Smoking Gods of the Mayas

The ancient Mayas were passionate smokers, and so were their gods. Even today the Lacandón Mayas, or

Caribes, as they prefer to call themselves (no relation to the Caribs), believe that the white-haired, bearded Lords of Rain and Thunder (Nucuch Yum Chacob) who inhabit the highest heaven[34] love to smoke cigars; the comets (or meteorites) passing through the dark-blue tropical skies are still-glowing cigars they have thrown away.[35] Old Yucatec legends also refer to smoking deities. They describe the Balams, the Gods of the Winds and the Four Directions, as heavy smokers who pound huge rocks together to create a spark to light their cigars. To mortals all of this appears as thunder and lightning.[36]

These ancient beliefs are reflected in archaeological finds, some of which are shown in this book. Many lordly figures smoking or holding cigars are shown in paintings in the codices and on ceramic vessels or in carvings on stone monuments. Whether some, most, or all of these figures represent supernatural beings is a question long debated and still unresolved. While the smoking personages in the codices are certainly deities, as in El Fumador on the jamb panel of the Temple of the Cross at Palenque, a number of painted ceramics show lords with human characteristics who probably represent rulers and their entourages.

Among the deities God L appears to be a heavy smoker, while Gods A, B, D, F, and N could be characterized as occasional indulgers. Among the mythological animals monkeys have a strong lead in smoking, with jaguars second and frogs a distant third. Detailed discussions of these examples are included in Part II of this book. The "flare gods" in general and God K in particular are dealt with in Chapter 6.

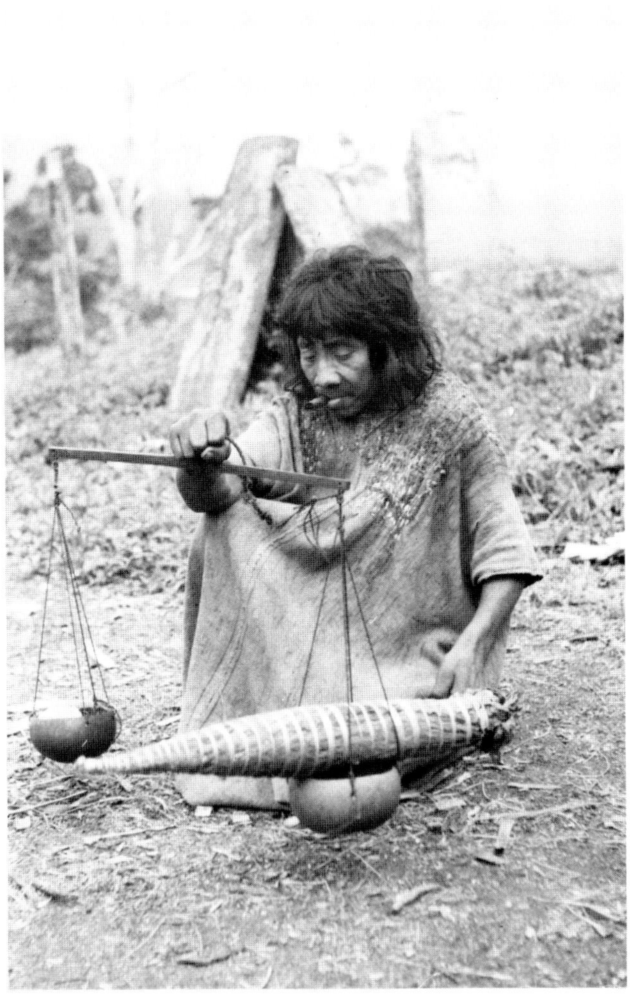

Figure 33. The weighing of the tobacco bundle (photography by Gertrude Duby Blom)

Tobacco in Maya Religion and Folklore

Even though our knowledge of ancient Maya rituals is incomplete, one is inclined to speculate that tobacco was used very early in religious ceremonies. Today among the Lacondones, a few hundred Maya Indians who preserve the ancient ways, tobacco plays an important part in their rituals. They grow their own tobacco, dry it on racks, and store it in large funnel-shaped bundles.[37] In the pre-Conquest *Codex Vindobonensis Mexicanus* similar bundles, presumably of tobacco, are shown being presented to 7 Flower, an important deity associated with sacred mushrooms, tobacco, and other precious substances.[38]

The Lacandón Mayas are heavy smokers, and they also use their wrapped cigars as trade articles in towns neighboring the Lacandón Forest.[39] At the season of tobacco harvest the Lacandón performs his tobacco ritual: He takes the first leaves of his crop, rolls a large cigar, and lights it with a "new fire" ignited by sun rays concentrated by a crystal lens. First the cigar is held in front of the sacred *olla,* and then it is offered to the god whose image decorates the incense burner ("god pot"). The god is usually pictured smoking and sitting on the sign for the earth *(caban).*[40] In the past the Lacandón Mayas of Cholti, a tribe with the same name but living in a different location, celebrated the Hicsión, the "Feast of Cigars." This ceremony has been described in a Spanish manuscript, *Testimonio de diferentes cartas y providencias desde el Pueblo de Nuestra Señora de los Dolores de Lacandones,* written in 1696 by three missionaries, Antonio Margil de Jesús, Lázaro de Mazariégos, and Blas Buillén. The manuscript not only gives the factual

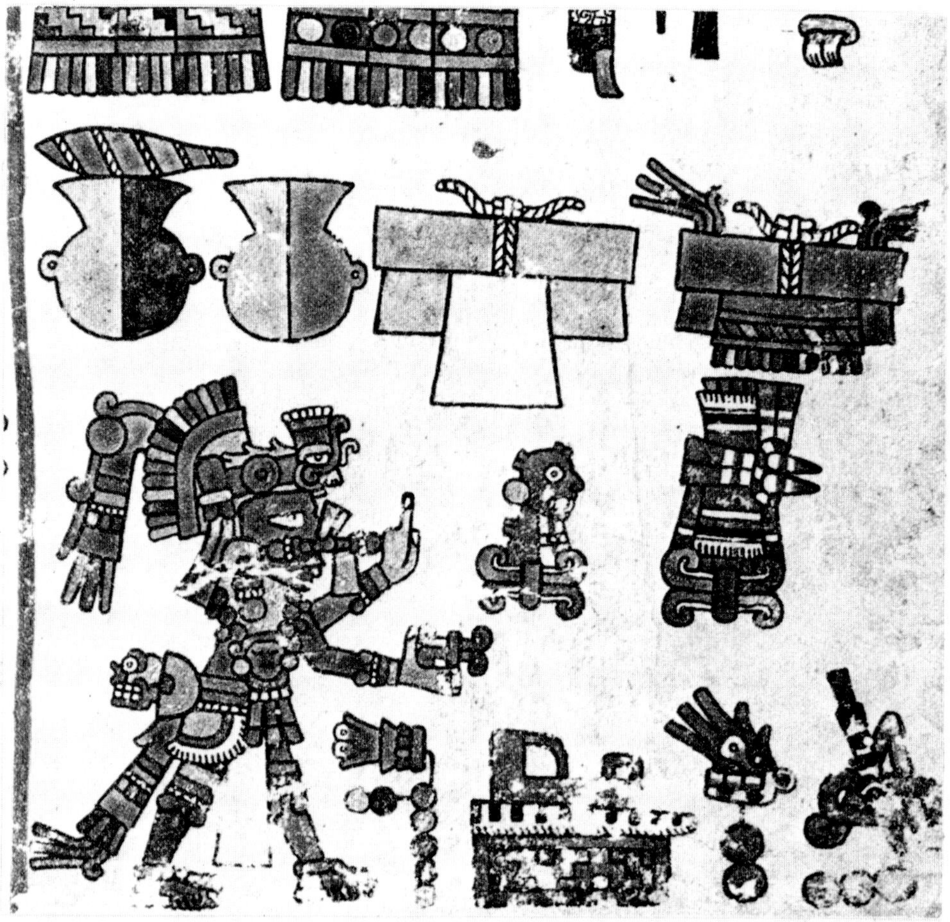

Figure 34. Detail in the Codex Vindobonensis Mexicanus *showing Goddess 7 Flower with offerings. The bundle atop the large jar strongly resembles—and, according to Jill Furst, is—a tobacco bundle*

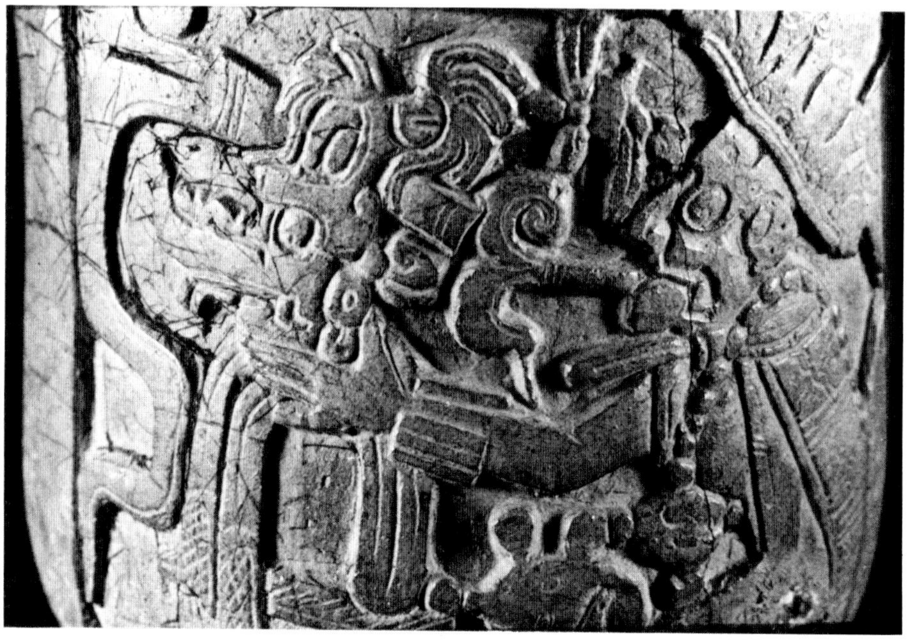

Figure 35. Detail on a Classic Maya vase, Chocolá style, showing God L, one of the principal deities of the Maya underworld, smoking

description of the festivities but also provides insight into a truly democratic aboriginal society:

> Regarding the point, whether they have a King, they say that they have never known what this office might be; only the caciques are their lords, kings, priests and confessors, without being paid any tribute or other profit. Simply as a token that they are their caciques, they celebrate a particular feast that they call Hicsión Feast or Feast of the Cigars. They spend twenty days making cigars in all the houses; each one makes as many as he wishes in his own house and then, when the twenty days are over, they go out to the woods, rivers and lakes, seeking deer, fish, honey, etc. And they come on the day appointed, and assemble in the house of the idols, and make gifts of all these things mentioned, and make their caciques intoxicated with the above-mentioned water. And they rub the snout of the idols with the fat of the animals, and they give them [the idols] ground cacao to drink, putting a little in the mouth. Each family gives its cacique a small bundle of cigars to signify that he is the master. They paint themselves, eat, drink, and dance and it is in great solemnity.[41]

Offering tobacco to the old gods has now largely disappeared from both the Lacandón and Yucatec *milpa* ceremonies.[42] In the modern Maya folklore, however, ceremonies such as the New Year celebration in the Tzeltal town Oxchuc, still include, among other things, the ritual offering of thirteen calabashes of tobacco.

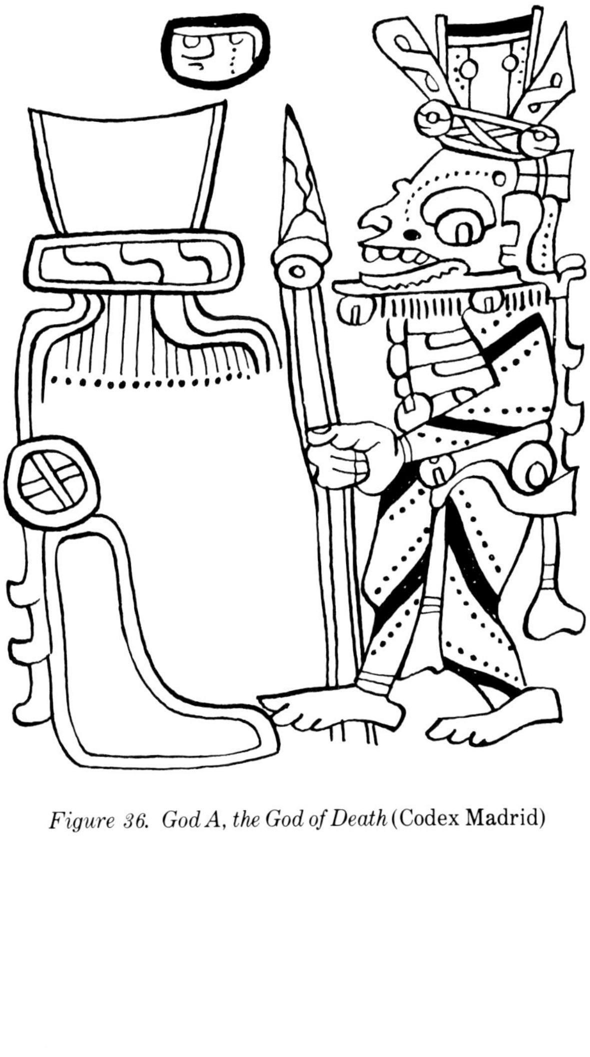

Figure 36. God A, the God of Death (Codex Madrid)

Native tobacco is considered a traditional defense against the supernatural forces of evil among the Tzotzils of Chenalho, Chamula, Husitán, Huitiupán, Pantelho, Venustiano, Carranza, and Zinacantán. The Tzotzils of Larrainzar (formerly San Andrés Istacostoc), still attribute magic powers to tobacco. They sprinkle powdered tobacco on the chest and face of the very ill to protect him from Pucuh, the Demon of Death.[43] The divine power of tobacco in the Tzotzil Mayas' folklore has also been confirmed by ethnologist Colixta Gutiéras-Holms, who conducted extensive interviews with them about their life and religion. Here are some of the answers he received:

> *Komel, ch' ulelal,* I even learned how to heal harm done by magic and I prayed always to heal, never to kill. If (harm) comes through the air, it is cured with tobacco....
>
> There are people, who chew *moy* (powdered tobacco mixed with lime) every day, from the time they get up in the morning, because the little gourd-bottle containing it is kept at the head of their beds, the way others smoke cigarettes. On hearing thunder, the people bring out their *moy* and keep it in their cheek, and in that way it will not thunder too loudly. When we die, the *pilico (moy)* defends us, we will rest....
>
> Lightning is *anhel,* and to avoid being killed by it, the people use *moy,* which is their protector....
>
> Even when death has taken place in a distance, in another house they are afraid. The ground leaves of the wild tobacco are rubbed on the back of the neck, on the forehead, and on the inside of the elbow and knee joints. It is this way the dead do not come to seize us.[44]

The Tzotzils' belief in the protective power of tobacco against the gods of the underworld is also recorded by Holland in his *Medicina Maya en los Altos de Chiapas:*

> These Gods—Ik'al, Kitsil Bak, Mukta Pishol, Walapotok, Shapinkte, Natijikol, Yalem Bekelt, Me'chamel, to name a few—favorite pastime is to haunt the crossroads at night. To be safe from them, the Indian always carries with him his chewing tobacco, the *mui,* as an absolute assurance against bad spirits, the gods of the Underworld.[45]

Similar beliefs are professed by the neighbors of the Tzotzils, the Tzeltal Mayas. The older, more conservative Tzeltal male of the Tenejapa area often carries a small, polished tobacco gourd *(cul bankilal)* to protect himself from demons and other evil spirits he may encounter on the trail. While in everyday usage the Tzeltal calls the tobacco simply *may,* in its ceremonial context he respectfully addresses it as *bankilal,* "elder brother," or *bankiltik,* "honored elder brother."[46]

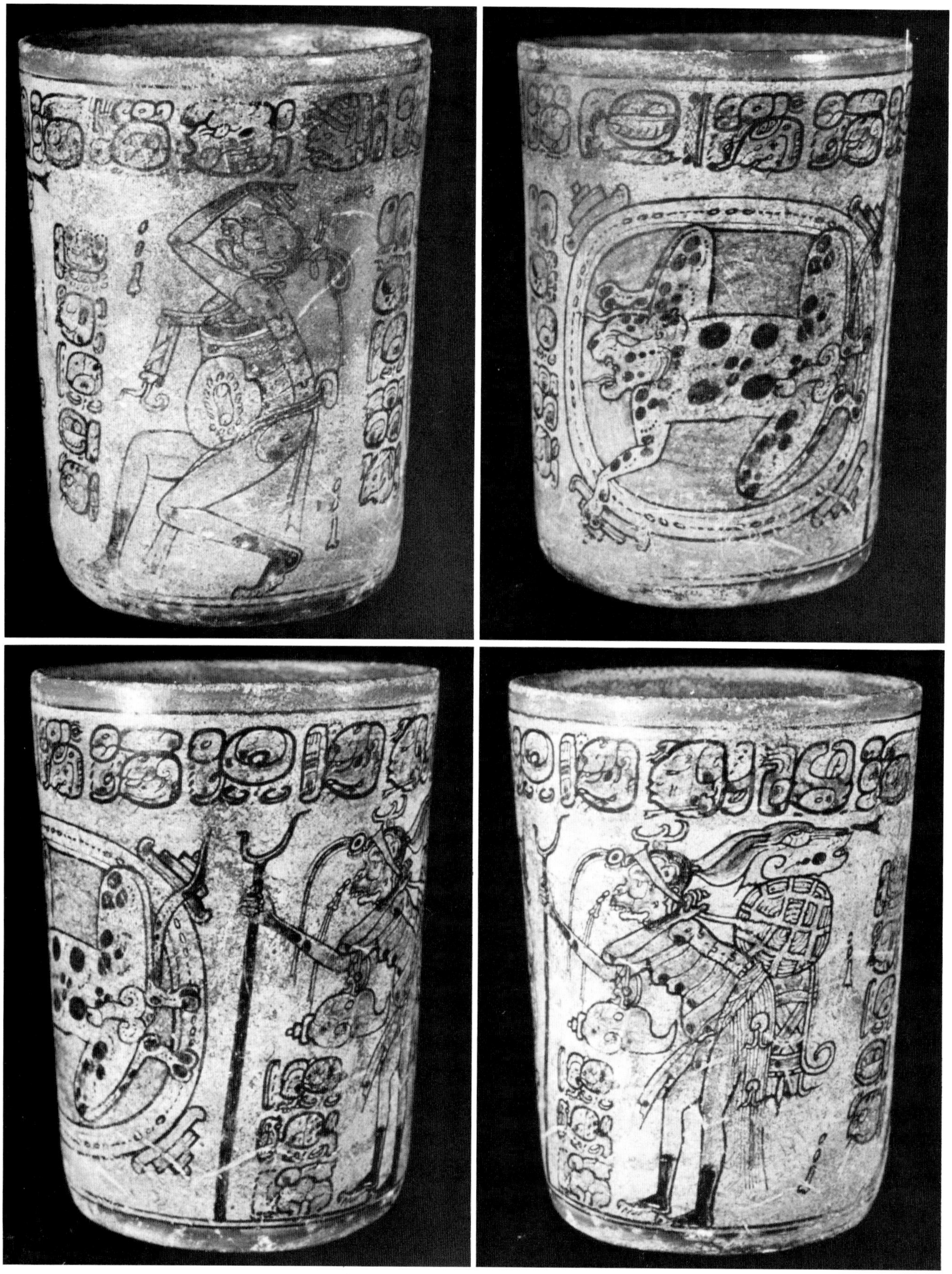

Figure 37. Death gods and a jaguar pictured on a Classic vase from Petén (November Collection)

The Chorti also holds his tobacco and tobacco utensils in great respect. When he dies, they are placed in his grave, as reported to ethnographer John G. Fought:

> In a place where I live, when people die, where I live, when they see that he is already expiring, the sick person who is lying dying, they lay out his sandals, they lay down his machete beside him, they lay out his gourd cup, his gourd bottle, and his tobacco, they lay down a pipe beside him.[47]

An unusual and certainly unappetizing application of tobacco in folk healing is performed by the Chorti Maya *ah huht*, translated literally, "spitter." The *ah huht* is called to cure seizures, ward off the evil eye, or treat "strong blood," especially in children, which makes him something of a pediatric shaman. The *ah huht* chews different herbs, such as rue, sage, artemisia, but most often tobacco, and spits it onto the patient from head to toe. If the illness is especially severe, he sprays his saliva-mixed tobacco in the form of a cross from head to crotch and shoulder to shoulder.[48] This treatment is repeated until the patient gets well or dies—or, I suppose simply gets tired of being spat upon. A similar custom exists among the Indians of the Amazon jungle. There the shaman healer drinks *natema*, a mixture of tobacco juice and another herbal extract called *piripiri*, which permits him to see into the body of the patient as "though it were glass." He then regurgitates over the body of his patient to rid him of the disease.[49]

Tobacco is also the magic tool of the Chorti diviner, the *ah q'in*. The *ah q'in* is a specialist in predicting future events, finding lost objects or animals, and determining the causes of sickness, either magical or natural in origin, and then recommending the best cure for it. He is considered a good and respectable man who receives his power and information directly from the benevolent Sun God (much as the oracle of Delphi received her supernatural power from the Greek god of the sun, Apollo). He is called upon to counteract malignant spells caused by sorcerers, to predict whether it will rain and how much, to foretell whether the crops will be good, and so on. To facilitate his work, the *ah q'in*, like Aladdin in the fairy tale, conveniently carries with him his personal genie, the *sahorin*. Unlike the Arabian genies, however, who prefer bottles and oil lamps as dwellings, the *sahorin* is housed in the right calf of the *ah q'in*. Ethnologist Charles Wisdom described how the *sahorin* is questioned:

> He [the *ah q'in*] goes to the house of his patient and darkens the interior, since diviners say they cannot work in strong light. He sits next to the bed, chews tobacco, rubs the saliva on his right leg, and in a few minutes begins his diagnosis. This consists of asking direct questions of the spirit concerning the probable cause of the sickness.
>
> The questions often run: What caused the sickness? Was it a evil wind? Was it a sorcerer? Did he appear in the guise of an animal? Did he work black magic? Did he stab a candle? Was it the *hijillo* of an apparition or a spirit of the dead? Was it lightning fright? etc.
>
> If the spirit in his calf twitches the calf, the answer is "Yes," while no twitching signifies "No." The whole procedure is carried out with great dignity and in dramatic manner. Each time the diviner gets an affirmative answer, he turns to the gathered family and triumphantly announces his find. He often says: Ha! Didn't I tell you so! It is as I thought it was. I knew it at once. Soon I will tell you what is ailing this man here, what he has in his body, etc.[50]

Similar data on the Chorti diviner were later recorded by John G. Fought, who translated a rich collection of legends, folk tales, and religious and medical texts. He wrote:

> The diviners, when there is someone who becomes ill, then someone who knows how to divine is brought. If it is night, and someone in the house falls ill, that same night someone goes to get him, so that he will divine and see in whose grasp that sick person is. . . . Then he asks for tobacco, chews it, and begins to divine. He inquires if it is an ordiniary disease, or a little disease which befalls anyone just like that, if it wasn't raised by someone else.
>
> When he is speaking with the calf of his leg he says this: Wise Diviner, knowing one (then he names different saints and angels), reveal to us today the punishment and bondage where it plays over the head of this little creature, where his angel and spirit are being threatened.[51]

A very similar method of conversing with spirits is used by the *ah q'in*'s counterpart, the *po si*, the ominous sorcerer, who uses the same calf-twitching method to ask his *sahorin* which method he should use on his unsuspecting victim. Since he, the *po si*, is a person of evil intentions, the *sahorin* lives in his left calf instead of his right. The *po si* crosses his left leg over his knee, rubs his calf with his palm, spits tobacco on it, and then puts his yes-or-no questions to the spirit:

> Is it necessary that I bury the image of so-and-so? Perhaps it should be of wax, perhaps a piece of cloth? Is it necessary that I send worms up his nose to kill him, to make him suffer? Will it be best to send a frog into his stomach, an earthworm into his intestines?[52]

A custom very similar to that of invoking the sahorin exists among the Chibcha and Puinavi Indians of Colombia. They, however, do not rub the tobacco on their bodies but eat it, and if certain joints of their body move, that is regarded as a sign that they will have good—or bad— luck.[53]

Tobacco is used by another member of the Chorti "medical academy," the *ah nir-om,* "curer", or *curandero* as he is called in Spanish. The *ah nir-om* treats frights (*aigres*) and expels poisonous foreign substances from his patient's body. The ailment is usually caused by sorcery, an enemy's curse, pursuit by an evil spirit, or contact with a ritual object. His most effective treatment is what is called "seizing." It makes use of live black turkeys, chickens, tobacco, and artemisia leaves. The seizing treatment is called *chucur,* a corruption of the Chorti Maya word *tcuk-ur,* "seizing" or "capturing." The *ah nir-om* who specializes in *chucur* is called a *chucurero.* The *chucurero* slowly passes different objects, especially tobacco leaves, over the body of his patient, at the same time calling upon the saints, particularly San Antonio del Monte, patron of medicine, to drive out the fright, *aigre,* or foreign matter, and also calling upon the invader to come out and enter the object he is holding in his hands:

> *Aigre,* come out of the body of so-and-so; Fright, come out of the body of so-and-so; Fire, come out, come out. See, here is a new place for you, a better place for you, come into this, quit the body of so-and-so.... Saint so-and-so [here he calls several saints by name], make the man's sickness leave, drive it into what I hold in my hand.

If the illness is a serious one, especially *hijillo,* "poisonous air," after the treatment the tobacco leaves are buried to prevent contamination. It is said that the leaves usually turn red as evidence that they contain the fire or fever of the patient. If the patient is not cured in eight days, the *chucurero* repeats his treatment, giving it four times in all if necessary.[54]

According to Maya popular belief, *hijillo* comes straight from the underworld, called in different dialects Metnal (a striking resemblance to the Aztec word Mictnal), Xibalba (*xib* meaning "fear," "terror," or "trembling with fright"), or Olontic "down below."[55] According to Landa it was the place to which the wicked were banished after death to be tormented "by the devils and by great extremities of hunger, cold, fatigue, and grief."[56] The entrance to Metnal was believed to be in the cave-filled region of Alta Verapaz, at the crossing of four roads where there was an unbelievable stench or corpses and blood (other sources place the entry to the Maya underworld at Palenque, Cham, Izapá, and other locations).[57] The diseases that afflicted men came from this hole, and the Maya curer had to send them back to the underworld, using incantations and magic herbs, such as tobacco.

In the Maya community of San Carlos tobacco is also used in a voodoolike casting of spells. If a sorcerer (*brujo*) wants to cast a spell (*envivida*), he obtains a hair or other small object from the body of his intended victim and rolls a cigarette, hiding the object in it. Then he gives the cigarette to a friend of the victim and, without letting him know what is in it, asks him to give it to the victim. After the person on whom the spell has been cast smokes the cigarette, he falls ill and may even die.[58] The Chortis also use tobacco to cast less malignant spells. They believe that a mixture of pounded tobacco leaves and *aguardiente* (a strong alcoholic drink like brandy) given to a woman will cause her to lose consciousness so that a man can have his way with her.[59] Anyone who has tasted *aguardiente* can vouch for its potency.

[Yucatán] is a land with a great quantity of medicinal plants of various properties. . . . If there were any person here who possessed a knowledge of them, it would be most useful and effective.

Diego de Landa

This plant, kuutz, *is the tobacco, or* picietl. *It is naturally healing and so medicinal that there is hardly a disease to which it is not applied.*

Ricardo Osado, *Libro de Judeo*

A paste of tobacco is to be pressed on the abdomen of a man. . . .

Ritual of the Bacabs

This is an herb of marvelous virtue against wounds, ulcers, noli me tangere, herpes, and all other things. . . .

Jean Nicot

4. Tobacco as Medicine

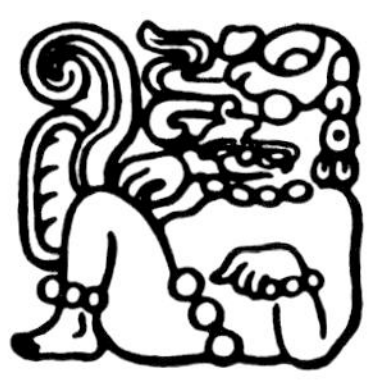

In the life of the ancient, and to some degree also of the contemporary, man, religion, superstition, and medicine cannot be completely dissociated. This principle certainly appears to be true with tobacco. Tobacco as medicine was identified by Fray Ramón Pané, the Italian monk who was a member of Borgia Pope Alexander's tribunal that burned Savonarola and who accompained Columbus on his second journey to the West Indies in 1493. He described *cogobia* (also spelled *cogobha* or *cohobba*), a powerful snuff of tobacco and perhaps other herbs, as not only an offering for religious rituals but also a medicine given to the sick to sniff, smoke, or eat to clear the body of evil humors.[1] Another early description of tobacco as a cure-all is given in the *Relación de Texcoco:*

> [The herb *picietl*] would relax you and prevent the feeling of pain that is caused by excessive work. It is dried, ground, wet, wrapped, and mixed with a little lime in the mouth. It is placed between the lips and gums, not more than the quantity which would fill a hazelnut. They do it when they work and also when they go to sleep. They say that smoking [of tobacco] clears the brain, purges rheumatism. The Spaniards, who suffer from this disease, admit that they use it to obtain cure and they like it very much, and they also use it to clear malaria, taking it as a suppository because it purges them. They also place toasted leaves [of tobacco] under their arms; when there is pain, this relieves it.[2]

Sahagún mentioned that the Aztecs regarded tobacco *(picietl)* as a sure remedy for various ailments. For an abscess for example, it was recommended that the sufferer "place a little lime mixed with the herb *picietl* in a lump or place in the form of a cross, and this will suck the material out of the abscess."[3] For head colds the following was prescribed: "Take the herb called in Indian language *yexucoton* or *picietl* and inhale it in a green, powdered form and with your finger rub it into your mouth to throw out the fever of the rheumatism."[4] Tobacco was also thought to be very effective against the bites of poisonous snakes and insects:

> The remedy for snakebites is the sucking of the place where it was bitten and to make an incicison and place a thin transparent piece of cloth made of maguey over it, expose the bite to fire, to warm it and to rub it with ground tobacco. . . . For scorpion bites it is recommended that the wound should be sucked and rubbed with ground *picietl.*[5]

Tobacco is also frequently mentioned as medicine in other early Spanish manuscripts. In the *Relación de Alahuistlán,* for example, it is written that "when [the Indians] have a cold they cure it by putting a herb called *piciete* in their nose."[6] The Cruz and Badianus Manuscript, an Aztec herbal now in the Vatican Museum, was written by two Aztec Indians, Martinus de la Cruz, a native physician, who composed his work in Aztec, and Joannes Badianus, who translated the text into Latin in

1552. The herbal also describes tobacco *(picietl)* as a sure remedy for "rumbling of the abdomen":

> For one whose bowels are murmuring because of diarrhea, make a potion, let him take it with an early clyster, of the leaves of the herb *tlatlanquaye* [*Iresine calea Standl?*], the bark of *quetzalaylin*, the leaves of *yztac ocoxochitl* and these herbs *tlanextixiuhtontli, elocacatl*, the tree *tlanextia quahuitl* ground in bitter water and ashes, a little honey, salt, pepper, and [stone] alectorium, and finally *picietl*.[7]

Another early description of Aztec healing with tobacco comes from the chronicler Durán. With remarkable insight Durán not only analyzed the pharmacological effects of the "Divine Medicine," the magical ointment of the Aztec priests, which was composed of tobacco, *ololiuhqui* (morning-glory seeds), and ground venomous insects, but also clearly recognized the psychological influence the priests exerted upon their subjects:

> This pitch also served a medicinal purpose: it was used to cure the sick and little children. Therefore, it was called the Divine Medicine, and people came from all parts to pay a visit to the dignitaries of the temple so as to be rubbed with the Divine Medicine. The pitch was applied to the ailing part of the body where pain was felt, and the result was a remarkable relief. The cause of this cure (according to my own opinion, and I have persuaded many of the natives of this) was their extraordinary faith and trust. The Indians use the ointment to cure their ailments, since its properties are like those of *picietl* [tobacco] and *ololiuhqui*, which calm and soothe. Once it is applied as a plaster, it deadens pain; this it does by itself. But its effect is stronger when it is mixed with so many kinds of poisons. It has been proved that when it is applied to any aching place it will diminish the pain and give relief. Thus these priests cured with the pitch or ointment, and [when they had] anointed the body with this painkiller, the strong effects of tobacco and the *ololiuhqui* seemed a sudden and miraculous thing to them.[8]

Tobacco was used extensively as a medicine outside Mesoamerica. The Roanoke Indians consumed tobacco to rid themselves of "bad humors."[9] The Cherokees used tobacco leaves fried in butter to promote the healing of wounds.[10] The Iroquois of New York placed wads of tobacco inside tooth cavities to ease pain.[11] The Jivaro, Conibo, Campa, and Piro tribes of Peru cured headache and dysentery with tobacco.[12] The natives of Venezuela treated rheumatism with it.[13]

These claims of the therapeutic effects of tobacco impressed the Spaniards so deeply that they too soon began using it as a remedy. The first record in a European tongue of tobacco as an analgesic is that by Oviedo y Valdés, an old acquaintance of Columbus who in 1514 became superintendent of the gold smelters of the West Indies. In his *Historia general* he wrote: "I am aware that some Christians have already adopted the habit [of smoking], especially those who have contracted syphilis, for they say that in the state of ecstasy caused by the smoke, they no longer feel their pain."[14] The Spaniards took tobacco home with them for their smoking pleasure and also as a remedy for a wide range of diseases. In 1543 tobacco was recommended for its healing properties by a certain professor at the University of Salamanca. A physician of Seville, Nicolai Monardes, wrote in 1571:

> This herb, commonly called tobacco, is a very efficient herb, known among the Indians, chiefly those of New Spain. After taking possession of these countries, our Spaniards, being instructed by the Indians, availed themselves of this herb in the wounds, which they have received in war, healing themselves with it to the great benefit for all.[15]

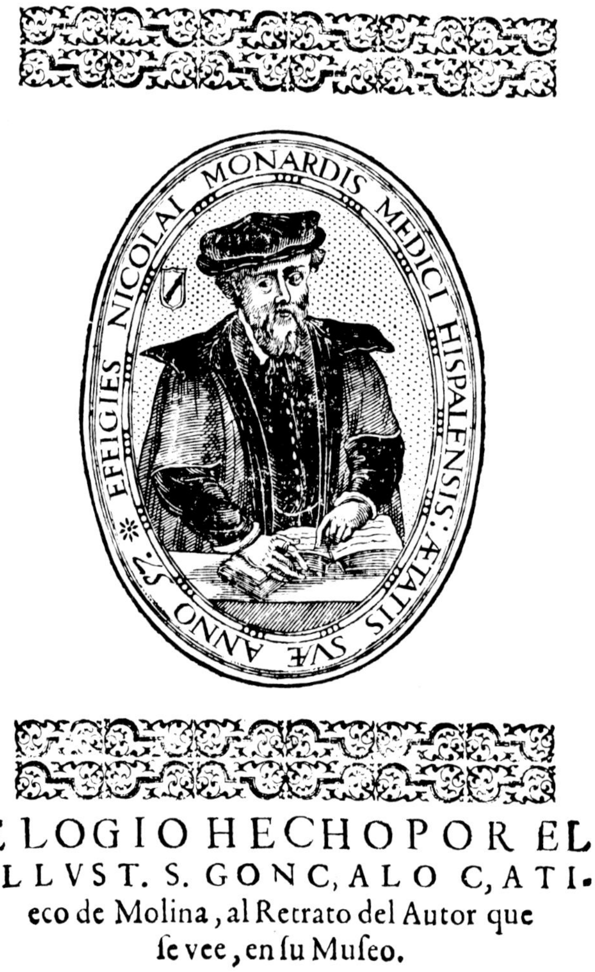

Figure 38. The title page of Nicolai Monardes' Historia

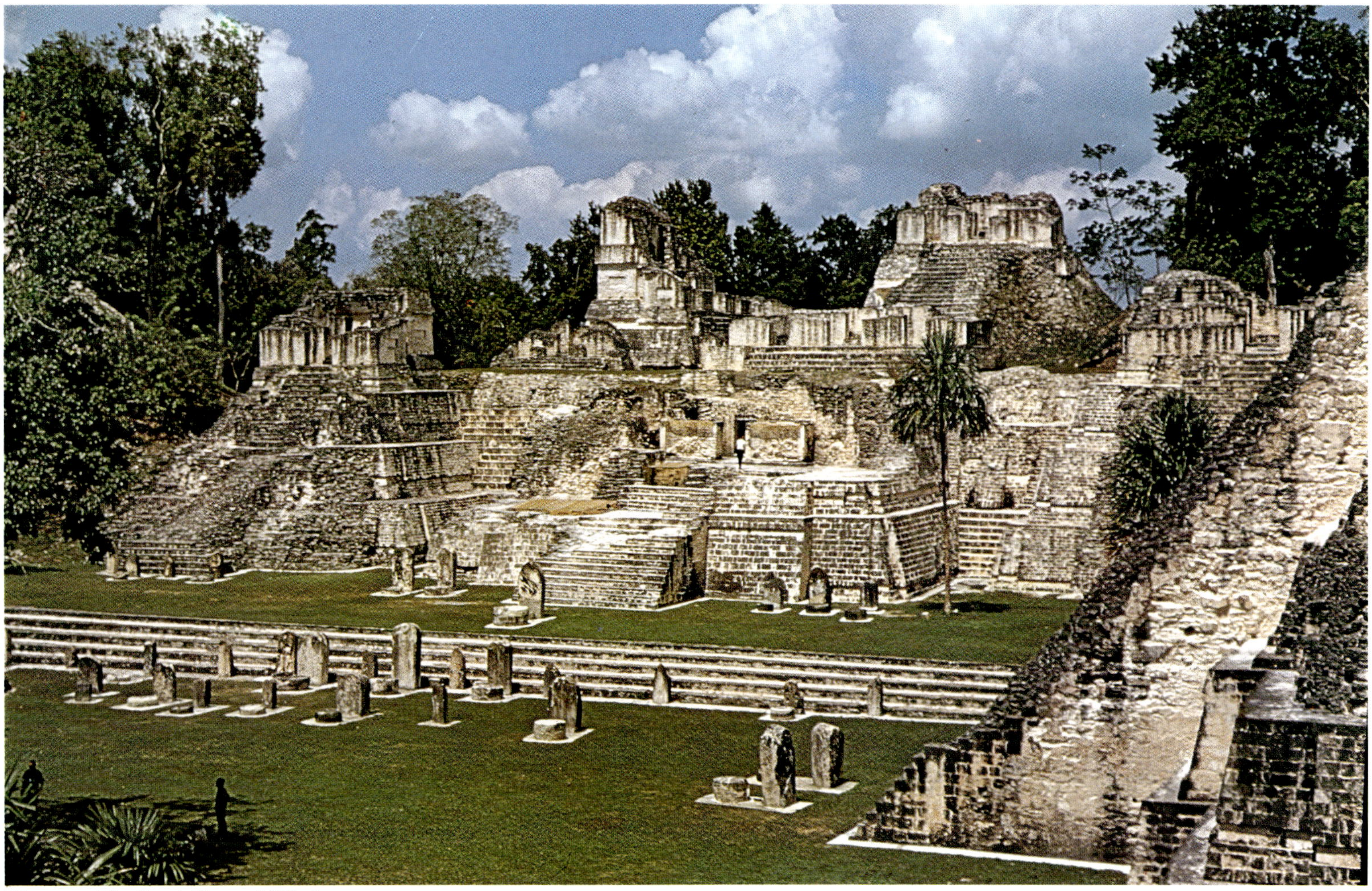

Plate 23. The Great Plaza at Tikal

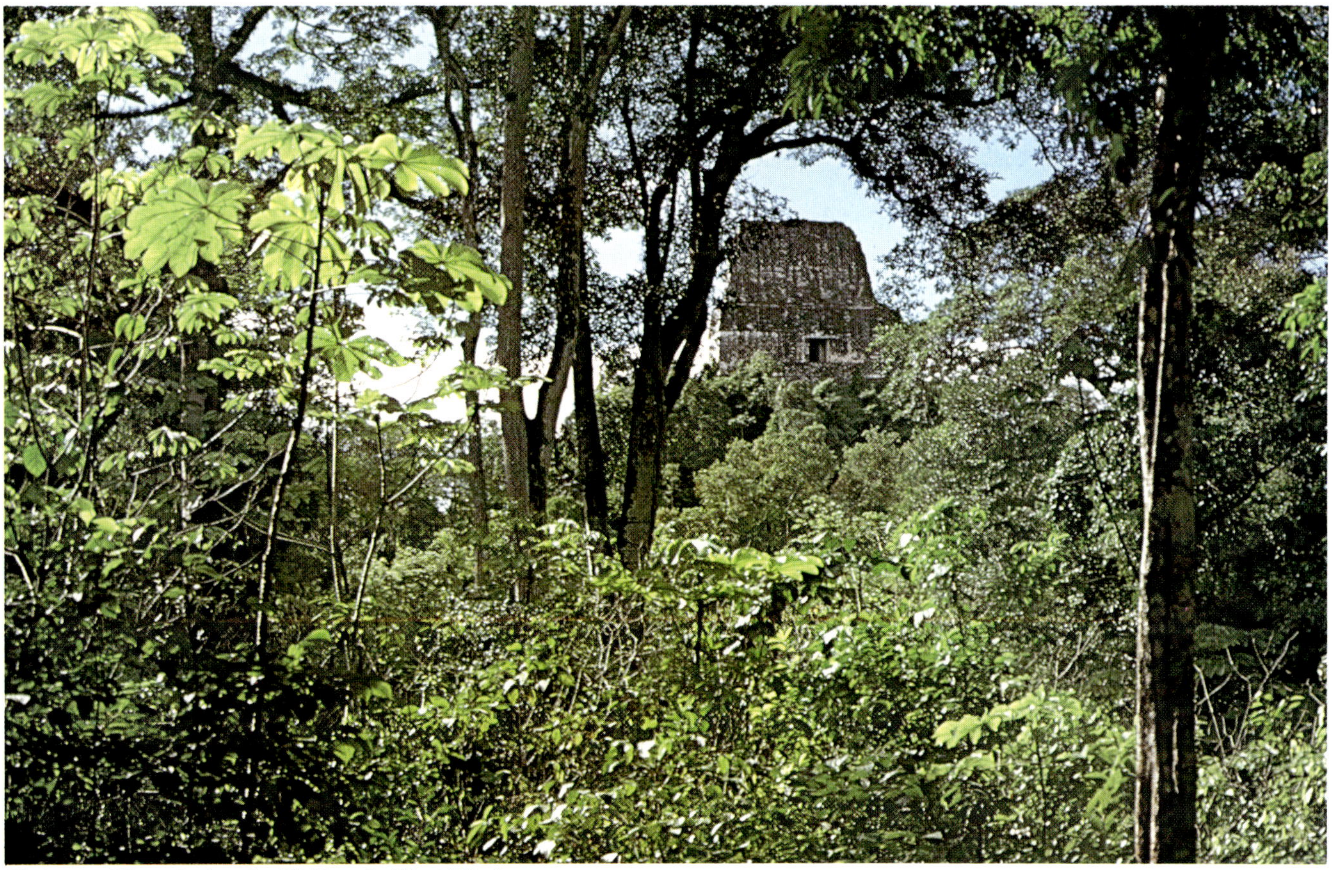

Plate 24. The rain forest with Temple IV at Tikal

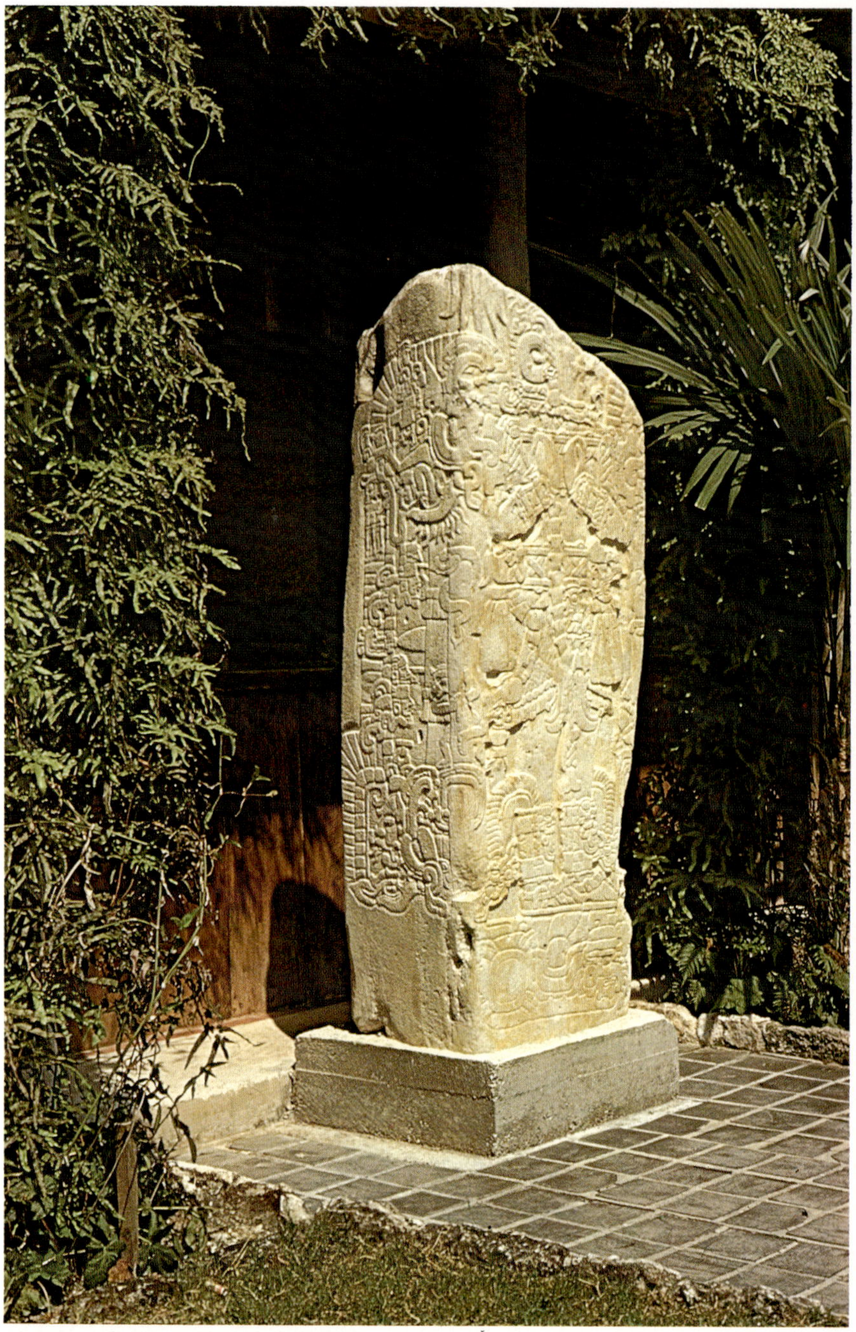

Plate 25. Stela 1 in the Tikal Museum

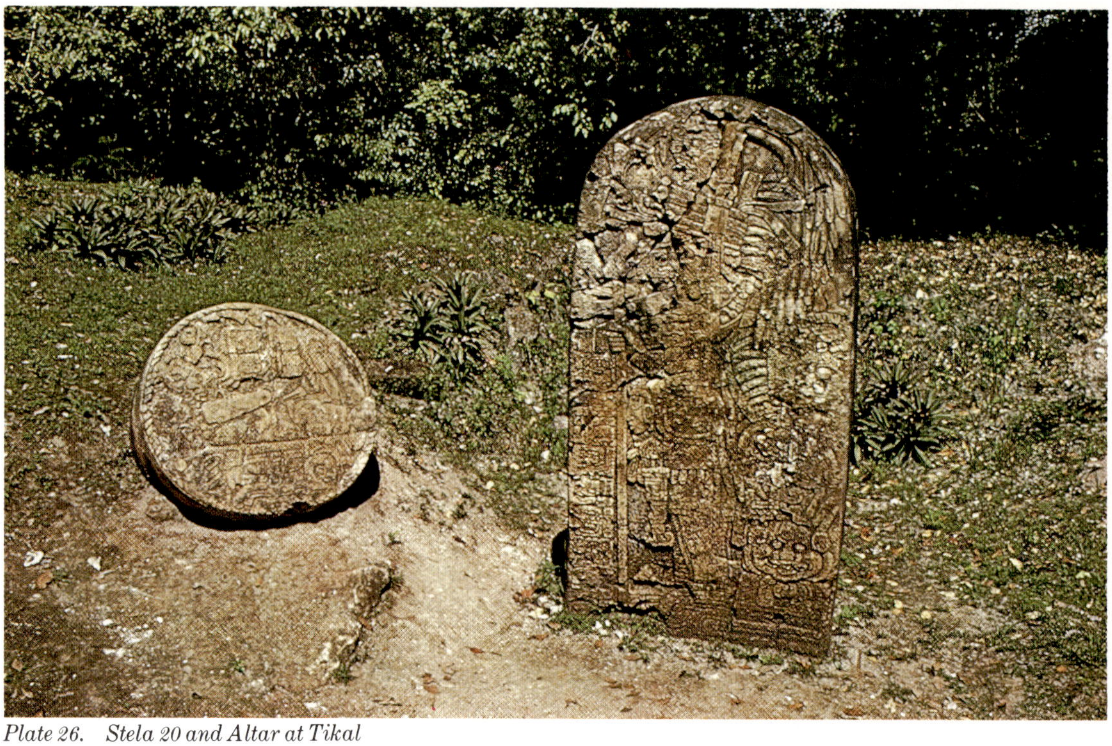

Plate 26. Stela 20 and Altar at Tikal

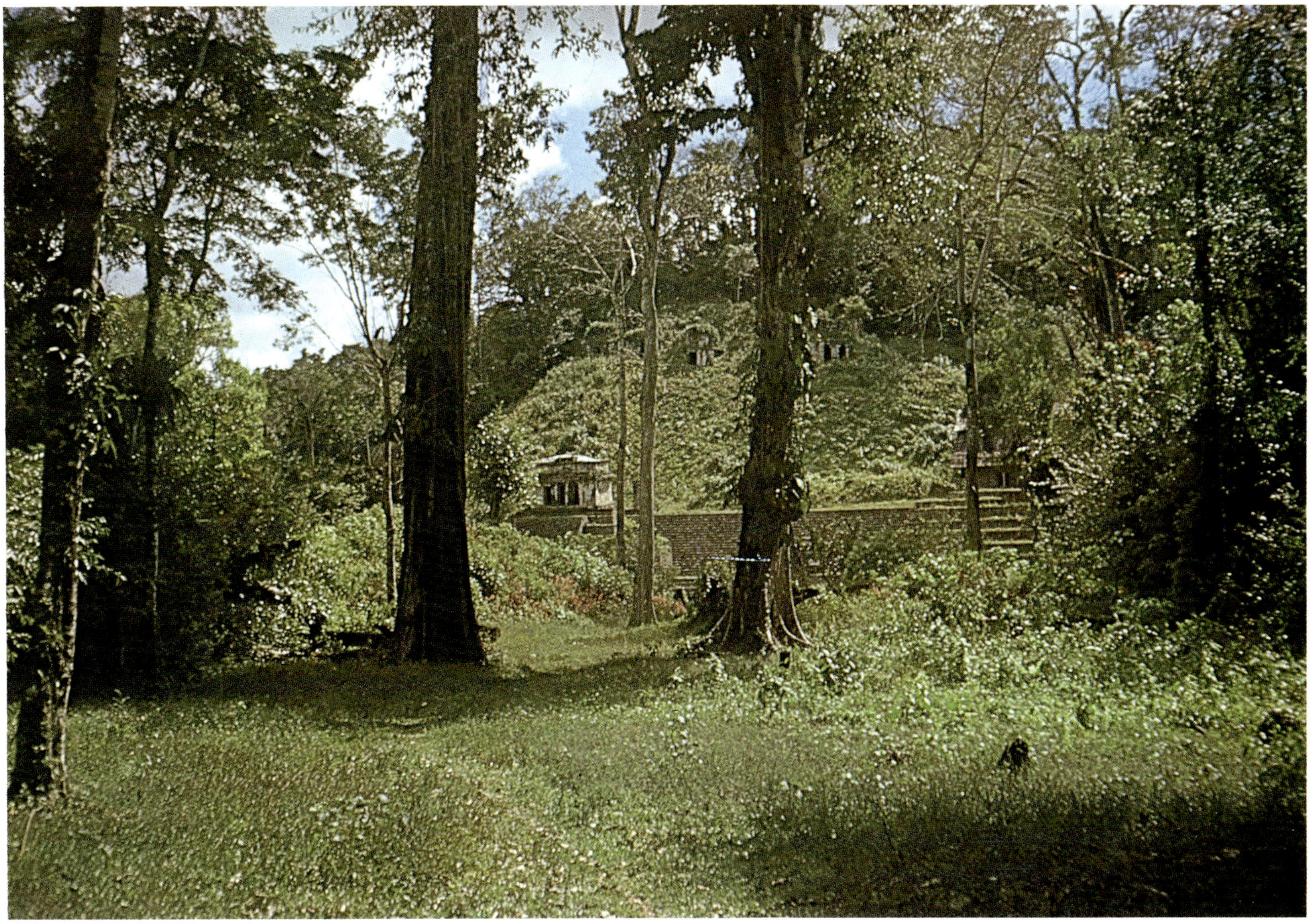

Plate 27. Bonampak

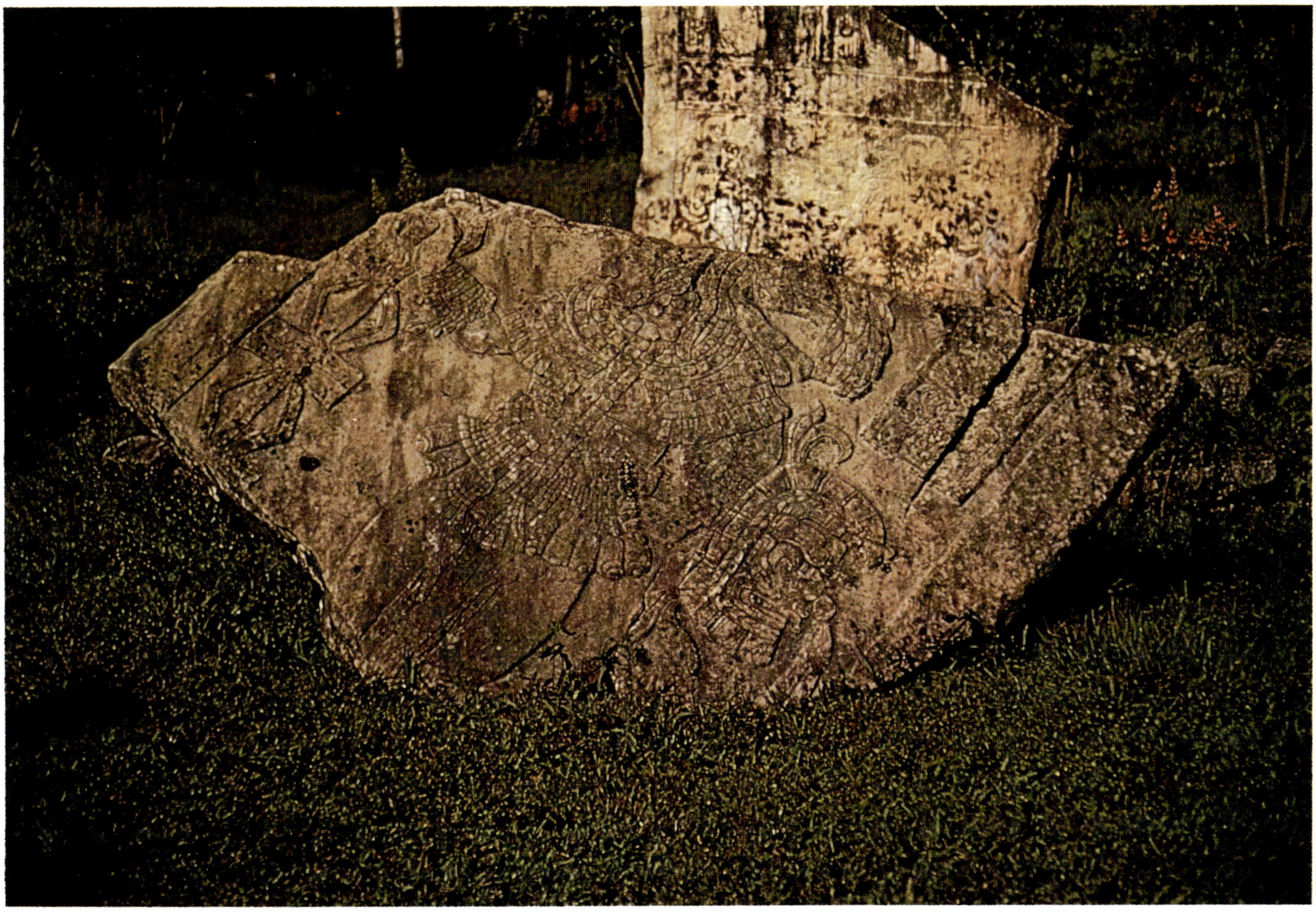

Plate 28. Stela 1 at Bonampak

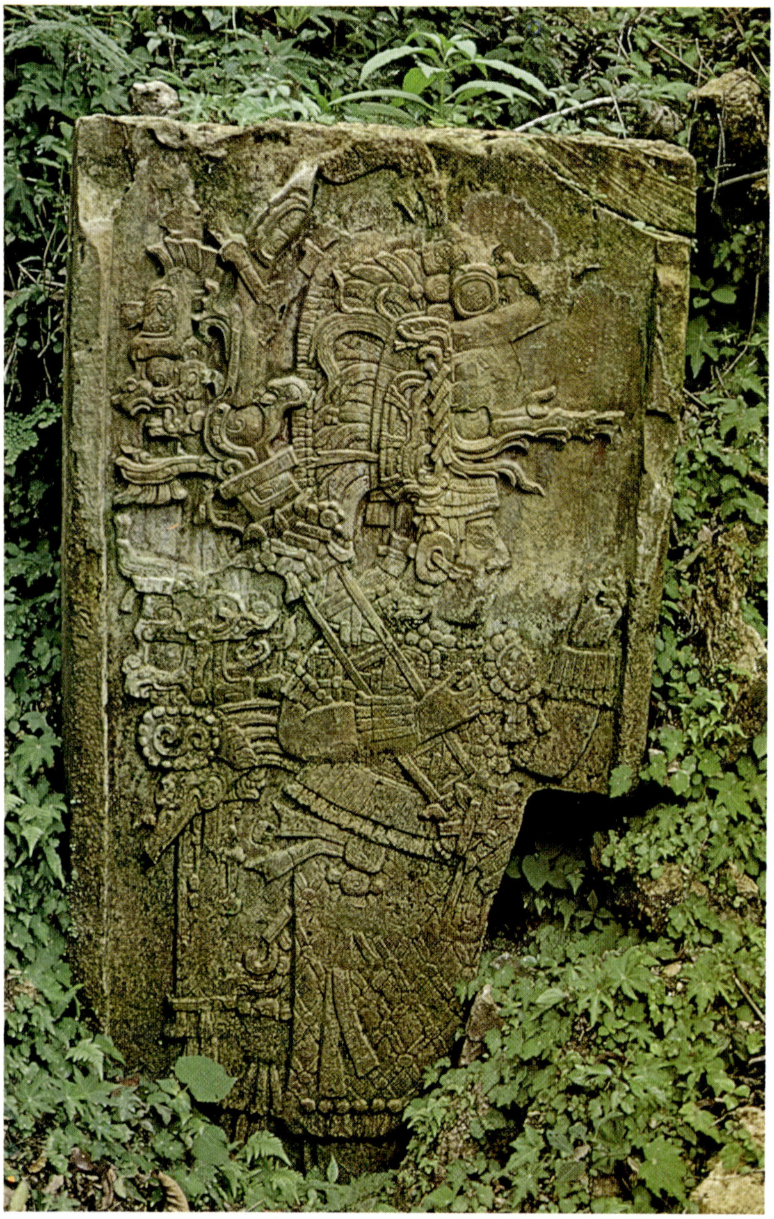

Plate 29. *Stela 16 at Dos Pilos*

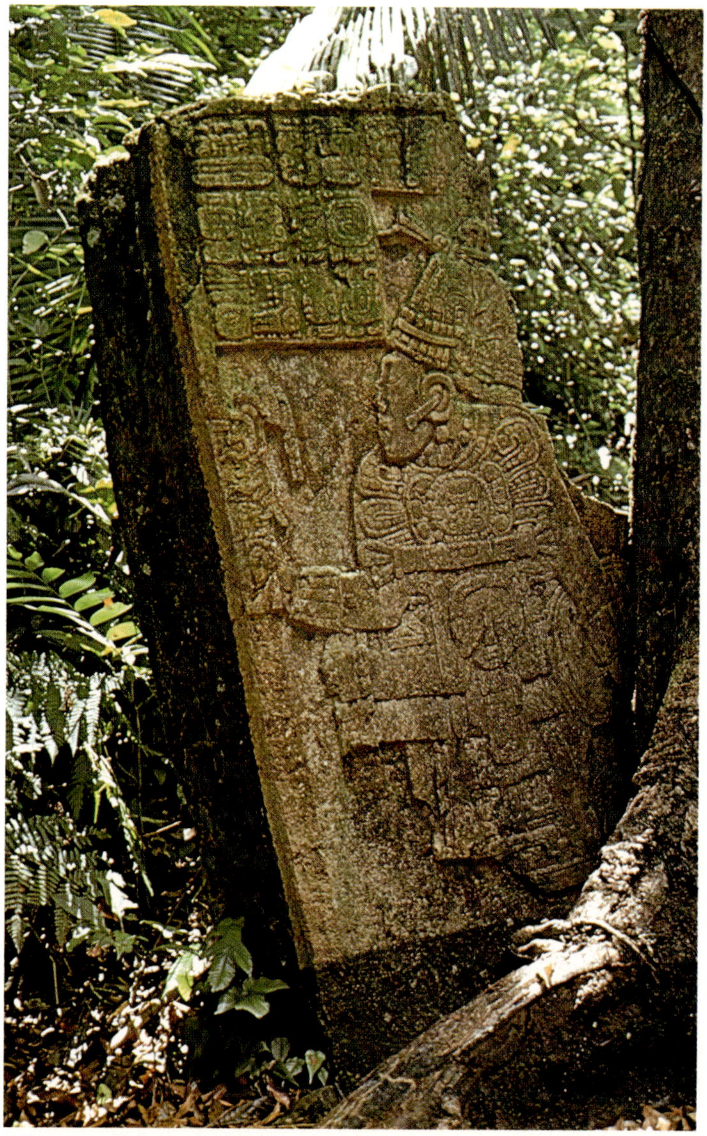

Plate 30. *Stela 3 at Aquateca*

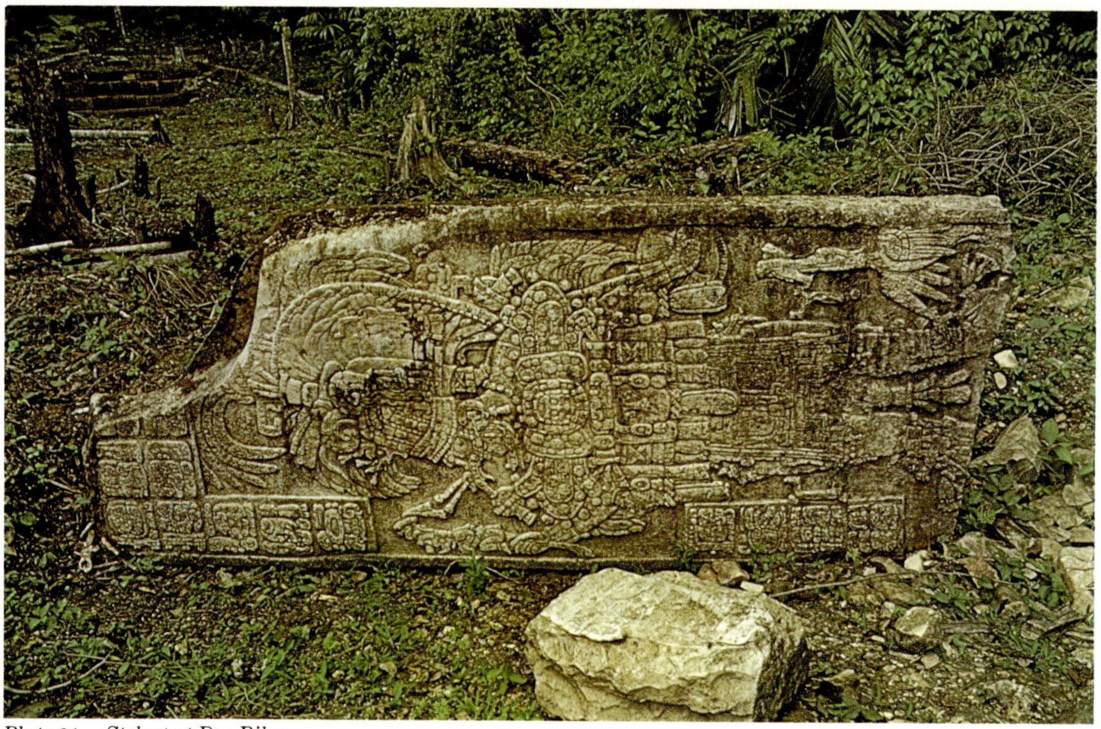

Plate 31. *Stela 1 at Dos Pilos*

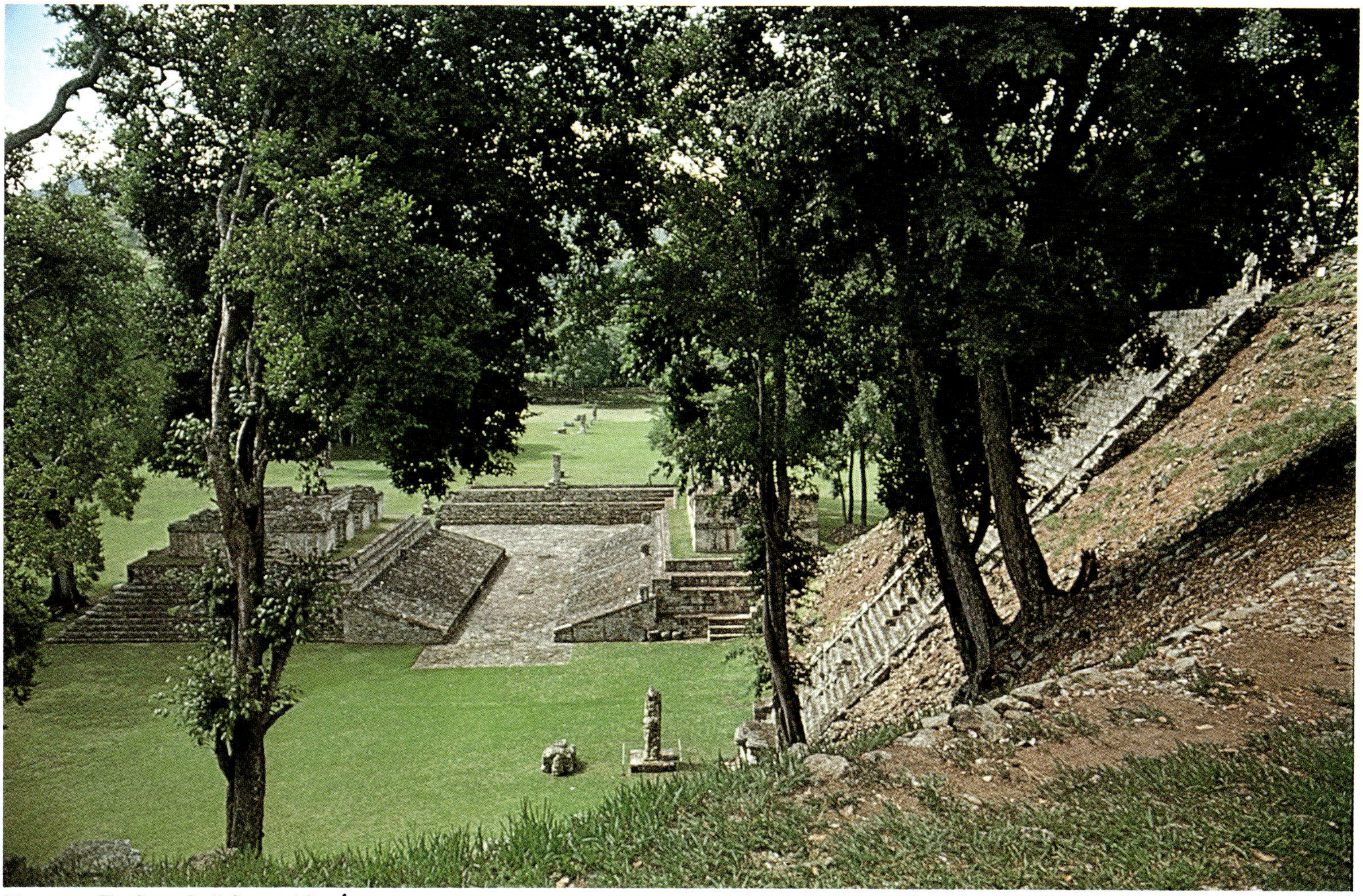

Plate 32. The Great Ball Court of Copán

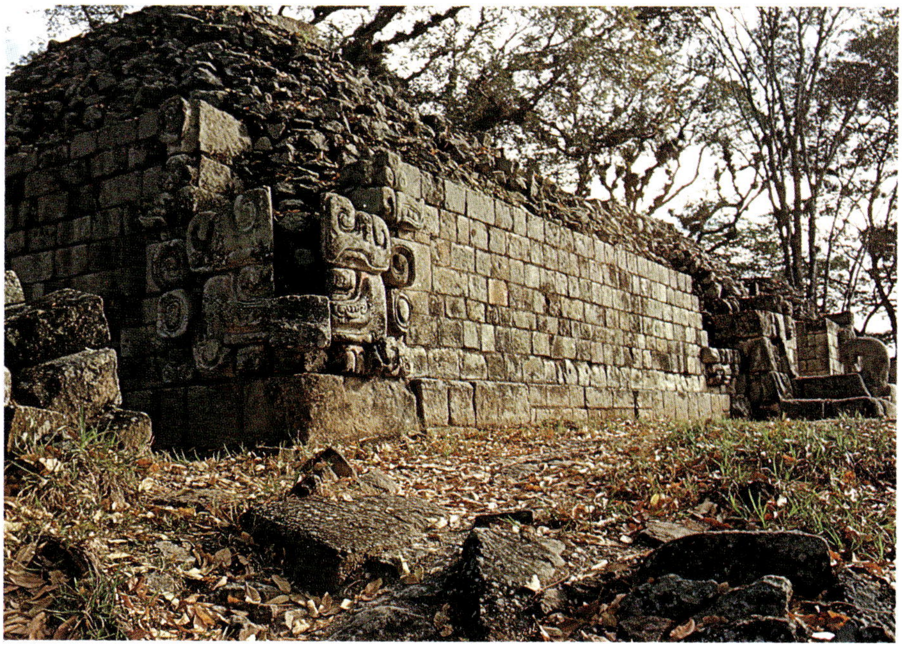

Plate 33. Outside wall with long-nosed mask, Temple 22, Copán

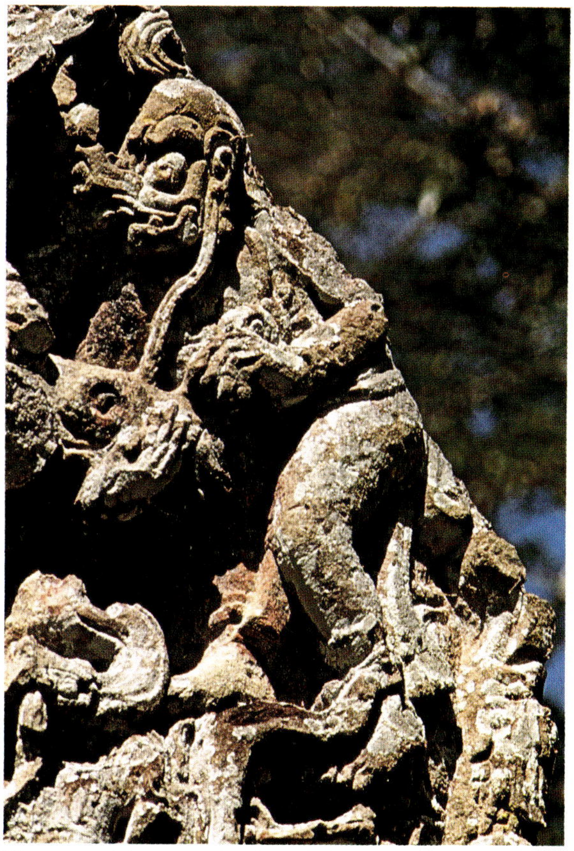

Plate 34. A long-nosed god, probably God K, on Stela D at Copán

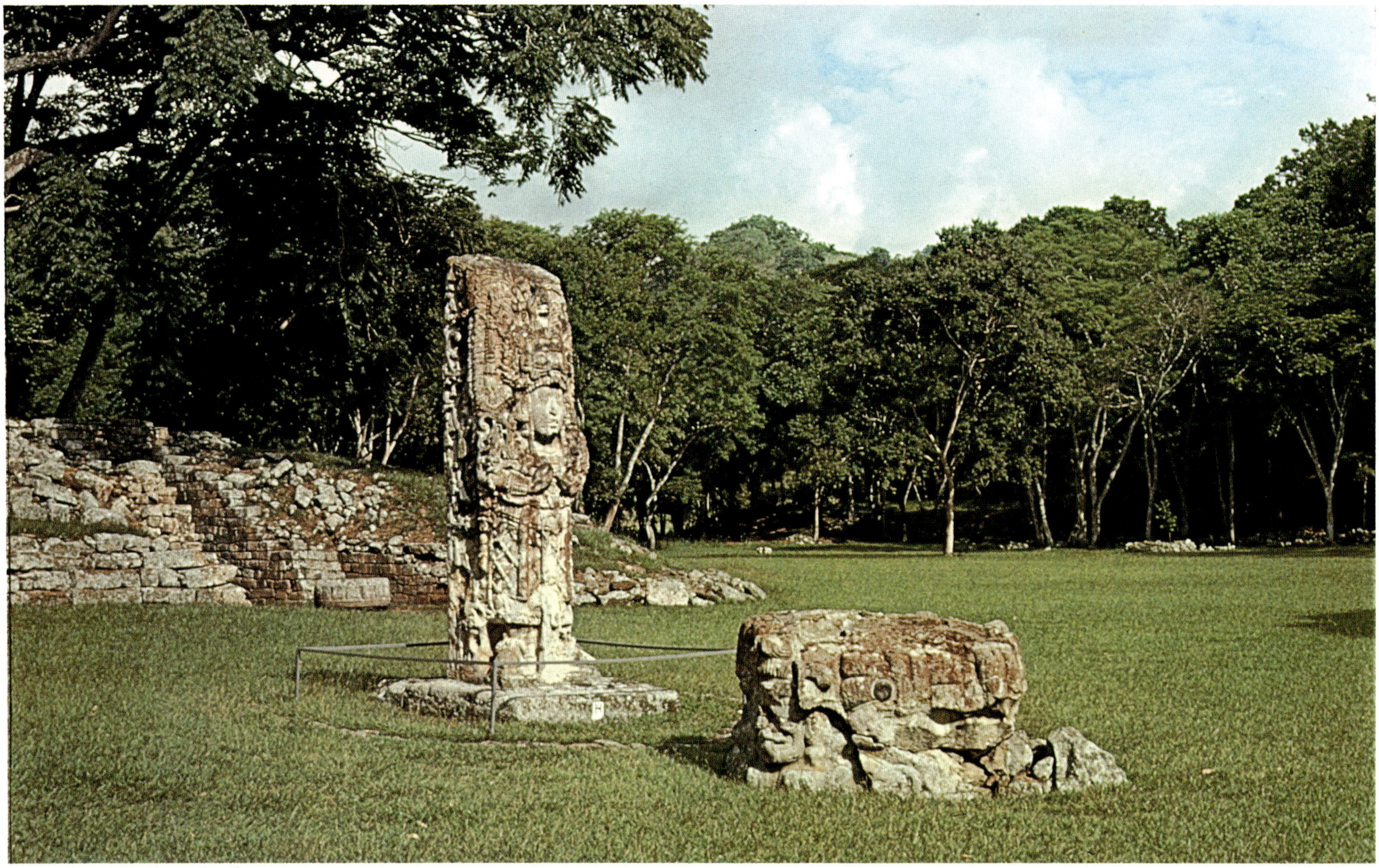
Plate 35. Stela H at Copán

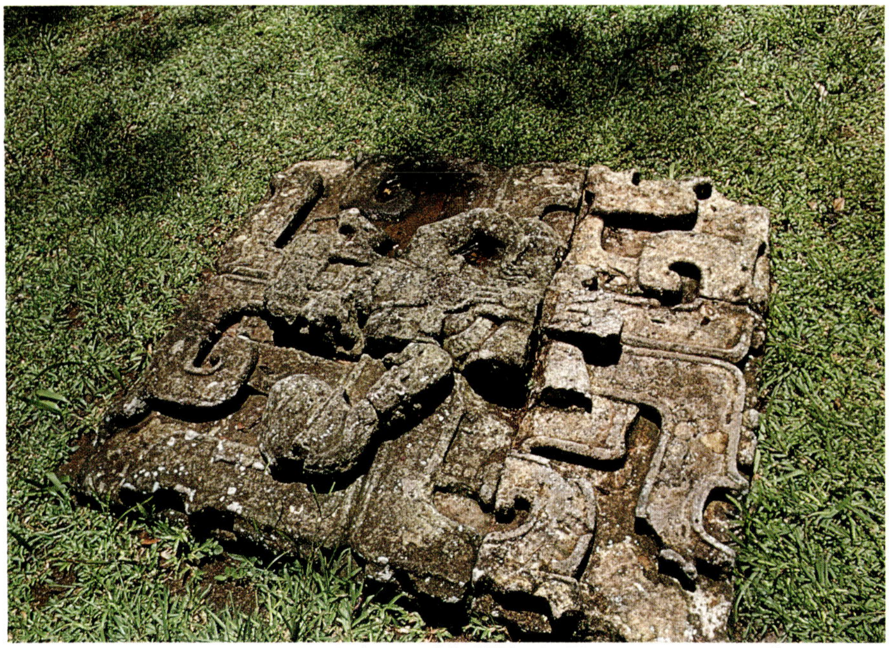
Plate 36. Flat altar carved with the image of God K in the Western Court of Copán

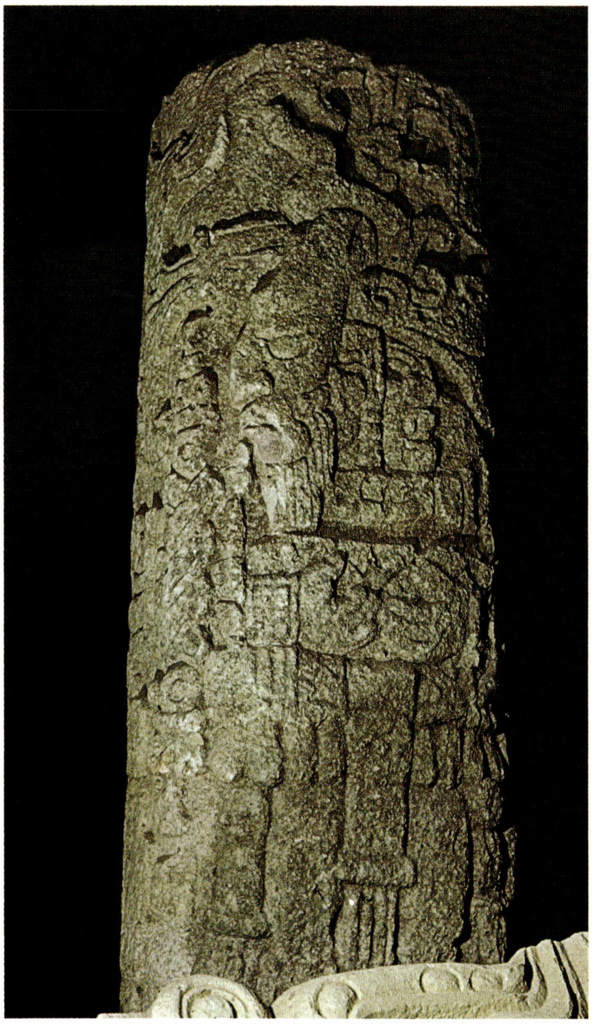
Plate 37. Stela 11 in the Copán Museum

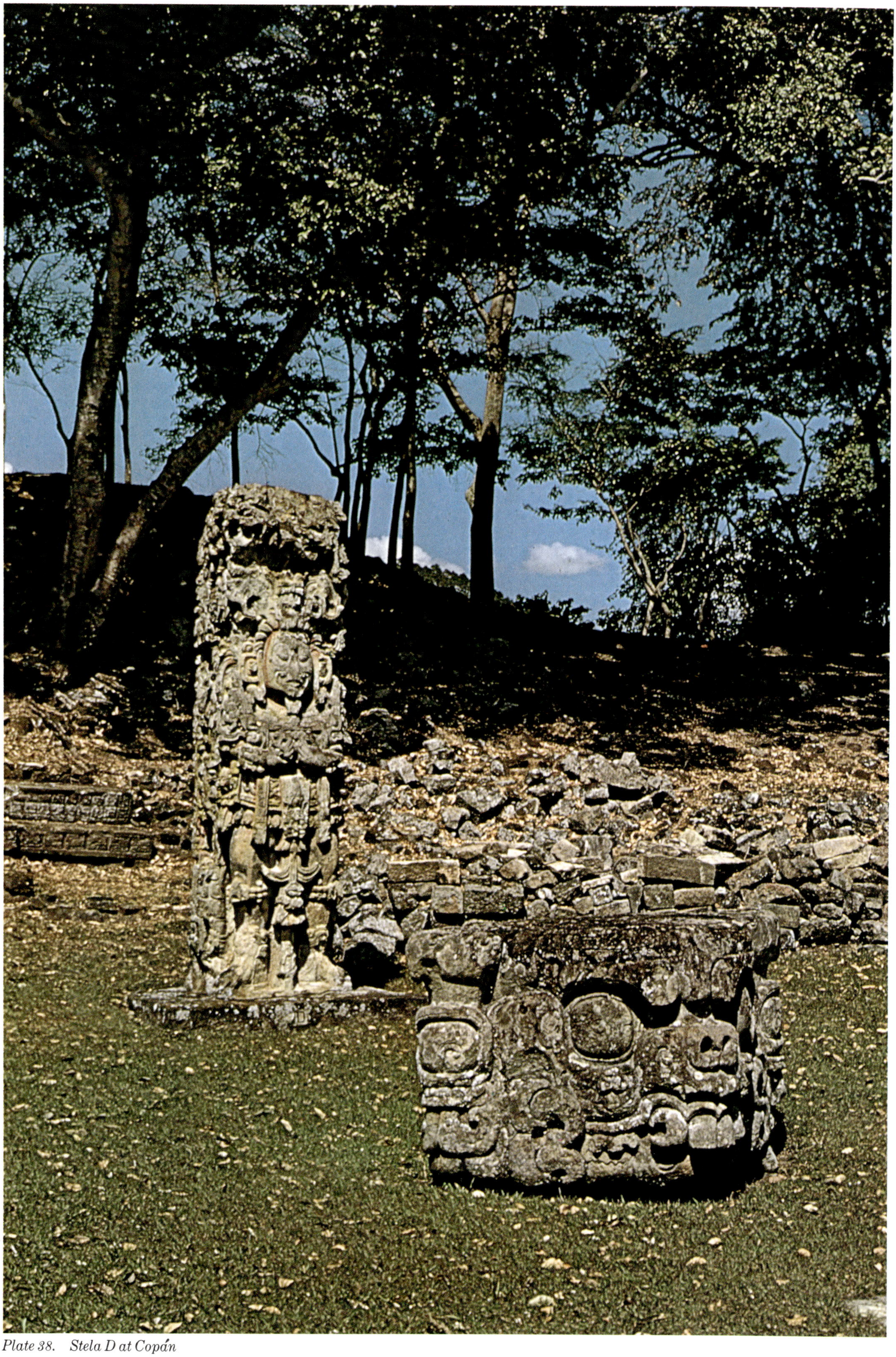

Plate 38. Stela D at Copán

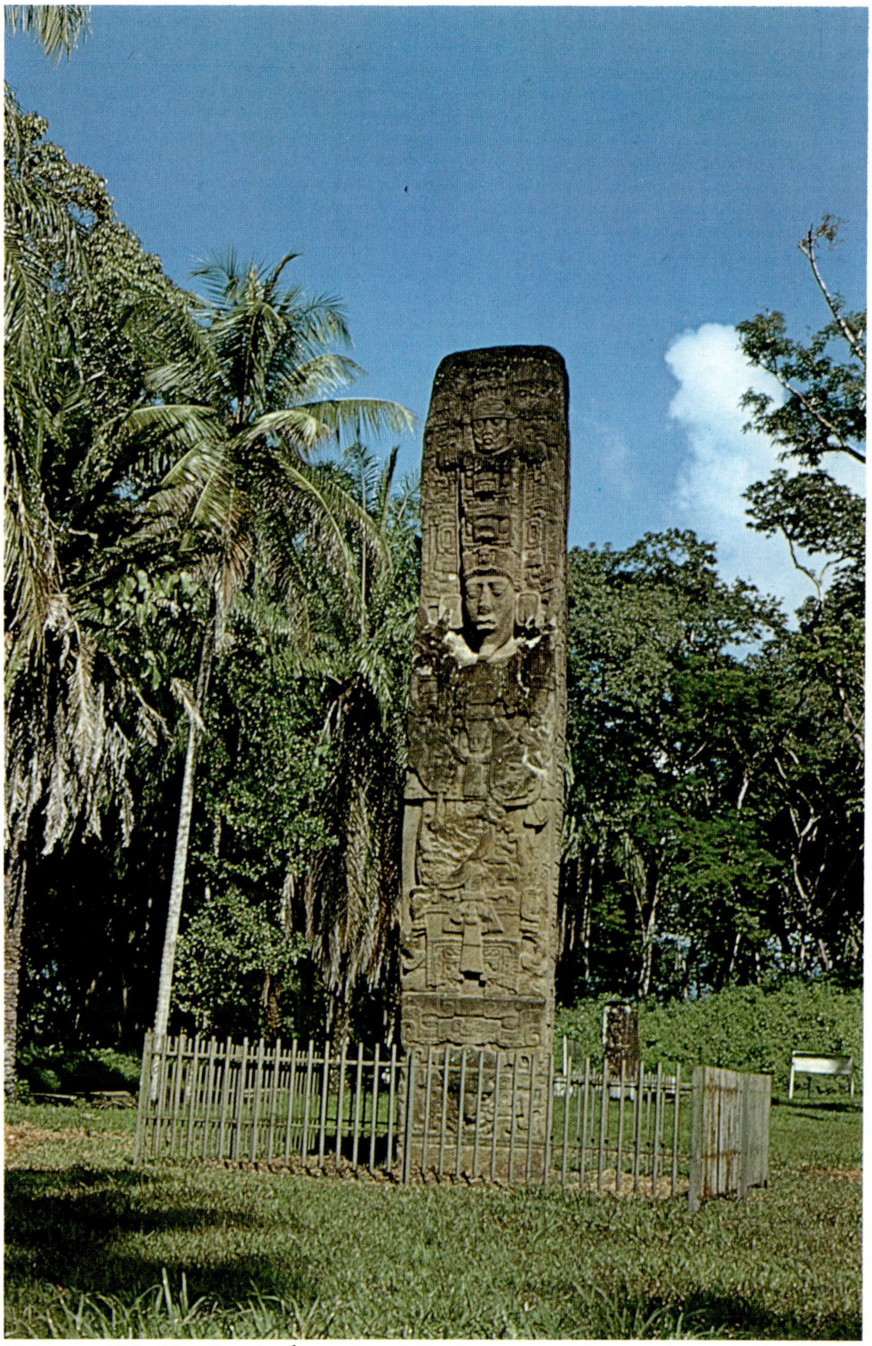

Plate 39. Stela E at Quiriguá

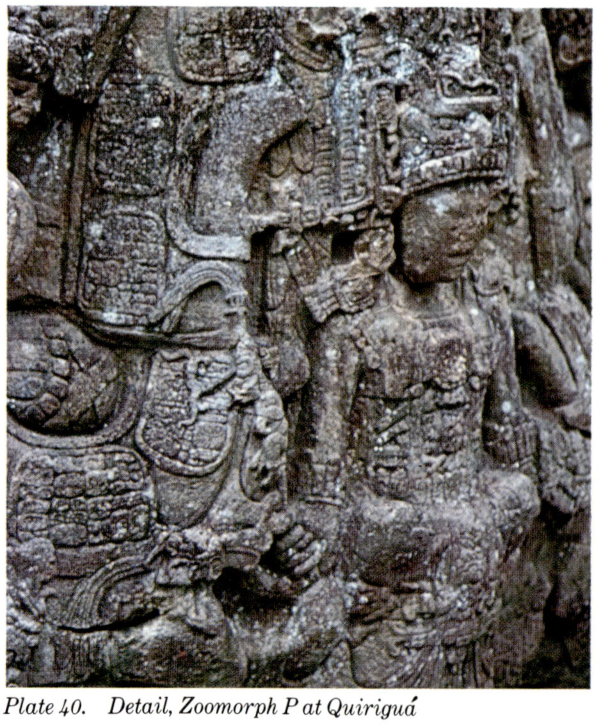

Plate 40. Detail, Zoomorph P at Quiriguá

The fame of tobacco as a healing herb rapidly spread from the Iberian Peninsula to France and from there to the rest of Europe. Jean Nicot's French-Latin dictionary, published in 1573, gives the following definition for the genus *Nicotiana*: "This in an herb of marvelous virtue against wounds, ulcers, noli me tangere [lupus of the face], herpes, and all other things."[16] Tobacco was especially popular as a medicine in England and in Holland. Jan Beverwijk (1594–1647), the famed "first parish doctor and professor at Utrecht," wrote that tobacco

> has a wonderful way of ridding the head of all moisture and phlegm; it has a tendency to promote dessication; . . . it turns the stomach and loosens the bowels. . . . It is an excellent remedy against the plague, and reliable persons told me that when a great many people in London died of the plague, those who frequented the tobacco houses never caught the disease.[17]

This view was generally accepted. It has been said that during the reign of Charles II the worst floggings the boys ever received at Eton were the ones they were given because they refused to smoke this "preventive medicine against the plague."[18]

The extravagant claims of tobacco enthusiasts soon aroused an attitude of satirical disparagement among the more enlightened, among whom Ben Jonson was the earliest to offer criticism in print.[19] James I of England also included a vitriolic chapter on tobacco therapeutics in his *Counterblaste*:

> It cures the Gaut in the feet, and (which is miraculous) in that very instant when the smoke thereof, as light, flies up to the head, the Vertue thereof, as heavie, runs down the little toe. It helps all sorts of Agues. It makes a man sober that was drunke. It refreshes a weary man, and yet makes the man hungry. Being taken when they goe to bed, is makes one sleepe soundly, and yet being taken when a man is sleepie and drowsie, it will, as they say, awake the braine, and quicken his understanding. As for curing the Pockes, it serves but among the pockie Indian slaves. Here in England it is refined, and will not deigne to cure heere any other than cleanly and gentlemanly diseases.[20]

In Mesoamerica tobacco was used not only to treat but also to prevent disease. Anguiano recorded that the Jicaque Indians chewed *yute* paste, "which is leaves of green tobacco mixed with lime," to protect them from contagious diseases.[21] Paso y Troncoso noted that chewing tobacco "strengthens the teeth."[22]

Curing with Tobacco Among the Mayas

When the Spaniards arrived in Yucatán, they found a rich medical lore among the Mayas. They reported:

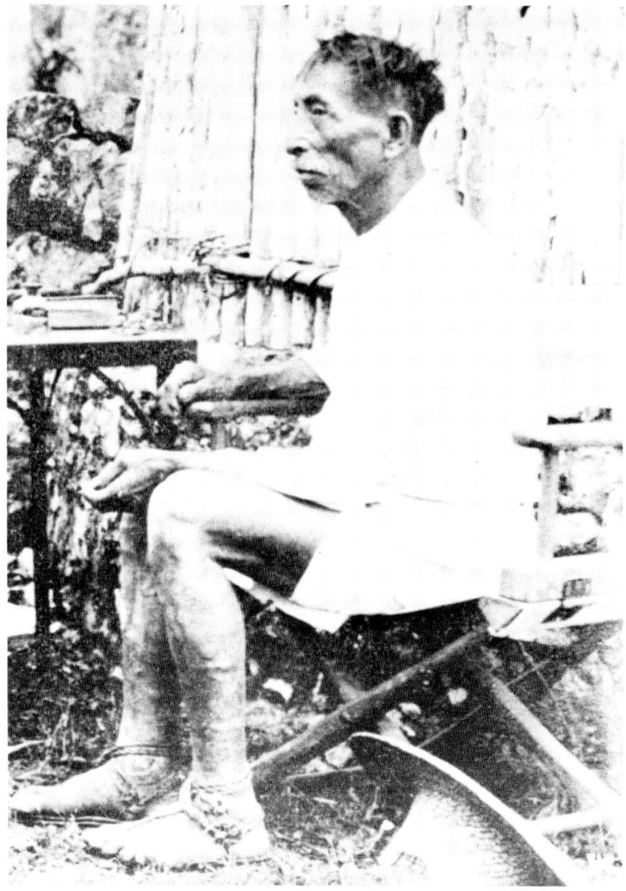

Figure 39. Yucatec herb doctor (from Roys, The Ritual of the Bacabs)

> There is in this land a great quantity of medicinal plants of various properties; and if there were a person in it who had knowledge of them, he would find them most useful and effective, for there is no disease to which the native Indians do not apply plants. However, if you ask them for an account of their properties, they cannot give any other than they are cold or hot and they are accustomed to employ them for the effects, for which they apply them. Nevertheless, there are many of great virtue for every sort of sickness and antidotes for poison; and on the other hand there are also some that are very poisonous and deadly.[23]

The Mayas believed that ailments were sent by supernatural forces as a punishment for their sins, and it was to the gods that they prayed for cure. This concept of divine protection as opposed to punishment is not peculiar to the Maya religion. It is also cleary expressed in the Old Testament as in this verse from Exodus:

> . . . if thou wilt diligently hearken to the voice of the Lord thy God, and wilt do that which is right in his sight, and wilt give ear to his commandments, and keep all his statutes, I will put none of these diseases upon thee, which I have brought upon the Egyptians: for I am the Lord that healeth thee. [Ex. 15:26]

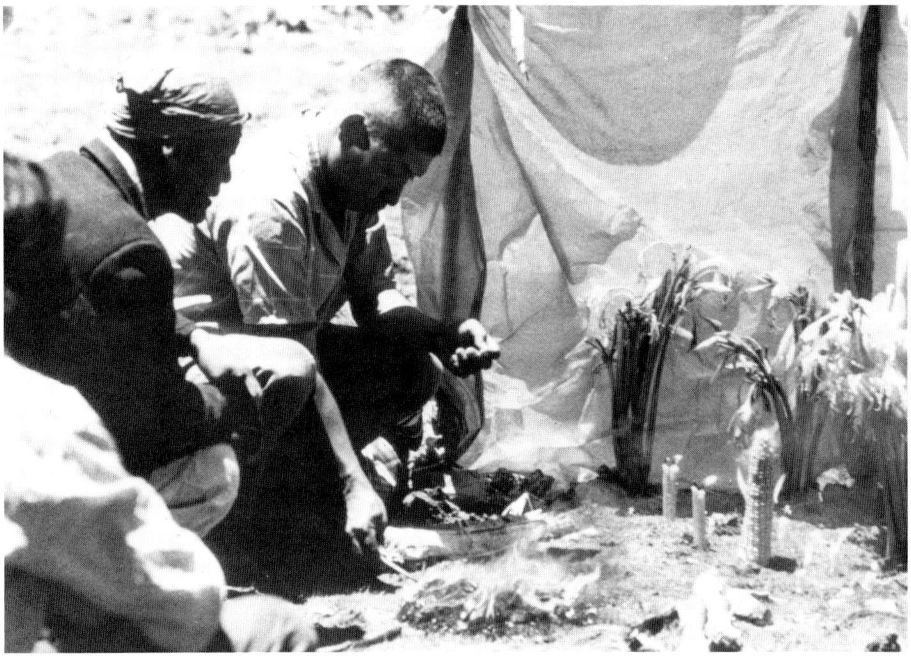

Figure 40. Curing ceremony among the Mam Mayas

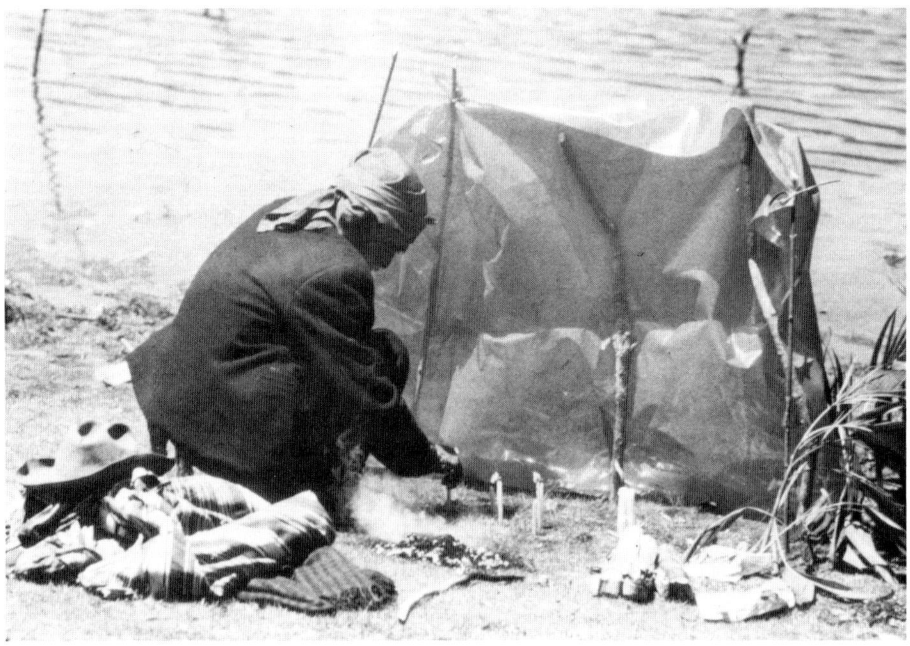

Figure 41. A Mam curandero

For the Mayas religion and medicine, priest and physician were inseparable. That is shown in Landa's *Relación de las cosas de Yucatán:*

. . . the priests, the physicians and sorcerers . . . are all the same things. . . .

On the following day the physicians and the sorcerers assembled in one of the houses with their wives, and the priests drove away the evil spirit. Which being done, they opened the bundles of their medicine, in which they kept many little trifles, each having his own little idols of the goddess of medicine, whom they called Ix Chel. And so they called this festival *Ihcil Ix Chel,* as well as some stones they called *am* [spider], of the kind they used for casting lots. . . . There were also surgeons, or, to be more accurate sorcerers, who cured with herbs and many superstitious rites.[24]

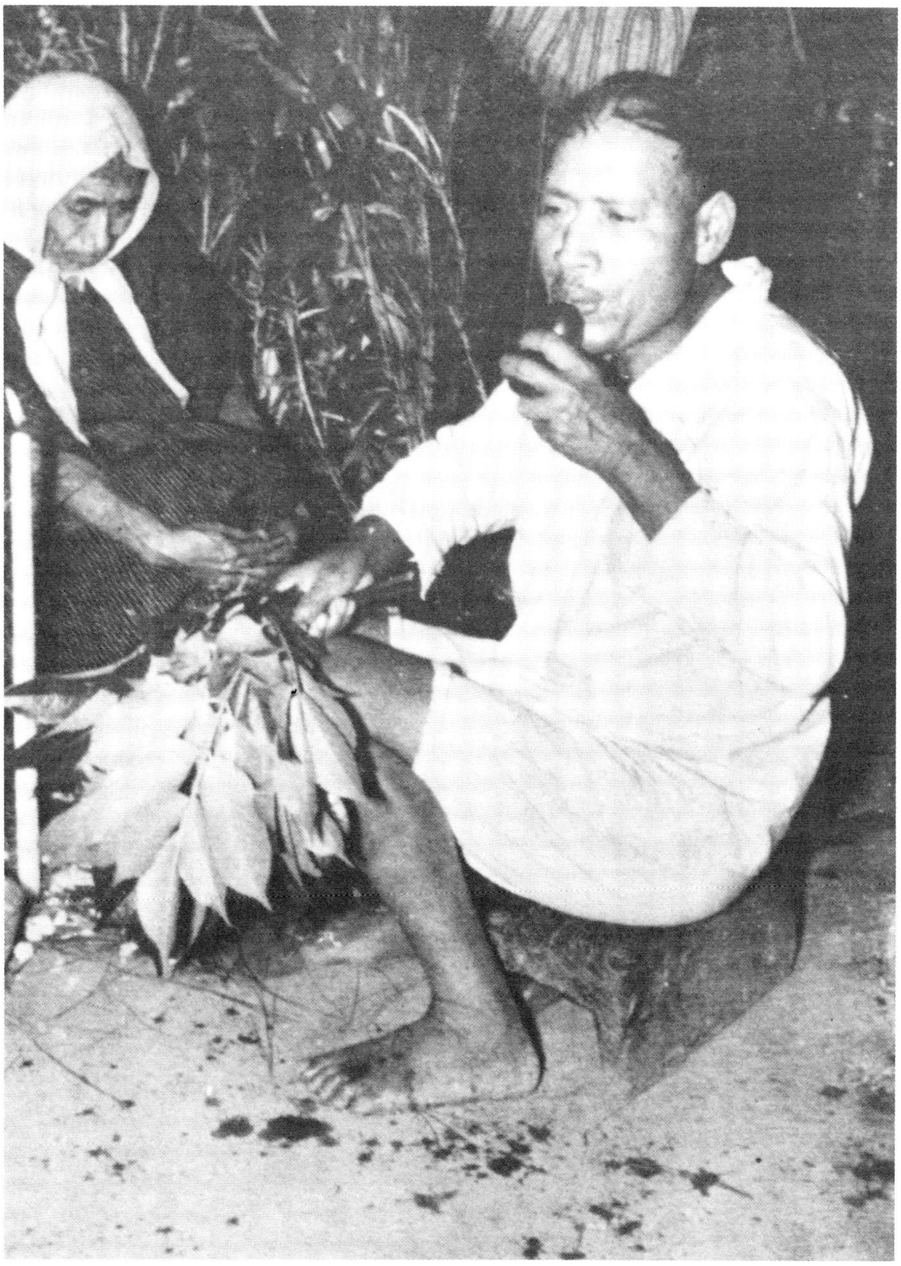

Figure 42. A Tzotzil curandero *(after Holland)*

The Mayas also believed that internal and external malignant forces caused ailments. Two mysterious powers dwelled inside the body, the *kinam* and the *tamcaz*. The *kinam* was a force of ambivalent character; sometimes it acted on behalf of its host, but more ofter it had evil intent. The effects of *tamcaz* were always malignant: stiffening of the body, spasms, epilepsy, frenzy, and nervous shock.[25] The Mayas attributed other diseases to external causes, such as spells cast by sorcerers, ill winds, and malevolent animals. In the sixteenth-century *Diccionario de Motul* are listed the names of persons capable of casting malignant spells (the names were formed by adding the phrase *ah pul*, "he who casts," to the name of the disease). Such persons were thought to be capable of making children scream or refuse the breast and of causing diarrhea, strangury, bloody stools, tapeworms, lung disease—indeed, any illness.[26]

Among the malevolent animals mentioned in ancient Maya medical scripts are snakes in the bowels and a venomous lizard that could cause a throbbing headache merely by touching one's garment or biting the shadow of one's head.[27]

Especially dangerous was the "purple *tancaz* parrot" pictured in the *Book of Chilam Balam of Kaua*. Yucatec villagers still believe that when the parrot flies over the house it vomits a substance that causes death if it falls into the mouth of a sleeping child (an explanation as logical as many I have read in the *Journal of Pediatrics* to explain the cause of crib death).[28]

A considerable body of medical literature was written by the Maya Indians. Unfortunately, the earliest extant manuscripts date only from the eighteenth century. There are indications, however, that these manuscripts were compiled from older, probably pre-Conquest, sources.[29] A comparison of the prescriptions of an eighteenth-century *yerbatero*, "herb doctor," with the data in Landa's *Relaciones* and the *Motul Dictionary* indicates that the pharmacological agents and the ways of administering them had changed very little since the sixteenth century, when Maya medicine was certainly still unaffected by Spanish influence.

Tobacco as a remedy for physical and mental ills is mentioned in the *Princeton Codex*, or, as it is better known, *Ritual of the Bacabs*.[30] This manuscript, which contains a mixture of incantations and straightforward medical prescriptions, was so named by William Gates because of the frequent mention of the Bacabs, benevolent deities of the Maya pantheon. While the manuscript was written in the latter half of the eighteenth century, it was certainly copied from a much earlier text. *Ritual of the Bacabs* recommends tobacco for toothaches; chills; lung, kidney, and eye diseases; and so on.[31] The ritual to dispel a seizure is as follows:

This is also a worm, in the tooth also

. . . Then, how? I drag in my red *tuncuy*-tree, my white *tuncuy*. This, then, is to kindle the fire, where I roast my red breath, my white breath. What is his bush? The red *chacah*-tree, the red *kutz*-tobacco, the white *kutz*.

This was his mother, the *ul*-snail, in a way, your father. You are the great substitute (or mask, *koh*). I bite you. I bite the back of your claw. Hun Ahau. I curse you, seizure! First scattered there, first scattered then![32]

Tobacco is also mentioned in the "Incantation for Placenta." Today among the Mayas of Chiapas a birth is attended by the husband and a midwife. If the placenta is retained after birth, a "man who knows" is called in. In addition to manual efforts he also combats the forces of death with his magical chant: "Chac Bolay! Where is my symbol, when I arise to lay my hand on the bowels of Itzam-cab? My supreme belly are you. . . . Bed-soiled are you. . . . Covered tobacco (*kutz*) are you then. The lust of the sire (*cit*), the lust of the darkness . . . are you."[33]

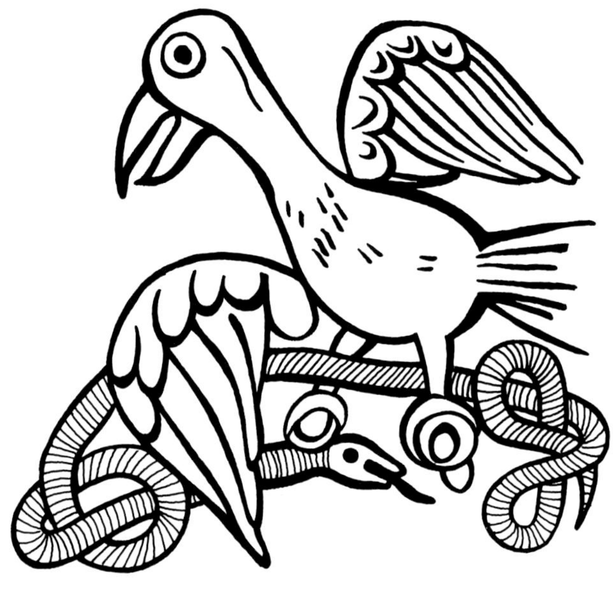

Figure 43. A spider-snake-macaw-wind seizure (after Roys)

The remedy for tapeworm infestation is the following mixture of folk remedy and chant:

Can-tippte [tapeworm] *in the abdomen of man. Very good.*

. . . A paste of *kutz* (tobacco) is to be pressed on the abdomen of the man. Then he is to be anointed with the juice of *kutz* on his abdomen. This is the drink given for snake-pulsation. Shortly ago, how? I created the *can-chac-che* shrub, its stalk. Shortly ago, how? I created the *kutz-aban* ("tobacco bush"), its wine. Shortly ago, how? the white *can-tiplah* [flower?]. It is to be drunk, its wine. Well steeped is its wine. Then I stretch you out, white snake-pulsation, red-snake-pulsation. Amen.[34]

For eruptions, fever, and seizures, "the *chacah* [*chacau*, chocolate?] is to be drunk with two peppers, a little honey, and a little tobacco juice. It is to be drunk." [35]

Another famous Maya medical treatise is the *Yerbas y hechicerías del Yucatán* (Herbs and Magic Spells of Yucatán), better known as *Libro del Judeo* (Book of the Jew), by Ricardo Osado, several manuscripts of which are extant.[36] In the *Yerbas* tobacco, especially its green leaves, is the ingredient in different concoctions recommended for such maladies as lack of strength, pain in the bones and all over the body, snakebite, abdominal pains, pain in the heart, recurring chills, convulsions, loss of speech, sore eye, "dark flint," buboes, various kinds of pox, and retention of urine.[37] Whiles Osado's treatise is a bona fide pharmacopoeia, containing neither ritual acts nor magic incantations, his exotic medicines include a wide range of plants, as well as roasted, fried, and baked animals from poisonous snakes to screaming hawks.

The *Descripción de San Bartolomé* also mentions tobacco as a remedy for snakebite:

> There is a great quantity of medicinal herbs and the natives use it as such to cure all kinds of diseases. The most commonly used and also abundant in the area is *picietl*, which in Spanish is known as tobacco. The natives use this herb fresh and dry and also use it crushed to cure snake bites and by soaking the herb and applying the extract to the wound itself, covering it with the same leaves to relieve pain.[38]

Tobacco was listed as a medicine to prevent miscarriage. It was mixed with lime and then smeared on the abdomen and massaged into it.[39]

Descripción de la Villa de Espíritu Santo describes how tobacco was used to treat chest diseases:

> In this province there is a great deal of tobacco which the natives use many different ways. They chew it, once it has been ground up, they mix it with lime and when the substance is mixed, they keep it in their gums. The natives contend it numbs their hunger and thirst. They also claim it numbs them. They are also known to put it in tubes [*cañuotos*] which are filled up with ground tobacco without lime. They claim this is good for asthma, painful chest, and helps them expel phlegm. It also alleviates rheumatism and helps with pain in other principal parts of the body with many diseases which plague them.[40]

Frowned upon by the church and persecuted by the civil authorities, the Maya *ah-men*, or *yerbatero*, still finds his professional services very much in demand. In his dual role as doctor and sorcerer he uses incantations as well as herbs, among them tobacco, both as medicine to cure disease and as charm to ward off evil spirits. It is reported that today, in the Oxchuc, Chiapas, area, where the men chew locally grown tobacco to "provide warmth in the cold days," tobacco is also applied to a patient's body during certain rituals while the healer vocally offers it as a gift to the supernatural in return for a lost soul.[41]

The Tzeltal Maya medicine man mixes tobacco (*mai*) with black pepper (*'asus*) and sulphur (*asupre*) to treat the illness known as *hel'ik'*, "wind remains." To cure bloating (*pumel*) the *yerbatero* tells his patient to eat tobacco leaves mixed with garlic cloves. Tobacco is also eaten to prevent constipation (*mak ca' ubel*) and to enhance the passage of gas (*mak cisubel*). *Mai* is also used as a drawing agent in a plaster over snakebites (*ti' ben can*).

The hybrid variety of *N. tabacum* (*towaco may*), which has very large leaves, is used for the treatment of the disease of the prostate called *hukulub* and to help alleviate difficulty in urination (*mak cusubel*). The medicine is made by browning the leaves in a griddle, grinding them into a fine powder, which is mixed with cane liquor and drunk.[42]

In the Maya community of Cham Kom, Redfield has reported the present-day use of tobacco in curing ceremonies. Tobacco smoke is passed over the body of the patient to counter the "evil winds" that caused his sickness.[43] Martínez also described tobacco as *muy medicinal* and said that it was used by the Yucatec Mayas to treat paralysis of the tongue, toothache, gout, and various inflammations.[44]

A widely practiced "curing" with tobacco can be seen by travelers in Petén. It is thought to be an excellent remedy for scorpion and tarantula bites to which it is applied locally, and it certainly does wonders to disengage the larva of a *chiclero* fly that has burrowed under the skin.

In this way they would inhale the smoke until they became unconscious and lay sprawling on the earth like men in drunken slumber.

Gonzalo Fernández de Oviedo y Valdés

Tobacco is not a hallucinogen, but in massive doses it is a powerful intoxicant.

Gerald Weiss

I do not wish to suggest that we can reduce all of Maya culture to a series of drug-induced inputs. Rather, I believe that anthropologists must become sensitive to the role of sacred plants in cultures around the world and not permit their own cultural blinders to prevent them recognizing their significance.

Marlene Dobkin de Ríos

5. Unexplained Effects of Tobacco

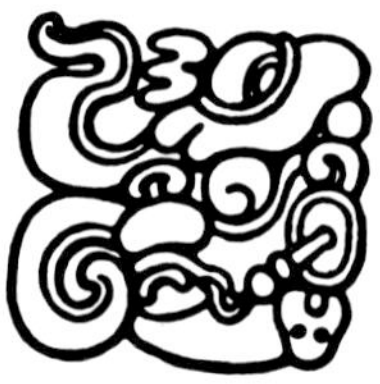

According to the noted Guatemalan historian Fuentes y Guzmán, the highland Mayas held tobacco in great respect:

They also worshiped and attributed deity to the herb they call *picietl*, which is tobacco, concerning which they had superstitions. Smoking it, and becoming intoxicated with it, they invoked the demon to learn future events and to consult on the requests and petitions of others with which they had been charged.[1]

The drunken behavior of those who smoked tobacco was described by other early observers. Jerónimo Mendieta, in his *Historia eclesiástica indiana*, wrote:

Others say that some regarded the herb called *picietl*, and the Spaniards tobacco, as the body of the goddess Ciuacoatl. And for this reason it has some medicinal effects. It must be handled cautiously because it is dangerous, it takes away the judgment from those who take it, and makes them behave foolishly and wildly.[2]

Another early traveler described the intoxicating effect of tobacco:

Then they set fire to one end [of the cigar] and, putting the other end into their mouth, they drew breath through it, and they fill themselves with this cruel smoke that they lose their reason; and some take so much of it they fall down as if they were dead and remain the greater part of the day or night stupefied.[3]

Another account, from 1526, says: ". . . in this way, they became unconscious and lay sprawling on the earth like men in a drunken slumber."[4]

Similar effects were mentioned by the very keen observer Sahagún:

The one who sells *piciete* first grinds its leaves, mixes it with some lime, and after it is mixed rubs it very well into the hands; some do it with incense of the earth, then put in their hands and mouth to relieve headache and provoke drunkenness.[5]

Nuñez Cabeza de Vaca, who traveled extensively above the Gulf of Mexico in 1527–28, also remarked upon the eagerness with which the Indians made themselves "drunk" with tobacco smoke.[6] The *Relación de Nuchistlán*, which describes the life of the Indians at the time of the Conquest in the area that is now the southern part of the Mexican state of Zacatecas, says of tobacco intoxication: "They locked themselves in a house to drink some of the [juice] of *pesietl*, perfuming themselves with the herb that provokes drunkenness; afterward they went out to take a bath for mystical reasons and to talk to their gods."[7] Muriel Porter quotes Solís y Rivadeneyra as saying that the Aztec priests of the Valley of Mexico "got angry and intoxicated with tobacco smoke, which they always used to lose consciousness to understand themselves and the demons."[8]

These records are puzzling. How could *N. tabacum*, which is smoked today by millions of

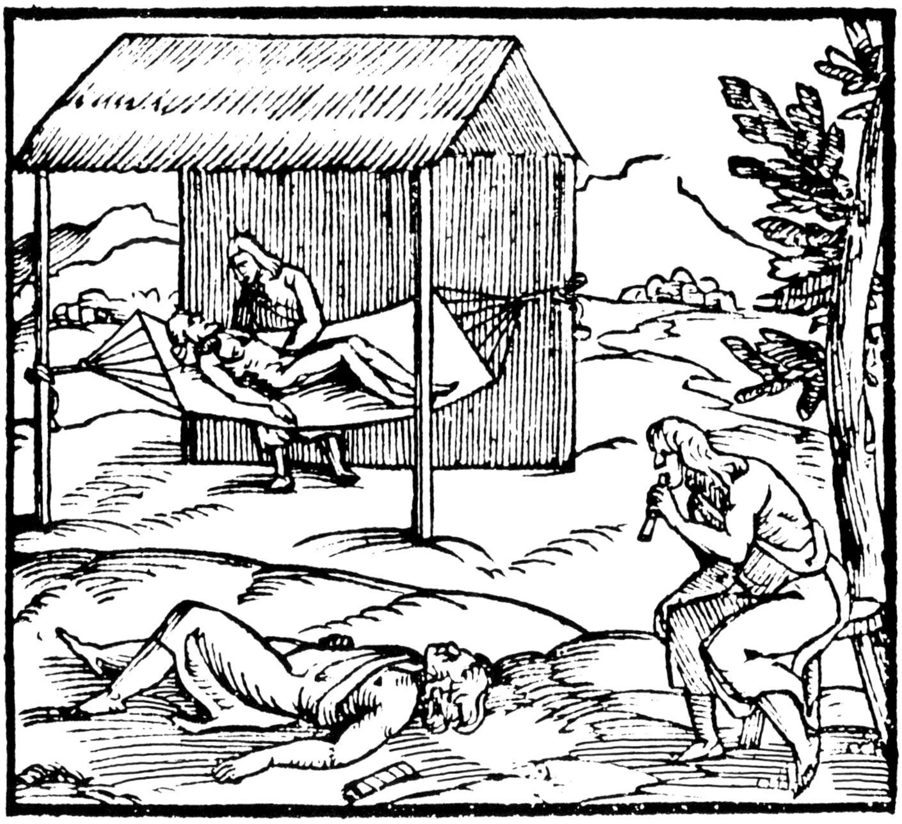

Figure 44. Tobacco narcosis (after Benzoni)

people without any of these effects, have made the Indians "intoxicated," behave "foolishly," "lose their judgment," "fall down as if they were dead," "invoke demons," and so on? The theory offered in the past that the Indians were more susceptible to the effects of tobacco certainly can be dismissed. What then is the explanation?

The paradox—that the tobacco the observers described is a substance allegedly possessing hallucinogenic and narcotic effects that are unknown today—is expressed by Johannes Wilbert:

> Tobacco has long been known to play a central role in North and South American shamanism, both in the achievement of shamanistic trance states and in purification and supernatural curing. Even if it is not one of the "true" hallucinogens from the botanist's or the pharmacologist's point of view, tobacco is often conceptually and functionally indistinguishable from them.[9]

To explain this apparent contradiction, one must consider the following possibilities:

1. The ancient Mayas may have smoked a species of tobacco more potent than the presently commercially available species.
2. The Mayas could have consumed the same species of tobacco that is smoked today but in much larger quantities.
3. The Mayas may have potentiated the effect of tobacco by nonpharmacological means.
4. The Mayas may have consumed herbs other than tobacco with psychotropic properties, in conjunction with or instead of tobacco.

Let us examine these possibilities one by one.

1. The first explanation of the observation that the Mayas became intoxicated while smoking is that, contrary to present belief, the tobacco the ancient Mayas smoked was not the *N. tabacum* that smokers today use but *N. rustica*, or weed tobacco, which has a much higher nicotine content. Nicotine ($C_{10}H_{14}N_2$), the principal agent of tobacco smoke, has been the subject of scientific investigation for many years. However, its effects on the central nervous system are just beginning to be explored. Nicotine may act upon the brain by changing the concentration of biogenic amines, primarily serotonin, and also by altering the vitamin B_6 metabolism.[10] While commercially grown *N. tabacum* seldom contains more than 2 to 3 percent nicotine, *N. rustica (noholki*

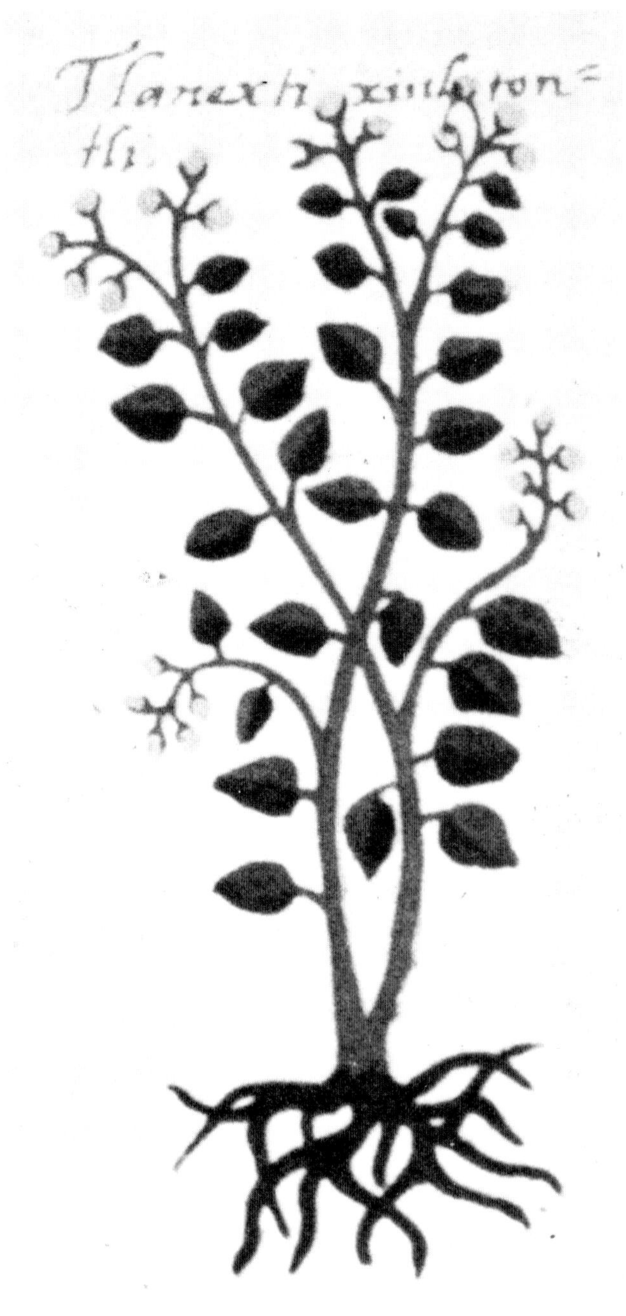

Figure 45. Nicotiana rustica (Codex Badiano)

'k' uits) a variety well known to the Mayas, has been grown with as much as 10 percent nicotine in the leaf.[11] The latter is smoked nowadays only in some rural areas of Mesoamerica, in water pipes in the Middle East, and in some areas of Russia, where air-cured *mahorka* tobacco is manufactured from it. It is most probable that during the pre-Columbian times *N. rustica* was more widely used and that it was the tobacco known as *picietl,* the sacred herb of Aztec rituals, the divine tobacco of the Indians of the eastern woodlands, and also the *petum* of the natives of Brazil. The narcotic effect of *N. rustica* was well known to the Natchez Indians, who gave it to their intended victims of sacrifice to stupefy them before the rite of strangulation.[12] In the United States today *N. rustica* is used only in the manufacture of insecticides because of its high nicotine content, and farmers keep a close lookout for it because it can be poisonous to cattle and sheep.

It also must be recognized that nicotine is by no means the only bioactive substance in the tobacco leaf. Recently alkaloids of the harmala group, harman and norharman, have been isolated from cured commercial tobaccos and their smoke.[13] They constitute a chemical group of B-carbolines, which include harmine, harmaline, tetrahydroharmine, and 6-methoxyharmine, all with hallucinogenic properties.[14] While to date no native varieties of tobacco have been analyzed for these substances, it is a reasonable supposition that their concentration may vary widely, depending upon the variety and growth, and that some of the native-grown tobaccos may contain a relatively high concentration of them.[15]

That the Indians in ancient times may have used more than one species of tobacco, some of them very powerful, is a view toward which several of the modern investigators lean. Peter T. Furst has written:

> We know that American Indians, who for the better or worse gave tobacco to the world, used it exclusively in their sacred and ritual enterprises and never just for recreation—at least not until the white man corrupted its original purpose. We also know that, as noted earlier, tobacco was widely employed by Indians to induce trance states kin to those triggered by "true" hallucinogens, such as *psilocybe* and fly agaric mushrooms, the *Banisteria* vine or *iboga*. Some authorities suspect—and it may yet be confirmed pharmacologically—that tobacco, which belongs to the same botanical family as the hallucinogenic *Datura* (e.g. Jimson-weed) does indeed contain "mind-altering" components, which may vary significantly from species to species.[16]

Although he does not include tobacco in his list of hallucinogens, Richard E. Schultes describes it as a "psychomimetic" herb:

> There are some forty-five extratropical specia (of tobacco) indigenous to North and South America. While there are hundreds of references to the use of tobacco by American Indians in religious, curing and intoxicating practices, writers of the past hardly began to explore the real role and significance of tobaccos in belief and ritual. Nevertheless, the literature from 1492 to the present leaves no doubt about the supreme ritualistic and mythological status accorded to these plants by the indigenous people of nearly all parts of the New World. There is also ample evidence that in many areas tobacco

has been employed to trigger ecstatic states very similar, perhaps even identical, to those induced by the "true hallucinogens." Whether or not some species of *Nicotiana* have chemical constituents with real hallucinogenic activity remains to be seen. The possibility exists, of course, that the solanaceous *Nicotiana* belonging to a family rich in hallucinogenic principle may, when fully investigated, yield interesting phyto-chemical data, supporting ethno-botanical literature that ascribes "hallucinogenic" effects to certain tobacco preparations.[17]

Johannes Wilbert, who conducted perhaps the most thorough examination of the tobacco-shaman complex among the natives of the New World, also leaned toward the theory that they may have used different species of tobacco for different purposes: "In many tribes the tobacco they themselves cultivate or collect in the wild state is reserved for ritual and ceremonial use, whereas white man's tobacco, or 'Virginia Tobacco,' a hybrid domesticate of *Nicotiana tabacum*, is freely smoked."[18] In his *Rerum medicarum Novae Hispaniae* of 1651, Francisco Hernández differentiated two kinds of tobacco: *picietl*, a cultivated small variety, and *quauhyetl*, a larger, wild variety.[19] Similar observations were made by N. Ross Crumrine among the Mayos, a tribe far from the Maya area. The Mayo shaman prays, chants, smokes his special native tobacco, and blows the smoke at the patient, while everyone else present smokes commercial cigarettes. The shaman sings the songs of various saints and hears them speaking to him and giving him advice.[20]

Scanning the older manuscripts with this hypothesis in mind, one finds references to several different kinds of tobacco used by the Mayas in ancient times. The *Ritual of the Bacabs* contains the following chant, which is supposed to cure asthma: "For four days he would drink the juice of red tobacco, the white tobacco, the black tobacco. Then he would be asleep, then he would be curled up."[21]

The *Descripción de San Bartolomé* also mentions two kinds of *tabaco:*

> There are two different types of this plant [tobacco]. One is known as *boniato*, which they use to calm swelling and cure chills by heating the leaves and applying them to appropriate parts of the body, to mitigate pain and take the chill away. The other variety is brighter and stronger and the natives call it "Mexican tobacco."[22]

That some potent *Nicotiana* preparations can induce toxic states is also evident in an article entitled "Herbal Intoxication," which appeared in the *Journal of the American Medical Association:*

> *Case 3.* A 20-year-old man came to the hospital after fainting after smoking one *Hare Rama Bidis* cigarette. He complained of nausea, headache and weakness. He became anxious and decided to leave the hospital after a urine sample was obtained.
>
> The *Hare Rama Bidis* cigarettes are also imported from India and are labeled "tobacco" and sold as *Class A* cigarettes. We have identified the ingredients as *Nicotiana* spp, and have assayed the nicotine levels at 6.42% to 9.03% (commercially available tobacco cigarettes [*N. tabacum*] contain 0.7% to 3.0% nicotine). Patient 3 thus consumed an extremely potent nicotine preparation, and his urine contained large amounts of nicotine (1.5 mg/ml).[23]

2. Another theory to explain why tobacco may have had hallucinogenic properties under certain circumstances is that, while tobacco is consumed in "normal" quantities is not a potent hallucinogen per se, in massive doses it could serve as a powerful one, and the Indians may have consumed enormous amounts during their ceremonies.[24] It is recorded, for example, that the Tenetehara Indians of Brazil regularly smoke very large amounts to induce trancelike states. Cigars thirty-four centimeters long are smoked by the Tukanos. The Waraos, who inhabit the Orinoco River delta in Venezuela, are known to smoke cigars fifty to seventy-five centimeters long for the same purpose.[25] Just such enormous, funnel-shaped cigars are pictured in the 1570 edition of Mathias l'Obel's *Stirpium adversaria nova* and in Thévet's work, which he rather amusingly entitled *Les singularités de la France Antartique, autrement nomée Amerique*, published in 1558.[26] Both of these books depict American Indians smoking cigars ("cornets") of startling dimensions. Although today we are not used to such huge cigars, it is worth mentioning that even in our age very large cigars are still made. In Andorra the men smoke cigars sixty centimeters long and 5 centimeters thick.

3. Another possibility that should be considered in the effort to explain the hallucinogenic effects of tobacco is that the natives may have used it only as a mild psychotropic agent but potentiated its effects by nonpharmacological means. It is well known that the use of psychedelic agents is only one of several ways of achieving a trancelike state conducive to the sense of seeing and contacting the supernatural. In many cultures other methods are used, such as chanting, suggestion, massage, fasting, self-torture, sensory deprivation, breathing exercises, Yoga meditation, ritual dancing, and drumming.[27]

There is ample ethnographic evidence that the Maya shamans used (and in some areas still use) reverberation, chants, dancing, and different forms of music either to induce or to potentiate trancelike states. Even more important are data indicating that in ancient times both fasting and penitential self-torture were widely practiced among the

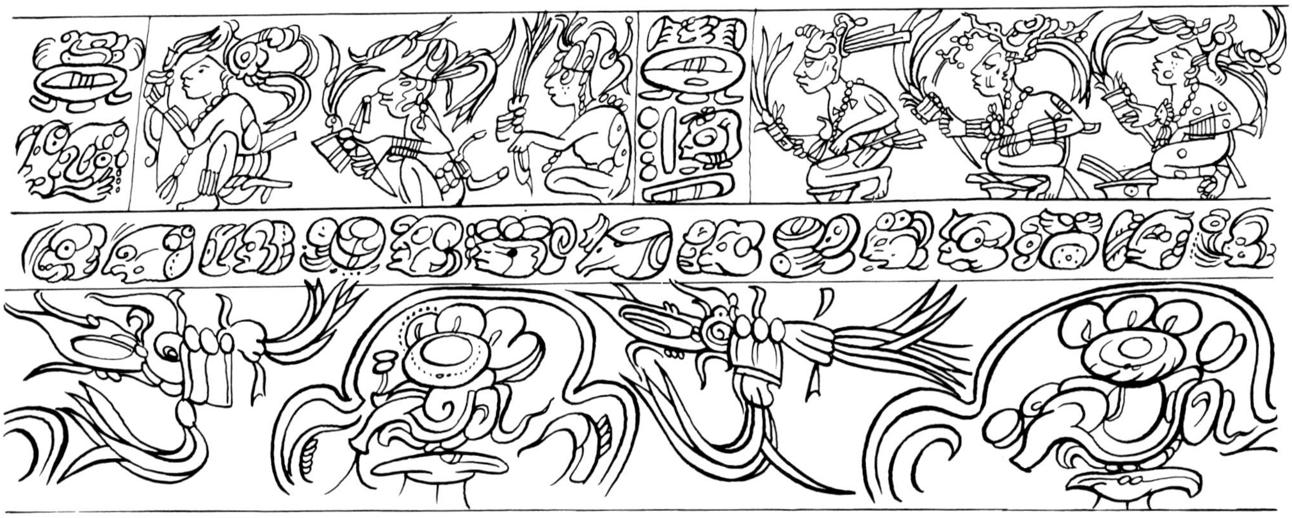

Figure 46. Line rendering of a polychrome vase from the Classic period, probably from Petén, showing a penitential rite (after Gordon)

Mayas, a matter that was studied extensively by Joralemon.[28] Some of these rites, such as the ones depicted on the lintels of Yaxchilán, were definitely associated with visions and hallucinations. A scene carved on Lintel 24 is especially interesting. It shows a kneeling woman who is drawing across her tongue a thick, twisted cord into which large thorns are woven. Facing her at the left stands a male figure, probably Shield Jaguar, the ruler of Yaxchilán holding over her a staff, that Benson describes as a "ceremonial torch."[29] Furst, however, suggests that it is probably not a torch but a long tubular cigar or smoking tube

of the kind still employed by the shamans of the Warao of Venezuela to place themselves into ecstatic trance states in which they travel to other worlds to confront the supernaturals. In the Warao case the "cigar" consists of a palm stipule wrapping around several tightly rolled leaves of black tobacco sprinkled with granules of fragrant resin. The tube is consumed along with the tobacco by the shaman who in the course of feeding the supernaturals with the smoke of as many as twenty or more of these cigars in a single seance saturates himself with an extraordinary high dose of nicotine and other alkaloids.[30]

The combination of self-torture and tobacco consumption has also been recorded among some of the other Mesoamerican nations, such as the Zapotecs and the Mixtecs. In *Relaciones geográficas*, written in 1579 at the request of Philip II of Spain, Francisco del Paso y Troncoso reported that the Zapotecs of Macuilxochtlil worshiped an idol named Coquebila. In honor of the god they held a fasting period of forty to eighty days, during which time they consumed only a quantity of tobacco and offered blood from their tongues and ears.[31] According to Ronald Spores the Mixtecs of Zacatepec worshiped two stone idols, Yahatujyuty and Nanahuconuhu. When the cacique of the city decided that there would be a fast, the priests would abstain for five or six days, like passionate poker players, "eating nothing but inhaling tobacco smoke." Violation of the fast was punished by death by stoning.[32]

4. The last and probably the most likely hypothesis is that the Indians who behaved in such peculiar ways were either not consuming tobacco at all or were using tobacco mixed with other herbs, some of which had psychotropic properties. The term *psychotropic* is generally used to describe substances that cause psychological change or modification of mental activity, serving as sedatives, stimulants, or "psychic deviators" (hallucinogens).[33] Such drugs, especially the hallucinogens, induce states of ecstasy like those described by the early observers. The ten general characteristics of such ecstasy states, mystical unions, and drug-induced alterations of mind have been delineated by Dobkin de Ríos and by Ludwig:

Altered thinking: varying disturbances in concentration, attention, memory and judgement.

Disturbed time sense: a sense of timelessness and the slowing down or the accelerating of time.

Loss of control, which may either be feared or, given particular cultural beliefs, willingly accepted in order to experience divinity or become a mouthpiece for the gods.

Change in emotional expression: expression of emotional extremes unlike those of normal waking hours, from ecstasy and orgiastic equivalents to profound fear, terror and depression.

Changes in body image: various distortions, often with a profound sense of depersonalization; at times, schism between the body and mind and feeling of dissolution of

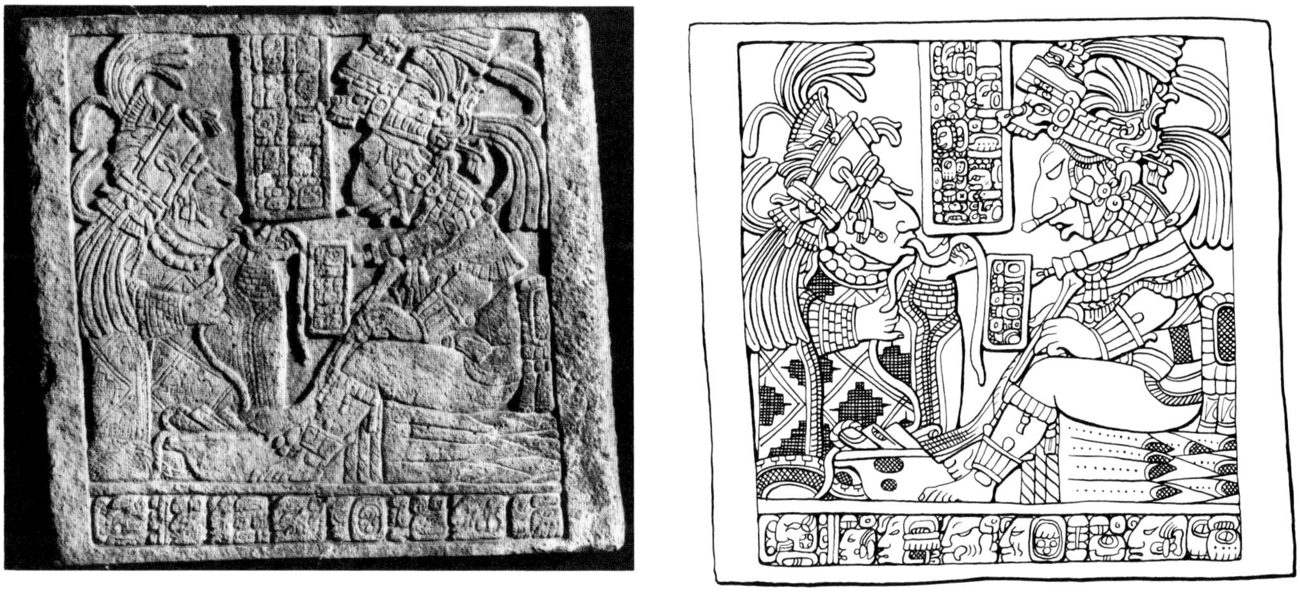

Figure 47. Lintel 17 at Yaxchilán. The scene shows two seated lords, one pulling a thorned cord across his tongue (photograph courtesy British Museum)

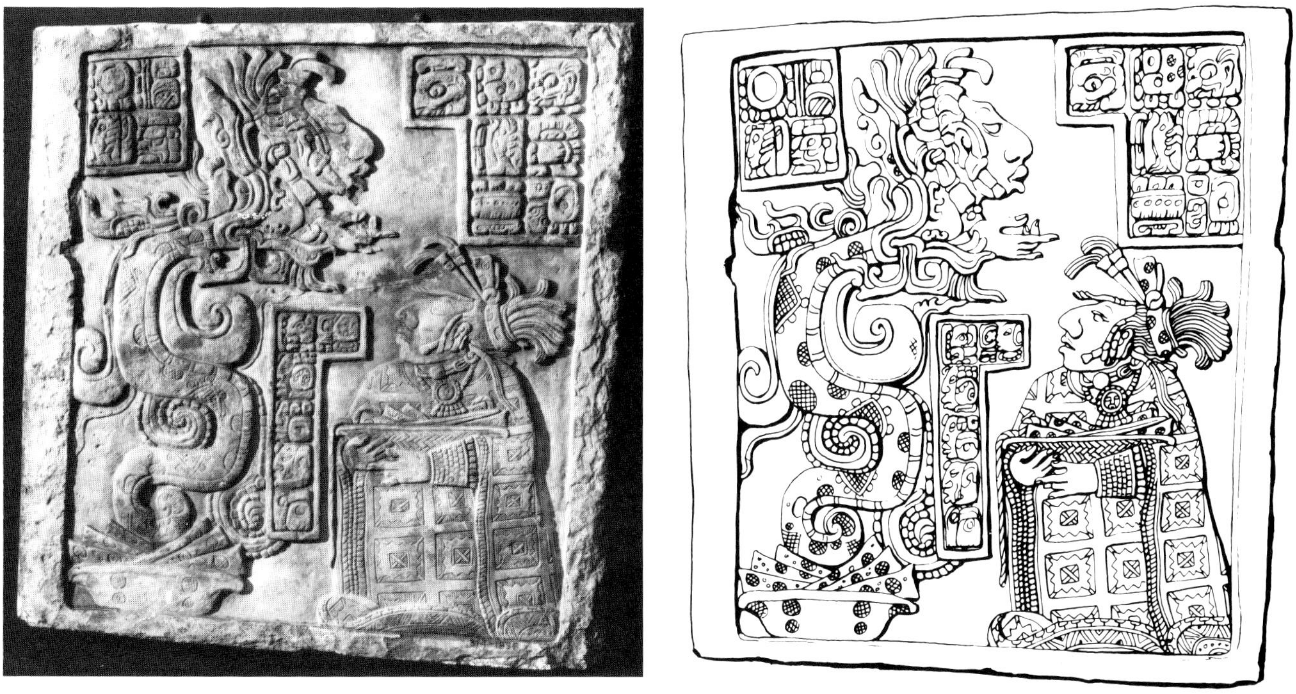

Figure 48. Lintel 15 at Yaxchilán, showing the figure of a kneeling person, perhaps a woman, holding blood-letting paraphernalia. In front of the figure, as if in a dream, a deity emerges from the throat of a serpent (photograph courtesy British Museum)

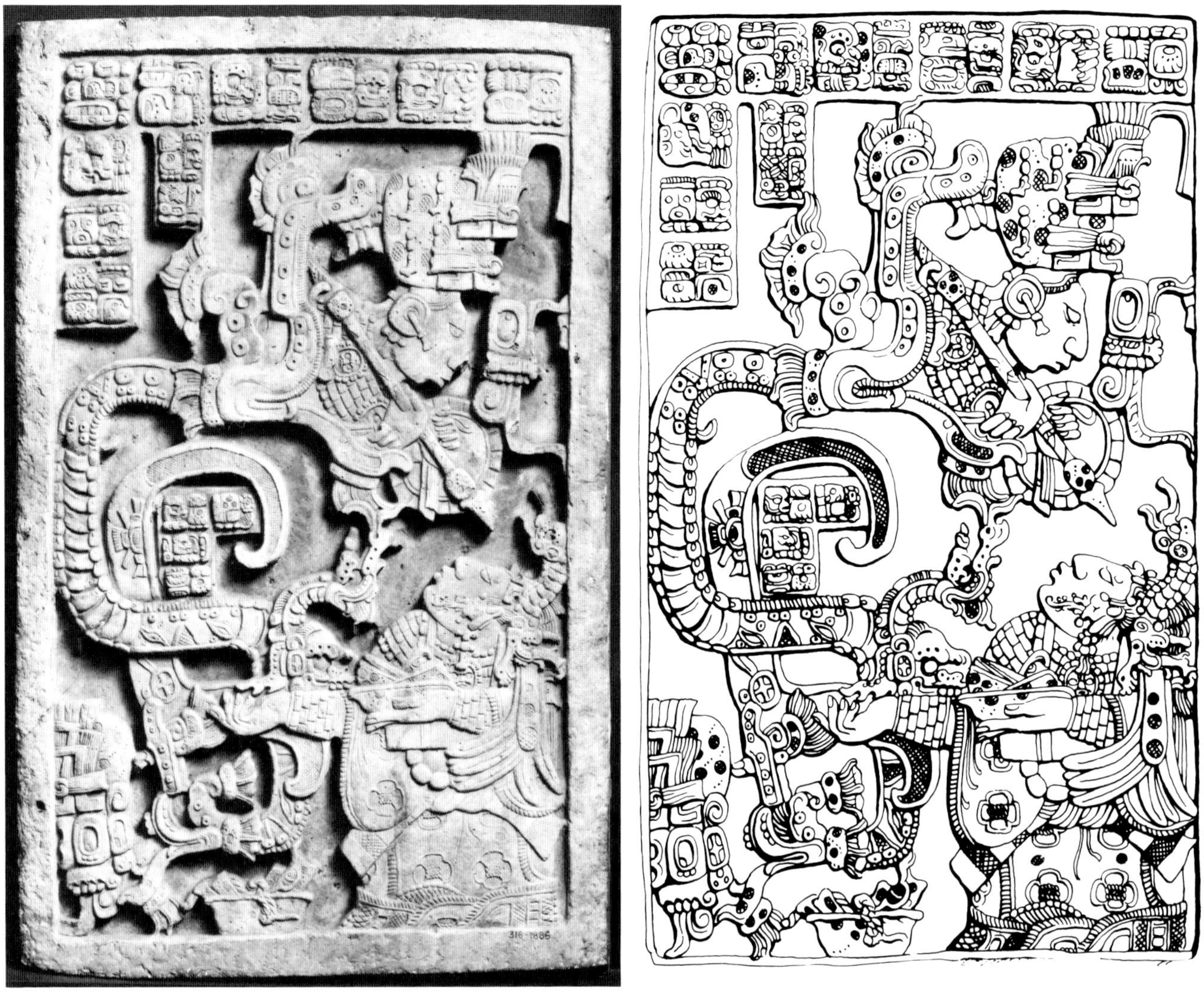

Figure 49. Lintel 25 at Yaxchilán, showing a scene similar to that on Lintel 15. A penitent nobleman (or noblewoman), holding a dish with ceremonial objects, looks at a deity emerging from the throat of a serpent (photograph courtesy British Museum)

boundaries between one's self and others, the world, or universe.

Perceptual distortions: increased visual imagery, hyperacuteness of perceptions, geometric patterns, synesthesias.

Change in meaning or significance: special significance attached to subjective experience with great insight or profound feelings.

Sense of the ineffable: difficulty in communicating the nature of such experience to someone uninitiated.

Feeling of rejuvenation.

Hypersuggestibility: increased propensity to accept or respond to specific statements of the person guiding the session.[34]

The indigenous peoples of the Americas are known to have practiced religious cults in which the supernatural was worshiped with the aid of hallucinogenic plants, such as peyote, morning glory, mescal bean, jimson weed, to name only a few. It is probable that the early Spaniards, being only vaguely aware of these herbs but having learned the use of tobacco on the Caribbean islands on which they first landed, thought that everything the Mayas smoked or chewed was tobacco and labeled it as such. They may also have witnessed ceremonies in which the Mayas consumed tobacco but "spiked" it with other herbs of strong narcotic or hallucinogenic properties. The latter herbs, however, eluded the untrained eyes of the European observers.

Not all the Spaniards mistook hallucinogens for tobacco. Gumilla, for example, in his *El Orinoco illustrado,* clearly recognized that a powder having the smell of tobacco was in reality *yupa,* a strong narcotic:

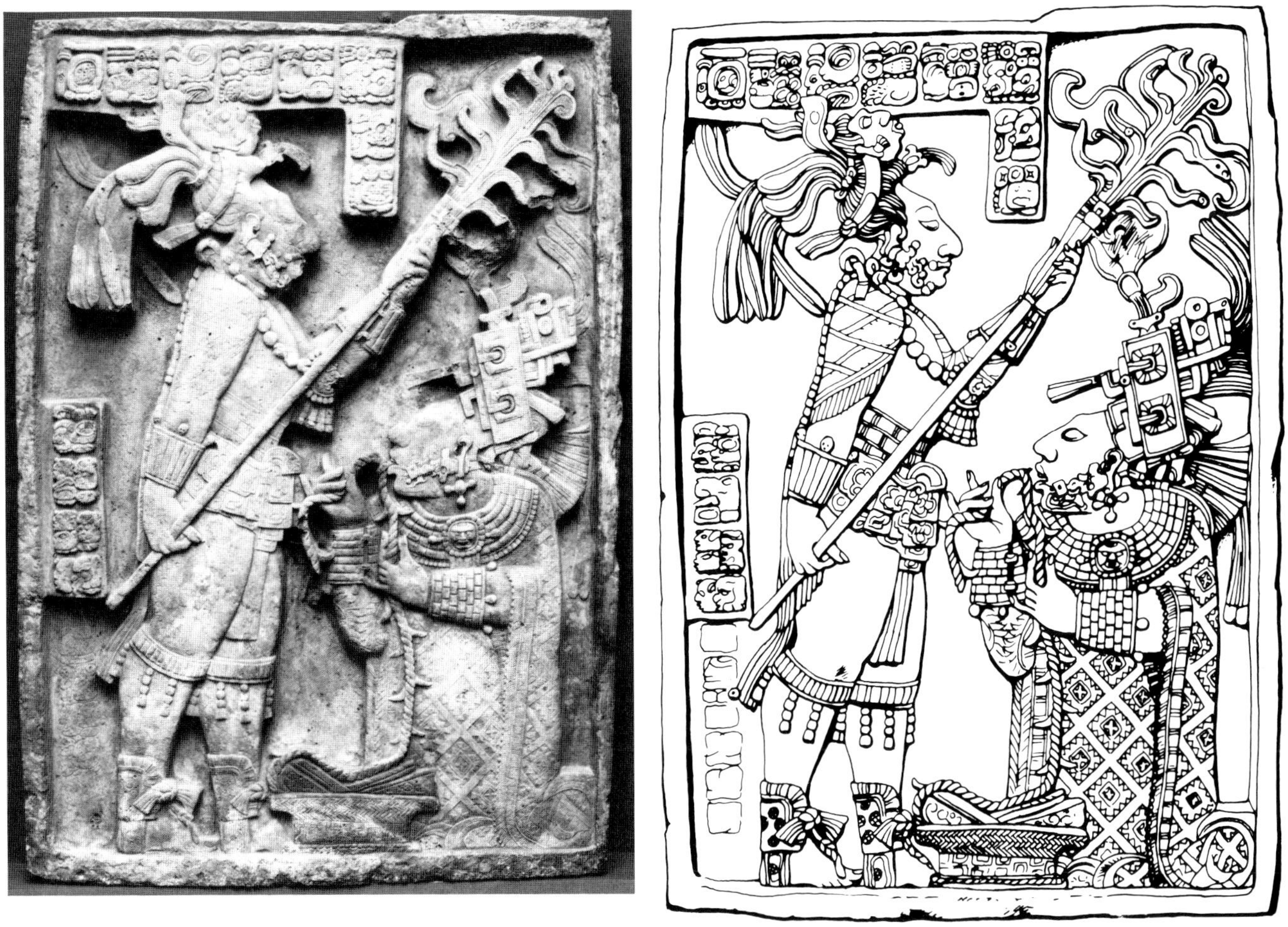

Figure 50. Lintel 24 at Yaxchilán, showing two lords, one in a kneeling position, pulling a thorned cord across his tongue, the other, standing, holding what appears to be a long, smoking "tobacco tube" or a foliated ceremonial staff (photograph courtesy British Museum)

> They have an evil habit of intoxicating themselves through the nostrils with certain malignant powders which they call *yupa*, which takes away their reason, and, furious, they grasp weapons. . . . . They prepare the powder from certain pods of *yupa*, . . . but the powder itself has the odor of strong tobacco.[35]

It is also possible that we are unjustly criticizing some of the early historians as poor observers when we should blame some contemporary scholars instead. Many of the modern translators of ancient Spanish manuscripts took considerable liberty in their interpretation. They freely translated such Spanish expressions as "smoking reed," "odoriferous herb," and "smoke" as "tobacco." This naturally gives the impression to all to whom the original manuscripts are unavailable that the natives smoked nothing but tobacco. We know, however, that in North America alone, the leaves of more than sixteen plants other than tobacco were smoked for pleasure or for ceremonial reasons. Their use was largely abandoned when tobacco became plentiful and was found to provide more agreeable smoke. This possibility was considered in 1899 by Joseph McGuire:

> In the early references to smoking a notable peculiarity is that the term employed very commonly is "herbs," which may be because of ignorance of the plant smoked, though it is certainly suggestive also of there being more than one, for it is known that certain of our Indians consider it essential to their ceremonial smokes or dances to have a mixture of different plants.[36]

Other sources indicate that it was the habit of the Indians to mix different, mostly aromatic, substances with their smoking tobacco.[37] Such admixtures have been mentioned repeatedly by students of North American Indian ethnology. Over a great part of eastern North America, for example, the natives smoked a concoction of herbs called kinnikinnick with or instead of tobacco. *Kinnikinnick*, a word derived from the Algonkian family of Indian languages, refers to "mixing" or "mixture" of smoking materials. Shrub sumac, bearberry, manzanita, and dogwood were among the plants the Indians used to make kinnikinnick.[38] Naturally there is a possibility that some of these mixtures contained hallucinogens. We know, for example, that the California tribes and also the Aztecs added jimson weed (Jamestown weed, or *Datura stramonium)*, a powerful narcotic, to their smoking tobacco. The herb's present name dates to its use in early Virginia as a potherb. In 1676 a troop of British soldiers sent to quell Nathaniel Bacon's rebellion aganist the colonial government of Virginia ate the plant in a salad. Their behavior was bizarre for several days afterward; they were observed "running naked and playing silly games, and after 11 days return'd to themselves again, not remembering anything that had pass'd."[39]

There are indications that not all the herbs the natives of Mesoamerica called *picietl* were tobacco plants but that they were similar-looking hallucinogenic herbs, possibly such as jimson weed or henbane (*Hyosciamus niger*). Juan Bautista Pomar wrote in the *Relación de Texcoco:* "The herb *picietl*, as it is called, is the same that in Spain is called henbane" (la yerba que llaman picietl, que según dicen es al misma que en España llaman beleño).[40] Sahagún, who used three different Aztec words, *picietl*, *yetl*, and *yecuxoton*, to define tobacco, in one chapter of his *Historia general* also identified *yetl* as "the henbane to Castilla" (que es como belenõs de Castilla).[41]

The *Relación de Ostomán* also gives an indirect clue that *pitziete* or *picietl* may not always designate tobacco. In a prescription for colds the manuscript clearly distinguishes between *pitziete* and *tabaco*:

> It is said that it was a healthy land where they lived, but sometimes they fell ill of colds and nose bleedings and when they had this cold they use the juice of *pitziete* through their noses and some other leaves they call tobacco [*tabaco*].[42]

Hallucinogenic kinnikinnick still exists in modern times, as shown in a report from the Department of Psychiatry and Pharmacology, Center for the Health Sciences, School of Medicine, University of California at Los Angeles:

> There are presently 192 distinct herbs commercially available and used as smoking substances. These substances are in bulk or in 18 different preparations consisting of rolled cigarettes or blended smoking mixtures. Such preparations contain numerous non-psychoactive ingredients, of which the most common are *yerba santa*, rosemary, thyme, *mullein* and spearmint. . . . 44.4% of these preparations contain plants with known psychoactive effects. . . . The use of these smoking substances has resulted in a number of intoxications.[43]

The article lists the following herbs with psychoactive ingredients that are today available commercially for smoking purposes in the Los Angeles area: *yohimbe* bark, Scotch broom, California poppy, catnip, cinnamon, *damiana*, hops, *hydrangea*, juniper, *kavakava*, kola nut, lobelia, passion flower, periwinkle, prickly poppy, snakeroot, thorn apple, valerian, wild lettuce, and wormwood.[44]

To know for a certainty that such hallucinogenic mixtures were used in America before the arrival of the Spaniards, one need only read carefully the account given by Ramón Pané, who was left on Haiti to minister to the natives and who wrote a detailed account of the Indian rites and superstitions—thus becoming the first ethnologist in the New World. Pané's observations were first published by Peter Martyr in 1511:

> Wherefore, when their chiefs consult the *zemes* [idols] about the outcome of the war, about the harvest or about their health, they go into the house dedicated to the *zeme*, and there snuffed *cohobba* into their nostrils (so they call the intoxicating plant by which the bovines also are thrown into frenzy) they say that the house is turned upside down, the roofs and floors become interchanged, and that men walk with their feet upward. Such is the strength of the powder *cohobba*.[45]

It is most surprising that both the contemporary and initially even the later writers regarded *cohobba* as the local term for the tobacco plant and considered the powerful snuff to be pure tobacco.[46] Later, however, several expressed the opinion that the snuff was either made of or mixed with *Anadenanthera peregrina*, a shrub with powerful narcotic effects.[47]

According to Durán's *Book of the Gods and Rites*, tobacco was used with another potent hallucinogen, *Rivea corymbosa* (morning glory), in the manufacture of the magical ointment of the Aztec priests, the "food of the gods":

> This was the divine food with which the priests, ministers of the temples, especially those with whom we are dealing, smeared themselves in ancient times. They took all these poisonous animals and burned them in the divine brazier which stood in the temple. After these had been burned, the ashes were placed within certain mortars, together with a great deal of tobacco; this herb is used by the Indians to relieve the body so as to calm the pains of toil. In this it is similar to Spanish henbane, which, when mixed with lime, loses its poisonous qualities, though it still causes faintness and is harmful to the stomach. This herb, then, is placed in the mortars together with scorpions, live spiders, and centipedes, and there they were ground, producing a diabolical, stinking, deadly ointment. After these had been crushed, a ground seed called *ololiuhqui* [morning glory, or as it is called in the Mexican markets today *Dondiego de día, Dompedro,*

maravilla, manto de María, or semilla de Virgen] was added, which the natives apply to their bodies and drink to see visions.... When the priests had been painted with this bitumen, they lost all fear. They slew men in sacrifice with the greatest of daring; still smeared, they went forth at night, alone, to the woods, to the dark caves, to somber and fearful cliffs, totally unafraid of injury or harm. But since they carried the divine food on their bodies, they sneered at wild beasts. Well they knew that lions, tigers, and other deadly animals of the woods would flee because of that Pitch of God—or, let us be clear, of the devil.[48]

In lowland South American and in Bolivian rituals tobacco was used with the powerful hallucinogen *ayahuasca* (*Banisteropsis caapi*). In Northern Peru the people still "cure" tobacco with a mixture of both liquid and powdered tobacco and the mescaline cactus called San Pedro (*Trichocereus pachonoi*).[49]

The association of tobacco with hallucinogens is especially apparent in the peyote paraphernalia of the Plains Indians, which includes not only bags of peyote, incense, and sage, an altar cloth, a water drum, and eagle feathers but also a generous supply of tobacco leaves. Tobacco smoke prayers are made at the peyote meetings of the Kiowa, Oto, Caddo, and Kickapoo Indians. The Osages are said to smoke traditional corn-shuck crook-leaf tobacco cigarettes during their peyote rites. Ritual tobacco smoking is an integral part of the peyote ceremony of the modern intertribal Native American Church.[50]

Sahagún mentioned that Mexicans mixed their tobacco with an "incense plant" (*yauhtli*). How this mixture was prepared and what it was used for are not known for certain. It was probably used as incense, and the *yauhtli* pellets were burned with pulverized tobacco as an offering to the gods.[51] We know also that the Mesoamerican lords regularly added resins and aromatic herbs to their smoking tobacco. These ingredients may have served only to make the mixture better smelling, but it is possible that some of them were indeed psychoactive. M. Fernández de Escheverría y Veytia, one of the early chroniclers, wrote on this subject:

They formed then a paste of these herbs with ground charcoal and filled with it the cane tubes which were lighted at one end and passed to the guests to hold in the hands and enjoy the fine odor; at the beginning, then it was only the perfume to delight the sense of smell; but subsequently they had another use, because applying fire at one end, they sucked at the other and swallowed the smoke. This not only served them as a delightful experience but also as a medicine.[52]

Clavijero noted:

After eating the lords were accustomed to conciliate sleep with tobacco smoke. They employed it in a paste form to be smoked, or in powder by the nostril. For smoking they put it in a tube or cane, or other finer material, the leaf with resin of liquid amber or with other odoriferous herbs. They received the smoke, pressing the tube with the mouth and covering the nose with the hand to the end that it may penetrate the lungs more quickly.[53]

Early records indicate that the Zapotecs smoked tobacco mixed with the leaves of other plants:

They [the inhabitants of Miahuatlán] also mix (tobacco) with the leaf of another plant, which they call *nantzi*, that is *madre*, which they burn and inhale the smoke and eject it through the nostrils. They say it eases headache and hay fever. They call this mixture *suchietl*, which is to say, "flower that smells."[54]

A very strange "additive" to tobacco was mentioned by the traveler Thomas Gage. He described a most potent drink of the Pocomán Mayas consisting of tobacco leaves, fermented cane, roots, and (God forbid!) a well-grown toad:

Among themselves they make drinks far stronger than wine. These they confection in those great jars that come from Spain. They put in them a little water and fill up the jar with some molasses or juice of the sugar cane, or some honey to sweeten it. Then, to strengthen it, they put roots and leaves of tobacco, with other kinds of roots which grow there and which they know to be strong in operation. Nay, to my knowledge, in some places they put a live toad, and closed up the jar for a fortnight or a month, till all they have put in be thoroughly steeped, the toad consumed, and the drink well strengthened. Then they open it and call their friends to drink it.[55]

Thompson, who dismissed Gage lightly, calling him a charming fraud, made the tongue-in-cheek remark that this was indeed a tale of uncommonly strong waters because even the witches cauldron of Shakespeare's *Macbeth* contained only a toe of a frog.[56]

Recent research indicates that Gage's anecdote should be taken seriously. Frogs and toads not only served as decorative motifs (if indeed there is such a thing in Maya art) but also had important iconographic, calendric, and glyphic symbolic meanings.[57] It was also their function to be the court musicians and guests of the Chacs, the Yucatec rain gods.[58] Besides being blessed with musical talent, the poisonous toad (*Bufo marinus*), which is rampant in the Maya area, also contains a potent indole substance with hallucinogenic properties: I-hydroxy-N-dimetiltriptomine, or *bufotenine*.[59] This miserable toad, which plagued my camps along the Usamacinta River many times, is an especially ravenous species and earned the title the "worst toad" in Felton and Fowler's *Best, Worst, and Most Unusual*. In the 1930's the Australian government

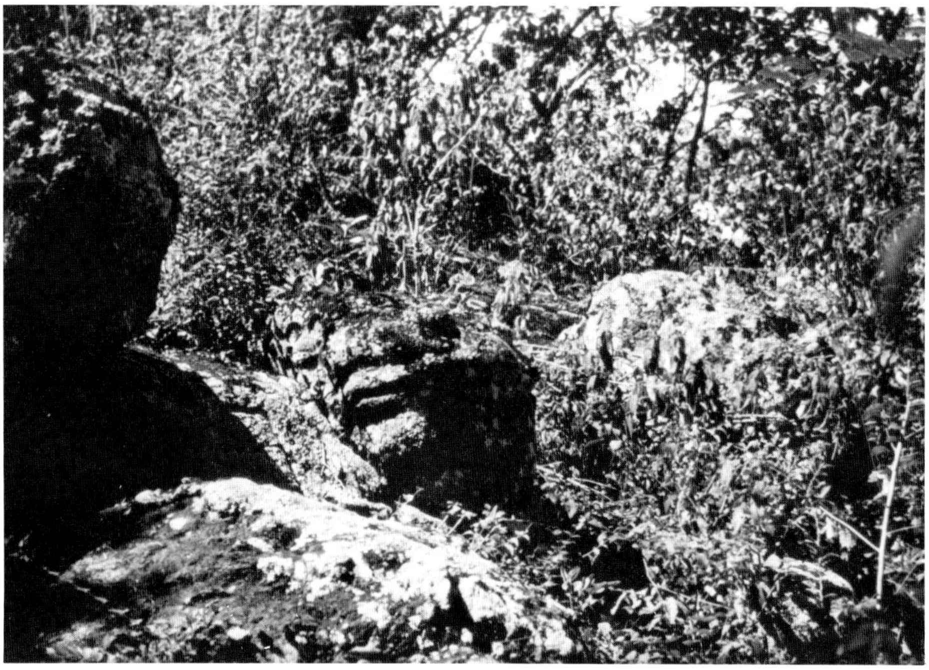

Figure 51. El Sapo, the Toad, a rock sculpture in the foothills of Copán.

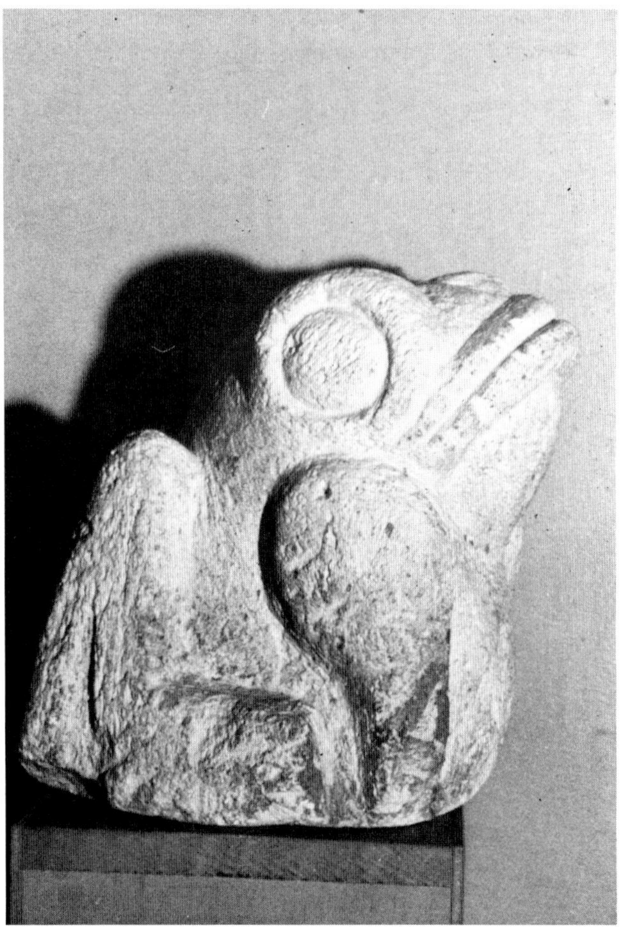

Figure 52. Sculpture of a toad or a frog (Copán Museum)

imported a platoon of *bufos* to Queensland to help eradicate the cane beetles that were destroying the crops. The Australians underestimated the toads; after the beetles were gone, the toads, now up to eight inches long and greatly increased in numbers, began devouring garbage, ping-pong balls, dogs, cats, and cattle, killing their prey with poisonous glandular secretions. By the early 1970's the *bufos* had become something of a biological nightmare and the Australian National Wildlife Department was offering a reward of thirty dollars for each toad brought in. "Wanted: Dead or Alive" posters proliferated throughout the continent, and as a public service at least one radio station was broadcasting tape-recorded mating calls of male *bufos* in the hope of luring love-starved females to destruction.[60]

Excavations by David Sedat at San Jerónimo, Baja Verapaz, Guatemala, uncovered artifacts suggesting that toads were used in some obscure way in Maya rituals.[61] Michael D. Coe found large quantities of *bufo* remains at the Olmec site at San Lorenzo.[62] They may have been used as food, but it is also possible that they were used as "fortifiers" for fermented ritual beverages, a custom still practiced among the Quiché Mayas.[63] The use of toad poison has also been reported among the Guiana Indians, the Amahuacas of Peru, and the *curanderos* in the Veracruz area.[64] Furst believes that the many bowls representing toads found in the Maya region may have been used by the shamans to store the exudate of these venomous creatures.[65] While it is possible

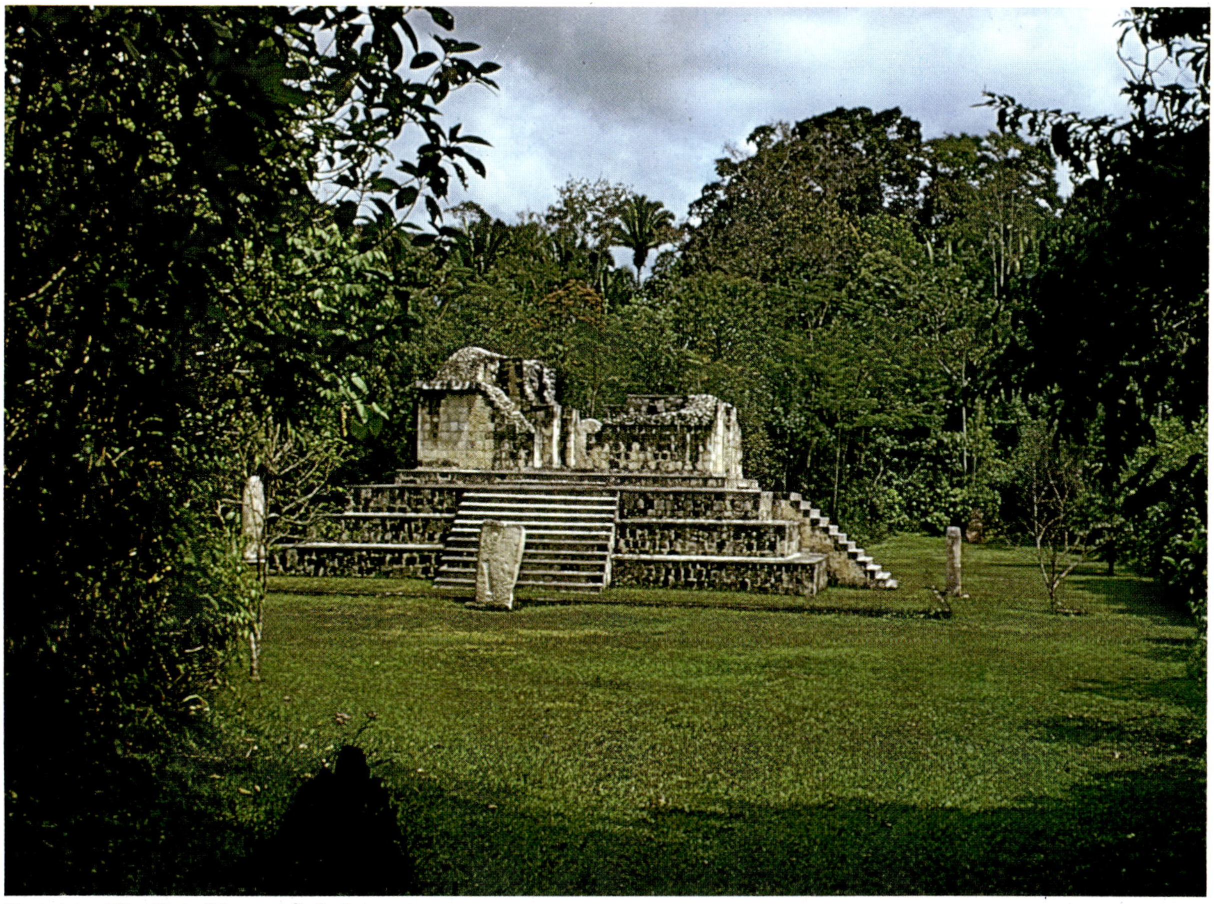

Plate 41. The Main Plaza at Seibal

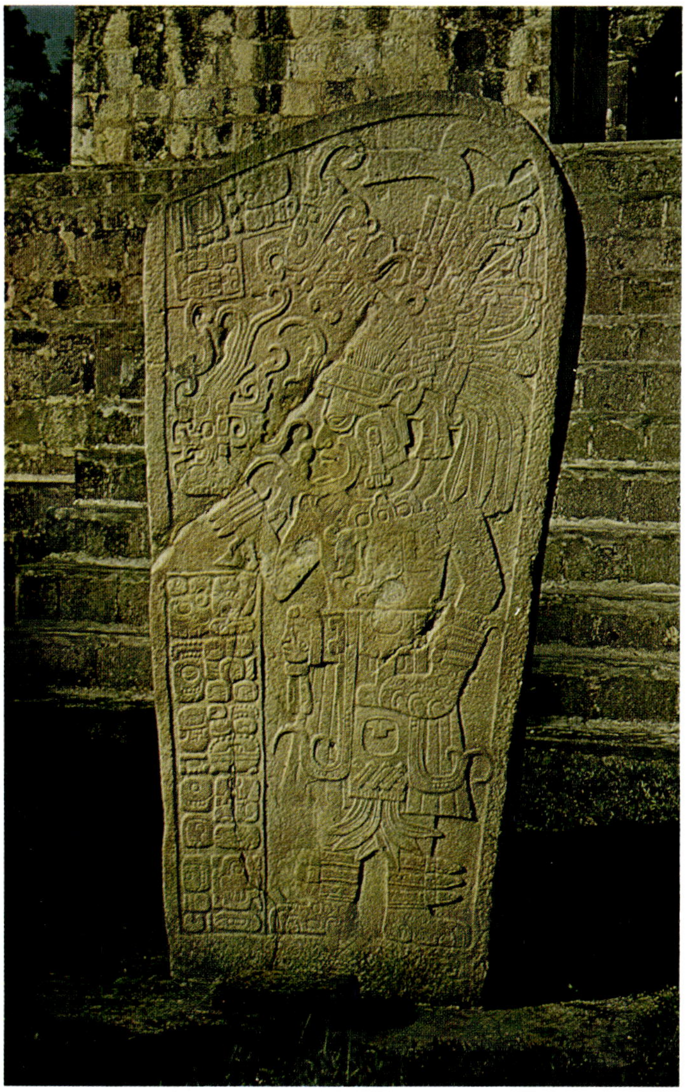

Plate 42. Stela 8 at Seibal

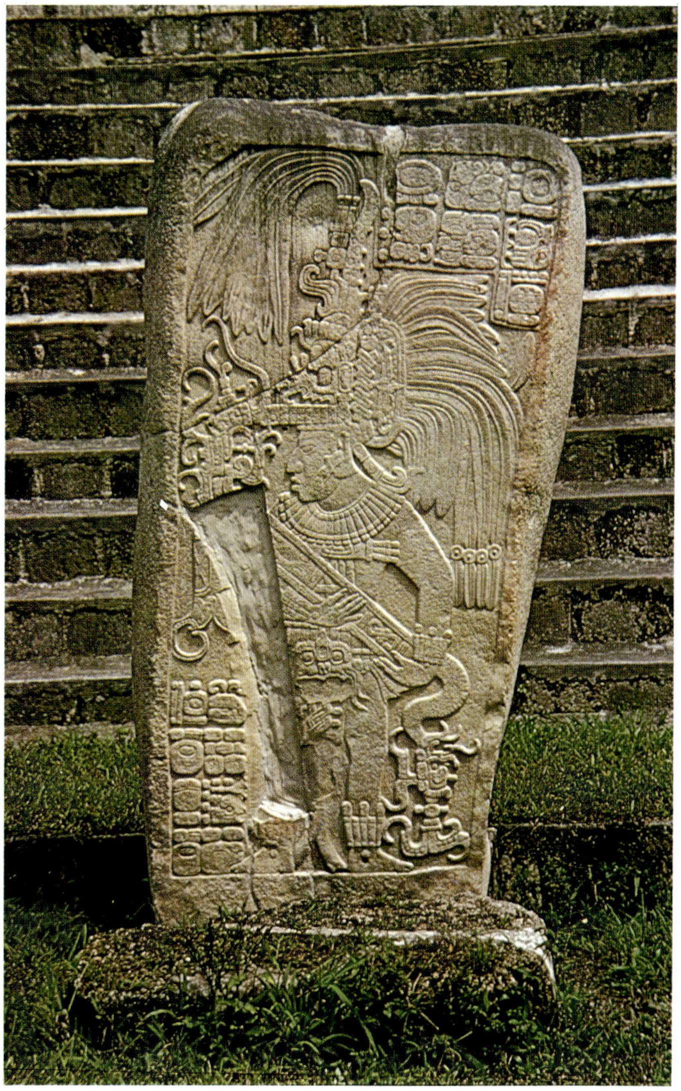

Plate 43. Stela 9 at Seibal

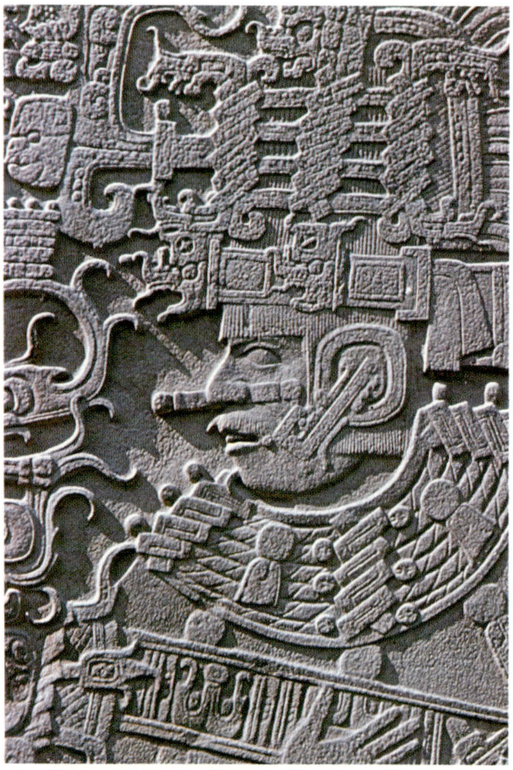

Plate 44. Detail, Stela 10 at Seibal

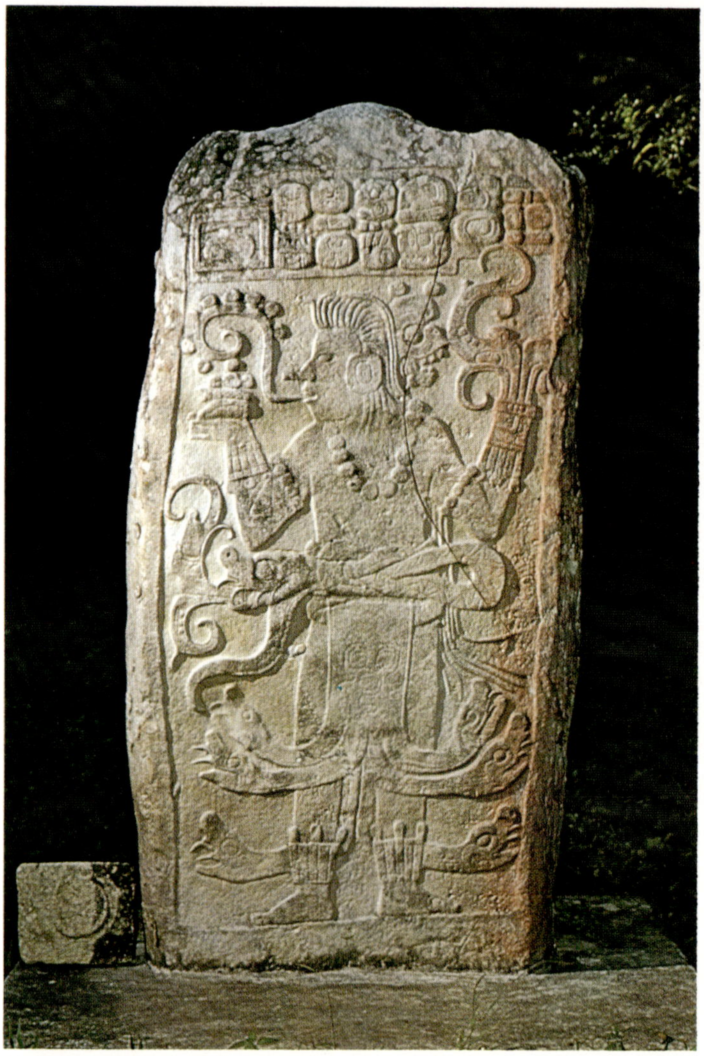

Plate 45. Stela 13 at Seibal

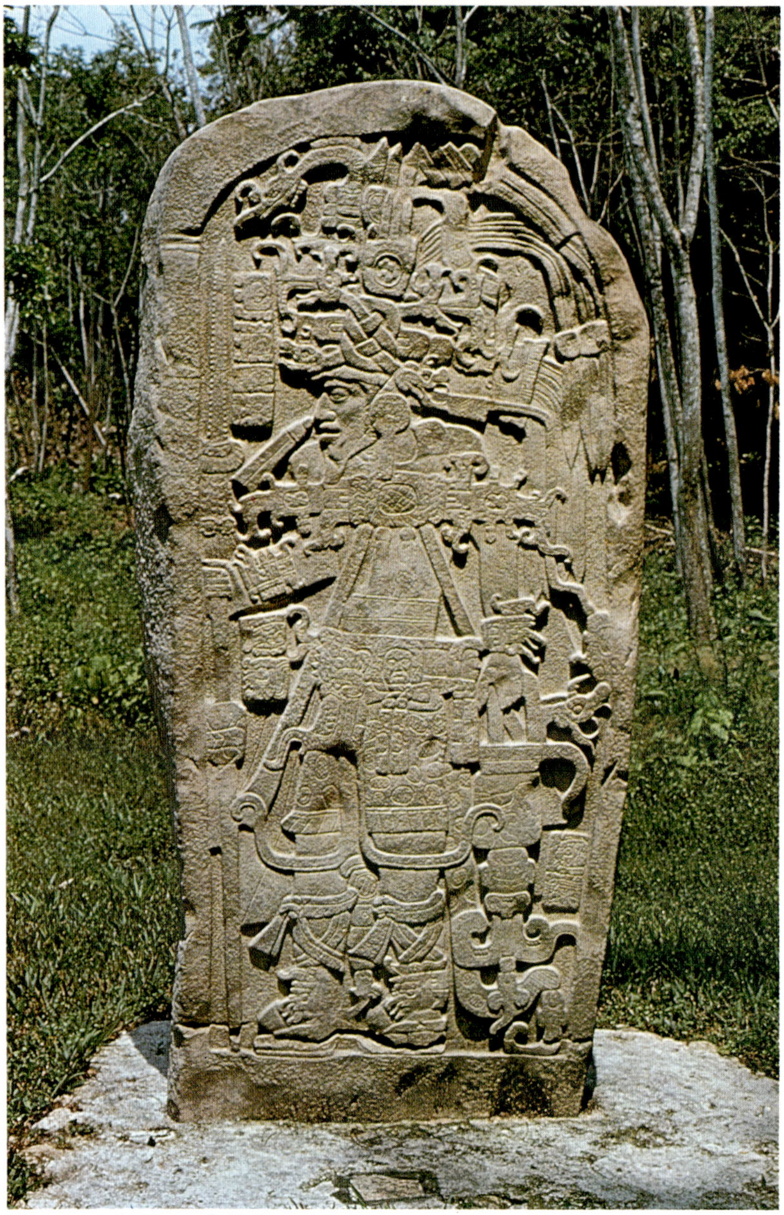

Plate 46. Stela 1 at Seibal

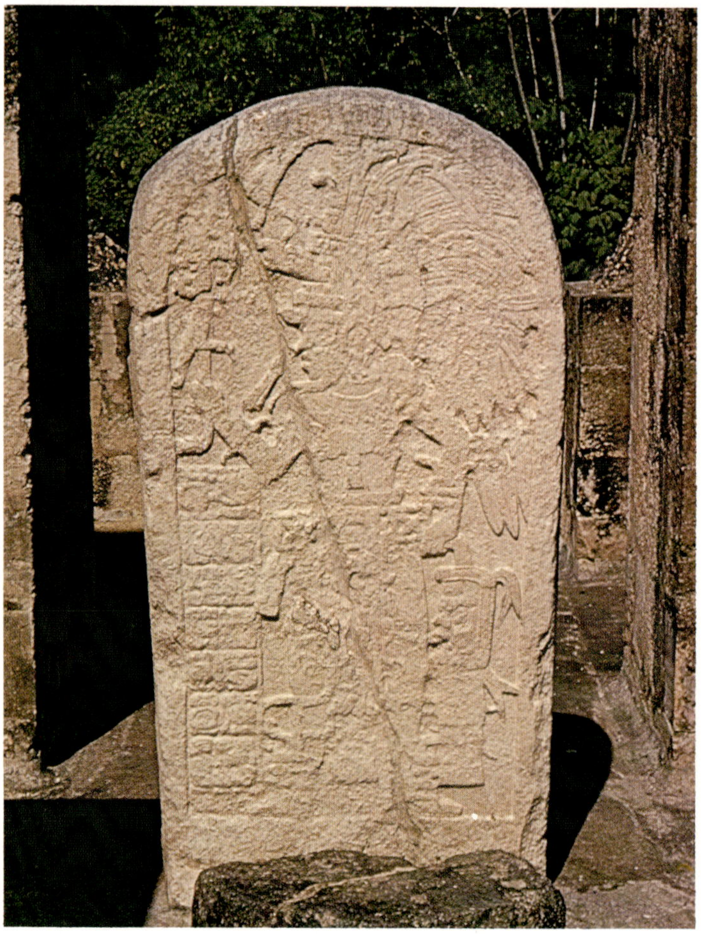

Plate 47. Stela 21 at Seibal

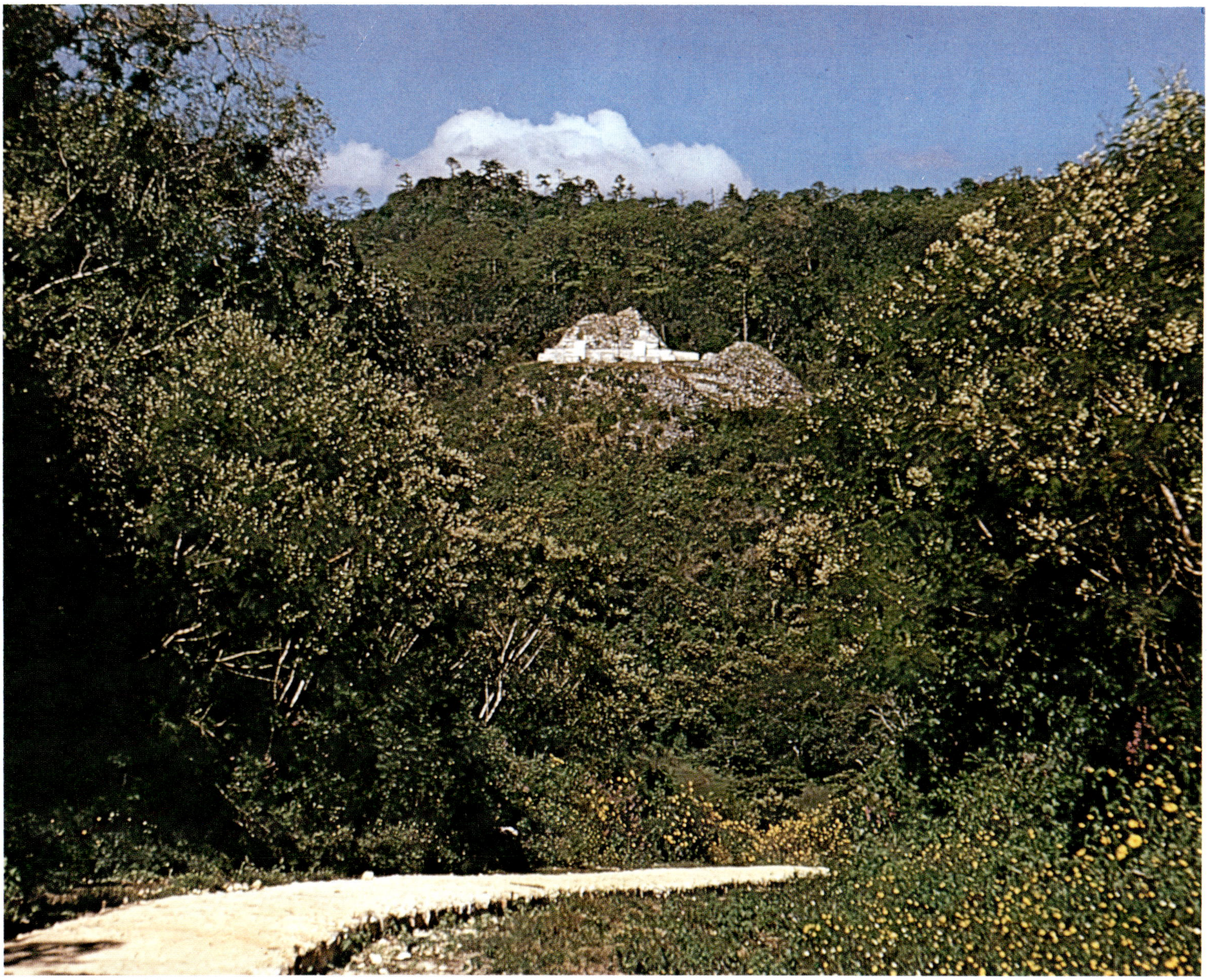

Plate 48. Chinkultik

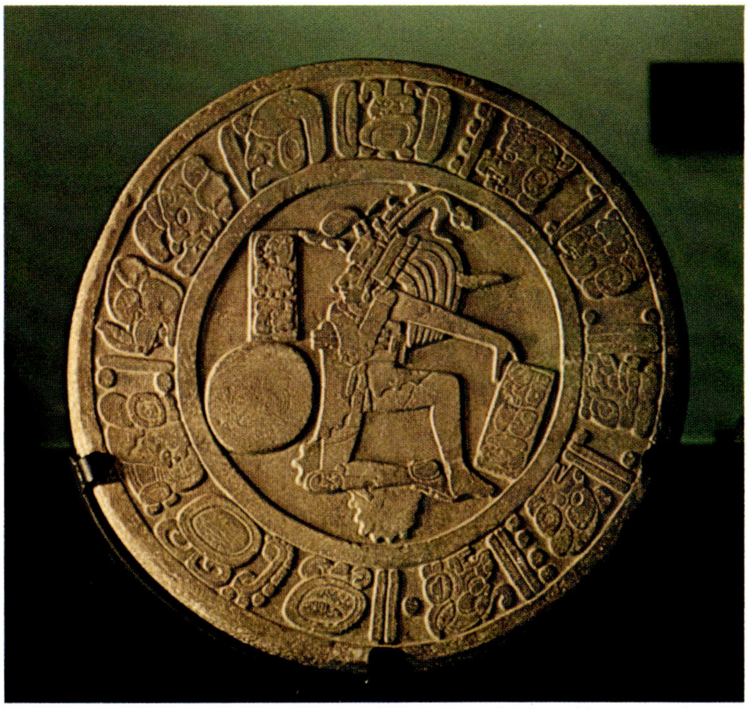

Plate 49. Ball court marker from Chinkultik (Museo Nacional de Antropologia e Historia, Mexico City)

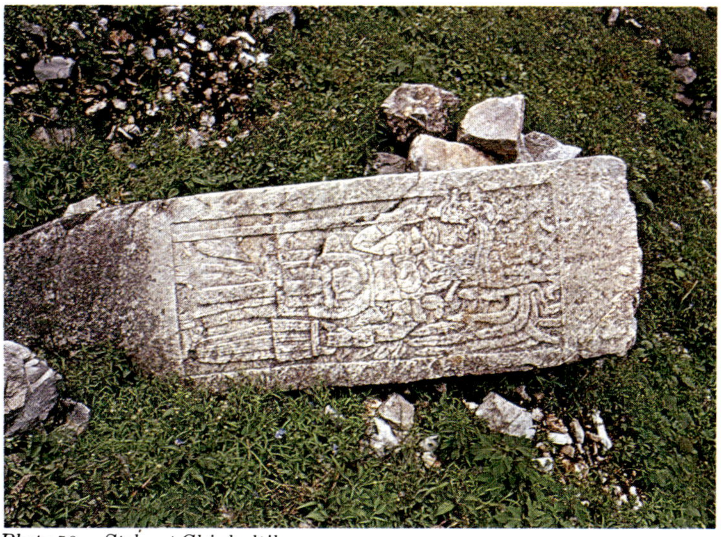

Plate 50. Stela at Chinkultik

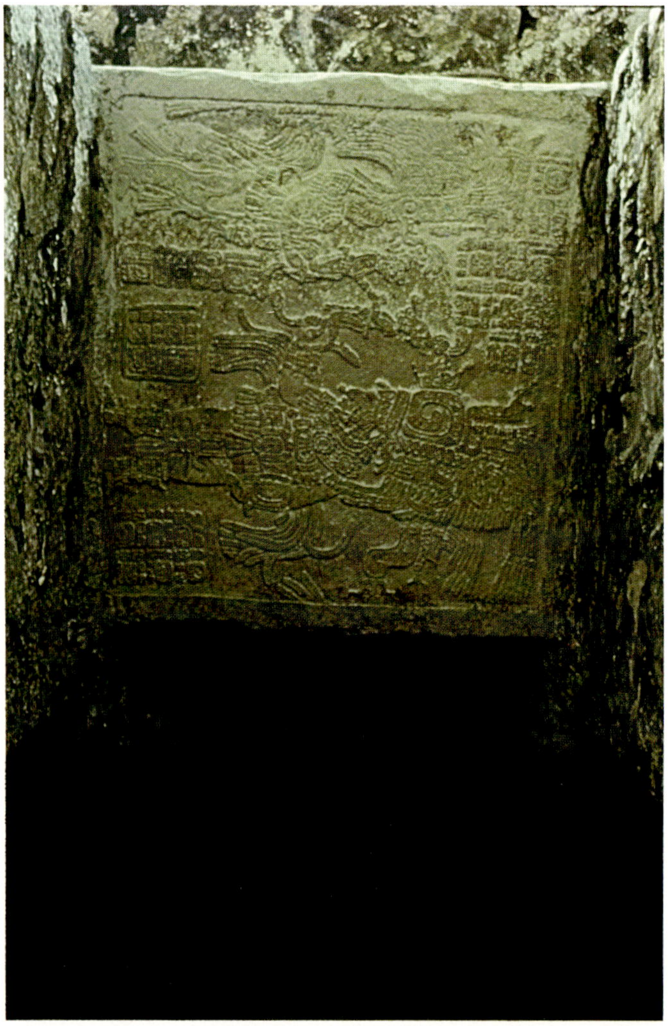

Plate 51. *Lintel 3 at Yaxchilán*

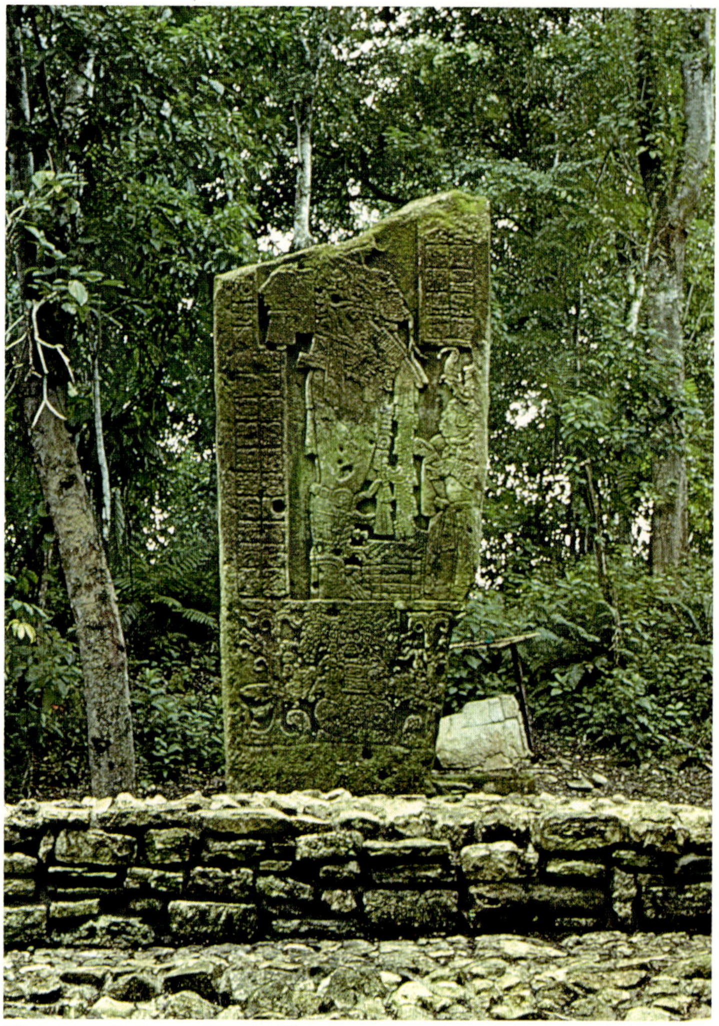

Plate 52. *Stela 1 at Yaxchilán*

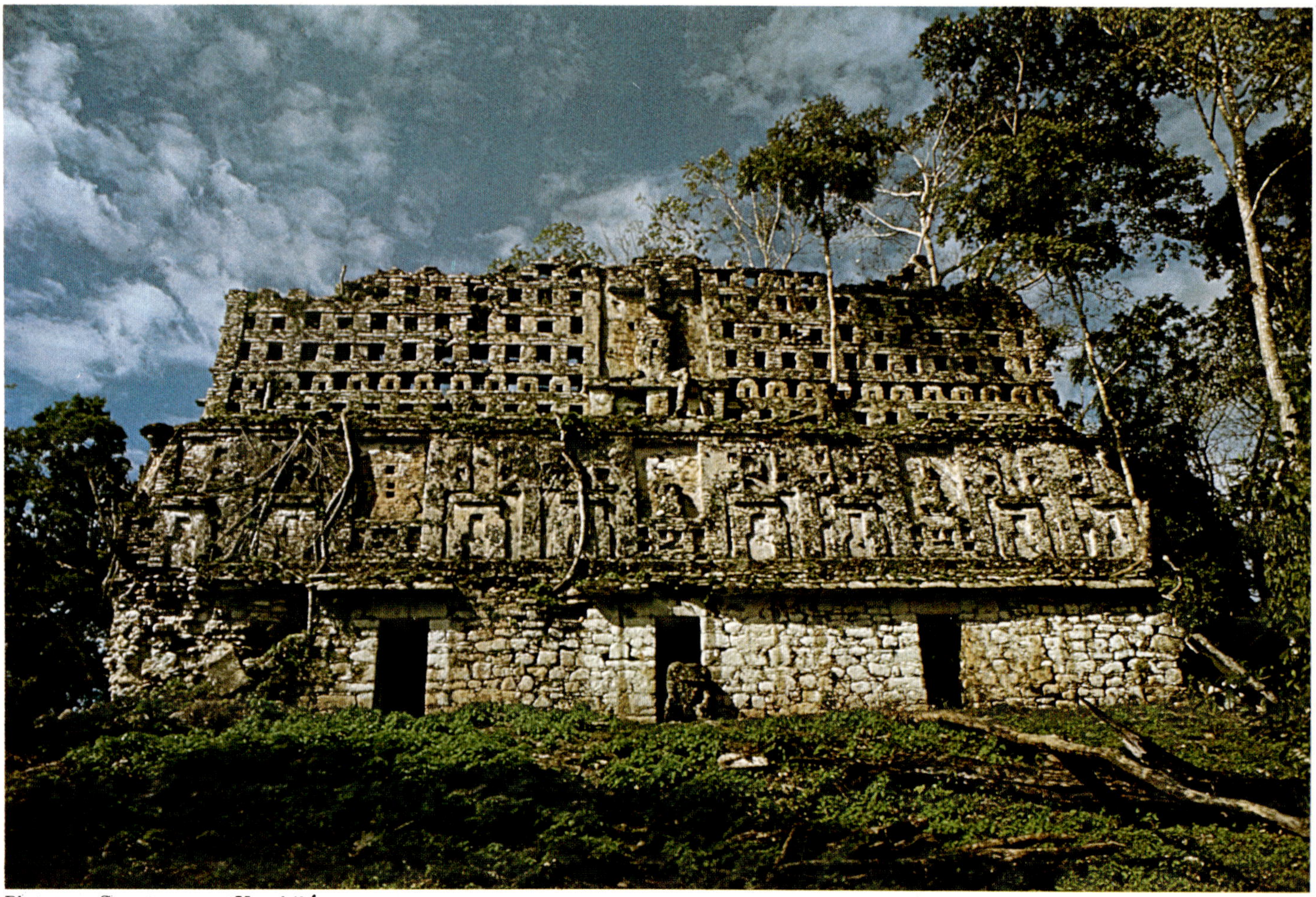

Plate 53. *Structure 33 at Yaxchilán*

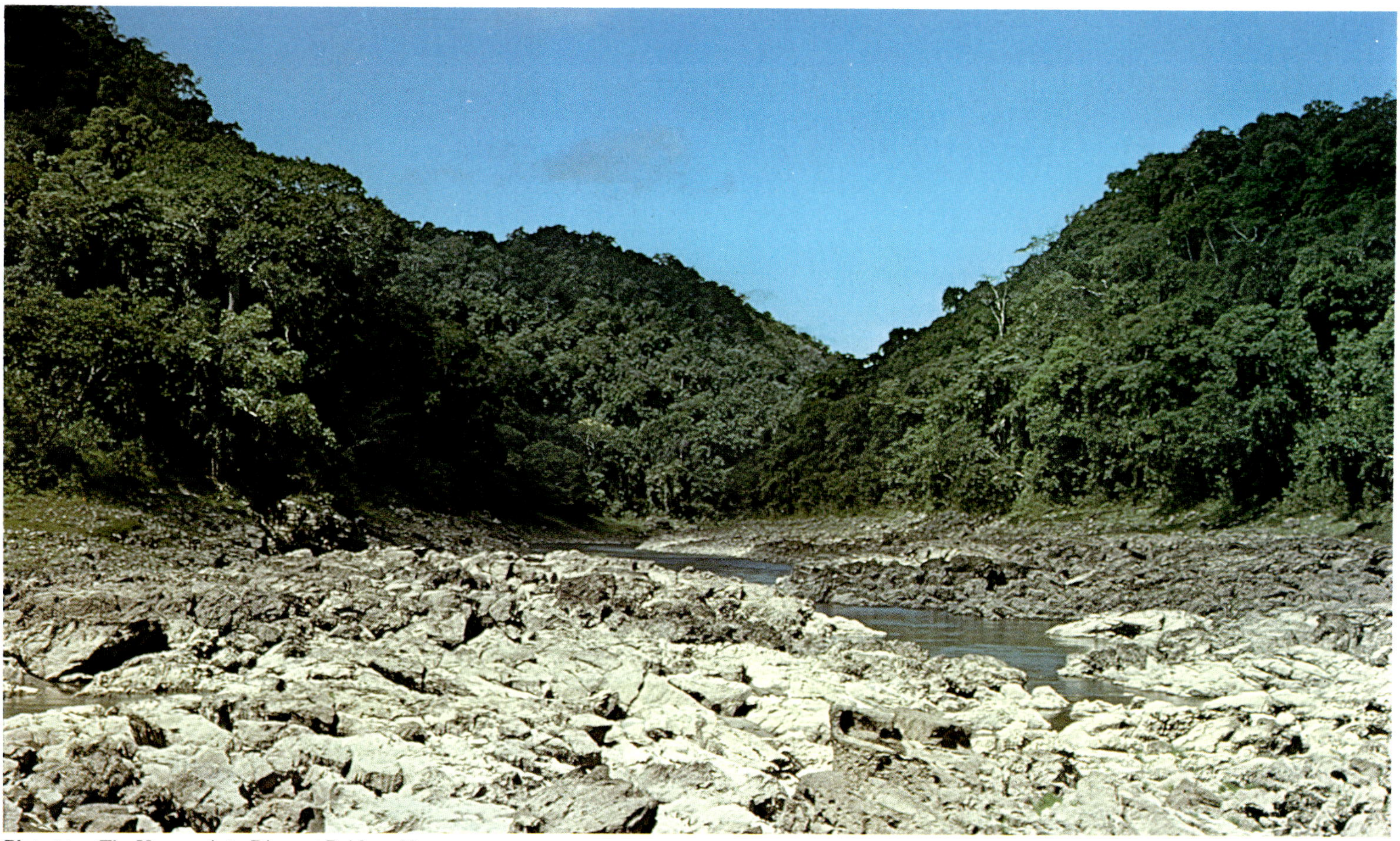

Plate 54. The Usumacinta River at Peidras Negras

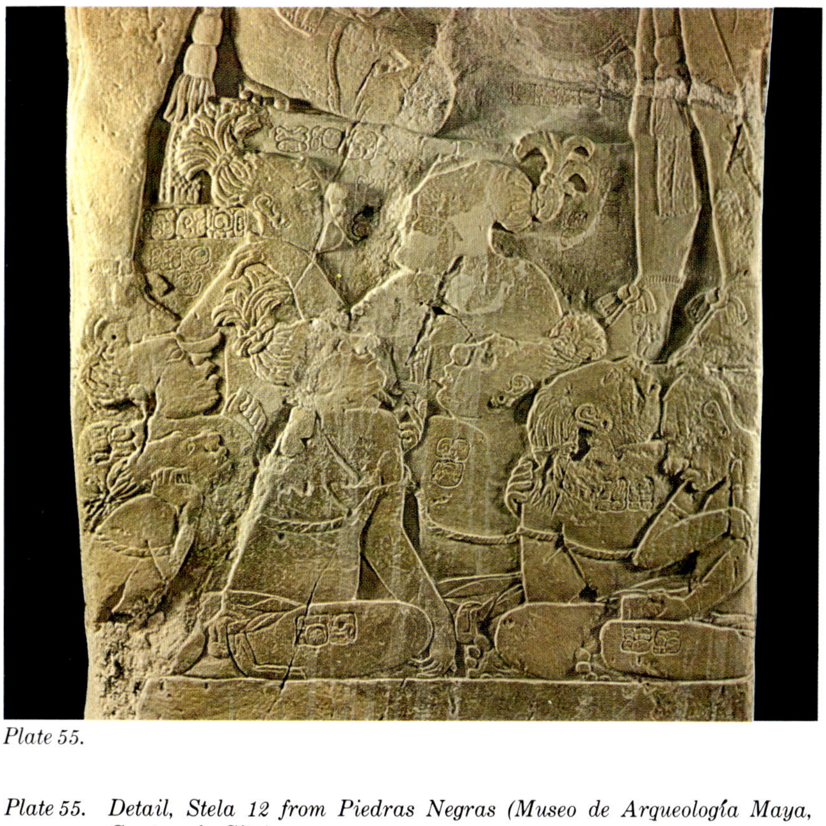

Plate 55.

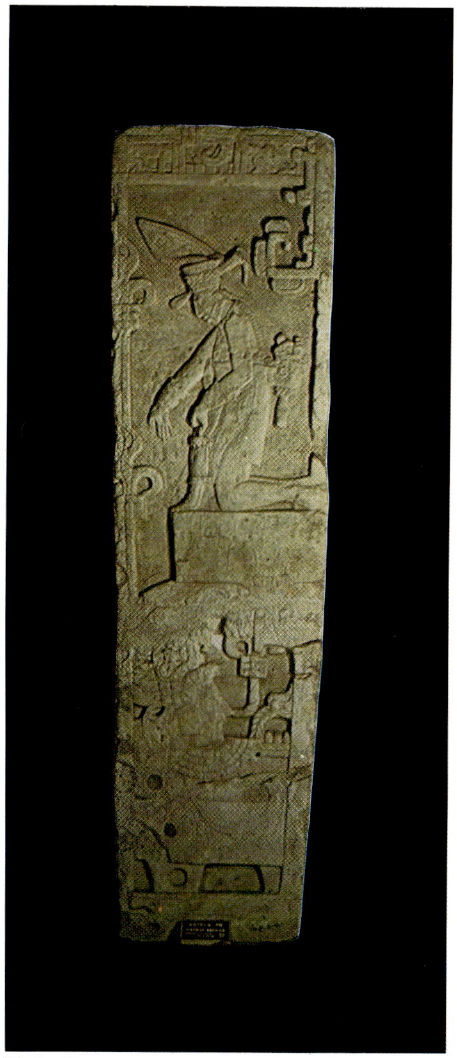

Plate 56.

Plate 55. Detail, Stela 12 from Piedras Negras (Museo de Arqueología Maya, Guatemala City)

Plate 56. Stela 10 at Piedras Negras (Museo de Arqueología x Maya, Guatemala City)

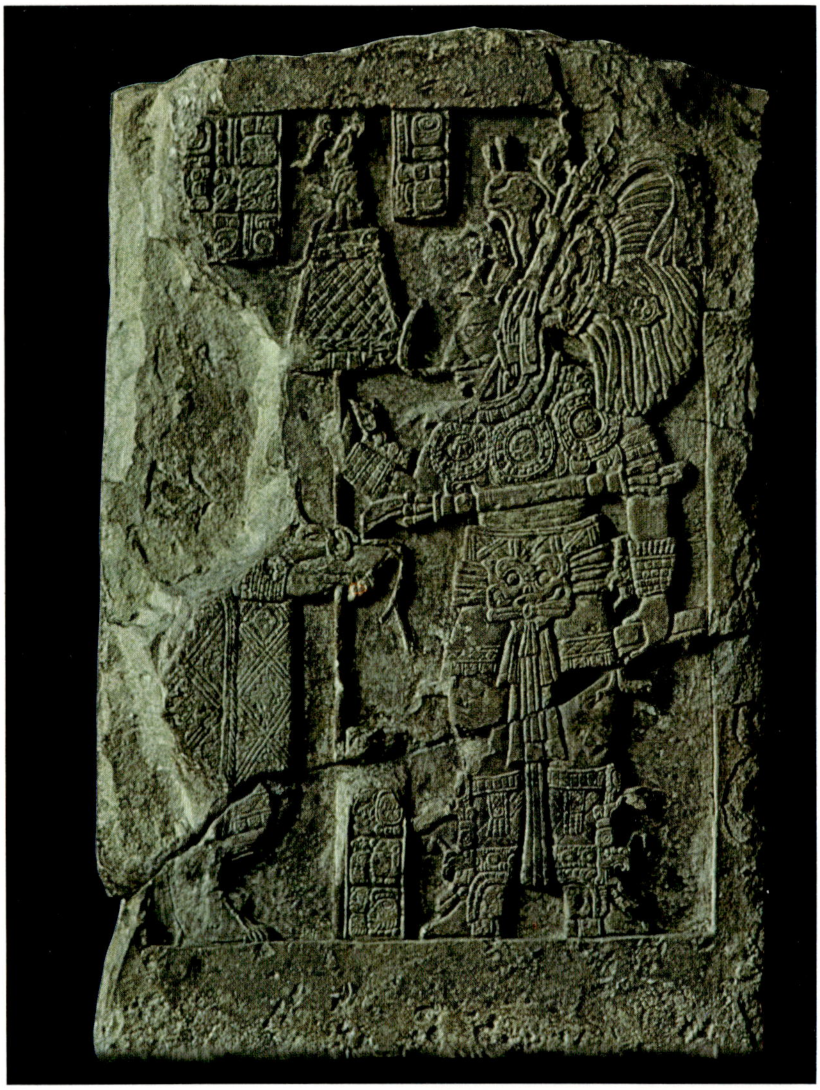

Plate 57.

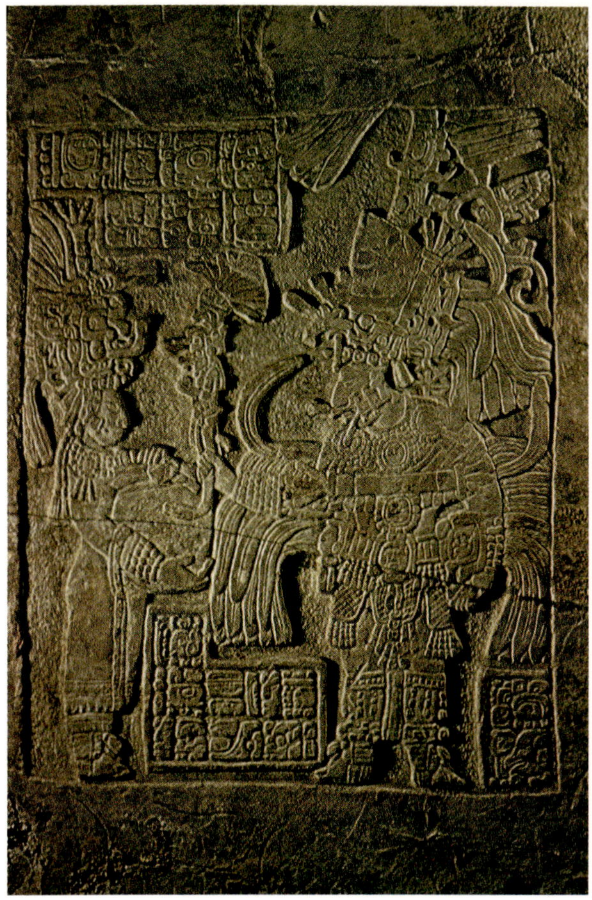

Plate 58.

Plate 57. *Lintel 43 from Yaxchilán (Museo Nacional de Antropologia e Historia, Mexico City)*

Plate 58. *Lintel 53 from Yaxchilán (Museo Nacional de Antropologia e Historia, Mexico City)*

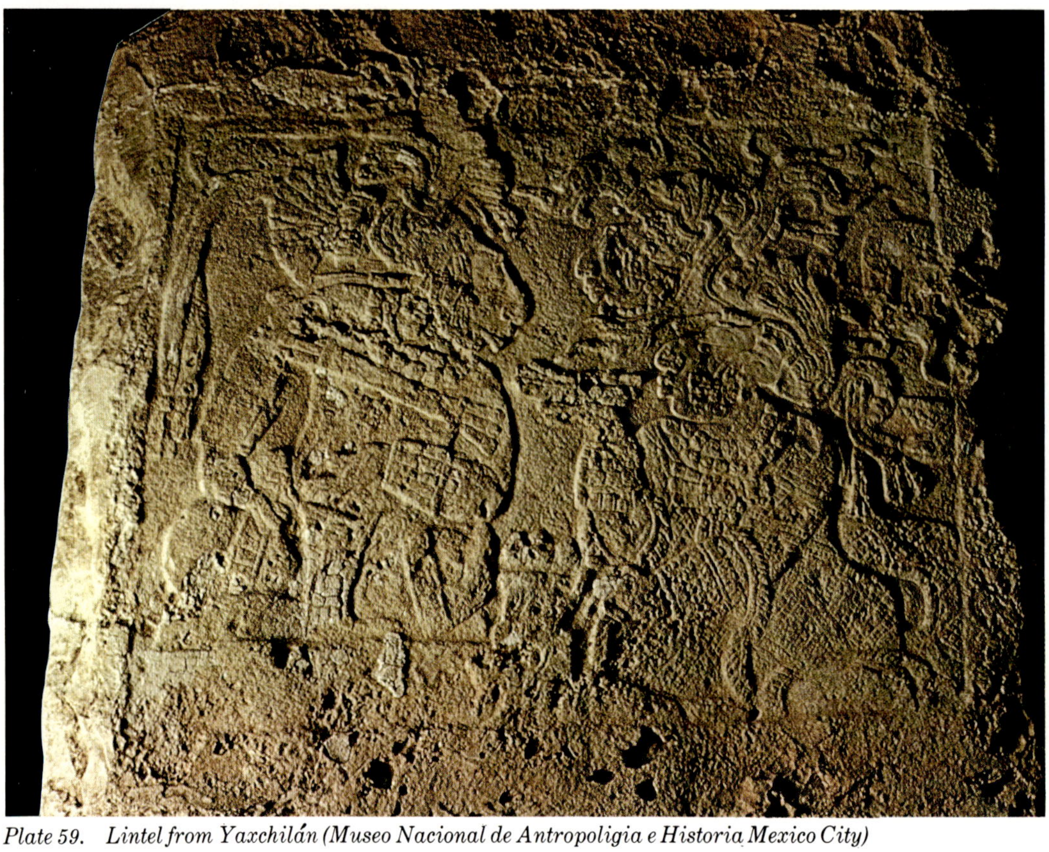

Plate 59. *Lintel from Yaxchilán (Museo Nacional de Antropoligia e Historia Mexico City)*

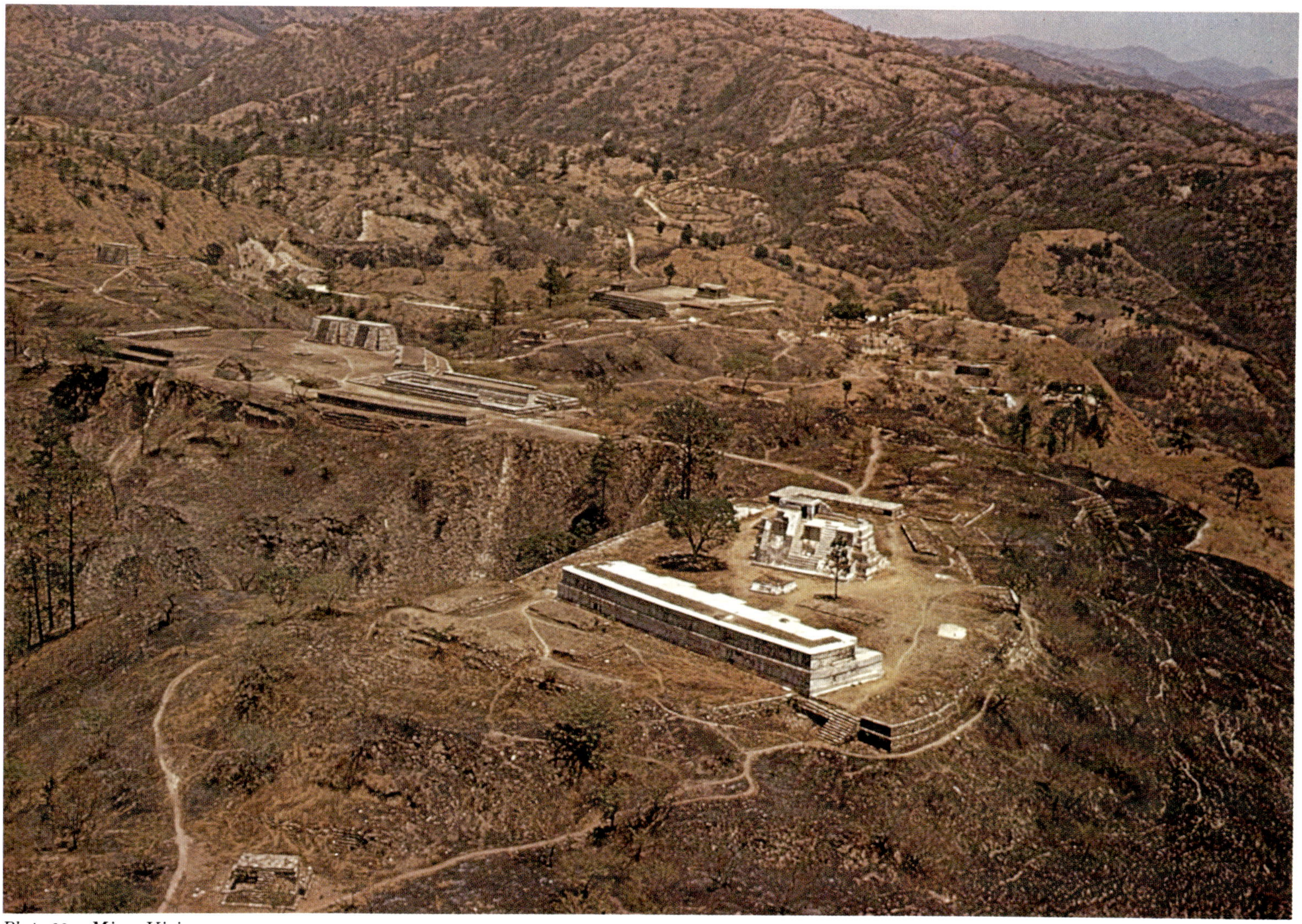
Plate 60. Mixco Viejo

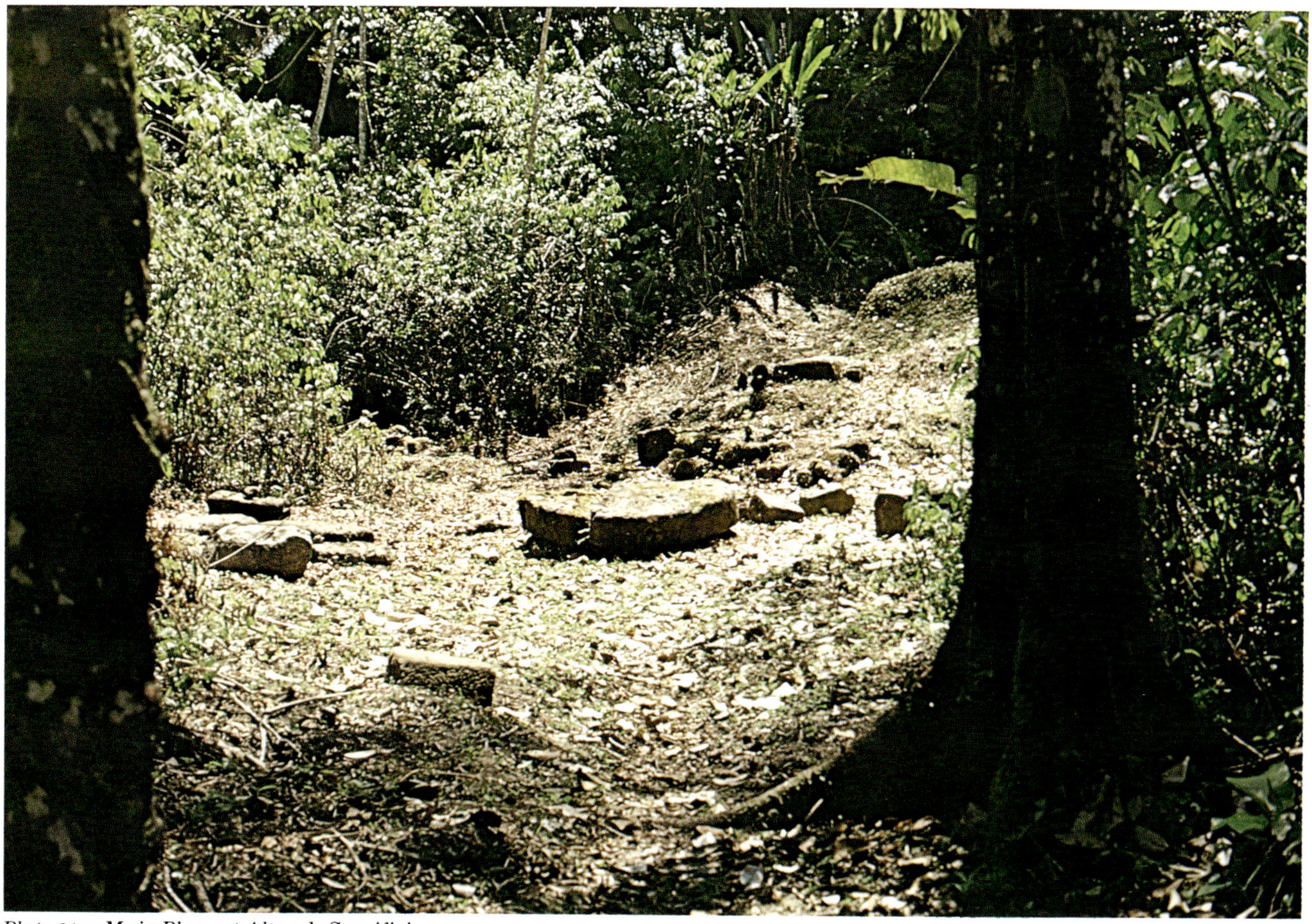
Plate 61. Main Plaza at Altar de Sacrificios

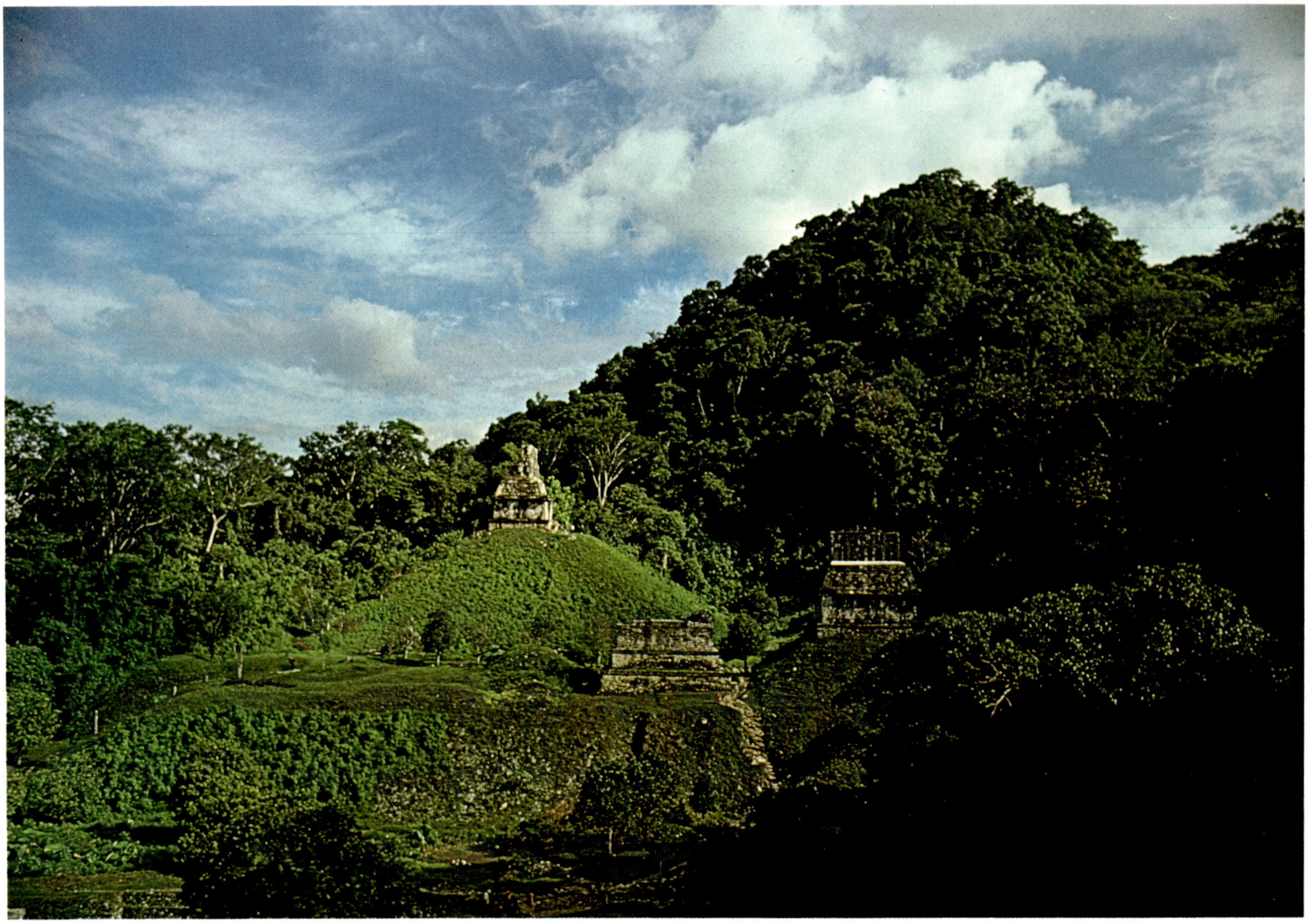
Plate 62. The Temple of the Cross Group at Palenque

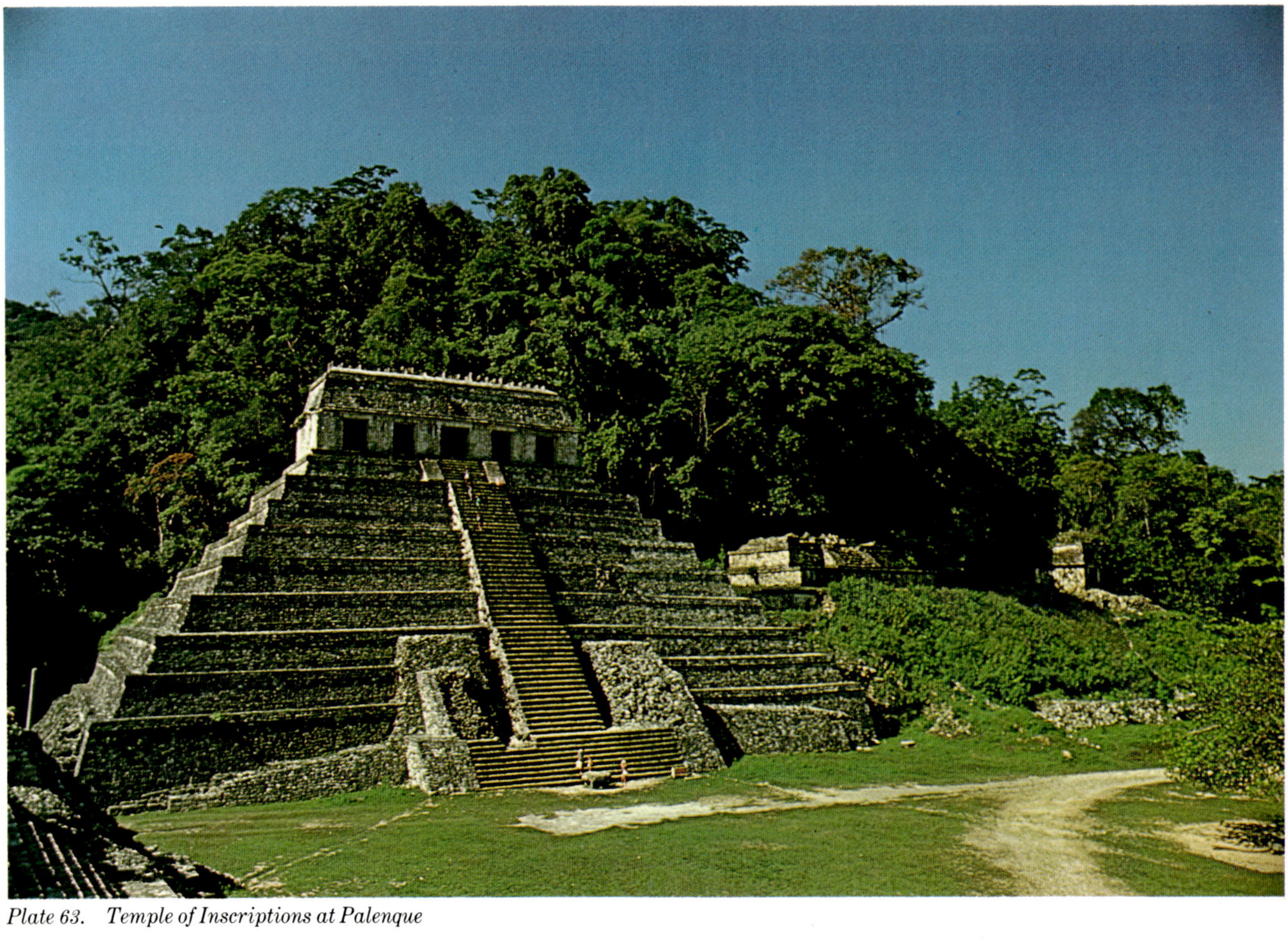
Plate 63. Temple of Inscriptions at Palenque

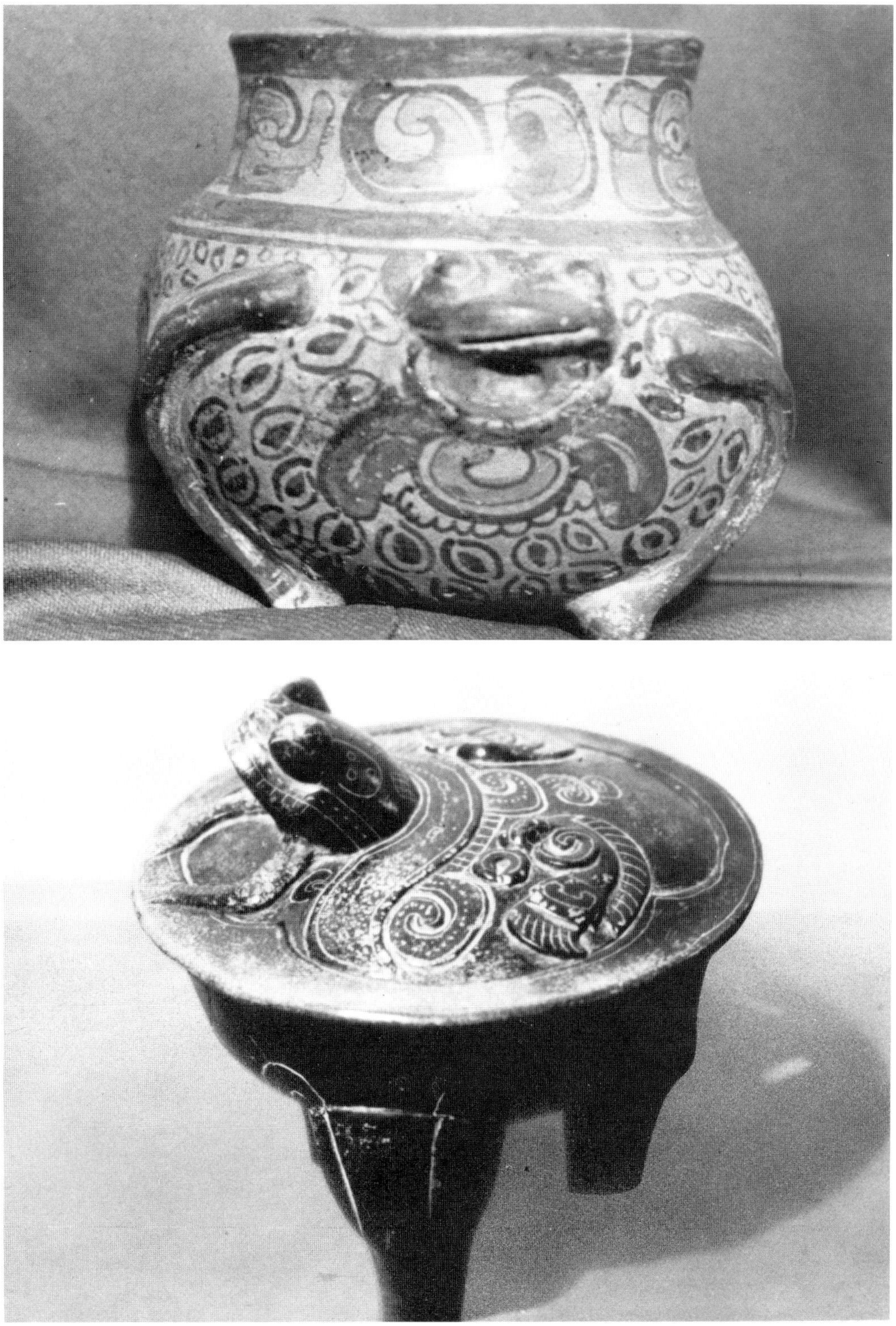

Figure 53. Ceramic bowls made into the image of toads: (a) *late Classic,* copador *style (Copán Museum);* (b) *early Classic, probably from northern Petén (private collection)*

that the Mayas used *bufotenine* as an additive to other psychotropic substances, such as tobacco, or consumed it by itself as a hallucinogen, most Maya scholars still express serious doubts about this theory.[66]

While the use of toads as a source of psychotropic material remains questionable, there is near-general agreement that the Mesoamerica nations, the Mayas among them, used hallucinogenic mushrooms. This was eloquently affirmed by the

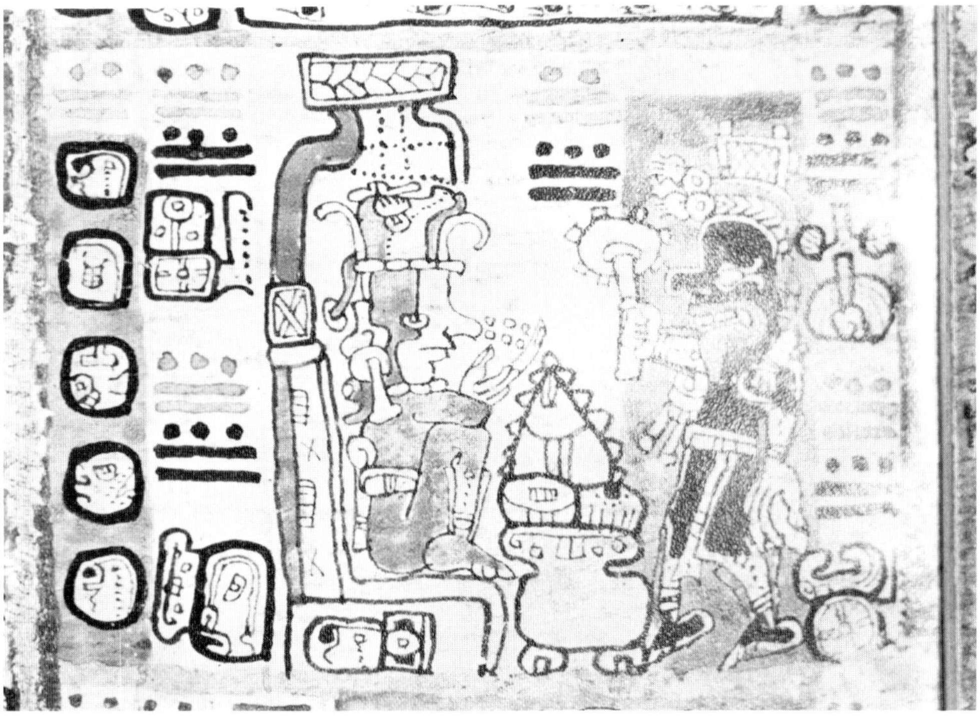

Figure 54. Encounter between God F and black God M. The latter is holding an object that may be a mushroom, a rattle, or a mace (Madrid Codex)

"anti-toad," Sir J. Eric S. Thompson: "All accept the archaeological evidence of hallucinatory mushroom cults because it is strongly supported by colonial sources and by present-day practices in Middle America. There is continuity from ancient to modern times."[67] Present-day Mesoamerica grows more than twenty species of mind-alterating mushrooms, including *Amanita Muscaria*, *Psilocybe*, and *Stropharia*.[68] These mushrooms contain a pharmacopoeia for different mind-alterating alkaloids, several of them known to be used in native rituals. From a pragmatic point of view, the use of mushrooms for hallucinogenic purposes could be very old, for, while many other psychotropic substances must be boiled, pulverized, and specially prepared before their alkaloid properties are released, psychotropic mushrooms can be used raw or dried, later reconstituted, and mixed with other materials.[69]

One of the early descriptions of a mushroom ceremony is given in the *Madrid Codex* (not to be confused with the Maya manuscript of the same name, also called the *Codex Tro-Cortesianus*), which contains the following account of the consumption of mushrooms at Tenochtitlán:

> Very soon the people were given mushrooms to eat. They ate them during the so-called flute-playing time. They had not eaten anything but had only drunk a little cacao at night. They ate the mushrooms with honey. When the mushrooms begin to affect them, they began dancing or crying. Some who still have their reason go to their place; they sit against the wall; they do not dance but remain with their heads down. One sees that he is going to die; he begins to weep; another that he is going to die in the war. Another that he will be eaten by wild animals. Another sees that he will become a prisoner of war. Another sees that he is going to be rich, to be happy and esteemed. Another sees that he must beg people, that he must be a slavemaster. Another sees that he will be an adulterer: that his head will be broken with stones, that he will be flattened with stones.
>
> Another sees that he will be a thief; he also will be stoned. Another sees that he will have stones thrown at his head, that he will be imprisoned. Another sees that he will die in the water. Another sees that he will have a quiet peaceful life and death. . . . And when the effect of the mushroom has worn off, they began to talk and tell each other what they have seen.[70]

According to early literary sources mushroom cults flourished among the Zapotecs, the Mixtecs, the Otomis, the Matlazincas, and the Tarascans.[71] More recent research indicates that ingestion of psychotropic fungi is still practiced by some of these nations despite powerful bans imposed by both ecclesiastical and government authorities.[72]

As far as the ancient Mayas are concerned, we are on somewhat more uncertain ground. The possibility that they engaged in mushroom rituals was doubtless first inferred from the large number of mushroom effigy stones discovered at Maya archaeological sites.[73] The view that these "mushroom stones" represent psychotropic fungi has been challenged by Kohler, who thinks that these specimens were used to manufacture culinary and utilitarian pottery objects, and by Rose, who believes that they were in some way connected with the Maya ball game and were perhaps employed in shaping the rubber balls.[74]

In the ethnographical examination of the subject five names are prominent: Stephen F. de Borhegyi, Peter T. Furst, Merle Greene Robertson, the mycologist Bernard Lowy, and Karl Herbert Meyer, the last of whom more or less acted as chronicler and critic of the controversy.[75] Lowy believes that the *Codex Madrid* and the *Dresden Codex* contain certain references to *Amanita muscaria*, a potent hallucinogenic fungus.[76] On page 51 of the *Madrid Codex*, for example, is painted a scene that is described by Villacorta as God E seated in a temple and being confronted by God M, who is holding in his hand a "calendar stone or maybe a rattle."[77] This scene was earlier described by Förstemann as the "Black God . . . swinging a rattle" because "he has been invited for a hunt."[78] This scene was interpreted by Lowy as the Black God "holding up a sacred fungus."[79] Mushrooms are also referred to in the *Popol Vuh* and *The Annals of the Cakchiquels* in association with religious activity.[80] At the time of her research in 1972, Merle Greene Robertson found that in some areas the Mayas had retained or readopted mushroom rituals. There were still some Lacandón Indians living around the lakes of the Metzabok region who had used hallucinogenic mushrooms in their ritual ceremonies apparently "as long as anybody can remember."[81] Similar observations were made by students of Michael D. Coe among the present-day Chol Mayas.[82] Greene Robertson identified two species of hallucinogenic mushrooms common to the Usamacinta River basin, *Panaeolus venenosus* and *Psylocybe cubensis*. The Lacandón Mayas placed the mushrooms "in the ritual bowls or on small stone altars in ancient Maya *oratorios*. The purpose of their use, according to the Lacandones, was to contact their gods."[83]

These mushrooms are found in abundance around Palenque, much to the frustration of the local chief of police, who cannot figure out what the herds of hippies are doing in the pastures digging up and carting away cattle manure. To grow, *sporae* of the *Psylocybe (Stropharia) cubensis* must pass through the alimentary tract of a ruminant—cow, sheep, or deer. Deer were the only such animals before the Spanish entry, and Furst suggests the very interesting possibility that the veneration of the deer in ancient Maya religion and mythology is connected with the growth of these miraculous mushrooms in the deer's droppings.

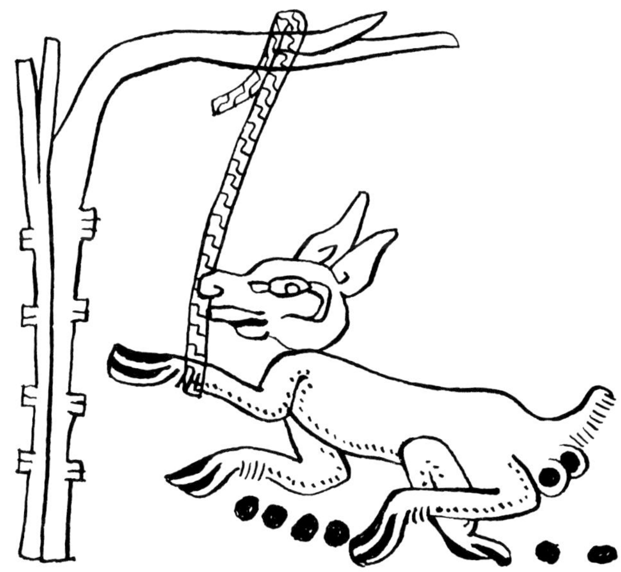

Figure 55. Snared deer (Madrid Codex)

Furst also examined the subject from the view of etymology and discovered entries in sixteenth- and seventeenth-century Quiché and Cakchiquel dictionaries relating to intoxicating fungi. These books mention *xibalbaj okox*, "mushroom of the underworld," and *muxan okox*, "mushroom that makes the eater crazy," as well as *k'aizalah okox*, "mushroom that makes one lose one's judgment."[84] Meyer, following the same path, found more literary references that led him to the conclusion that "at least the historic Cakchiquel Maya in the Guatemalan Highlands knew the properties of hallucinogenic fungi and ingested them."[85]

Using these results of ethnological research in interpreting scenes on the monuments at Yaxchilán, Greene Robertson theorizes that some of them are indeed related to the ingestion of hallucinogens and that even the famous "bundles" of Yaxchilán, which were thought to contain barter, may instead have held, among other ritual paraphernalia, hallucinogenic herbs or mushrooms.[86] The scene on Lintel 25, for example, is interpreted by Greene Robertson as follows:

> The sacrifice and the hallucinogenic effects may be expressed in symbolic apparitions. On Lintel 25, this vision takes the form of a giant serpent-god emerging from the ritual bowl, and the head of a man wearing a year-sign emblem on his headdress emerges from the jaws of the serpent. He is speaking to the penitent person, a woman, who has performed the [self-sacrificial bloodletting] rite and is now holding the bowl in her left hand, and in the right hand is offering blood to the gods. . . . If one is to

> consider [the depictions of] serpents rising from sacrificial vessels . . . as apparitions from the use of hallucinogenic drugs and idolatrous incantations, then it is required that one investigate the availability of hallucinogens that could have been used to enable one to undergo such obvious self-torture, and one that would have produced the required hallucinogenic effects.[87]

It thus appears highly probable that some of the early records erred in identifying the herb the Indians used to induce narcosis and hallucinations as ordinary tobacco, and Thompson may have accepted their interpretations too readily.[88] The plant or plants the Mayas used to induce hallucinations were either not tobacco or were tobacco mixed with other substances. The question, of course, is why they were consistently called tobacco. The explanation may be that at the time of the Conquest the Arawak word *tobaco,* the Nahuatl *picietl,* and the Maya *zigar* designated not just tobacco plants and objects made of tobacco but anything smokable and even the act of smoking itself. The Spaniards, unaware of any other plant that was used for smoking, associated these names with tobacco. An analogy exists in American slang, in which a "smoke" means a tobacco cigarette. And in Cuba, to this day a cigar is *un tabaco.*[89]

There is ample linguistic evidence to support the theory that in the early times these words were used in a much looser context. Fray Ramón Pané, the monk who accompanied Columbus on his second voyage in 1493, noted that not only the plant but also the tube through which the natives smoked was called *tabaco.*[90] Oviedo y Valdés, by royal decree official chronicler of the Indies from 1532, also used the term *tobaco* inconsistently, sometimes using it to mean the act of smoking, at other times the smoking instrument, and at still other times the herb itself:

> The Indians of this island, among other of their vices, know one very evil thing, to indulge in smoking, which they call *tobaco,* in order to go out of their senses. They do this with the smoke of a certain herb, which, as we have been able to gather, is the quality of henbane.[91]

In another chapter of his book he wrote:

> The cacique employed a tube, shaped like a *Y,* inserting the forked extremities in their nostrils and the tube itself in the lighted weed. In this way they would inhale the smoke. . . . Those who could not procure the right sort of wood took their smoke through a hollow reed [*cañuela*]; it is this the Indians call *tobaco,* not the weed, nor its effects, as some have supposed (que es dicho llaman los indios tabaco, e no a la hierba o sueño que les toma como pensaban algunos).[92]

In the revised edition of his book, however, he wrote: "There is an herb they call *tobaco,* which is a kind of plant, the stalk of which is as tall as the chest of a man."[93] Las Casas, who was an excellent linguist, also used the word *tobaco* to refer to cigars ("muskets"), not to the plant" (estos mosquetes, o como los llamaremos, llaman ellos tabacos).[94] As late as 1721, Francisco Ximénez, in his *Historias del origen,* used the words *tobacos* and *cigarros* alternately.[95] The chronicler Hernández further complicated the matter with his statement that the reed cigarettes in Mexico were called *tavacos* or *tabacos.*[96]

Similarly, the root words *ciq, ziq,* or *cii,* which occur in several Maya dialects,[97] sometimes denote tobacco cigars and at other times mean anything that is smokable or inflammable. Brasseur de Bourbourg, who in 1861 first translated the *Popol Vuh,* noted that the Maya word *ciq* may be translated not only as "cigar" or "tobacco" but also as "smoking," "perfume," or even "to make noises," "to cry," and so on ("Ciq signifie aussi parfum, voix cri, lamentable, cigar, fumer, c'est evidement l'origin de l'espagnol et de notre cigare").[98] Adrián Recinos, the other noted translator of the *Popul Vuh,* also defines *zig* or *zikar* as "to smoke" (*fumar*).[99] Seler, analyzing the same Maya words, translated them as "honey"—more probably a root with narcotic properties that was added to drinks. The word *ci* or *cii,* he translated as "good-tasting," or "good-smelling."[100] He also believes that the Mayas used *ci* for the maguey plant, as well as the strong alcoholic drink made from it. In the Ixil Maya dialect *cib* also means "smoking," but not necessarily "tobacco."[101] The custom of using the words for "tobacco," "to smoke," and "pipe" interchangeably has also been recorded among other native tribes of the New World.[102]

The evolution of the words "tobacco" and "cigar" now appears to be clear. In pre-Columbian and early post-Conquest times, when the natives smoked the plant that we now call tobacco, as well as possibly some other herbs with more potent narcotic properties, these words simply meant "to smoke"—anything. After the Spanish invasion, when hallucinogens gradually ceased to be used in rituals and tobacco smoking became for the most part a pastime, the words *tabaco* and *zigar* acquired the meanings that they have today: the tobacco plant and the tobacco cigar. In contemporary culture where hallucinogens appear to have made a comeback, there seems to be an etymological process in the opposite direction. When a decade ago you asked for a "smoke," you knew that you would be given a tobacco cigarette. Nowadays, you cannot be so sure.

Emphasis has now shifted to present tobacco, as seen through native eyes, with corresponding stress on spiritual roles played by the plant.

Sir J. Eric S. Thompson

The most important and solemn feast in which the most splendid ceremonies took place was that of the god known as Tezcatlipoca.

Fray Diego Durán

Coe has identified God K as a Mayan equivalent of Aztec Tezcatlipoca, a view which is clearly correct.

David Kelley

6. God K, the Deity with the Smoking Cigar

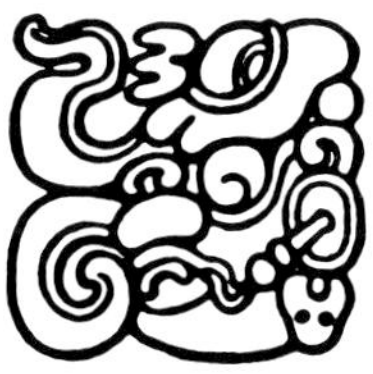

Peculiar groups of small gods on different Classic stone monuments were documented in detail by Alfred Maudslay,[1] but not until 1913 did they receive special attention, in the work of Spinden. The effigies are small anthropomorphic spirits with long, upturned muzzles, wide-open mouths, prominent incisors, and shorter lower jaws. They are most commonly seen on the rulers' ceremonial scepters (hence their old name Gods of the Manikin Scepter), which during late Classic times in some centers replaced the double-headed serpent ceremonial bar as a symbol of authority.[2] Such effigy gods can be seen on the monuments at Palenque, Machaquilá, Calakmul, Yaxchilán, Tikal, and Etzná, as well as on many ceramic vases of the Classic epoch.[3] Linda Schele's recent classification divides these effigy gods into three subgroups: jester gods, shield bearers, and flare gods.[4] The last group is probably the most significant and is of special interest in respect to tobacco research.

The Flare God is Schellhas' God K, one of the principal deities of Maya mythology, whose exact nature long puzzled Maya specialists. Besides Flare God and God of the Manikin Scepter, God K is also known under several other names: God with the Ornamented Nose, Zimmermann's God G8, Ah Bolon Dzacab, God of the 819-Day Cycle, and God GII of the Palenque Triad. He is generally regarded as a benign deity, as are Gods B and D, to whom he bears an obscure relationship.[5]

Schellhas' first identification of God K in 1894 was based primarily on the effigy's proboscis.[6] Recently, however, interest has focused on the tube-and-coils

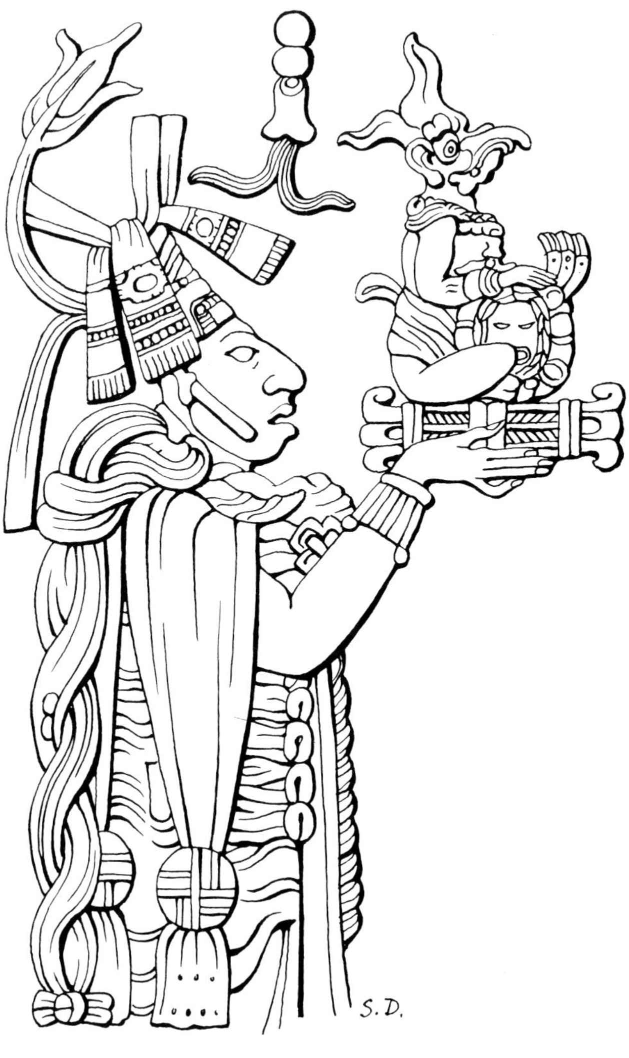

Figure 56. Dignitary holding the shield-bearing god, Sanctuary Tablet, Temple of the Sun, Palenque (drawn after Maudslay)

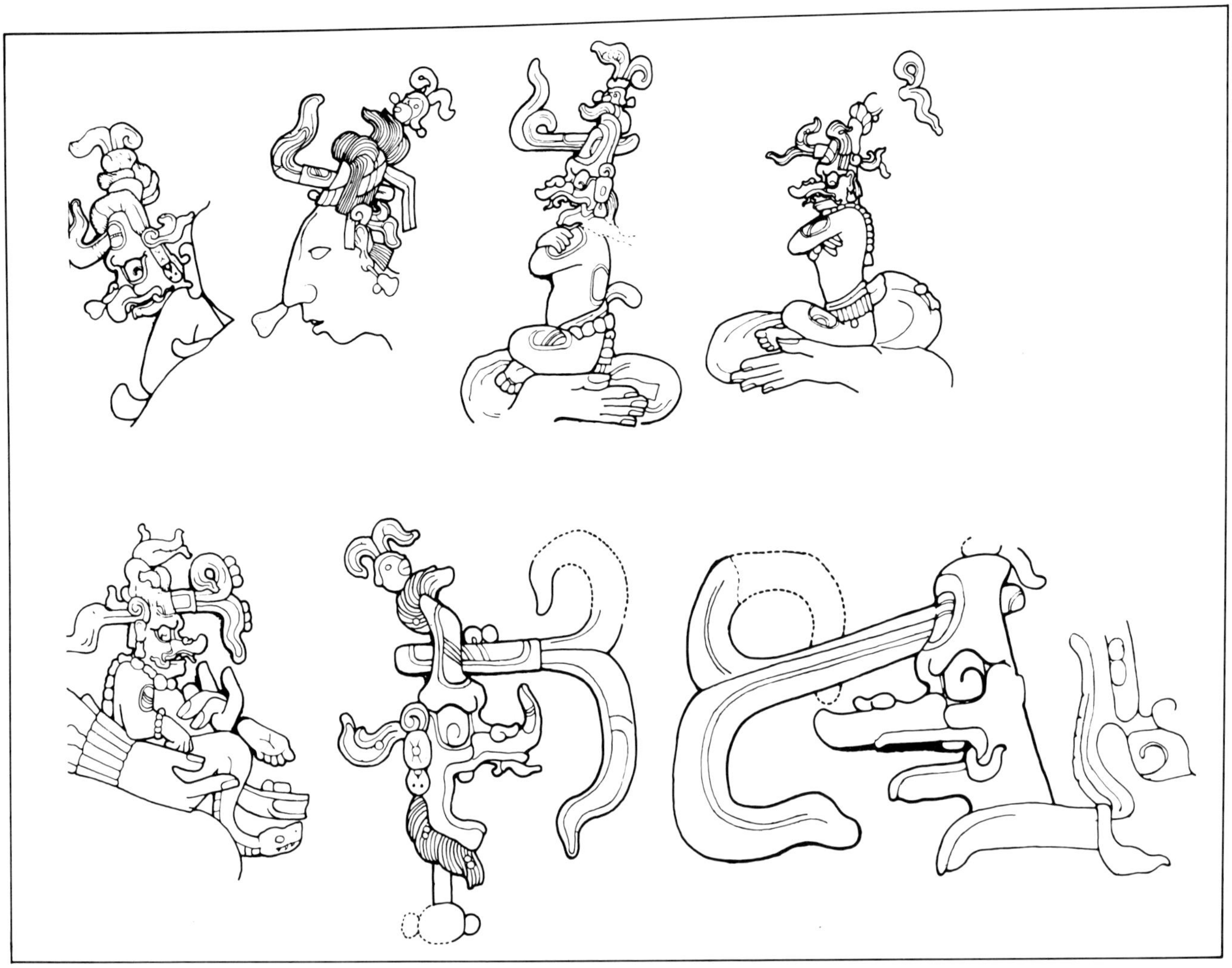

Figure 57. Flare Gods at Palenque. Left to right: Sarcophagus Lid, Temple of Inscriptions; Temple XIV; Temple of the Sun; Dumbarton Oaks Tablet (probably Palenque region); Palace House A; Palace North Structure; Tier 3 (after Linda Schele)

forehead ornament, which was identified by Spence as foliation and by Thompson as *bil*, a vegetable element, the sign of floral growth.[7] Michael D. Coe initially also accepted this explanation.[8] Later, however, he changed his views because the forehead ornament "often appears in both codices and monuments over the glyph 'fire' (T563), and, therefore, it surely represents smoke." He identified the forehead tube at different periods as a bone tube, a pipe, and finally a cigar, and the flamelike appendage rising from the tube as smoke, an iconographic complex indicating power and divinity.[9] Schele agrees that the object probably represents a smoking tobacco cigar.[10] This theory has been reinforced by David Kelley on a glyphic basis:

As a separate glyph, the tubular pipe occurs only once, prefixed to "star" glyph (Tikal St. 5.B.10). . . . I have suggested that the combination is to be read *budz ek*, "smoking star," a known Yucatec term for comets. The inscription suggests that the pipe has been infixed onto the head of God K, though the infix has been so conventionalized in the codices that it is no longer recognizable as a separate element.[11]

A disturbing feature of this very attractive cigar theory is the great variety in shape and size of the ornament as it is portrayed on monuments and ceramics. Some of the examples are tubular and closely resemble a modern-day cigar. Others, however, are funnel-shaped, and still others are rhomboid. The variation that probably least resembles a cigar is represented in the celt-shaped objects piercing the foreheads of the stucco-covered wooden statuettes of God K found in Burial 195 by William R. Coe at Tikal.[12] The ax shape of these implements are especially well defined because they are shown in three dimensions. They closely resemble the jade celts found in abundance at Olmec

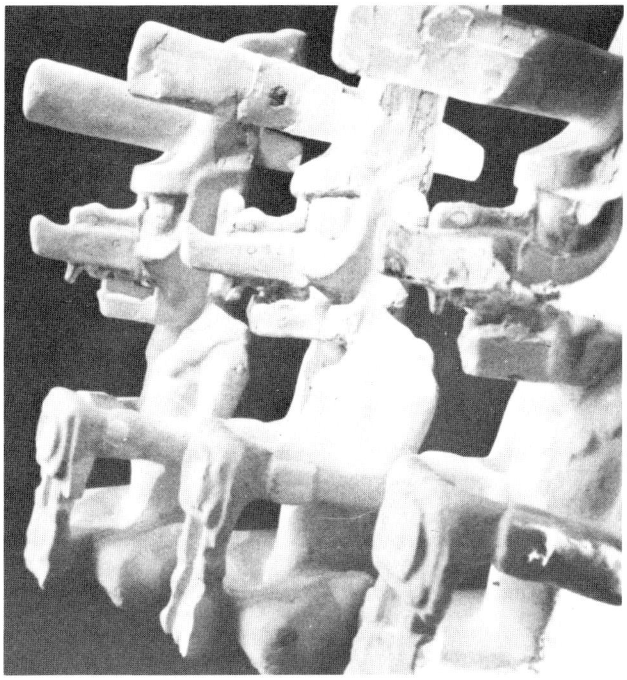

Figure 58. Three stuccoed wooden statuettes found in Burial 195 at Tikal. The long-nosed deities with the celt-shaped tubes across their foreheads probably represent God K (after William R. Coe)

sites and the "ax gods" of the Nicoyas and the Guanacastes. Such celts, made of different kinds of stone, were also found at several Maya sites, such as Tzabajob, Chichén Itzá, Tzinzuntzán, Zaculeu, and Pomona.[13] One can only guess whether they were made only for utilitarian or also for ceremonial purposes. According to Sedat, the Mayas in the Guatemalan highlands believe that they are from the heavens and were thrown to the earth by lightning.[14] Hellmuth photographed such celt-shaped object piercing the head of God K on a carved early Classic bowl from the Tikal area, and I photographed one on a polychrome plate from northern Petén.[15] To confuse the issue further, on both of these artifacts the front end of the object emerges not from the forehead but from the mouth of the deity.

The troublesome feature is that the "handle" of this "cigar" protrudes from the nape of the deity. Other very important archaeological finds suggest that there was such a deified ax. The first one is a polychrome plate made during the Classic period of Petén, now in the official collection of the Guatemalan government. The plate shows the seated figure of an anthropomorphic jaguar who is holding an ax in his right hand. A snake serves as the handle of the weapon. The ax blade is of conventional shape, and from it emanate two typical coils of smoke. It strongly resembles the object piercing the forehead of God K in some of his portrayals (a manikin scepter sans manikin!). Another example of a snake-handled ax is seen on the Dumbarton Oaks Tablet, wielded by the dancing God GI. The latter ax has neither manikin nor smoke curls, but it has a beautifully executed serpent body and head.

Further examples of snake-handle axes can be seen in the Temple of the Warriors at Chichén Itzá.

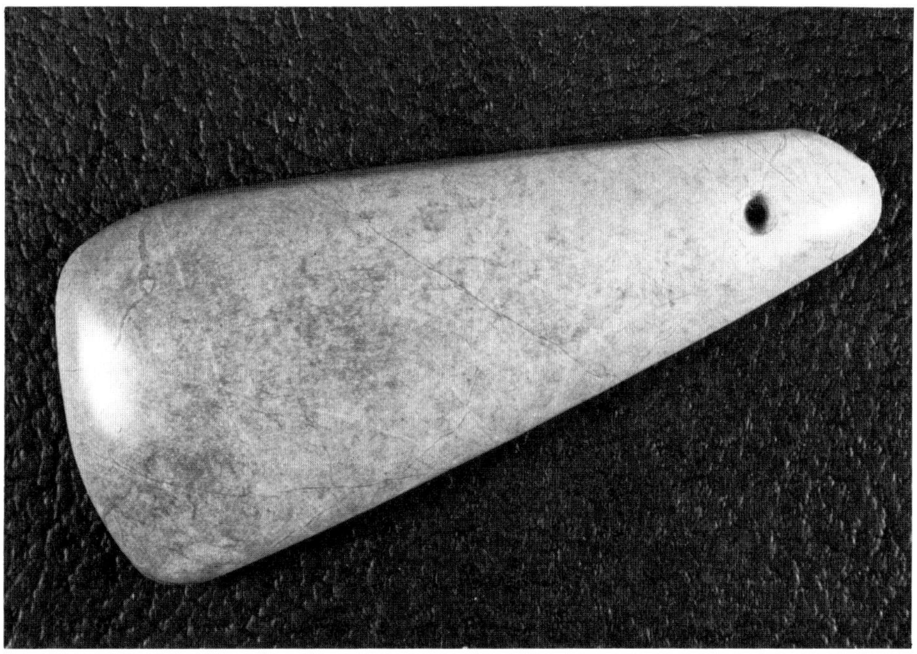

Figure 59. Maya jade celt

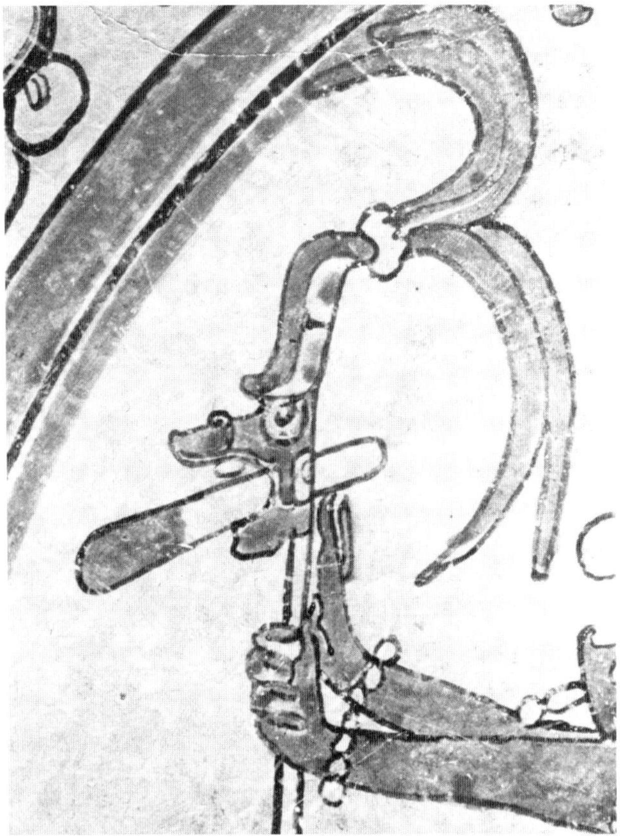

Figure 60. Detail of a polychrome plate, Classic period, probably from Petén. It shows God K as a manikin scepter. Note that the tube emerges from the mouth instead of the forehead

Warriors dressed in Toltec-Maya combat gear carrying serpent-handle battle axes are portrayed on both the columns and the frescoes of the Northern Colonnade.[16]

Undoubtedly not all these examples are simply representations of a god. Like the serpent spears and the jaguar shields on the Sanctuary Tablet of the Temple of the Sun at Palenque and on Stela 5 at Yaxchilán, they are deified weapons. These finds suggest the possibility that, while most "forehead flares" probably do represent cigars, some of the celt-shaped objects piercing the heads of jester gods may be just what they appear to be: celts. One could go even further and ask: Is it possible that a deity such as God K may have had various implements at his disposal, and, depending on the occasion, he placed in his forehead short cigars, long cigars, smoking cigars, torches, or even celts or other objects? Is it conceivable that the shape, size, and identity of the forehead ornament reflect the particular role the god is playing? Does God K with a short, round cigar mean one thing and the same figure with a large nonsmoking celt-shaped object mean something—or someone—else? My answer to these questions is a cautious yes, which agrees with the line of thinking of Kelley, who long suspected that the varied headgears in Maya iconography may be ciphers for the names of different gods and lords.[17]

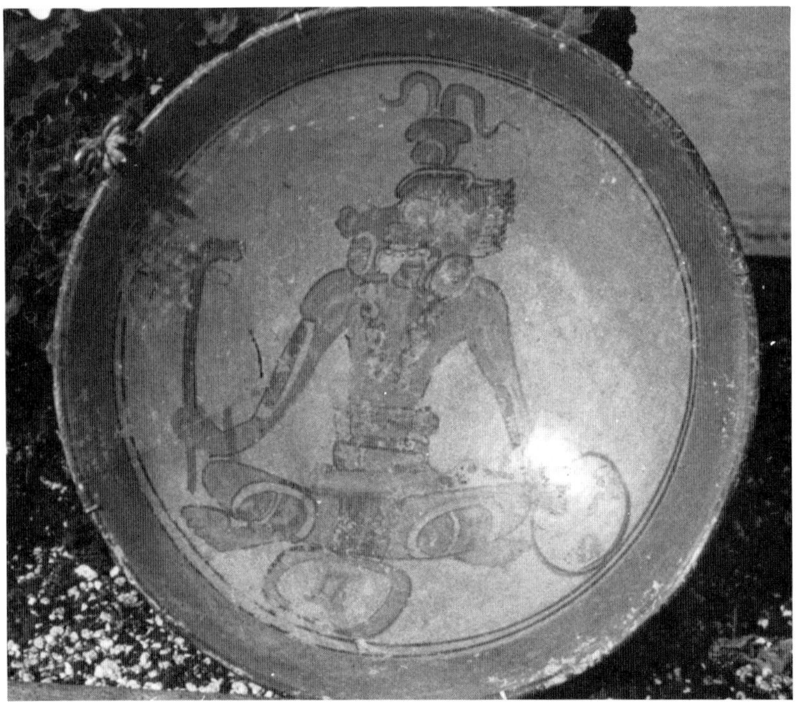

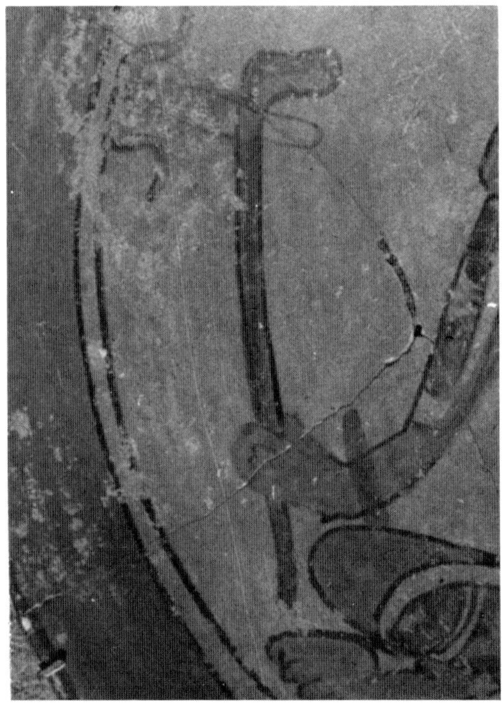

Figure 61. Polychrome plate and detail showing a seated jaguar holding an ax with a smoking blade, late Classic period, probably northern Petén (Official Collection of the Guatemalan Government)

The study of two particular vases, D-29 and D-33, discussed in detail later in this book, certainly indicates that the theory is valid. The first vessel is painted with the figures of two similar monster-headed deities exhibiting typical features of God K. While one of the gods has a large forehead "cigar," the other sports a huge floral ornament emerging from his forehead.

On the second vase appear four typical God K figures. The deities are paired, the pairs facing each other—an arrangement indicating that we are dealing not with repetitious portraits but with at least two and probably four gods. From the forehead of the first god quetzal feathers emerge, the second god has a smoking forehead tube and from the foreheads of the third and fourth gods only smoke is rising. There are very few "coincidental findings" in Maya iconography.

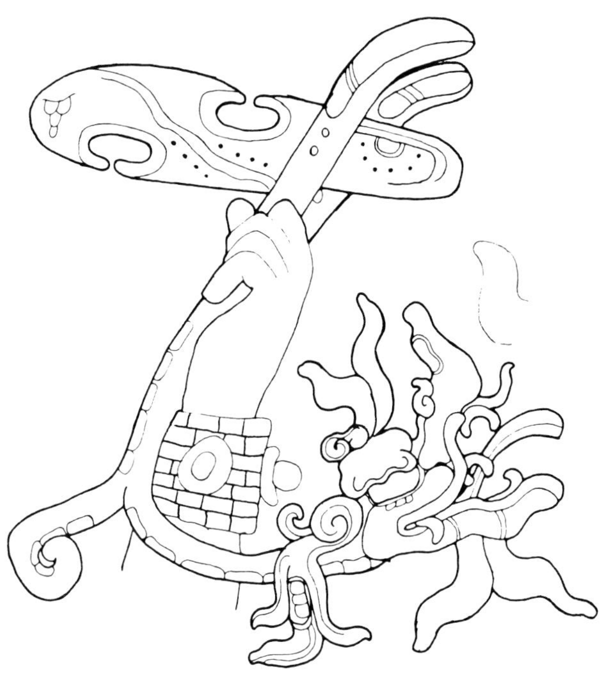

Figure 62. Ceremonial ax with a snake handle, Dumbarton Oaks Tablet (after Linda Schele)

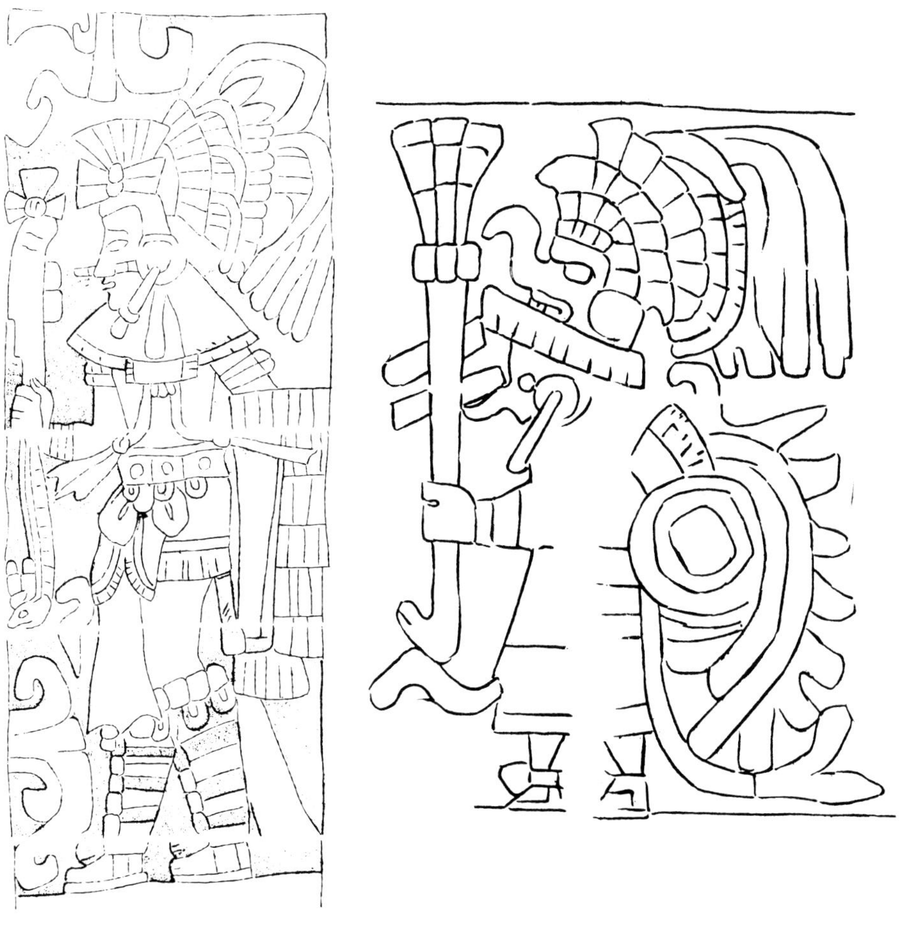

Figure 63. Toltec Maya warriors holding snake-handled battle axes on the (a) *columns and* (b) *frescoes of the Northern Colonnade, Temple of the Warriors, Chichén Itzá (after Morris)*

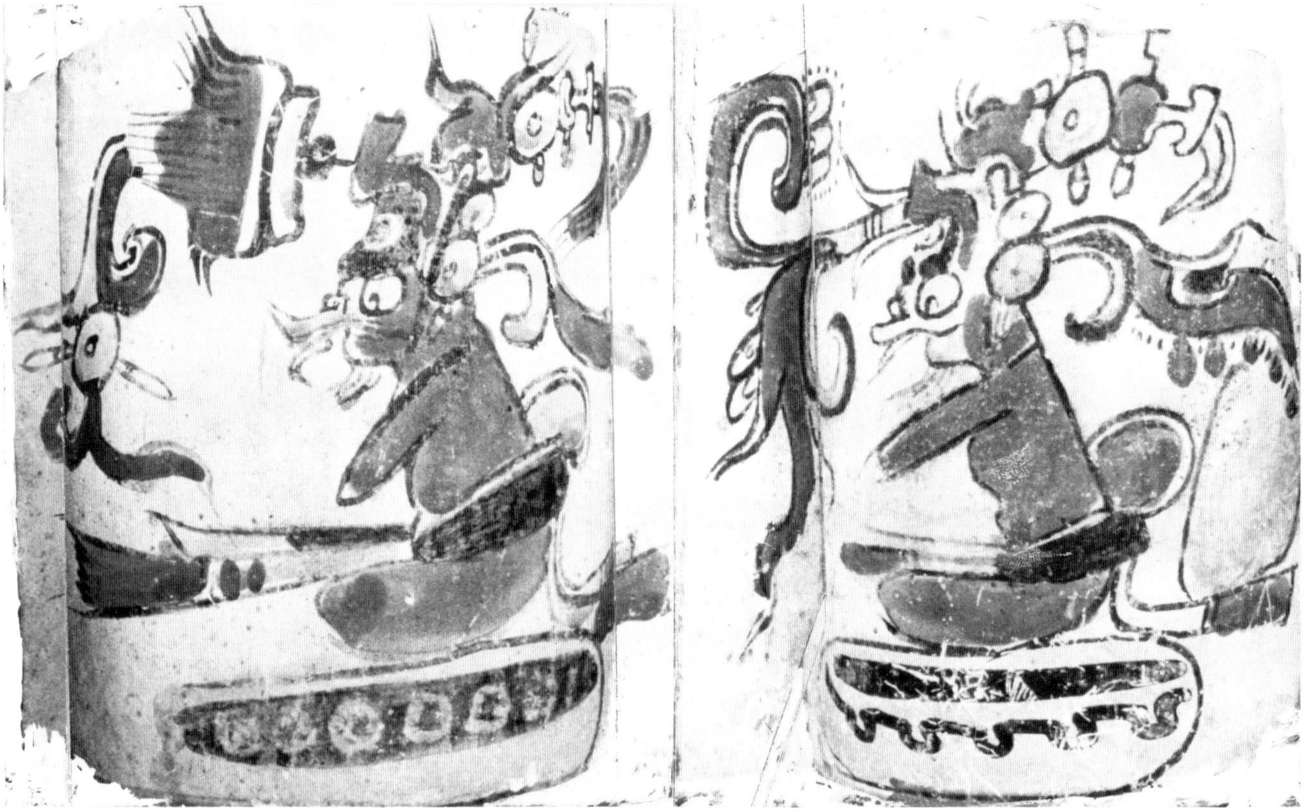

Figure 64. Two figures with God K characteristics on a polychrome vase from the late Classic period of Petén. One figure has a smoking forehead tube, the other a forehead flower ornament

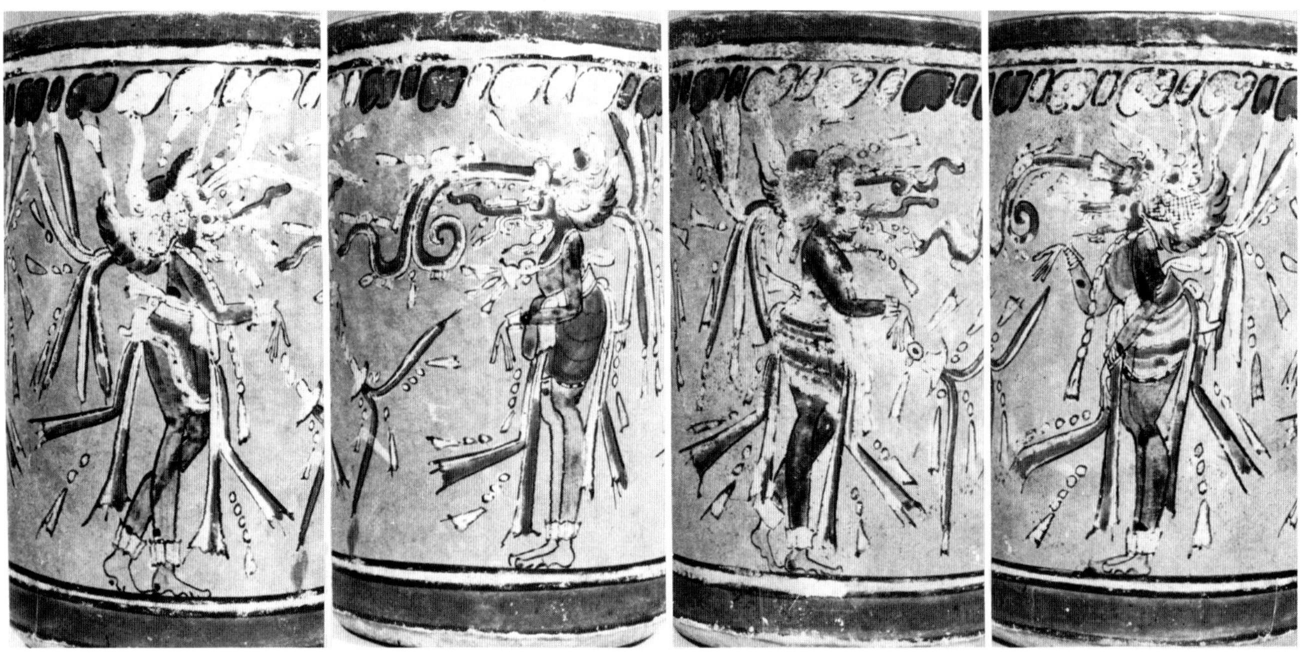

Figure 65. Four figures, probably all of God K, on a polychrome vase, late Classic period, Petén. The first god's forehead has feathers, the fourth god has a smoking forehead tube, and the second and third gods have smoke coils only (photograph by Nicholas M. Hellmuth, courtesy of Foundation for Latin American Anthropological Research, St. Louis, Missouri)

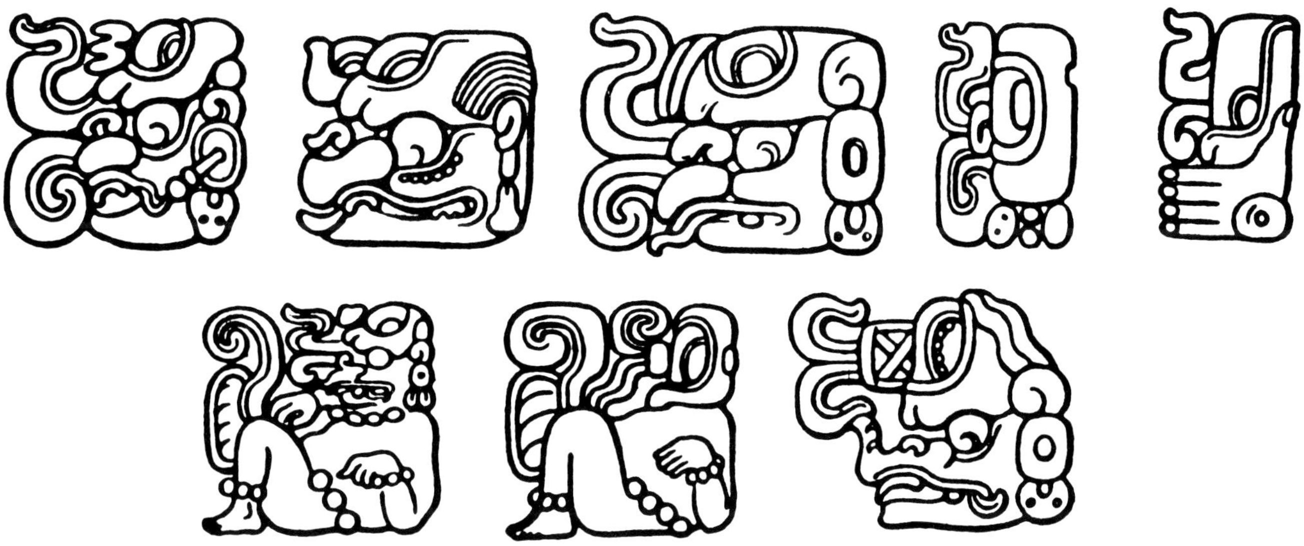

Figure 66. Hieroglyphs of God K (after J. E. S. Thompson)

The Mythological Identity of God K

The first to consider the identity and nature of God K was Brinton, who regarded him as simply a special manifestation of God B. Schellhas, who recognized the astronomical significance of this deity, thought that he might represent a star. This theory was based on a study of the initial pages of the *Madrid Codex*, where a row of repetitions of God K's name glyph is painted just below the signs of the cardinal points and parallel to a series of the signs of God C, thought to be the Lord of the Polar Star. Schellhas, and later Lounsbury, regarded God K as the patron of the ninth day (Muluc).[18] Förstemann, Schellhas' contemporary, associated God K with the date 13 Oc and considered him to be a weather god because "his ornamental nose, according to the conventional mode of the Central American peoples, is intended to represent the blast of the storm."[19] Barthel in a 1952 article on the Venus cult also regarded God K as one of the storm gods.[20] Seler, following the same line of reasoning, associated God K with the old Rain Serpent and believed that the long-nosed masks ornamenting so many of the ancient Maya buildings are portraits of this deity.[21] Kelley, reaching deeper into the possible calendric associations of God K after examining the pertinent portions of the Palenque temple texts, expressed the view that the birthday of this deity (an event that is possibly commemorated on the piers of the Temple of Inscriptions) is 1 Ahau 13 Mac and that God K is also the mythical regent of the year Ben.[22]

Probably the most important calendric association of God K is with the so-called 819-Day Cycle. In 1943, Thompson published his observations on five Maya texts from Yaxchilán and Quiriguá, in which the month position is separated by a regular set of clauses.[23] He further demonstrated that these clauses occur at intervals of 819 days and that they are composed of six fairly constant explanatory glyphs, the fifth of which is described as a "grotesque head with a branching flamelike element emerging from the forehead"—evidently God K. Thompson went to great lengths to relate this cycle to the grouping of the lords of the skies and the underworld and to astronomical events but could not come to specific conclusions.[24]

The 819-Day Cycle was reexamined by Berlin and Kelley about the decade later, when they noted that the clauses strongly refer to both colors and directions.[25] In his *Deciphering the Maya Script*, published in 1976, Kelley presented fourteen texts with the 819-Day Count, most of them including the name glyph of God K in the fifth position of order.[26] From these observations several important conclusions were drawn by Kelley and by other Mayanists regarding the nomers for colors and directions. Kelley believes that the presence of God K in these texts probably indicates that they refer to the birth and descent of important personalities.[27]

A very important step in the clarification of this deity's astronomical aspect was made by Seler while he was studying the time tables of the *Dresden Codex*, where God K is pictured six times. He recognized God K as an apotheosized personification of Ah Bolon Dzacab (Nine of Many Generations or Nine Matrilineages), regent of one of the four groups of years, to whom reference was made both

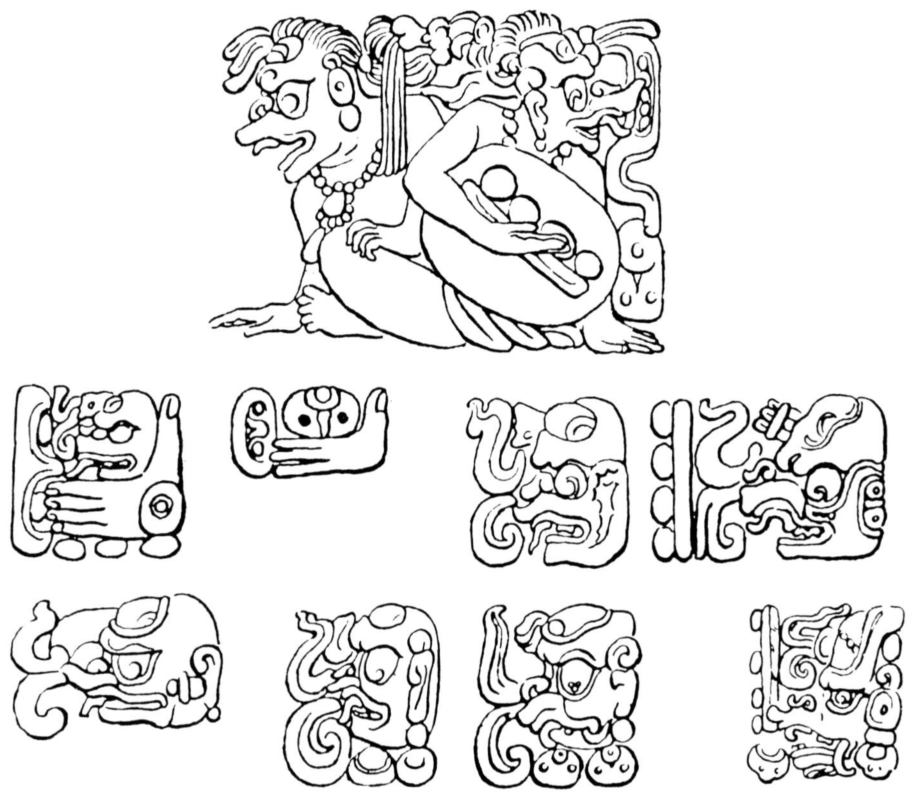

Figure 67. Hieroglyphs of God K in the inscriptions at Copán (after Seler)

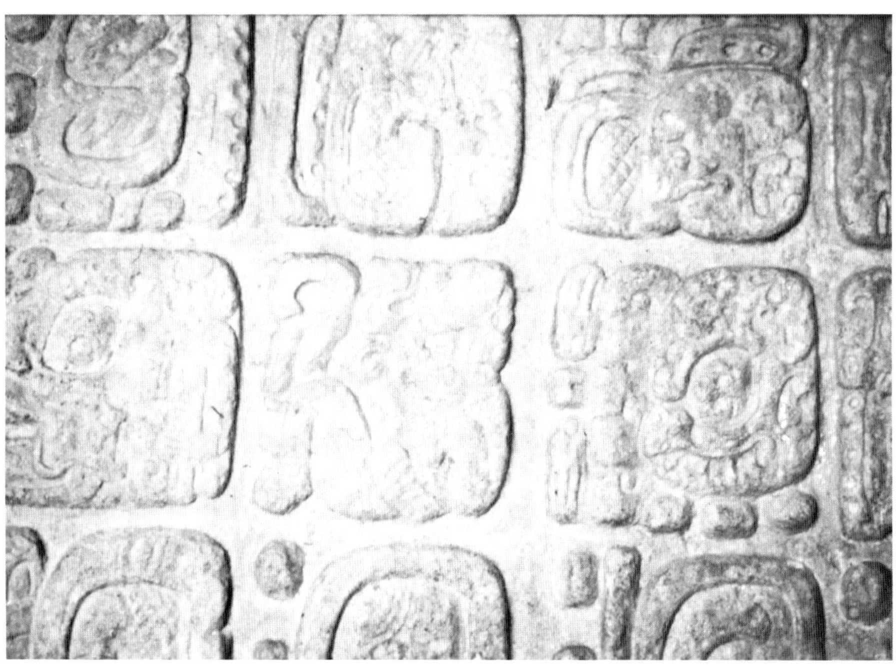

Figure 68. The name glyph of God K on the Sanctuary Tablet, Temple of the Foliated Cross, Palenque

in *The Book of Chilam Balam* and by Landa.[28] Seler's identification has been accepted by later students of Maya archaeology, including Kelley, who, like Seler, found the number 9 (*bolon*) prefixed to the glyph of God K on inscriptions at Copán, Palenque, and Chichén Itzá.[29]

Thompson further elaborated on the hypothesis that God K is Bolon Dzacab, but he revived the theory of Lewis Spence that this deity also symbolizes the terrestrial and hence vegetal personification of the omnipotent Yucatec god Itzamná.[30] Through the whole of his reasoning Thompson continued to emphasize God K's aspect as a deity of vegetation and floral growth, probably because of his interpretation of the forehead ornament as being "seed and sprouting leaves."[31] For the same reason he correlated God K with Centeotl, the Aztec God of Maize, whose birthday, 1 Flower, corresponds with the birthday (day of emergence) 1 Ahau of Bolon Dzacab.[32]

Kelley accepts God K as Ah Bolon Dzacab, but he also identifies him with God GII, one of the deities of the important triad of gods of Palenque, who appears in both the iconography and the glyphic texts of the Temple Group of the Cross and the Temple of Inscriptions.[33] He also recognized the other two members of the triad: God GI as Kukulcán, the Feathered Serpent, and God GIII, who has both jaguar and sun characteristics as the God of Death.[34] Following the lead of Seler and Berlin, who suggested that the Cross Group of Temples in general and their Sanctuary Tablets in particular were dedicated to specific deities, Kelley believes that God K, or Bolon Dzacab, has especially strong associations with the Temple of the Foliated Cross, where his name glyph is carved three times.[35] Kelley further believes that the Tablet of the Foliated Cross records not only the date but also the event of the birth of the deity:

> The figure of a deity resembling God K, holding a maize stalk with the head of the maize deity in it, is shown emerging from the shell. It seems quite possible that the event referred to is his emergence. . . . If the lower glyph block on the shell, then its use on the exterior table of the Temple of the Cross must refer to the emergence of Berlin's God GII on the date 1 Ahau 13 Mac.[36]

A somewhat weak point in Kelley's deductions is that, while all other representations of God K at Palenque are distinguished by the smoking forehead cigar, the jester god crawling out of the shell on the Foliated Cross Tablet has a "non-smoking" forehead tube, which looks like a bone rather than a cigar.

Thompson, like Lounsbury and Kelley, attributed to God K strong agricultural aspects, but Michael D.

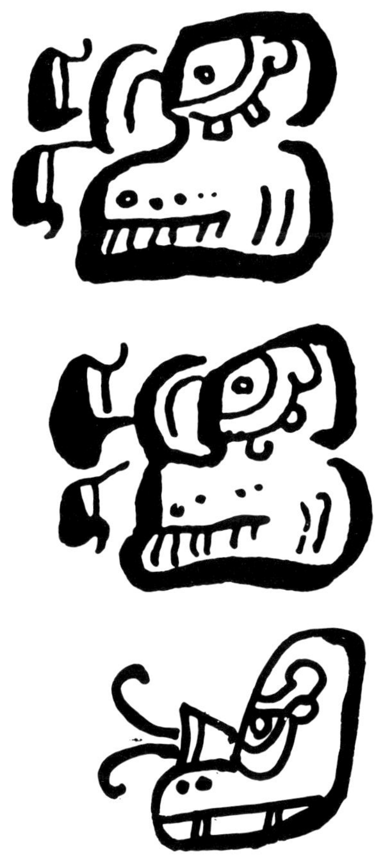

Figure 69. Name glyphs of God K in the Dresden Codex *(upper two) and in the Codex Madrid (lower)*

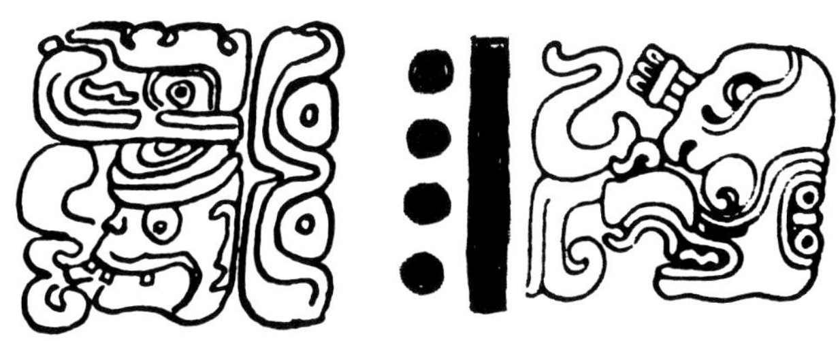

Figure 70. Glyph "Smoking Star Comet" on Stela 5 at Tikal (after Kelley), and hieroglyph of God K with the number 9 (bolon) *prefixed (after Seler)*

Coe and Schele now emphasize that God K's name Bolon Dzacab (Nine of Many Generations) is associated with ancestral blood-letting rites and is portrayed in scenes showing royal power, and they infer that he is the protector of noble lineage and ancestry, even royal ancestry (the *Titulargottheit* of Seler). In this respect, Schele concludes, God K on the Sanctuary Tablet of the Temple of the Foliated Cross is "not emphasizing the cycle of life in the local sense of crops, but in the general sense of life itself."[37]

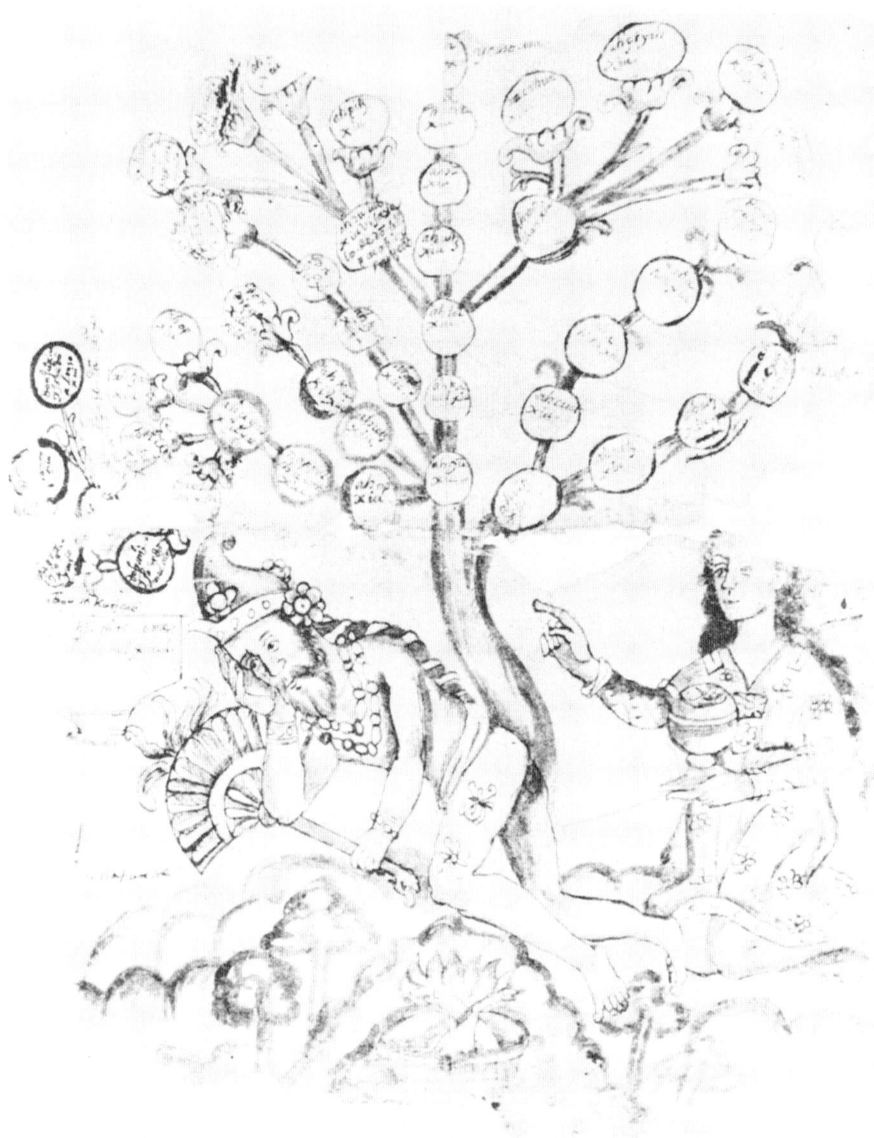

Figure 71. The family tree of the Xiu family, rulers of Uxmal (after Morley). Note the manikin scepter held by the kneeling dignitary

That the Maya lords regarded themselves as having divine ancestry was recognized by Thompson (who seems continuously to have emerged on both sides of any argument). He stated that "the prestige of hereditary chiefdomship is a necessary adjunct to claims of divine descent or divine right of kings."[38] The same conclusion was reached by Kubler, who found that Classic inscriptions often give legendary, even remote dates, implying that some principal figures of Maya dynastic history boasted mythological ancestries flourishing in remote times, possibly even occupying another cosmological state.[39] Evidence that the cult of God K as an "ancestry god" survived into colonial times is the painting depicting the genealogical tree of the Xiu family of Mani, former rulers of Uxmal. It shows the founding father of the Xius holding a snake-footed manikin scepter.[40] The role of God K as patron of noble ancestry is especially evident at Palenque, where the rulers apparently regarded themselves as being of the lineage of God K, the snake-footed deity with the smoking cigar emerging from his forehead. On Piers B, C, D, and E of the Temple of Inscriptions a lord is portrayed holding an infant. This first pictorial recording of these stucco tablets was made by Annie Hunter, who under the supervision of Maudslay put a "gratuitous foot on each babylike figure, ignoring the fact that the second leg of each was made in the form of a serpent."[41] Spinden in 1913 recognized the error of the otherwise most meticulous Englishman and described the infant as having instead of a foot an

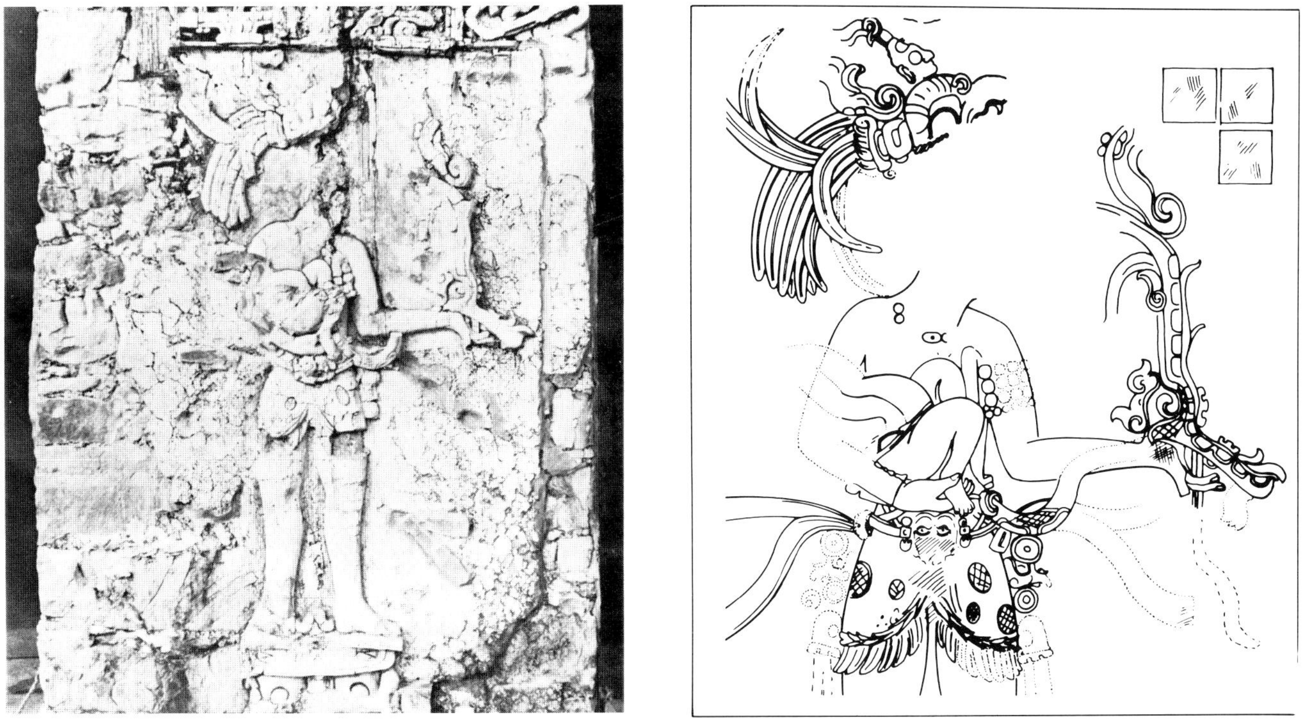

Figure 72. Stucco panel and detail on Pier B, Temple of Inscriptions, Palenque. The composition shows a serpent-footed infant held by a standing dignitary (after Merle Greene Robertson)

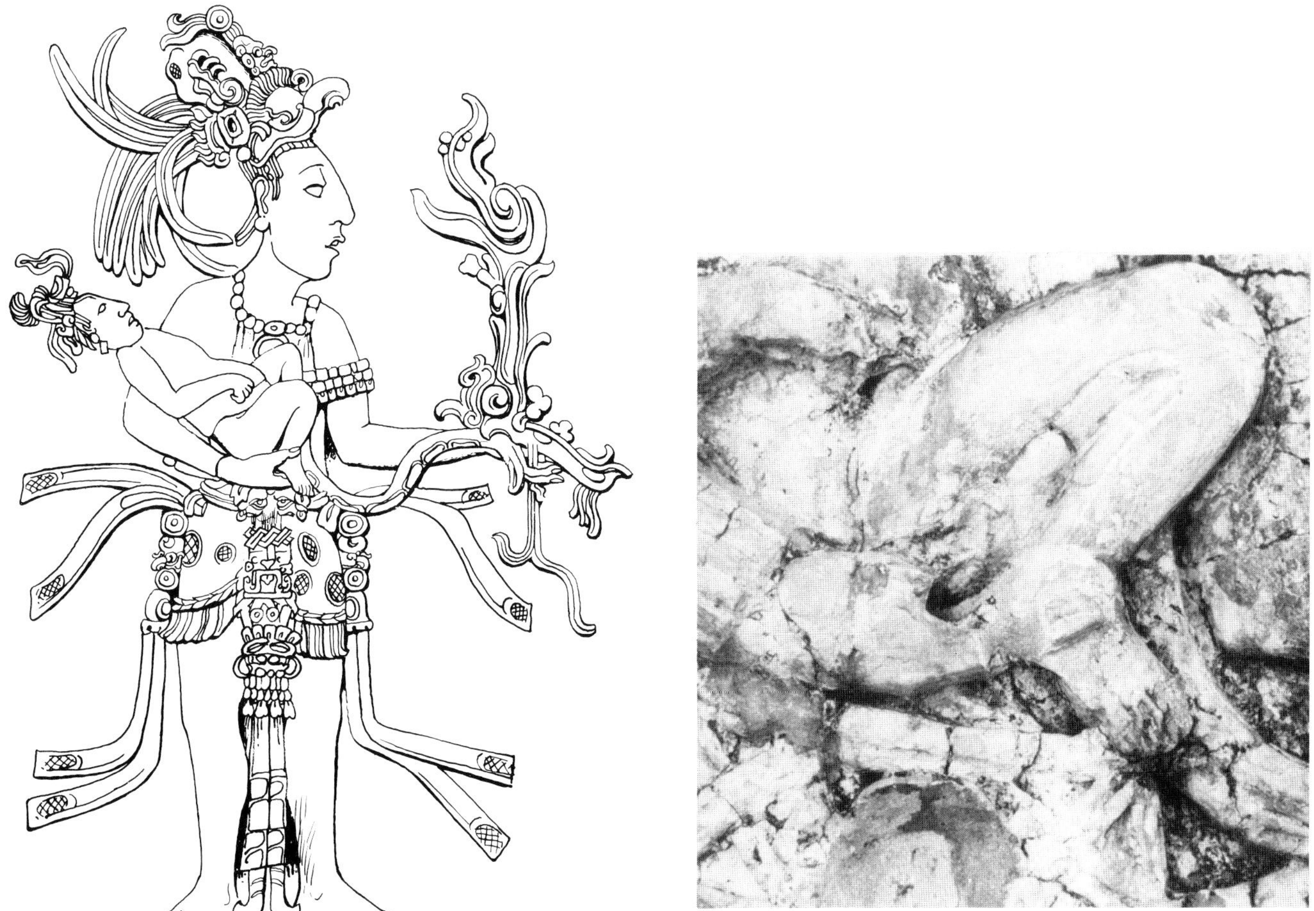

(a) (b)

Figure 73. Reconstruction drawing of the stucco panel shown in Fig. 72. showing how the monument may have looked before damage (a) and detail of the stucco panel (b)

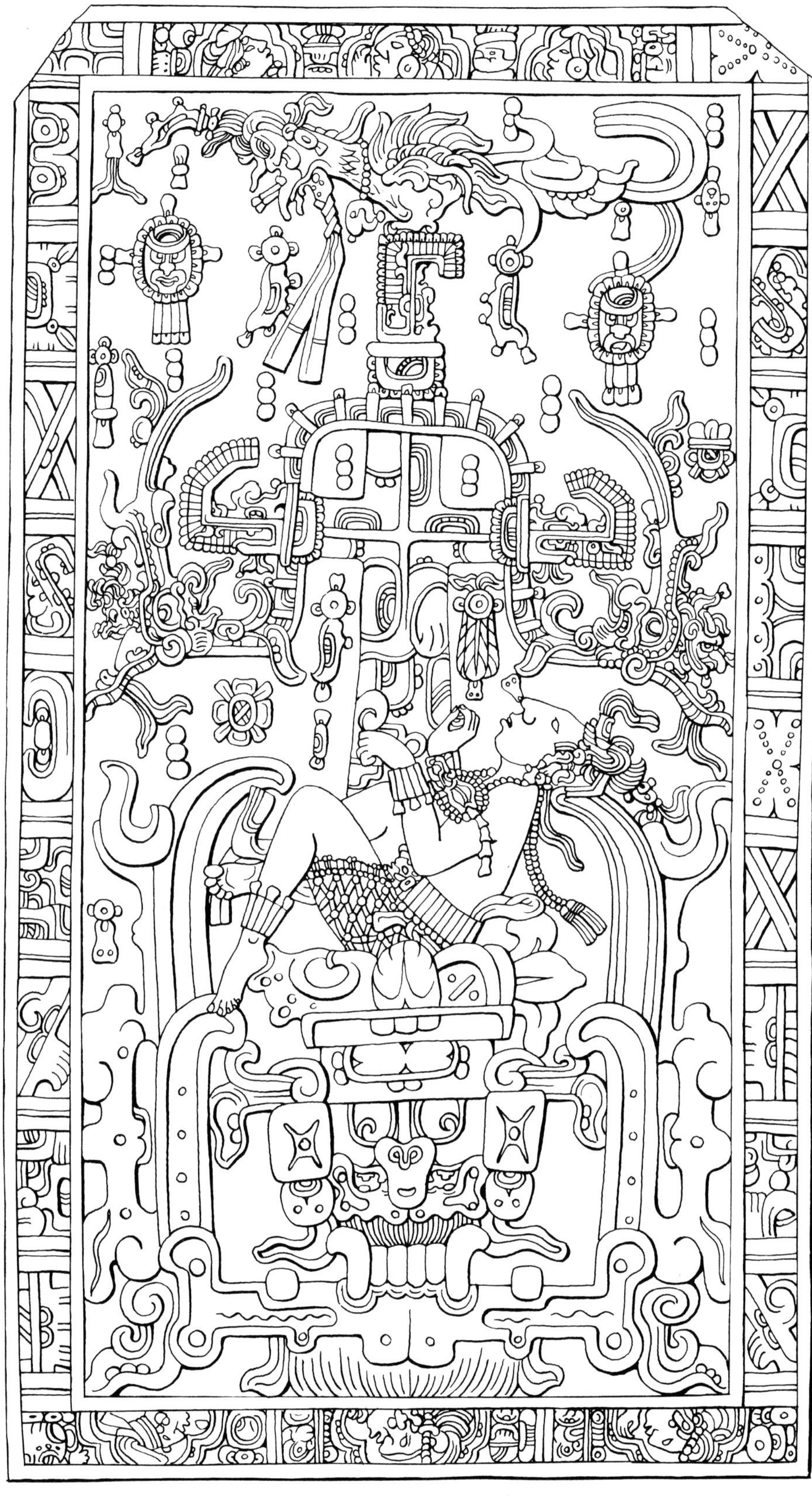

Figure 74. Line rendering of the Sarcophagus Lid, Temple of Inscriptions, Palenque. The stone slab is carved with the image of Lord Shield Pacal, ruler of Palenque. He is shown with the smoking forehead tube, the insignia of God K, falling into the underworld. The head of God K is also seen emerging from the jaws of the skeletal double-headed serpent bar

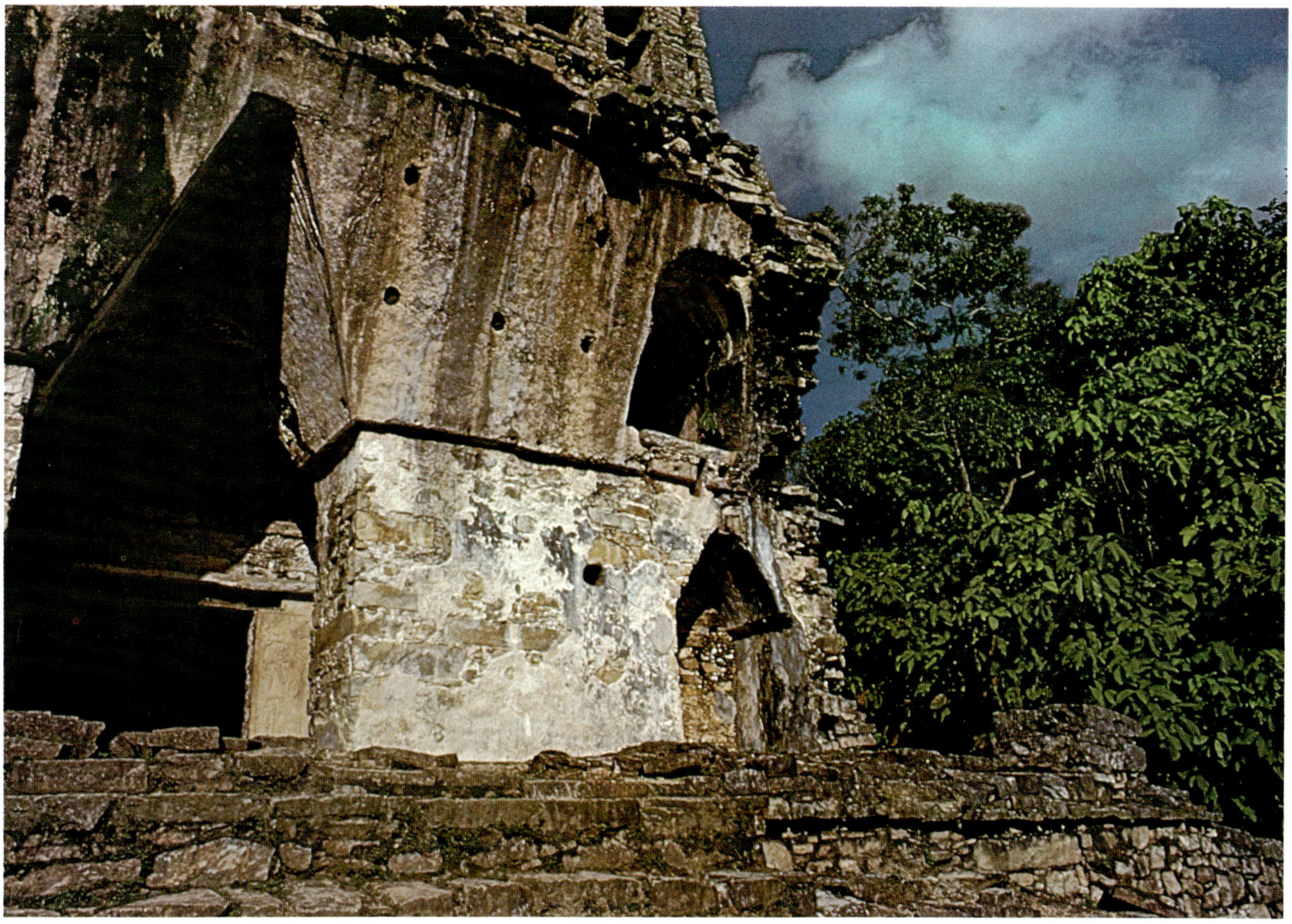

Plate 64. The Temple of the Cross, Palenque

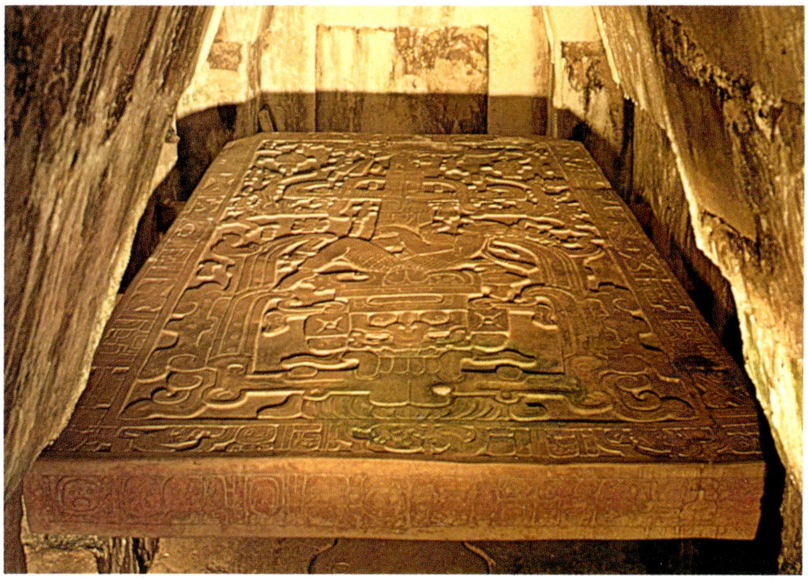

Plate 65. Sepulchral Chamber, Temple of Inscriptions, Palenque

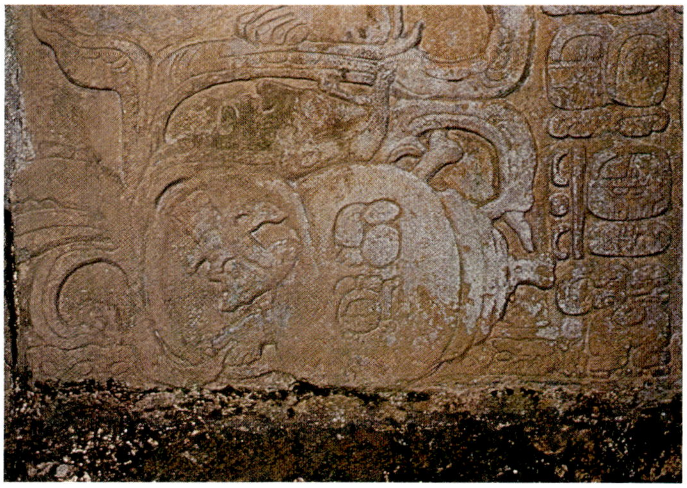

Plate 66. Detail, Sanctuary Tablet, Temple of the Foliated Cross, Palenque

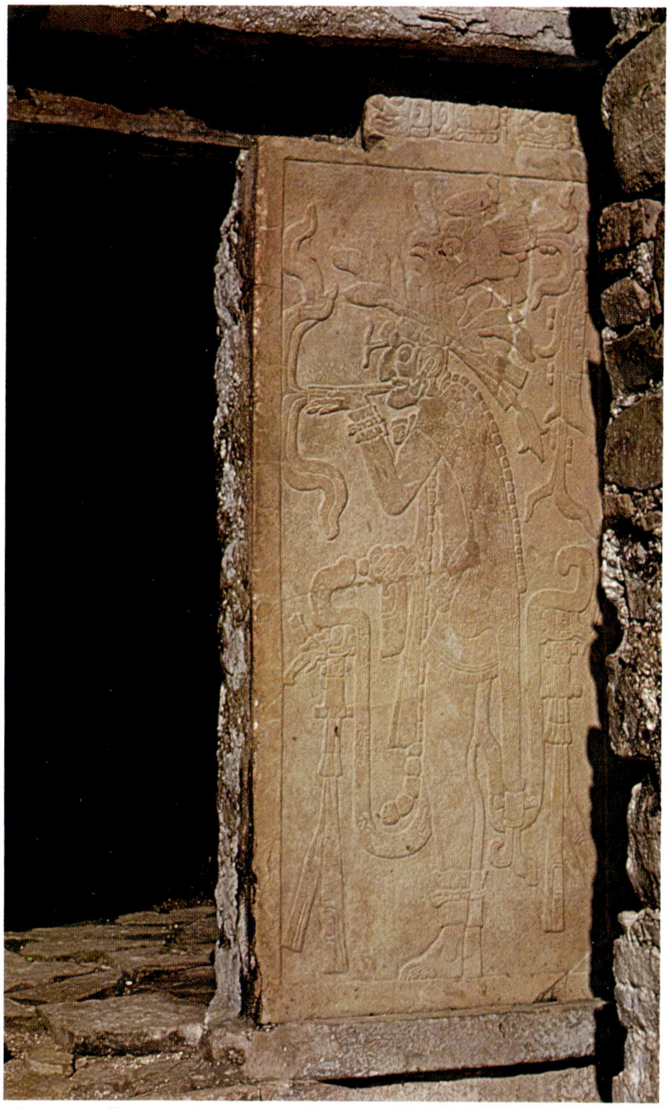

Plate 67. Eastern Door Jamb, Temple of the Cross, Palenque

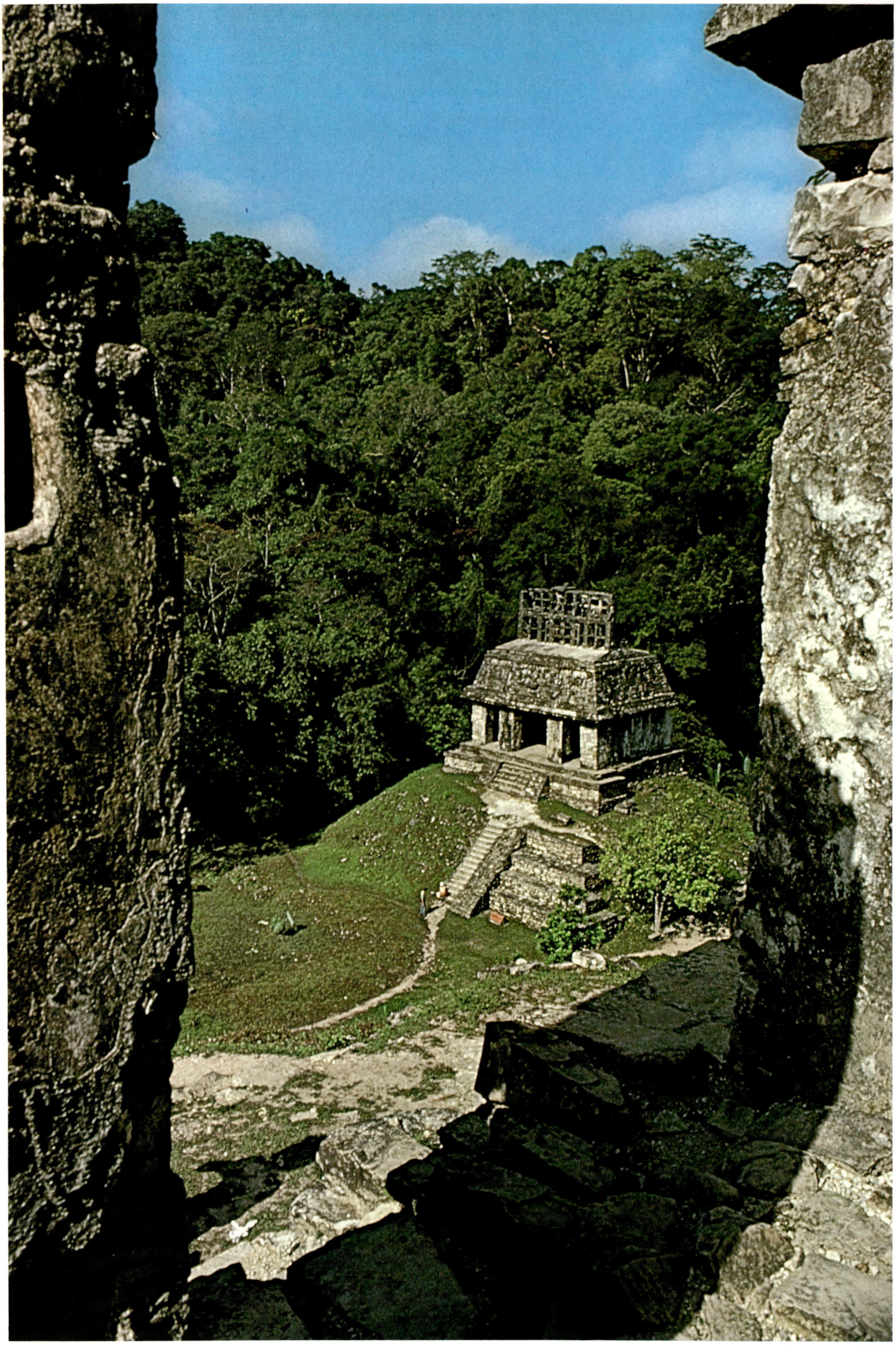

Plate 68. Temple of the Sun, Palenque

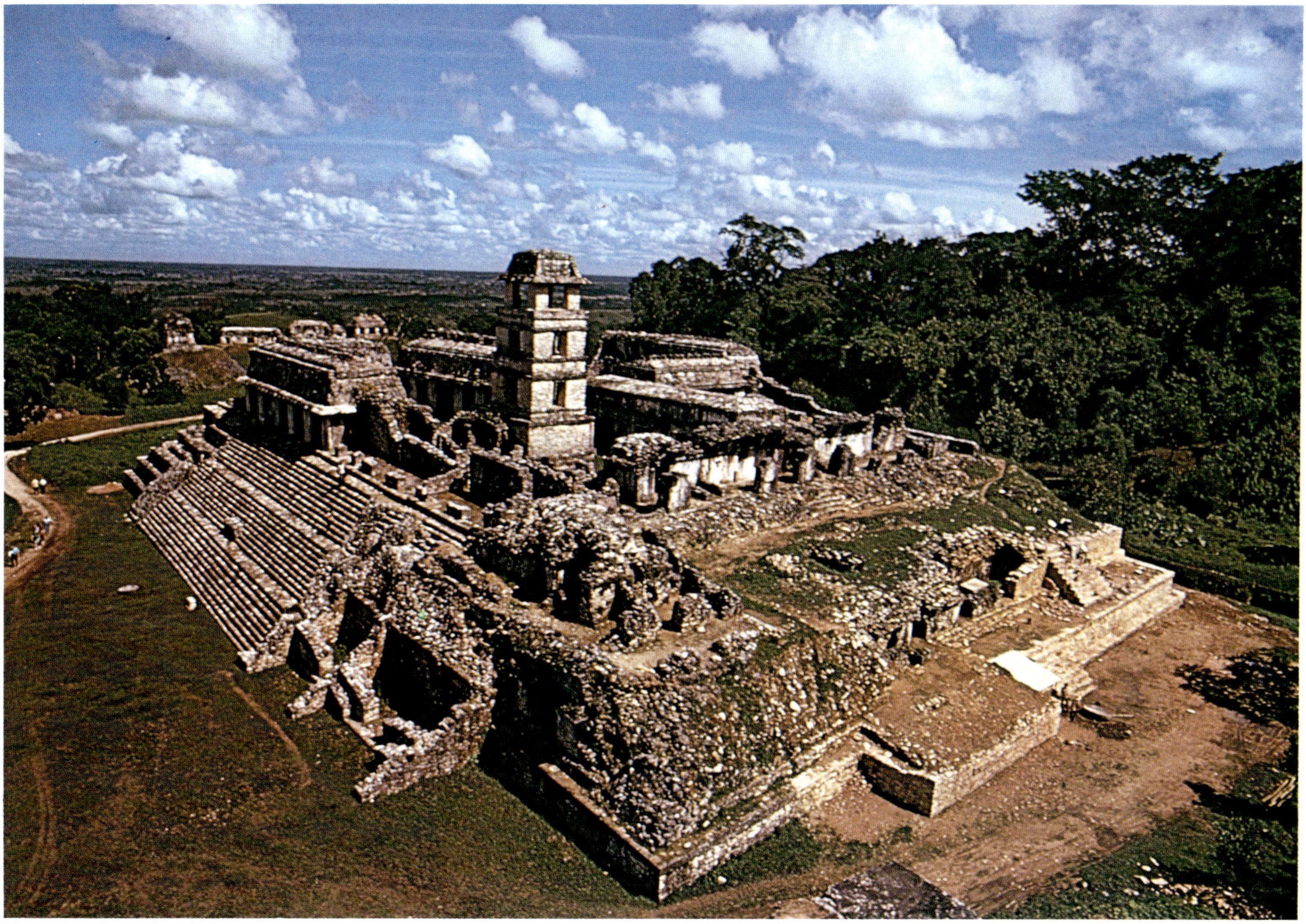

Plate 69. The Palace, Palenque

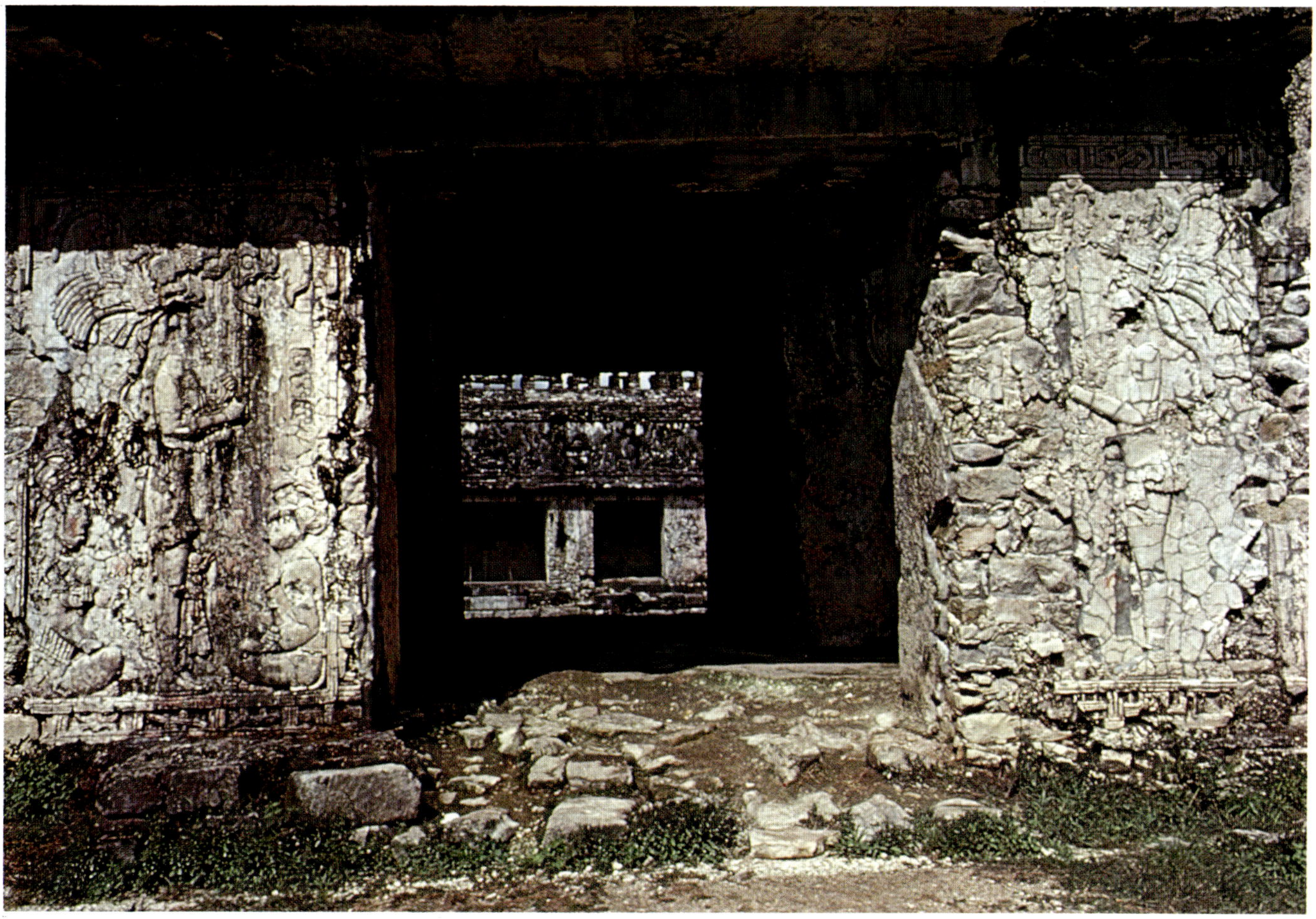

Plate 70. Eastern Piers, Palace House A, Palenque

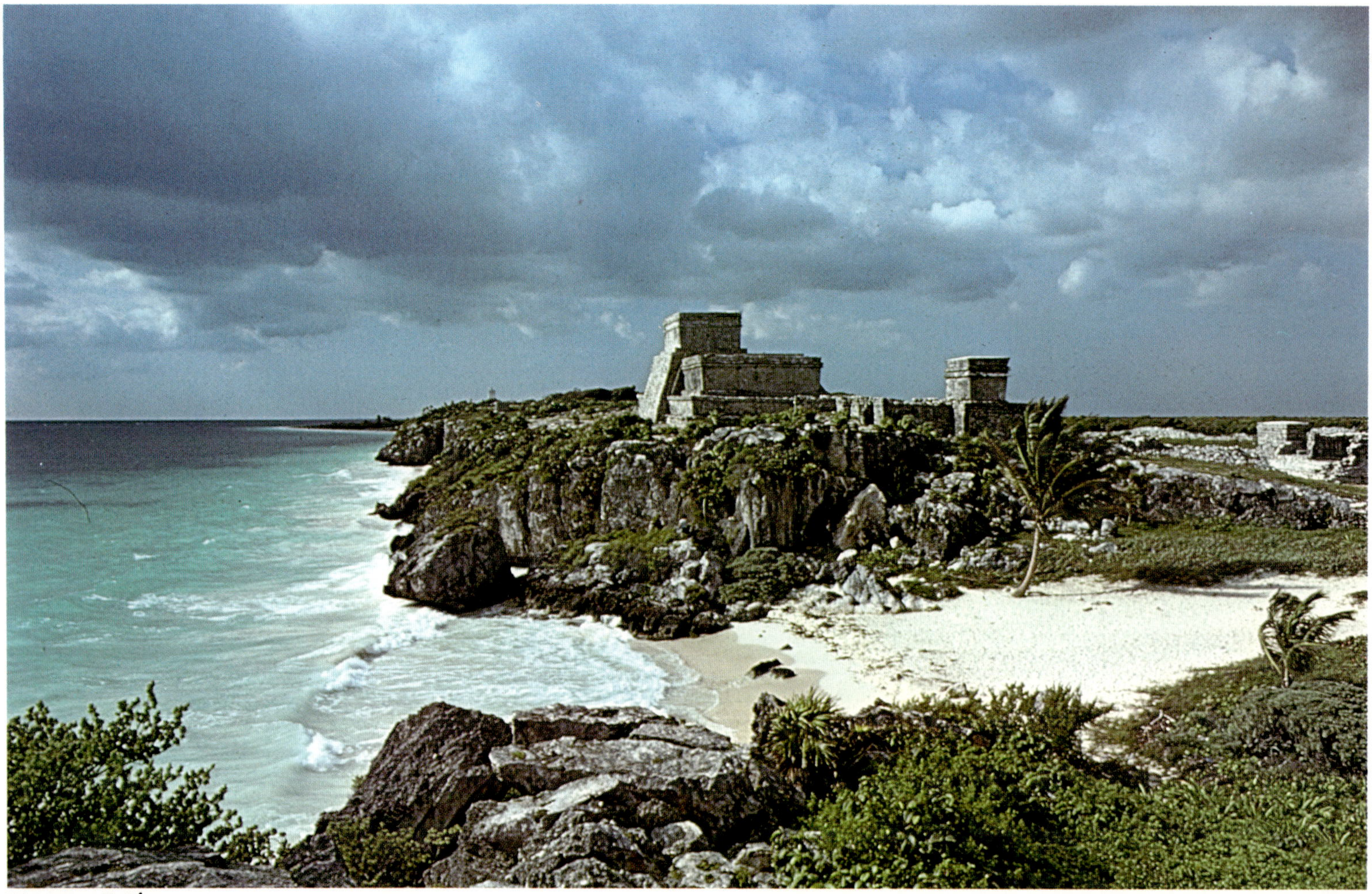

Plate 71. Tulúm

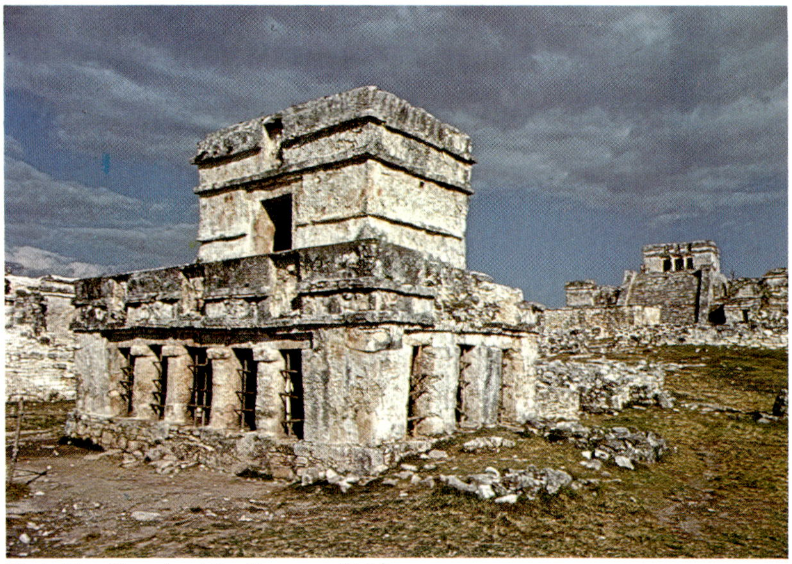

Plate 72. Temple of the Frescoes, Tulúm

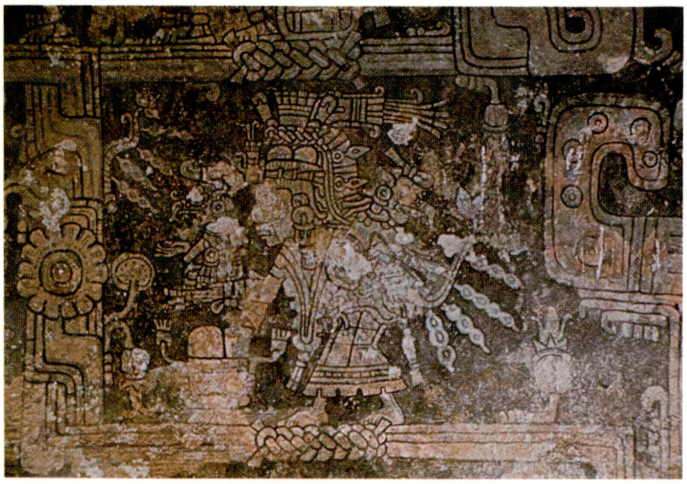

Plate 73. Detail of mural painting, Temple of the Frescoes, Tulúm

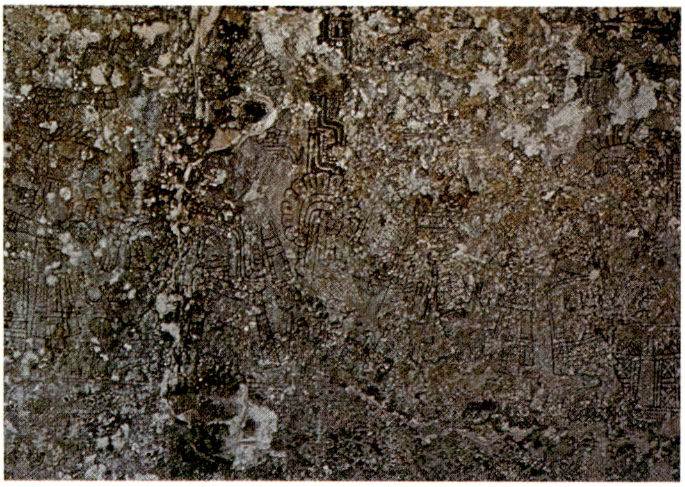

Plate 74. Detail of mural painting, Temple of the Diving God, Tulúm

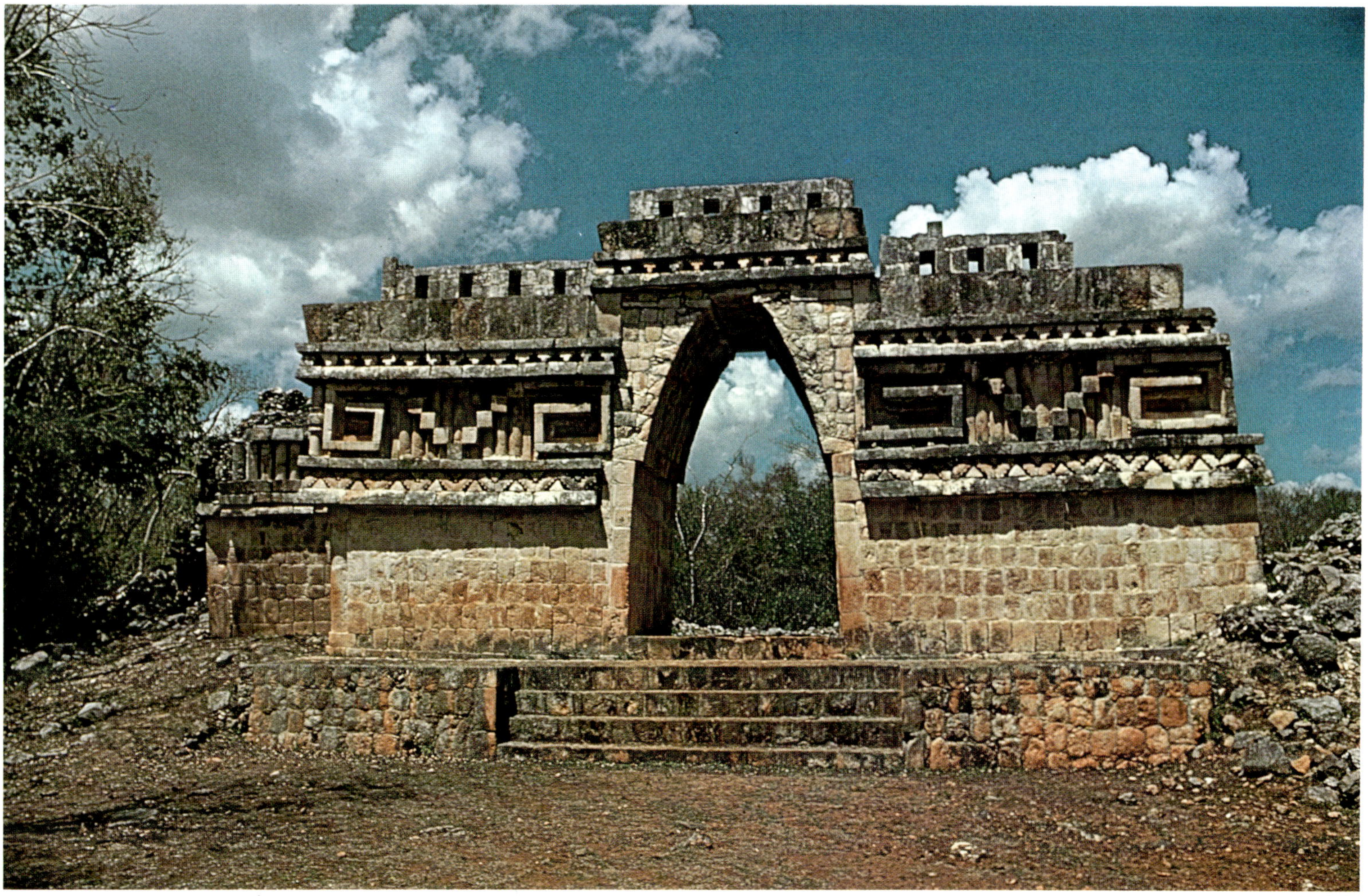

Plate 75. The Arch of Labná

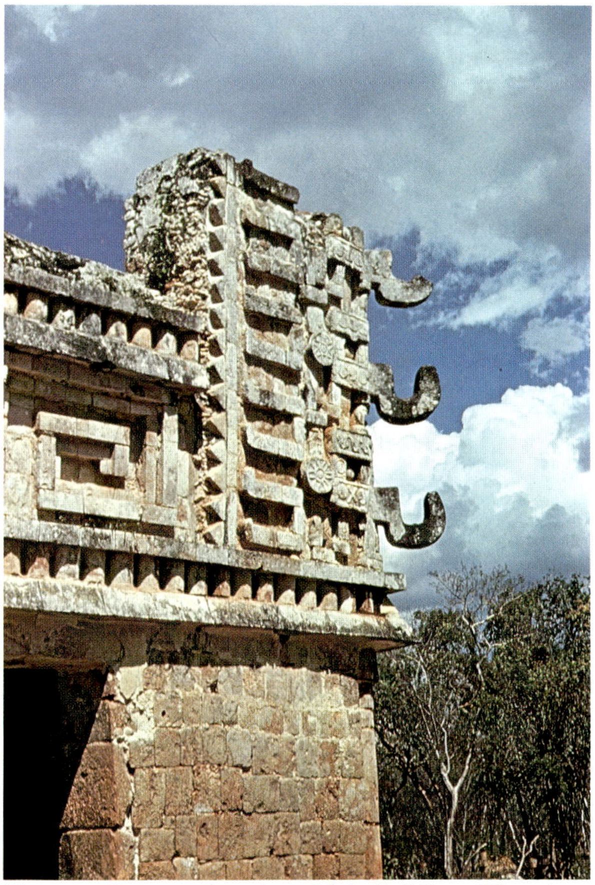

Plate 76. Long-nosed masks on Main Building of Xlapak

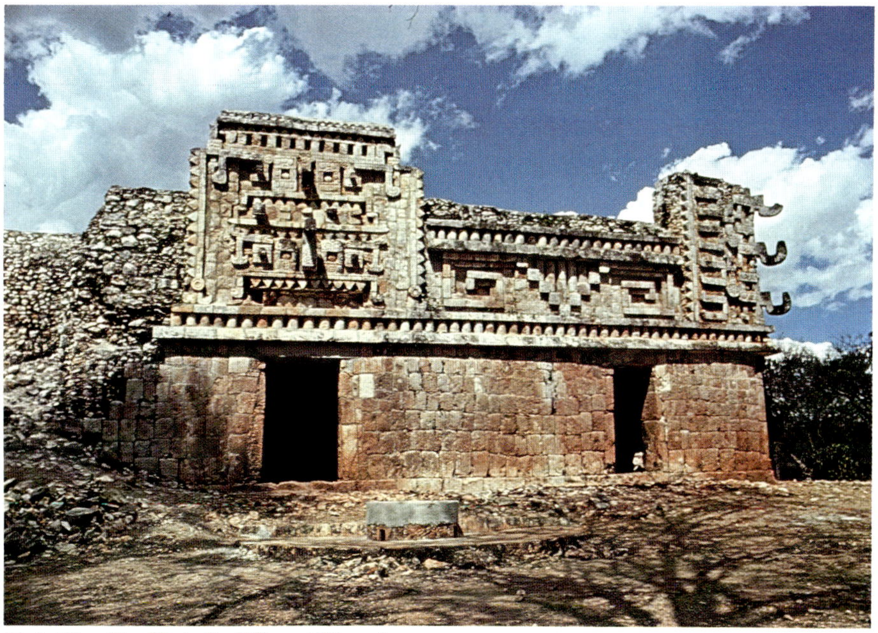

Plate 77. The Main Building at Xlapak

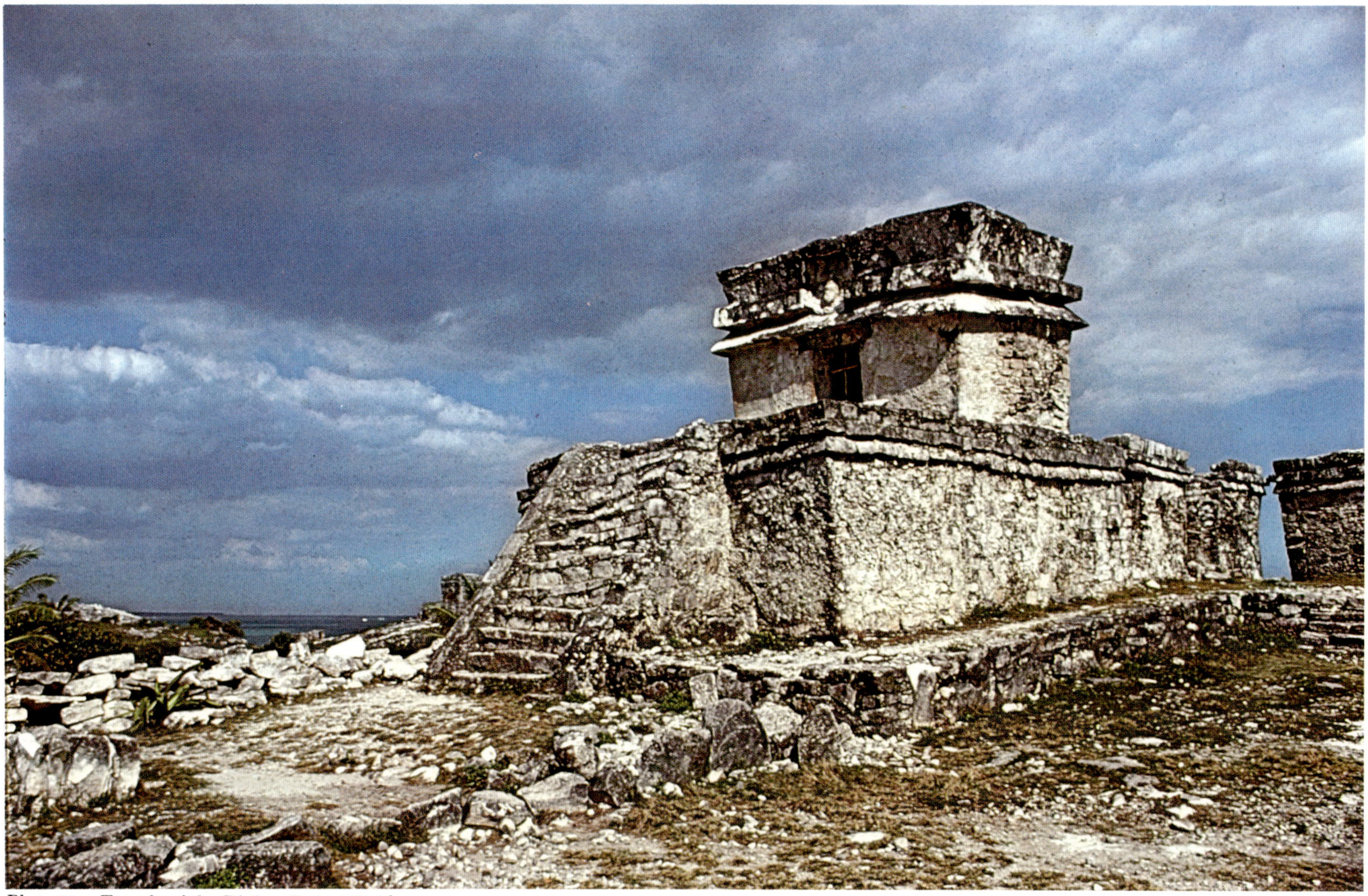

Plate 78. Temple of the Diving God at Tulúm

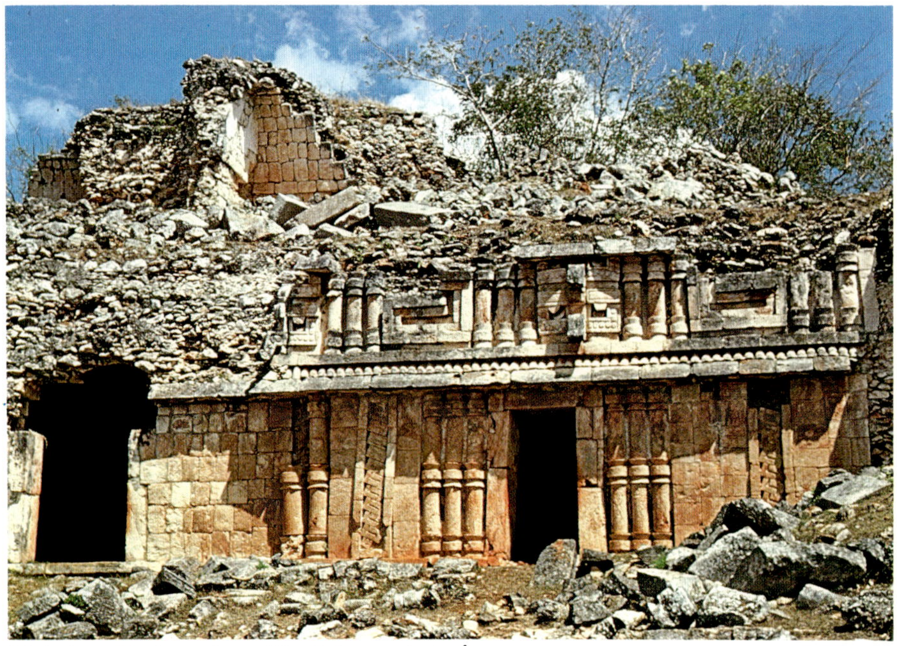

Plate 79. Eastern wing of the Palace at Labná

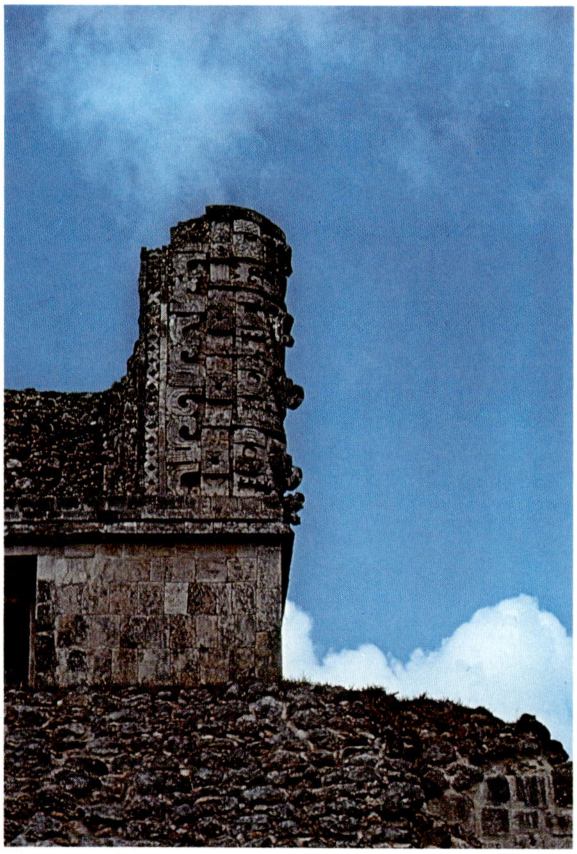

Plate 80. Long-nosed masks in the Nunnery, Uxmal

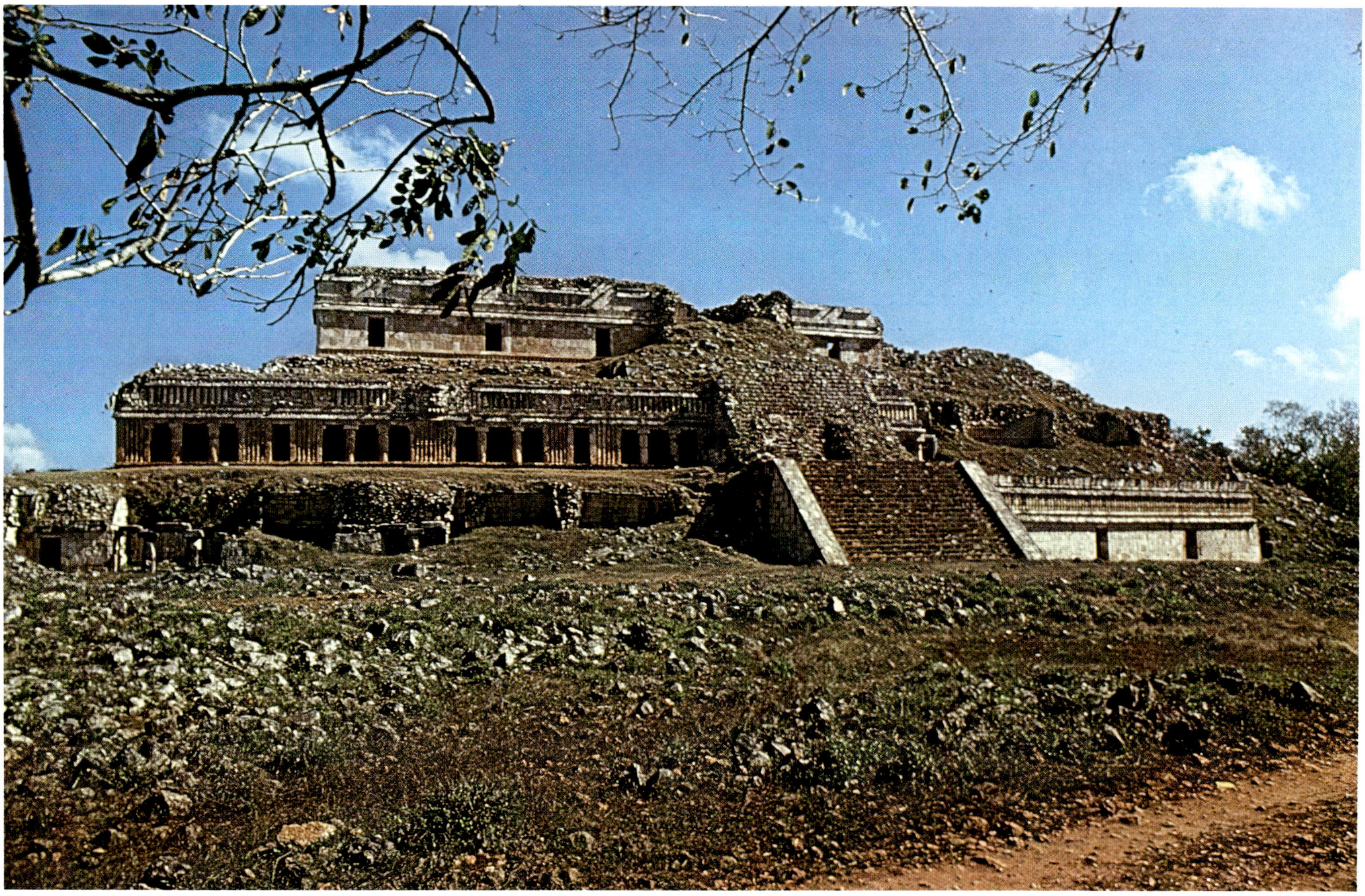

Plate 81. The Palace at Sayíl

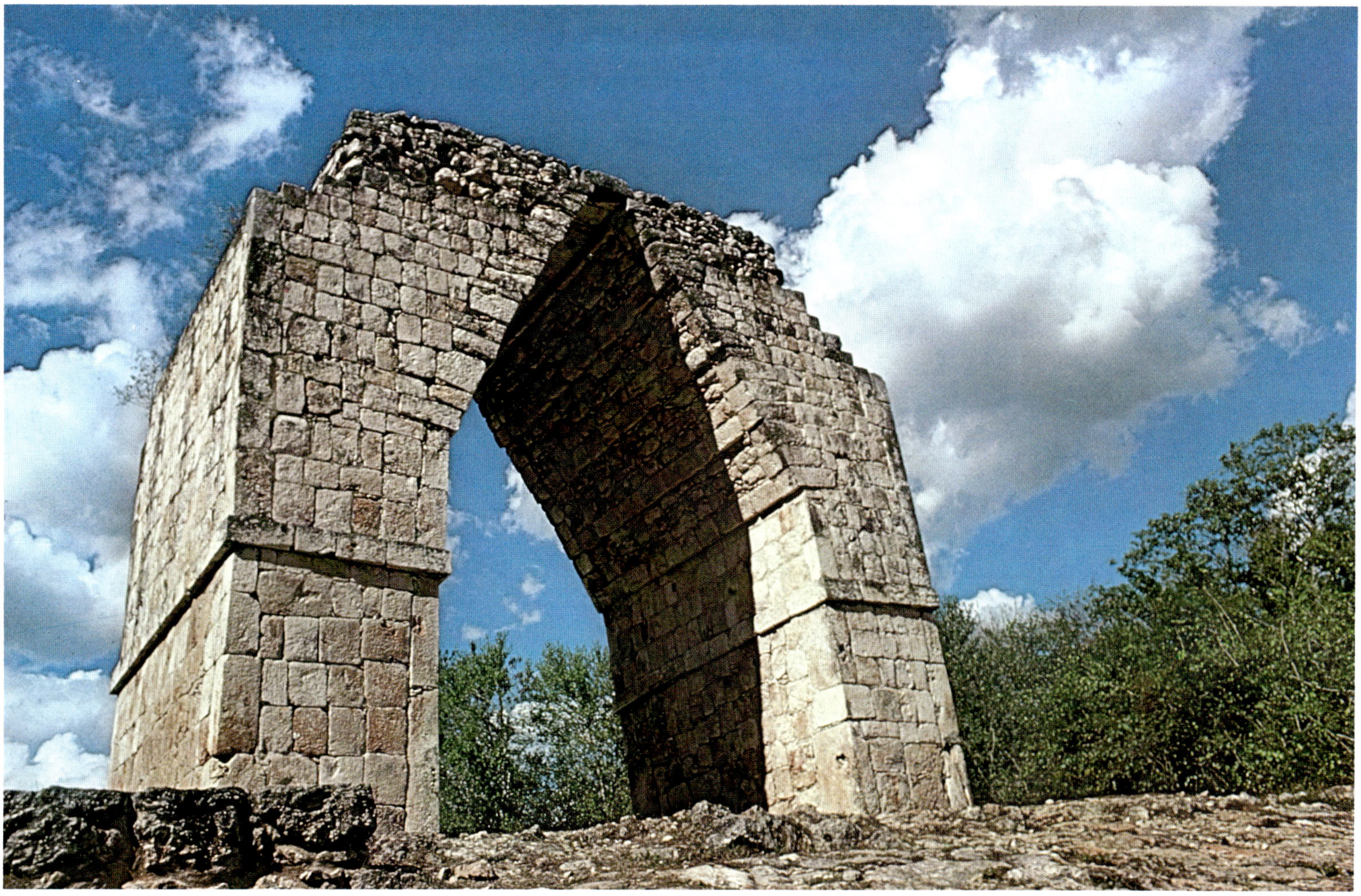

Plate 82. The Arch of Kabah

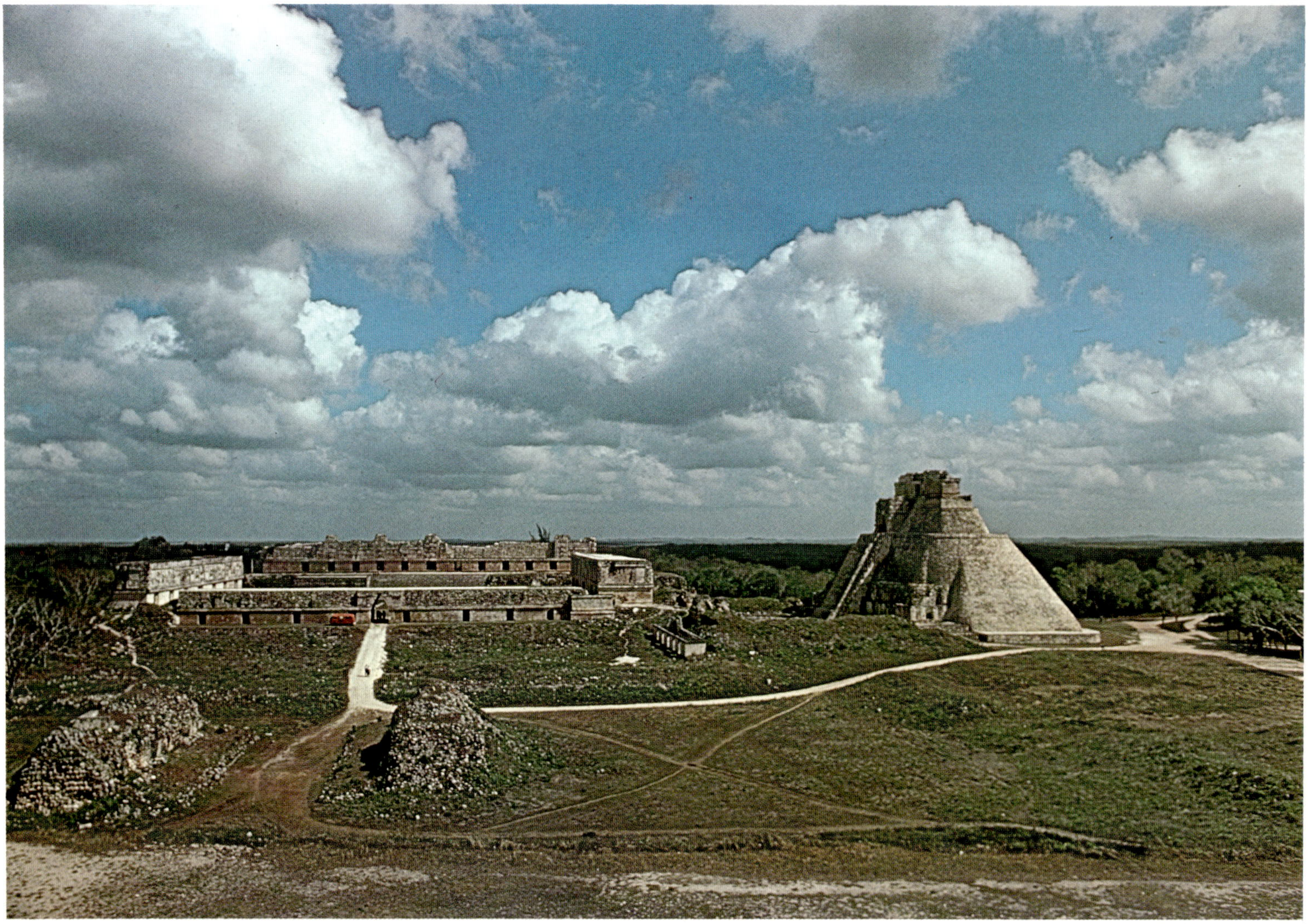

Plate 83. Uxmal

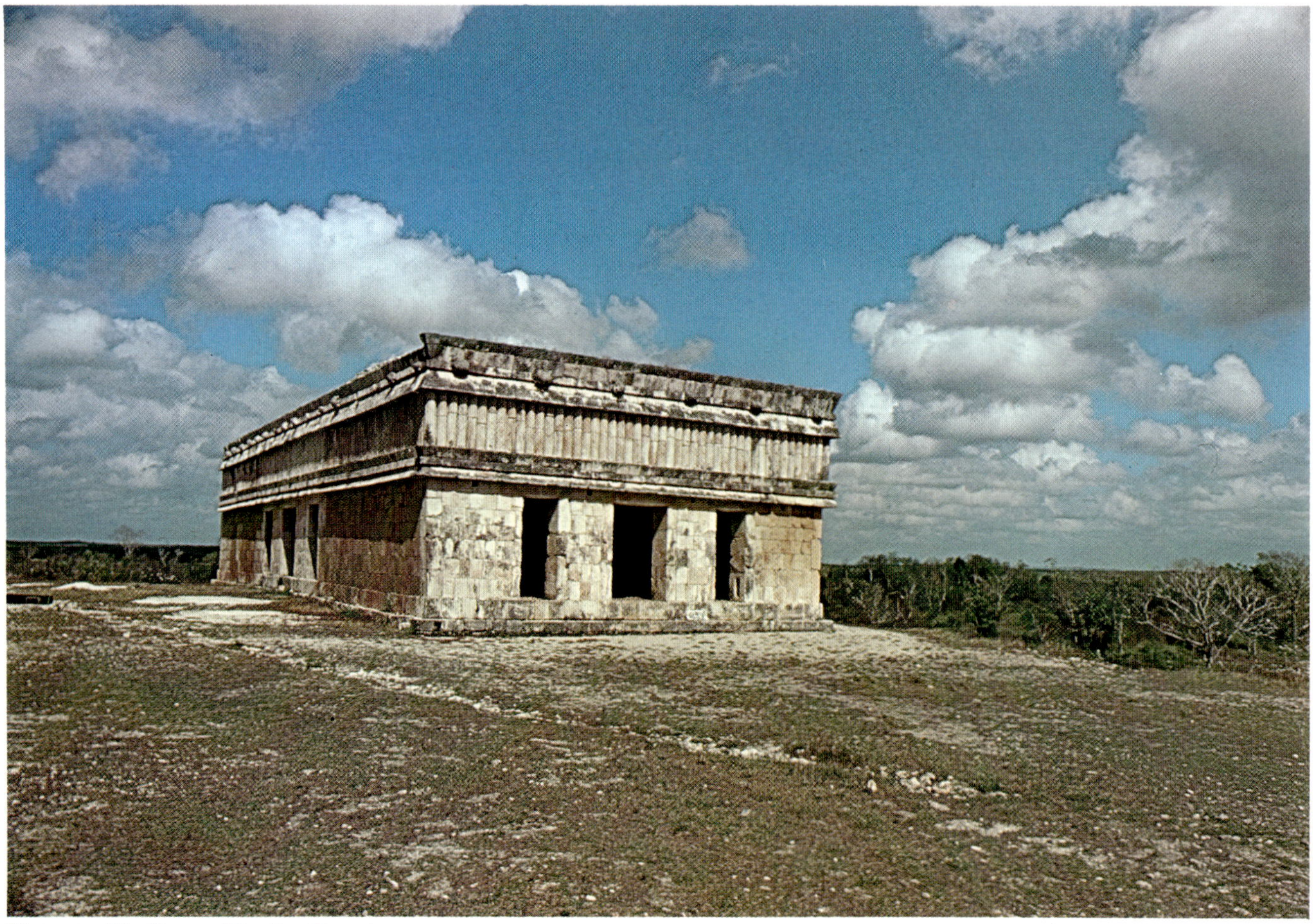

Plate 84. The House of Turtles, Uxmal

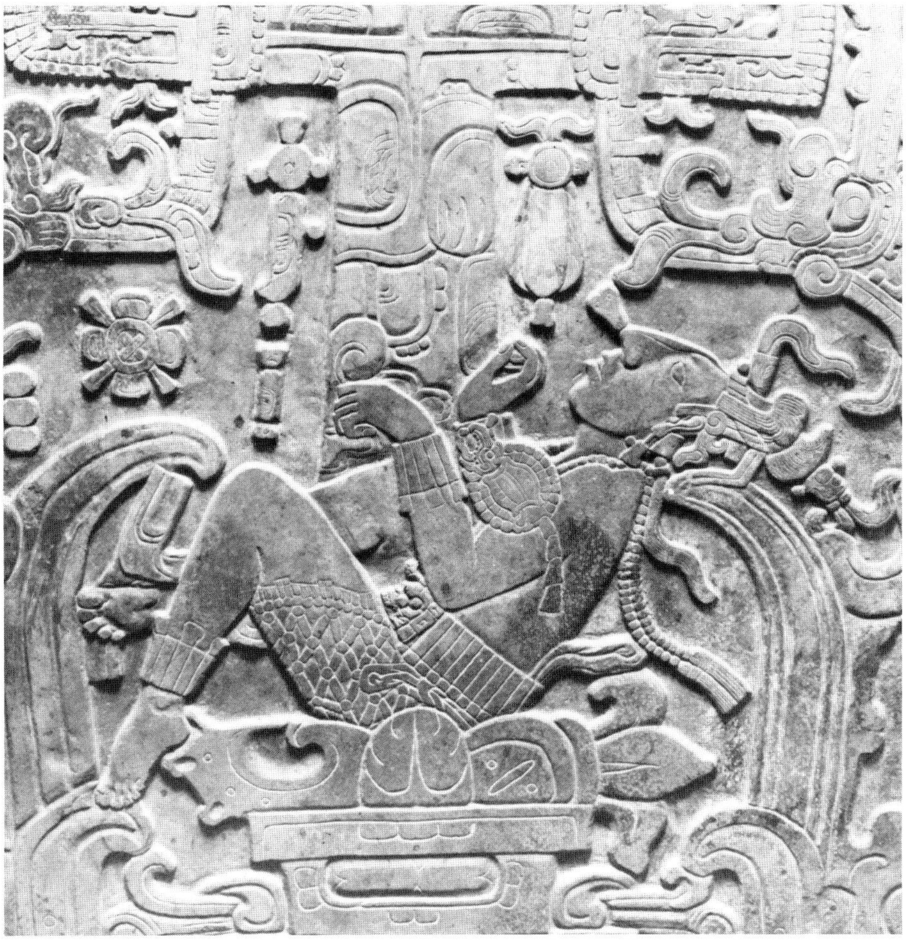

Figure 75. Detail, Sarcophagus Lid, Temple of Inscriptions, Palenque. It shows Lord Shield Pacal as God K, the god with the smoking forehead tube (after Merle Greene Robertson)

"ophidian [serpentine] appendage," concluding therefore, that the infant was the serpent-footed God of the Manikin Scepter.[42] By his birth defect (the child has six toes on his right foot) Greene Robertson further identifies him as the historical figure Chan Bahlum, who ruled Palenque between the Maya dates 9.12.11.12.10 and 9.13.10.1.5. The stucco tablets of the piers are damaged to the degree that the head of the child can no longer be seen, but Greene Robertson theorizes that if the heads of these infants were intact, they would be complete with smoke curls and cigar and that they represent Chan Bahlum, not as the human offspring of Lord Shield Pacal and Lady Ahpo Hel but as the incarnation of the important deity God K.[43] According to Schele the divine origin of Shield Pacal is also specifically stated on the Sarcophagus Lid of the Temple of Inscriptions, where

> Pacal is shown falling in death into the maw of the Underworld. He is attired in the expected Palenque tradition except for the cigar penetrating his forehead and the *ahau*-foliation topknot. The cigar is clearly and carefully shown. It was important to differentiate the actual cigar penetration from those instances where the lord wears a headdress or other attributes of divinity. The Palencanos wanted to specify that at death Pacal was a divine being. Thus at birth (TI piers) and at death (TI Sarc. Lid) the ruler of Palenque is shown to be a living incarnation of God K. The person of the Palenque ruler was defined as a living divinity on earth.[44]

Apparently the apotheosis of Shield Pacal as God K was generally observed in the Maya realm. After his death (9.12.11.5.18) the idea that the serpent-footed god with the cigar in the forehead was one of the main symbols of divine rulership became accepted in several of the major centers, especially at Yaxchilán. Before the date of Pacal's death there are only three known instances where an artist used the image of God K in monumental iconography: Stela 19 at Dzibilchaltun, Stela 12 at Etzná, and Stela 1 at Tikal. After Pacal's death, however, the serpent-footed manikin scepter became a common motif.[45] The question naturally arises: Why did the ruler of Palenque choose to be identified with this particular deity? A reasonable explanation again comes from Greene Robertson, who in her extensive studies of the sculpture of that city found that both Pacal and his son, Chan Bahlum, had congenital foot deformities: Pacal had a clubfoot; Chan Bahlum, polydactilia. Instead of trying to hide the deformities, they used them to their advantage—the malformed feet became "snakes," and the rulers identified themselves with the serpent-footed God K.[46]

Unlike Greene Robertson, David Kelley believes that Pacal's relation to God K was calendric rather than morphological. The rule of Shield Pacal was associated with a major calendric cycle so that the tropical year and the Maya year were in the same relationship that they had at the date of the "birth of the gods" at Palenque. This coincidence furnishes a direct association between Pacal and God K and suggests that time connection between the ruler and the deity is responsible for the sudden emergence of God K as one of the three most important gods of that epoch.[47]

The next question is: How did the cigar come to appear in the foreheads of all these distinguished individuals, gods and mortals alike? The answer lies hidden in the long-forgotten labyrinths of Maya mythology. It is possible, however, that the early Maya lords, who, to judge from their surviving portraits, were scantily dressed and certainly had no pockets on the loincloths, carried their cigars stuck in their hair or helmets. Having to produce

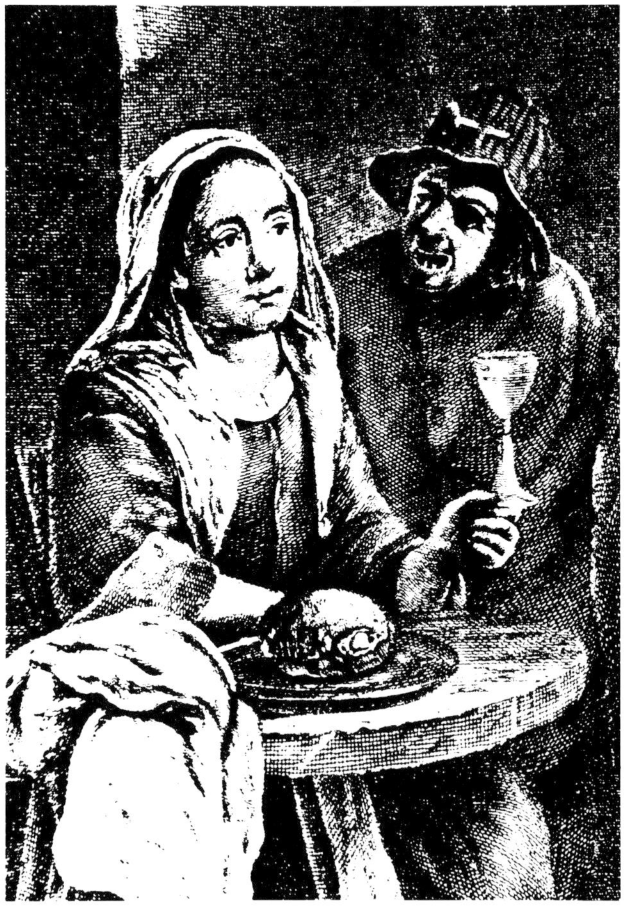

Figure 76. Detail from the engraving Le retour imprévu, *by Egidius van Tilborgh, showing a man and a woman, the man with a pipe stuck in his hat (after Brongers)*

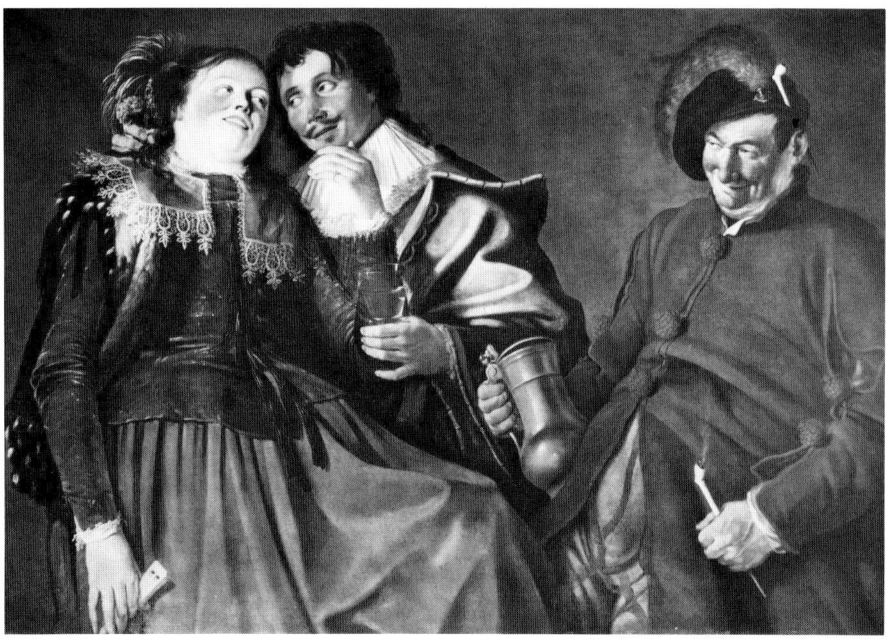

Figure 77. The Enticement, *by Hendrik Gerritzs (Museum Boymans van Beuningen Collection)*

fire and keep it burning, they may have kept their cigars lighted, especially when traveling. The lighted cigar would have provided them not only with a continuously available "smoke" but also with a source of fire for other purposes. And since, as we believe, in ancient times smoking was limited to members of the ruling caste who claimed to be descendants of gods or even gods themselves, it is possible that the forehead cigar may have become one of the insignia of divinity.

It must be admitted that this theory is farfetched, but it is not without historical "post-cedent." It is known that Dutchmen carried their smoking paraphernalia stuck through the front of their hats during the sixteenth and early part of the seventeenth centuries. Examples of this custom can be seen in a painting by Hendrik Gerritze (1585-1657) in the Boymans van Beuningen Museum of Rotterdam, on an engraving by the Flemish artist Egidius van Tilborgh (1625-1678) called *Le Retour Imprévu* and in the painting *Strolling Fiddlers*, by Christian Dietrich.[48]

God K in Maya Iconography

The representations of God K in Maya iconography can be divided into two principal categories: true portrayals and impersonations. While in his true portrayals the god is shown with all or most of his divine facial characteristics, God K impersonators are lordly personages bearing only the smoking forehead tubes. They may be rulers imitating the long-nosed god, or they may represent God K himself in human disguise. Of both these possibilities there are plently of examples in the mythology of other ancient cultures. In Egyptian, Hindu, Greek, and Roman legends gods often appear in the guise of mortals, and rulers are not infrequently pictured carrying divine paraphernalia.

Recognizing God K

The typical representations of God K are easy to recognize. Occasionally, however, some or most of his characteristic features are missing, which can make it difficult to identify him among the array of long-nosed Maya deities. The characteristic features of true portrayals of God K may be subdivided into diagnostic features (those present only on God K) and nonspecific features (those also present on the other deities), as well as constant (always present) and frequent (often present).

True Portrayals

Diagnostic—Constant. This feature is always present in a true portrayal. Without it the figure cannot be identified as God K.

1. *Proboscis.* At the beginning of the twentieth century, it was this feature by which the Cyrano de Bergerac of the Maya pantheon was first identified. The snout is not only elongated (like the olfactory organs of his colleagues, Gods B and D of the codices and various Yucatec rain gods) but upturned (like the fret of Zotz, the ominous bat messenger of the Maya underworld). That it is elongated *and* upturned makes the muzzle of God K both unique and distinguishable. The nose is often ornamented with small frills and curlicues in a baroque fashion.

Staring at the nose of God K as much as I have, I have been impelled to the conclusion that his power, like that of Samson in his hair, was seated in his majestic nozzle. It could have tenderly blown a fragrant breath at those he favored and with a mighty sneeze blasted the undeserving into the depth of Mitlán, the underworld of the ancients. God K without his snout is as unimaginable as Neptune without his trident or Saint John of Nepomuk without his halo. Identifying any creature as God K who lacks this majestic snout, whether on pots, pans, monuments, or codices, would be a grave mistake.

Diagnostic—Frequent. The following features are often but not always present on true portrayals of God K. The presence of any one of them can serve as a mark of identification.

1. *Forehead tube thought to represent a cigar.* This is a fairly constant "trademark" of this deity. The identification as God K of any portrait lacking the forehead tube is suspect. It is nearly always present on ceramic representations and on stone carvings, but is usually absent from paintings in the codices. The object may be tubular or funnel-shaped, or it may resemble a celt. Sometimes it is undecorated, but more often it is striated, dotted, or marked with oval symbols. It also varies greatly in size and, if painted, in color.

As a rule the tube emerges from the forehead; however, in two paintings, both of them on Petén ceramics, it protrudes from the mouth. On most portrayals the "handle" of the tube is sunk into the head and is not visible; on others it emerges at the nape. As discussed earlier, these tubes probably represent cigars, but the possibility that (on some portrayals) they may represent torches or celts cannot be excluded.

2. *Scrolls emerging from the forehead tube, thought to represent coils of smoke.* To paraphrase an old saying, "If there is smoke, there is God K." These curls are usually represented on the images of this deity. Coils and scrolls are commnon motifs in Maya iconography, however, and they should be considered diagnostic of God K only if they emanate

from the tip of the forehead tube. Exceptions to this rule are a few examples on Classic ceramics and hieroglyphs, where the smoke coils emerge directly from the forehead. It should be mentioned too that on some of the stone monuments the coils rise from the "handle" of the tube at the nape of the deity, and there are also a few instances where the scrolls emanate from both ends of the tube, and even from the top of the head of the god.

3. *Snake-foot appendage.* Another common diagnostic feature of God K is the snake foot. The mythological basis for this curious appendage, which replaces one of the deity's legs (usually the left), is unknown, but it is probably associated with the serpent's aspect as emblem of rulership and by its rain-bringing and fertility associations. Floyd Lounsbury, who believes that there are two God K's, an old one and another born at a later date (a matter discussed later in this chapter), thinks that the old God K may not have had a snake leg but that the young god did.[49] The discovery of several ceramic vessels from the very early Classic era portraying God K as snake-footed makes this hypothesis somewhat difficult to defend. A possible reference to the emergence of the snake foot in Mesoamerican mythology can be found in the *Codex Tellerianus Remensis,* where the Mexican War God Huitzilopochtli, one of the several manifestations of the Aztec Tezcatlipoca, is mentioned as "having no snake legs because he did not sin."[50] To find a mythological analogy to the snake leg of God K, David Kelley reaches as far as the apocryphal scripts of the Hellenized Jews of the late centuries B.C.; where Jehovah appears as the snake-footed sun god, holding a whip and a shield, sometimes with the head of a rooster or lion.[51]

Snake-footed images on true portrayals are most common on manikin scepters; the serpent leg serving as the scepter itself and the body of the god with the contralateral leg bent at the knee as the manikin. There are several notable exceptions to this, such as Stela 1 at Tikal and the Dumbarton Oaks Tablet, where the full-figure portrait of God K is shown with one leg replaced with the body of a serpent.

Nonspecific—Constant. Some of the body characteristics and attire of God K are uniform in all his portrayals; however, he also shares them with other members of the Maya pantheon, especially with Gods B and D, and to a lesser degree with older old gods and monster-headed deities.

1. *Shape of the head.* The head is usually somewhat oblong and reptilian on the full-figure representations, but it is more angular and stylized where it is shown without the body. In such instances cosmic signs and glyphic symbols may be interwoven in the composition.

2. *Elevated forehead.* The high forehead, which closely resembles the T617a *nen* ("mirror") grapheme, is also a trait of God K but is shared with other Maya divinities. On some of the portrayals the foreheads are very high, resembling cossack helmets; on others they barely exceed normal dimensions.

3. *Face.* The face of God K is always presented in profile. The eye is the usual "saucer" type seen in representations of most Maya deities. The pupils are central and scroll-shaped. The lower jaw is either recessed or missing.

4. *Body.* The body of God K is anthropomorphic and is scantily clothed with a belt or loincloth with front and /or back hangers. Body jewelry is modest and usually consists of earplugs, a bead necklace, wristlets, and gaiters. Exceptions to this attire are the portrayals in the codices, where God K is often shown fully clad.

5. *Posture.* As a rule God K is shown in a cross-legged sitting posture with the body either in profile or *en face,* the head always looking sideways. Standing representations can be seen only on manikin scepters, on a few Classic ceramics, on three lintels at Sayíl, and in the codices.

Nonspecific—Frequent. These features are common on the representations of God K, but are also shared by other deities.

1. *Mouth curlicues.* Like Gods B and D, God K is often shown with curlicues in the angles of his mouth. They may represent snakes.

2. *Body markings.* On some of his portraits the divine nature of God K is reinforced with oval saurian body markings or, less frequently, with jaguar spots. Sometimes the body markings bear "mirror" infixes.

3. *Arms and hands.* On ceramics one arm is usually extended with the palm elevated as if in greeting. The other hand either rests in the God's lap or on the ground beside him. On stone monuments, especially at Palenque, God K is often shown with his arms crossed.[52]

4. *Colors.* If shown in color, God K is painted either red or deep orange. I know of only two representations in black, both on late-Classic vases, and of one portrait in blue, in the *Dresden Codex.*

5. *Glyphic symbols.* In his pictorial representations God K is often related to specific glyphic symbols, such as the glyph *nen,* the *kan* cross, and the capped Ahau.

Impersonations

The term *impersonations* is used in this description to denote some of the lordly figures on stone monuments who lack the grotesque facial

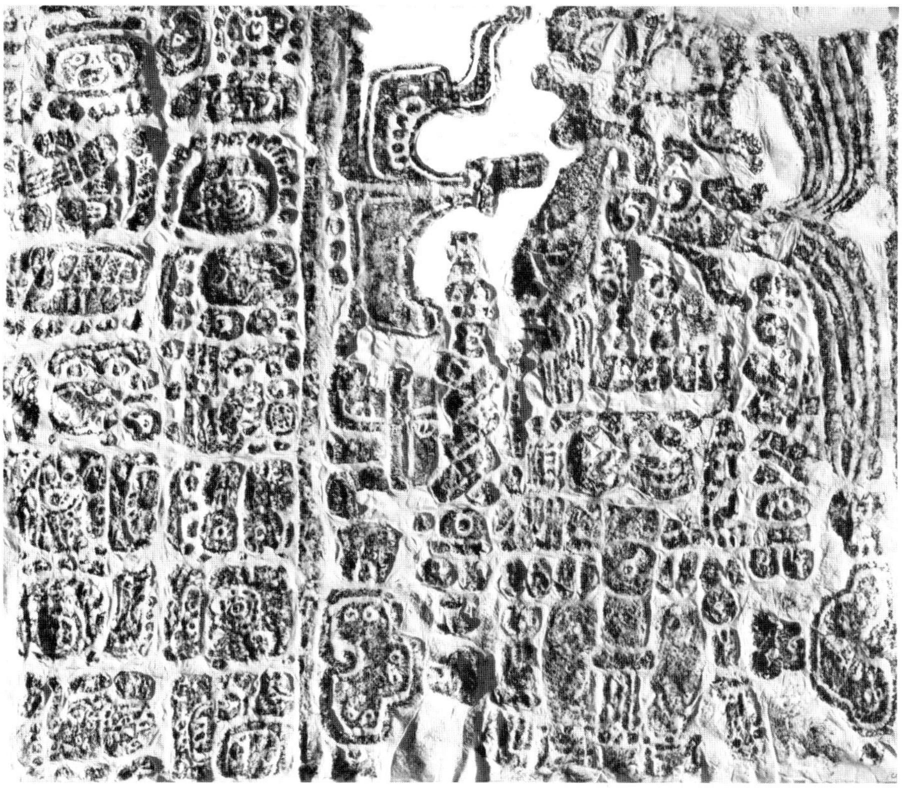

Figure 78. Stela 11 at Copán, showing a standing dignitary. The forehead flare indicates that he is either God K himself or a ruler impersonating him (rubbing by Merle Greene Robertson)

characteristics of God K but are distinguished by the forehead tube and the smoke coils. As has been mentioned, the possibility that these personages represent God K in human guise cannot be completely excluded because the Maya gods were polymorphous—able not only to assume the aspects of the animals or plants they represented but also to metamorphosize into human forms. When, however, we consider the very important written historical data on these monuments, it is more probable that they represented rulers who accorded themselves these divine features and claimed to be either the deity himself or his descendants. Such figures are shown on Stela 11 at Copán, on the Sarcophagus Lid of the Temple of Inscriptions at Palenque, and on a few Classic ceramic vases from Petén. Persons with normal human features but forehead flares are also depicted on some of the "eccentric" flints found in the Motagua River valley. A unique instance of God K impersonation can be seen at Palenque, where the Stucco Panels of the Temple of Inscriptions show the ruler Chan Bahlum as a snake-footed infant.

The Head of God K

The bodiless head of God K is also a common motif in Maya art and iconography. It was painted and carved on many pots and vases and decorates many stone monuments. Exactly what these isolated heads signify is difficult to determine. In recent times, as some of the violent aspects of Classic Maya life and rituals previously regarded as "peaceful and harmonious" are recognized, some students of Maya anthropology cry "decapitation sacrifice" every time they see a head without a body on an object. This conclusion is as wrong as it is to suppose that the profile of the British monarch on the ten-shilling piece implies that Elizabeth II had her head chopped off by subjects infuriated by the incompetence of the British National Health System.

Naturally, there are several instances where bodiless heads do indicate decapitation, especially when they occur along with portrayals of violent acts or armed individuals or when they are shown on plates or in baskets. We know that to such atrocities even the deities, God K among them, were not immune. Witness to this are the Venus Tables of the *Dresden Codex*, which show God K lying wounded by the spear of the God of the Evening Star. The (severed?) head of God K is held by the lord on Stela 9 at Seibal, is seen suspended on the ceremonial bar on Stela 8 at the same center, and is shown on plates on two ceramic vases. Deities holding God K's head on plates are also shown on several pages of the *Codex*

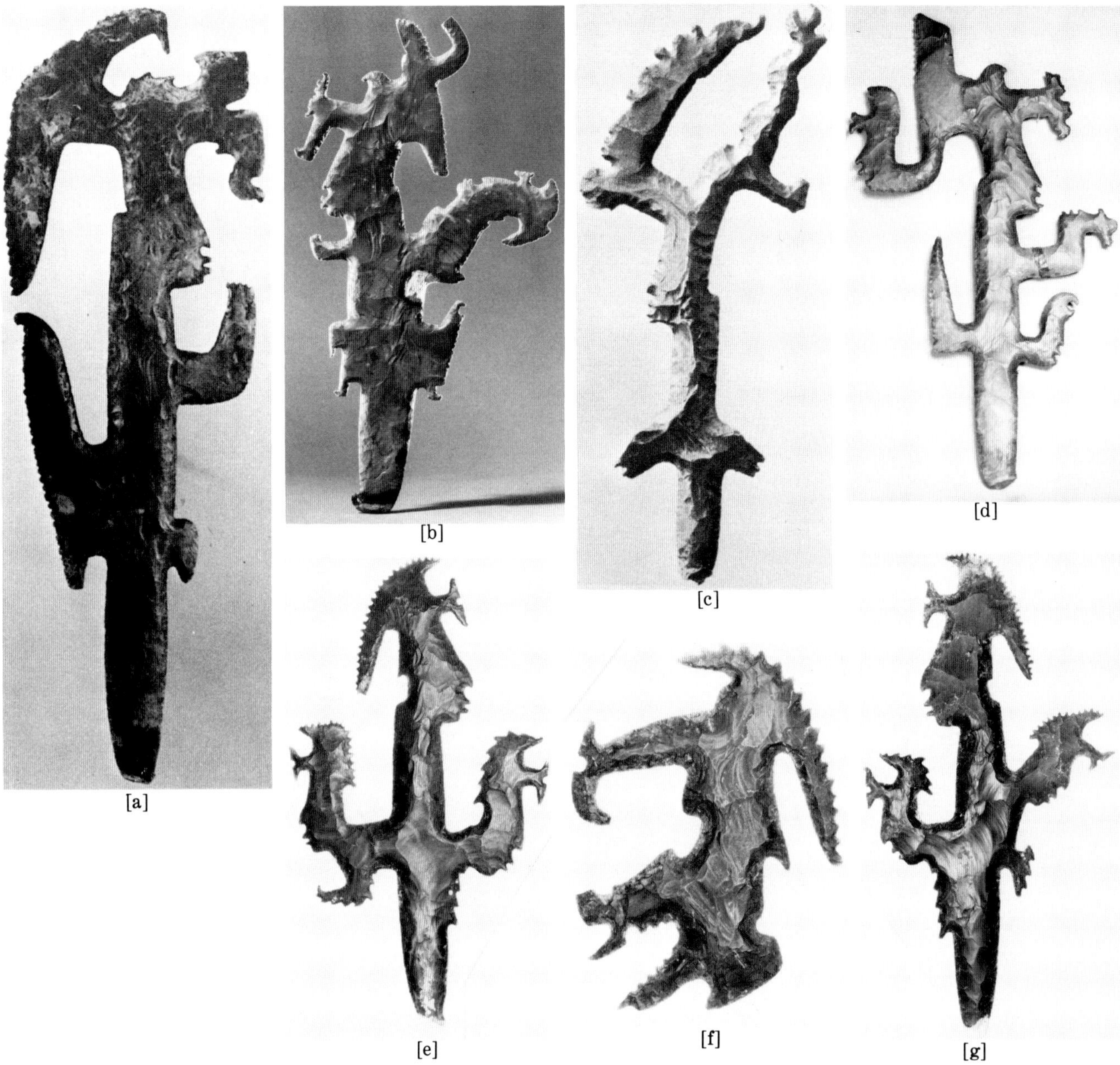

Figure 79. Eccentric flints from the Quiriguá region made into shape of standing human beings with forehead flares ([a] *Collection of Mr. and Mrs. Samuel Josefowitz,* [b] *Metropolitan Museum of Art after Easby and Scott,* [c] *after Michael D. Coe,* [d] *Cleveland Museum of Art,* [e–g] *Dumbarton Oaks Collection)*

Paris. Most of these examples indicate that the head of the "god with the forehead cigar" was cut off by something or someone. There are, however, several monuments, such as some of the stelae at Copán, a Ball Court marker at Yaxchilán, and Altar 1 at Iximché (Quiché), as well as several ceramic vessels, where the isolated head of God K appears in a peaceful context and probably indicates his presence or influence rather than his fall by beheading.

Peculiar representations of God K are the lordly figures shown wearing the head of the deity as a ceremonial helmet. Probably they indicate that the lords are under the protection of God K. Such portrayals are seen on Stelae 20 and 30 at Tikal, Stela 22 at Piedras Negras, Stela 3 at Santa Rosa Xtampak, on a lintel from the Usumacinta River valley, on Stela 1 at Itzán, and on Stelae 5, 6, and 7 at Seibal. Three examples of such "God K helmets" also occur in the *Dresden Codex* and one on a carved vase from Petén (D-34).

The God K group

The great diversity in size, shape, and regalia in the artistic presentations of God K is certainly puzzling, especially since the appearance of some of the other Maya divinities, such as God L or God A, is fairly uniform. There is similarly a great variety in name glyphs and affixes of the deity.[53] For this chameleonlike nature of God K one may seek an explanation in different regional trends or in artistic liberty. It should be repeated, however, that in Maya iconography very few things can be considered coincidental, and few changes were made at the whim of the artist. What, then, is the explanation?

The answer may be that in the representations of God K we may be dealing not with a single deity but with a group of supernaturals with similar but not necessarily identical functions, a possibility to which Kelley referred but never elaborated upon.[54] With this hypothesis in mind, I attempted to categorize the many examples of deities of the God K group presented on ancient Maya artifacts. I divided them into seven distinct subgroups.

1. *God K1, or God K of the Temple Slabs*

This deity is shown as a small, free-moving spirit held in the hands, on laps, on pillows, or on mat supports by human beings, some of them known historical personages. God K_1 has a small well-proportioned human body, which makes one wonder if we are dealing with an elf god or with a child deity. He is always shown with both the forehead tube and the smoke coils, but he may or may not be snake-footed. Examples of God K_1 can be seen on the Sanctuary Tablets of Temple XIV and the Temple of the Cross at Palenque, on the Dumbarton Oaks Tablet, on a lintel from the Usumacinta Basin (now in the Cleveland Museum of Art), and on Lintels 6 and 43 at Yaxchilán.

2. *God K2, or God K of the Serpent Bar*

In these representations God K is again shown either alone or in double portraiture as a small god with the forehead tube and coils but in the jaws of the double-headed ceremonial serpent bar. One may suppose that he is either being swallowed by or emerging from the throat of the sacred snake, the latter possibility being the more likely one. Because of the prominent serpent stigmata of God K, it is not impossible that these scenes represent his birth from the throat of the divine Double-headed Serpent. Such examples can be seen on the following monuments: the lintel of Structure 6 at Bonampak; Stela 16 at Dos Pilos; Stelae 12 and 14 at Naranjo; the Sarcophagus Lid, Temple of Inscriptions at Palenque; a Ball Court marker at Yaxchilán; Stela 12 at Machaquilá; Stela H at Copán; and a somewhat unconventional presentation on Stela 1 at Tikal. On the Dumbarton Oaks Lintel from Kuna Lacañha heads of God K form the two ends of the bar.

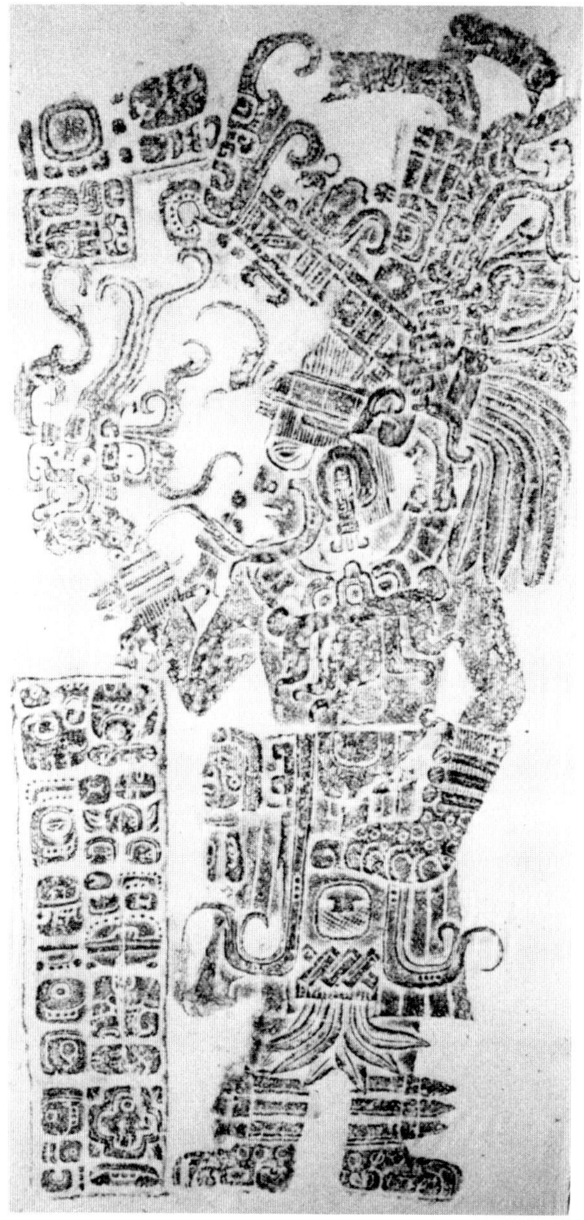

Figure 80. Stela 8 at Seibal showing a jaguar-gloved lord holding the head of God K (rubbing by Joan Wholey Patten)

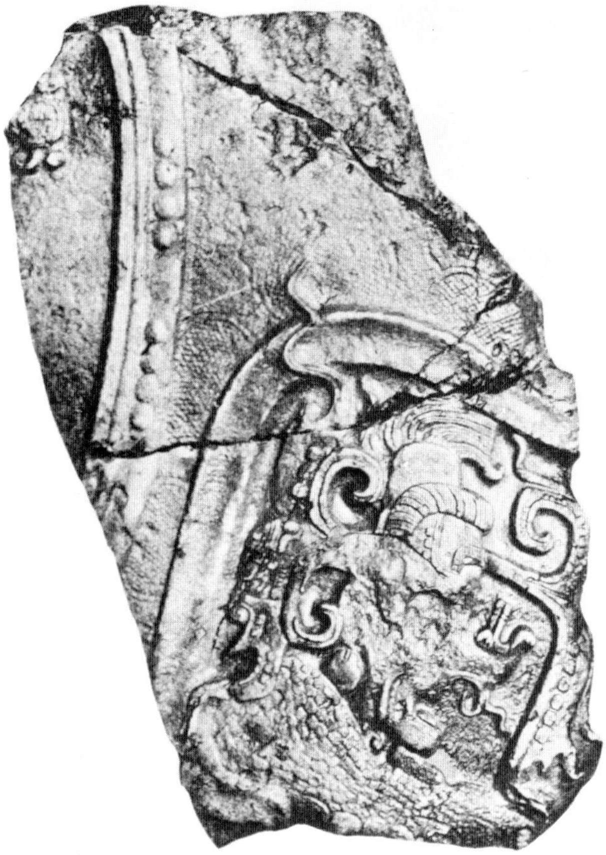

(a)

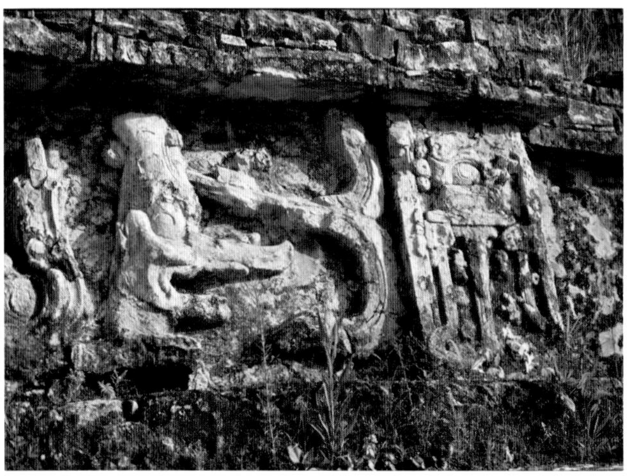

(b)

Figure 81. Head of God K on a fragment of Altar 1 at Itzimté (after Morley) (a) *and on the stepped platform of the Palace at Palenque* (b)

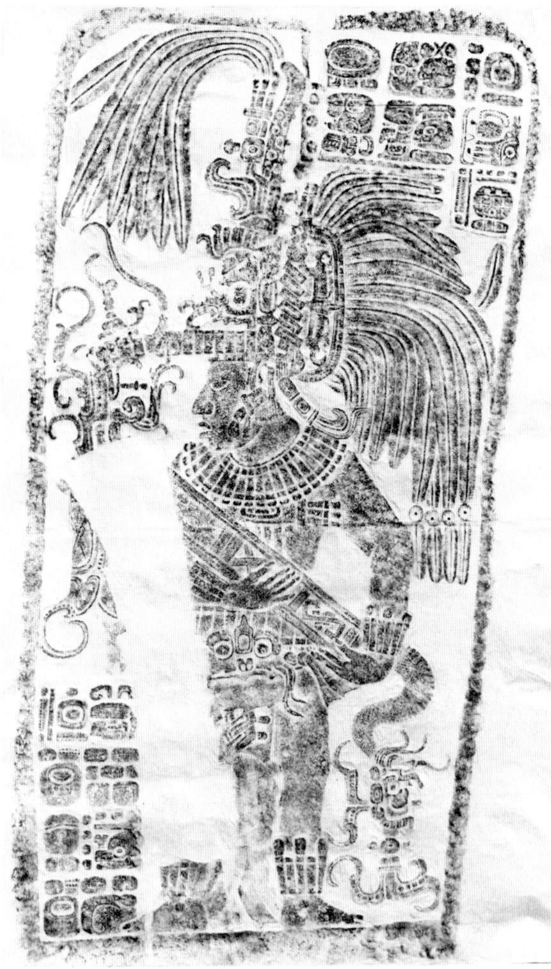

Figure 82. Stela 9 at Seibal showing the image of a ruler holding a ceremonial bar from which the head of God K is suspended upside down. Note that the forehead flare is also in a "reverse" position (rubbing by Merle Greene Robertson)

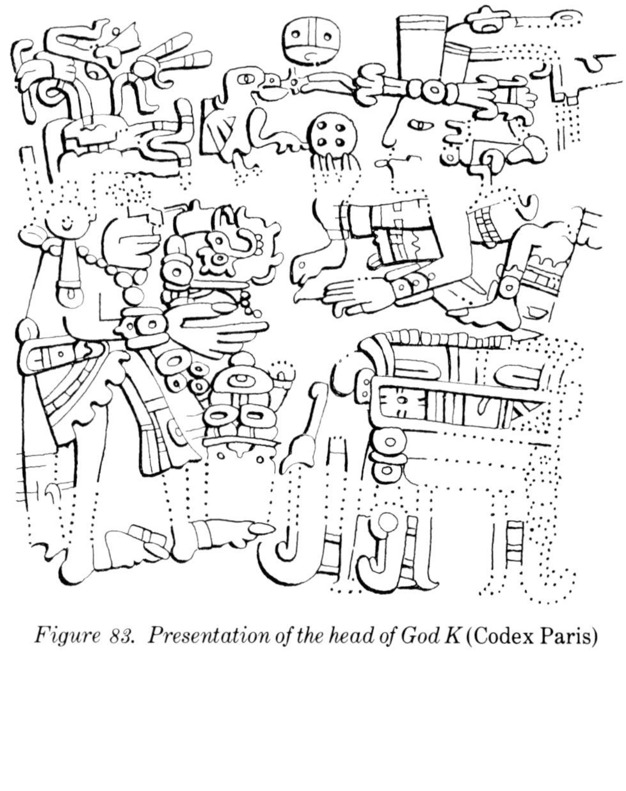

Figure 83. Presentation of the head of God K (Codex Paris)

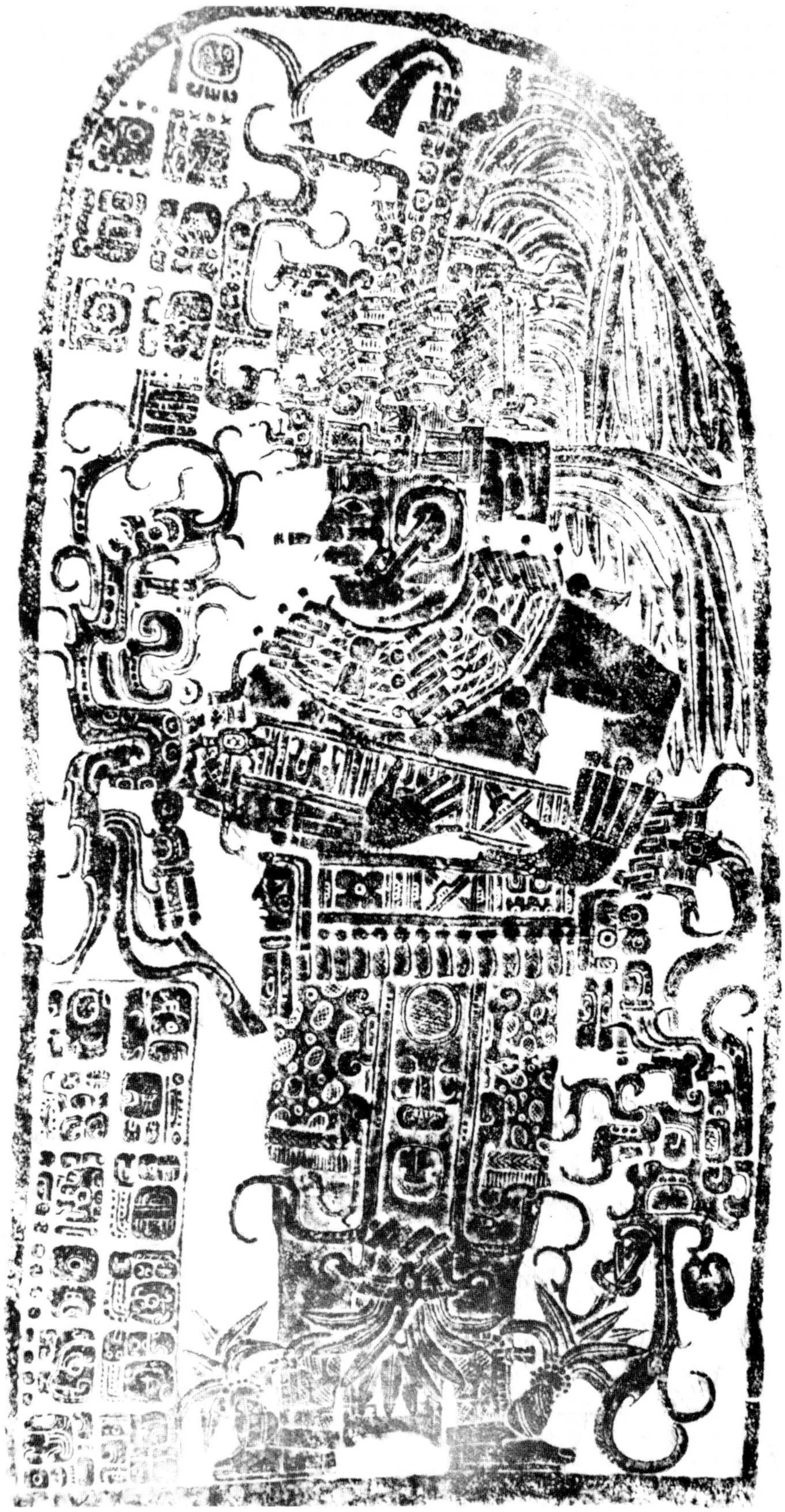

Figure 84. *Stela 10 at Seibal, showing a lord with a large ceremonial bar. He has a forehead (helmet?) flare and holds God K's head suspended from the low end of the bar (Rubbing by Merle Greene Robertson)*

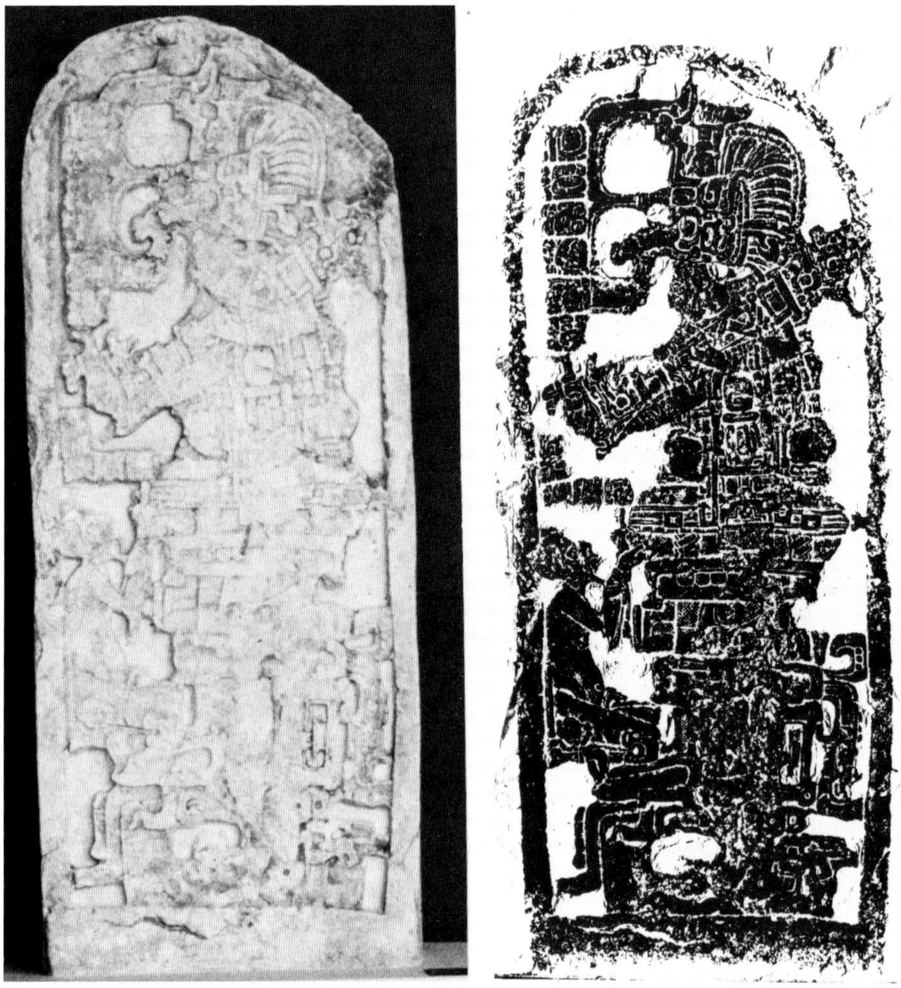

Figure 85. Stela 22 at Naranjo, carved with the majestic figure of a dignitary wearing the head of God K as a helmet (rubbing by Merle Greene Robertson)

3. *God K3, or God K of the Manikin Scepter*

The God of the Manikin Scepter is probably the most common presentation of this deity. The scepter is an implement that strongly resembles the object carried by the court jesters of medieval Europe. God K_3 is shown with a small anthropomorphic body with one leg transformed into a snake, which also serves as the handle of the scepter of the kind held by a lord. On some of the scepters God K_3 has both forehead tube and smoke coils (Stelae 2, 3, 6, 7, and 10 at Machaquilá; Zoomorph P, Stelae D, E, and F at Quiriguá; Stela 13 at Naranjo; Stela 7 at La Florida; the Wooden Lintel of Structure 10 and Stela 5 at Tikal; Stela 5 at Sayíl; and Stela 65 at Calakmul). On other scepters God K_3 is presented with forehead tubes only: Lintels 1, 3, 7, 21, 32, 42, and 53 at Yaxchilán; Stela 3 at Aguateca; Stela 4 at Itzimté; Stela 15 at Calakmul; Stela 13 at La Muñeca; the Wooden Lintel at Temple IV and Stela 5 at Tikal; and Stela 65 at Calakmul (Campeche).

4. *God K4, or God K of the Manikin Staff*

God K of the Manikin Staff is portrayed on a number of stucco tablets in the Sepulchral Chamber of the Temple of Inscriptions and in Palace House A at Palenque, as well as on a few ceramic vessels of the Classic period. On this representation dignitaries, probably human, are shown holding long staffs, each topped with the long-nosed head of God K. The staff could be ceremonial and intended to indicate the presence of the deity or to show that the carrier is protected by God K, but there is also the possibility that it may also have served as an ax-headed ceremonial weapon similar to the *fokos* of Hungary.

5. *God K5, or God K of the Ceramic Vases*

With a very few exceptions, on ceramic artifacts God K is portrayed either as a seated, often enthroned individual of normal size and proportions or with only his head depicted. Forehead tubes are

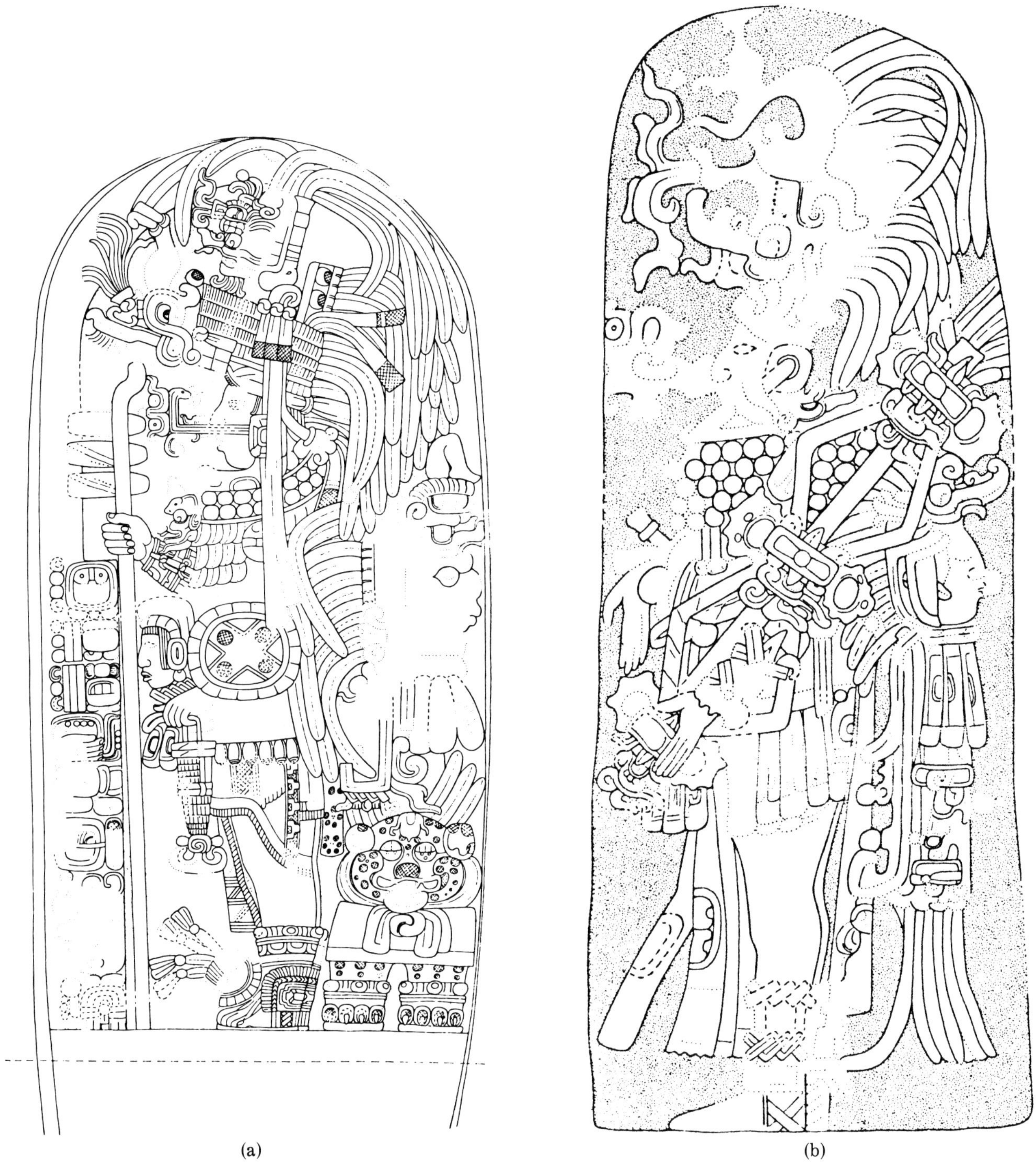

(a) (b)

Figure 86. Dignitaries with God K headdresses: (a) *Stela 20 and* (b) *Stela 30 at Tikal (after William R. Coe)*

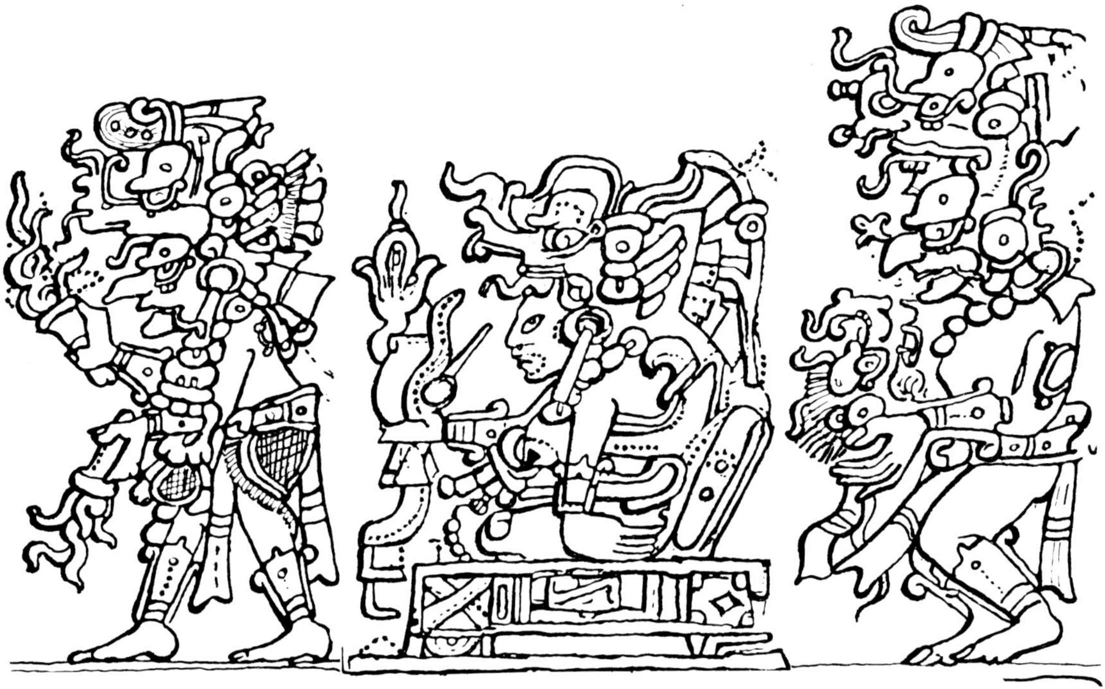

Figure 87. Deities wearing God K headdresses. Note that the one on the right holds the severed head of God K (Dresden Codex)

shown on most and smoke coils on nearly all representations. The scrolls are usually painted red, and they are very large and often frame the body of the deity. On these vases God K_5 is presented either in a twin portrait or in the company of other divinities and, with the possible exception of one vase, never in the company of persons who could be regarded as mortals.

6. *God K6, or God K of the Codices*

Representations of God K in the *Dresden Codex* and in the *Codex Paris* have neither forehead tubes nor smoke curls, and can be identified only by their name glyphs and noses. Unlike portrayals on monuments and ceramics, where the scantily dressed figures of God K are nearly always in sitting postures, the images of God K of the codices are usually fully dressed and often engaged in such activities as walking, offering sacrifices, and burning incense or lying wounded.

God K is portrayed several times in the *Dresden Codex*. His hieroglyph without his picture occurs on pages 10b, 49, 58, and 69.[55] On page 3, God K is shown seated in a posture identical to that seen on ceramics. A somewhat similar portrait occurs on page 7. On page 12 he is seen holding a dish filled with round objects identified by the glyphic text as cacao beans. On page 46 we see the deity lying wounded by a spear. Pages 34, 49, and 65 show portraits of other deities wearing the head of God K as a helmet. On page 67, God B is shown holding God K in a pouch. On page 25, God K is shown seated in a house, and on page 26 he is shown holding a beheaded turkey. In the *Codex Paris Manuscript*, God K also occurs several times; generally only his head is shown. Especially well preserved are the representations on pages 3 and 6, where his head is being presented to other deities by Gods E and B.

7. *God K7, or God K of the Temple Walls.*

It was Seler who in 1904 recognized God K in the ornament so common on the temple ruins of the Maya area. He thought that the peculiar "conventionalized face with the projecting proboscis-shaped nose, which applied chiefly to the corners of temple walls, displays unquestionably the features of God K."[56] Some of the best examples of such architectural ornamentations are found on Temple 22 at Copán, the Main Structure at Xlapak, the Temple of the Warriors and Las Monjas at Chichén Itzá, the Nunnery and the Palace of the Governor at Uxmal, and especially on the intricate Codz Pop temple façade of Labná, which is composed of more than one hundred interwoven long-nosed faces. If this theory is true, then it would certainly seem that God K was one of the principal gods of post-Classic Yucatán. That appears to be highly doubtful, however. It is more probable that these long-nosed masks represent the ancient Rain Serpent Dragon to whom God K is related but not identical.[57]

(Text continues on page one hundred and four)

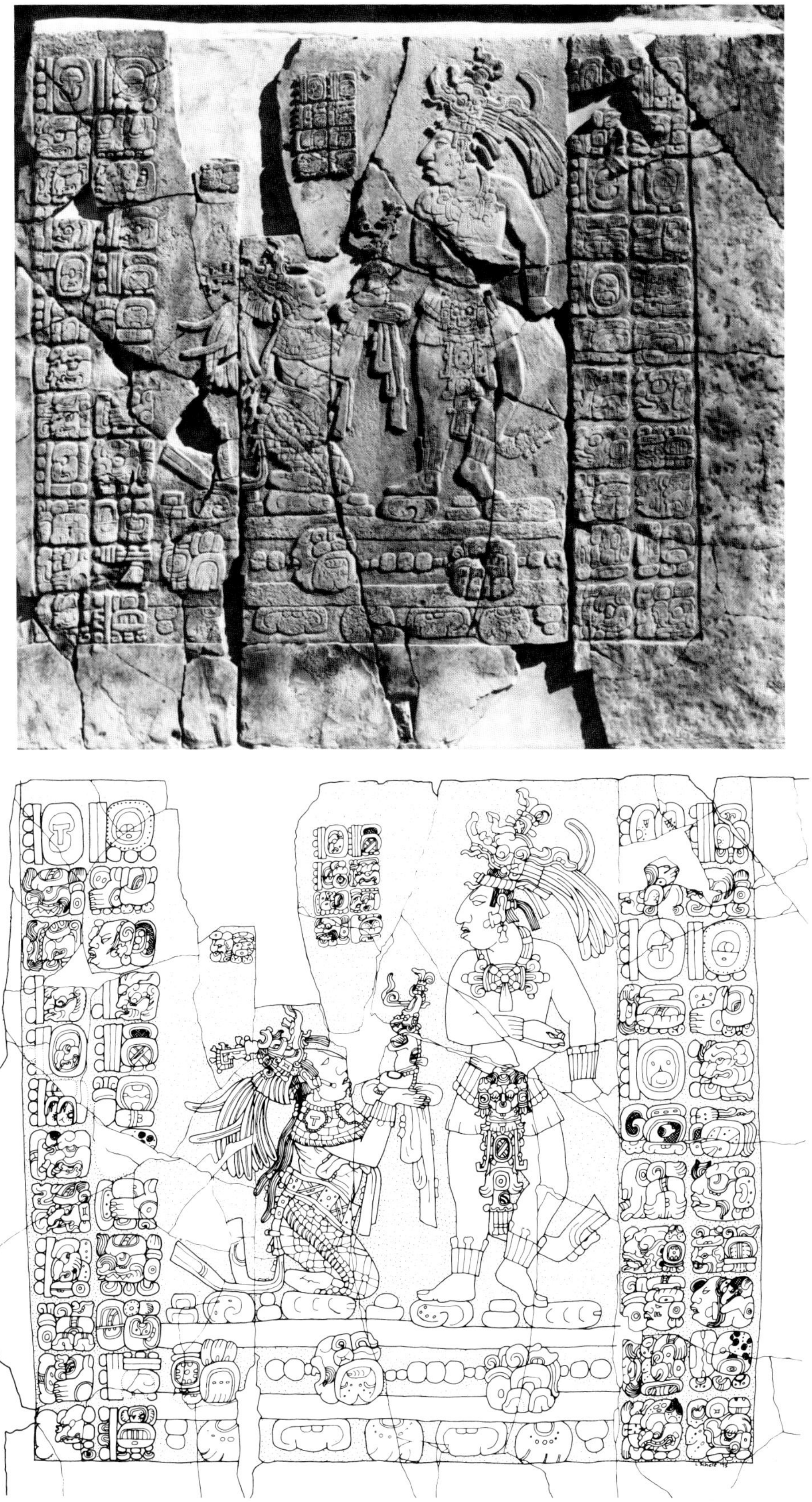

Figure 88. Sanctuary Tablet, Temple XIV, Palenque. The dancing lord is wearing a God K helmet; the kneeling lady is holding the small figure of God K on a folded cloth (photography by Merle Greene Robertson, drawing by Linda Schele)

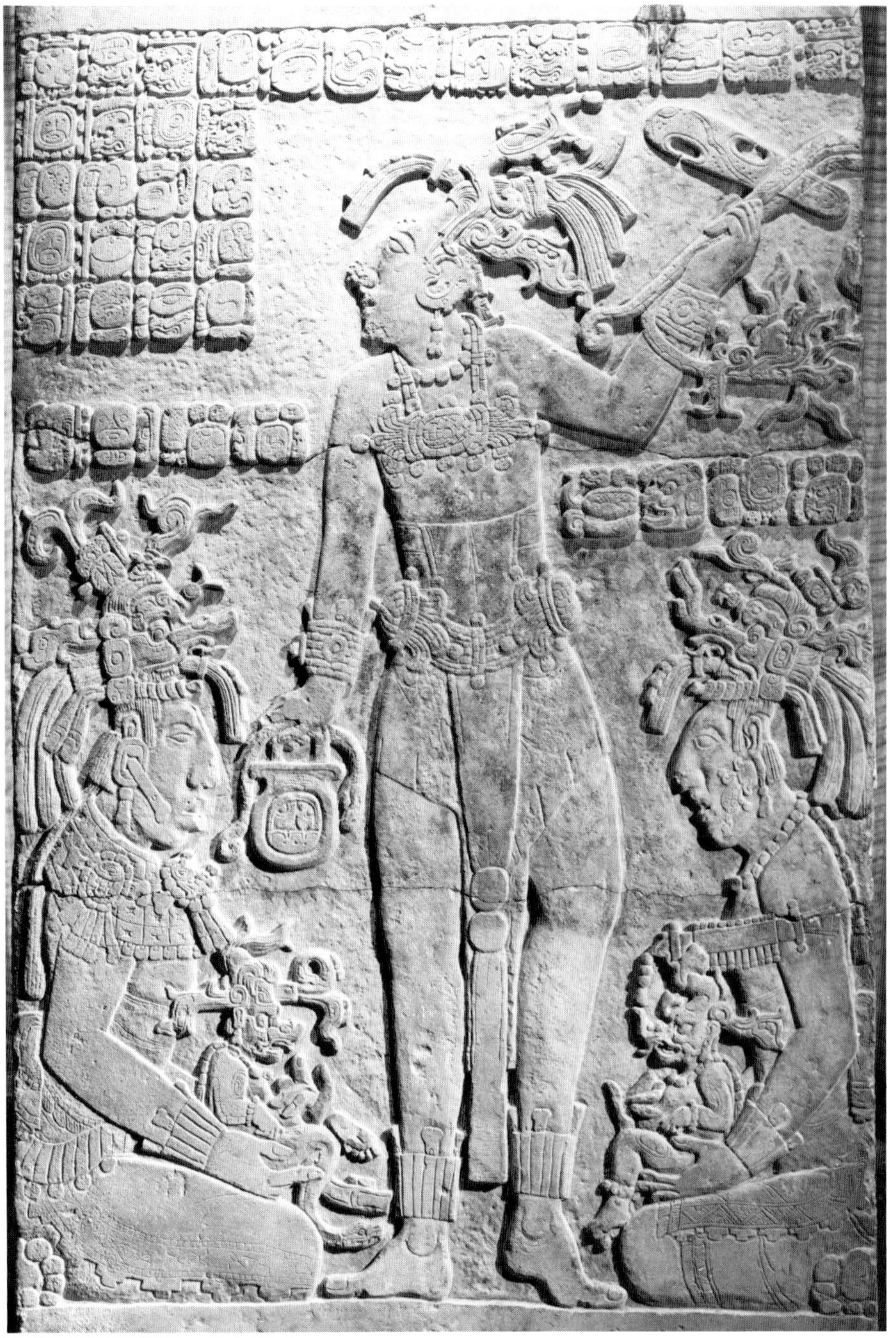

Figure 89. The Dumbarton Oaks Tablet, showing the ax-wielding God GI dancing. Two seated attendants are holding jester gods; the left one is the snake-footed God K (after Michael D. Coe and Elizabeth P. Benson)

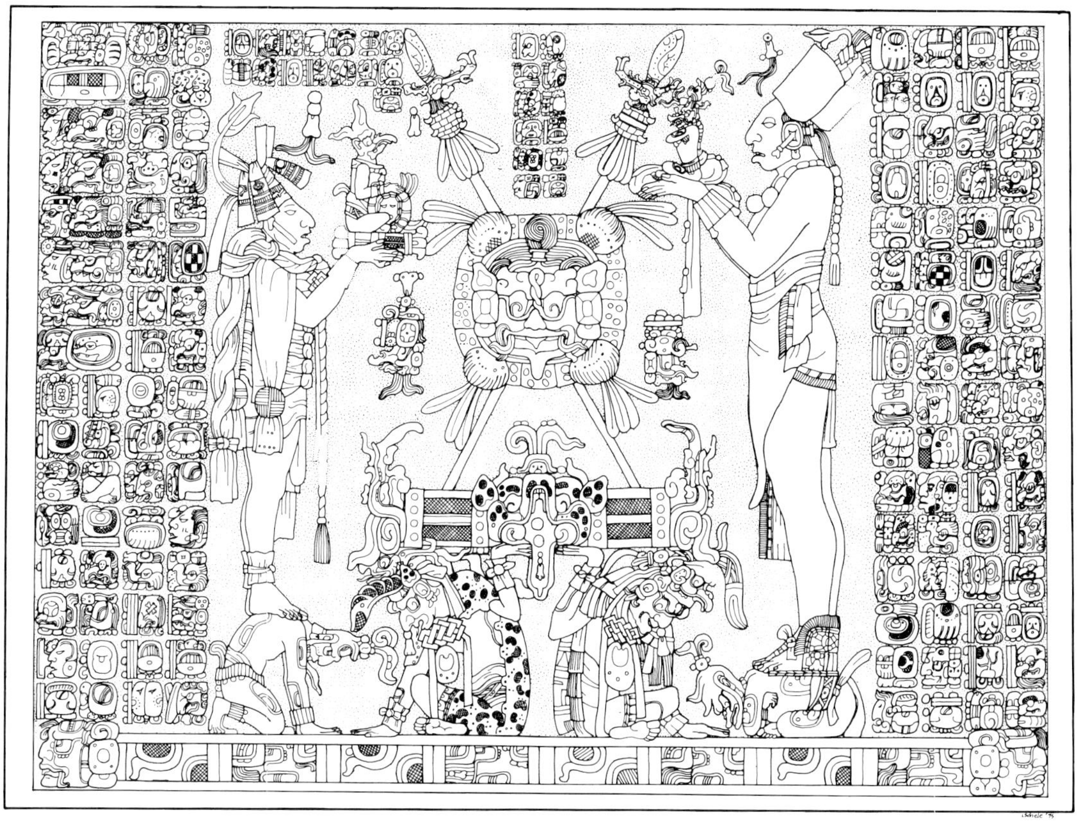

Figure 90. Sanctuary Tablet, Temple of the Cross, Palenque (drawing by Linda Schele)

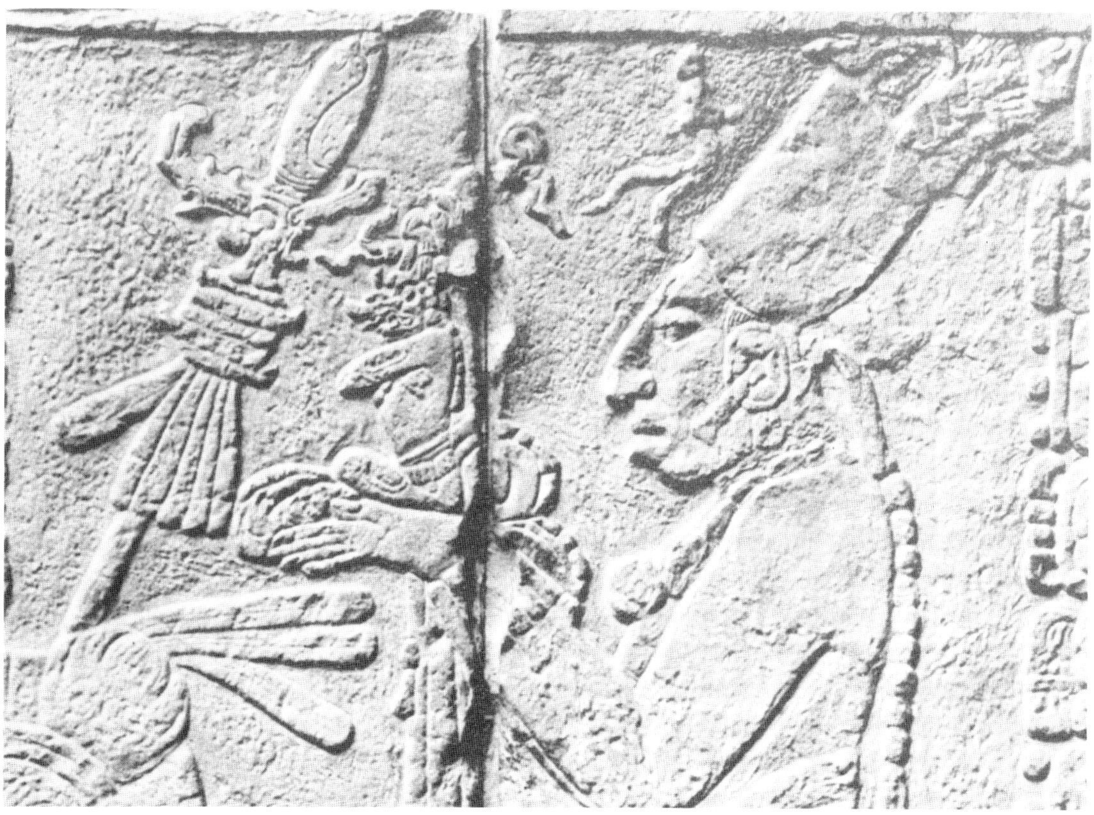

Figure 91. Detail, Sanctuary Tablet, Temple of the Cross, Palenque, showing Chan Bahlum, former ruler of that city, holding the small figure of God K (after Maudslay)

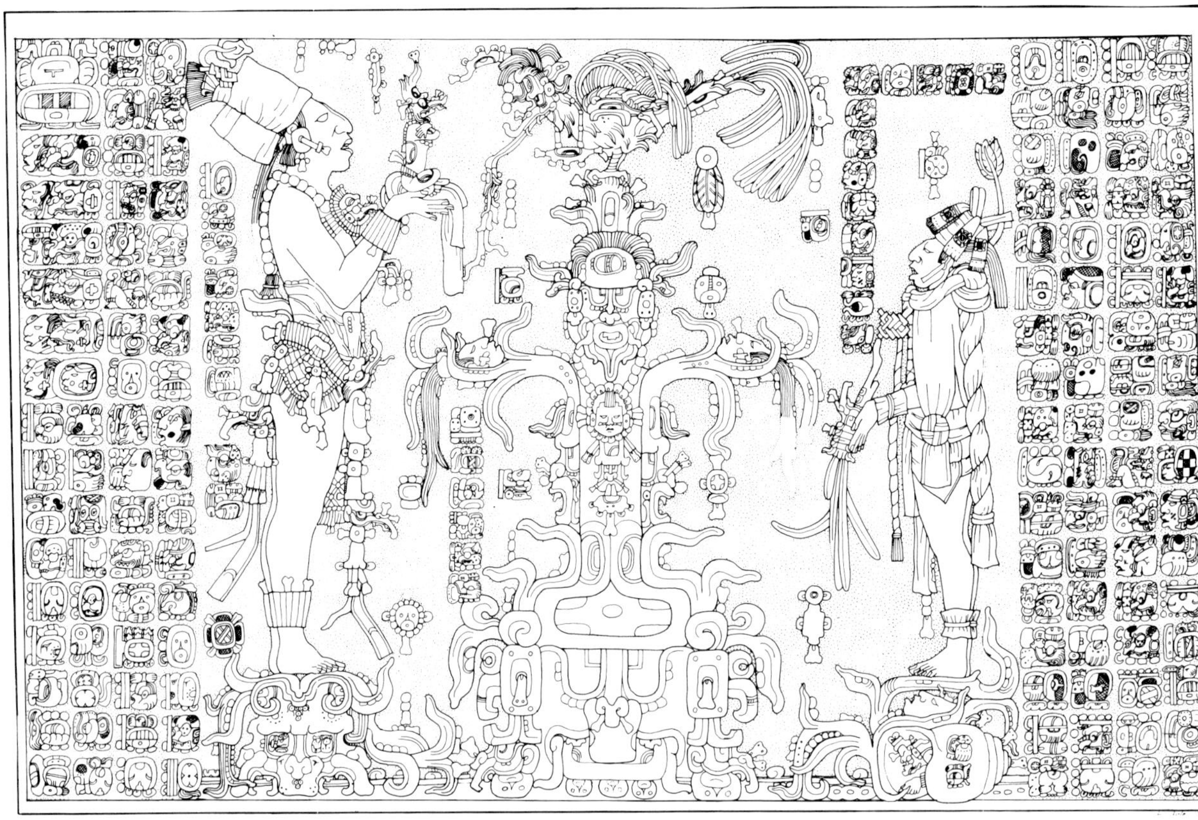

Figure 92. Sanctuary Tablet, Temple of the Foliated Cross, Palenque, the guardian of which is believed to be God K. His name glyph occurs three times on the monument. Also note his figure emerging from the conch shell in the lower right corner (drawing by Linda Schele)

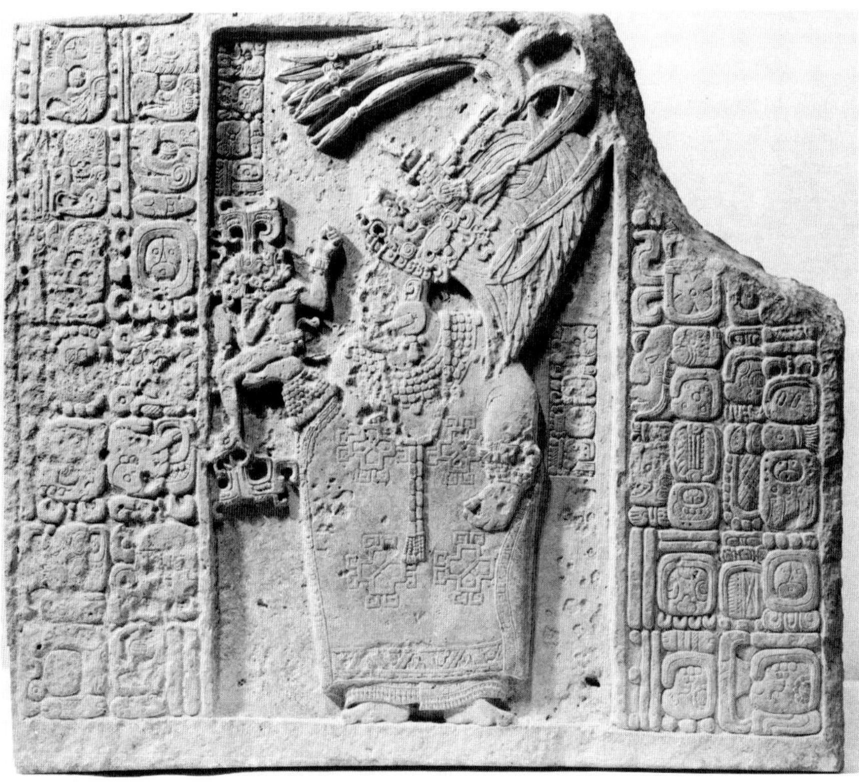

Figure 93. Stone lintel from the Usumacinta River valley. The monument shows a nobleman (or noblewoman) holding the snake-footed figure of God K (Cleveland Museum of Art)

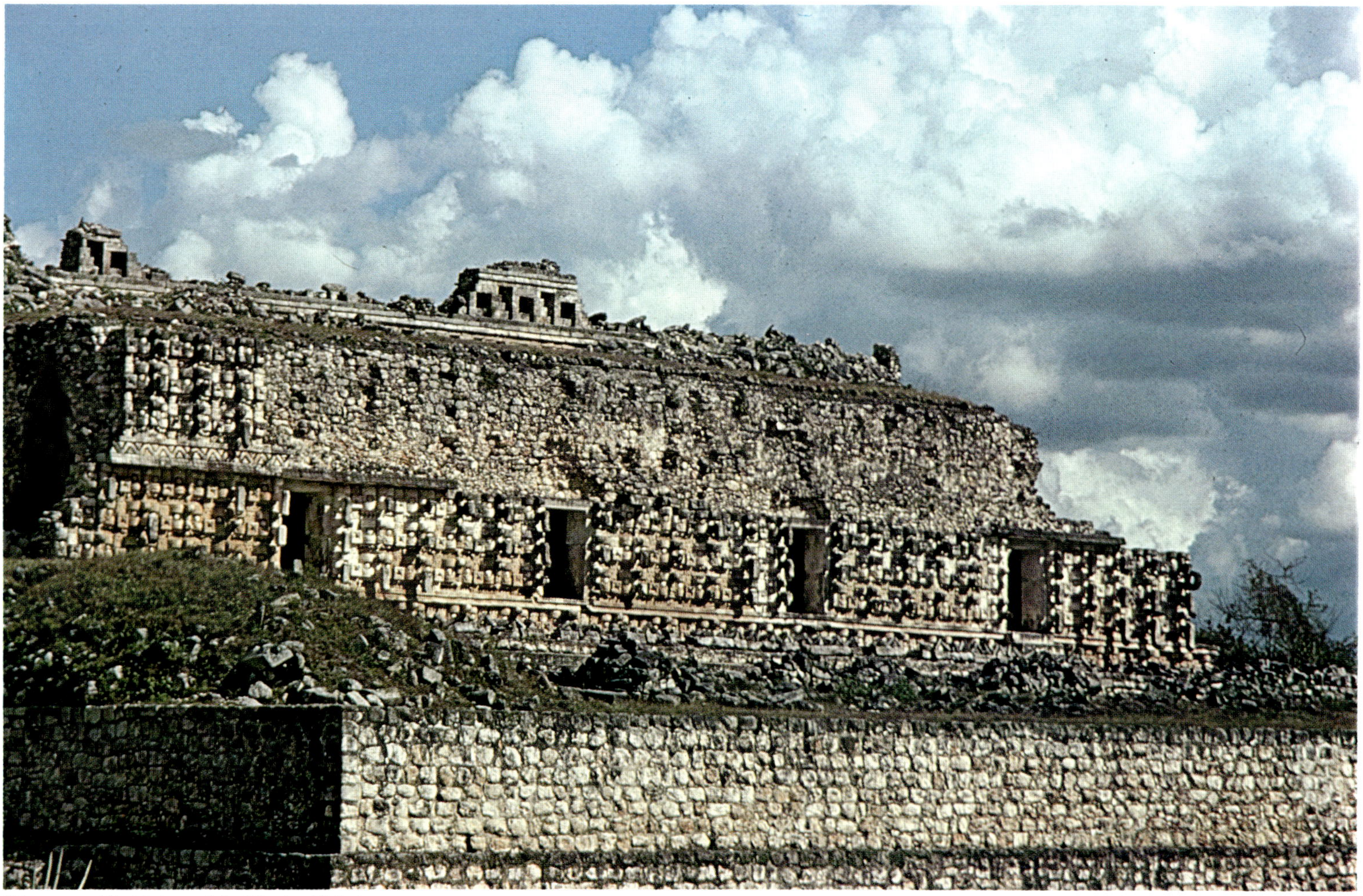

Plate 85. *The Palace of the Masks, Kabah*

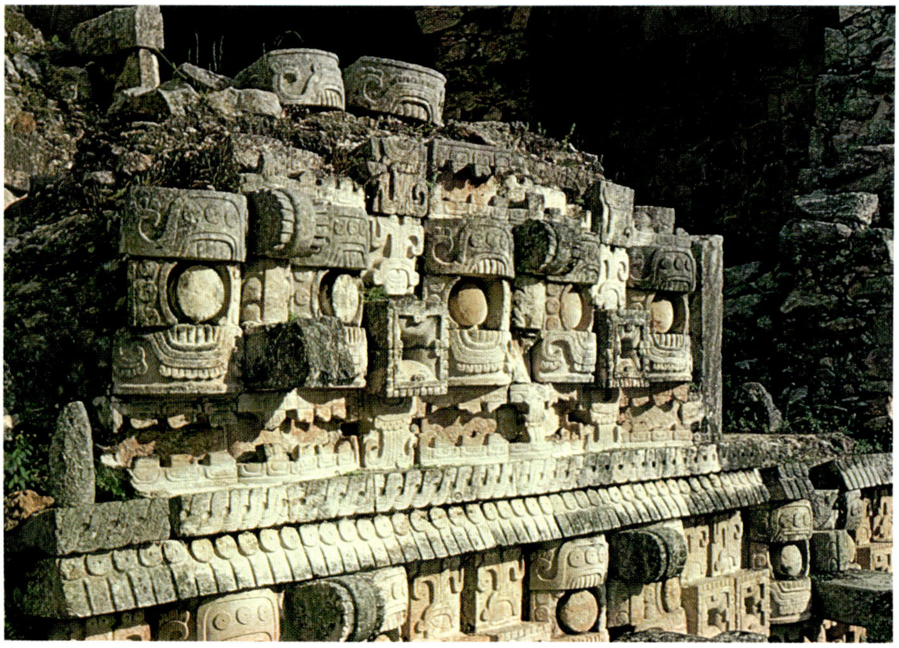

Plate 86. *Detail, Palace of the Masks, Kabah*

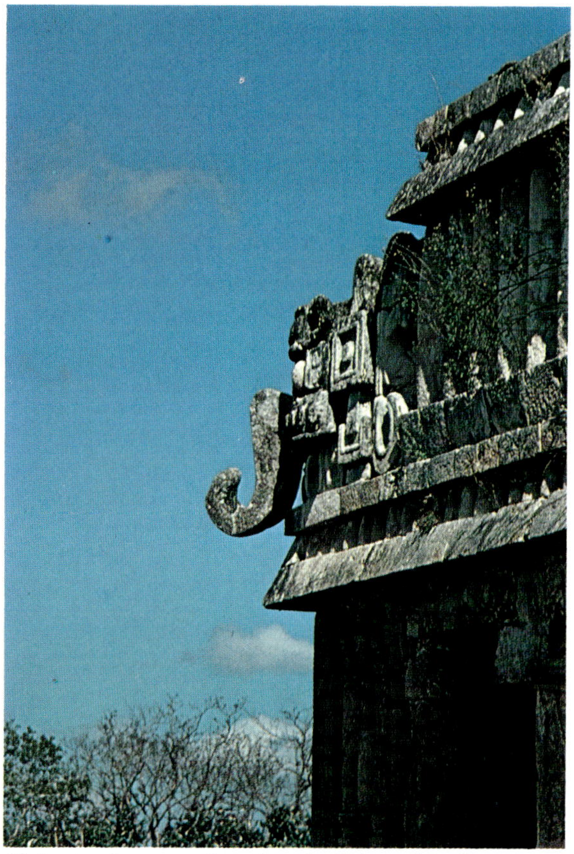

Plate 87. *Long-nosed mask on the Palace of the Governor, Uxmal*

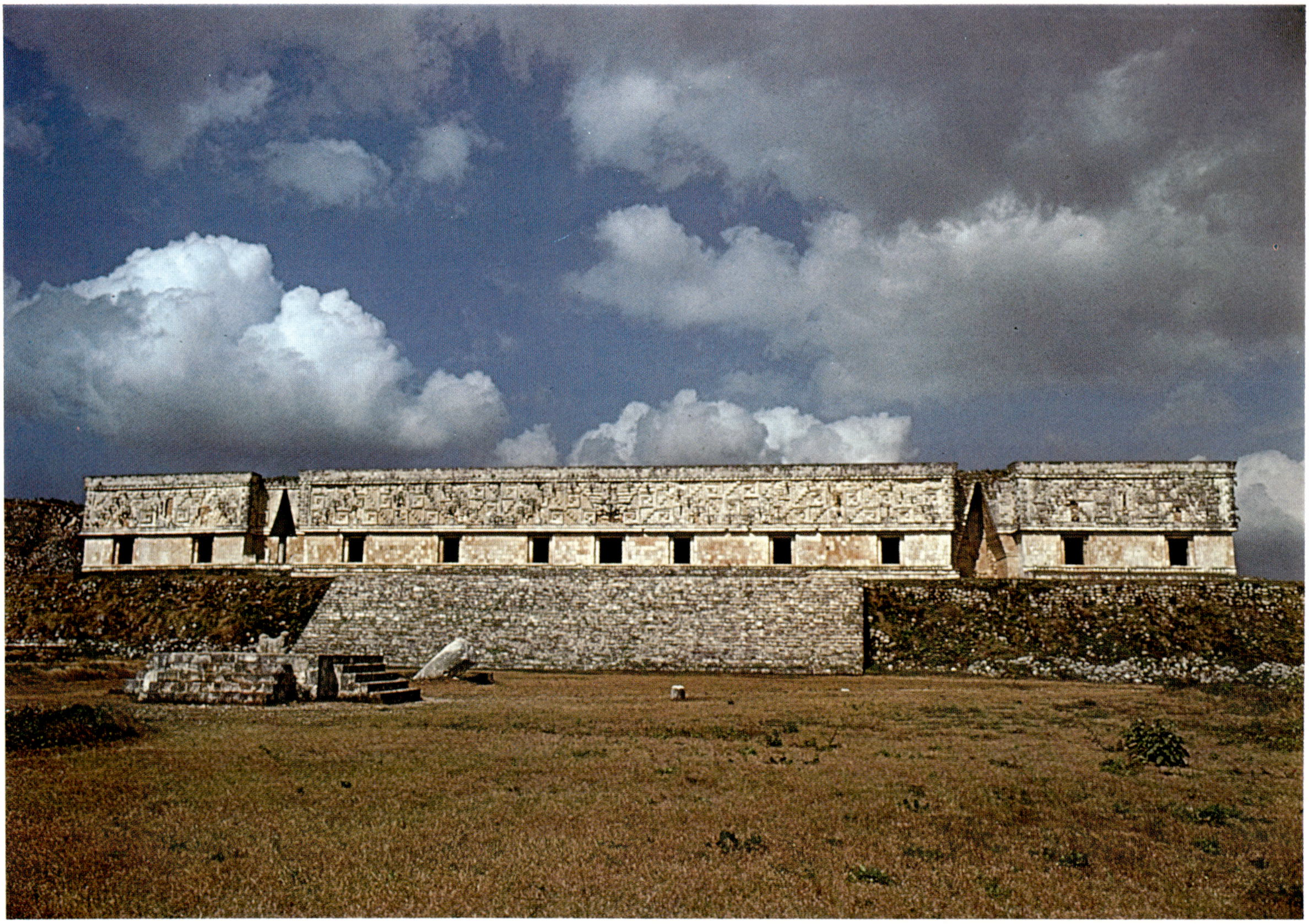

Plate 88. Palace of the Governor, Uxmal

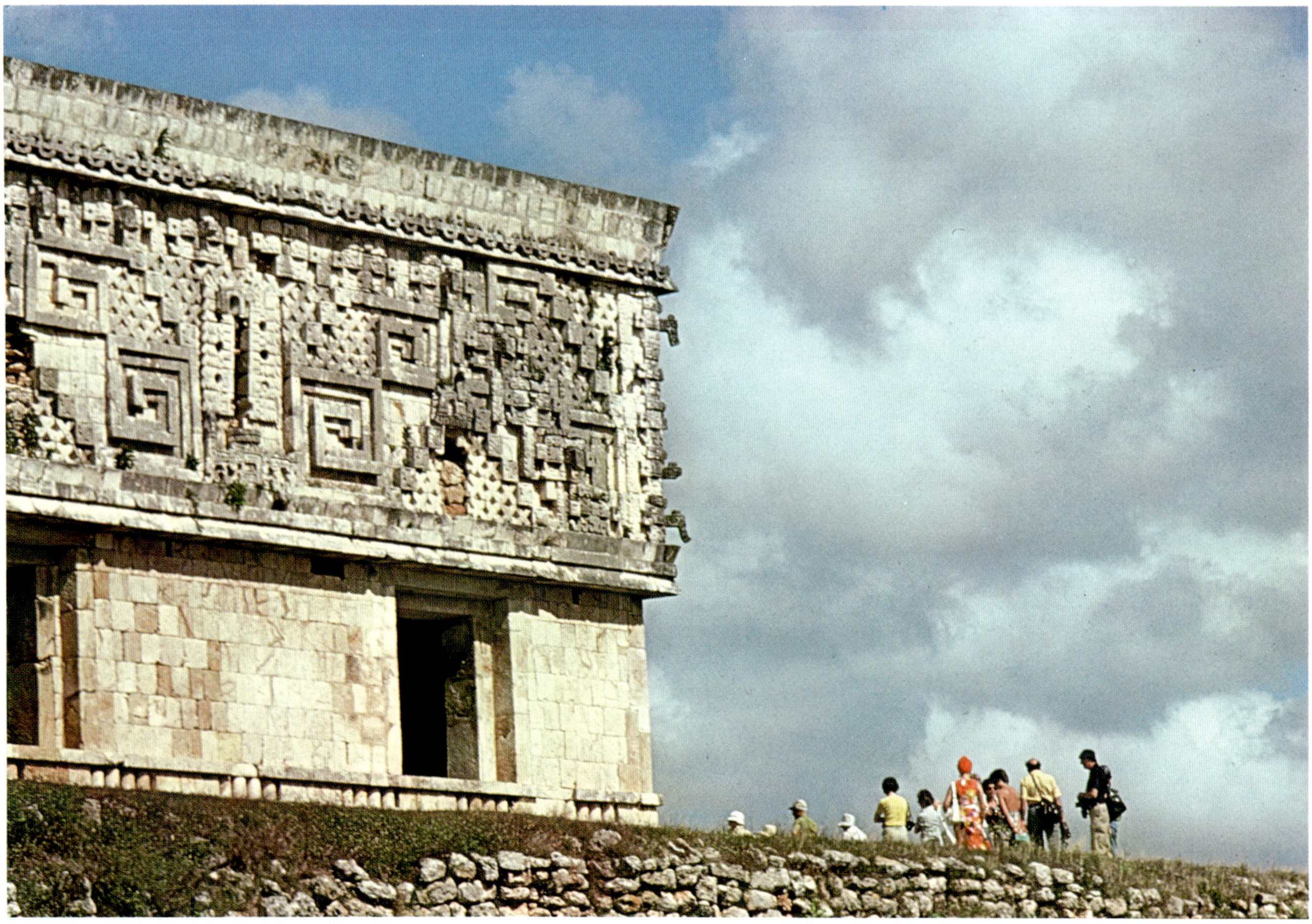

Plate 89. Palace of the Governor, Uxmal

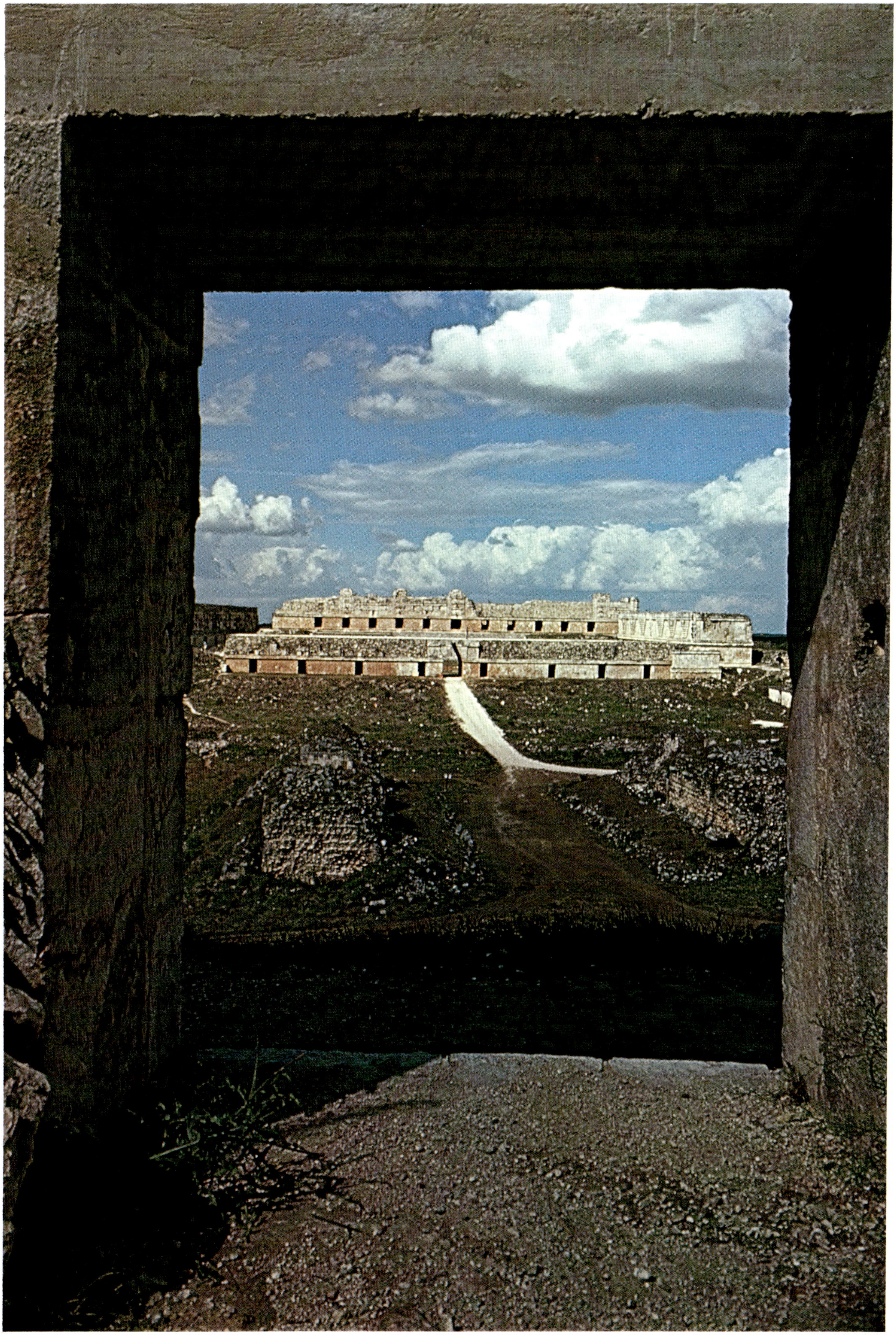

Plate 90. View at the Nunnery, Uxmal

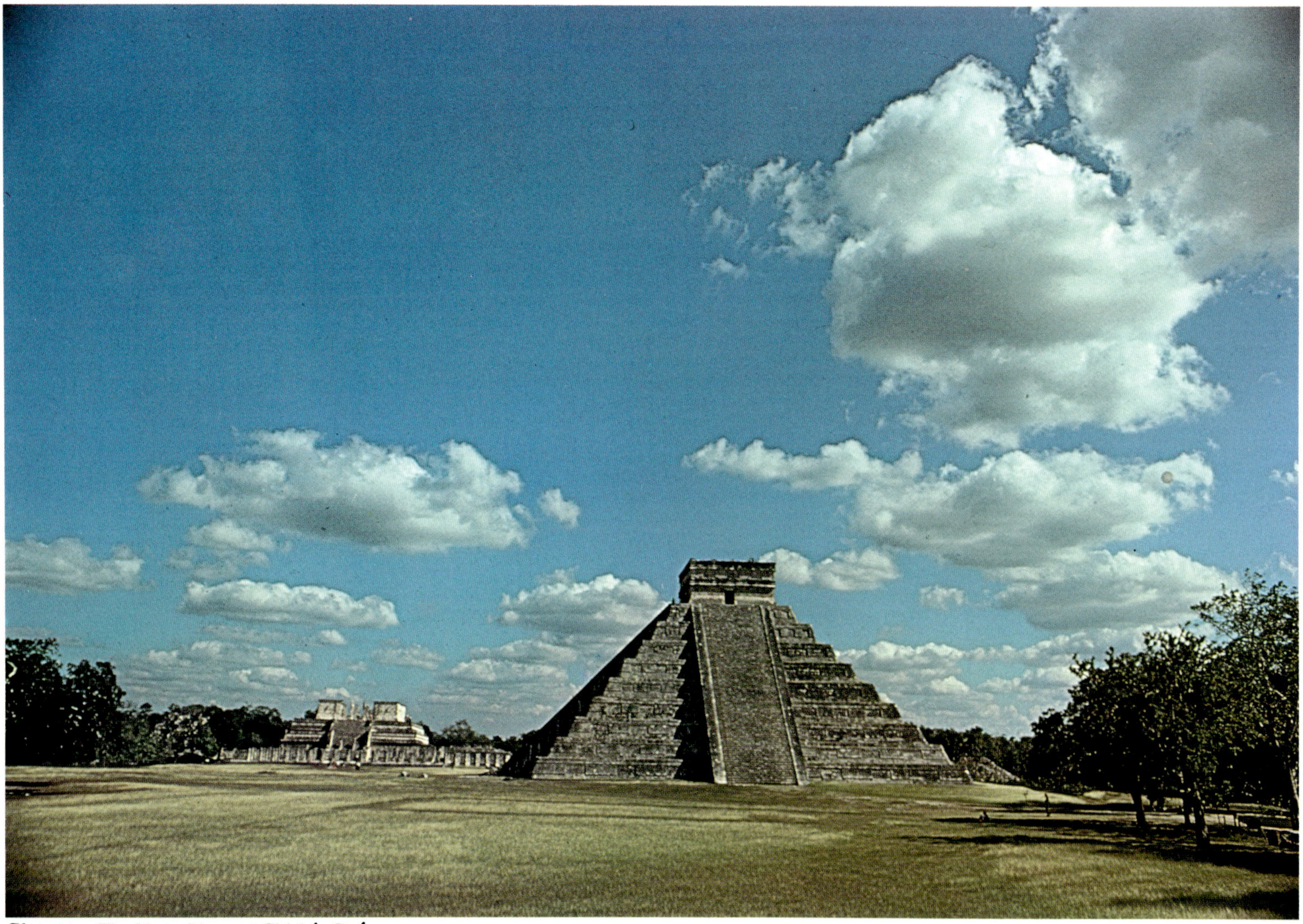

Plate 91. The Main Plaza at Chichén Itzá

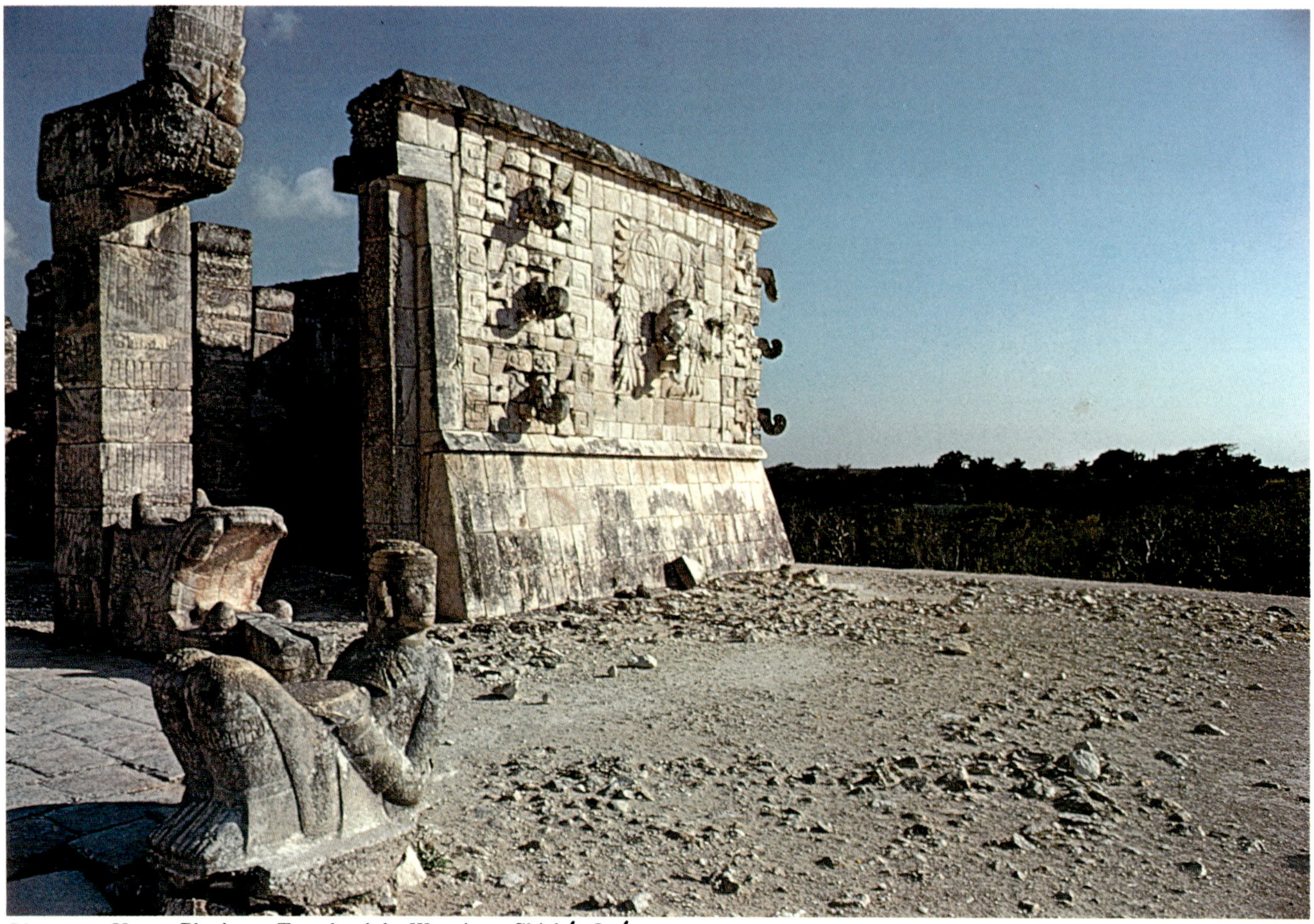

Plate 92. Upper Platform, Temple of the Warriors, Chichén Itzá

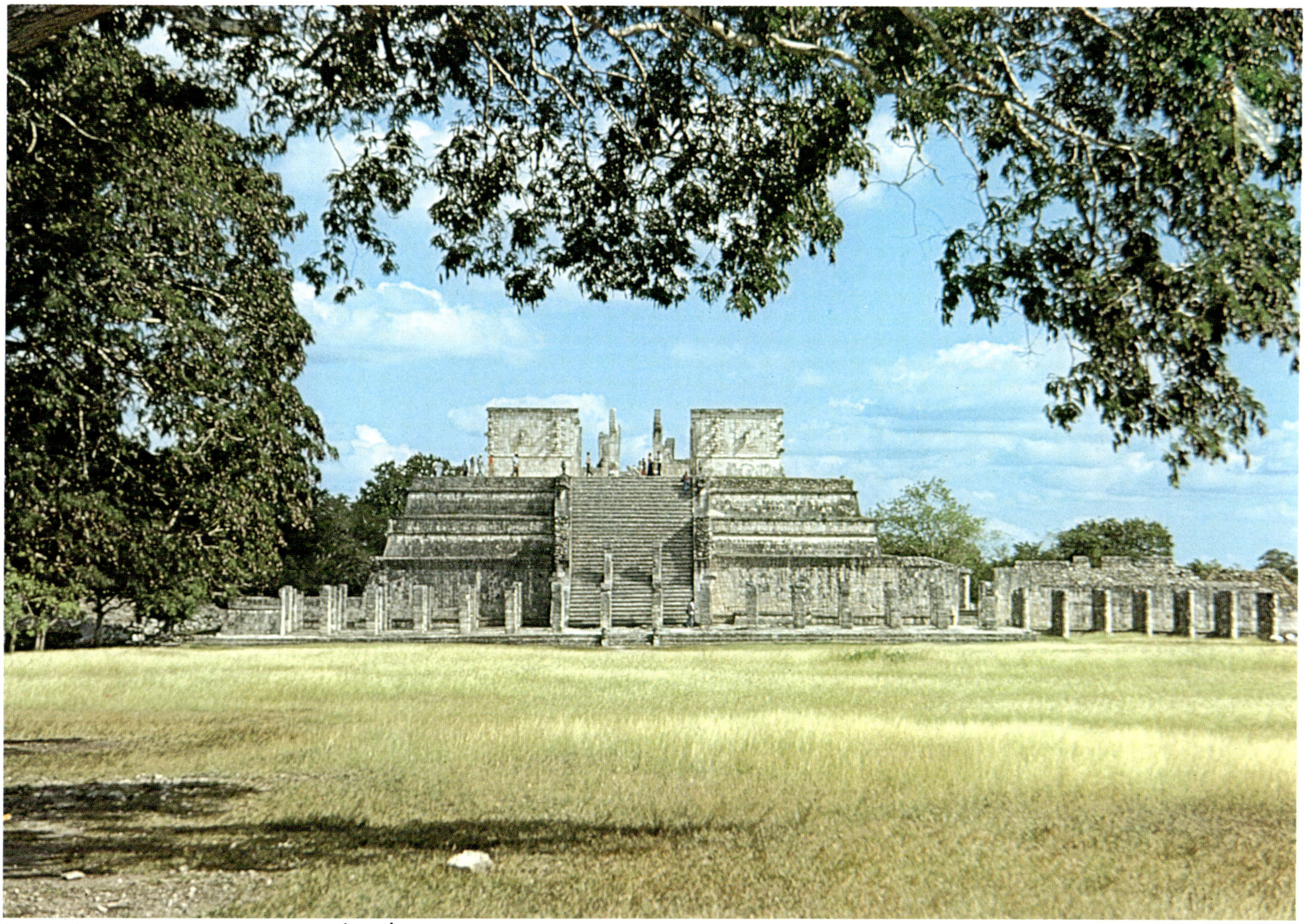

Plate 93. Temple of the Warriors, Chichén Itzá

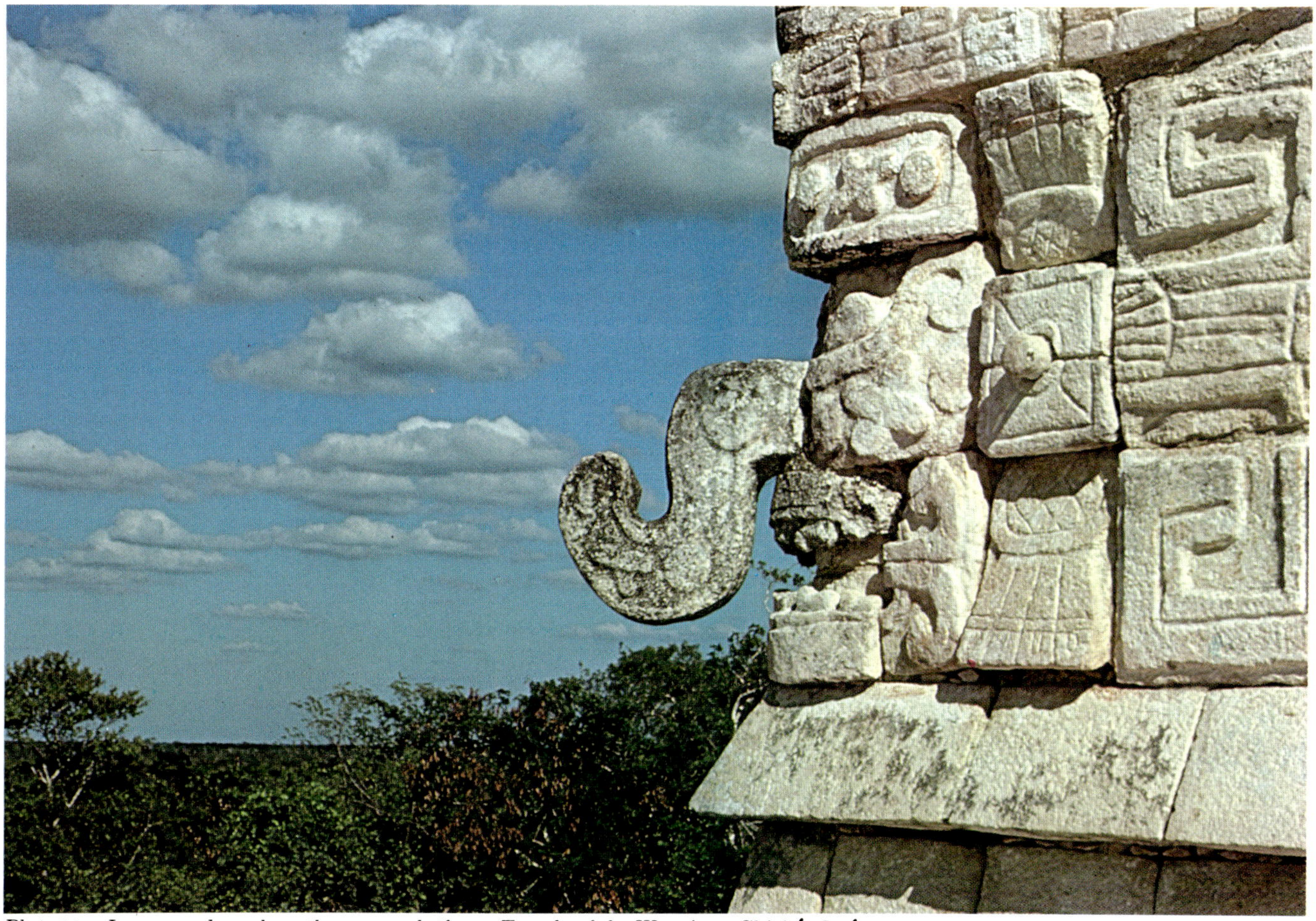

Plate 94. Long-nosed mask on the upper platform, Temple of the Warriors, Chichén Itzá

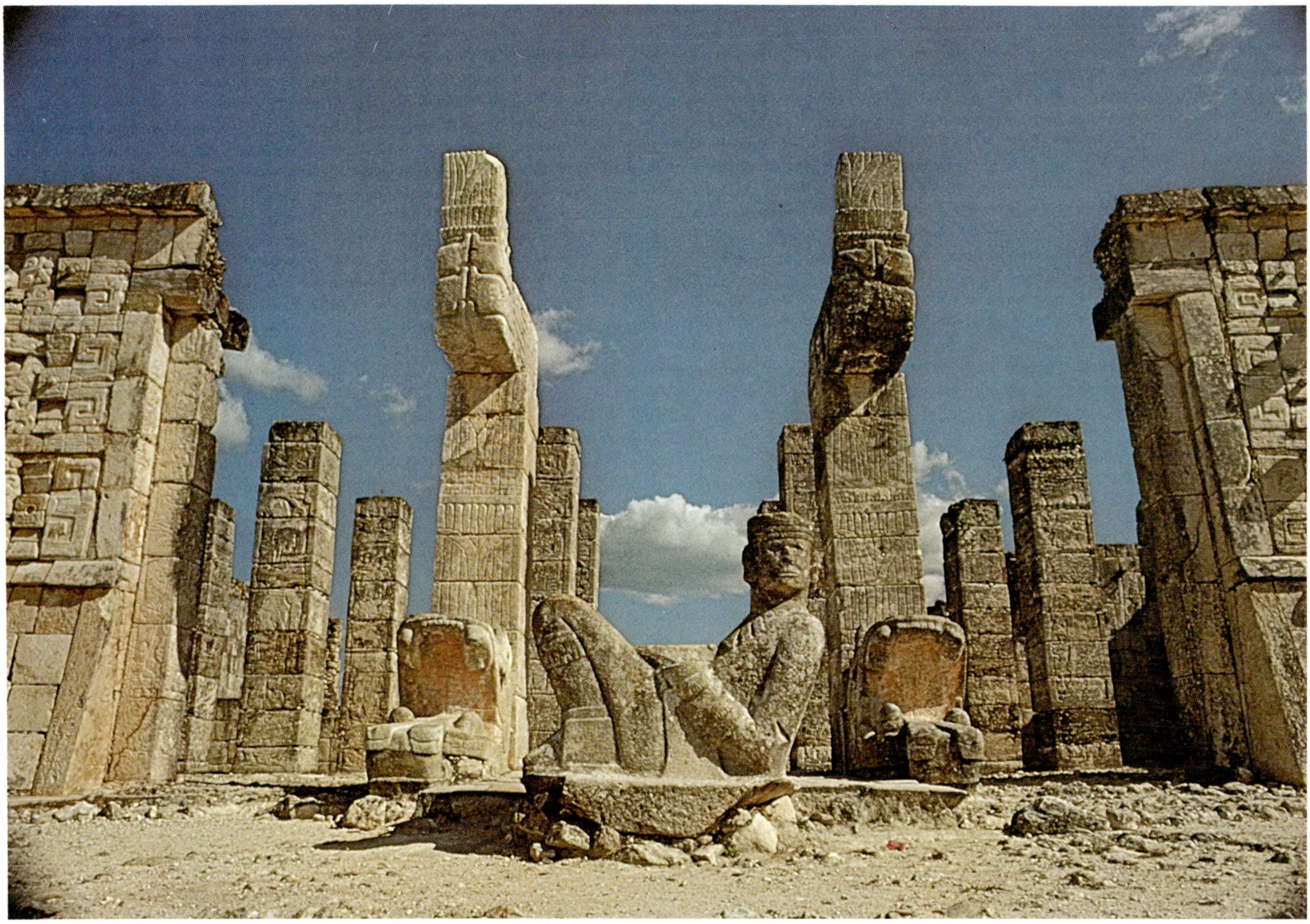

Plate 95. *Upper platform, Temple of the Warriors, Chichén Itzá*

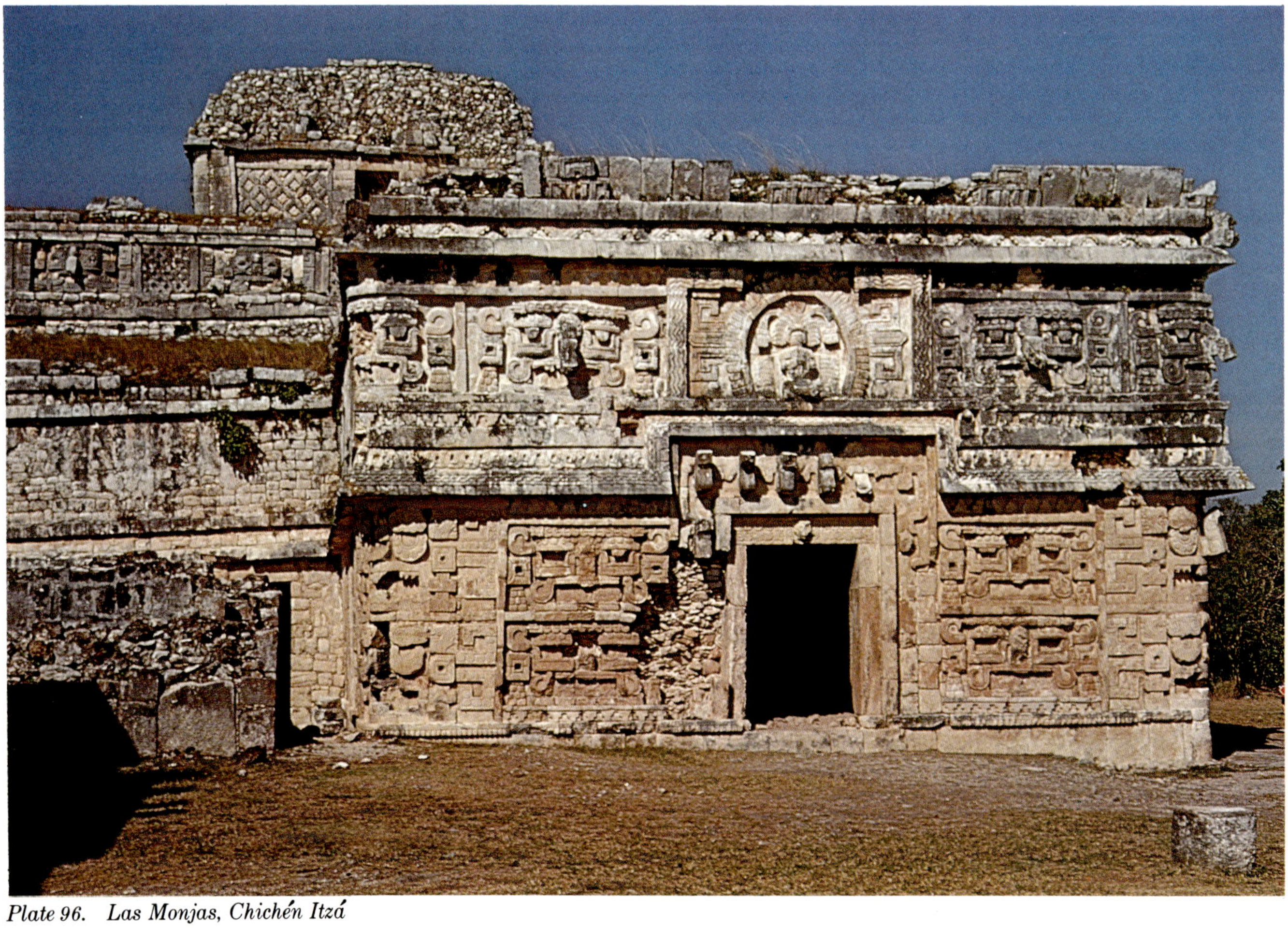

Plate 96. *Las Monjas, Chichén Itzá*

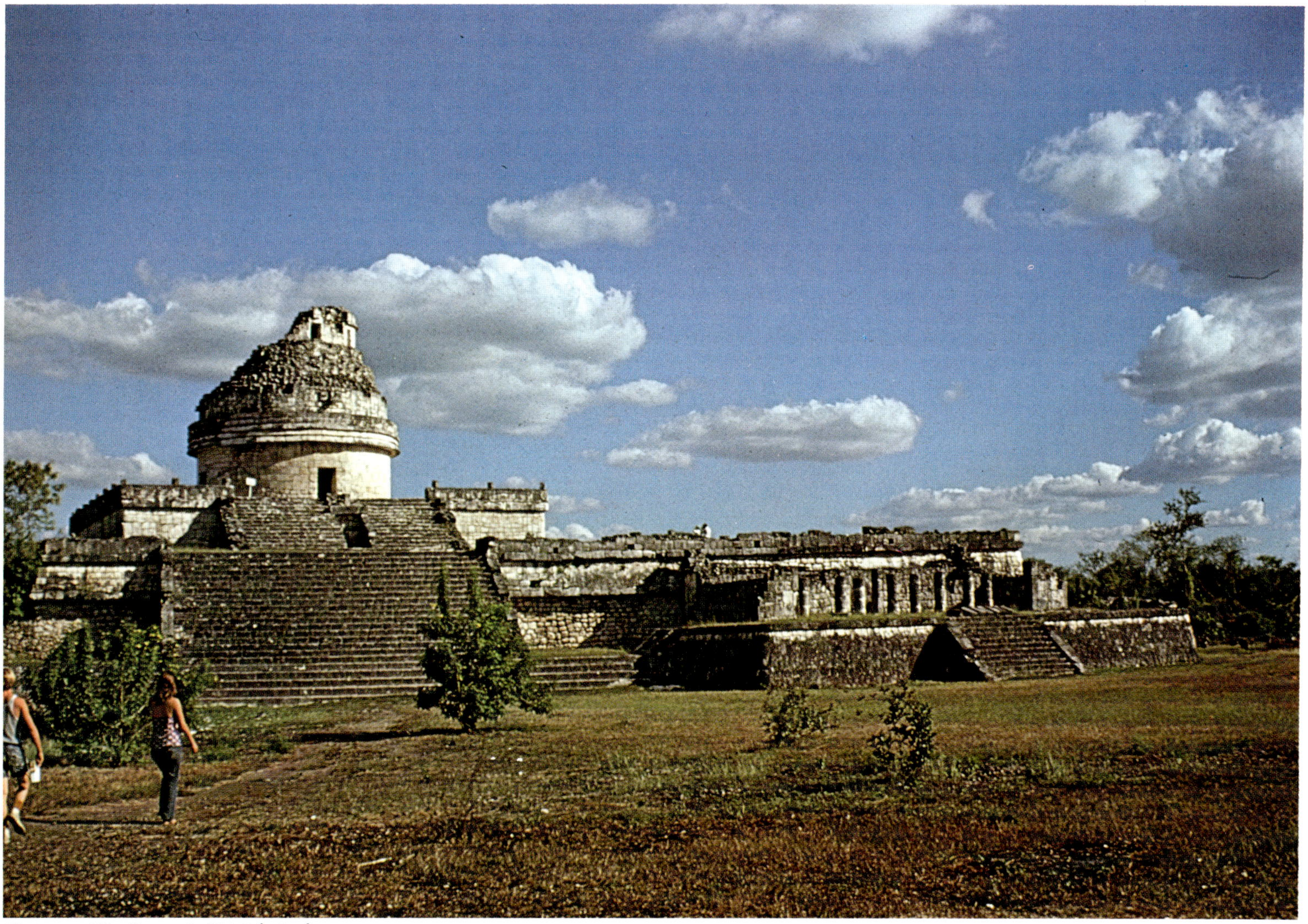

Plate 97. The Caracol, Chichén Itzá

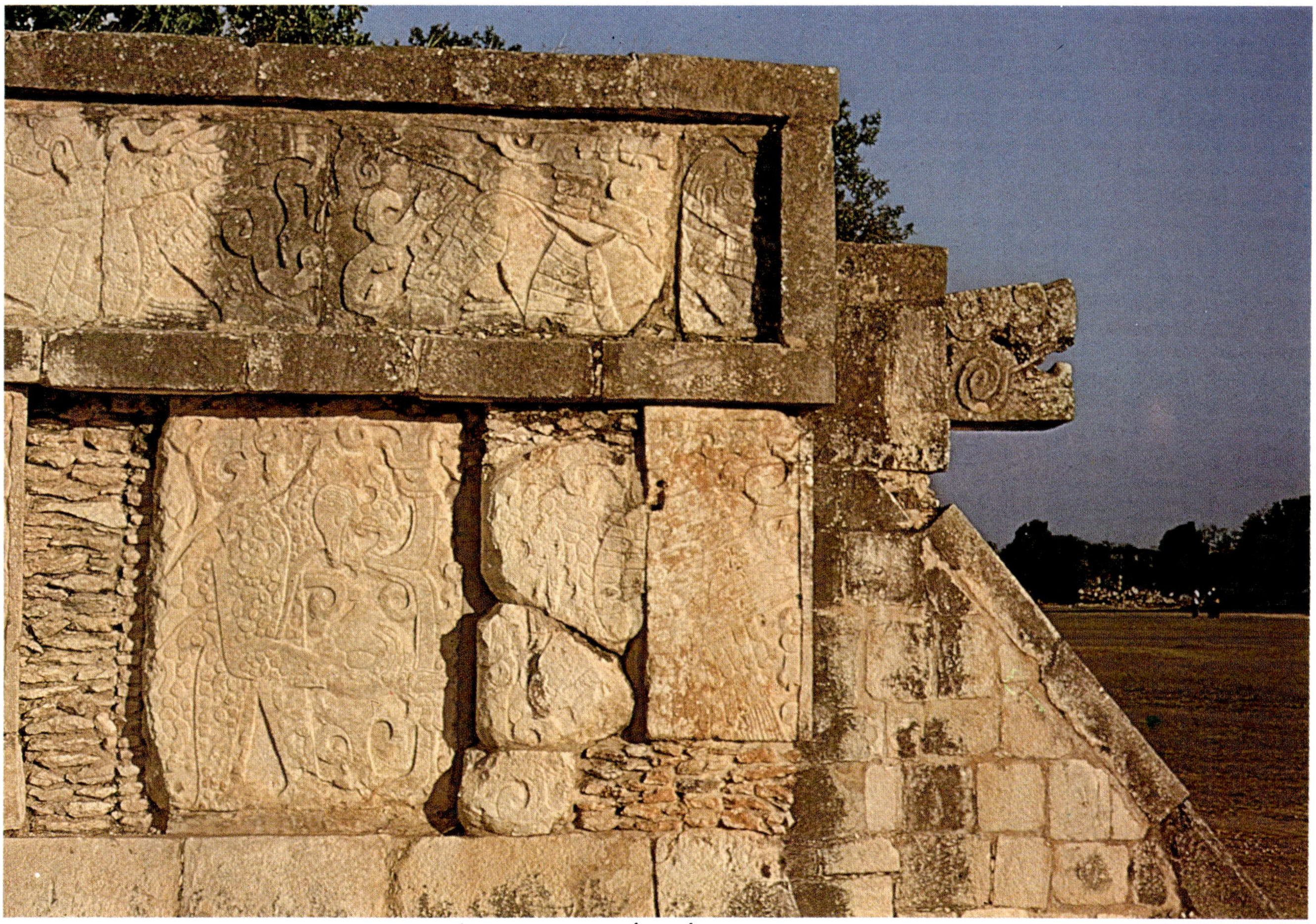

Plate 98. Fire breathing jaguar, Platform of the Jaguars, Chichén Itzá

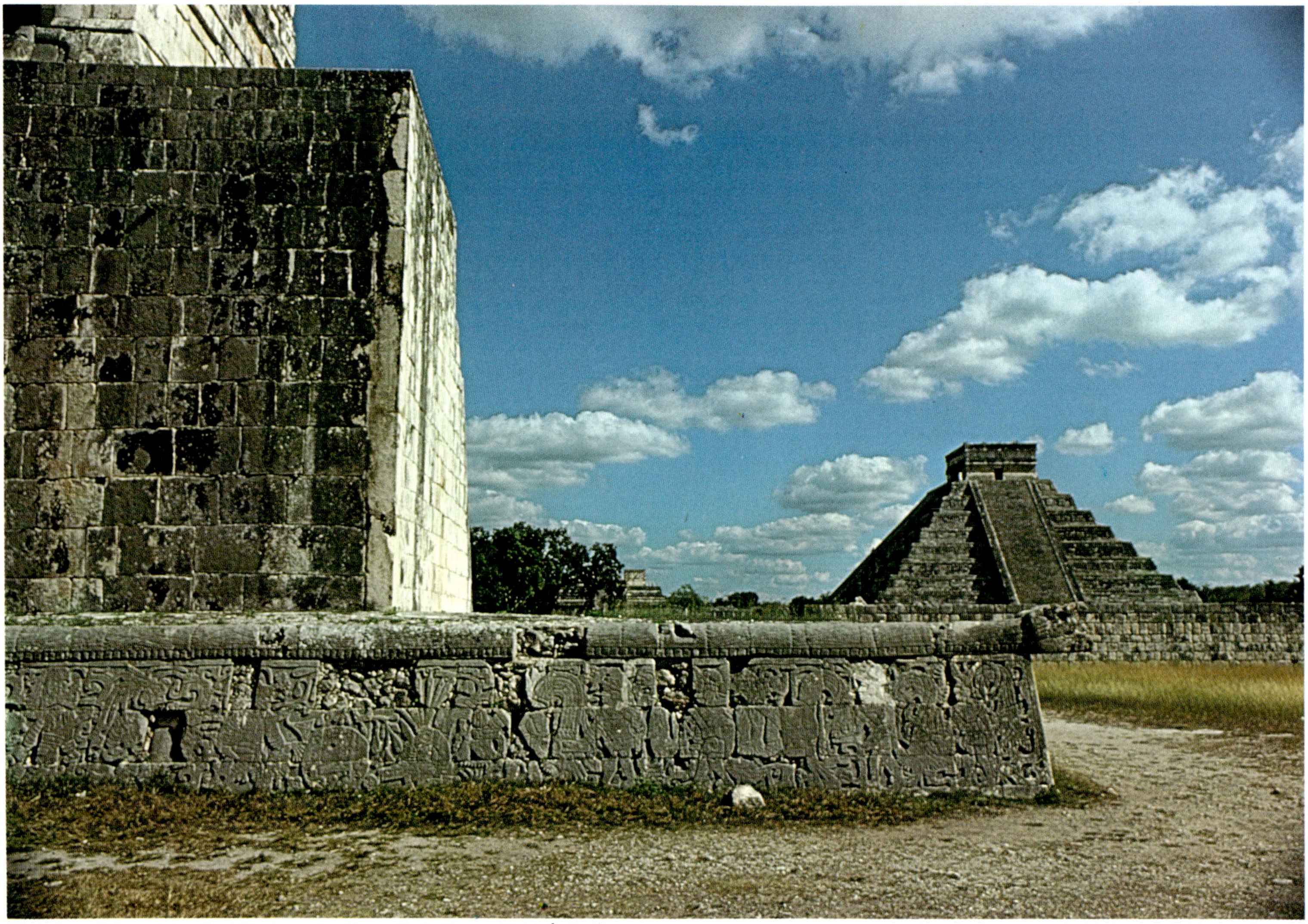
Plate 99. *View at the Castillo from the Great Ball Court, Chichén Itzá*

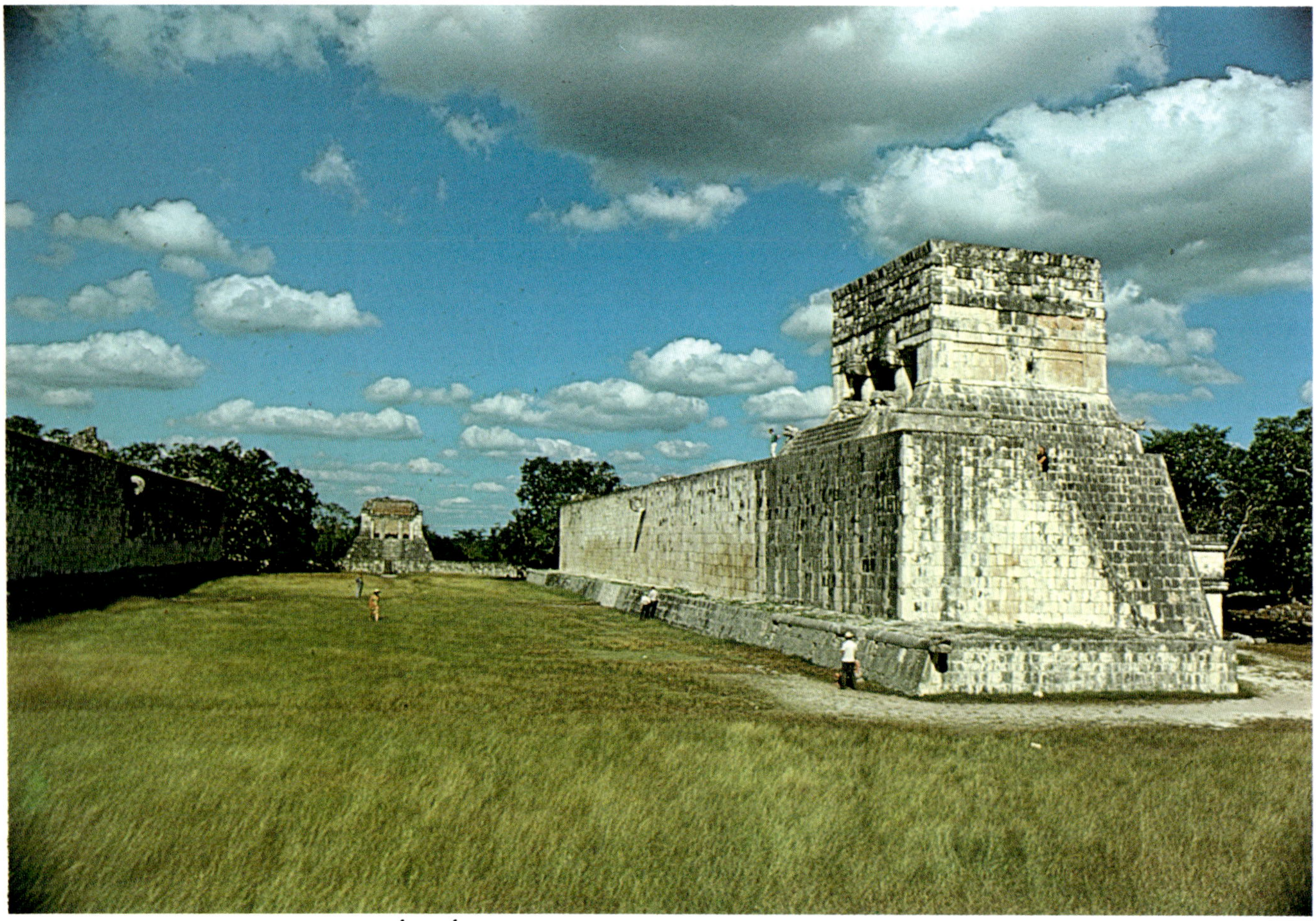
Plate 100. *The Great Ball Court, Chichén Itzá*

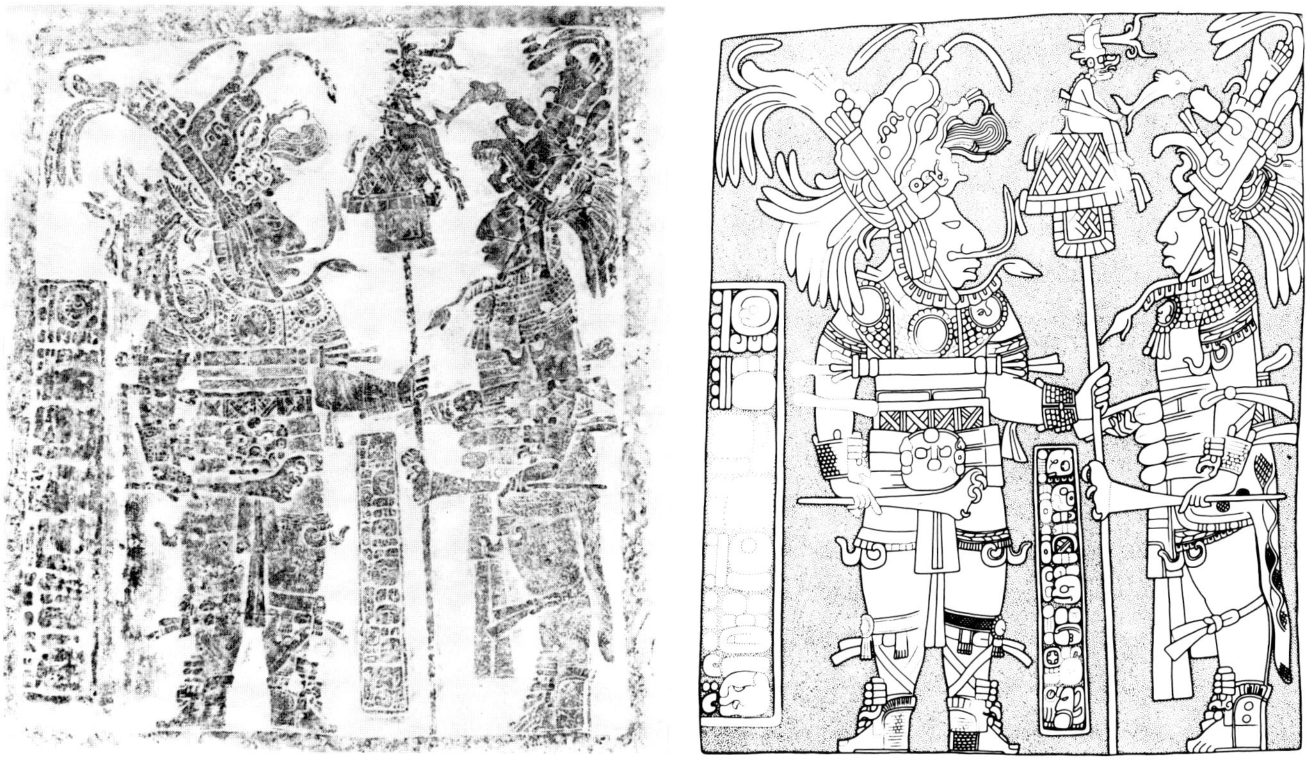

Figure 94. Lintel 6 at Yaxchilán, showing a lord with a male attendant. The former is holding a ceremonial staff topped by a mat cupola upon which God K is seated (rubbing by Merle Greene Robertson)

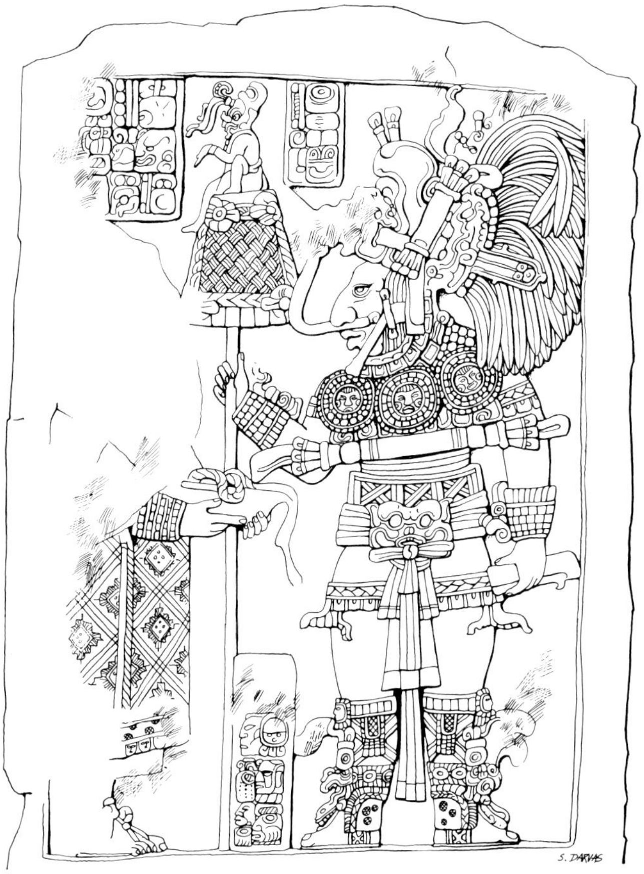

Figure 95. Lintel 43 from Yaxchilán (now in the Museo Nacional de Arqueologia y Historia, Mexico City) presents a composition very similar to that of Lintel 6 (Fig. 94). We see probably the same lord with the staff, the mat cupola, and with God K—this time, however, attended by a woman holding a dish

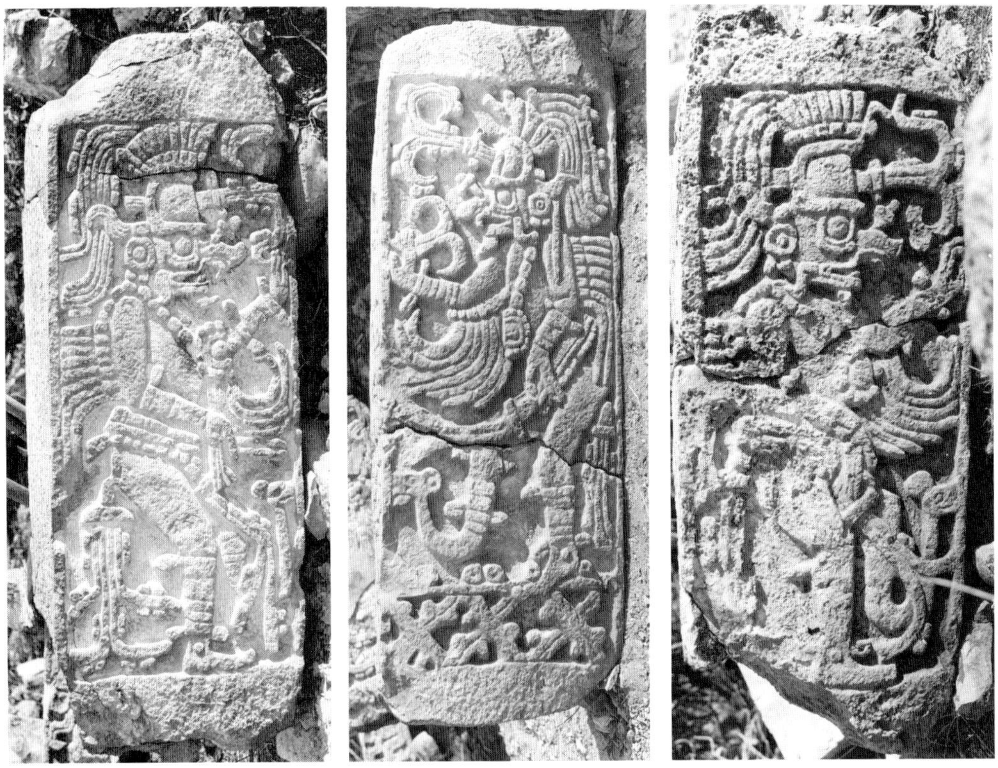

Figure 96. *God K on the lintels of Structure 4 B1 at Sayíl (after Proskouriakoff; photographs courtesy Peabody Museum, Harvard University)*

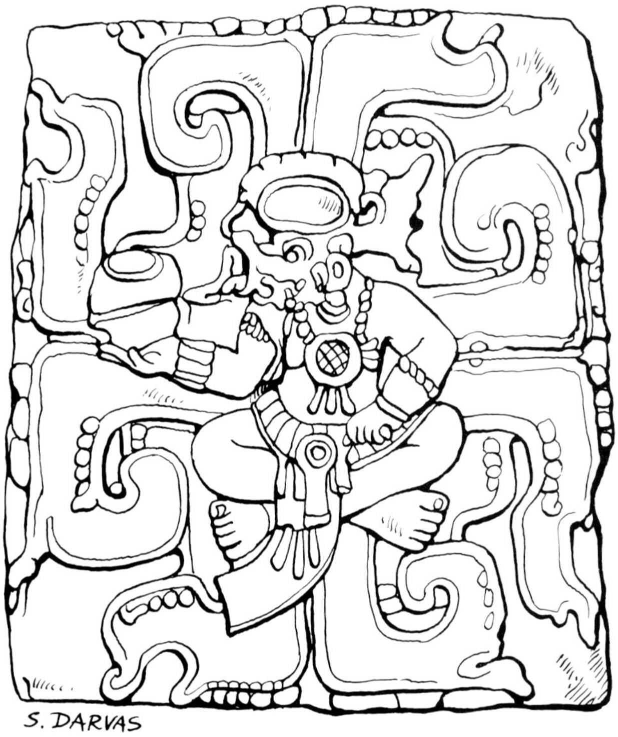

Figure 97. *Cross-legged figure believed to be God K on a flat stone slab sunk into the floor of the Western Court at Copán (drawn after Maudslay)*

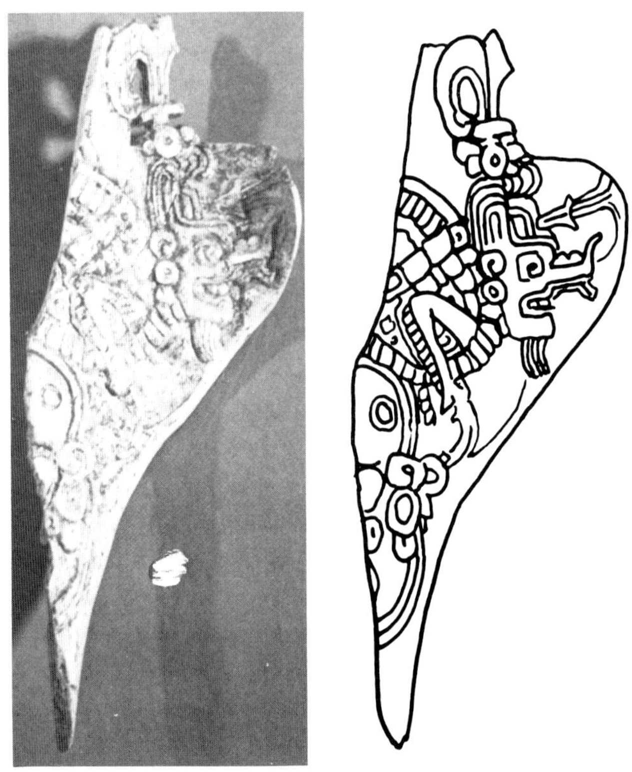

Figure 98. *God K on a shell carving found at Copán (Peabody Museum, Harvard University; drawing after Spinden)*

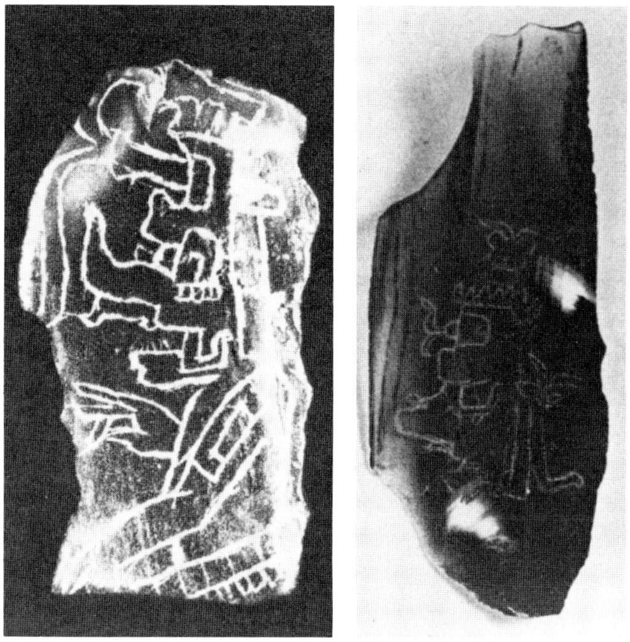

Figure 99. Obsidian fragments carved with the images of God K (Museo Nacional de Arqueología y Etnología de Guatemala)

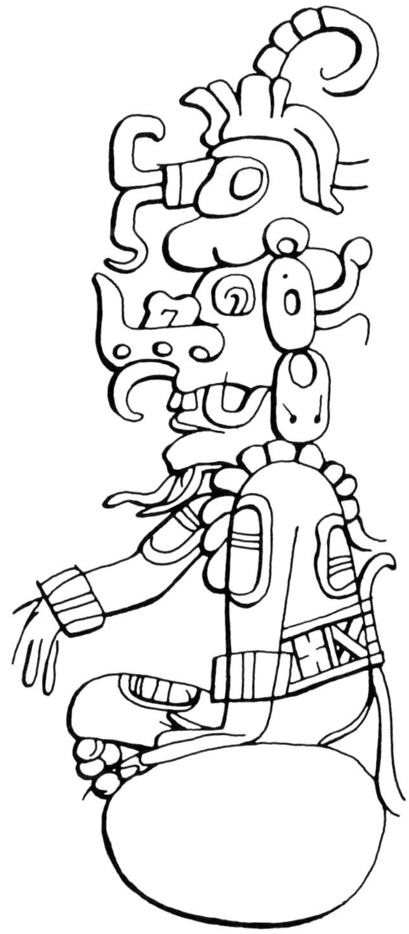

Figure 100. God K on an obsidian fragment found at Tikal (after William R. Coe)

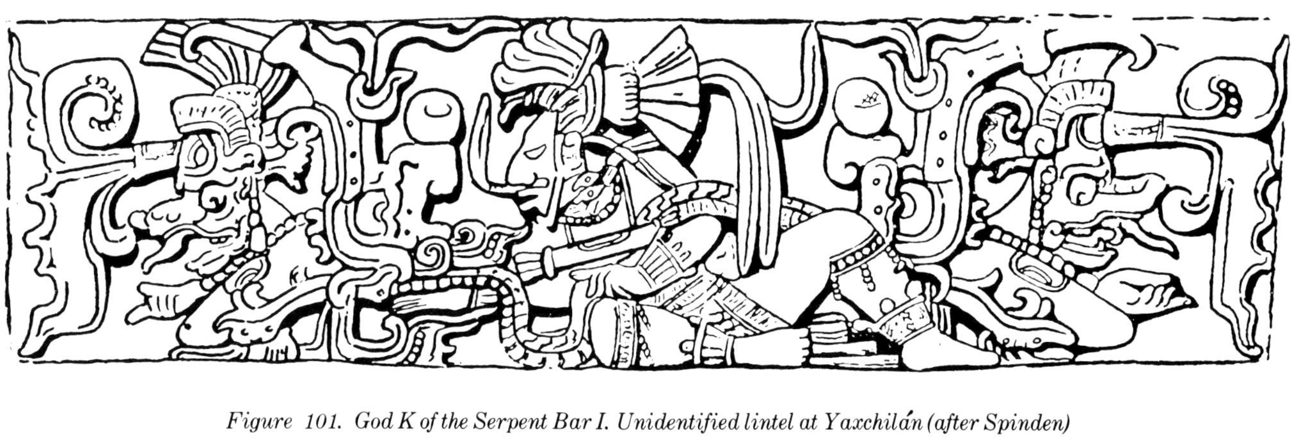

Figure 101. God K of the Serpent Bar I. Unidentified lintel at Yaxchilán (after Spinden)

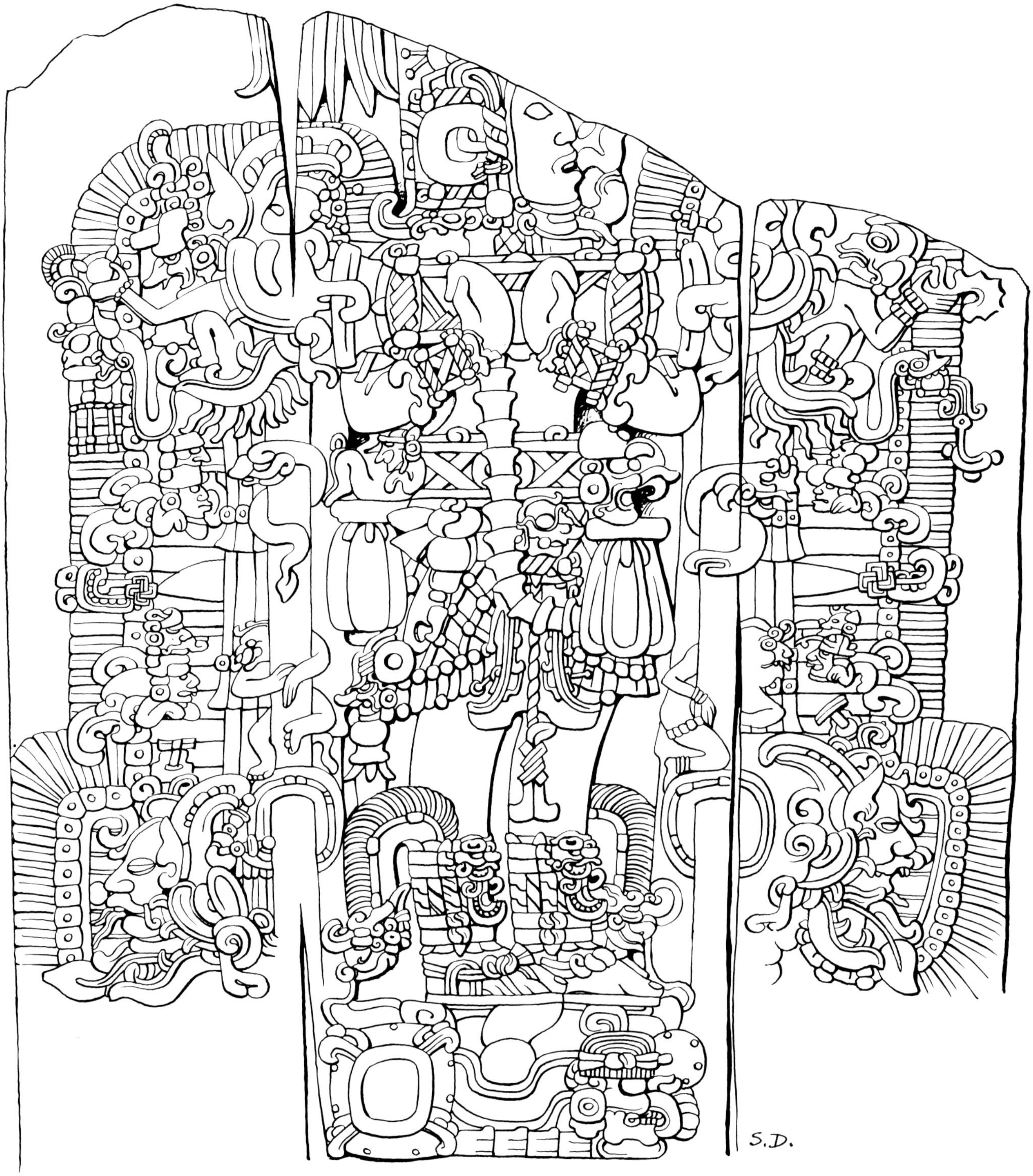

Figure 102. God K of the Serpent Bar II. Stela 1 at Tikal (drawn after William R. Coe)

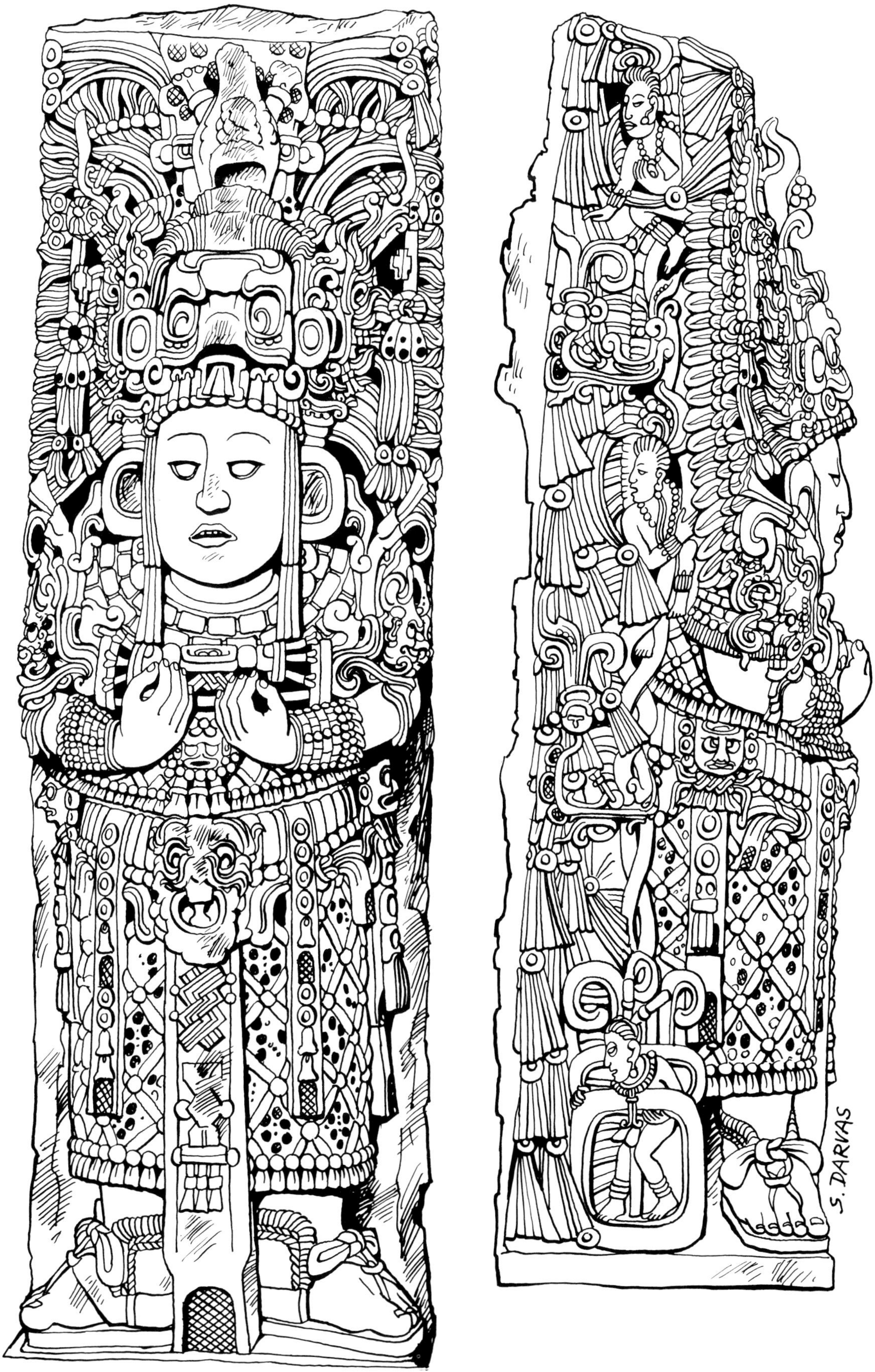

Figure 103. God K of the Serpent Bar III. Stela H at Copán (drawn after Maudslay)

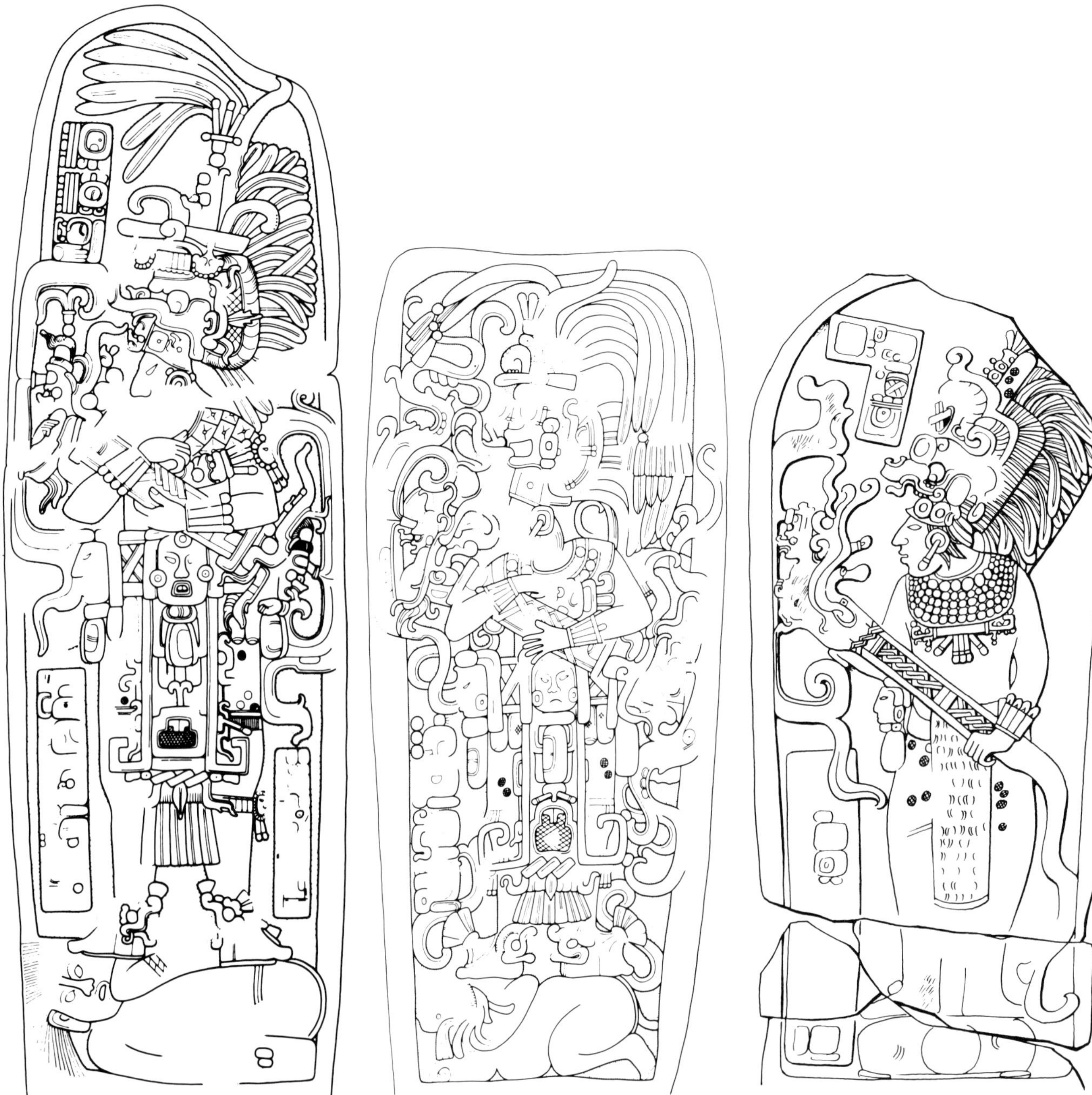

(a) (b) (c)

Figure 104. God K of the Serpent Bar IV. (a) *Stela 14 at Naranjo (after Graham and von Euw),* (b) *Stela 12 at Naranjo (after Graham and von Euw),* and (c) *Stela 12 at Machaquilá (drawn after Graham)*

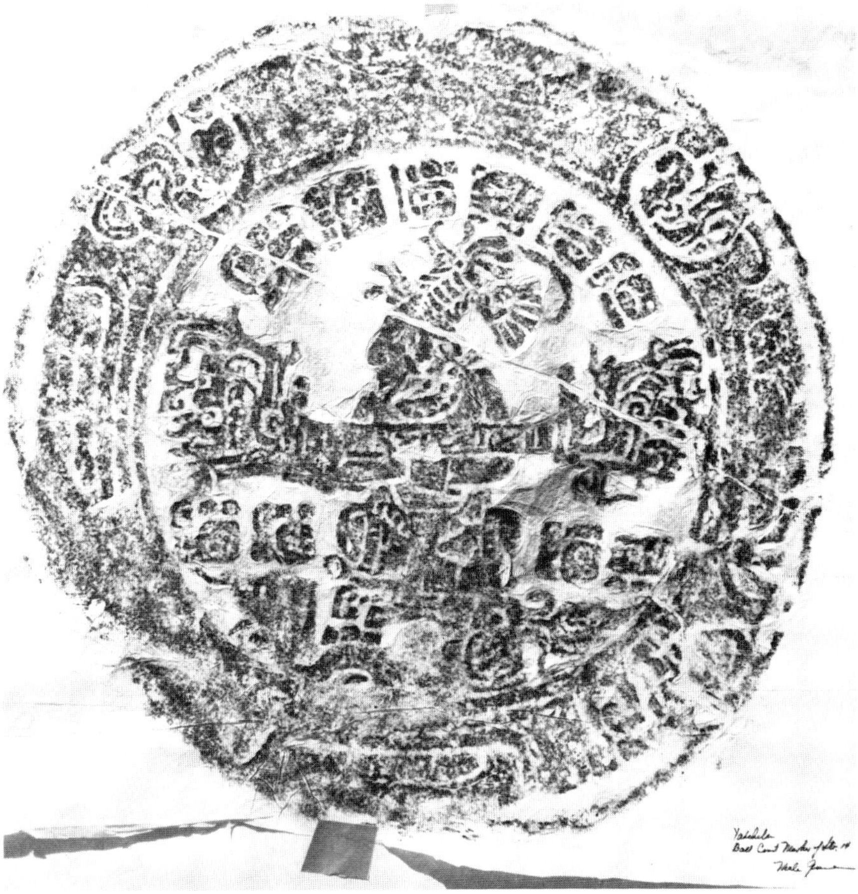

Figure 105. God K of the Serpent Bar V. Ball Court Marker at Yaxchilán (rubbing by Merle Greene Robertson)

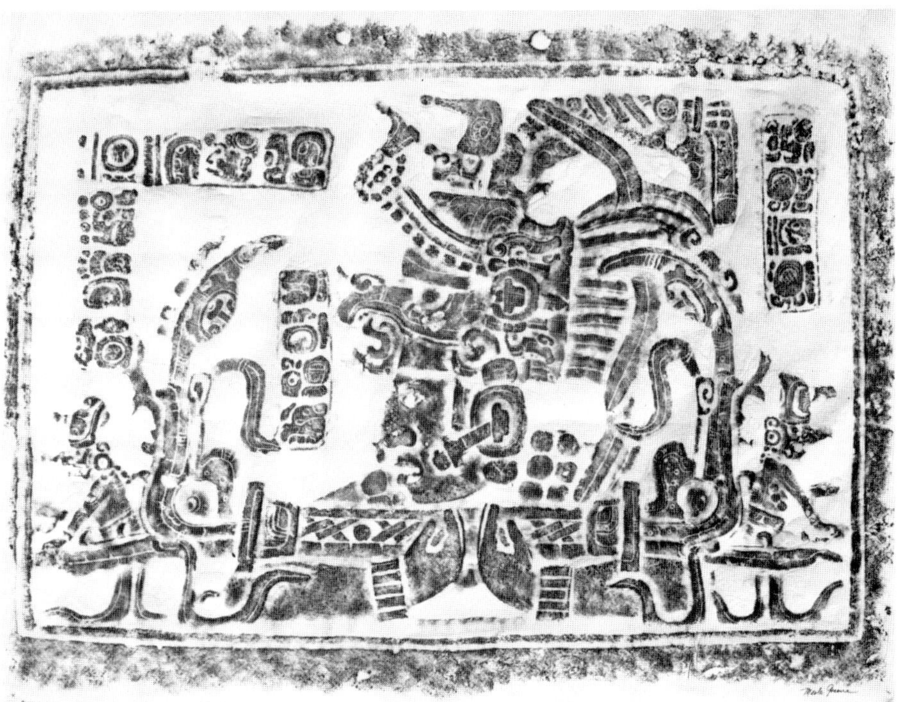

Figure 106. God K of the Serpent Bar VI. Lintel of Structure 6 at Bonampak (rubbing by Merle Greene Robertson)

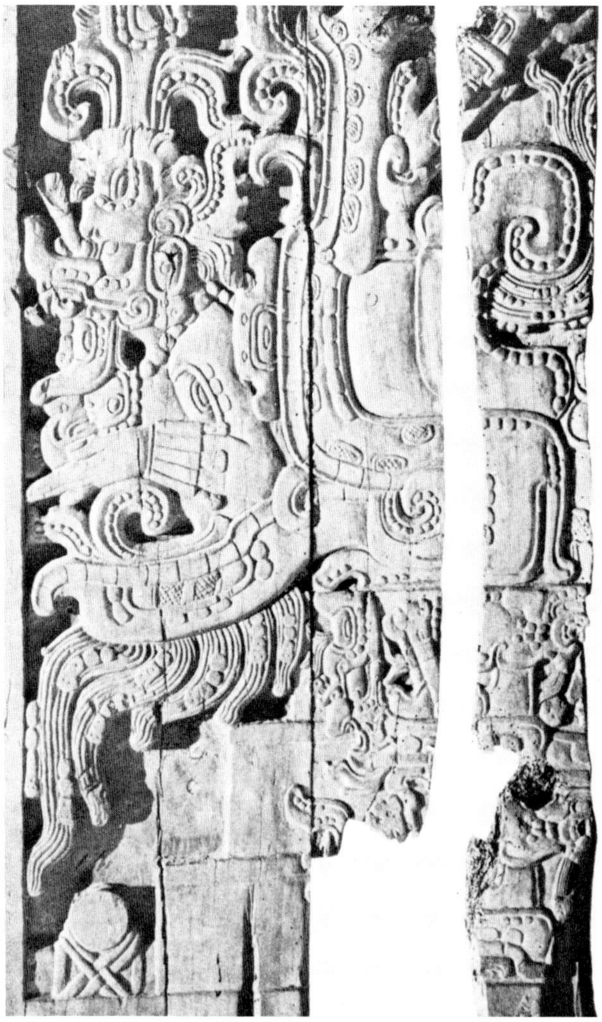

Figure 107. God K of the Serpent Bar VII. Detail of Lintel 3, Temple IV at Tikal (after Maudslay)

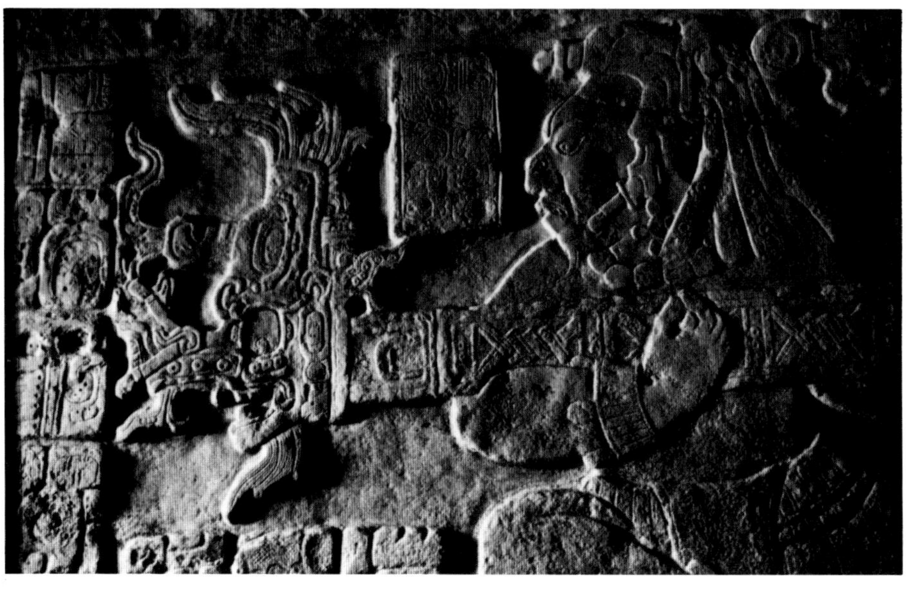

Figure 108. God K of the Serpent Bar VIII. Detail of lintel from Kuna Lacañha (Dumbarton Oaks Collection)

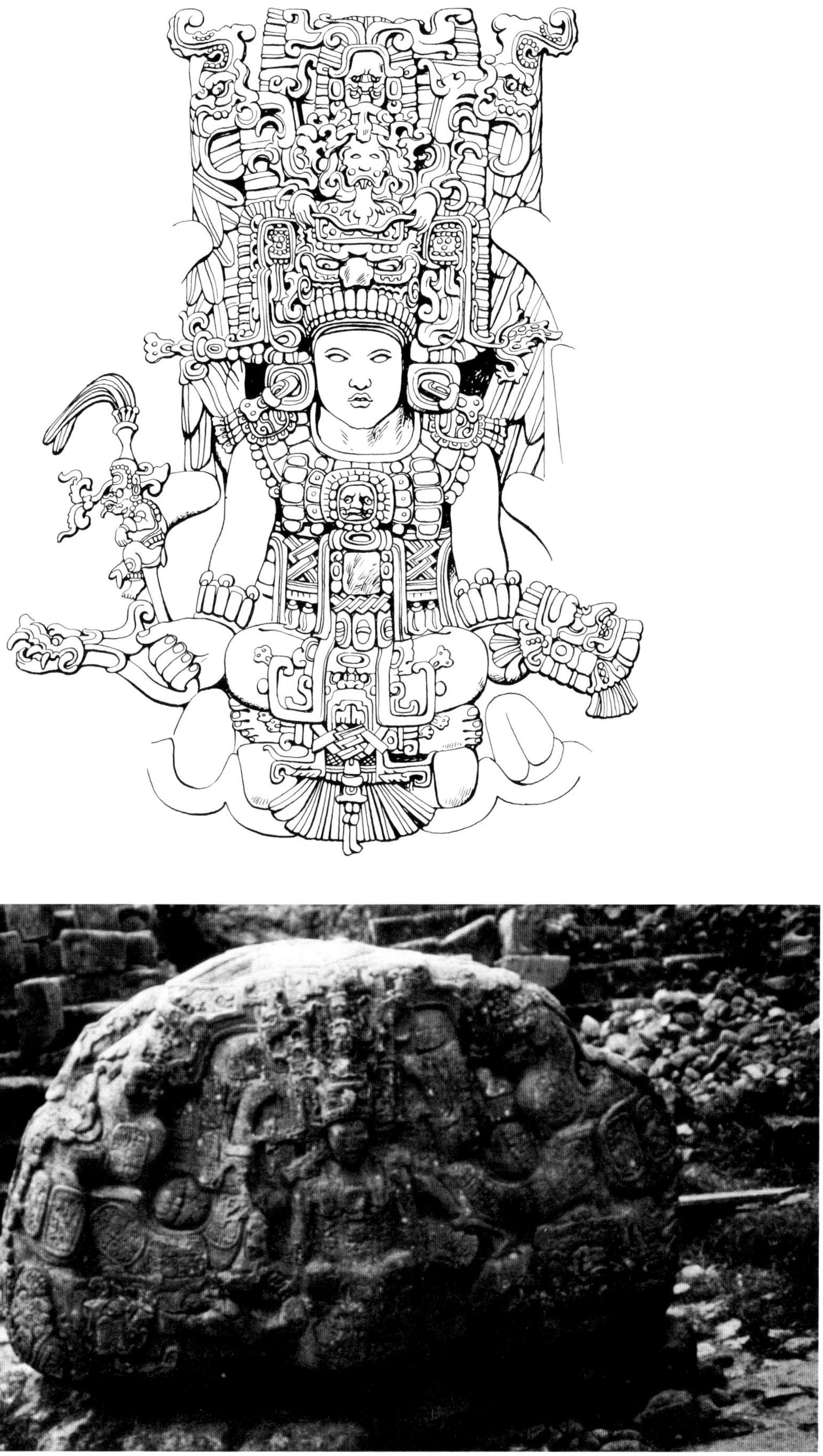

Figure 109. God K of the Manikin Scepter I. Zoomorph P and detail at Quiriguá

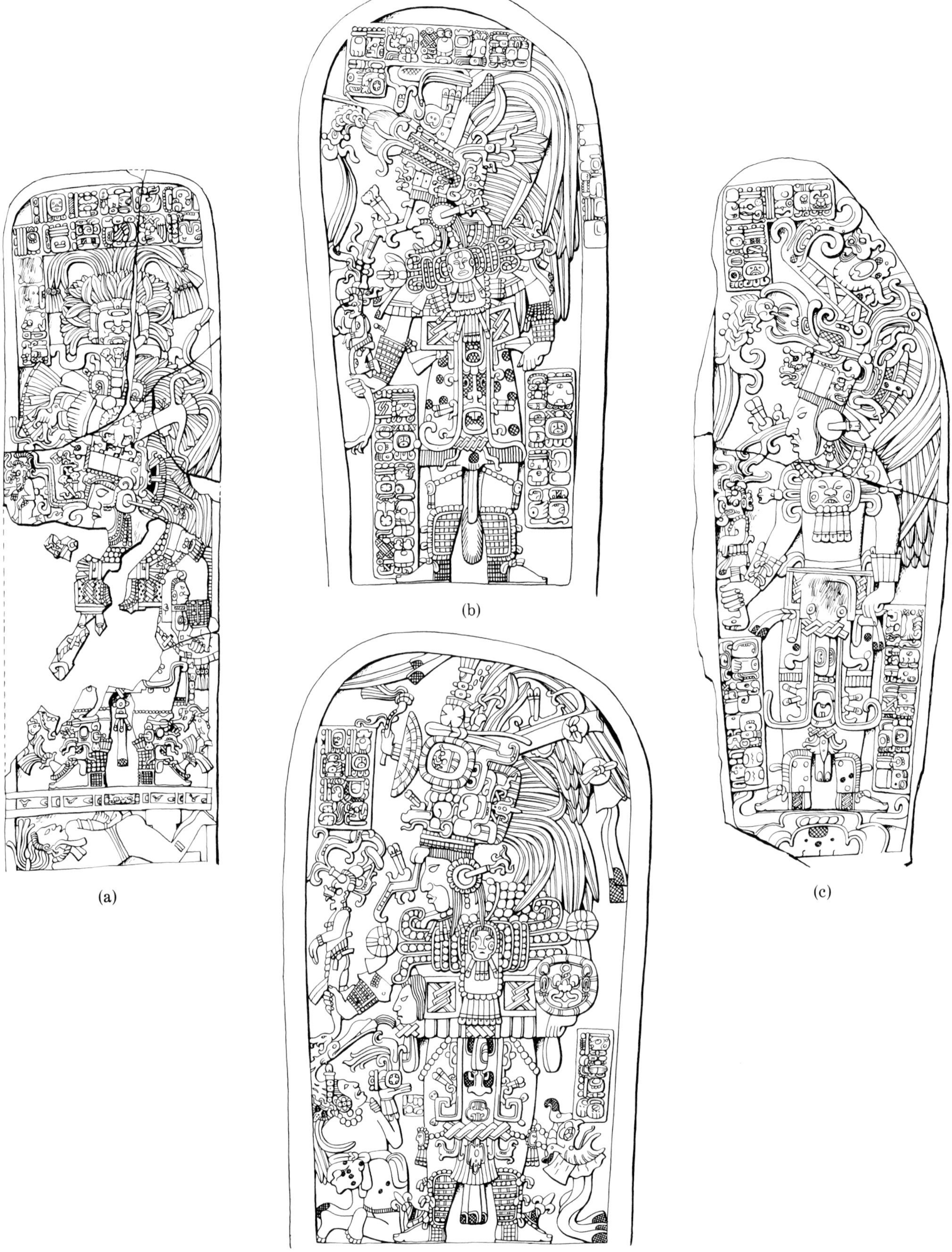

Figure 110. *God K of the Manikin Scepter II.* (a) *Stela 7 at Aguateca,* (b) *Stela 3 at Machaquilá,* (c) *Stela 2 at Machaquilá, and* (d) *Stela 7 at Machaquilá (drawn after Graham)*

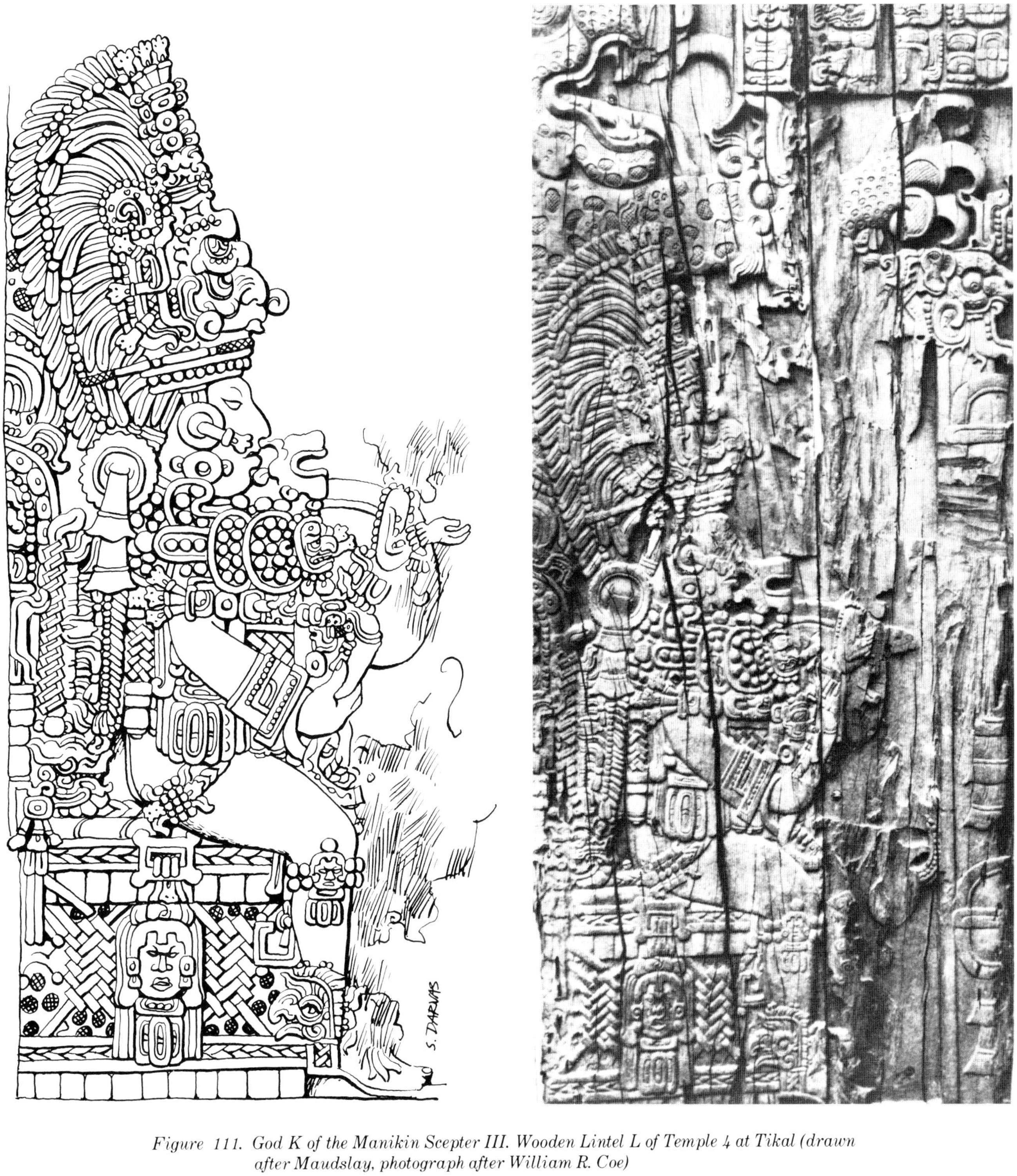

Figure 111. God K of the Manikin Scepter III. Wooden Lintel L of Temple 4 at Tikal (drawn after Maudslay, photograph after William R. Coe)

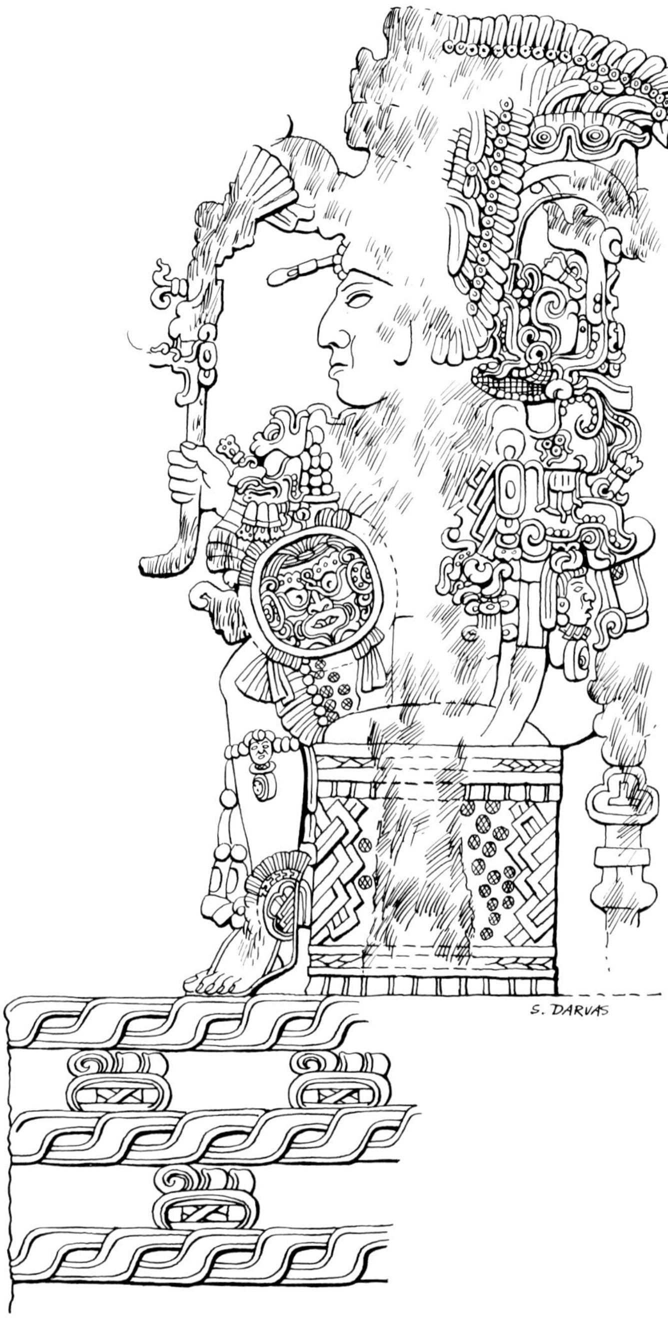

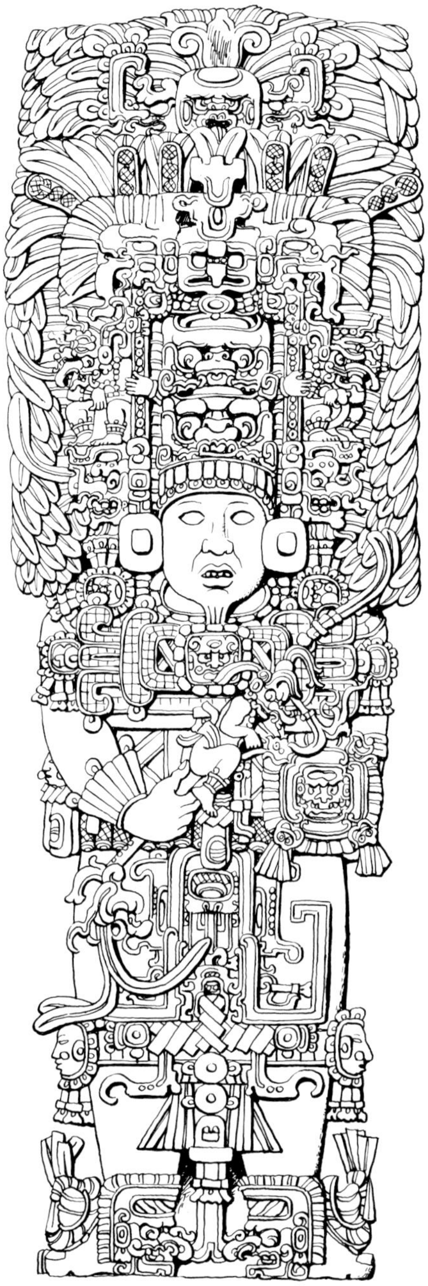

Figure 112. God K of the Manikin Scepter IV. Wooden lintel in Temple IV at Tikal (drawn after Maudslay)

Figure 113. God K of the Manikin Scepter V. Stela F at Quiriguá (drawn after Maudslay)

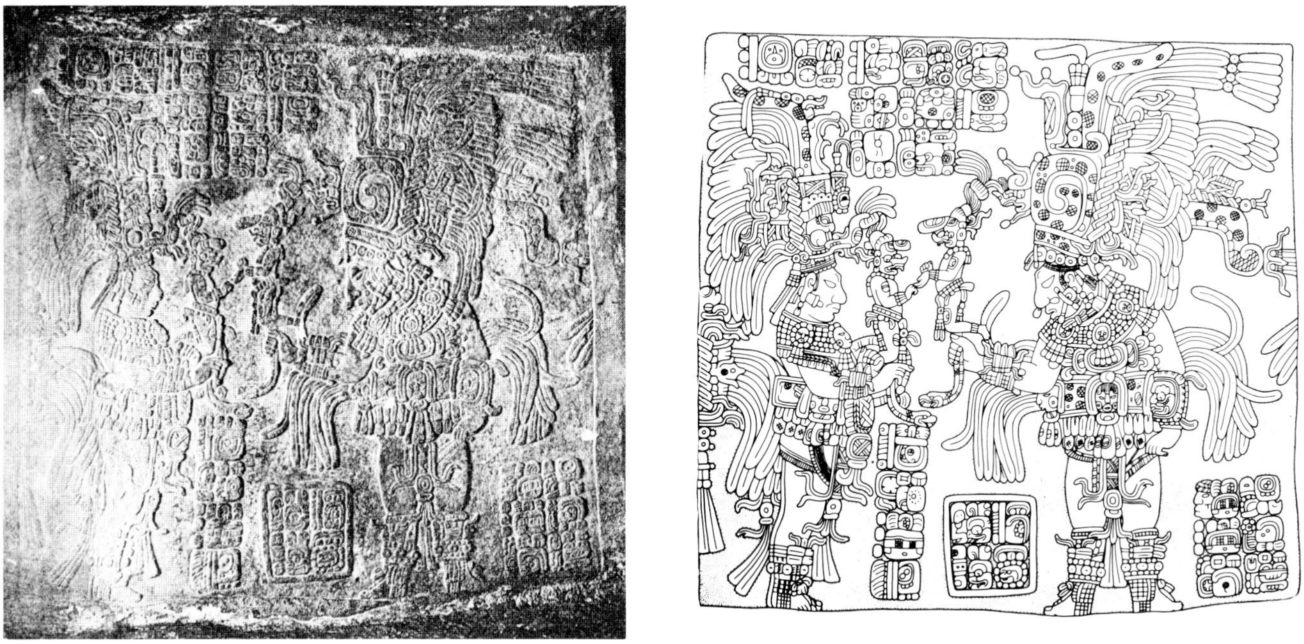

Figure 114. God K of the Manikin Scepter VI. Lintel 3 at Yaxchilán (after Graham and von Euw)

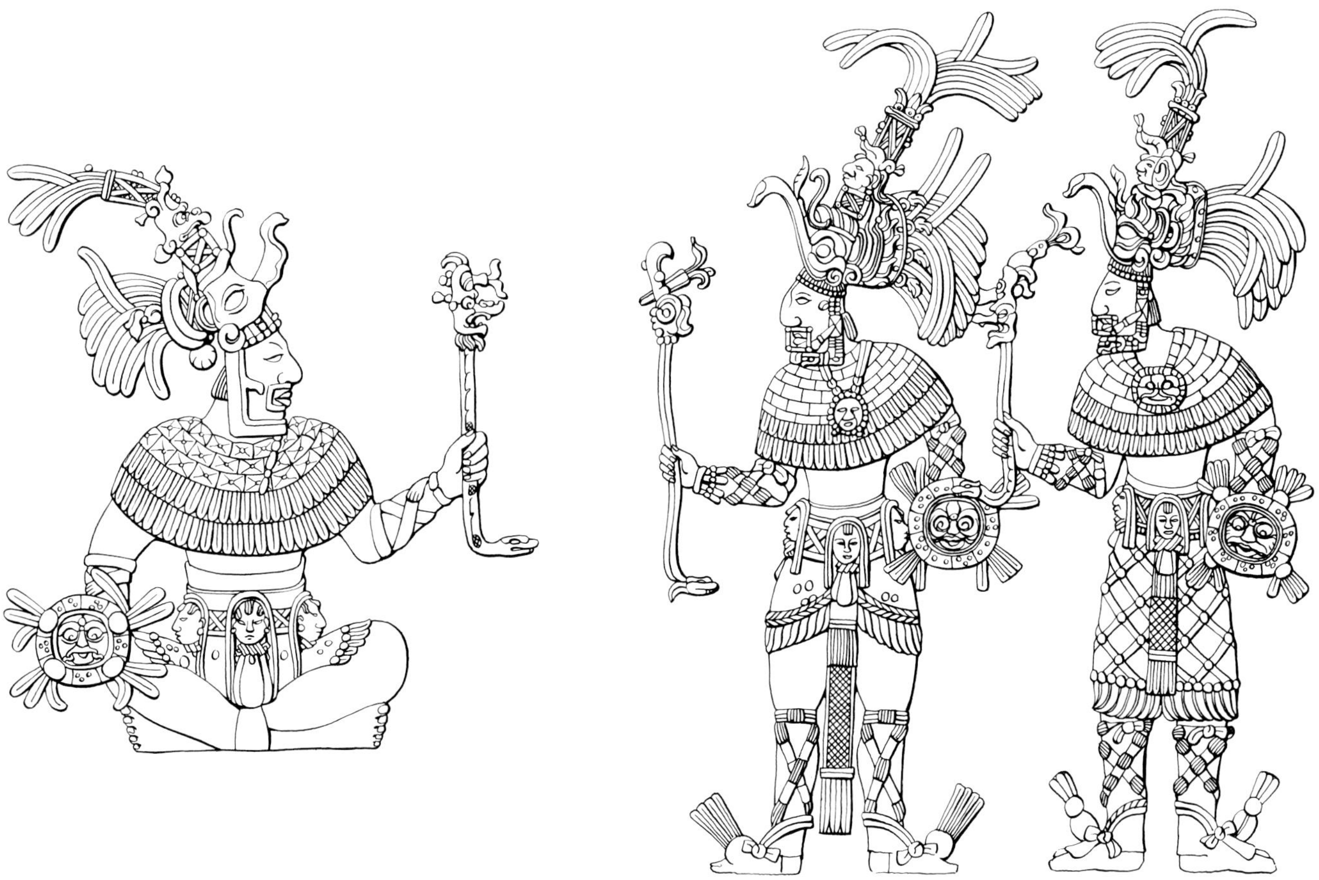

Figure 115. God K of the Manikin Staff I. Sepulchral Chamber, Temple of Inscriptions, Palenque (drawn after Rhuz-Lhuillier)

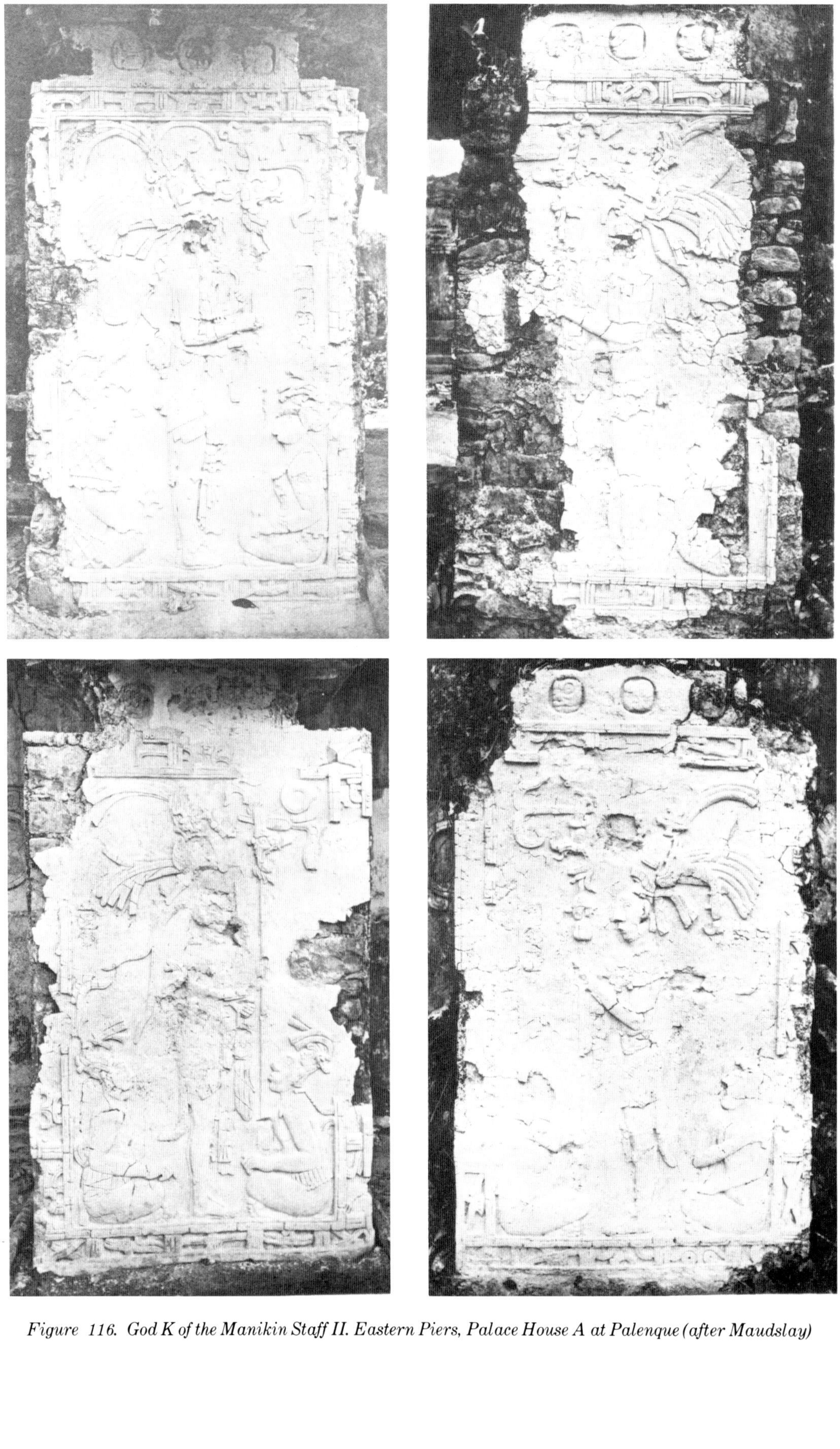

Figure 116. God K of the Manikin Staff II. Eastern Piers, Palace House A at Palenque (after Maudslay)

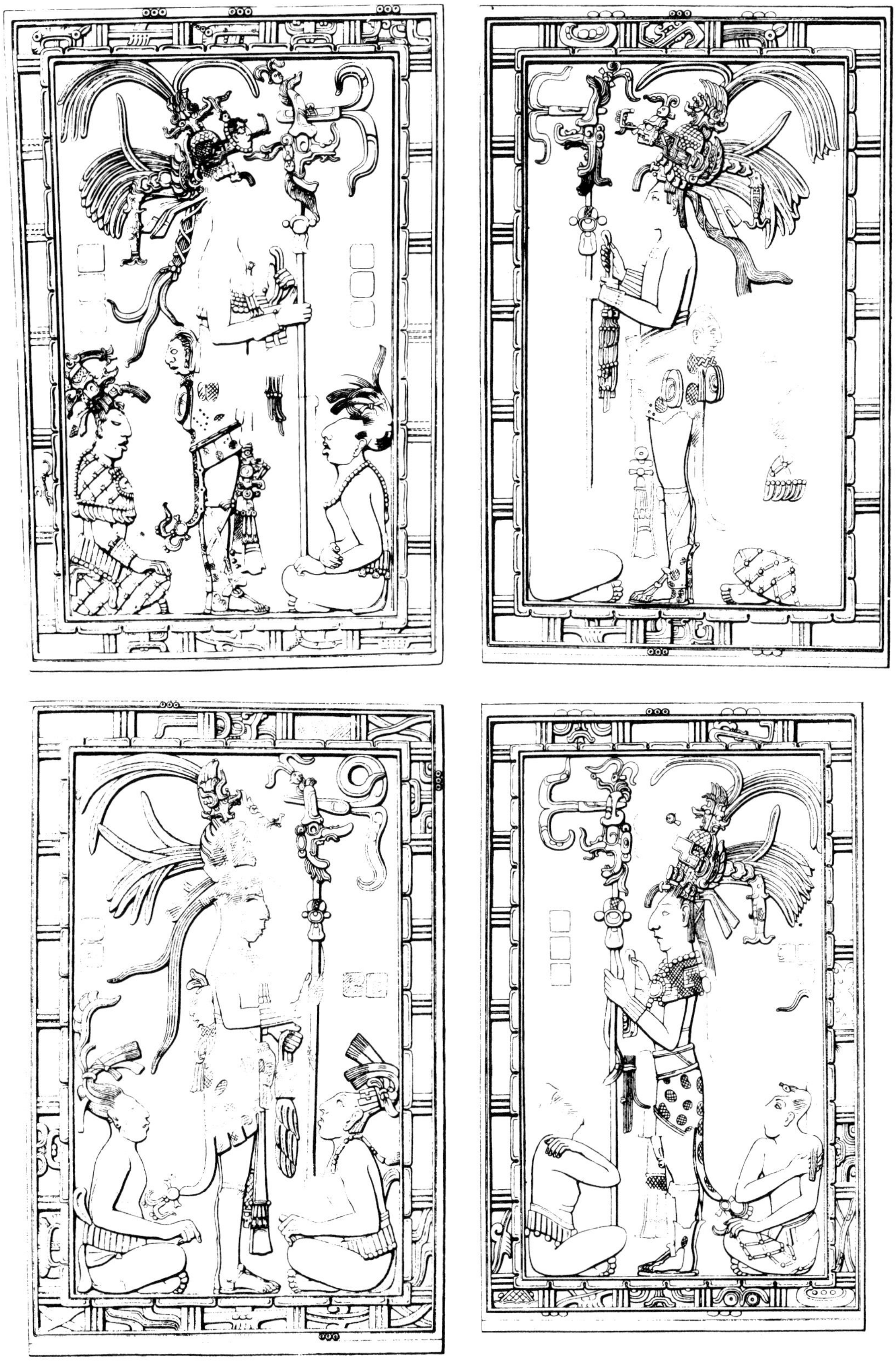

Figure 116. God K of the Manikin Staff II. Eastern Piers, Palace House A at Palenque (after Maudslay)

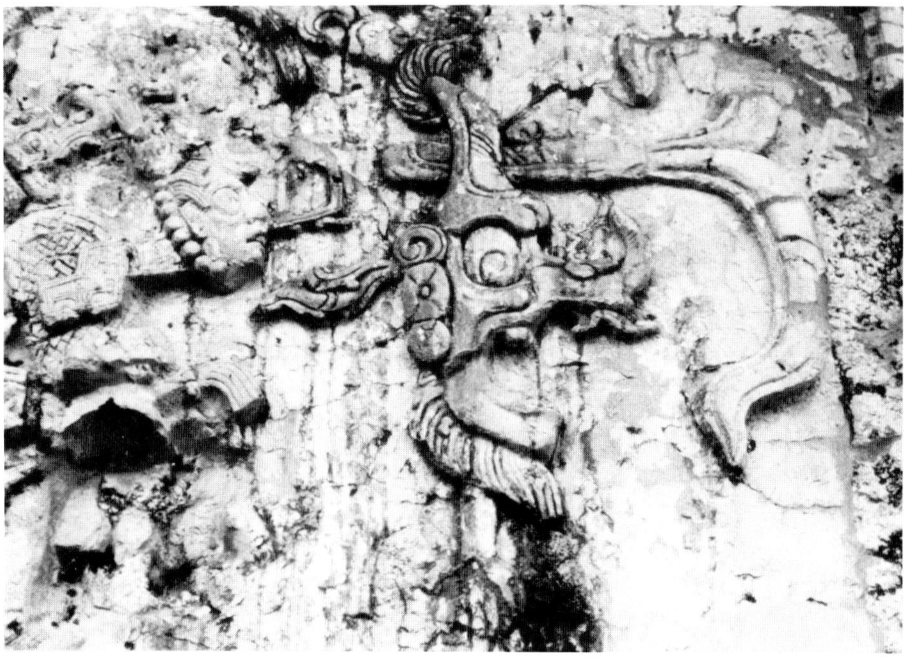

Figure 117. God K of the Manikin Staff III. Detail on Eastern Pier One, Palace House A at Palenque

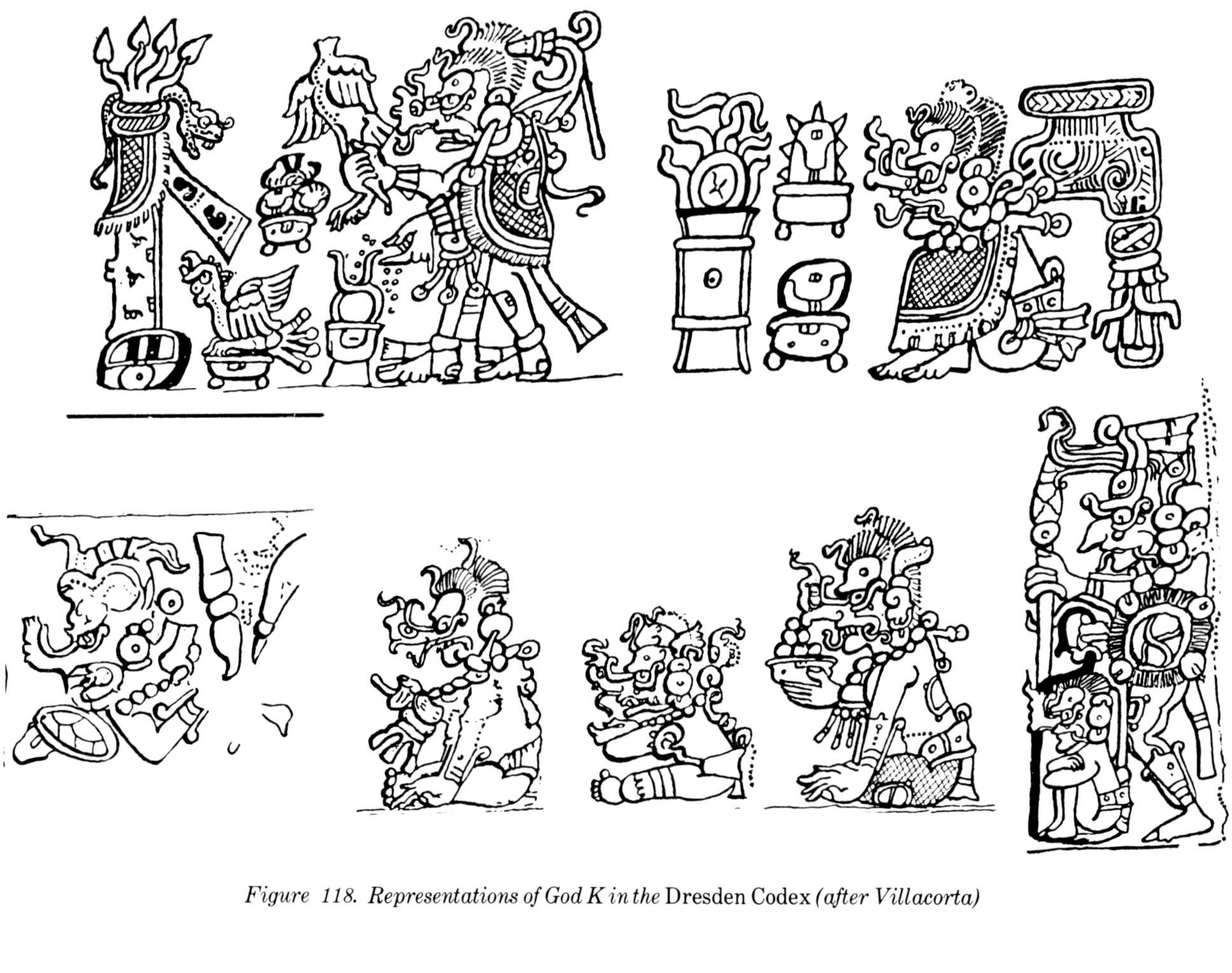

Figure 118. Representations of God K in the Dresden Codex *(after Villacorta)*

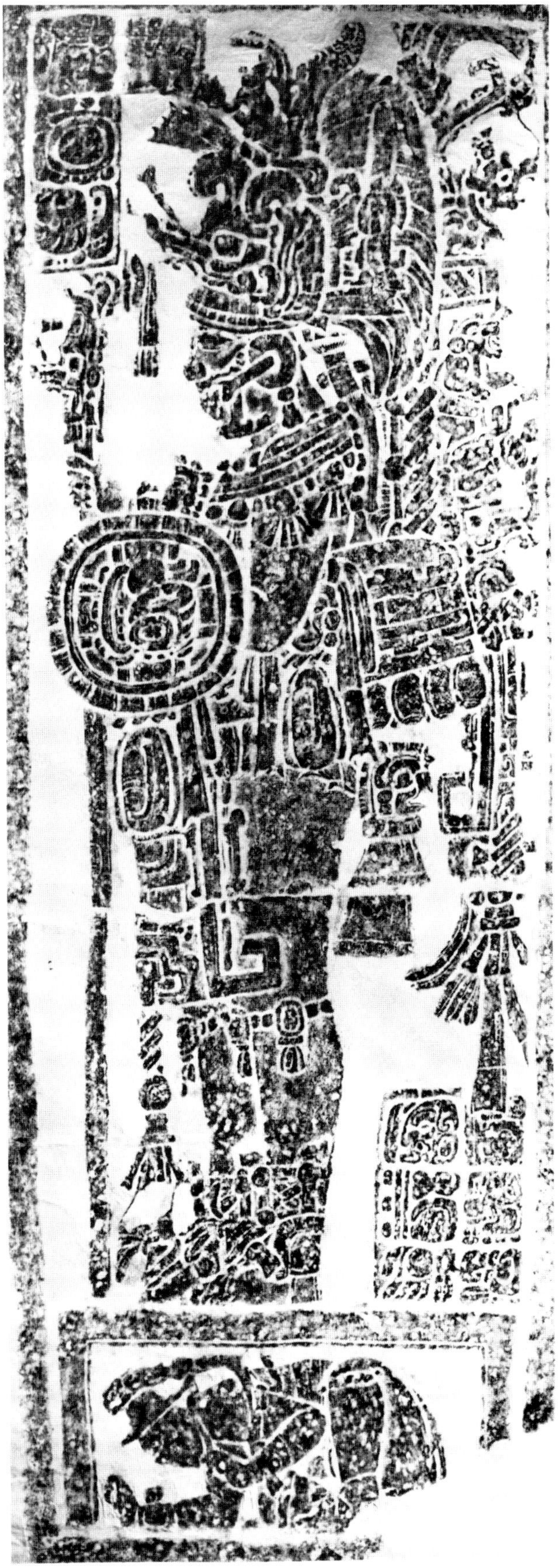

Figure 119. God K of the Manikin Staff IV. Stela 17 at Dos Pilos (rubbing by Merle Greene Robertson)

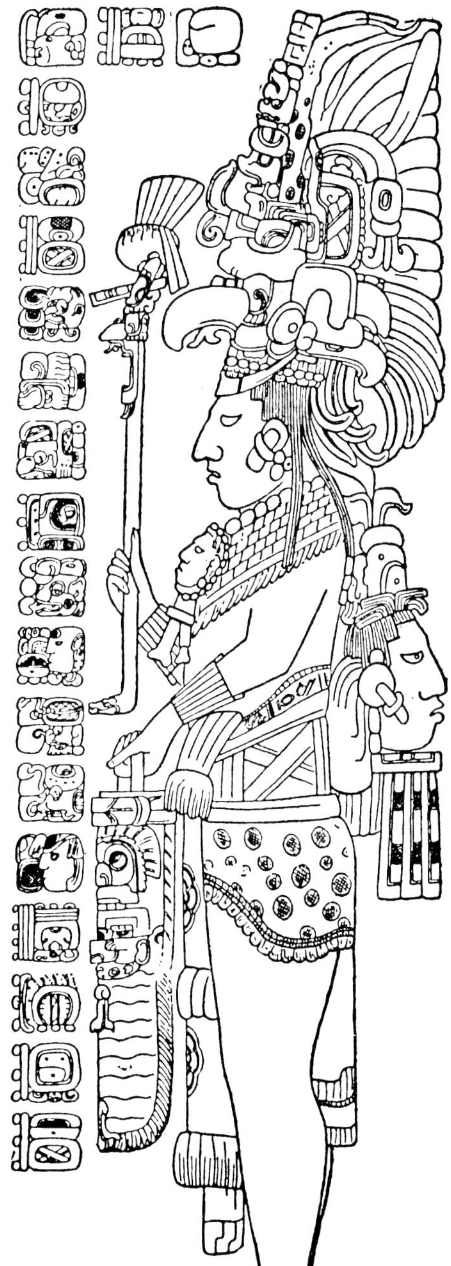

Figure 120. God K of the Manikin Staff V. Stela at Tzendales (after Spinden)

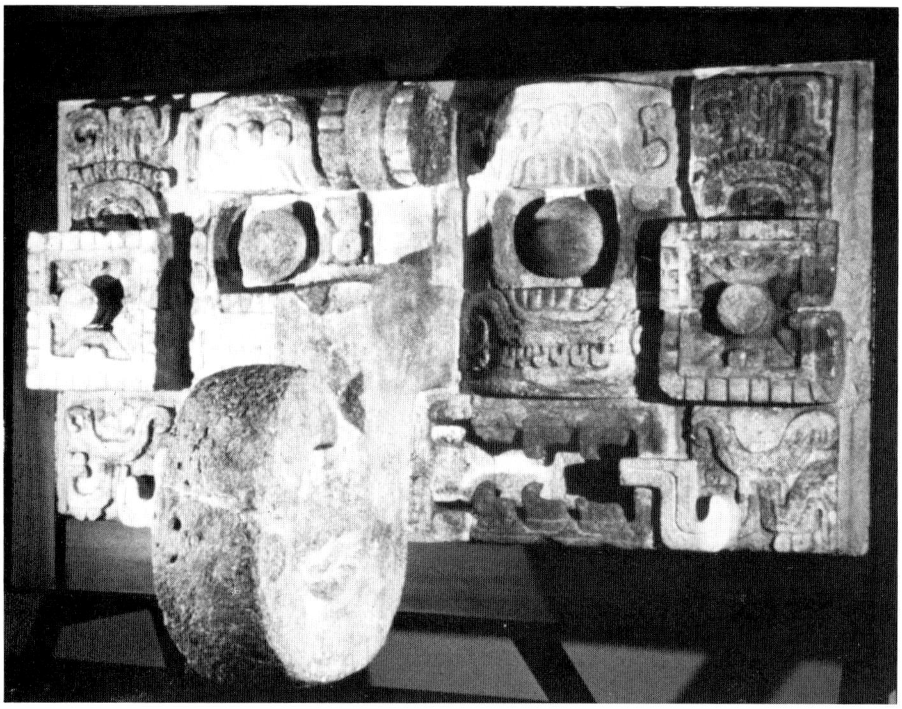

Figure 121. Long-nosed mask from Labná (Museo Nacional de Antropología y Historia, Mexico City)

(Text continued from page eighty two)

Recent Research

The research on the identity and function of God K is far from ended. The newest and most interesting theory is that offered by Michael D.Coe: that God K is both iconographically and functionally identical to the great deity Tezcatlipoca (Smoking Mirror) of central Mexico, whose cult was still very much alive at the time of the Conquest.[58] Tezcatlipoca, a dark, dangerous, multifaceted god, the antagonist of Quetzalcoatl, the benevolent Feathered Serpent, may have represented the "shadow" side of our human personality. His worship was an essential feature of the dualistic Aztec religion.[59] He was the patron of sorcerers, young men, princes, warriors, and highwaymen; but at the same time he was the eternally young Telpochtli, the enemy Yaotl, and the warrior knight Huitzilopochtli. He was also identified with Xipe Totec, God of Sunset and Sacrificial Pain.[60] His symbol was the obsidian mirror into which the Aztec shamans gazed until they fell into trances and in which they saw pictures revealing the future and the will of the gods.[61] In Aztec mythology Tezcatlipoca lost his right foot in the struggle in which he drew the Earth Monster from the waters. His single footprint was seen as the group of stars we call the Great Bear. In the codices he is identified by a round mirror on his temple and sometimes by another mirror he wears in place of his lost right foot. These mirrors emit coils of smoke, similar to that of the cigar in the forehead of God K of the Mayas. Being a nocturnal deity, Tezcatlipoca is usually painted black, but his face is striped with yellow bars. He bears the arms of the Aztec warrior knight: the shield (*chimalli*), the spear thrower (*atlatl*), and the spears (*tlachoctli*), and he wears the *aztaxelli*, the ornament of heron feathers.

As the legends go, originally there were four Tezcatlipocas, all offsprings of the old Father God and Mother Goddess. Each of them was assigned to a color and a direction. The first two created the death gods and the water gods, the third was the Feathered Serpent, and the fourth was the sun god. The four Tezcatlipocas initiated a series of "suns," or creations, each of which ended in collapse and disaster. After the end of the Fourth Sun, which was destroyed by floods, the sky fell to the earth, only to be raised again by the Tezcatlipocas, who used the four world trees to hold it up. To this myth there is also a possible reference in the Yucatec *Book of Chilam Balam of Chumayel*, where the Tezcatlipocas are identified with the four Bacabs of Maya mythology (a theory with which Barthel disagrees).[62]

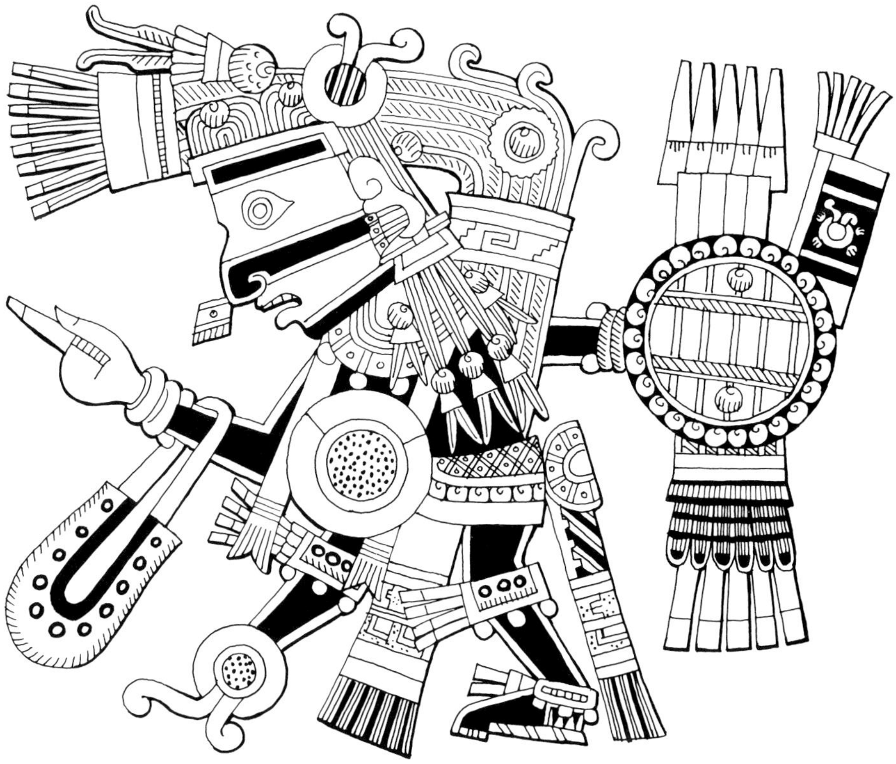

Figure 122. Tezcatlipoca, Smoking Mirror (Codex Borgia)

Michael D. Coe believes that the purpose of this myth is to confirm the divine nature of the royal lines and to identify the rulers as the descendants of the Fire God and Tezcatlipoca.[63] His view that God K (Smoking Cigar) and Tezcatlipoca (Smoking Mirror) are the same is certainly one of the boldest and most interesting theories in recent Maya research and undoubtedly will enrich the thought of Maya scholars for many years to come.

An example of renewed interest in the mythological role of God K is the discussion by Linda Schele and Jeffrey Miller of glyph T617a, "mirror" (*nen*). The glyph appears prominently in the texts of Palenque, is infixed in the foreheads of the cross deities—and is, in fact, one of the "god markings" of Maya iconography. The T617a mirror element is especially common in representations of God K. It is associated with the smoking-cigar element, and is infixed on the cigar itself, on the "cossack" helmet often worn by the god, or directly on his forehead. Schele proposes that this mirror symbol implies in both glyphic and iconographic context the process of royal accession and mythological defense of royal power.[64]

Another interesting essay on mirror-iconography is that of John B. Carlson. Based on his intensive

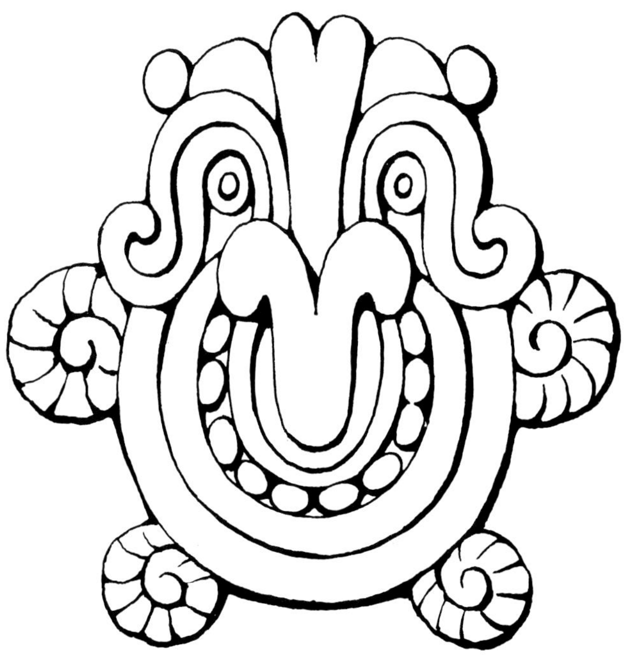

Figure 123. The symbol of Smoking Mirror

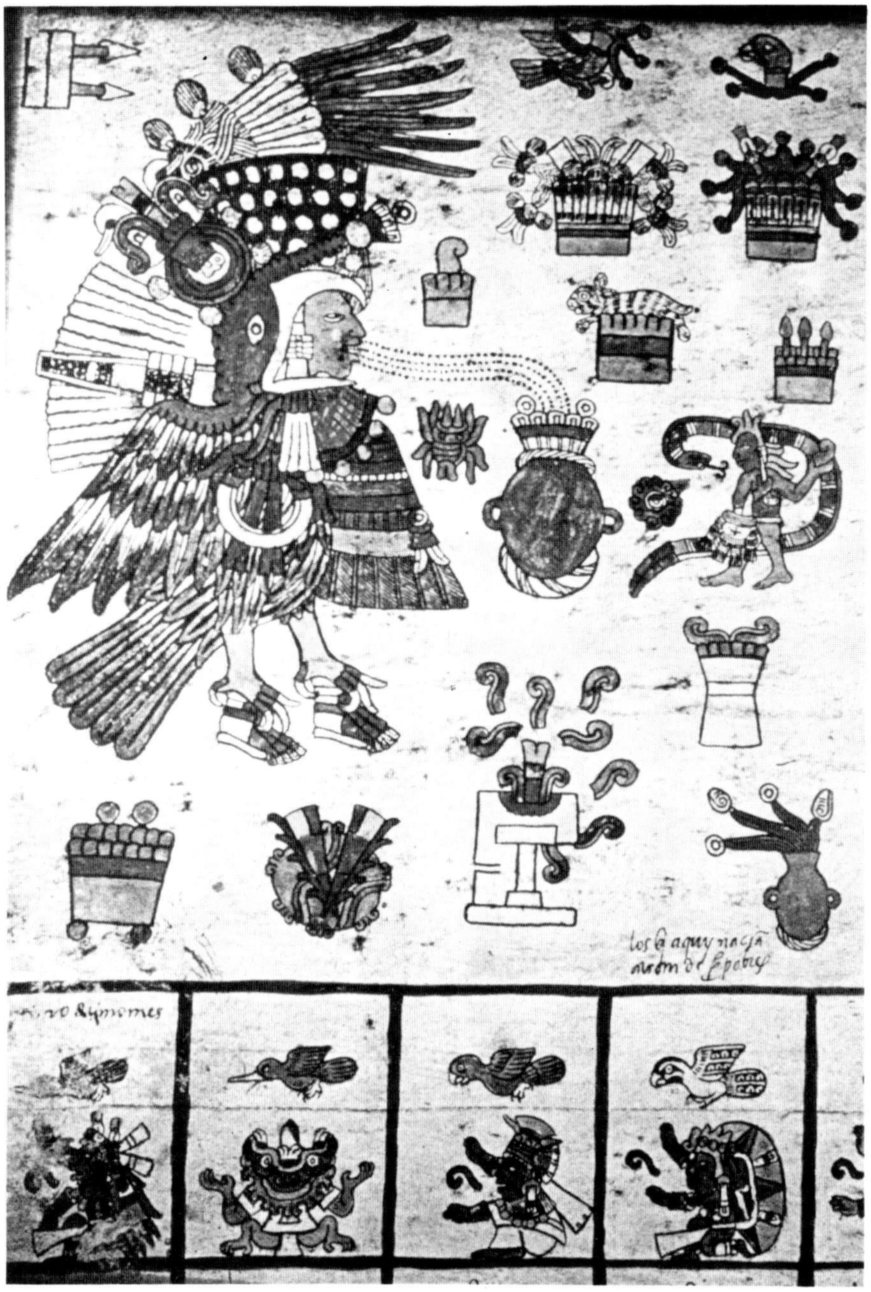

Figure 124. Tezcatlipoca as a bird (Codex Borbonicus)

study of Olmec concave iron ore mirrors, Carlson came to the conclusion, that these implements which have not only unique aesthetic design but also high optical quality were not only high status objects but as originally Ekholm suggested[65] were also used as "burning mirrors" to generate fire. Combining this hypothesis with Michael D. Coe's view on Tezcatlipoca, Carlson advocates that the Olmec concave mirrors were at the root of the general Mesoamerican mirror tradition and they were associated with the Smoking Mirror deity and a system of deified royal lineage and rulership. Elaborating further on Jeffrey Miller's *nen*/mirror hypothesis, Carlson looks at the "mirror" infix on the forehead of God K as an additional proof of this god's close affinity to Tezcatlipoca and at the same time as a representation of the mythological concave mirror. He, however, admits that "there is as yet no evidence that the Maya manufactured concave mirrors."[66] While only the future will determine whether Carlson's views are correct, whatever it is worth, we were able to find several figurines from the Island of

Jaina, which represent different dignitaries and ball players with concave mirrors hanging from their necks!

Another interesting development in the search for the mythological identity of God K is David Kelley's theory regarding the birth (emergence) of this deity. As a result of the analysis of recently deciphered calendric data, he is now convinced that there were at least two God K's during the evolution of Maya mythology: an "old" God K, who was primarily associated with the 819-Day Cycle, and a "young" one (perhaps one or more sons of the old god), whose birth occurred in A.D. 753 during the winter solstice.[67] Kelley further believes that this god is in several ways closely related to, if not identical with, Hunahpú, one of the Hero Twins of *Popol Vuh.*[68]

Lounsbury, another eminent Maya scholar, also entered the arena of speculation concerning the identity of the "deity with the forehead cigar." Working from an etymological rather than a calendric basis, he identifies God K with the highland rain god Toj, Tojil, or Hun Toj, who is also the patron of the ninth day, Muluc and whose role as the "Fire Bringer" has been most recently discussed in the excellent monography of Thomas Barthel.[69] (See discussion of Vase C-3 on page). Lounsbury raises the question whether the forehead tube of the deity may be a torch (*toj*) rather than a cigar, and the combination of the "smoking torch" and the "mirror" may be pictographic representations of something like the *tzotzil ch'ul toj ch'ul nen,* where the smoking torch is a rebus for the other nomen *toj* ("black flint") from which the mirror is made. He agrees with Kelley that there must have been two God K's, an ancient one who appeared long before 1.18.5.4.0, and a second one born on the latter date. Both gods have forehead tubes and mirror devices, but the latter has more. Both gods can also be named by the glyph complex T758:110, but the first god has a numerical prefix 1, while the second does not. In fact, the second god has a *ca* sign and an extra symbol, a swirl or spiral.[70] Lounsbury considers that in ancient Maya the cigar may have been referred to at least in ritual language as a "torch."

According to Miriam Beatriz Ríos Meneses, the cult of God K is still not completely forgotten. The Yucatec pueblos Santa Elena and Dzitas honor God K as the Wind God (Díos del Viento) on every January 21 with a special celebration during which turkeys are killed in his honor.[71] The custom recalls the portrait of the deity sacrificing a turkey in the *Dresden Codex.*

I must admit that even this lengthy analysis has not answered to my absolute satisfaction the final question whether or not we are dealing with a single deity (God K), a group of gods (a God K group), or an even greater group of deities (manikin or jester gods), some of whom may temporarily adopt and display that symbol of special significance and function, the smoking cigar.

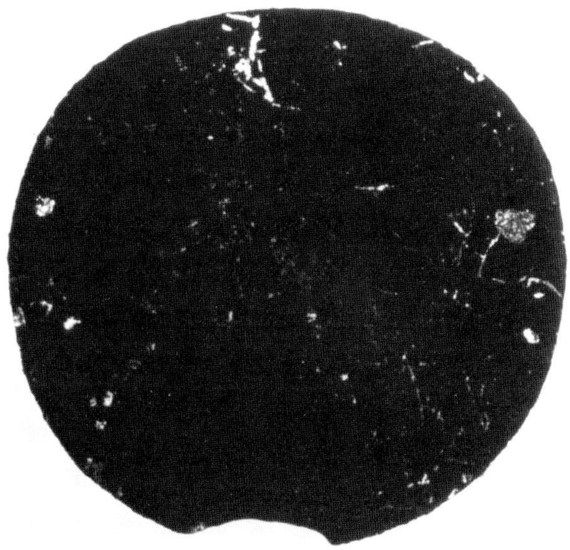

Figure 125. Obsidian mirror found in northwestern Mexico

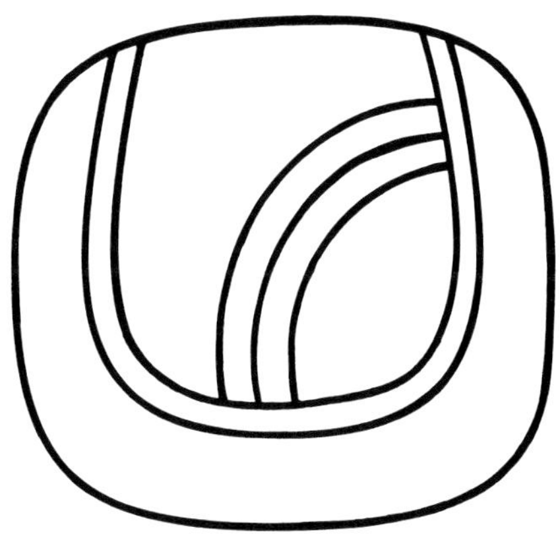

Figure 126. The nen, *"mirror," glyph*

PART TWO

Archaeological Material

Unlike an engineer, a tinker attacking a problem has an excellent chance of arriving at a variety of solutions.

François Jacobs

Tobacco was used with other substances on their altars, . . . the conjurers cast tobacco, copper, etc., on stormy waters to appease the angry gods, and the tobacco was among the cherished objects provided for their kings and priests in the abode beyond life.

James Smith

The visual images of Maya art are subject to strict control much like verbal language.

Linda Schele

[*If something*] *does not rest on logical inference, neither concurrence nor refutation is possible.*

Tatiana Proskouriakoff

7. The Classification of Archaeological Material

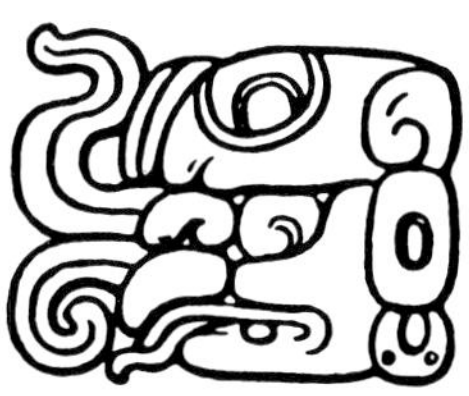

Of course, the only way to state with absolute certainty that the ancient Mayas indulged in tobacco smoking would be to discover tobacco cigars or cigarettes or pipes stuffed with tobacco (preferably still smoking) in Classic graves. Unfortunately, the only bona fide discovery of tobacco at a Maya site was the cache of cigars discovered by C. Rudy Larios in Group H at Tikal.[1] This chapter, however well supported, is, therefore, speculative. The only thing we know for certain is that the Mayas smoked something. This something was most likely tobacco, a conclusion based on archaeological material found at Classic and post-Classic sites, stone monuments, ceramic artifacts, and codices. While this material has proved very valuable in providing data on mythology and the lifeways of the ruling class, it revealed nothing about the smoking habits of the common people. As Thomas Carlyle wrote 150 years ago, "History is the essence of innumerable biographies."[2] For the Mayas the lives of ordinary people were not considered important enough to record and preserve for posterity.

The "smoking" scenes shown in the codices, on stone monuments, and on painted and carved ceramics have been grouped as follows, according to the degree of certainty with which I believe that they portray the act of smoking:

1. Scenes *almost certainly* showing the act of smoking, that is, portraying individuals with smoking objects resembling cigars or cigarettes in their mouths (Chapter 8)
2. Scenes *probably* showing the act of smoking, that is, depicting individuals with nonsmoking objects resembling cigars or cigarettes in their mouths (Chapter 9)
3. Scenes showing objects that *may* be large smoking cigars either held in the hands of individuals or depicted any other way in the composition (Chapter 10)
4. The "flare gods," who remain a subject of speculation and debate (Chapter 11)

The objects that the archaeological artifacts show the ancient Mayas smoking can be divided into two groups: cigarettes and cigars. It should be emphasized, however, that this division is not based upon the general definition of these objects but merely reflects their relative sizes. In the archaeological material I have studied, I was able to distinguish the kinds of "smoking materials" illustrated below:

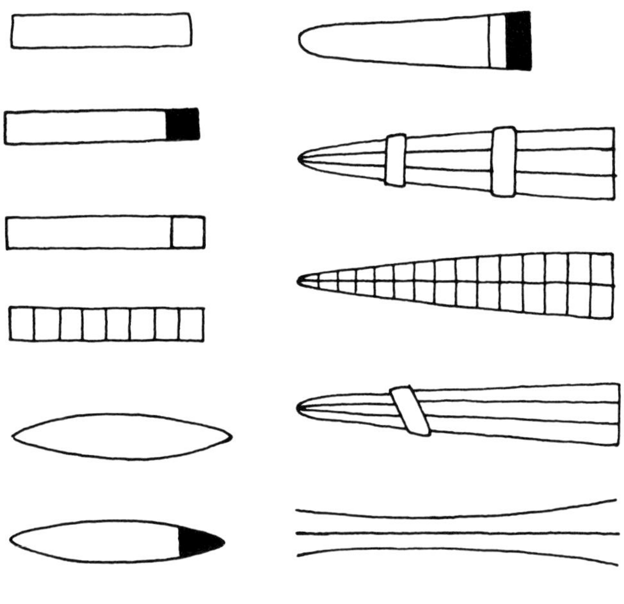

Figure 127. Different shapes of cigars and cigarettes depicted on ancient Maya ceramics

There is also considerable variation in the ways in which smoke is depicted:

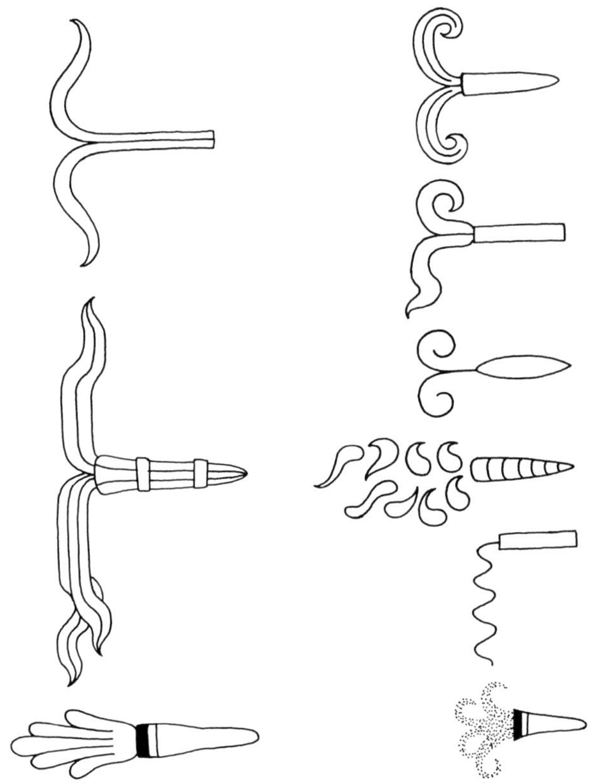

Figure 128. Portrayal of smoke on ancient Maya ceramics

In most representations of the act of smoking the smoke is shown as converging scrolls or coils. While most of the time it can be stated with reasonable probability that these scrolls represent smoke, sometimes this identification is not so easy. The scroll is a common motif in Maya iconography, and it is used to represent not only smoke but also foliage, breath, speech, stench, fire, and so on.

On most of the ceramic artifacts discussed in the following chapters, the pictorial representation is accompanied by a hieroglyphic inscription. It was a logical assumption that the text and the image are complementary and that, while the scenes painted or carved on the bowls and vases gave us pictorial evidence of smoking, the hieroglyphic text presents a written explanation of the act. Unfortunately, this assumption—with the possible exception of the *Madrid Codex*—was not accurate. I asked two renowned scholars in the field, Michael D. Coe and Linda Schele, to express their views on this subject. Coe replied as follows:

> In my monograph, *The Maya Scribe and His World*, I presented evidence that ran counter to orthodox Mayanist views on the nature of the hieroglyphic texts found on pictorial Maya ceramics, whether painted or carved. Following as usual their mentor, Eric Thompson, Mayanists believed that these extraordinary works of art were fashioned by illiterate peasant craftsmen, who used Maya glyphs and perhaps pieces of texts in a decorative fashion only, without understanding the meaning of what they were writing. This blind acceptance of established authority, so typical of Maya studies, meant that, heretofore, no one studied Maya pottery texts in a serious fashion.
>
> I found that pottery texts were almost surely meaningful, even when only one or two glyphs were repeated over and over, for these may have represented repetitive chants or cult names. In my scheme, there are basically two kinds of texts, both of which can be found on vases illustrated in this book. The first is what I have called "Primary Standard," occurring either in the band around the rim or in a prominently placed vertical or diagonal band on the vessel wall. A Primary Standard Text is virtually the same on all pots on which it occurs, with the proviso that the individual artist or scribe selected from the complete text (which may have run to nearly 100 glyphs or words) just those glyphs which interested him or his patron; however, whatever the glyphs selected, they always occur in the same order that they were found in the complete text. What did the Primary Standard Text say? We still do not know, but both I and the Soviet scholar Yuri Knorosov believe we are dealing with a funerary chant, describing the gods of the Underworld whom the deceased person (for whom the pot was designed) was to meet after his posthumous trip to the nether regions. For instance, the second glyph in the series seems to stand for the stairs down which he went on his fateful journey, and the third glyph is that of God N, the old shell god whom I have shown is one of the two chief deities of Xibalbá, the Maya version of hell.

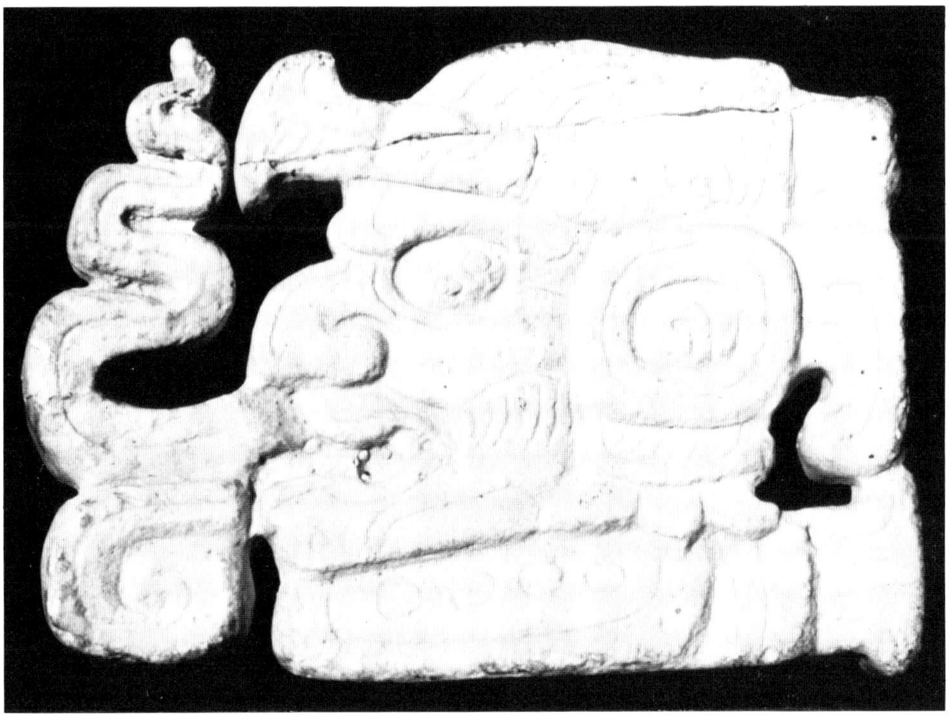

Figure 129. A breath (?) scroll on a glyphic head from Tikal (mold by Joan Wholey Patten)

The second kind of text is subsidiary to the first, and seems to refer to specific individuals and actions shown on the vase or bowl. I have called this the Secondary Text. Names of gods and perhaps deceased humans occur here, although the latter may also be referred to in the Primary Standard Text, following the chant. I would like to say that I have identified verbal glyphs referring to the act of smoking, but such is not the case, although such glyphs do occur in the scenes with gods smoking in the *Madrid Codex.*

There is still a great deal to be learned about pottery texts. For instance, not all Primary Texts around the rims are of the standard sort, and these are puzzling; perhaps many of them also refer to sequences of gods, but derived from other mythic originals. We do know that a few Secondary Texts contain historical material and even the Emblem Glyphs associated with particular Maya sites, but these also are probably posthumous references to events in the life of the defunct lord who had commissioned the vase for his own grave or tomb. There will be no easy solutions to these problems, but by recognizing that this rich collection of hieroglyphic material is as worthy of study as that on the carved stone monuments, the way will be opened to decipherment by future generations of scholars.[3]

In similar research Linda Schele also failed to find any glyphs on the ceramics illustrated in this book that refer directly either to tobacco in general or to the act of smoking in particular.

I was advised by some that, because there is no apparent relationship between the act of smoking and the text accompanying the scenes in which smoking occurs, I should omit remarks directed to the hieroglyphic inscriptions. I decided against that course for three reasons. First, while we did not find any connection between the act of smoking portrayed on these vessels and the inscriptions, someone may do so at some future time. Second, in the complex, esoteric scenes depicted on these vessels, smoking is not necessarily the only or even the most important activity, and disregarding the accompanying text might be compared to omitting the captions on the photographs of the participants of the Yalta Conference just because they do not mention Churchill's bulky cigar. Third, it simply seemed improper to comment in detail on half the message these vessels convey to us (the iconography) and completely disregard the other half (the glyphic text).

For writing comments on the glyphic texts, I am much indebted to Linda Schele, whose profound knowledge of Maya hieroglyphic inscriptions filled the gaps in my knowledge of this subject.

Archaeological evidence for Maya use of tobacco is scant.
Sir J. Eric S. Thompson

Limitations imposed by the lack of written documents force the student of this art to deal with the art objects directly. In spite of limitations, the challenge attracts.
Jacinto Quirarte

Mesoamerican studies often impress the readers as a Sahara of guesses where travelers crazed with a thirst for certainty suffer various mirages.
George Kubler

8. Scenes Almost Certainly Representing the Act of Smoking

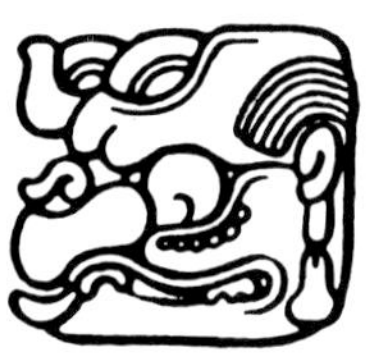

Scenes that almost certainly can be identified as representing the act of smoking appear in the Maya codices, on at least one fresco, and on monuments and ceramic objects throughout the Maya world. Those scenes are described in the pages that follow.

The Codices

The Mayas, being a literary people, produced a number of books concerning their history, astronomy, and religion. These books were written on fiber of the native *copo* plant backed with bark or deerskin coated with lime. These written records would have been the most reliable source of information for us. Unfortunately, all but a handful were destroyed by overzealous missionaries.

In the surviving codices are many scenes showing various deities, historical figures, animals, and so on, holding objects that are difficult to identify or engaged in activities associated with burning, incensing, or fumigating. If one had a one-track mind, turned his imagination loose, and did not hesitate to modify, rearrange, or change colors, as Ariel Baynes did in his two manuscripts "The Mayas Had Tobacco" and "Tobacco Comes to the Ancient Mexicans," he could interpret a large number of such scenes as acts of smoking, cultivating, or adorning tobacco.[1] Approaching the matter with a more critical attitude, however, one must limit the "smoking scenes" to the following examples.

One of the surviving codices, the *Codex Tro-Cortesianus*, or *Codex Madrid*, presents on page 79b a 260-day sacred almanac. The almanac is divided into five parts of 52 days each.[2] The subdivisions are presided over by Gods D, A, and E, respectively (interpreted by Thompson as gods of rain, death, and maize). In the center of the page, in a scene somewhat resembling a Chinese opium den, the three gods are lolling on objects identified by *caban* signs (representing the earth), and puffing large funnel-shaped cigars. From each cigar rises a group of black dots, undoubtedly depicting smoke.

On pages 86b and 87b is another 260-day almanac, again accompanied by a picture of the protector gods for the period. On page 86b God D appears, described by Villacorta as "sitting peacefully smoking a cigar (*sentado tranquilamente fumado un cigarro*)."[3] In the center of the following page, 87b, Gods D and H are pictured puffing on cigars of similar shape and size. Red and black dots at the ends of the cigars stand for sparks and smoke.[4] On page 88, God D is shown seated on a throne, the bottom of which is distinguished again by the *caban* sign, and which thus probably symbolizes earth. He is leaning comfortably against the woven back of the throne, which is decorated with the symbol *pop*, the Maya mat, sign of authority and overlordship. In the mouth of the deity there is a large, flared cigar with a cascade of black and red dots.[5]

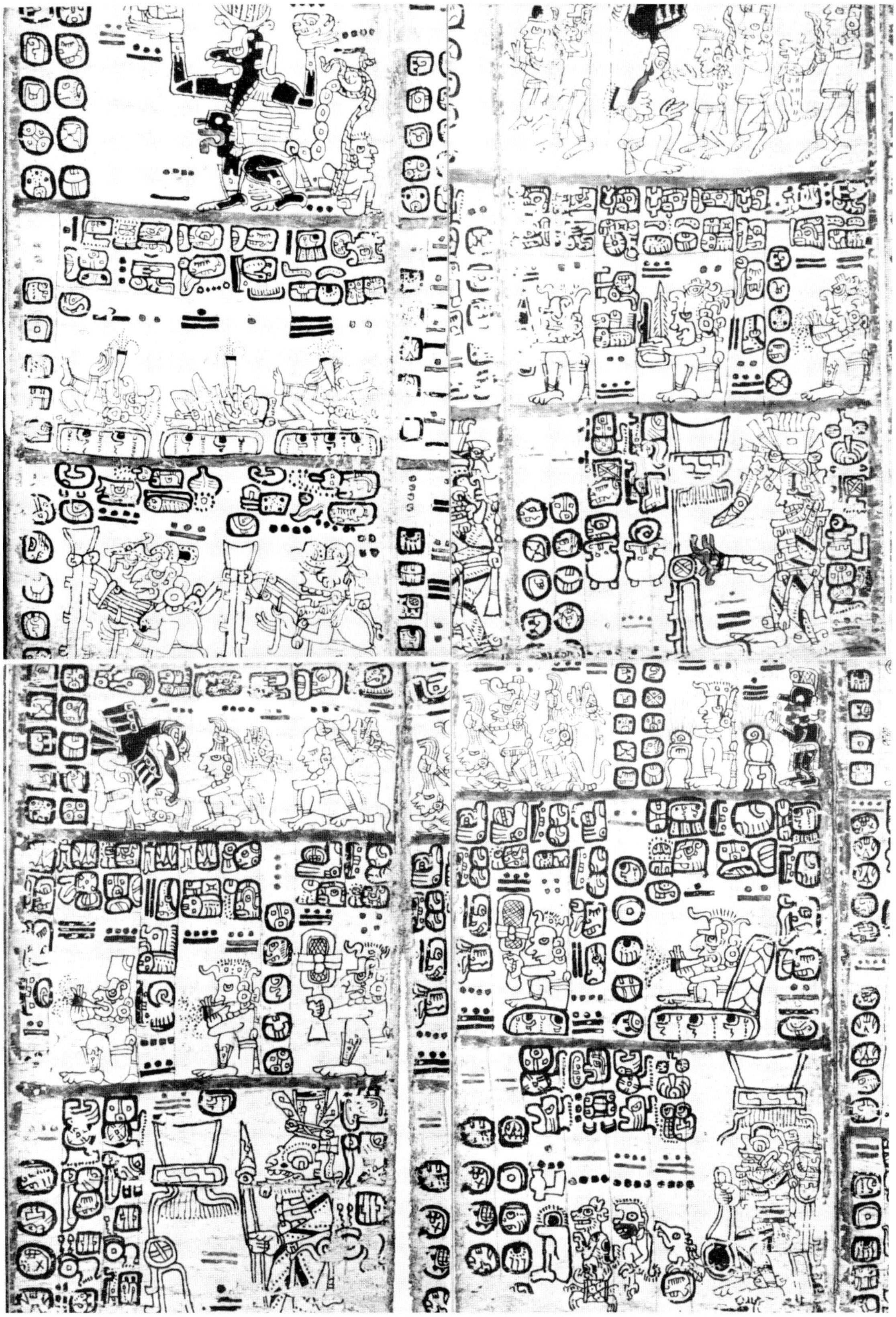

Figure 130. Pages from the Codex Madrid *on which the act of smoking is portrayed*

Glyphic Text

Pages 86b and 87b of the *Madrid Codex* are especially important not only because they pictorially present the act of smoking, but also because there is a definite possibility that smoking is also described in the accompanying glyphic text. Seler pointed out first that the initial glyph compound in the accompanying script T227:504 probably meant "smoking," a hypothesis echoed by David Kelley.[6]

There are five texts accompanying these scenes, each made of four glyph compounds. Glyph T227:504 occurs in the first position as the verbal compound in each case. The second position in each text is the nominal compound of a deity who is not pictured in the scenes. This fact favors an identification of the second compound as the object of the verb. The subject of the verb and the portrayed deity are named in the third and fourth positions.

The *tzolkin* subdivision M86b–87b is apparently part of a series of related sections that begin on page M83b and conclude on page 88b. In two of the sections, M85b and M87b–88b, the verb is related ideographically to the act portrayed. In the other sections the relationship between the verbs and the scenes is not yet understood, but it is possible that they are related to the act of smoking. It should be noted, however, that the hypothetical identification of T227:504 compound as the glyph for smoking has not been proved by testing the proposed reading of the graphemes in other contexts.

The smoking scene in subdivision 88b is of four glyphs each. The first compound (T110.526:251) reads *X-cab*. It is possible that the verbal compound in the first position refers to an activity associated with the *caban* (earth) throne. The glyph in the third position names the subject of the verb and the person participating in the scene. It is probable that the verb in this section of the *Madrid Codex* refers to the throne and the activity associated with it rather than the act of smoking itself.

Stone Monuments

The only Maya stone monument that shows the act of smoking is the well-known sculptured panel from the east side of the doorway leading into the Sanctuary of the Temple of the Cross at Palenque.[7] It portrays a standing aged man in profile. His age is indicated by his sunken cheeks and missing teeth.[8] His large Roman nose, wide "scroll" eye, with central pupil, and his elaborate headdress made of the feathers of the Moan bird (the ominous messenger of the underworld), which also appears to be perched on top of the headdress, identify him as God L, one of the most important Maya deities. Some believe that the wide fillet on his head represents tobacco leaves, rather than feathers.[9] In front and back curious taillike pieces of what appear to be cloth run from under his loincloth to the tops of suspended heads on either side. On this cloth are footprints, which may refer to the Road of Death.[10] Similar footprints are seen on Stelae 6, 11, 14, and 25

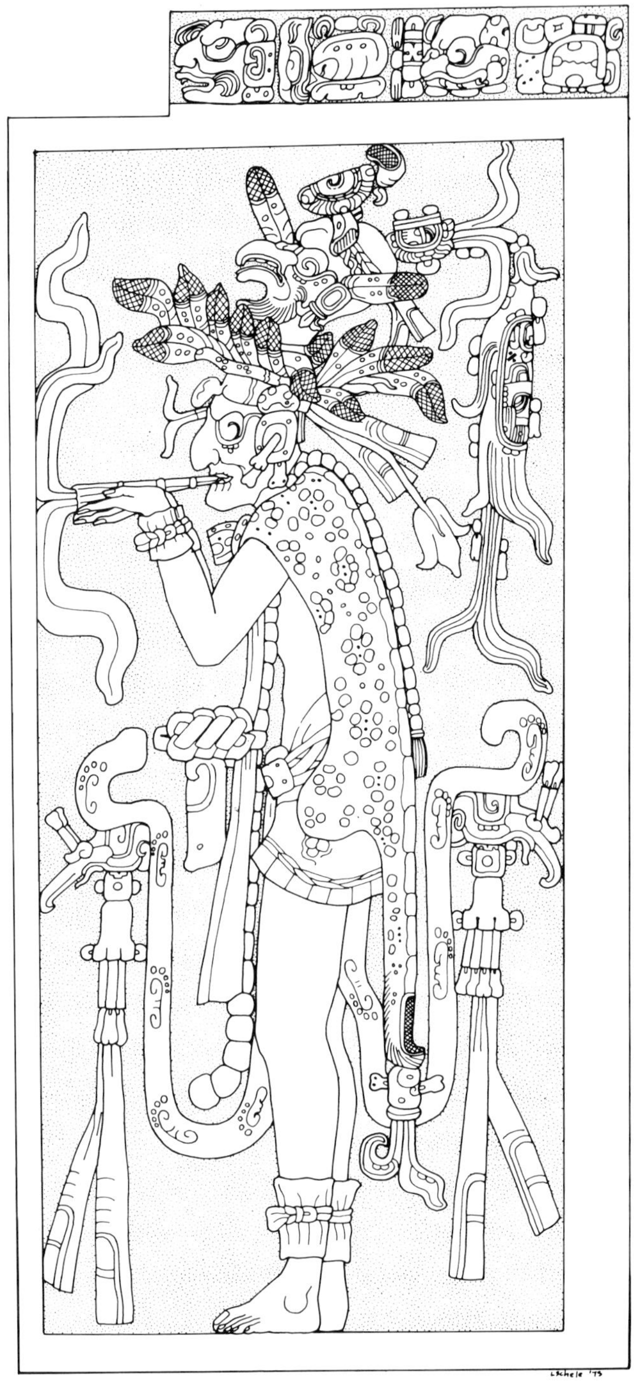

Figure 131. God L smoking a large cigar, door jamb, Temple of the Cross, Palenque (drawing by Linda Schele)

at Piedras Negras. In Benson's view they may represent bloody footprints. A long beaded lace shirt covers his chest and back, and a jaguar-hide cloak, with the tail of the beast still attached, hangs from his shoulder. A jaguar ear also appears over his own. The wide wristlets and anklets complement his wide belt buckle, which is shaped into a mat symbol, the sign of overlordship and authority.[11] In his hands, in a peculiar palm-upward position, he holds a large funnel-shaped cigar, which earned him his popular name, "El Fumador."[12] He is blowing smoke from the cigar, which resembles those still smoked by some of the primitive South American tribes.[13] (Cigars of identical shape were also popular in Europe in the sixteenth century and were depicted in Juan Bautista Muñoz' *Historia del Nuevo Mundo*.[14]) A small flare with two curls of smoke, just like that of God K, emerges from the forehead of the deity.

The iconographic significance of the figure, whom Spinden disrespectfully called "an aged and decrepit smoker," is uncertain.[15] Thompson believed that his smoking is an act analogous to the North American Indians' ritual of blowing smoke to the four directions of the earth.[16] Such customs have been practiced by the Hopis, who blow smoke to the four winds during their ceremonial dances and by the Zuñis, by whom "at many points in ceremonies, tobacco smoke is blown to the six cardinal points that the rain makers may not withhold their misty breath."[17] While Arthur Miller relates El Fumador to the cigar-smoking incident in *Popol Vuh*, recent observations by Schele attribute astrological significance to this monument.[18] She and Joralemon verified that at the time of the winter solstice the last rays of light of sunset fall on the panel on which God L stands smoking his cigar. Yet it must be admitted that, even after many years of research, we are not much farther along than John Lloyd Stephens was a century ago. Stephens, who was among the first explorers to see this tablet, said: "The leopard skin on the back, the beak and eyes of the bird on the headdress . . . all mystery, silent, defying the most scrutinizing gaze and reach of intellect."[19]

Ceramic Objects

One of the earliest portrayals of smoking in ancient Maya art is the brownish-gray ceramic container **Box of God K and God L Smoking** (A-1), executed in the early Classic style of northern Petén (now in a private collection). A detailed discussion of the box is offered by Michael D. Coe, and my remarks here will be limmited to the figures of God L and God K.[20]

The box not only offers an exceptionally fine example of smoking but also is one of the rare surviving ceramics on which God K and God L are shown standing and in full regalia. Moreover, it establishes a very important link between the religious cults and cultural styles of early Classic northern Petén and the iconographic program of Palenque. The box is carved on three sides; the back is undecorated. The front has two panels divided by a vertical row of hieroglyphs. In the left-hand panel appears probably the earliest known standing presentation of the serpent-footed God K (Plate 101). He is wearing a narrow belt with a back hanger, a round earplug, and a collar. In his left hand he is holding 7 Black Kan, a glyph that appears in the script of the Temple of the Sun at Palenque and on many other monuments of the Classic period that are often paired with another glyph carrying the numerical coefficient of the number 9. Foliage, possibly of maize, is falling from the glyph. God K appears in his full iconographic program, consisting of serpent leg, "mirror" body markings, characteristic facial features, a smoking capped Ahau on his head, a smoking forehead cigar, and an *akbal* symbol infixed in his forehead. A similar motif can be seen on Stela 31 at Tikal dated 0.0.10.0.0.

In the panel opposing God K, holding a javelin in his left hand, stands God L, the deity of the East Door Jamb of the Temple of the Cross at Palenque (Plate 102). A screech owl, regarded as the Moan bird, perches on his headdress. Behind the bird is a sky (*caan*) glyph with scrolls, perhaps maize foliage. While the Temple of the Cross example has the symbol for 12 Sky on his headdress, here the glyph has no numerical coefficient. The face of the God is aged, and he stands in the hump-backed pose of an old man. He has a short beard and facial scarification (perhaps a tattoo) and is smoking a large cigar. He wears a loincloth, a jaguar cloak, a collar from which a mat sign hangs, cloth wristlets and gaiters, and a cloth back hanger with footprints and smoke motifs. The figure, his attire, and the entire composition of the panel shows remarkable similarity to the Palenque door panel, demonstrating that the iconography of the Temple of the Cross group is deeply rooted in the very ancient tradition of northern Petén.

For a discussion of the hieroglyphic text and additional notes on iconography see Michael D. Coe.[21]

Another artist's representation of smoking dating from the early Classic period of northern Petén is the **Plate of the Smoking God (A-2),** first photographed by Donald M. Hales (now in a private collection). The outside rim of the plate is incised with six glyphs, and the bottom of the ceramic displays the head of a cigar-smoking supernatural personage etched in fine lines (Plate 103).

The head has a fleshy face with a skeletal but bearded jaw. The face has a *cimi* "death" sign, and the forehead is infixed with an *akbal*, "darkness," symbol. The teeth of the deity are bared, an infrequent feature of early Classic art, and he is puffing on a long, slim cigar. His forehead is crossed by a bone-shaped tube from which coils of smoke emerge. He wears what appears to be a shell ear

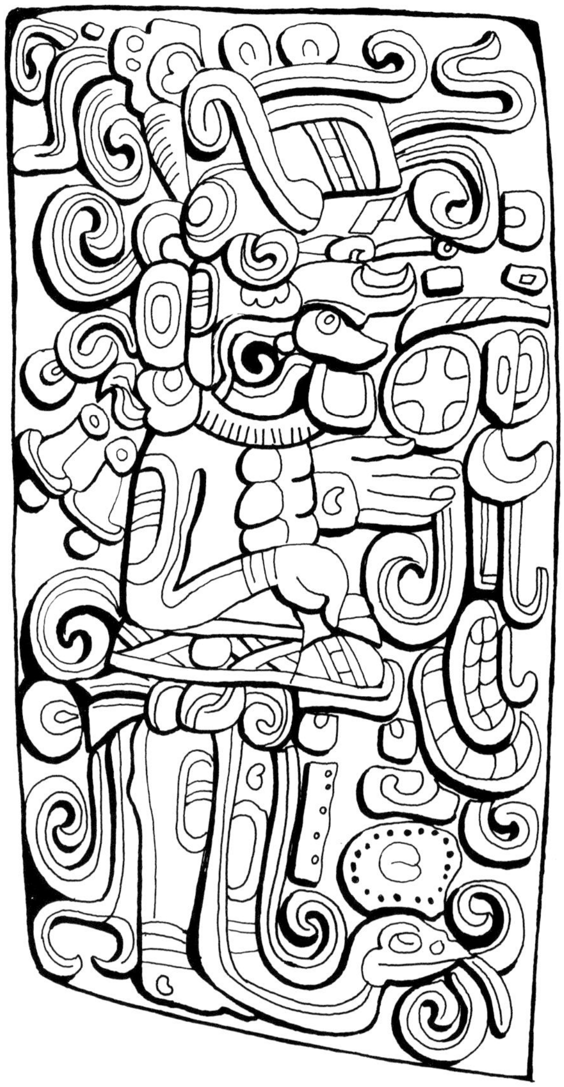

Figure 132. Detail, Box of God K and God L Smoking (A-1)

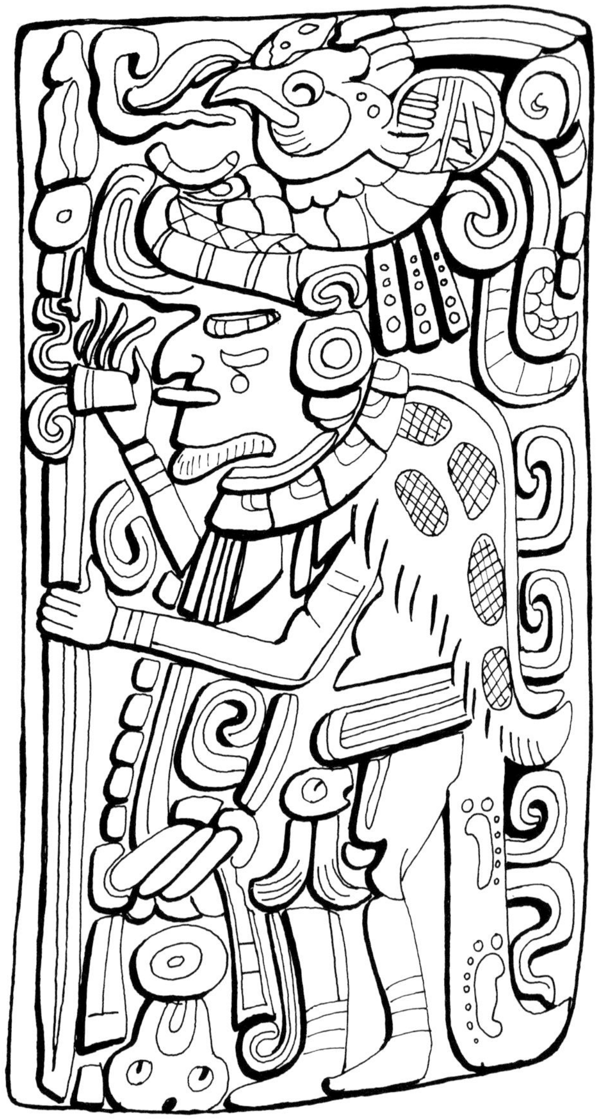

Figure 133. Detail, Box of God K and God L Smoking (A-1)

ornament and a beaded nosepiece, and his hair falls in front of his face.

The head seems to be emerging from the throat of a monster. It is also possible, however, that the god is wearing the head of a monster as a helmet, a representation seen on some of the stelae at Copán. The monster is undoubtedly the Bearded Dragon, a creature that is neither human nor beast and can be classified only as an iconomorph. Thompson identified it with Itzamná. Michael D. Coe has related it to the Mexican deities Xiuhcoatl and Ometeotl.[22] As on this plate, the Bearded Dragon is often shown on both monuments and ceramics emitting other deities from his gaping throat. From his forehead rises a complex of a "decorated Ahau" like the one on Princeton Vase 3. It has a long-fanged lower jaw and elongated ear extension like the ones seen on Princeton Vases 3 and 5.[23] The crown of the bent-back head of the dragon is topped with a *kan* sign.

The muzzle of the dragon swells near the end. A shell sign appears in the swollen area; on Lintel 3 of Temple IV at Tikal a *yax* (T16) sign appears on the dragon's muzzle. On the same lintel the dragon has the body of a feathered serpent. A skeletal head variant of T1016a is attached to the tail. A trifoil motif is attached to the rear head; the same motif appears at the top of the dragon's head.

Smoking is portrayed on a very important late Classic ceramic, the **Vase of the Dancing Lord** (A-3), provenance unknown (now in a private collection), which is decorated with a scene undoubtedly from the court of one of the Maya rulers (Plates 104–107). The vase has the usual flared rim and red

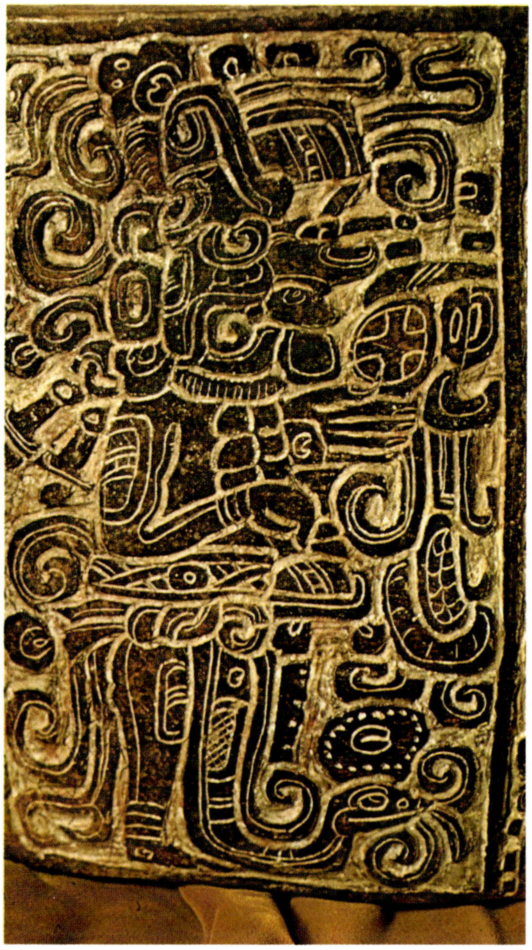

Plate 101

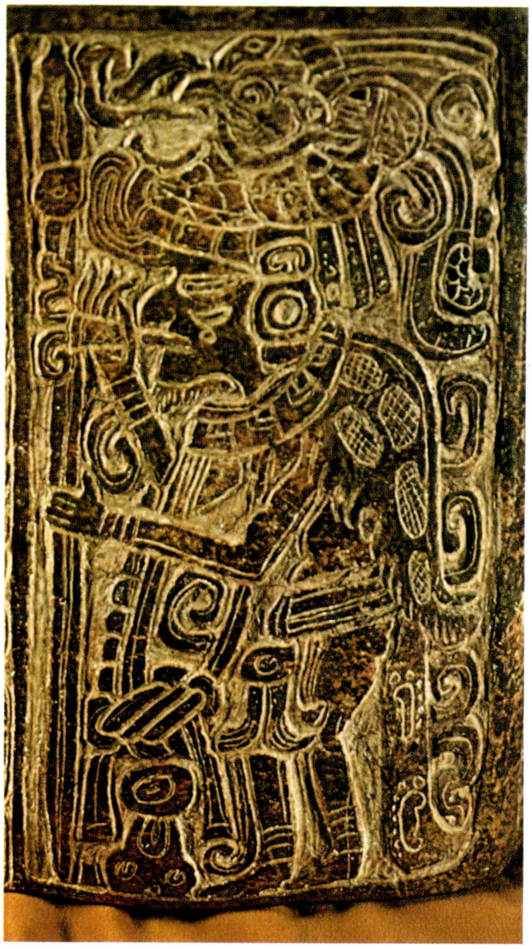

Plate 102

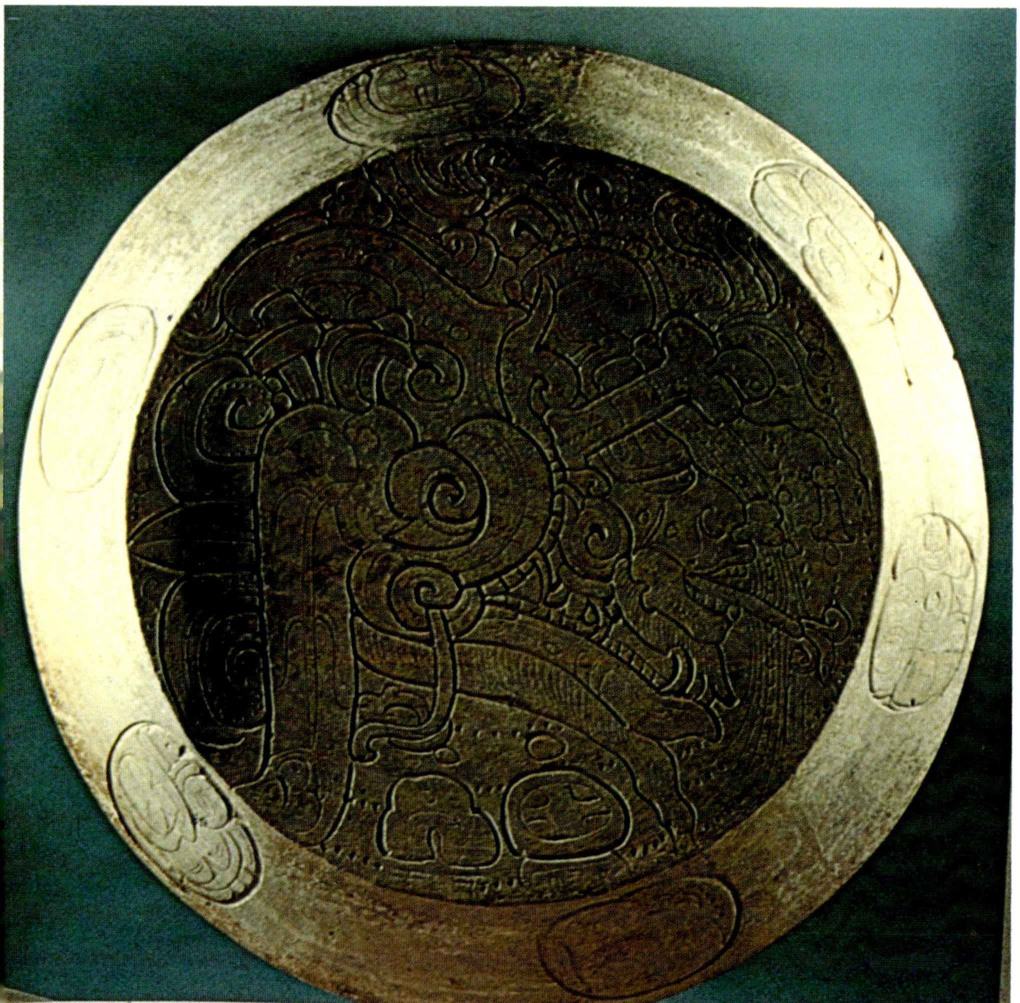

Plate 103

Plates 101-102. Box of God K and God L Smoking (A-1)
Plate 103. Plate of the Smoking God (A-2)/Photograph courtesy Donald M. Hales/

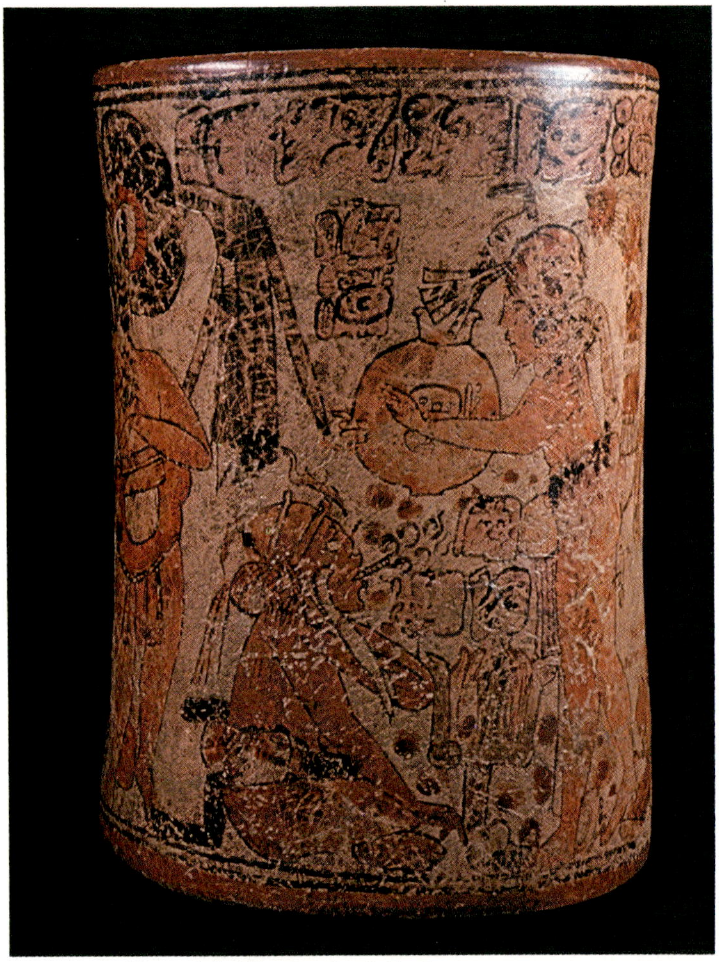

Plate 104. *Vase of the Dancing Lord (A-3)*

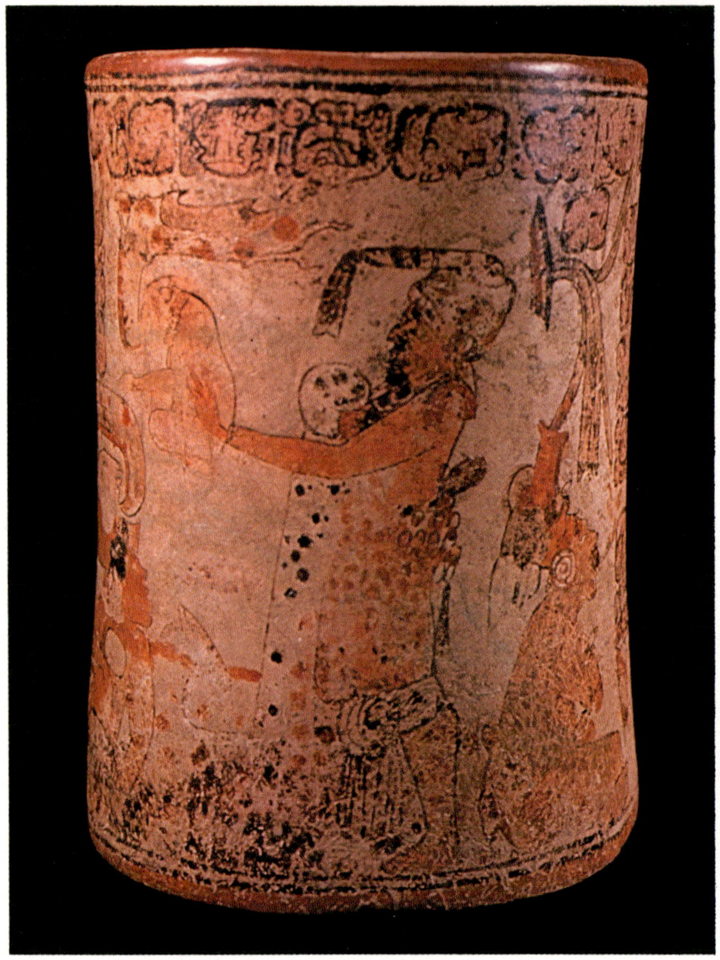

Plate 105. *Vase of the Dancing Lord (A-3)*

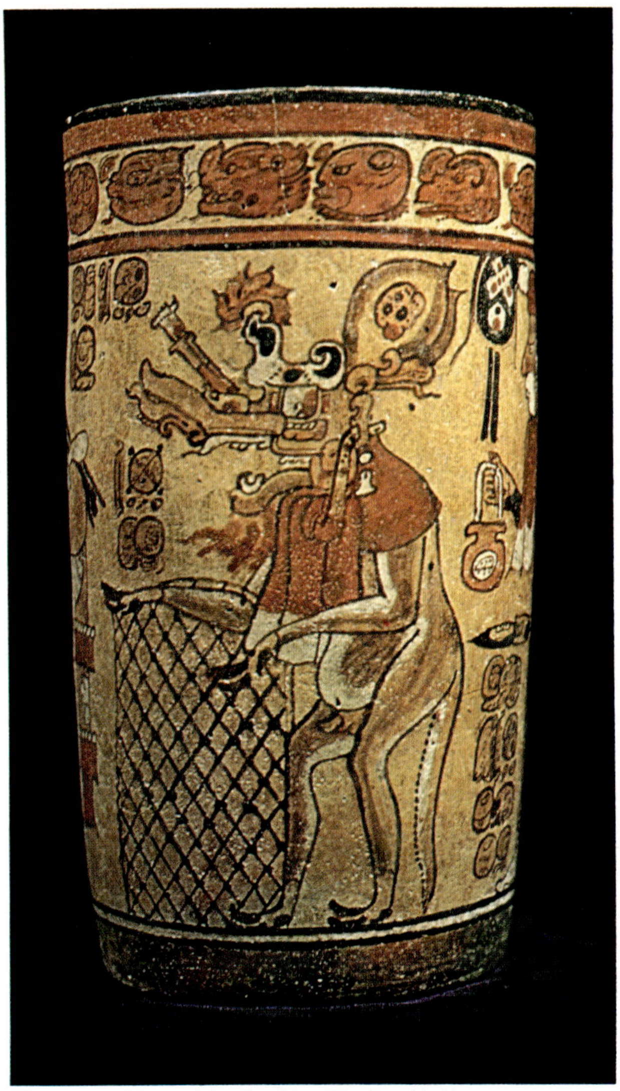

Plate 108. *Vase of the Smoking Monkey (A-4)*

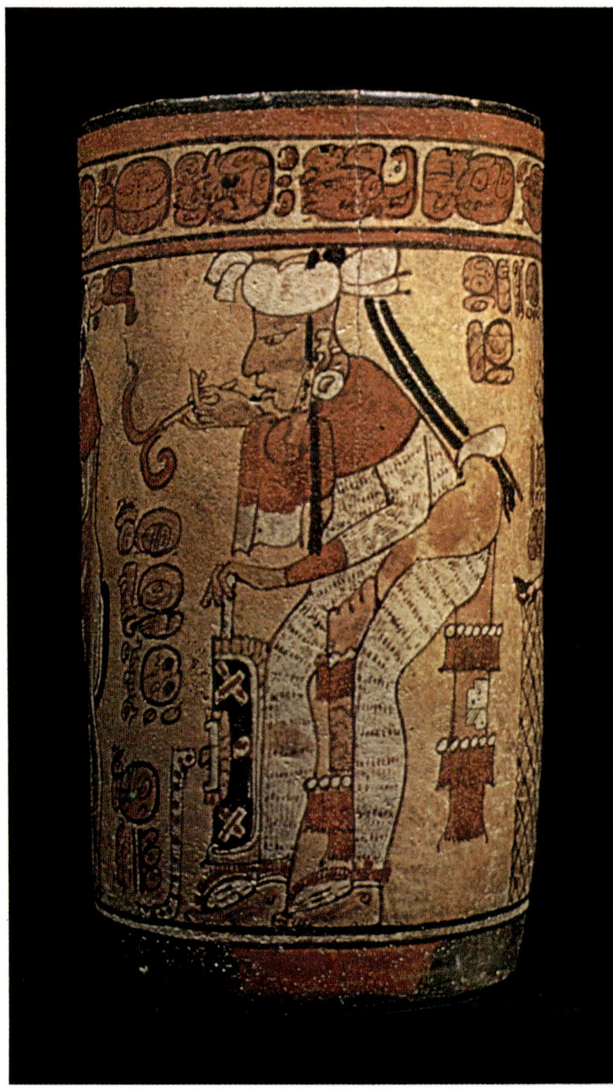

Plate 109. *Vase of the Smoking Monkey (A-4)*

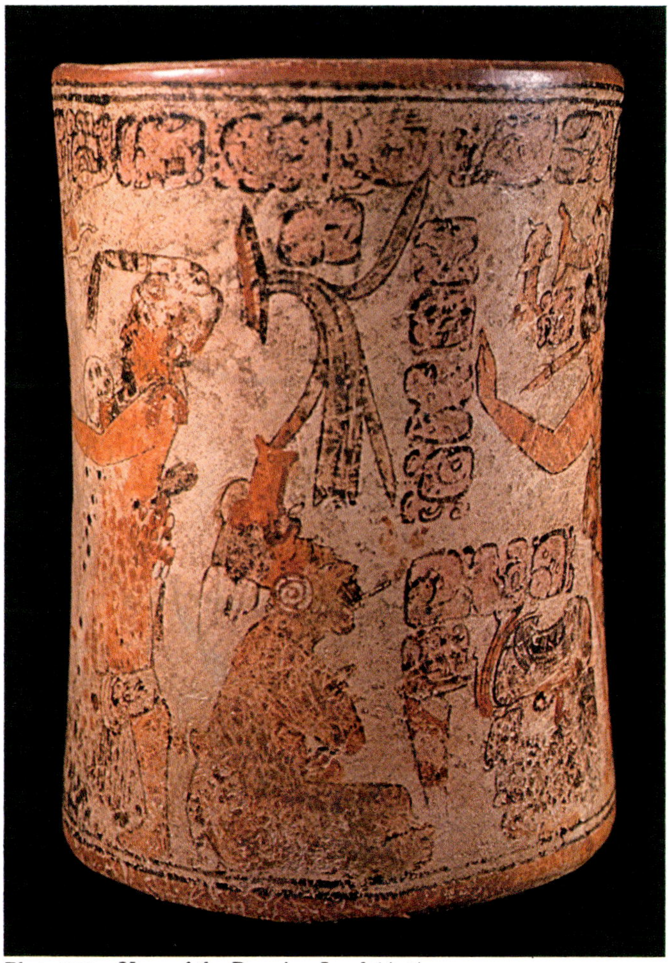

Plate 106. Vase of the Dancing Lord (A-3)

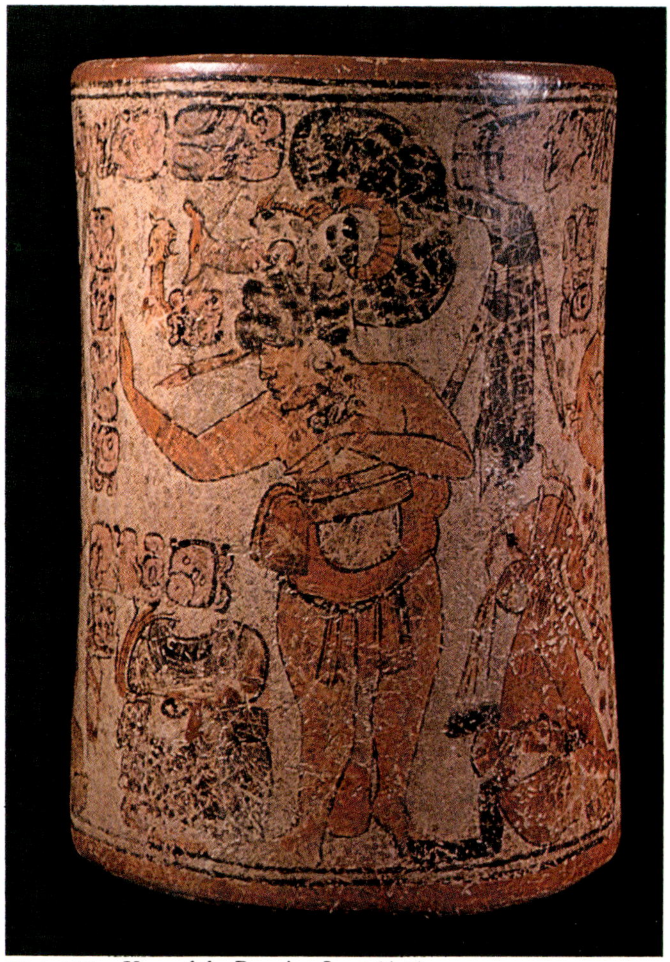

Plate 107. Vase of the Dancing Lord (A-3)

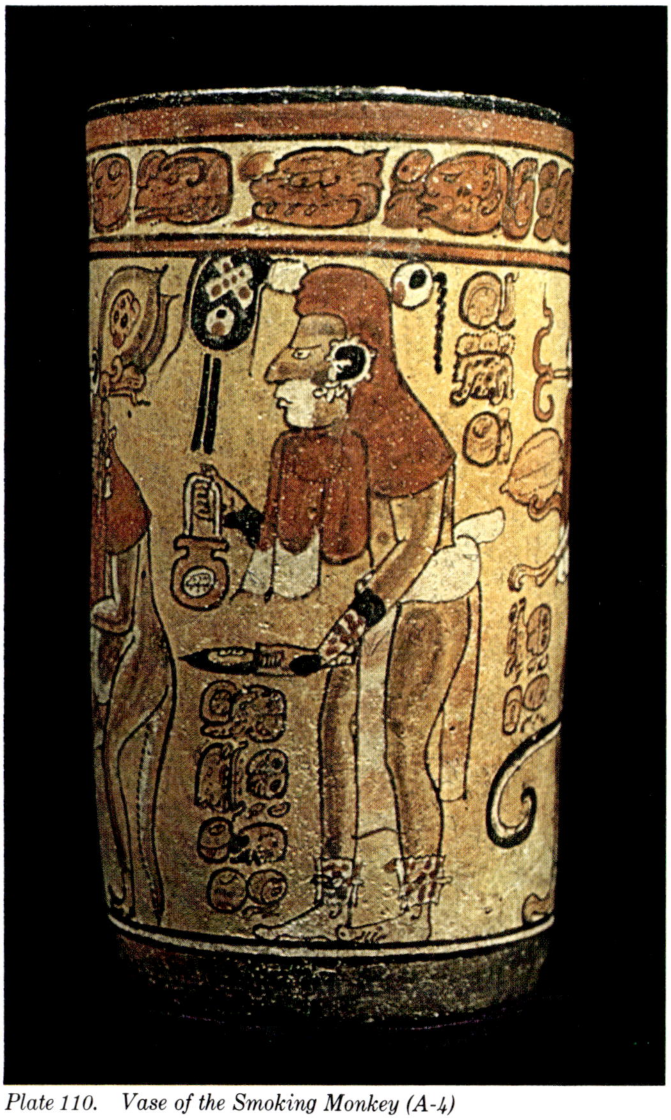

Plate 110. Vase of the Smoking Monkey (A-4)

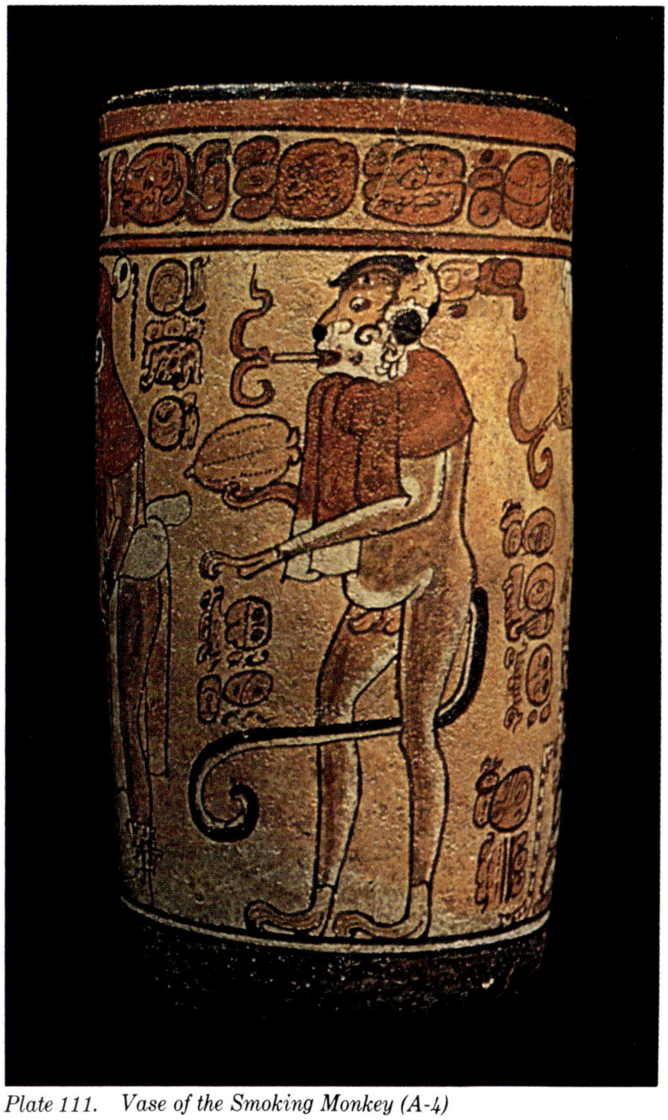

Plate 111. Vase of the Smoking Monkey (A-4)

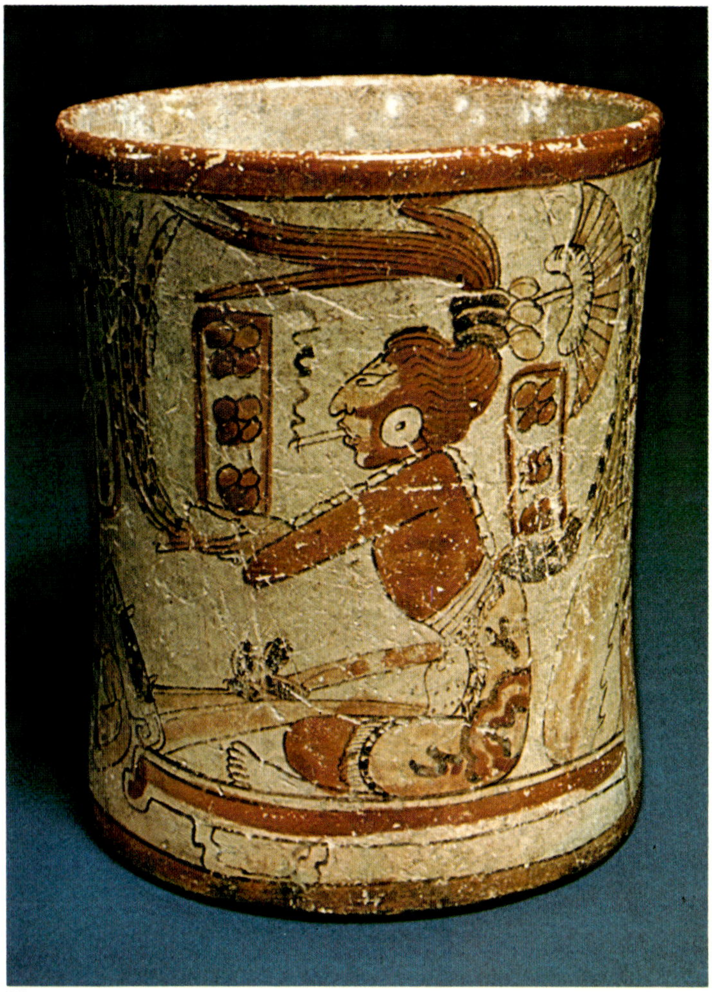

Plate 112

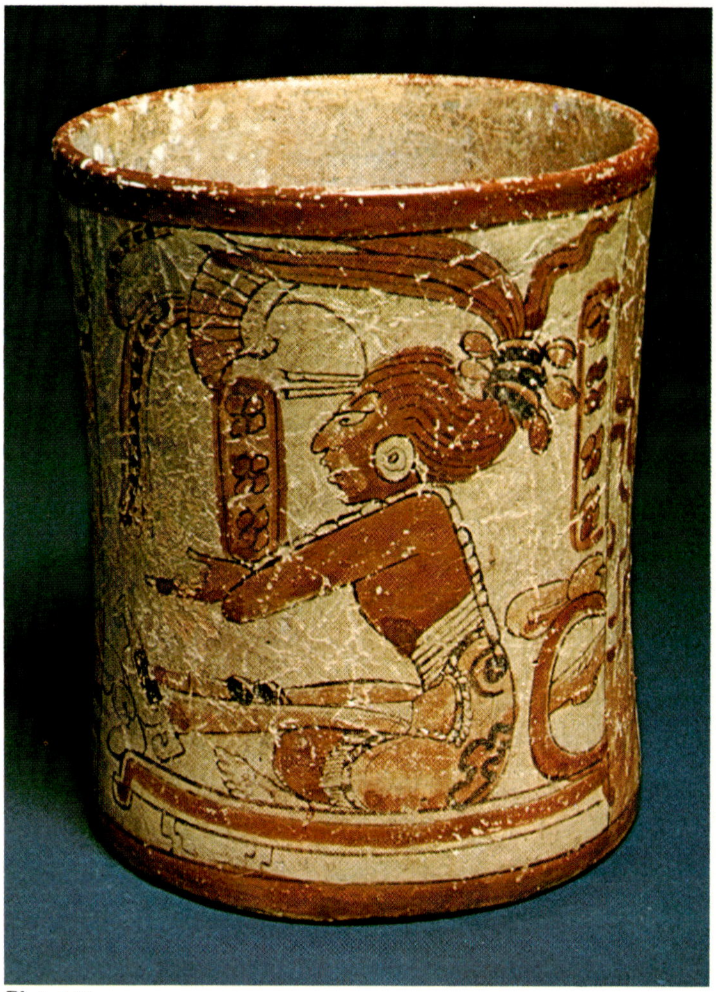

Plate 113

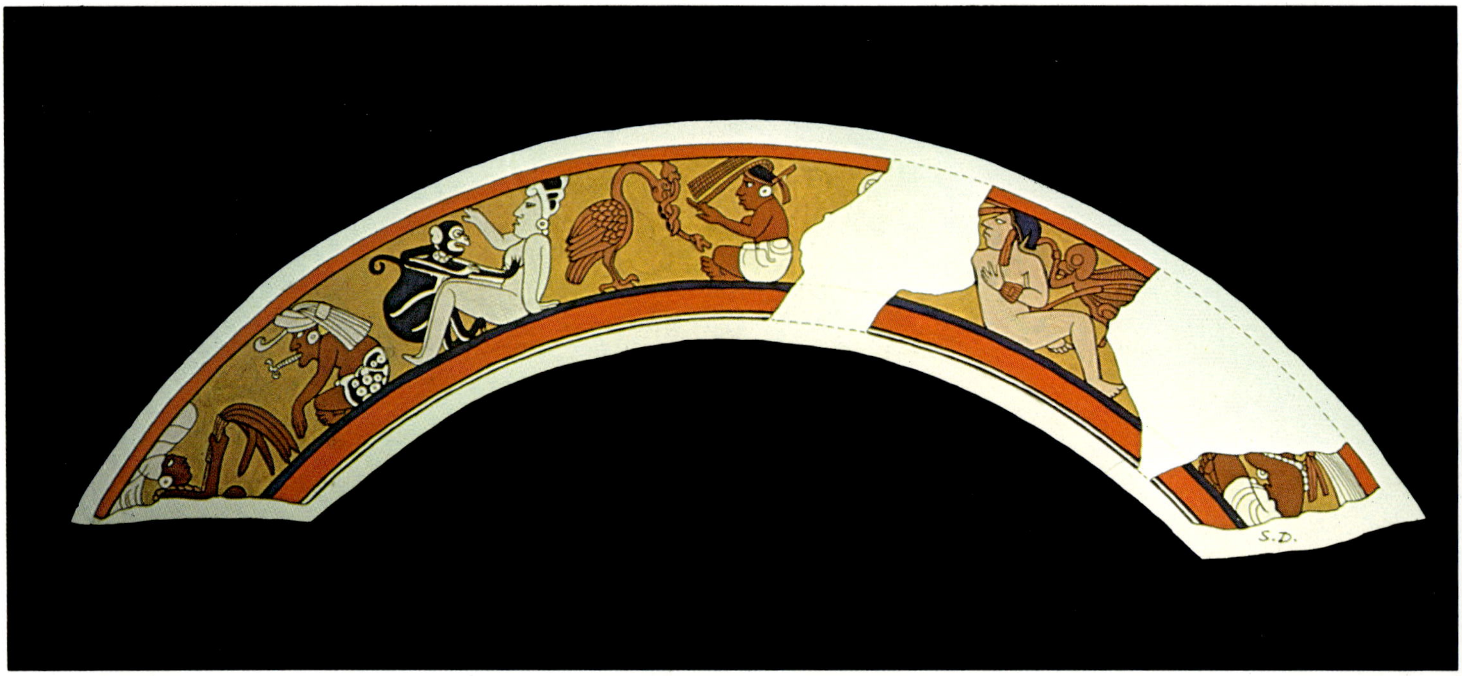

Plate 114

Plates 112-113. The Reynolds Vase (A-4)/Photographs courtesy Museum of the American Indian, Heye Foundation/

Plate 114. Plate of the Smoking God, the Woman and the Monkey (A-15)

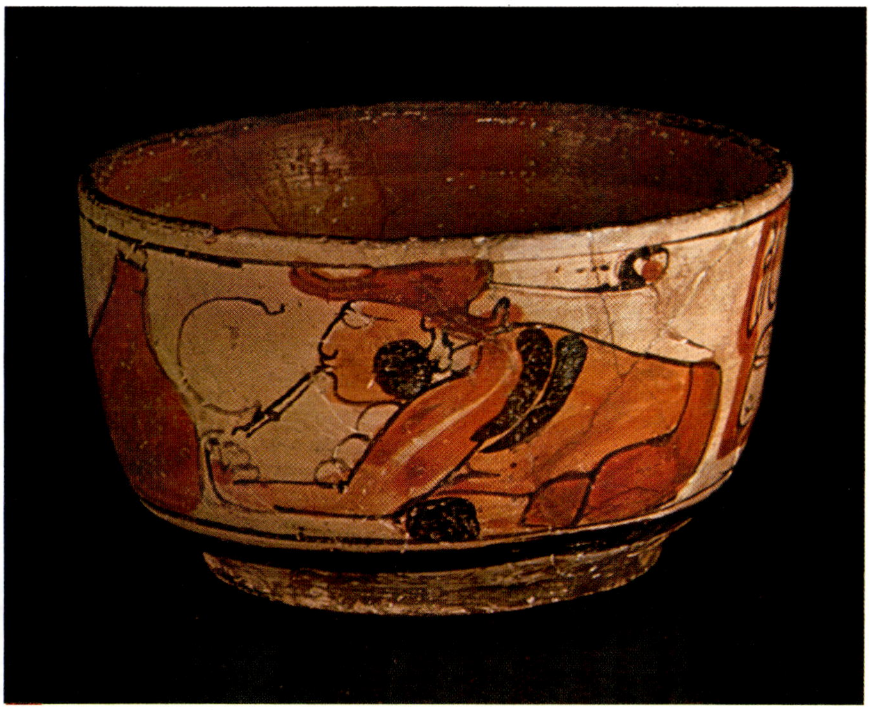

Plate 115

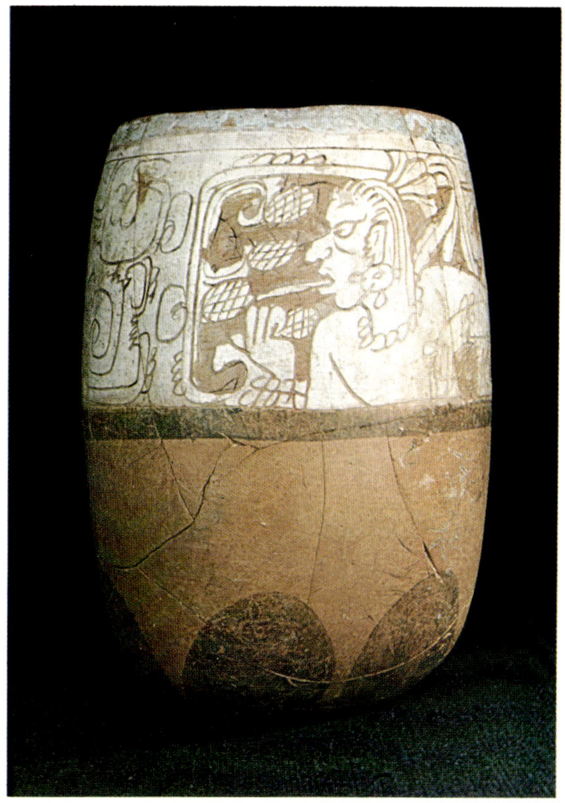

Plate 117

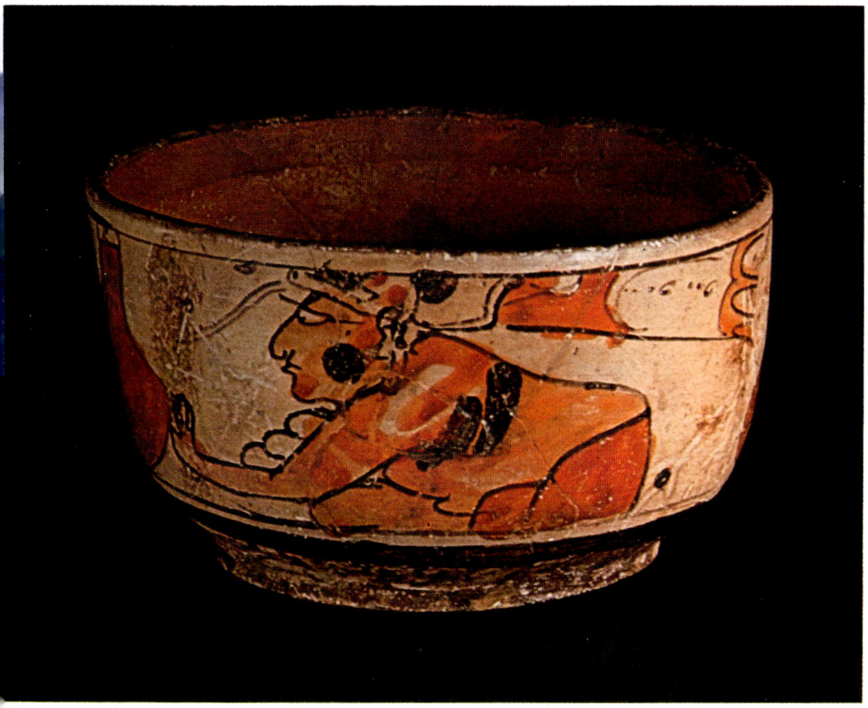

Plate 116

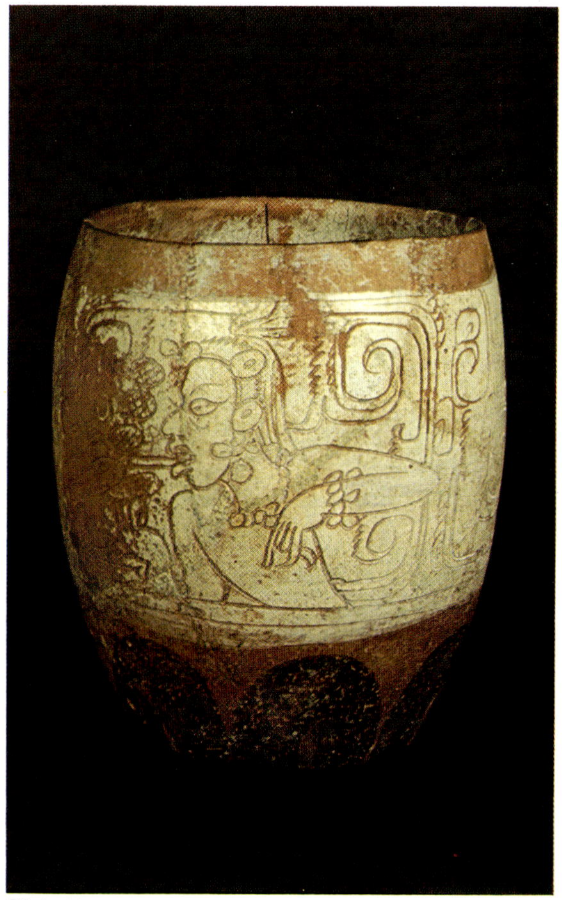

Plate 118

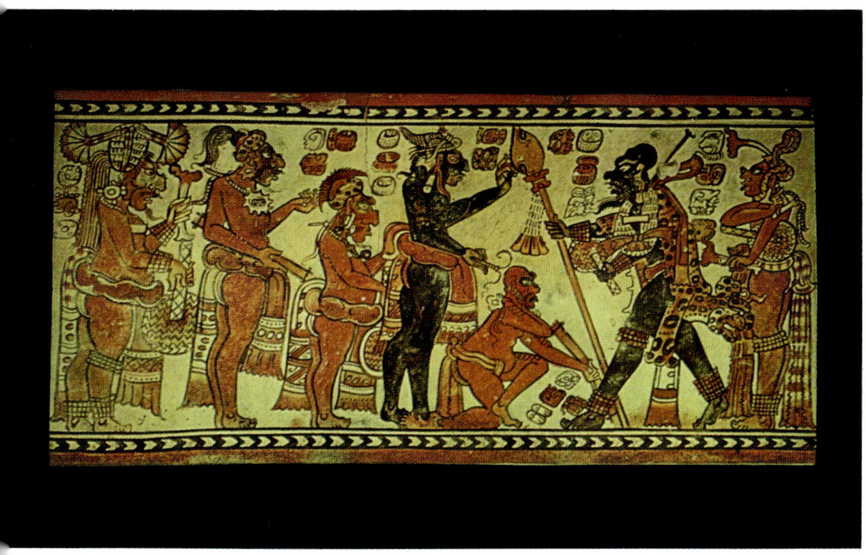

Plate 119

Plates 115-116. Bowl of the Crouching Lords (A-23)

Plate 117. Carved Vase, Lord Smoking I (A-7)

Plate 118. Carved Vase, Lord Smoking II (A-8) /Photograph Courtesy of Gillett G. Griffin/

Plate 119. Vase of the Black Warrior (C-8)/After Gordon/

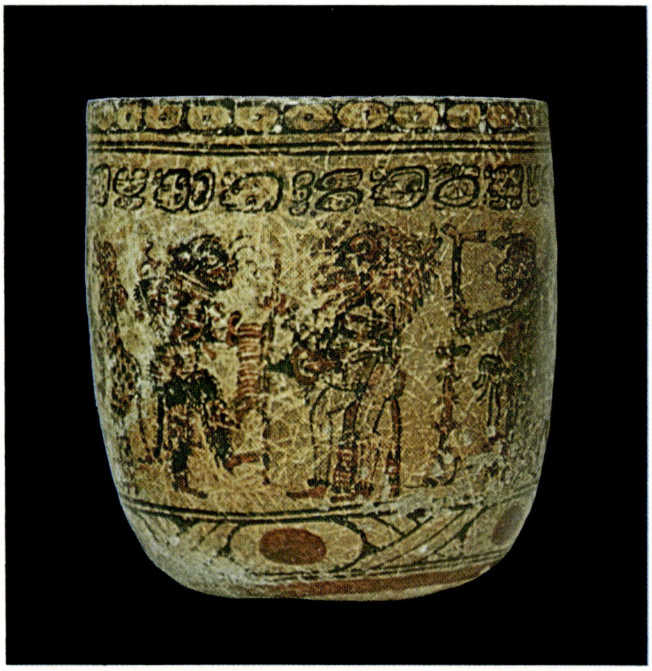

Plate 120. Vase of the Eleven Gods (A-16)

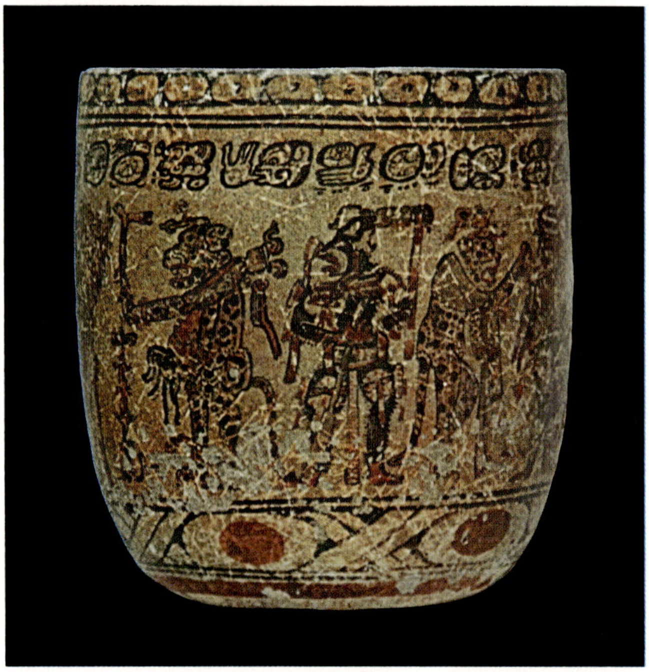

Plate 121. Vase of the Eleven Gods (A-16)

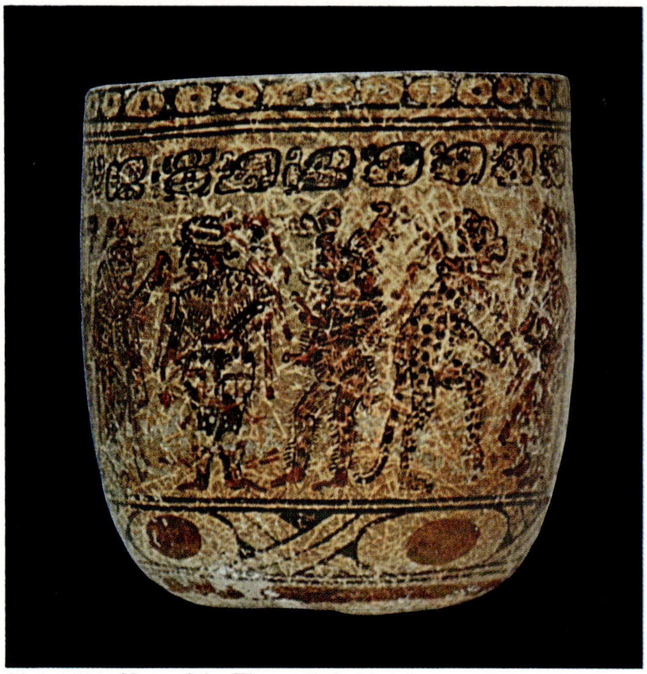

Plate 122. Vase of the Eleven Gods (A-16)

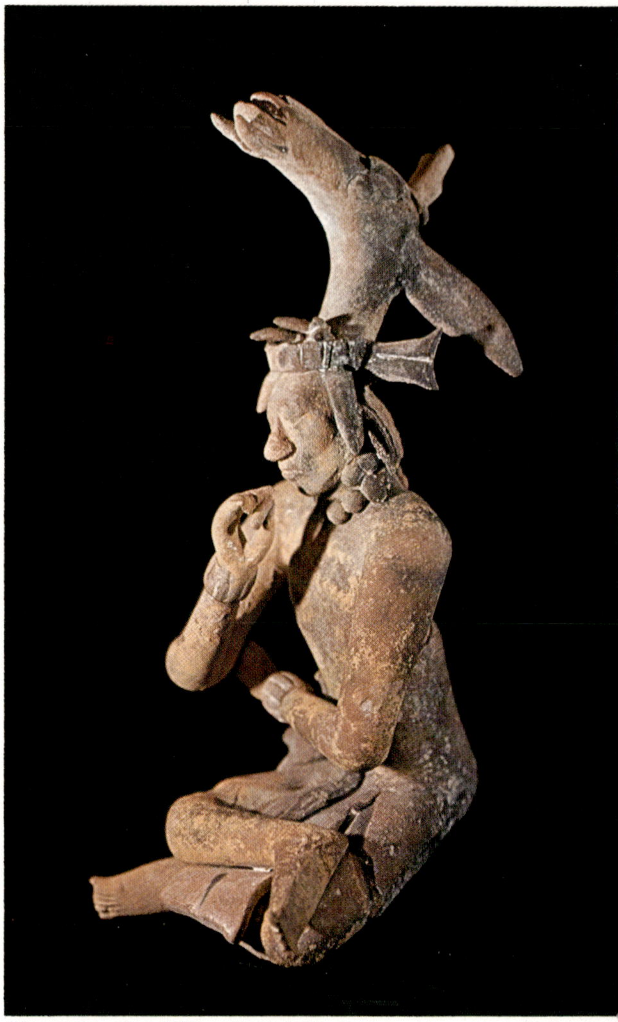

Plate 123. Lord Deer Seated (B-6)

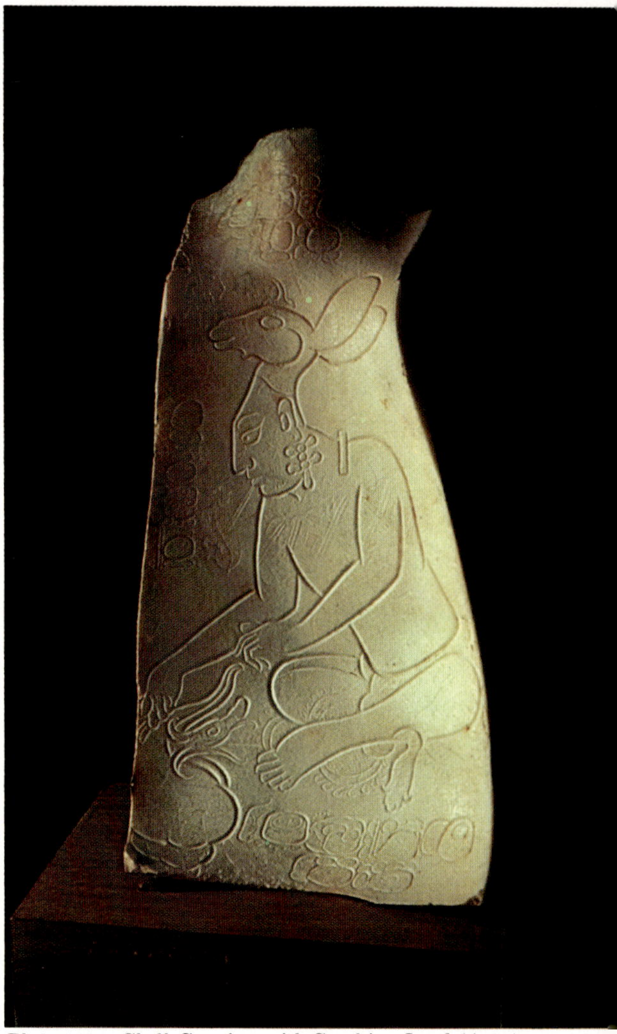

Plate 124. Shell Carving with Smoking Lord (A-21)

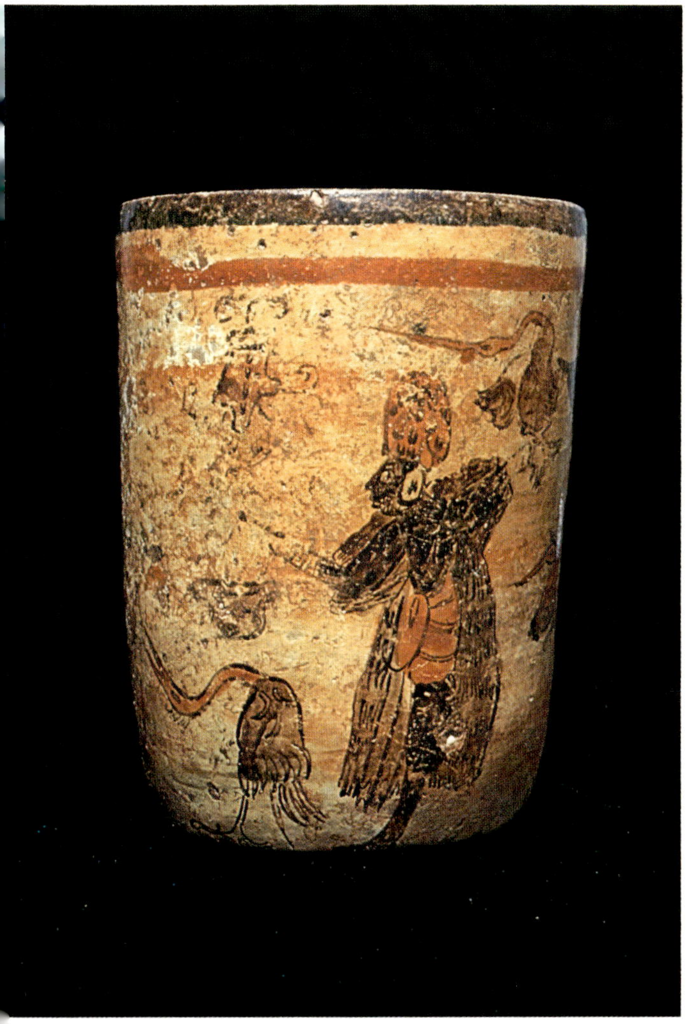

Plate 125. *Vase of the Water Lily Jaguar (A-11)*

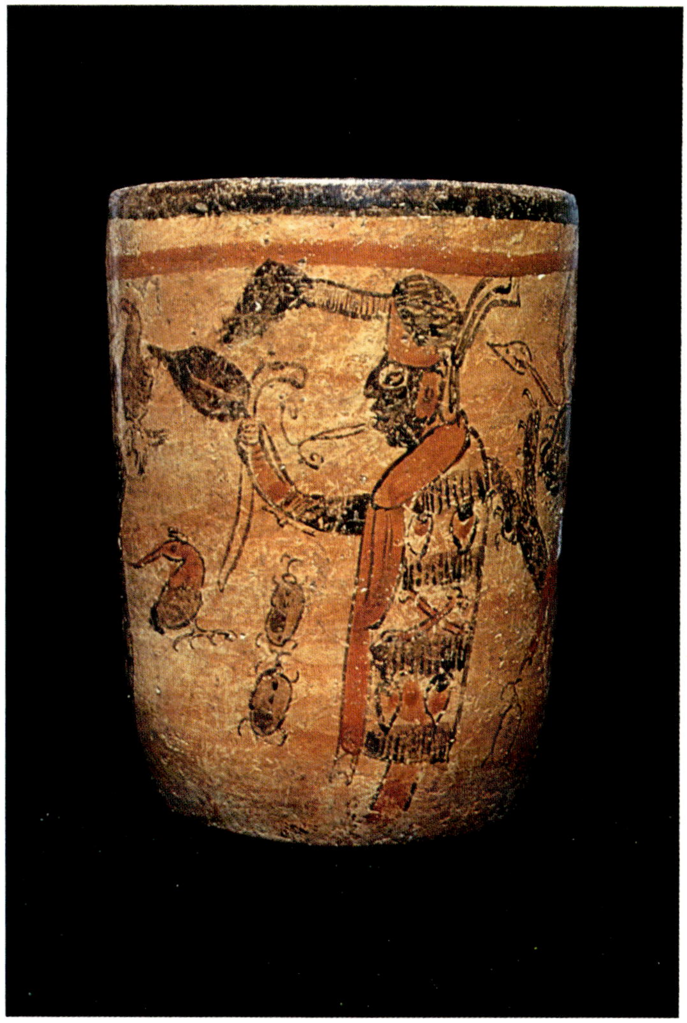

Plate 126. *Vase of the Water Lily Jaguar (A-11)*

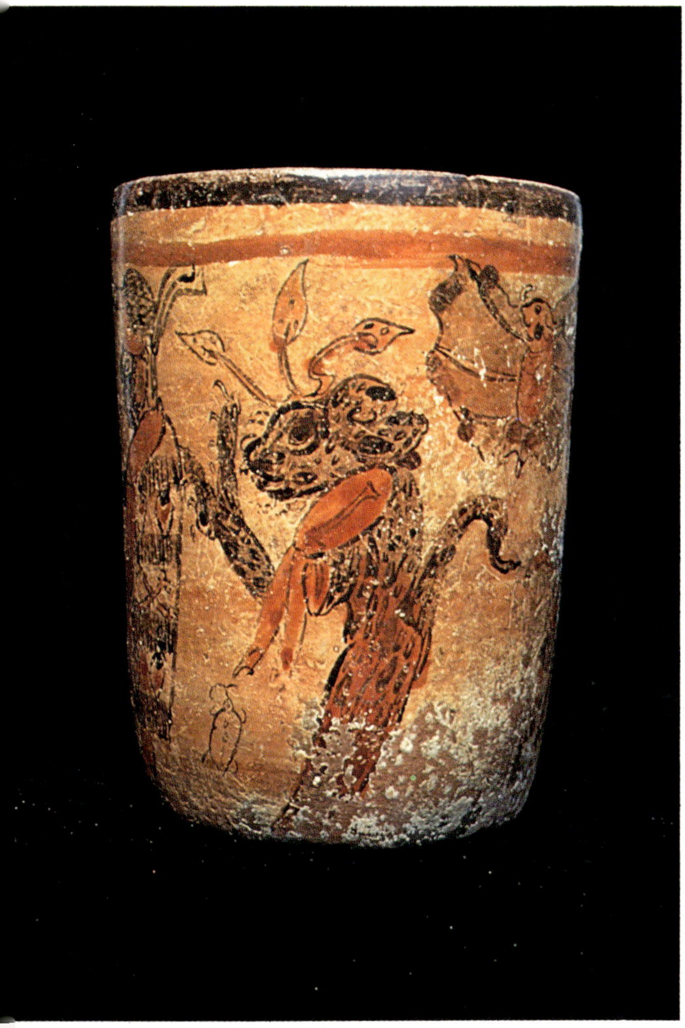

late 127. *Vase of the Water Lily Jaguar (A-11)*

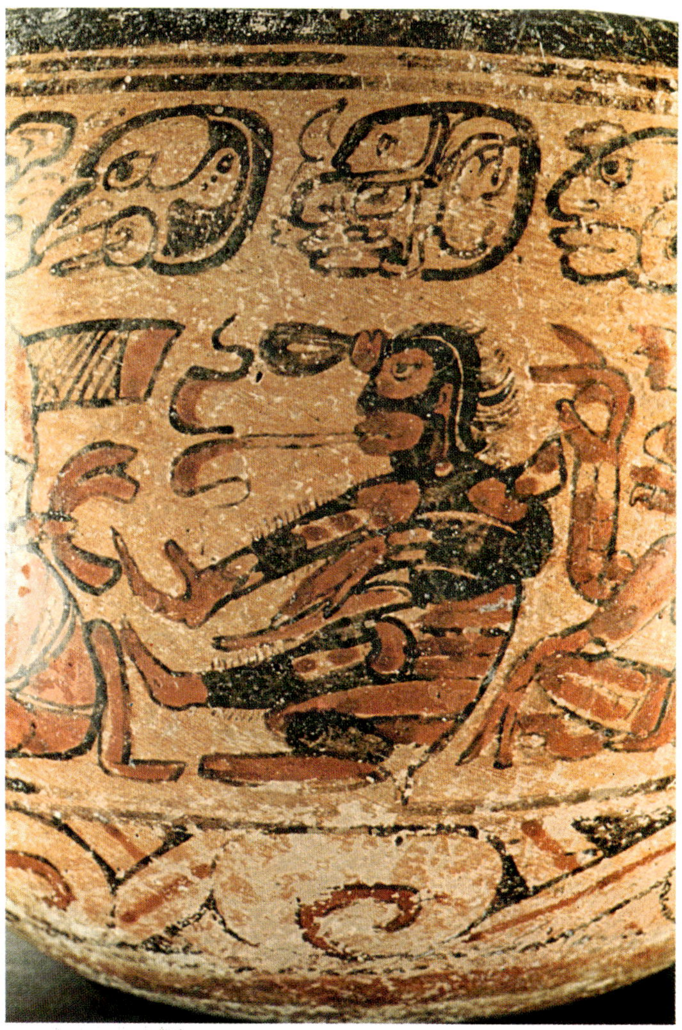

Plate 128. *Detail bowl of the Eleven Deities Five of them Smoking (A-13)*

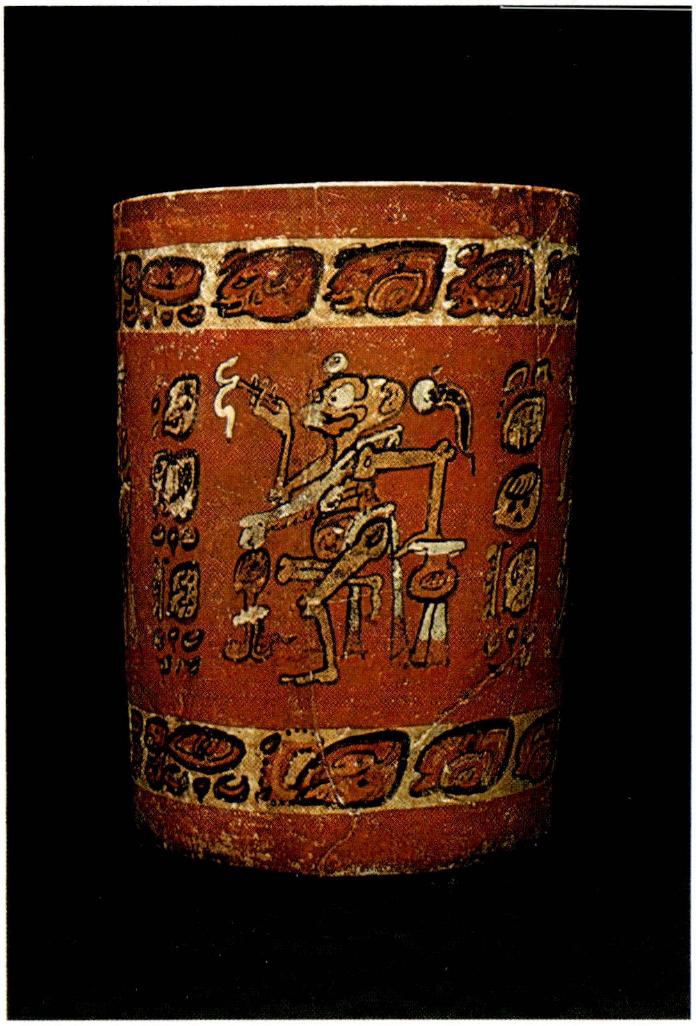

Plate 129

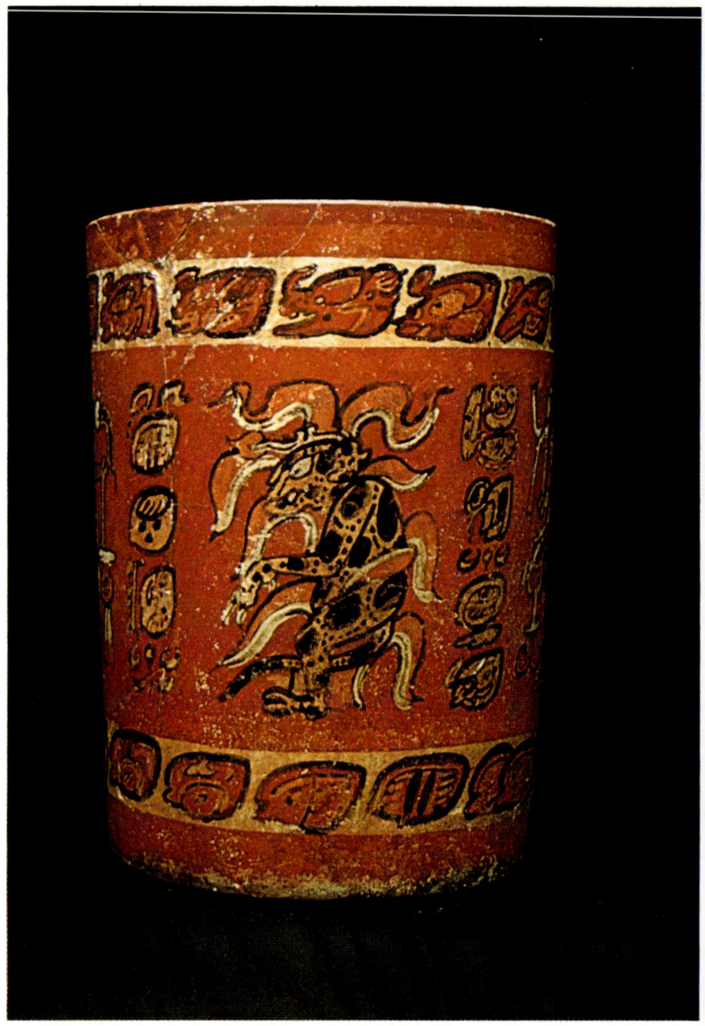

Plate 130

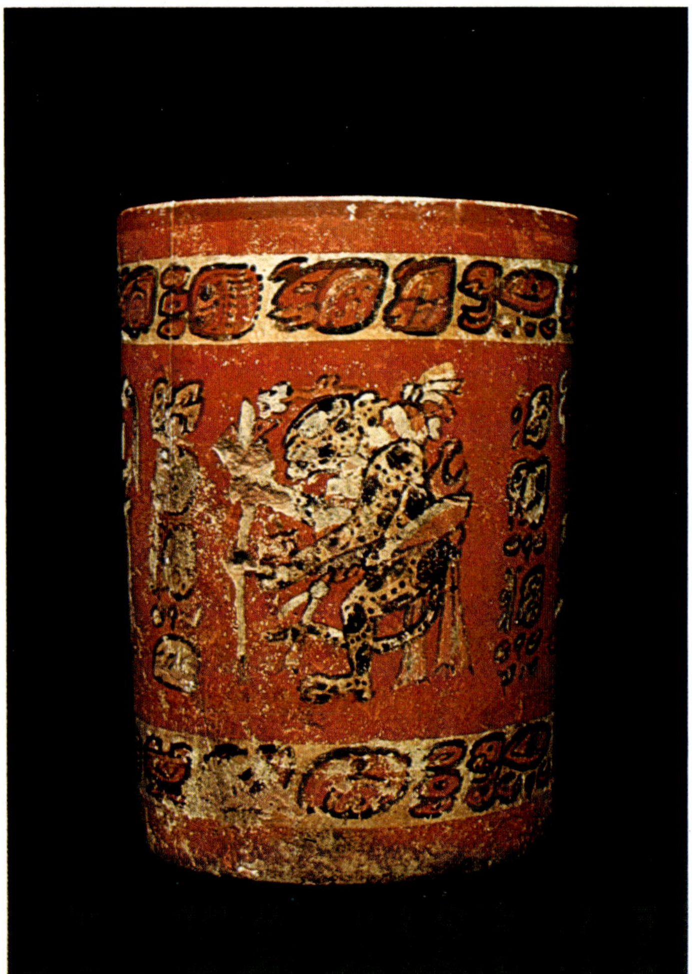

Plate 131

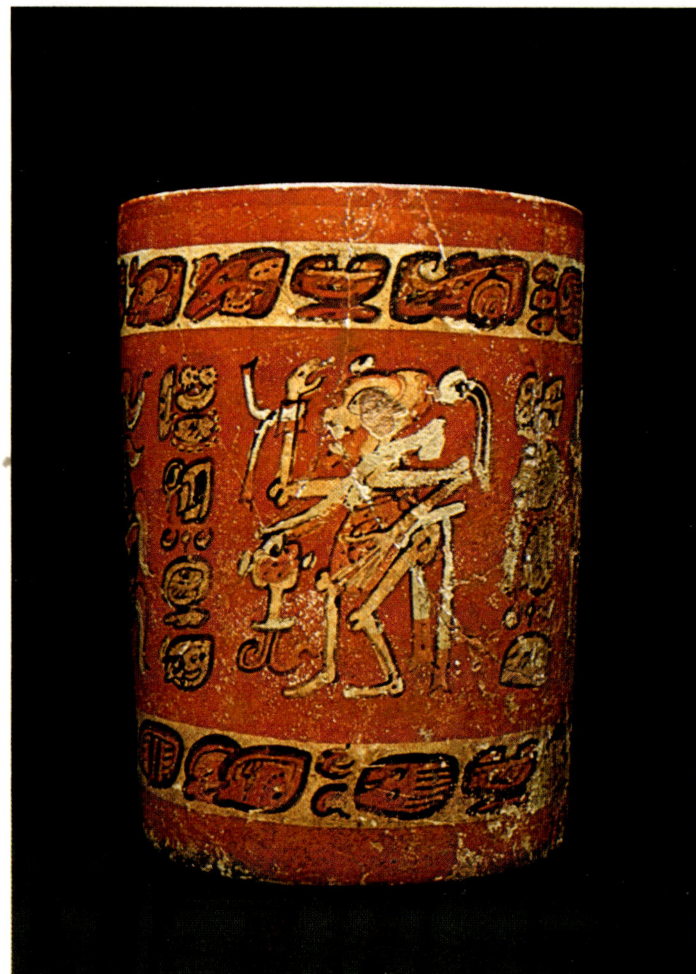

Plate 132

Plates 129-132. Vase of the Danse Macâbre (A-14)

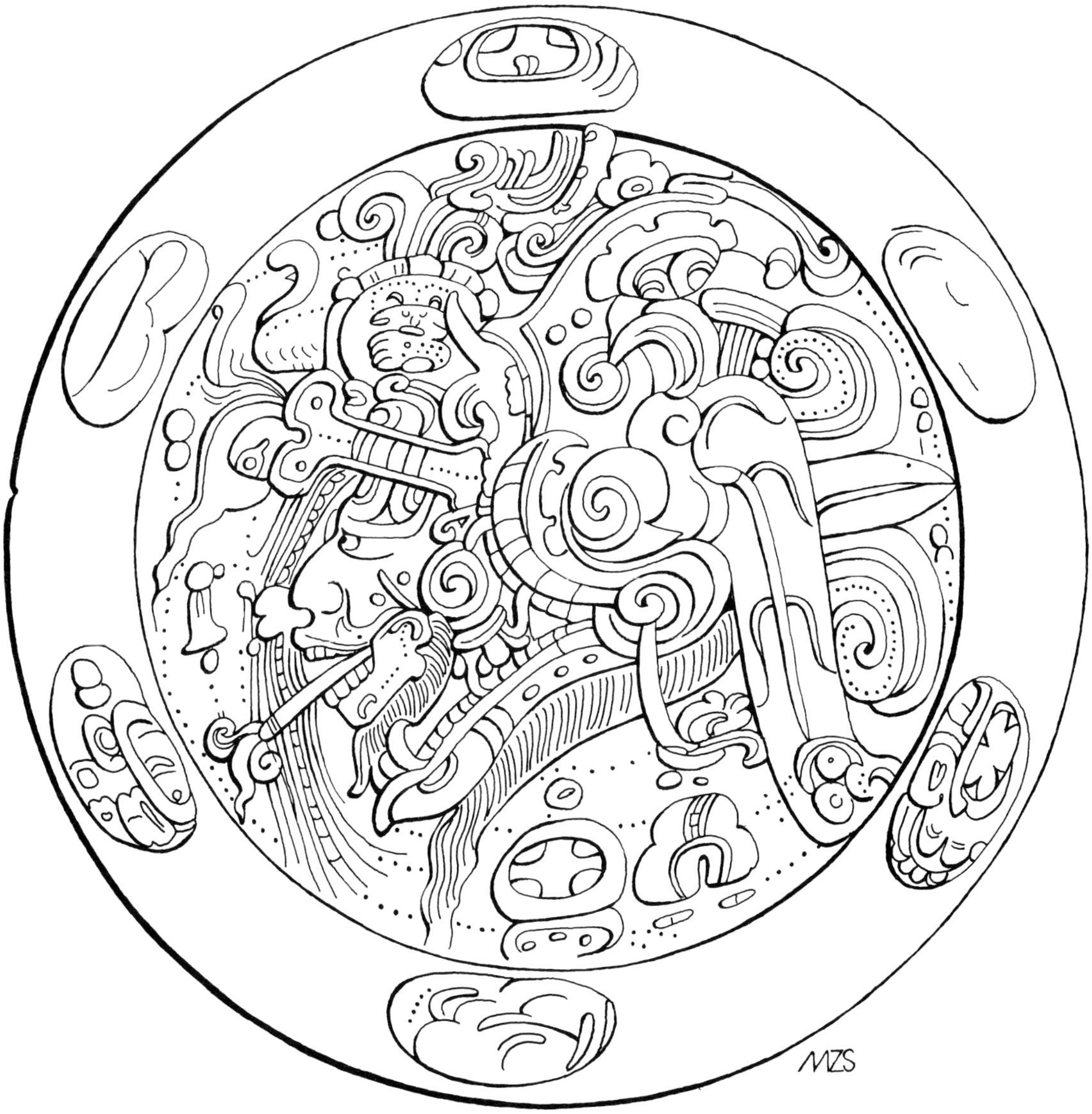

Figure 134. Plate of Smoking God (A-2)

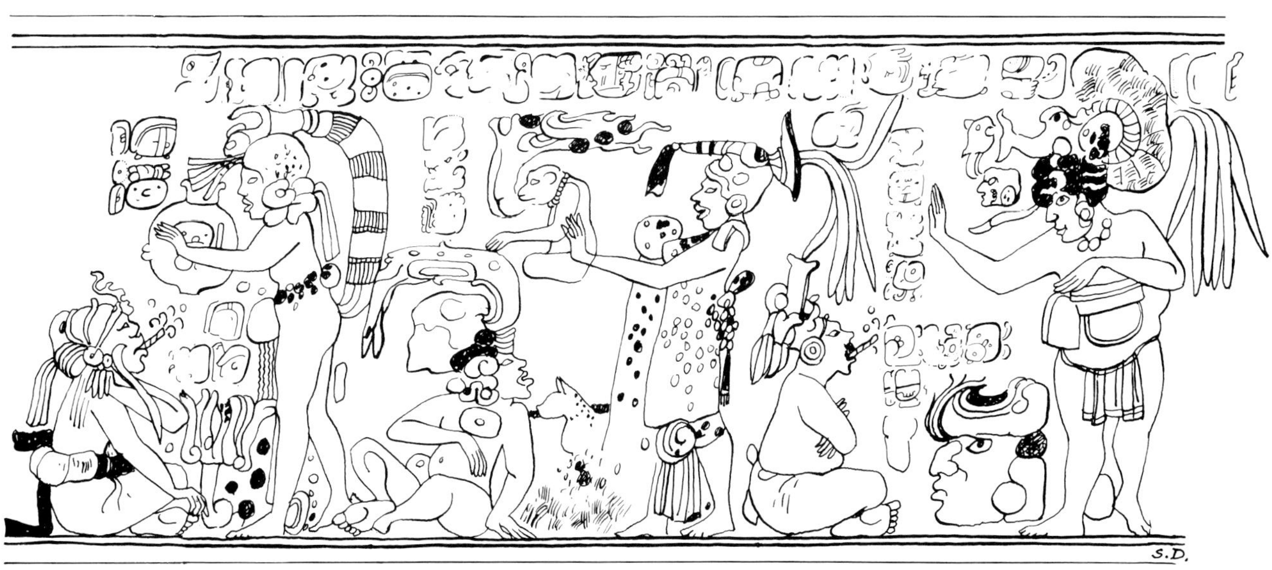

Figure 135. Vase of the Dancing Lord (A-3)

upper and lower margins, and is painted in bright orange and brownish green. The compostion of this scene is masterfully organized to present three pairs of human figures, one always standing on the right and one always seated on the left. It is not known for certain whether these pairs represent the same two persons three times or six different individuals. It should be noted, however, that each pair is closed inward by the following characteristics:

1. They face each other, and no elements in the composition are employed to separate them.
2. On two of the three pairs characteristic marks of identification are transferred from the standing main figure to the seated one. In the first pair the shared element is the similar headdress; in the second, the T757 "jog" animal.
3. The secondary texts are located within the intermediate space between the paired figures.

Each pair is dominated by the heavy, muscular figure of a standing lord. On the right-hand side of the scene he is in a graceful pose as if dancing. His right arm is extended and slightly raised; his left arm is held in front of his body in a palm-downward position. The lord's attire consists of a wide breechcloth with apron and a belly guard resembling the *sporran* of the Scots. He wears a necklace of what appears to be jade beads and a tall, truncated monster helmet decorated with the feathers of the quetzal bird. A small fish nibbles on the monster's nose.

Because it portrays an animal rare in the catalogue of Maya zoomorphs, the headdress is of particular interest to the iconographer. The nose is certainly that of a tapir, but the rest of the head reminds one of a rabbit or a deer. The animal head resembles some of the features of the beast used to identify the month Xul (T748). Since Xul is the generic root for "animal" (in Kekchi it is called *xul*), one may speculate that the purpose of this odd combination of features was to ensure that the grapheme would be recognized as "animal" (*xul*), rather than as a particular species, such as rabbit, deer, or tapir. The same combination of characters appears as a warrior headdress in Room 2 at Bonampak and in the name compound of Lord Kan-Xul II (Hok) of Palenque.

The dancing lord seems to be looking down at a large monster head painted green, which may relate to the zoomorphic pedestals on the fronts of the stelae at Copán and Quiriguá. At the left of the head sits a light-skinned attendant wearing conch-shell ear ornaments and an animal headdress similar to that of the dancing lord. He is smoking a slender white cigarette.

The standing figure at the left of this first pair is dressed in a skirt-length shirt. He (or she) is wearing a bulky "tie" of spotted cloth. He holds a small animal at arm's length, the T757 "jog," identified as a combination jaguar and dog by Thompson and as a rodent by Proskouriakoff.[24] From the mouth of the animal emerges an immense, elaborate tongue, a motif similar to the one on the Hieroglyphic Stairway at Copán. The symbol of the animal also appears on the headdress on the figure at the left of the standing personage. The pose of this figure is semireclining and reminds one of the tormented prisoners of the Banampak murals.

Turning the vase counterclockwise, we see another pair of figures, again one standing, the other seated. The standing lord is simply dressed, has a turbanlike headdress with a cantilever rear ornament, of perhaps feathers, and holds an inscribed *olla*. A large water-lily plant emerges from behind his legs. Facing him is a dark-skinned individual seated Turkish fashion, wearing a ceremonial cap and plain loincloth and holding his left hand on his right shoulder in a submissive gesture. Wearing a rather grim facial expression (as though annoyed because he could not get his favorite blend), he is smoking a white cigarette.

Glyphic Text

The glyphs on the rim are badly eroded, but enough compounds can be read to identify them as the Primary Standard Sequence or one of its variants. The introductory glyph of the text (T617a:713a.181) is located immediately to the left of the headdress of the standing dancer. The Secondary Texts are damaged beyond identification. It is probable that they contained descriptions of the activity portrayed, and perhaps would have provided us with clues to the identification of the participants.

Another important representation of smoking in Classic ceramic art is the **Reynolds vase** (A-4), probably from the Usumacinta region (now in the collection of the Museum of the American Indian, Heye Foundation). The vessel is painted bone white with upper and lower narrow red rims (Plates 112 and 113). The composition is divided into four panels. Two of the panels are decorated with a repetitive shamrocklike design, a common symbol in scene dividers on ceramics. The other two panels are occupied by the portraits of probably the same young lord seated Turkish fashion. His body and part of his face are brownish red. He is wearing a yellow kilt imprinted with a symbol identical to that on the side panels, a tightly strapped belt, a long loincloth tied to a symbolic design resembling affix T60 and to another unidentified object.[25] Beaded lace reaches from his arms to his back, and in his ear is a flared plug. The two portraits are very similar; however, there are some subtle differences, as are

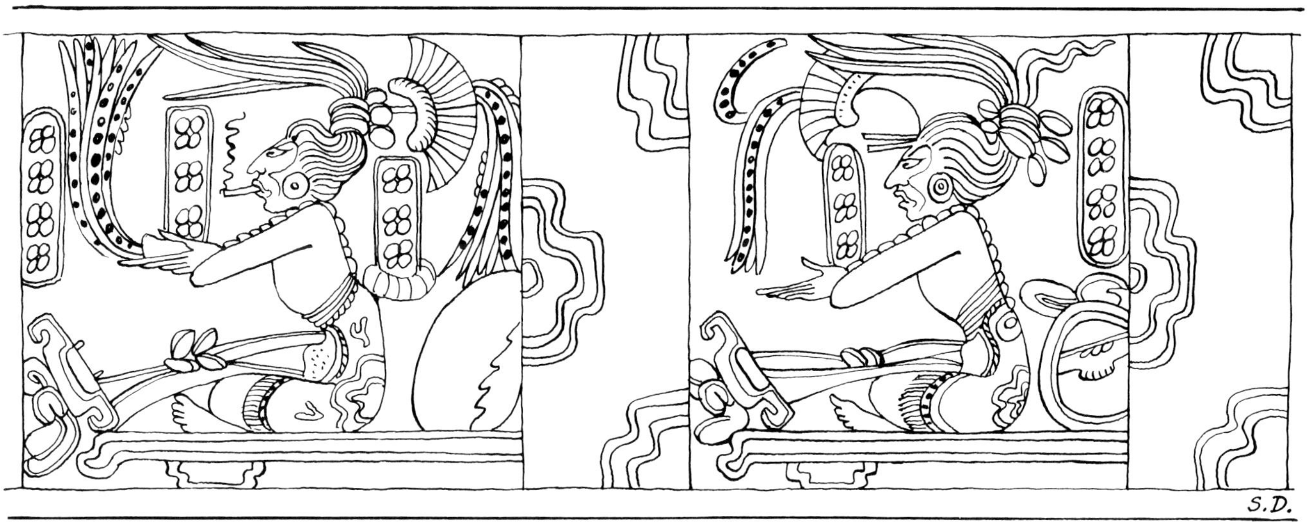

Figure 136. Reynolds Vase (A-4)

usually seen in repetitive portraits. In the first portrait the lord's long, manelike hair is pulled together in a distinctive *chignon* with three bands and four globular ornaments; in the other there are only three such ornaments. In the second portrait the hair is also arranged in a slightly different style.

In both portraits there is a fanlike ornament with floaters of what may be pheasant feathers. In the first portrait the ornament is in front of the lord, as though suspended in the air; in the second portrait it forms part of his headgear. There is a ceremonial "bundle"[26] in each tableau. In the first it resembles a large skyglyph (T561) tied into a bundle with a "knotted" affix (T60). In contrast to jaguar-hide bundles this kind of bundle is often shown on the ground rather than on a throne or on a dais.[27] In the second portrait the bundle has the appearance of a large cushion, stitched together in the middle. The Mayas portrayed many such bundles on both stone monuments and painted ceramics. Their exact significance is unknown. They may be but simple cushions, or they may be containers for important ceremonial objects.[28] Another difference between the two portraits is that in the first the lord is tight-lipped while in the second he is holding a white cigarette in his mouth from which a coil of smoke rises. Completing the scene are four short panels of glyphoids, two on each portrait, containing identical symbols of four small adjoining circles.

In interpreting this vase we must consider the possibility it shows not two portraits of the same person but pictures of two different people; namely, the Hero Twins of the *Popol Vuh*, Hunahpú and Xbalanque, who wandered through the realm of the underworld like Orpheus of Greek mythology and, after conquering hell itself, rose up to the sky to become heavenly bodies. The Hero Twins are portrayed on many Classic Maya pottery artifacts. Michael D. Coe calls the subject the "Theme of the Young Lords":

> [They] are a pair of identical or near identical young princes, richly garbed, with headdress to the front of which is usually fixed a jawless head. . . . Since they often exhibit the peculiar "god markings" (similar to Glyph 7 or 504) which indicate supernaturals in both codices and monuments, it may be assumed that the young Lords are also gods. I think it may be also assumed that they are the Hero Twins whose journey to the Underworld must be repeated by the dead person in whose grave the vase was placed. Occasionally another pair of youthful personages is encountered on vessels, but these are marked with black death spots and wear headbands. I have labeled these the Headband Gods and suggest that they are Hunahpú and Xbalanque during their Underworld sojourn.[29]

These "Headband Gods" are often shown smoking cigars.

The approach to deciphering some of the riddles of Maya ceramics by using the surviving legends of the Mayas has been developed by Michael D. Coe over the past decade. This methodology closely resembles that of Heinrich Schliemann, who followed the path outlined by Homer's *Iliad* to discover the ruins of Troy. Coe took the characters of the deities and demigods of the *Popol Vuh* to breathe life into the puzzling array of human and zoomorphic creatures of the ancient ceramic art. His approach has been criticized primarily on the basis that, while the most important *objets d'art* of Maya ceramics were painted between the third and ninth centuries

A.D., the extant copies of the *Popol Vuh* were not written until about seven hundred years later. If, however, one considers again the analogy of the blind Greek poet, whose work was based upon collections of ancient legends, folklore, and historical fragments, some of them reaching far back into prehistory, one finds Coe's theory both plausible and attractive.[30]

Smoking is also depicted on the **Vase of the Smoking Lords** (A-6), a polychrome vessel from the late Classic period, probably from Petén (now in the Museo de Popol Vuh, Francisco Marroquín University, Jorge and Ela Castillo collection). The tall ceramic cylinder is painted red and black on a white base. It has the traditional upper and lower rims and an upper circular glyph band. Two vertical columns of large T544 *kin* glyphs divide the composition into two tableaus, each showing the figure of a noble dressed only in loincloth, seated cross-legged on a low dais, and puffing billowing curls and rings of smoke from a large cigar. The manelike hairdos of the two lords are held together with simple, bandlike headdresses and, like the figures on the Reynolds vase, with four round pompom ornaments. Again the question arises whether or not two different individuals are shown, such as the Hero Twins, as Coe theorizes, or different aspects of the same person. Jacinto Quirarte, like most contemporary art historians, shies away from Coe's hypothesis, believing that vases painted with such "duplicate" portraits depict the same historical figure, not necessarily in different actions but in different aspects and maybe even in different dimensions.[36] Because such duplicate portraits are always painted on opposite sides of the vases, and thus only one of them can be viewed at the same time, Quirarte's theory certainly also appears to be a plausible one.

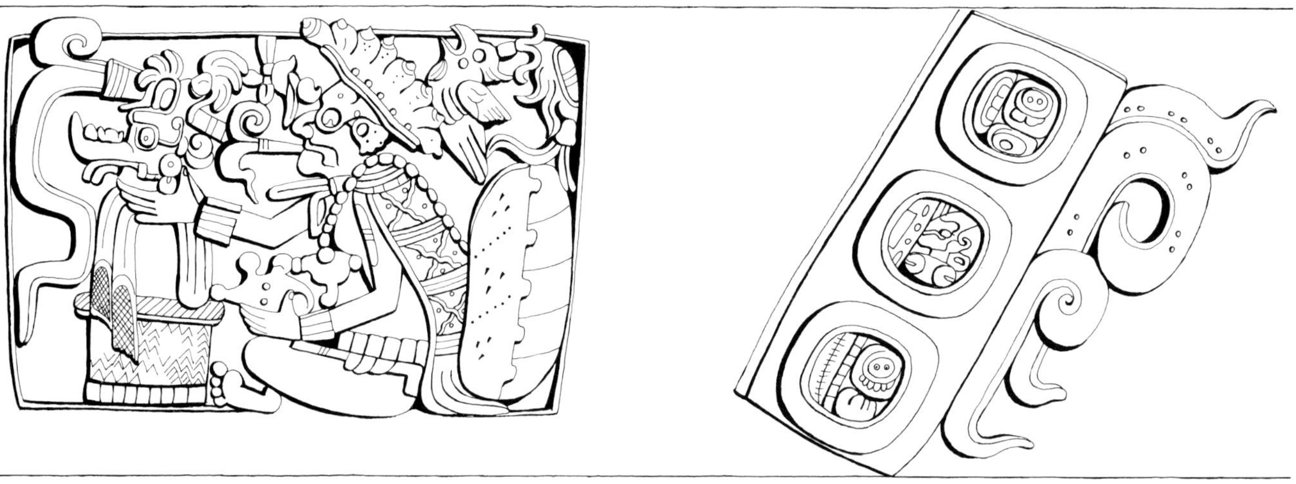

Figure 137. Carved Vase Chocolá Style, God L Seated (A-5)

A "prize find" for one studying representations of smoking in Maya art is the carved vessel Coe identifies as **Carved Vase Chocolá Style and God L Seated** (A-5).[31] The vessel (now in a private collection) is from the late Classic period of Chocolá, Yucatán. It shows the seated figure of God L, one of the two principal gods of the Maya underworld.[32] According to Matthews, God L is the underworld counterpart of Itzamná, the supreme Maya deity.[33] He is, considering his aged features, a remarkably active deity who, Coe says, "lives in a sumptuous palace, has sexual relations with one or more female deities, smokes cigars, rules over various divinities."[34] His identity is revealed by his large eyes, aquiline nose, and feather-and-Moan bird headdress. There is a large smoking cigar in his mouth, and in his hands, like Judith holding the head of Holofernes, he is grasping the head of God K, possibly in the act of placing it in or lifting it from a woven basket. The forehead of God K is pierced with what appears to be a large cigar emitting coils of smoke. The back of the vase is carved with a cartouche of three hieroglyphs flanked by elaborate scrolls, possibly smoke.

For a discussion of the glyphic text see Michael D. Coe.[35]

Smoking is shown on the near-identical **Carved Vase of Lords Smoking I** (A-7, Plate 117) and **Carved Vase of Lords Smoking II** (A-8, Plate

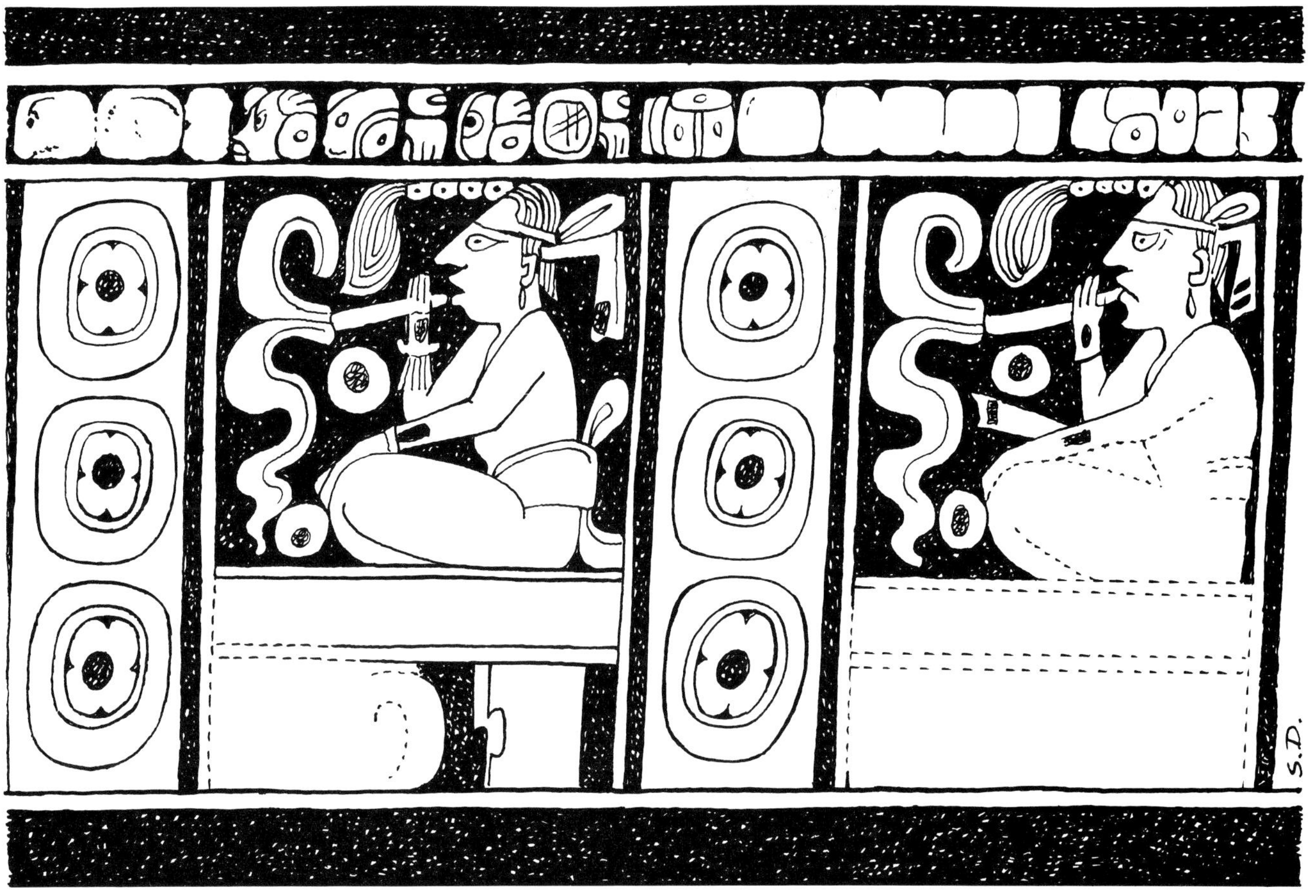

Figure 138. Vase of the Smoking Lords (A-6)

118), probably from the late Classic period, provenance unknown, both undoubtedly made by the same artist (one of the vases is in the collection of the Montreal Museum of Art, the other in a private collection). The vessels are burnished gray with an upper and middle black band and large black semicircles on the base. The upper half of each vase is rather primitively carved (evidently after firing) with the torso of a noble bent slightly forward. His hair is gathered, and his headdress appears to be made of feathers. He wears a circular ornament suspended on his pointed ear and necklace and wrist beads, probably jade. In his mouth is a long, slender cigar. Around the tip of the cigar are several crosshatched circles, probably representing puffs of smoke. The entire figure is enclosed in a tableau flanked by glyphoids.

The smoking lord is also portrayed on the **Shallow Bowl** (A-9).[37] The vessel was undoubtedly made by the same artist who made Vases A-7 and A-8 and bears near-identical decoration.

Gods and mythological animals are portrayed on the very important **Vase of the Smoking Monkey** (A-10), a ceramic vessel of the late Classic period, probably from the Maya lowlands (now in a private collection) (Plates 108–11). The vase has red and black rims and an upper glyph band, and shows four standing figures on a bright-orange base. The first figure is that of a man bent slightly forward with a smoking cigarette in his right hand. He wears a white turban, a short red cloak, a tricotlike body cover that strongly resembles "longjohns," and narrow decorated front and back aprons. With his left hand he is leaning on what could be an ornate blue fan or a back cover, which is often seen on backs of deer, a favorite animal of Maya mythology.

The human figure is followed by a peculiar beast with the body of a roebuck and the face of a monster with huge eyes and a truncated nose. A glyph, identified by Schele as T542a and by Barthel as T542b, is tattooed on the animal's ear.[38] The same glyph also appears on the name compound of one of Palenque's rulers (Tablet of the Cross, P10–Q16).

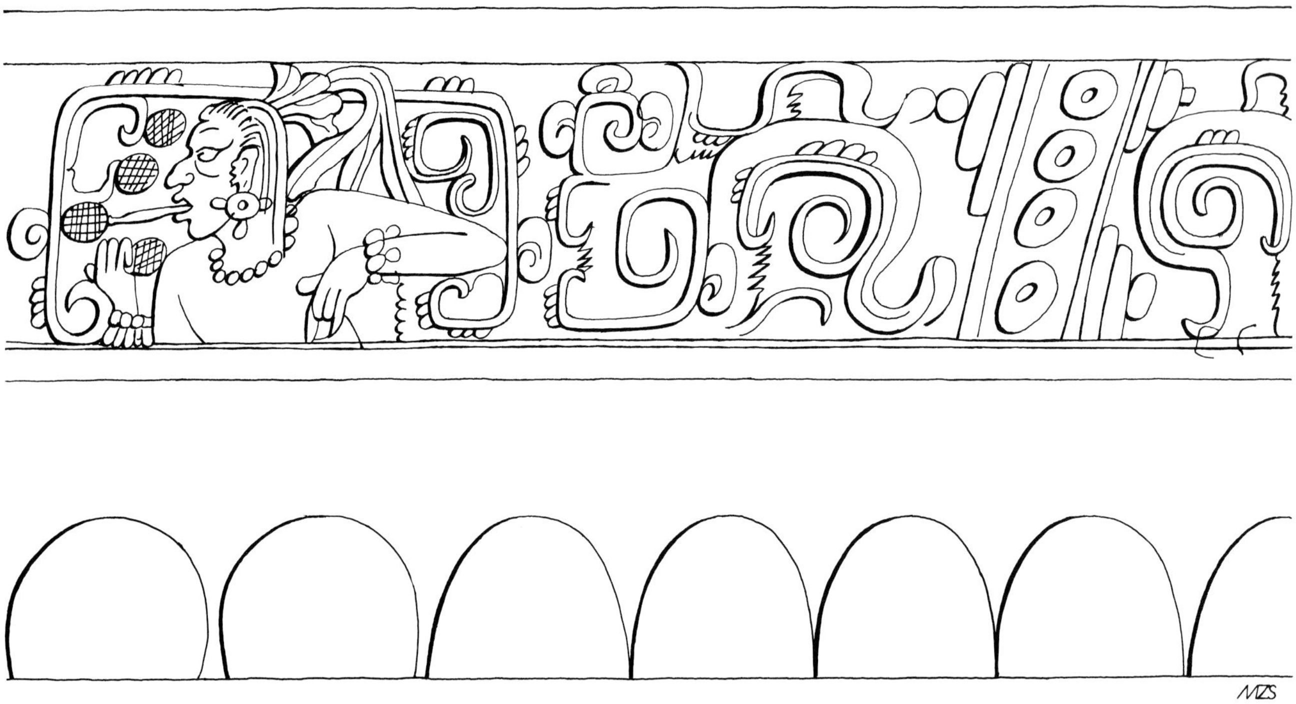

Figure 139. Carved Vase of Lord Smoking I (A-7)

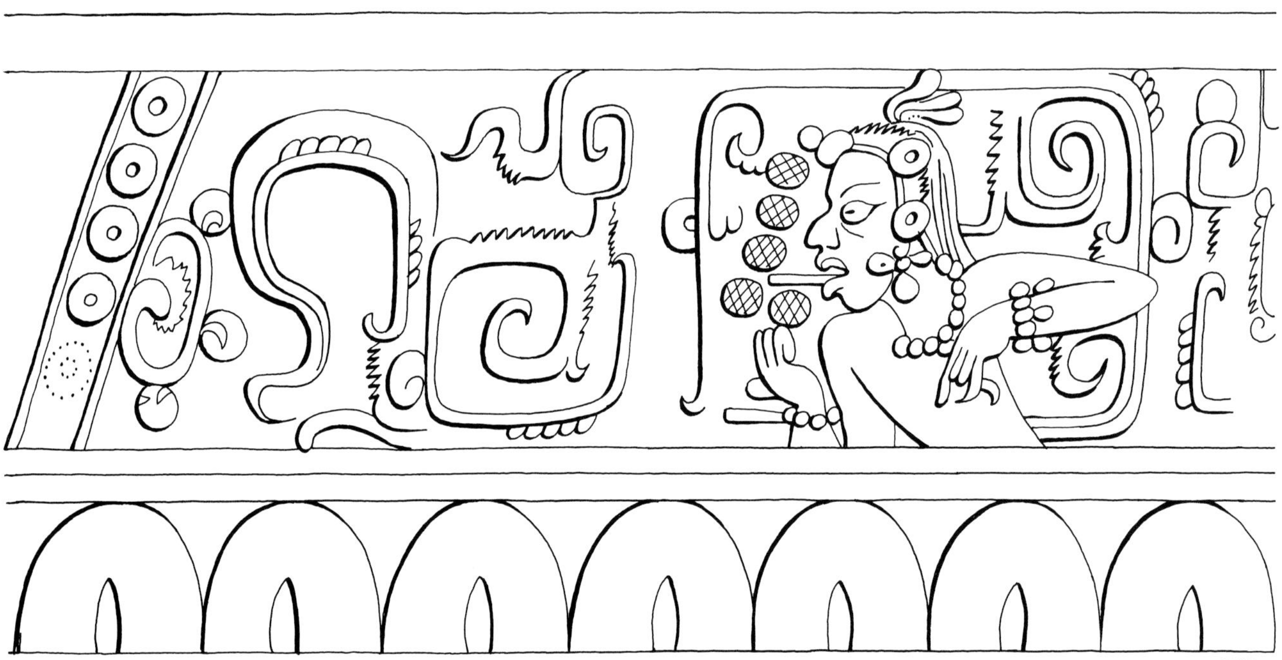

Figure 140. Carved Vase of Lord Smoking II (A-8)

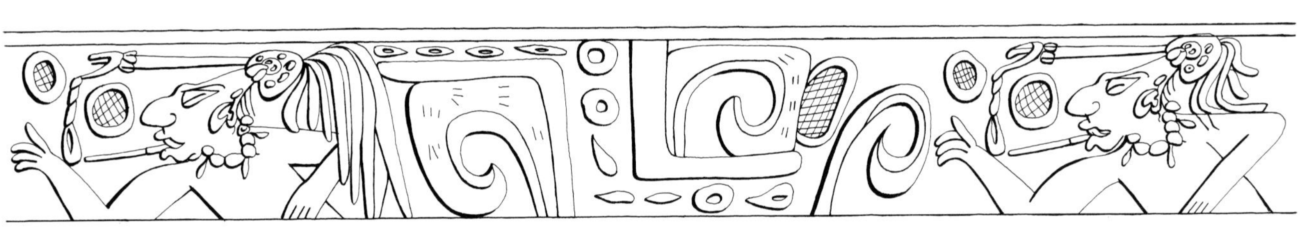

Figure 141. Carved Bowl of Lord Smoking (A-9)

The forelegs of the monster deer are caught in a large net. Like the other three figures in the scene the buck is wearing a short shoulder cape. He is followed by a second man who wears a white loincloth, a cape, which also covers the back of his head, and gaiters, and in his right hand he holds a vase decorated with what appears to be an *akbal* glyph. In his left hand is a flint knife. In my opinion this deity is the first member of the triad of gods identified by Berlin as GI, GII, and GIII at Palenque. GI is a multifaceted deity who not only is portrayed as the protector of the Temple of the Cross but also appears on Stela 11 at Yaxchilán and on Stela I at Copán, is shown in some of the fishing scenes on the carved bones of Tikal, and is pictured on several painted ceramics of the Classic period. It is probable that some of these representations of GI were intended to portray not the god himself but rulers impersonating him, just as the emperors of Rome and the kings of medieval France fancied having themselves portrayed as gods of Greek and Roman mythology. The characteristic features of God GI include shell earplugs, long gathered hair, and an eccentric flint knife, and he is often shown carrying an *akbal* jar. Kelley identifies GI as Kukulcán, GII as Bolon Dzacab, and GIII as the Death God (Lord of Xibalbá).[39] Coe has also related GI to the central figure of the Dumbarton Oaks Tablet, who is probably Lord Hok, one of the rulers of Palenque.[40] Last in the row, standing with his tail curling forward between his legs, is the most self-confident-appearing male monkey shown on known Maya ceramics. He too is wearing a shoulder cape. He is holding a large cacao pod in his right hand and is smoking a white cigarette. The animal is a spider monkey, as evidenced by the lack of the fifth digits from his hands and feet. His head is very similar to the head of God C (T1016), which supports the theory that the God C grapheme derives from the head of this animal.[41] The first and third figures in the tableau have blood-stained wristlets and gaiters made of paper or cloth. Such wristlets are usually signs of supernatural beings.

The figures in this scene, the monkey and GI, are also portrayed on another polychrome vase in the collection of the Museo de Popol Vuh, undoubtedly painted by an artist of the same school as that of the artist above. This artist was less talented, however.

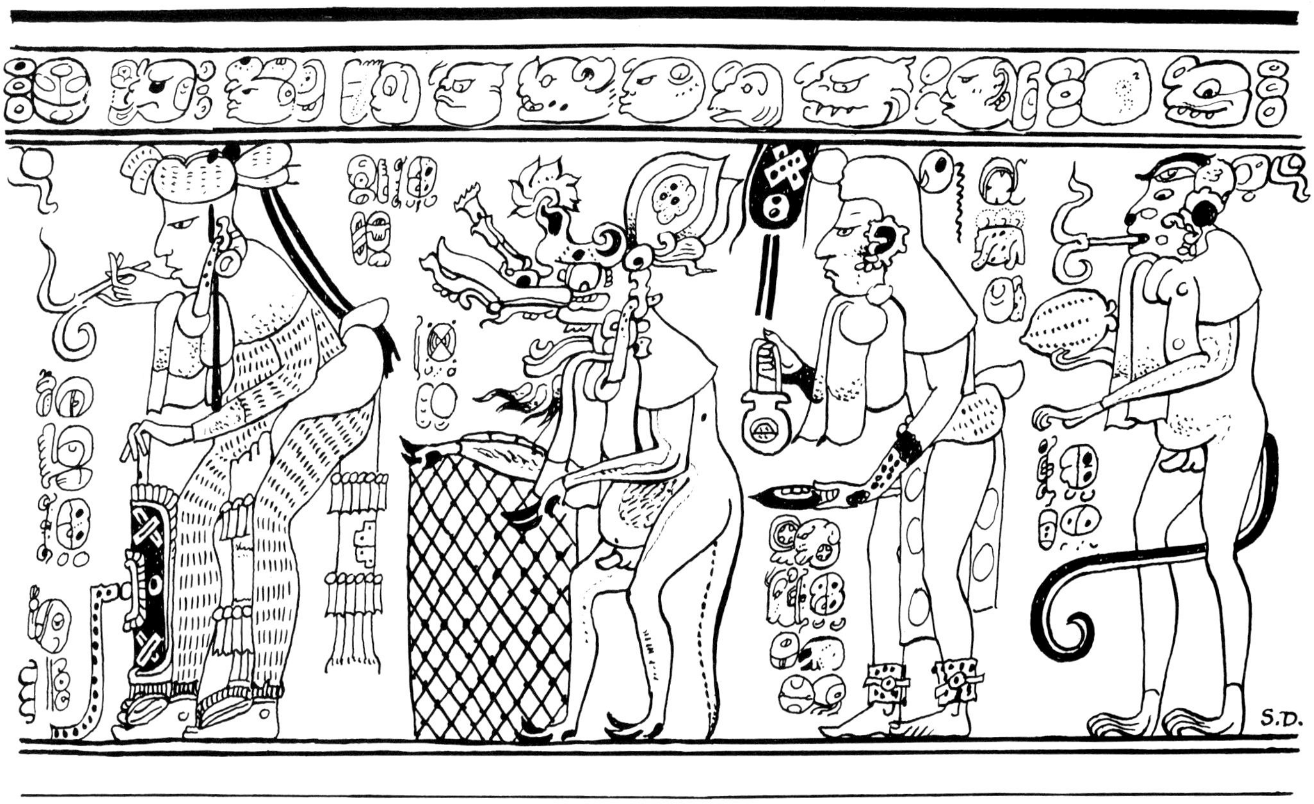

Figure 142. Vase of the Smoking Monkey (A-10)

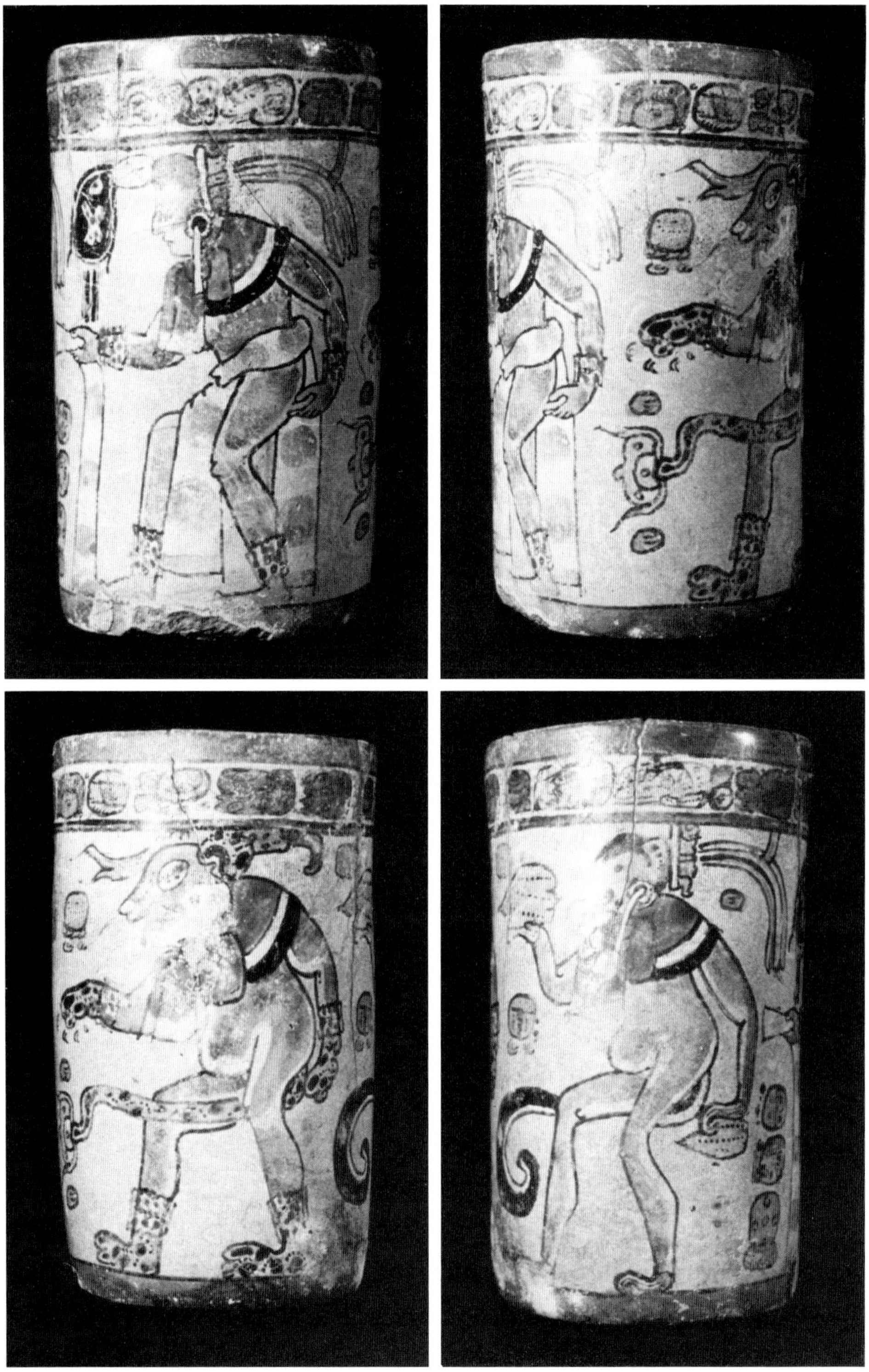

Figure 143. Polychrome vase of the same style and provenance as vase A-10, but probably executed by a less-talented follower of the artist. It shows essentially the same characters, but the deer is in jaguar disguise, and the monkey is not smoking (Museo de Popol Vuh, Francisco Marroquín University, Jorge and Ela Castillo Collection)

Glyphic Text

The glyphic text is a processional form of the Primary Standard Text. The introductory glyph is a zoomorphic variant of GI (T1011) and is placed over the front part of the headdress worn by the first figure. The Secondary Texts are very interesting. Each consists of four or five glyphs adjacent to the four figures. After every series of glyphs comes a jaguar-spotted *ahau* (T3.539), which in our opinion may be an indicator of a nominal phrase. This glyph is followed in each case by two other compounds.

There is evidence gained from the analysis of other processional pots that T3.539 functions in Secondary texts as a nominal indicator. If this hypothesis is correct, the secondary phrases on this vessel are of particular importance. The phrase adjacent to GI, the third figure in the scene, is composed of four glyphs with the nominal indicator in the second position. The compound is made of T281:23, which reads as *kan (na)* and T757. These two glyphs reoccur in the same relationship in the name compound of Kan-Xul II (Lord Hok), of Palenque, who is shown dancing as GI on the Dumbarton Oaks Tablet.[42] On this vessel a deity is recognized iconographically as GI of the Palenque Triad and identified on this vase by the glyphs T281:23.757. Since rulers other than Kan-Xul also have been portrayed as GI, we cannot attribute the GI = Kan-Xul II association at Palenque solely to the relationship of the probable names, but rather we accept the possibility that the Palenque Tablet which shows Kan-Xul as GI may have been a play on the relationship of the names. Since the glyph imprinted on the ear of the dragon-deer was also used as a ruler's name at Palenque, it is possible that at least some of the royal names of the Classic Maya were drawn from the names of gods.

The monkey is accompanied by a text of five glyphs. The nominal indicator occurs in the fourth position so that the name may be composed of the first three compounds. The first compound (T544.116) is the *kin* glyph. The second compound (T44:832) is a headless jaguar. A close parallel to this phrase is found in the Tablet of the Sun, C2–D2, in the nominal compound of GIII of the Palenque Triad, who is regarded as the Jaguar God of the Underworld and the Night Sun. Another close parallel is found on Lintel 47 at Yaxchilán in the apparent nominal phrase of an early Classic ruler. It is difficult to explain the association of the *kin* and headless jaguar graphemes with a spider monkey, especially since the graphemes are so closely associated with GIII, a jaguar deity. The association may be partly involved with a possible relation to the complex of sun deities.

The remaining nominal phrases have no recognized associations. The dragon-deer has a T561 *caan* (sky) compound, but the other compounds do not appear in other examples that we have so far recognized.

Smoking is also depicted on the **Vase of the Water-Lily Jaguar** (A-11), a polychrome vessel from northern Petén (now in a private collection) (Plates 125–27). The vase has a bright-orange base with red and black rims but no glyph bands. The scene shows four figures: three nobles painted black, signs of both war and underworld, an association reinforced by the presence of an anthropomorphic "water-lily jaguar," one of the rulers of the nether land. Above his head hovers a fret-nosed bat, perhaps Zotz, the Bat God. One of the nobles wears a tall red turban, a black cloak, flared earplugs, and a wide apron. He appears to be pursuing a long-necked water bird, probably a heron. The water-lily jaguar is easily identified

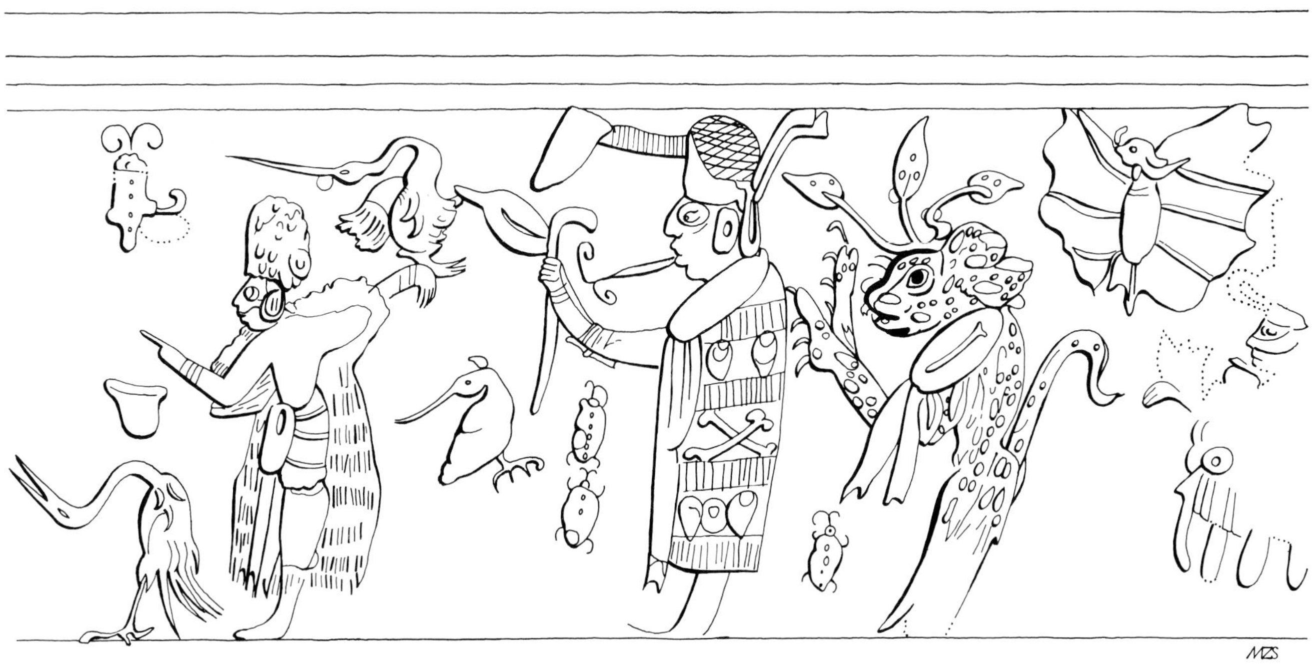

Figure 144. Vase of the Water-Lily Jaguar (A-11)

from the vegetation growing from the crown of his head.[43] There is a "hump" on his back, but whether or not it represents part of his attire cannot be determined from the painting, which is damaged.

He is followed by a waterfowl and two large insects. After them comes another lord, dressed in a turban, a shawl, and a long ornamental caftan. He is holding an object, probably a ceremonial ax, and is smoking a long white cigarette. He is followed by a standing jaguar, who probably has a sore throat because he is also wearing a red cloth around his neck. Above his head is a triad of water-lily leaves. The water-lily jaguar is probably the most commonly portrayed animal in Maya art. He is one member of the family of Maya jaguar deities symbolizing the night and the nether regions. He is usually portrayed with water-lily vegetation either on his head or surrounding him in a cartouche. He appears with his name glyph on page 8a of the *Dresden Codex.*[44] The portion of the vessel on which the third noble is painted is so badly eroded that only the outlines of the figure are visible.

The **Vase of Decapitation Sacrifice** (A-12) is a white incised ceramic vessel, probably from the middle Classic period of Petén (now in a private collection), shown in Nicholas M. Hellmuth's *Classic Maya Art Showing Dances after Decapitation Sacrifice.*[45] It is carved with an upper circular glyph band and shows a complex, esoteric scene. Parts of the composition and some of the glyphs are damaged beyond recognition. The scene is apparently centered around the figure of a deity seated on an ornate throne in a pose similar to that of the gods portrayed on the wooden lintels of Tikal. His hand and feet are shaped like jaguar heads. He is wearing a wide belt with an *ahau* belt buckle and a tall helmet with a floater of quetzal feathers. He appears to be looking at a kneeling attendant on his right, while the figure of another courtier appears to be emerging from behind his throne. Above their heads is a bird in flight, probably an eagle. At the left of the throne is a standing figure of a dwarf wearing a jaguar-hide belt and a cap decorated with quetzal feathers. In the mouth of the dwarf is an object that probably represents a cigar. Dwarfs served as court jesters in Mesoamerica even at the time of the Conquest. It was reported that Montezuma sometimes had in attendance at his meals hump-backed dwarfs who were his jesters. Above the head of the dwarf floats an animal, probably an armadillo. Turning the vase clockwise, one sees the prostrated figure of a young man leaning against a skull-supported bundle. Standing with its right leg against the abdomen of the youth is a water-lily jaguar with a sacrificial knife in his left hand. From the neck of the jaguar emerges a snakelike appendage, which is held by the left hand of the leaning figure. Above the head of the jaguar is the figure of another dwarf or a monkey, holding a huge smoking cigar in his mouth and blowing the smoke in the face of an anthropomorphic deer with the tail of a monkey. The procession of these eerie figures continues with a portrait of the Death God, who has a bleached face and a protruding belly. He is standing, resting his left hand on an object that appears to be a paddle. Still another figure is shown in this complicated scene; however, it is damaged beyond recognition.

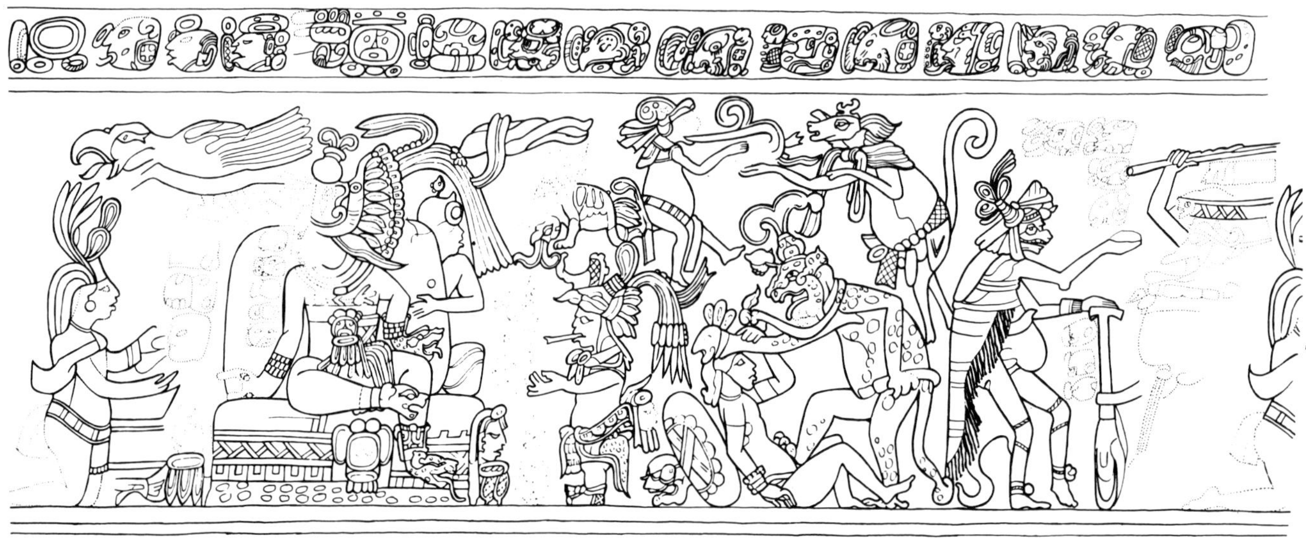

Figure 145. Vase of Decapitation Sacrifice (A-12) (after Nicholas M. Hellmuth, courtesy of Foundation for Latin American Anthropological Research, St. Louis, Missouri)

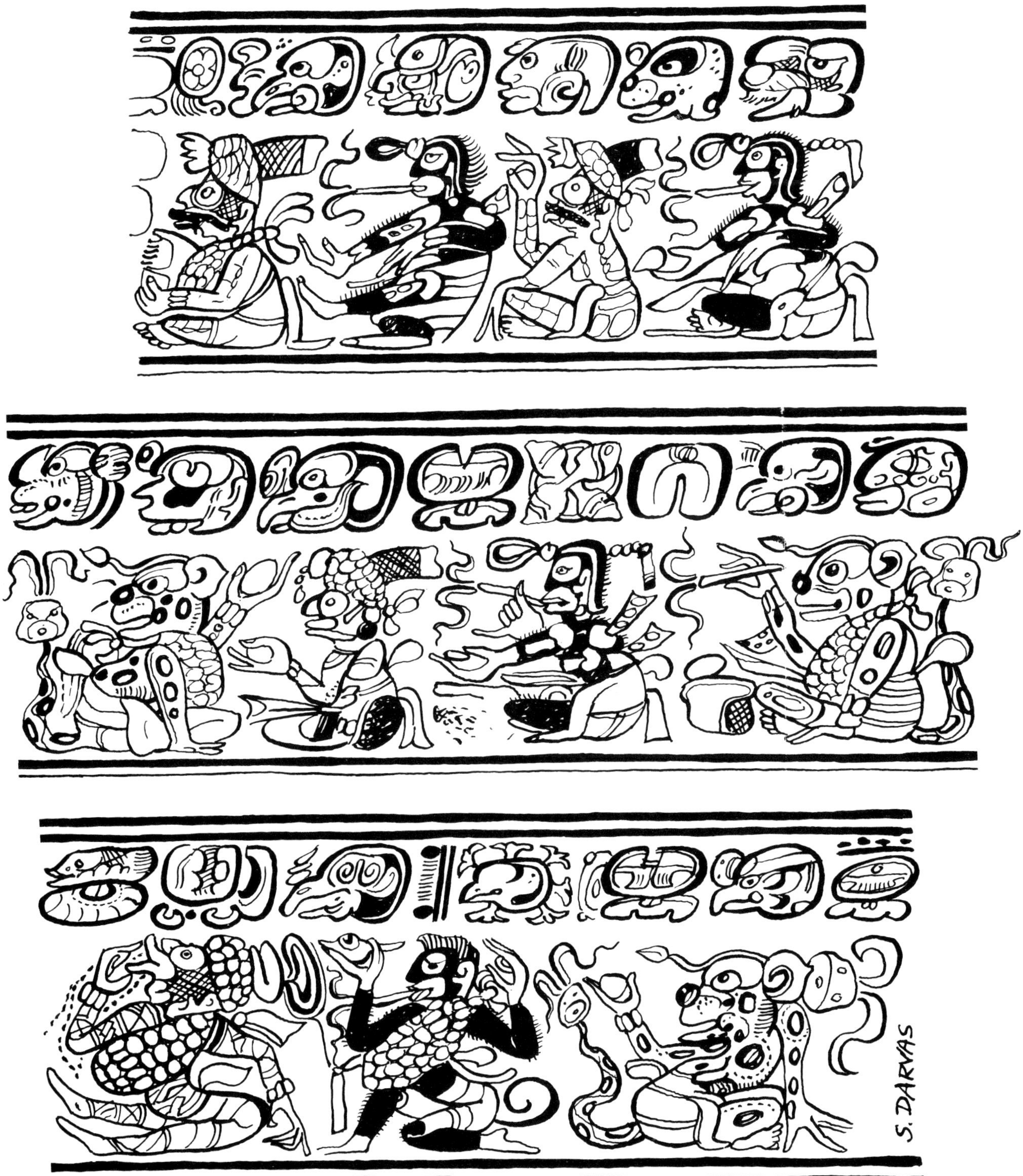

Figure 146. Bowl of Eleven Deities, Five of Them Smoking (A-13)

Glyphic Text

The interpretation of the texts must be based on the drawing without reference to photographs or the original work. The configuration of the details of some glyphs is suspect. The text does not appear to be within any of the sequences of the known Primary Standard Texts. There are no readily identifiable verbal compounds. We suspect the text may be composed of a series of nominal phrases. A *tzolkin* day 8 Ahau appears, but it is not specified with an *haab* or long-count position. If the 8 Ahau is a date, it must be assumed that it refers to the repeating cycle of 260 days rather than to a specific point in time.

The Secondary Texts may have referred to the specific events portrayed, but none of the glyphs survive in readable condition.

A remarkable example of the Maya potters' art is the **Bowl of the Eleven Deities, Five of Them Smoking** (A-13), a large, deep polychrome vessel, probably from the late Classic period of northern Petén (now in a private collection) (Plate 128). The bowl has an orange base and, in red and black, a "portrait gallery" of eleven seated figures belonging to three characters shown in repeated groups. The first group shows four dignitaries, their bodies painted red. They have peculiar large, round "bug" eyes and are wearing folded napkinlike headdresses, probably part of the ritual cloth complex described by Benson.[46] Two of them have netted bags across their shoulders. Hellmuth believes that they are "enema bibs," but I think that they are in some way related to the iconography of God N, who regularly wears a netted element in his headdress, hence his Yucatec name Pauoh (Net Bag) Tun.[47] The lips of this character are elongated in a peculiar duck-bill shape. I have seen this deity ("Hot Lips") on several other polychrome vases. His identity is unknown; however, his features are similar to the Nine Winds God of the *Vienna Codex*.[48] Such creatures (*bocones*) are also often seen on pre-Classic stone carvings found in the Motagua Valley area. The four figures in the second group are realistically painted jaguars. All of them have tadpolelike attachments on their tails, representing the iconic version of the Smoking Capped Ahau. On the forehead of each is a slender undulating appendage that may be a young water lily. One of the jaguars is holding in his right paw an object that may be a cup. Another is holding a slim smoking cigarette. The four figures in the third group are anthropomorphic spider monkeys or monkey impersonators, all of them smoking slender, white cigarettes. Their faces are painted black, and they seem to be wearing tight garments made of black (monkey?) fur. The first of them also has the monkey's curly tail and is holding in his right hand what could be an enema tray. The heads of three have looping attachments, and the fourth has the crewcut hairdo usually associated with spider monkeys. It is possible that this character, who also has a net bag hanging across his chest, is not the same "species" as the other three. It is evident that the scene depicts a procession, though, oddly, the participants are in seated positions.

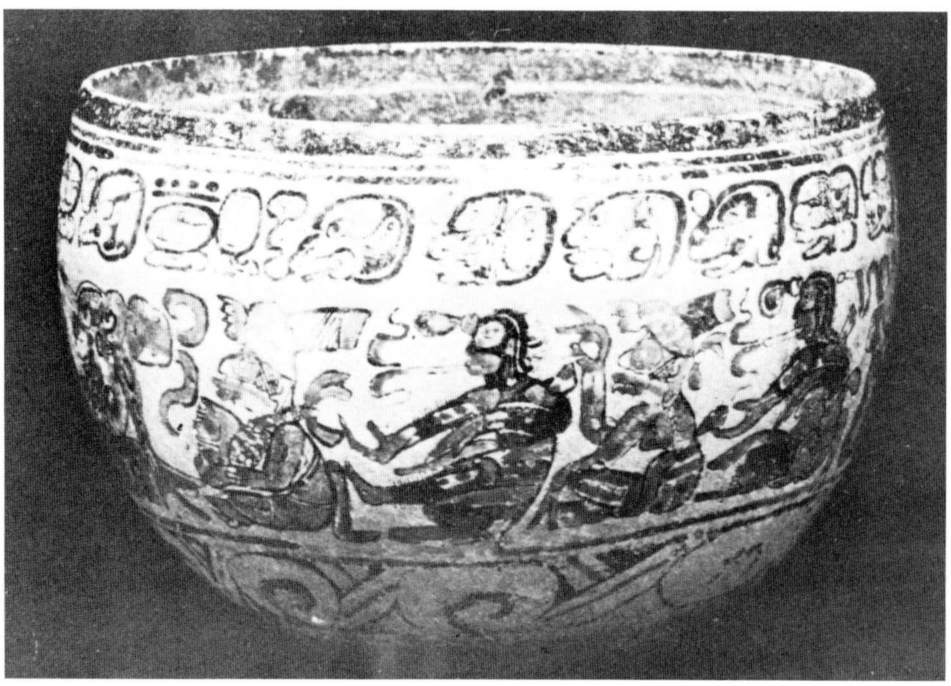

Figure 147. The bowl of the Eleven Deities, Five of them smoking.

Glyphic Text

The rim text is a variant of the Primary Standard which usually occurs on ceramics painted with this type of processional scene. It has no usual introductory glyph; however, it is possible that the glyph over the first monkey is a variant of T1011 and serves as an introduction.

A very powerful example of smoking in Classic Maya art is the **Vase of the Danse Macâbre** (A-14), a painted vessel from Petén (now in a private collection) (Plates 129–32). The vase is slightly flared and is painted in bold black and white on a rich red base with four figures and upper and a lower glyph bands. The surface of the vessel is further divided by vertical glyph columns into four panels, each occupied by one figure: two skeletal images and two jaguars. The dancing skeletons almost certainly represent God A, the Death God of the codices, or Cizin, as he is called among the Mayas (the name Cizin is from the root *ciz*, "stench" or "flatulence").[49] Cizin is often depicted in the codices and also on pottery, as on the vase in the collection of the Metropolitan Museum of Art in New York.[50] In these representations he is seen often dancing, "almost jolly," as Thompson remarks.[51] The skeletal figures on this vase are also in dancing positions, with right legs bent at ankles and knees. The first skeleton holds a smoking cigarette in his right hand. In his left hand is a pouch or an opened jar, a motif consistently associated with underworld scenes. The infix on the jar is *akbal*, the sign of darkness. The other three figures on the vase carry similar pouches with infixes representing either *akbal* or T585 (*be* or *bix*). Spinden believes that such pouches or jars may relate to tobacco iconography and were used to carry tobacco or a mixture of tobacco and other substances.[52]

On the two panels flanking the dancing figure are two jaguars, one on each panel, their tails between their legs, realistically painted down to their spots. Behind the one on the left are red scrolls, probably representing flames. The jaguar on the right is wearing a loincloth and appears to be holding a scepter. On the opposite side of the vase is a second skeleton. Either this one is a nonsmoker or the painted cigarette has been eroded by time and the moisture of the tropical forest. Both skeletal figures wear death collars from which disembodied eyes dangle with nerve stalks still attached—a symbol long known to relate to the God of Death.[53] These vivid pictures of *danse macâbre* are reminiscent of the legend of Kitsil Bac, "Noisy Bones," a favorite bedtime story of the Tzotzils. Kitsil Bac, one of the dozen or so Death Gods of that Maya nation, wanders about at night and loiters at the crossroads to prey on unsuspecting travelers. On one occasion his plans were foiled by a man's shrewdness and the magic power of tobacco. The story has been recorded by Holland:

> [Kitsil Bac] one night stripped off his flesh and wandered off as a skeleton. A man chanced upon the pile of flesh, and suspecting that it belonged to one of the malignant gods, urinated on it and threw wild powdered tobacco [tabaco silvestre molido] on it. When Kitsil Bac returned, he was unable to reclothe himself in his flesh because of what the man had done.[54]

In Yucatán the Death God was known as Cumhau or Hunhau and was identified in the *Diccionario*

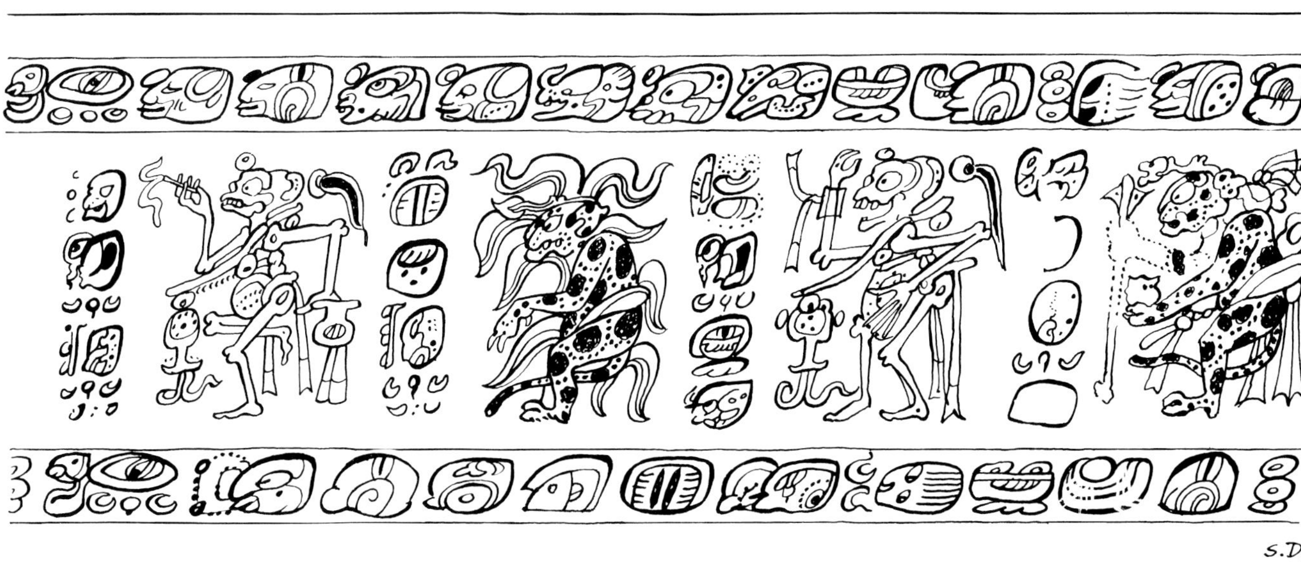

Figure 148. Vase of the Danse Macâbre (A-14)

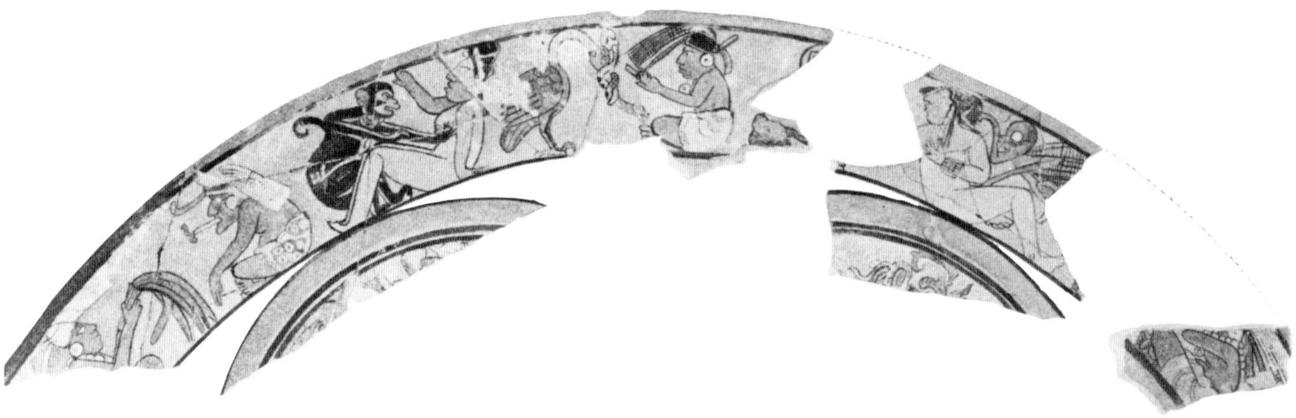

Figure 149. Plate of the Smoking God, the Woman, and the Monkey (A-15) (after Robert E. Smith)

Motul as "Lucifer, prince of hell." In his aspect as the regent of certain unlucky years, he was also known as Uac-Mitun-Ahau, or Lord of the Sixth Hell. In manuscripts and on ceramics he is usually pictured as a skeletal figure, wearing a death collar, a mantle, and crossbones in his headdress. The Mayas informed the Spaniards that the reason they sacrificed men and animals to the Death God was so that "their lives might be prolonged and they would not go soon to suffer the pains of hell." It has been also recorded that they performed a dance in his honor, called the "dance of the devil," probably also pictured on this vase.[55]

Glyphic Text

The rim and basal texts appear to be variants of Coe's Primary Standard Text. This particular variation appears to be peculiar to ceramic vases which present a series of [anthropomorphic, zoomorphic, and skeletal] figures. In most of the examples we have examined these figures can be iconographically specified as Underworld personages.

The Secondary Texts of this vase are extremely important. Each phrase consists of three or four compounds located in front of the four figures portrayed. In three of the four texts, the last or next to last grapheme is composed of T3, one of the variants of the third person possessive and the jaguar-cover *ahau* (T539). Figure 2 shows a jaguar walking in front of an arch of flames. The first two compounds of the adjacent text are composed of the "fire" glyph (T122:563a) and the Ix (Jaguar) glyph (T524). Thus a figure of a jaguar with fire is specified glyphically as *Kak Ix*, Fire Jaguar.

The jaguar-spotted *ahau* should be a part of the nominals of the figure. Since Maya grammar calls for the structure of verb-subject, it is unlikely that the jaguar-spotted *ahau* compound is verbal. Because it is used by three of the four nominal phrases, it is likely that it represents a generic title which in some way specifies a quality, location, or other common characteristic of the figures. Tentatively the jaguar-spotted *ahau* can be viewed as a signal for one kind (but not all kinds) of nominal phrase on pottery texts. The Vase of the Smoking Monkey uses the T563a nominal indicator in the Secondary Texts in a context identical to this text.

Jacinto Quirarte has noticed the occurrence of this compound in a number of similar scenes; he therefore regards it as a designation of jaguar lineage.[56] David Kelley, however, has proposed a simpler explanation.[57] T539 is recorded as the name of GIII of the Palenque Triad in his birth phrase (Tablet of the Sun, D1) at Palenque. GIII is now also known to be the Jaguar God of the Underworld and probably the Night Sun in the Underworld. *Balam*, the root for "jaguar" is identical with the root *baalan*, glossed as *"escondido tras de algo"* (hidden behind something),[58] and *balaan* glossed as *"cosa escondido o encubierta"* (a thing hidden).[59] It is possible then that the jaguar-spotted *ahau* compound designates the names for *balan*, *ahau*, or the "hidden lord" deities of the Underworld.

Our friend the monkey appears again on the very important **Plate of the Smoking God, the Woman, and the Monkey** (A-15) of the late Classic period. Robert E. Smith found the plate while working with the Carnegie Expedition at Uaxactún (it is now in the National Museum of Guatemala).[60] It has an interior orange background and on the outside a red rim and a band of eight figures. Six of them are human; two are animals, a monkey and a heron. The first figure from the left is a seated man holding in his hand a ceremonial bouquet of quetzal feathers. The second figure is an elderly, cigarette-smoking man sitting cross-legged and wearing a cap with a neck guard and a jaguar-hide kilt. The white

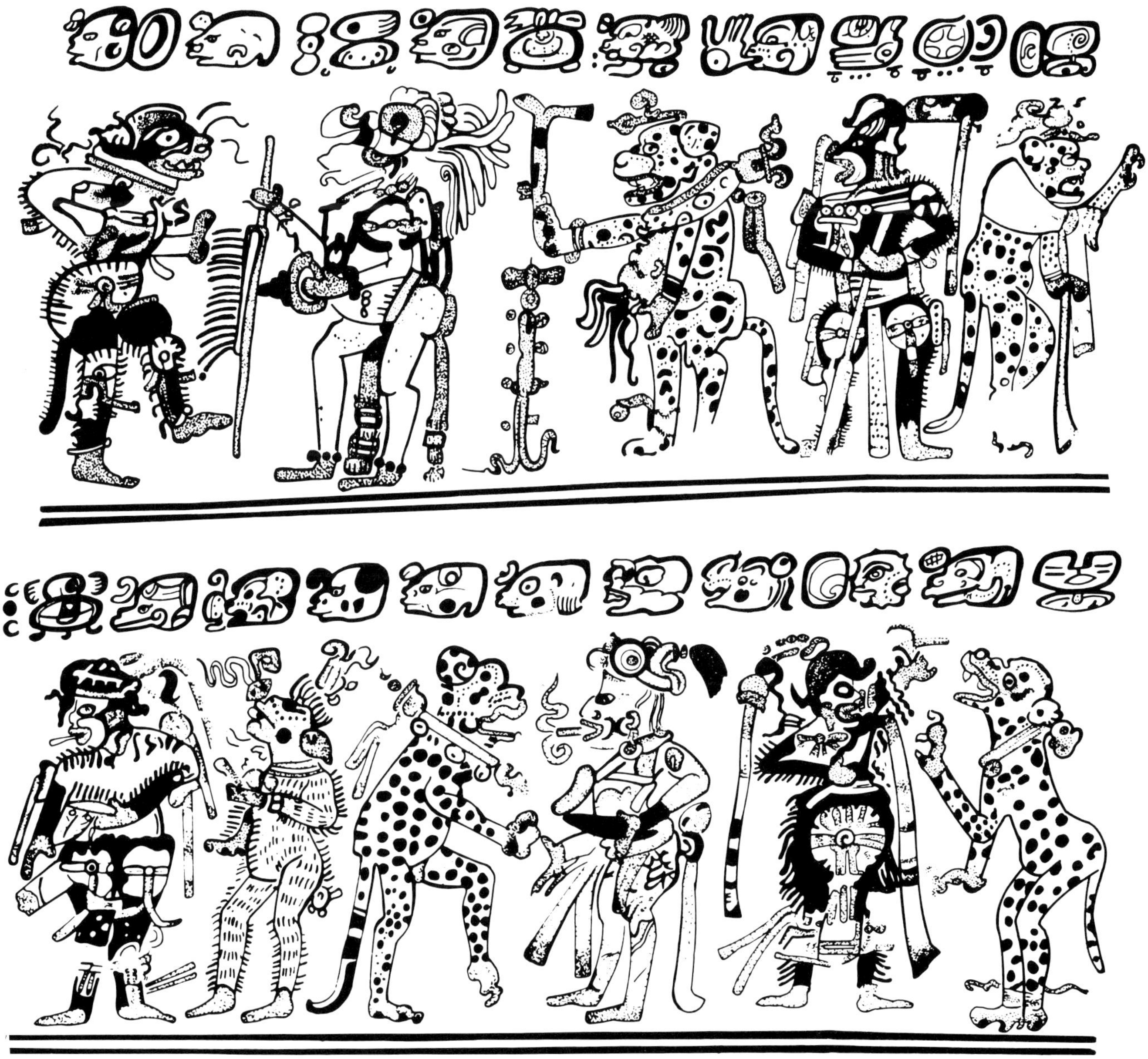

Figure 150. Vase of the Eleven Gods (A-16) (after Michael D. Coe)

cigarette between his lips sends a thin puff of smoke into the air. Next to him is the figure of a reclining nude woman with a large spider monkey. The monkey is squatting next to her and appears to be fondling her breasts. This strange couple is followed by a bird of prey with a snake in its beak. The serpent appears to be snapping at the next figure, a seated man in a white kilt who is holding a small object, perhaps an unlighted cigarette, in his left hand. The plate is broken here, and several fragments are missing. Two additional figures, another reclining nude woman (a rare subject in Maya art), and a seated man with his arms crossed, are clearly outlined.

It is difficult to explain this idyllic scene, which could have inspired Henri Matisse's lyric painting *Women and Monkeys*. Thompson wrote of the mythological role of the monkey:

> Sometimes the monkey serves as the symbol of the sun; a particularly fine example is to be seen in the full-figure variant on Yaxchilán L48. For a connection between the monkey and sun, we must turn to Mexico. In Mexico the sun god as patron of the arts of singing, music, and

entertainment was Xochipilli ("Flower Prince"), and one of the guises was that of the monkey. Furthermore, the day Ozomatli, "Monkey," is under his rule. In the Maya area the sun god was also patron of poetry and music, and was likewise connected with flowers (the plumiera, symbol of procreation). The monkey also symbolized arts and crafts as well as licentiousness and the fact of procreation, but the sun god was the first person to cohabit. With these various threads forming a definite warp and woof, it is entirely appropriate that the monkey should substitute for the more usual representations of the sun.[61]

This monkey-sun relationship also permeates the iconographic synthesis of Merle Greene Robertson,[62] and was used by Marvin Cohodas to explain the scene on this plate as the "mythic copulation of the Sun God and the Moon Goddess, which causes her descent, the coming of the rains, and the conception of new maize."[63]

While this idea appears somewhat far-fetched, we must take into consideration the cycle of Maya legends built around the deities of the sun and moon. The Sun God and the Moon Goddess, before they occupied their position in the sky, were the first to inhabit the earth. Sun was the patron of music and poetry and was a famous hunter; Moon was the goddess of childbirth and weaving. With the aid of the deer, who made the Moon's sexual organs, they were also the first to have intercourse. Moon, however, soon became unfaithful to her husband, and her name became synonymous with sexual license.[64] The scene portrayed on the plate may depict this love making between Sun and Moon, with the Sun God in monkey disguise. It recalls the tales of Leda and the Swan or Europa and the Bull from Greek mythology. If it can be done with a swan, it certainly can be done with a monkey. Naturally, it is also possible that the seated god represents the Sun, and while he enjoys his cigarette, his wife is "monkeying" about behind his back.

Scanning Maya ceramic collections for further representations of smoking, one finds in Coe's *The Maya Scribe and His World* the Vase of the Eleven

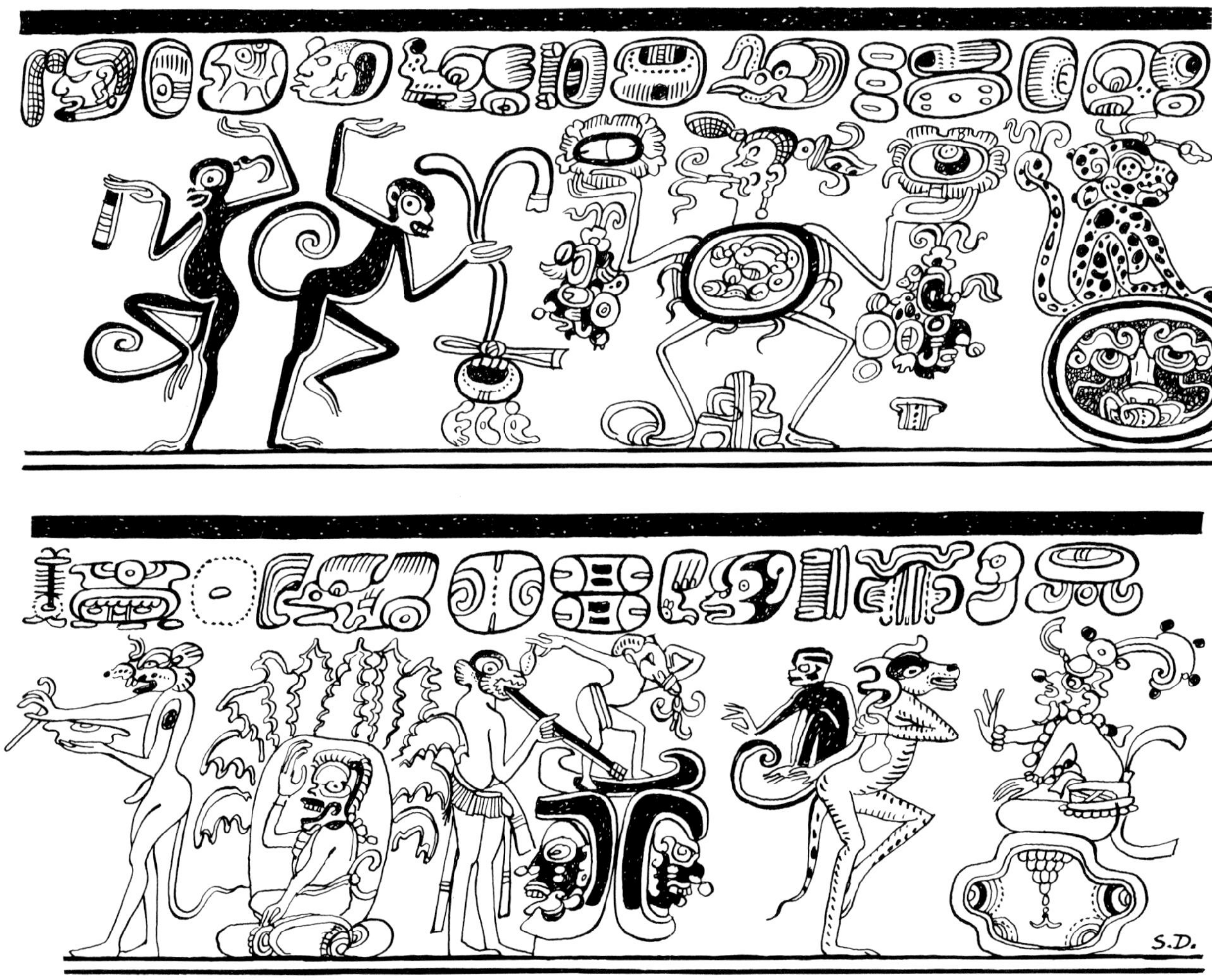

Figure 151. Bowl of God N (A-17)

Gods (A-16).[65] This Classic ceramic vessel (Plates 120–22) is of unknown provenance (now in the British Museum). A lidded bowl, it is painted red and black on a light-orange base. There is a repetitive circular (Muluc?) design on the upper rim followed by (top to bottom) a glyphic band, a circular row of eleven figures, and a base band composed of circles and crossed bands. The figures are apparently participating in a sacrificial dance in which jaguars are sacrificed by anthropomorphic gods. The felines wear death collars; the deities brandish sacrificial weapons. Coe's (abridged) description of the figures is as follows:

> (1) Feline deity with one leg raised in dance; the strange spiky attachments to the body are similar to those of Figures 4, 6, and 10. This being wears a death collar. (2) This grossly misshapen deity of basically human form, brandishing a weapon (or an eccentric flint?) in one hand and a feathered staff in the other, seemingly intent upon sacrificing his opposite number. (3) Water-lily jaguar, distinguished from other feline or near-feline deities by his pure jaguar character (excluding of course the anomaly of standing upright) and by the vegetation curling over the head. A severed human head hangs down from his death collar, and a banner-like object, perhaps of paper, with a long pendant, is carried in one hand. (4) A black god holding a paddle-like object on his shoulders are circular devices with tabs. Spikes or hairs cover the body. (5) Another jaguar god with an elaborate headdress and a long garment pendant from one shoulder. (6) Another black god smoking a cigar and grasping a sacrificial knife. (7) This hairy animal seems to be "Shaggy Dog," one of the canoe passengers on an incised bone from Tikal. While "hands" are those of a carnivore, the feet are human. (8) The jaguar on this vase wears a death collar and perhaps may be separated from the others by the oddly formed mouth. (9) This deity is human in shape, smokes a cigar, and also holds a knife; around his neck is a curved pendant like those worn on pottery tomb figures from Colima. (10) We have another spiky or hairy black god; this one has circles around his eyes and carries a knife. (11) The fourth jaguar god is peculiar in its low forehead with the ears set well back; the shortness of the tail might also be distinctive. The lid has four pendant devices which Thompson identifies as water symbols.[66]

Because the glyphic text begins with a 1 *katun* expression, it is believed that the vase may have been painted for someone who died during his or her first *katun* (a period of slightly less than twenty years).[67]

For a discussion of the glyphic text see Michael D. Coe.[68]

The Bowl of God N (A-17) is a late Classic polychrome vessel, probably of northern Petén (now in a private collection). It is orange with a lower red rim and an upper circular glyph band, and it is decorated in red, black, and yellow.

The scene on the bowl shows a charming pair of dancing spider monkeys, one of them facing an aged enthroned god, the other is turned toward another deity, God N. Each of the monkeys is holding a piece of cloth; the one held by the second monkey is longer and resembles the glyphic compound the T60 knot above a T501 *imix* glyph.

The scene recalls the legend of 1 Chuen and 1 Batz, half brothers of the Hero Twins. According to the *Popol Vuh,* 1 Chuen and 1 Batz were great sculptors, dancers, and pipers, and they aroused the jealousy of the Hero Twins, who persuaded them to climb a tree and then changed them into monkeys:

> They once became animals
> And turned into monkeys
> Because they just boasted
> And mistreated their younger brothers.[69]

From then on, 1 Chuen and 1 Batz, the two funny little spider monkeys, spent their time piping and dancing with their

> Appearance funny, all
> Potbellied at the bottom
> With edges of their stomach bulging.[70]

Next to the monkeys stands the figure of God N smoking a cigarette. God N, or Pauoh Tun, is, with God L, one of the two principal gods of Xibalbá, the Maya underworld. To date, however, the two gods have never been found on the same vessel.[71] God N's body is painted in a manner similar to his portrait in the Lower Temple of the Jaguars at Chichén Itzá. A glyph is infixed in the shell-like body it is not, however, his usual name glyph. He holds a cartouche of an infixed *kin* (sun) glyph in his right hand and an infixed moon glyph in his left. Seemingly suspended below these cartouches are the heads of two black deities, in an arrangement similar to that shown on the Tablet of the Sun at Palenque. An element resembling the T687 (pyramid) glyph rests on the ground line between his legs. God N is followed by the well-painted figure of a water-lily jaguar seated on top of a shield motif. The iconographic complex at first glance resembles the ominous Jaguar God of the underworld. It has the appropriate serrated eyebrows and anthropomorphic nose, but instead of the conventional mouth it has a T528 *cauac* glyph infixed into a T60 knot.

Next to the jaguar stands a mysterious anthropomorphic carnivore, which could be either a coatimundi or the Fox God, a deity rarely seen on Maya ceramics. Besides this portrait (if indeed it is one) only three other representations are known.[72] He is holding a paint pot and a brush. The Fox God or coatimundi is followed by the figure of God A, Ruler of the Dead, seated on a throne backed with a glyphic pillow. The glyph is the same as that infixed in the mouth of the head, which serves as the jaguar's seat. Behind the Death God are fan-shaped jagged (plant?) motifs.

The God of Death is followed by additional anthropomorphic animals. The first has a human body and is dressed in only a short skirt and a loincloth. The leaflike element above the nose suggests that he may be either a bat or Cama Zotz, the Bat God himself. He is holding a long, tubular object, perhaps a blowgun, to his mouth. In front of this figure is a canoe-shaped T683 lunar grapheme on which stands a creature with a human body and the head of an armadillo. The grapheme rests on two lunar symbols standing vertically and turned base to base. The heads encompassed in these symbols appear to be identical, but one of them is fleshy, while the other skeletal. Next to the armadillo prances a deer with a spider monkey on its back. The deer appears to be looking into the eyes of an aged deity seated on a *cauac* quadrifoil and wearing a headdress resembling an elaborate shell element.

The identity of this god is difficult to determine. I believe that he is the same one portrayed on the Vase of the Dancing Lord (described earlier) and on the Vase of the Black Line (described below). The discernible similar characteristics of these three examples are as follows:

1. Toothless, aged appearance
2. Simple dress with bead necklace
3. Cantilevered element with streamers on the forehead (on two of the three examples)
4. *Spondylus* shell element (in side view) at the crown of the head (on two of the three examples)

He is apparently one of the major gods and should be distinguished from other aged deities (Gods B, D, L, M, and N).

Glyphic Text

The inscription on the bowl is Primary Standard Text. The introductory glyph is T238.617a:126 and is placed above the head of the old god.

The **Vase of the Black Line** (A-18) is a tall, white, cylindrical vessel, probably from the middle Classic period at Petén (now in a private collection). The vase was undoubtedly painted by the same artist (or one of his students) who created the magnificent Vase 42 in Coe's *The Maya Scribe and His World.*[73] It has the identical elegant whiplash style and delicate lines equaled only by the best of the Japanese scroll painters. A narrow black upper rim and an upper band of beautifully painted hieroglyphs complete the composition. The text and the scene begin at the line that divides the scene. The first figure is that of God GI of the Palenque Triad, dancing. He is identified by his "god eyes," filed teeth, cross-sectioned *Spondylus* shell diadem, knotted pectoral belt, prominent basses at the hips, and long, gathered hair. He has a snout instead of a nose and an upturned upper lip. The shell earplug he wears in most of his other portraits is missing. The crosshatched motif at his nape closely resembles the details used as date indicators on GI's zoomorphic representation on Stela 6 at Tortugero. As on the Dumbarton Oaks Tablet, the Tablet of Creation at Palenque, and Vase 45 in *The Maya Scribe,* God GI is wielding a trilobed flint in one hand and an ax in the other.[74] According to Kubler, such trilobed objects belong in the archaeological classification of "eccentric flints," which diffused to the Maya area from Teotihuacán probably before the Classic period.[75] The hand hole and the tridentate shape are believed to represent a jaguar paw and may have been used in ritual killings as an "imitation of the jaguar method."[76] There is not, however, any surviving Maya representation showing such an act, and other possibilities should also be examined.[77] The postion of the god's body is most unusual. It is well above ground line in a jumping position that may very well represent dancing, but it is not like other dancing positions in Maya iconography. It may simply indicate motion rather than dancing.

Between the god's legs hangs a serpent or a loincloth with a serpent-head appendage, which may be a phallic symbol. From the mouth of the snake come large coils of smoke. This iconographic and glyphic motif is characteristically associated with cigars, cigarettes, and fire but here may simply signify breath or a bad odor. A similar motif appears on the "sister vase" of this vessel (Coe's Vase 42).[78]

Behind GI stands the dwarflike figure of an anthropomorphic Uinal Frog deity, identified not only by the configuration of his eye and face but also by the three dots at the back of his head, which correspond to the bone visible below the ear membranes of lizards and amphibians.[79] The Uinal Frog is distinguished not only by being the patron of the *uinal* period (20 days) but also by being called "friend" by Michael D. Coe.[80] The frog wears the "drum-major" headdress common on the monuments of Yaxchilán and Palenque, but to my knowledge not seen elsewhere on ancient Maya ceramic artifacts. He is also decorated with a bar pectoral and wears a simple cloth belt with shield motif at the sides and a back rack of feathers. In his left hand he is casually holding a flexible shield like the shields in the "capture" scenes of Yaxchilán. In his hand is a ceremonial spear aimed at the figure of

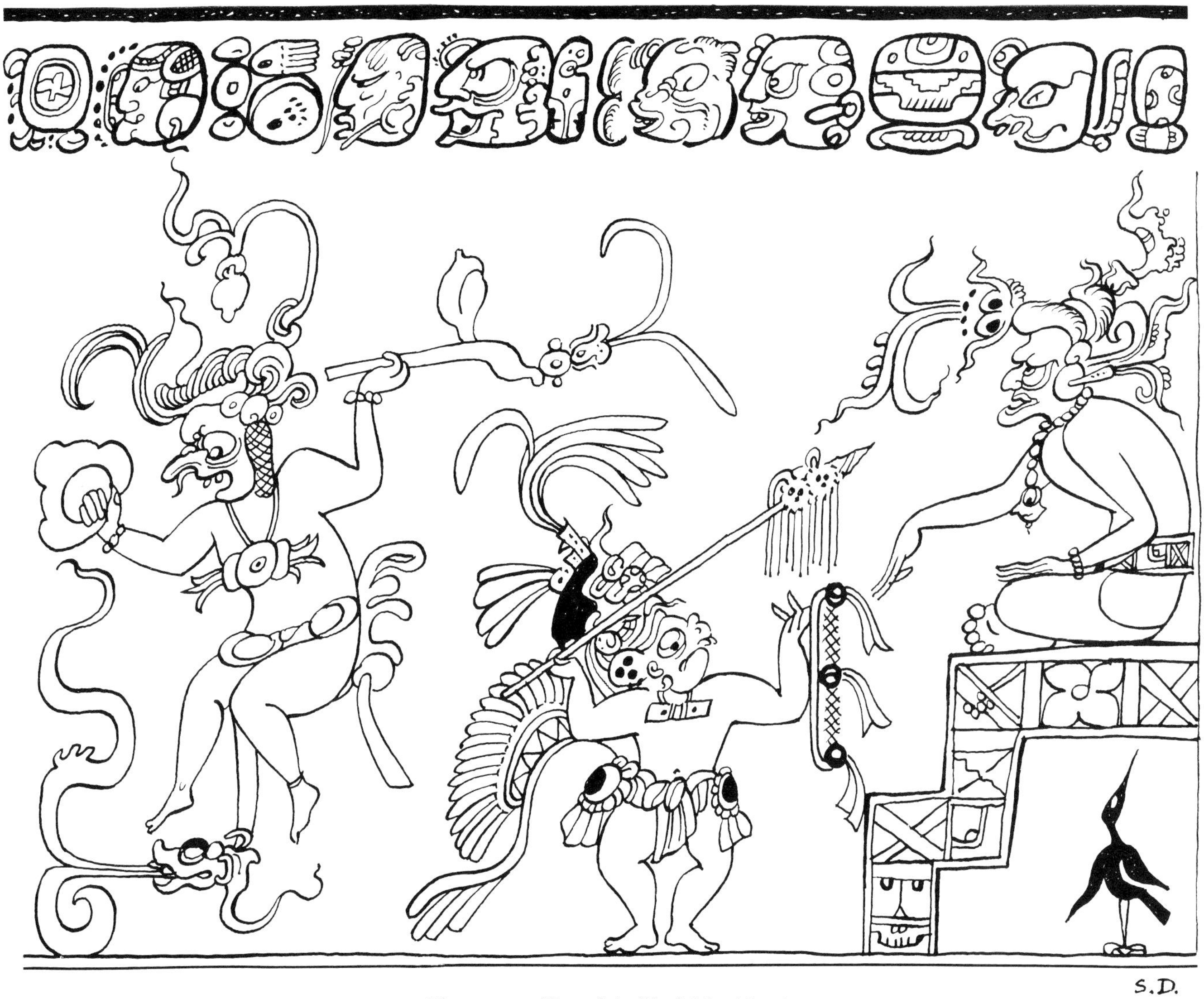

Figure 152. Vase of the Black Line (A-18)

a scantily dressed aged deity seated cross-legged on a stepped sky-band platform, similar to those seen in the *Dresden Codex.* The aged god cannot be identified by iconographic characteristics. His pectoral resembles those worn by the Serpent Bird on the Tablet of the Cross and the Sarcophagus Lid at Palenque and on Stela H at Copán. He is wearing a sky-band belt, a short kilt, a headdress with a cantilevered front element, a shell (?) motif at the crown of his head, and foliage at the back, the last as part of the earplug assembly. Michael D. Coe believes that the aged god may be related to God N or is God N himself.[81] Under the platform stands a black bird, probably a cormorant, with upturned beak.

Glyphic Text

The hieroglyphic text on the rim of the vessel is Primary Standard. Two of the compounds are especially interesting. The introductory compound, which is located above the boundary line dividing the composition, appears at the first glance to be T38.617a:126, an introductory glyph compound often seen in Primary Standard Texts. On closer inspection, however, it is evident that the infix in the main element is not what one may expect in the mirror grapheme, but it closely resembles the counterweight element in the earplug motif. Because the quality and the craft of calligraphy precludes any error by the scribe, we have to find another explanation. Both the earplug and the mirror graphemes occur as infixes in the water group (T36-41) prefix to emblem glyphs. One, therefore, may accept that it was this common attribute which allowed the interchange in the emblem glyph and which may explain their appearance as introductory glyphs in Primary Standard Texts.

The second very important compound on the vase is in the fourth position.

The third compound in the text was designated by Michael D. Coe as the "winged" *quincunx* compound of the Primary Standard Text.[82] Usually it consists of Affix T61, the T76 "wing" grapheme, and T585a (*quincunx*). In this case, *quincunx* has been replaced by a cartouche with infixed footprint.

In 1950 Thompson read T585 as the numerical classifier *bix* used exclusively with numbers 5 and 7.[83] Later Lounsbury proposed that T585a be read as being based upon the identification of T585a as Landa's second *b*. Landa's first *b* was apparently the grapheme for *be* in Yucatec, root for "road." Landa's drawing represents a footprint on a road. Lounsbury proposed that Landa's second *b* is a misdrawn *quincunx* grapheme with its center circle missing. The winged *quincunx* in our text confirms Lounsbury's hypothesis. While in all other cases the main grapheme is compound T585a or its head variant, here the footprint Landa's first *b* substitutes for the *quincunx*, Landa's second *b*. It should be noted, however, that, while this substitution supports the reading of T585a as *be* and indentifies it as Landa's second *b*, the proposed reading does not eliminate the correctness of the *bix* reading. We believe that the T585a grapheme may be polyvalent.

The next artifact to be described is the **Bowl of the Smoking Black Lord** (A-19), a deep polychrome vessel with an orange base made during the Classic epoch in the Maya lowlands (now in the November Collection). It has a most unusual scene, which depicts not only the act of smoking but also the paraphernalia of ritual (hallucinogenic?) enema. The composition is centered around a large *olla*, which probably contains the fluid to be infused. A small human figure is either painted on the *olla* itself or is seated in front of it. The small circles and the curved dotted lines at the mouth of the *olla* indicate that it is filled with some kind of fermented liquid, probably *balché*. In a fashion one can see on medieval coat of arms, the *olla* is flanked by a large black bird, probably a king vulture, and an armadillo. The rest of the figures, five human and two animal, are positioned around the great jar in a symmetrical manner. Turning the bowl counterclockwise, one sees the figure of a lord, his body painted red, his face around the eyes and the forehead black. He is holding an object with a nozzle, which may be a clyster instrument. Above him is a second, smaller *olla*, also containing fermented material. Gazing at this vessel is another lord, his body painted black, his face colored black and red. He is dressed in a loincloth and smoking what appears to be a dark cigar. As the vase is turned, another noble becomes visible. His red body is in a kneeling position. He wears a kilt and a cloth headdress and is facing another dignitary similarly painted and dressed. Between the two figures is another "enema tray," with fluid pouring from its snout. This is followed by a seated figure painted red and still another vertical enema set. The similarities in body painting raise the possibility that some of the characters painted on the bowl represent the same person in different situations.

The **Vase of the Seated Processional** (A-20) is a tall ceramic vase of the Classic period, probably from northern Petén (now in the November Collection). The vase has a conventional upper and lower red margin and an upper glyphic band and is painted on a white slip surface in red and black. The principal figure in the scene is u snout-nosed seated lord. Most of his body is painted red. He is wearing a short monster helmet, elaborate earplugs, wide wristlets, a rear ornament, a kilt, and a peculiar nose decoration resembling a bone. His left palm is turned downward; his right hand is raised as if gesticulating. He is seated Turkish fashion on a low cushion resting in the paw of an *imix* monster supported by four scantily dressed attendants. A fish nibbles at a flower emerging from the monster's forehead. The creature's earplug has a *kan* cross, and crossed-band motifs are attached to the top of his head. His mouth is open at a ninety-degree angle. His head rests in a nose-up position on the hands of four small bearers with human bodies and zoomorphic heads. The lord is being addressed by a tall individual who stands before him in a long white tunic, a tall cap with a feather floater, a bow-tie-shaped neck ornament, a pendant belt, and elaborate sandals. The divine nature of this figure is revealed by his animal ears. The same character appears in vomiting scenes on other pots. The left hand of the standing lord is raised in a gesture of salute. In his right hand he holds a cloth and a ceremonial vessel.

A line divides the rest of the composition into an upper and a lower area, the latter probably representing the underworld. There is a procession of seated figures on each level, four elegantly dressed nobles on the "first floor" and four deities with skeletal and one with zoomorphic features in the "basement." The lords wear what Michael D. Coe calls "napkin" headdresses, kilts, and necklaces.

The lords on the upper level seem to compose some kind of orchestra. The first one is holding a pair of rattles, the second a ceramic drum, the third a turtle carapace and deer horn. The fourth sits empty-handed. The second lord also appears to be smoking a cigarette. All the deities on the lower level have underworld characteristics: the first three and the fifth have skeletal heads and torsos. The fourth is a

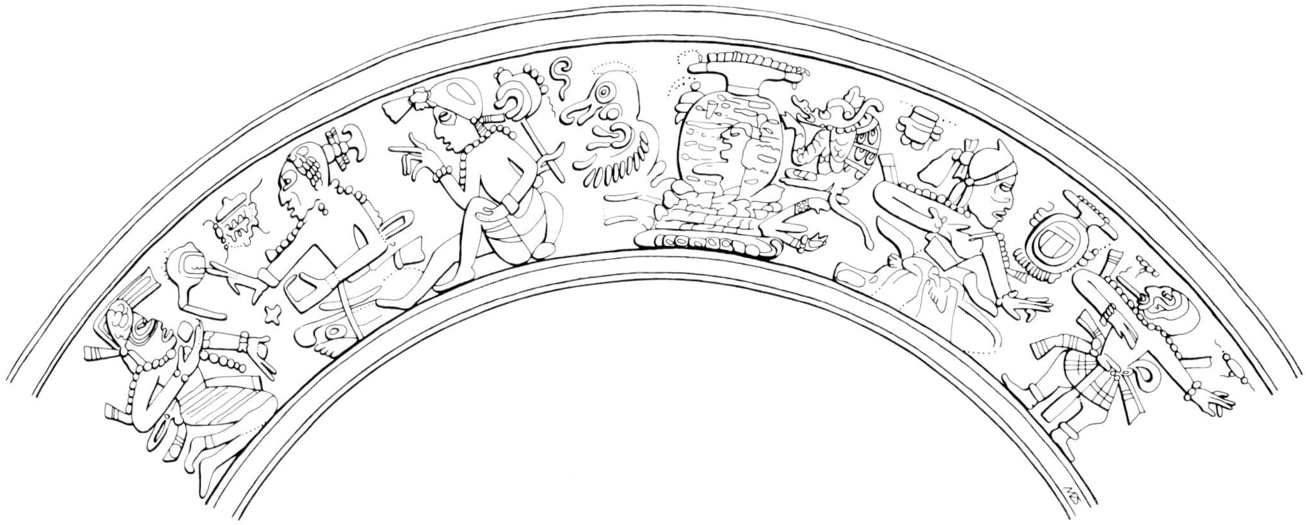

Figure 153. Bowl of the Smoking Black Lord (A-19)

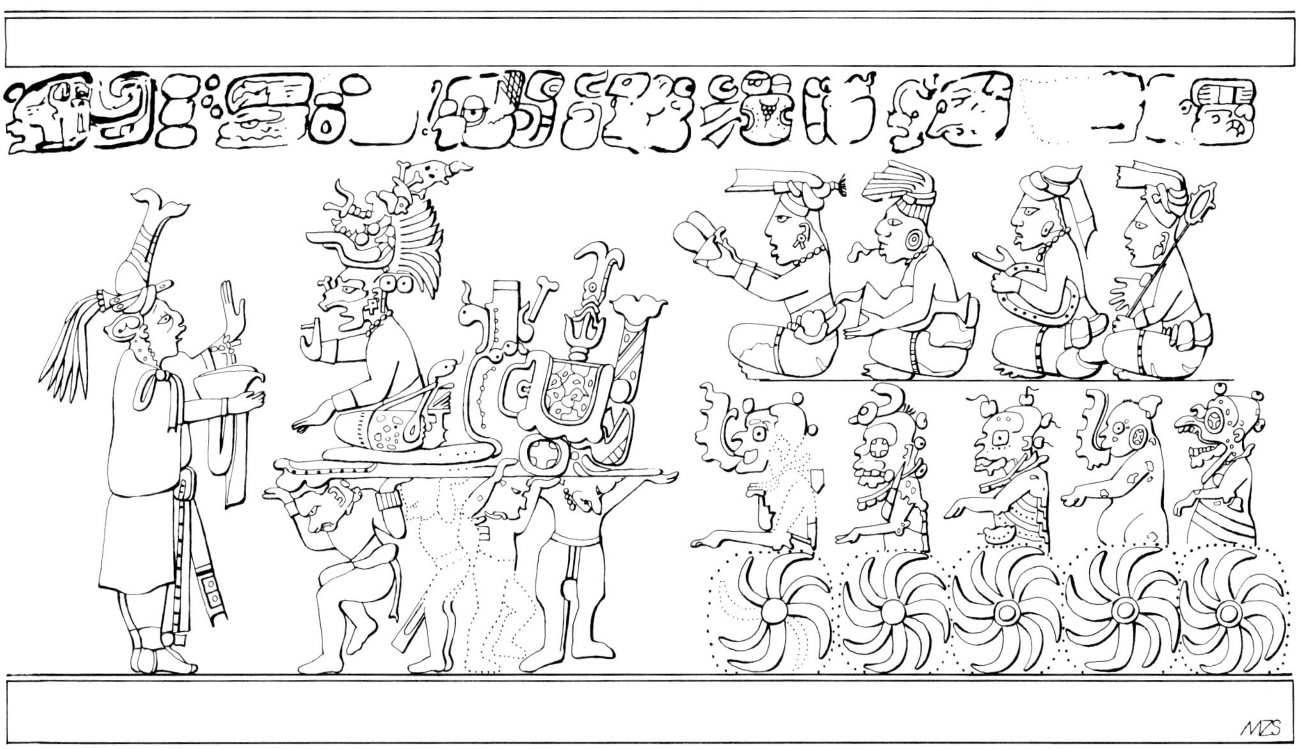

Figure 154. Vase of the Seated Processional (A-20)

rabbit, a full-figure version of the T757 jog glyph. The first and fifth Death Gods also have peculiar truncated noses, and the eye of the second Death God bears a *kan* infix. All the skeletal deities are wearing headdresses decorated with round symbols closely resembling the "death eyes" usually seen on collars and pectorals. The jog rabbit has a long, scroll-like tongue. The five underworld gods are shown with only the upper parts of their bodies visible, the lower portion being covered (or replaced) with large, whirling wheels reminiscent of glyph T627b.

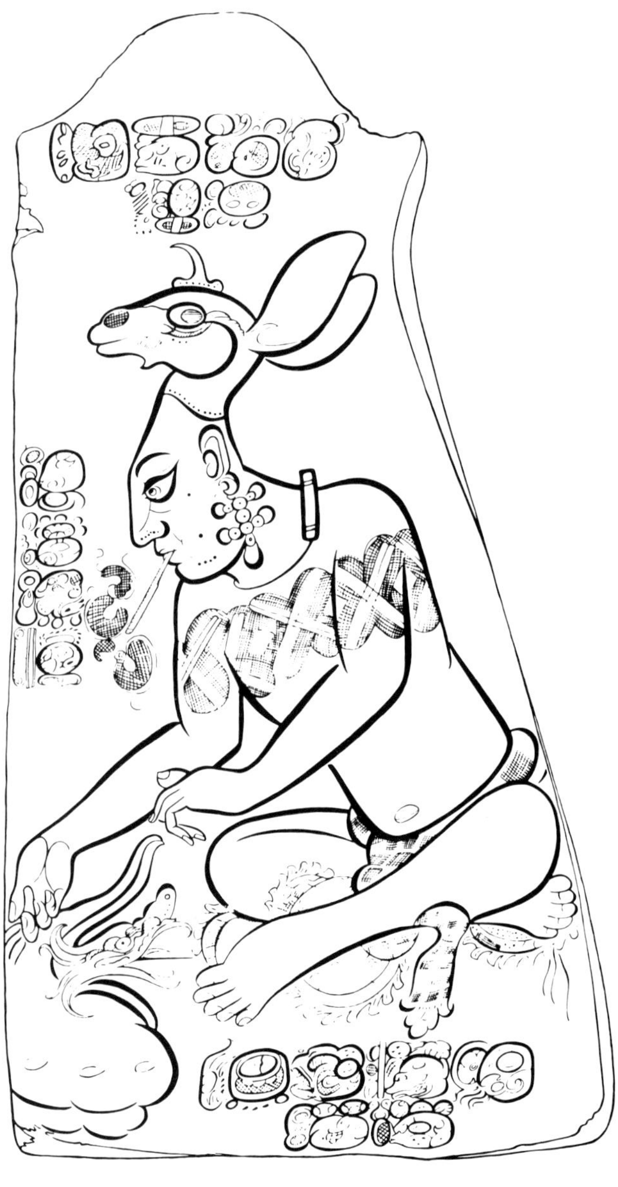

Figure 155. Shell Carving (A-21) (drawing courtesy Cleveland Museum of Art)

Smoking is also depicted on an artistic **shell carving** (A-21), late Classic period, probably from Chiapas (Norweb Collection, Cleveland Museum of Art, Plate 124). The design shows a scantily clad male figure seated cross-legged on a cushion. His high rank is indicated by his deer-head helmet, ornamental ear pendants, and face and body tattoos. He is leaning forward to touch an unidentified animal, perhaps a serpent. In the mouth of the lord is a slim cigar or a cigarette, from which smoke emanates. Three short columns of glyphs complete the composition. The carving shows several similarities to a ceramic sculpture from Jaina, described in Chapter 9.

Glyphic Text

The text on this shell is composed of three sets of glyphs which appear to have been designed to read continuously. It begins with the set of glyphs above the head of the figure and appears to read completely across each horizontal row. The event glyphs of the inscription are located in this first section, but they are not clearly recognizable and do not relate to other identified event glyphs.

The text continues to the section located on the right of the figure's face. The last two glyphs—a jaguar and God C group—are clearly nominal and name both the subject of the event(s) recorded in section one and presumably the person on the shell. His personal name is very likely the jaguar glyph in position 3. The jaguar head has a counterweight from an earplug placed in front of it. The God C glyph group below the "Jade-Jaguar" is a title compound which appears in Maya royal names throughout the last Classic period. The text below the figure appears to record further titles and/or names of the protagonist. None of the event or name glyphs can be associated with the act of smoking shown on the shell.

Cigarette smoking is also portrayed on the very important **Vase of the Palace** (A-22), a polychrome vessel probably made by an artist of northern Petén during the late Classic period (now in a private collection). The vase is painted in dark red, white, black, and blue on an orange base. The scene shows an enthroned ruler and four attendants. Their similar attire and identical blue-striped double neck pendants may indicate that they are members of the same group, probably of the same clan. The vase has a conventional red rim, a circular upper glyph band, and four vertical and one angular glyph columns.

The composition is centered around the figure of a light-skinned lord seated cross-legged on a dais. He is dressed in an elegant white kilt and wears a napkin headgear. He shares his throne with a cloth-tied bundle composed of seven flat white objects topped with a bouquet of quetzal feathers. This bundle may represent a folded codex or a tribute of cloth. A second bundle decorated with three black

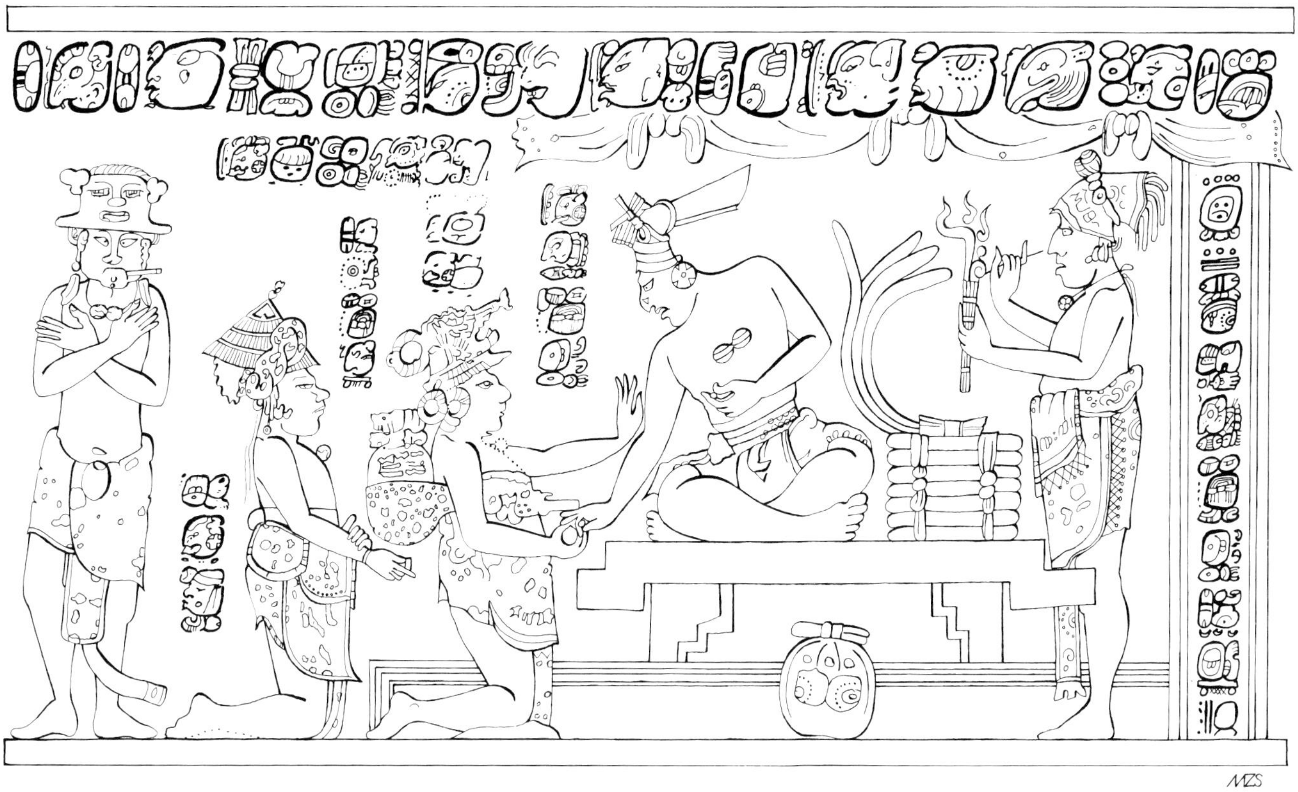

Figure 156. Vase of the Palace (A-22)

dots and a glyph compound is on the ground in front of the throne. A white swag curtain is tied above the dais, identifying the location as a palace interior.

Around the seated ruler are gathered three attendants of noble rank. The one on his right is an erect, dark-skinned man dressed in a blue kilt and matching cap. He holds in his right hand a smoking object similar to the large cigar smoked by El Fumador of the Temple of the Cross at Palenque. In his left hand is a sharp stylus with which he is touching the burning end of the flare. He is either putting copal to the torch or lighting a cigar to be offered to his king.

On the left side of the enthroned king are two courtiers paying homage to their master. The first is a dark-skinned man wearing an animal-head helmet. He appears to be touching the hand and arm of the ruler. He carries a cargo-band package that contains a tied bundle, perhaps a tribute. The second figure is kneeling. His headdress resembles a Chinese coolie hat bound with a twist of spotted cloth.

The most interesting person in the composition is undoubtedly the sloe-eyed, dark-skinned individual who is standing alone, sulking, between two glyph columns. He is wearing a white skirt, a blue loincloth, and a most appealing flowered pink hat on which a human face is painted. From the corner of his mouth hangs a smoking white cigarette. The figure is one of the very rare *en face* representations of the human head in Maya vase painting.

Glyphic Text

The text is one of a number of texts now known which mention Ruler IIIb of Naranjo. The rim text consists of the Primary Standard Sequence and includes the name glyph of this Naranjo lord. The introductory glyph of the sequence is located above the short text to the immediate left of the seated ruler. The first ten glyphs of the rim follow the order of the Primary Standard Text.

The historical addendum begins over the face of the left kneeling figure. This and the glyph to the right appear to record the verb expressing the event. The name of the Naranjo ruler is to the right of the verbal expression and to the left of the Introductory Glyph. The same name is found on Stelae 19 and 33 at Naranjo and on the Naranjo vase published by Michael D. Coe.

The vertical text which serves as the boundary of the scene, records a Calendar Round date, verb and subject. The calendar round, 4 Ahau 13 Ceh, is the *hotun* ending 9.18.5.0.0. This date is in the center of the reign of Naranjo Ruler IIIb, whose accession was 9.17.13.4.3.5. Akbal 5 Pop (Stela 14) and whose latest date is 9.19.0.3.0 (Stelae 7 and 10). The verbal expression is noted in the third glyph block. It appears to be the same as or closely related to the verbal glyph in the Primary Standard Text immediately before the ruler's name.

The protagonist of the verb and date is named in the fourth position. The first half of the compound is the Rodent Bone (T758:110). The same glyph compound occurs in the short text immediately in front of the seated ruler in the scene. In both texts, the personal name is followed by the Tikal emblem glyph and the Anterior Date Indicator. The vertical texts include three more glyphs which along with the Anterior Date Indicator apparently record titles to the protagonist.

The internal texts in the scene function as nominal phrases of the nearby persons. The problem arises, however, in deciding to which persons the texts refer. The longer L-shaped text includes some of the glyphic compounds which appear on the Naranjo painted pot mentioned above. It is likely to give general information about the scene. The remaining three texts definitely name three of the four persons. The text between the ruler and kneeling figures names the same person as the vertical text. Since both of the adjacent persons face this text, it may name either of them. If the text names the seated ruler, it is likely that the other two texts refer to the persons next in line, i.e., the two kneeling persons. However, each of the texts is located behind the kneeling person and each is located spatially closer to the figure which faces it. Spatially, at least, the short nominal texts make more sense if they refer to persons who face them.

If each figure is named by the text he faces, the standing person with the strange hat would then be named by the text at his left leg. This text begins with his personal name. The third and last glyph names him as Ahau or "Lord." The first kneeling person would be named by the text in front of his face. The first two glyphs are his personal name. The third is the Tikal emblem glyph. The last glyph is apparently a title. The second kneeling figure (in front of the seated ruler) would then be named by the text in front of his head. The first glyph is the T1.60:757 introductory glyph. The personal name is located in the second position, the Tikal emblem glyph is in the third, and the anterior date indicator closes the text.

This scheme produces the fewest problems. The two kneeling figures are identified as persons from Tikal. The seated figure presumably is the Naranjo ruler named in the PSS. In all other occurrences of his name he carries the Naranjo emblem glyph. If the text in front of his face names him, the presence of the Tikal emblem glyph in his nominal phrases would produce very awkward problems.

The vertical text names the event and its date, 9.18.5.0.0. 4 Ahau 13 Ceh. The protagonist of the event is an important person from Tikal. He is accompanied by another Tikal person. Both kneel in front of the Naranjo ruler, and the principal Tikal person carries a bundle in a back strap. The person standing to the left of the Tikal visitors is named without an emblem glyph. The lack of an emblem glyph suggests that he was local. The Naranjo ruler is identified in the rim text as Ruler IIIb. The date is a *hotun* ending; the event may have been the celebration of the period ending. The visitors are apparently of important rank, but that rank was low enough for them to assume subordinate positions to the ruler of Naranjo. This conclusion seems to be supported by the fact that neither of the names appears on the monuments of Tikal.

Smoking is also portrayed on the **Bowl of the Crouching Lords** (A-23), a shallow polychrome vessel probably from the late Classic period of Yucatán (now in the Mint Museum of Art) (Plates 115 and 116). The ceramic portrays two lords in the cross-legged, crouching position often seen on late Classic Yucatec pottery and also on some of the earlier Motagua pottery. Both lords are unclad except for their short kilts and napkin headdresses. They have round and bar-shaped black marks on their faces and bodies, probably signifying that their portraits are posthumous. Their arms are extended to touch with their hands the large brown jars in front of them. One of the characters is distinguished by a peculiar forehead ornament: the other, by a long, slender cigarette in his mouth, which emits a thin coil of smoke (the rather grim expression of the nonsmoker could indicate that he is out of cigarettes). The composition is divided in two by an odd-shaped glyph block.

Another example of smoking is seen on the **Plate of the Smoking Husband** (A-24), a polychrome ceramic, probably made during the late Classic period of Yucatán (now in a private collection). It depicts an enduring family scene, which has changed very little during the centuries: a matronly woman preparing dough on a low *metate*, her spouse sitting next to her comfortably smoking a slim cigar—only the newspaper is missing from his hands. Across from the *metate* is a large bowl, probably to hold corn. There are also two glyphs, too

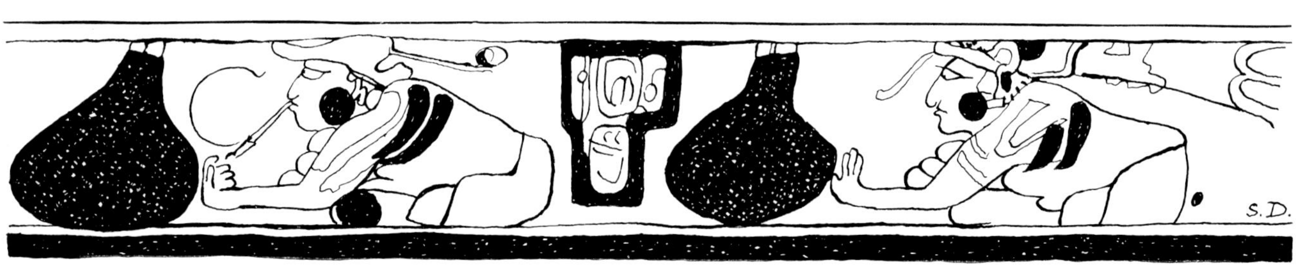

Figure 157. Bowl of the Crouching Lords (A-23)

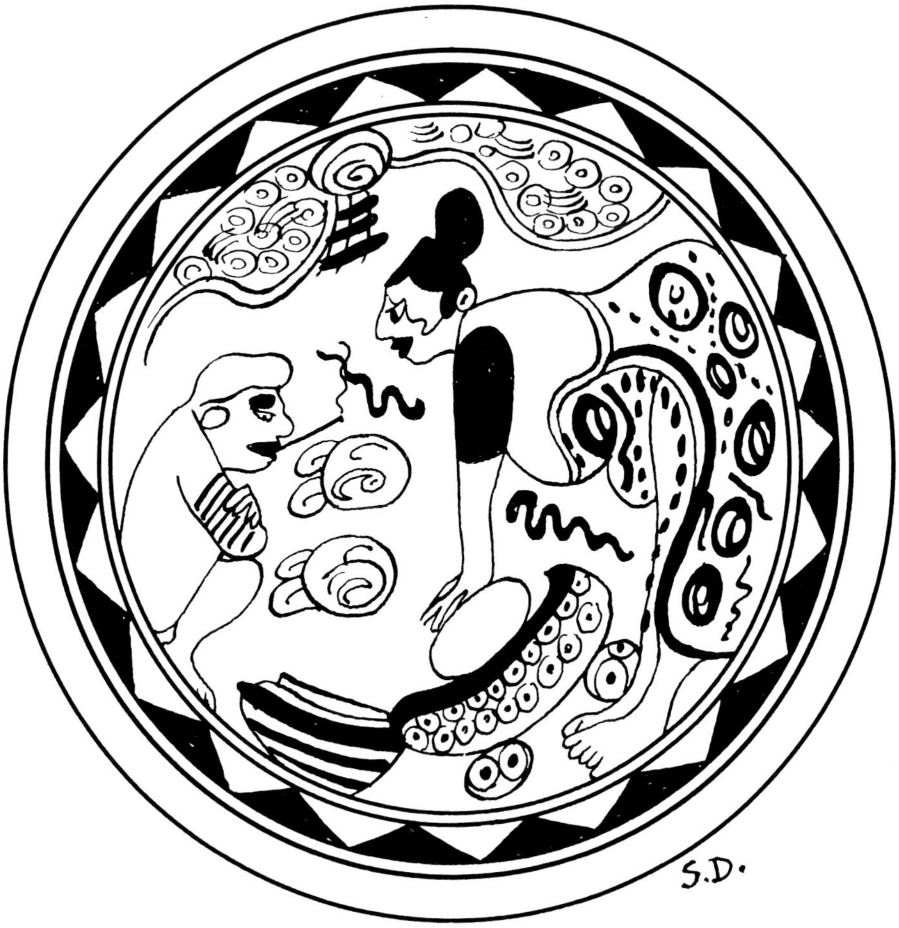

Figure 158. Plate of the Smoking Husband (A-24)

freely drawn to be identified, floating between the two participants of the scene. The picture is topped by a canopy, indicating that the idyl is taking place in the couple's home.

A smoking deity is presented on the plate of the **Plate of Black God L Smoking** (A-25). The shallow polychrome plate shows God L identified by his flat rimmed sombrero decorated with feathers, in a dancing position. His body is painted black, has a jaguar tail and except for a narrow belt and round ear ornament he is unclad. He is holding a featherwisk in his right hand and carries a wrapped bundle on his back. His eyes and mouth are rather simply sketched and there is a jaunty smoking cigarette in his mouth. In front of the god as well as on top of his headdress there is a bird of unidentifiable species, probably a water fowl. The upper bird seems to be holding (swallowing?) a snake or a jaguar tail. The rim of the plate is decorated by a triangular black-and-red design. Across the face of the Black God runs a zig-zag white line.

Fresco

The only known ancient Maya wall painting that depicts smoking is the fresco on the inside wall surface of the Temple of the Diving God (Structure 5) at Tulúm. The painting is separated into upper and lower levels by a horizontal band of jaguar skin and is also partly enframed by the body of a celestial serpent stretched above the participating figures. The intertwining bodies of six slender serpents further subdivide the upper level into four and the lower level into three tableaus.

The body of the celestial band is unusual because it is composed of mat symbols alternating with bare (or perhaps eroded) areas. Just below and probably also part of the body of the sky monster is a wide strip of jaguar hide with U- and V-shaped motifs and four iconographic complexes, representing, according to Arthur G. Miller, the chest and two wings of Xux Ek (Wasp Star), the malevolent God of the Planet Venus, who is also pictured on pages 46 to 50 of the *Dresden Codex.*[84] The fresco, on the east

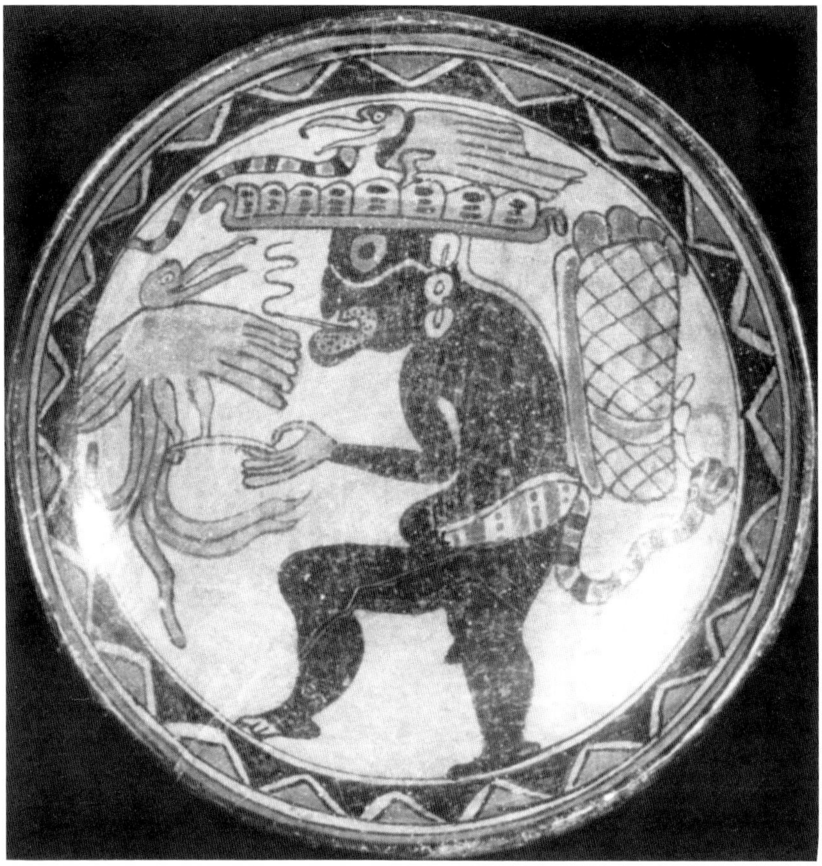

Figure 159. Plate of Black God L Smoking (A-25)/ (photography by Nicholas M. Hellmuth, courtesy of Foundation for Latin American Anthropological Research, St. Louis, Missouri)

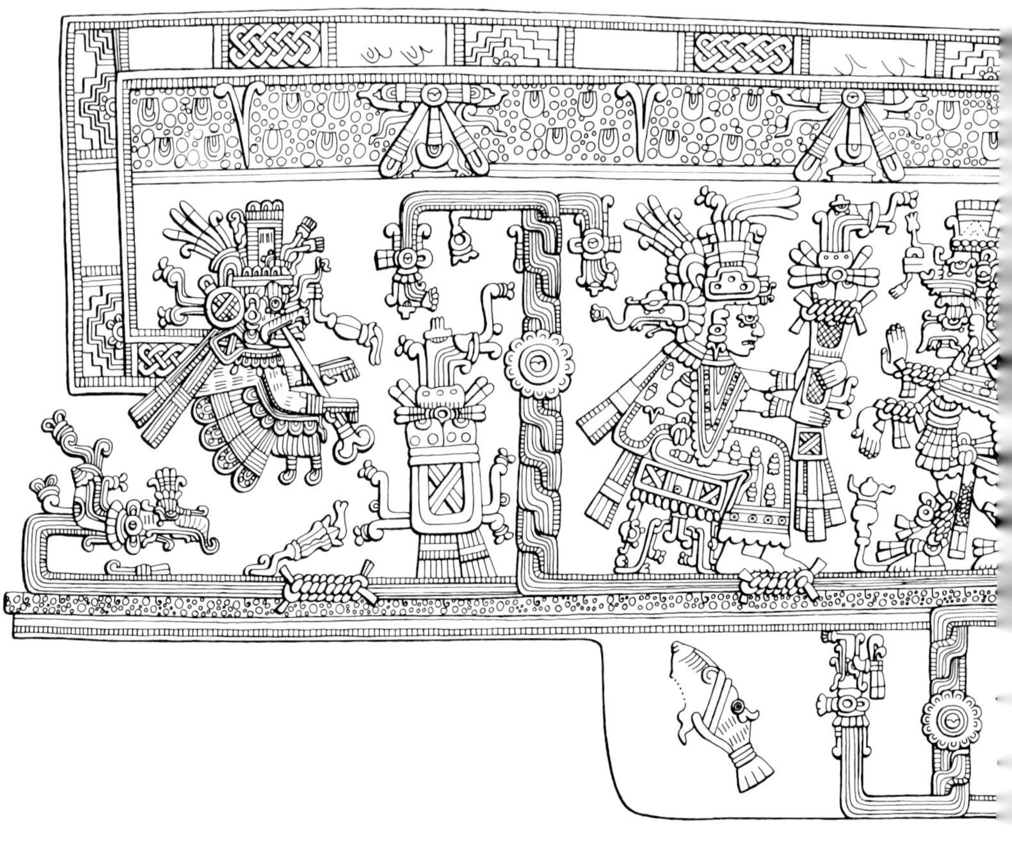

Figure 160. Reconstruction drawing of the mural pair

interior wall of the temple, may simulate the actual appearance of Xux Ek low on the eastern horizon for the 5 unlucky days of every 582 days (the Venus cycle).[85] At each end of the celestial monster is a creature whose upturned nose is very similar to that of God K, but the claws and arm flaps suggest that it may be a bird. Both creatures are wearing ornamental headdresses, necklaces, and wristlets and have large smoking cigars in their mouths. In front of the left celestial bird monster is a tall ceremonial object, probably an *incensario*. The identity of these avimorphs is puzzling. Probably they belong to the Principal Bird Deity Group defined by Bardawil, a flock of ominous supernaturals who are commonly seen on the monuments of Copán, Tikal, Palenque, and Izapá.[86] While the Moan bird perched on the hat of God L on the Palenque panel is worn by a smoker, the examples under discussion are the only ones where a bird god himself is puffing a cigar.

The two central tableaus are occupied by four personages, two in each panel, engaged in giving and receiving a complicated ceremonial object composed of tassels, scrolls, a mat symbol, and the head of a snake, that appears to be spouting fire. On the left-hand panel the object is offered by an elaborately dressed seated lady (?) to a long-nosed deity, probably God B, who wears the head of God K as a helmet. On the right-hand panel another seated female in royal dress and regalia is being presented with the same object by a standing knight, who is also wearing a God K headdress. Certainly one of the personages and probably all four represent supernaturals.

The lower part of the fresco is an underwater scene. The three compartments, divided by serpent bodies, are painted with the images of floating fish, God N (?) emerging from a turtlelike creature and a snake, respectively. The painting is apparently the allegory of the Sky, the Earth, and the Waters and bears a strong stylistic similarity to the frescoes of Santa Rita.[87]

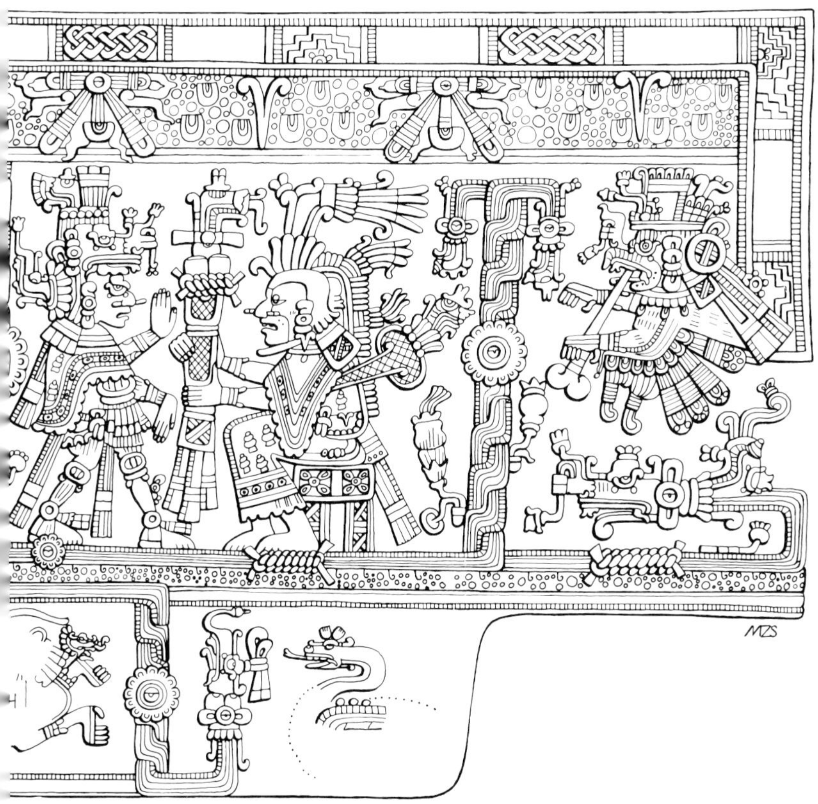

God, Tulúm (drawn after Dávilos)

History is little more than the register of the crimes, follies, and misfortunes of mankind.

Edward Gibbon

One of the problems lies in the difficulty that untrained people have in seeing Maya forms . . .

Linda Schele

Intuition plays an important role in science, but it should be strictly a private role.

Tatiana Proskouriakoff

9. Scenes Probably Representing the Act of Smoking

The **Vase of the Seven Gods** (B-1) is a tall white cylinder of the late Classic period, probably from northern Petén (now in a private collection).[1] It shows a hierarchically arranged palace scene with seven deities, each identified by his own name glyph. It has an upper circular row of glyphs, a planetary band and a vertical glyph column.[2] The principal god, who appears to be smoking a cigarette, has been described by Michael D. Coe:

> On the right is Figure 1, seated upon a jaguar throne under a swagged curtain and definitely the most important deity in the group. The cigar stuck at a jaunty angle in his mouth, his aged appearance, and the Moan-bird headdress show him to be God L. This sinister bird has its own headdress mainly consisting of the sign Ix-Sky (Ix-561, Bolon Caan). Covering the ear is a device which probably is significant in God L's identification while above is affixed a jaguar ear, shared with Figure 2 and indicating the Underworld affiliation of God L. Certain costume features are held in common with the other six figures: (1) a device like a bird's wing at the brow; (2) a long nose tassel; (3) a shoulder cape with long fringe; (4) a heavy belt, suspiciously similar to the yokes worn by ball players, decorated with the crossed-bands motif; (5) kilt of spotted jaguar skin.
>
> Behind God L is one of three bundles which play a prominent role in the drama. Partly obscured by God L's belt is the well-known *imix*-Comb-*imix* combination which Kelley believes to be a title and reads *ba-ca-b(a)* but which Proskouriakoff feels is more likely to indicate the end of the statement. In this particular example, the Kelley interpretation seems more appropriate.
>
> That God L rules the Underworld is indicated by the crossed bones and death eyes which appear in the upper left; by the weird peccary-like animal with death collar above him, possibly a constellation; and by the vertebrae which support the throne. At the right of him and below are four heads of the jawless reptilian Cauac Monster, a rain deity.[3]

The vertical column of glyphs divides the other six gods from God L. These deities are arranged in two groups, one above the other, separated by a horizontal band of death eyes. They all wear jawless monster-head headdresses with a spray of quetzal feathers, and all but the fifth have aquiline noses, large eyes, and filed incisors. God 2 also has jaguar spots around his mouth and the feline ears that identify him as the Old Jaguar God. Apparently he is a nonconformist, because he is the only one in the tableau who wears his hair long and whose headdress is topped by an *ahau* sign. Gods 3 and 4 are portrayed as jawless creatures with quill-like tongue extensions. Figure 4 is distinguished by a curving tongue, and God 5 by his monstrous face. Figure 6, who has barblike whiskers around the

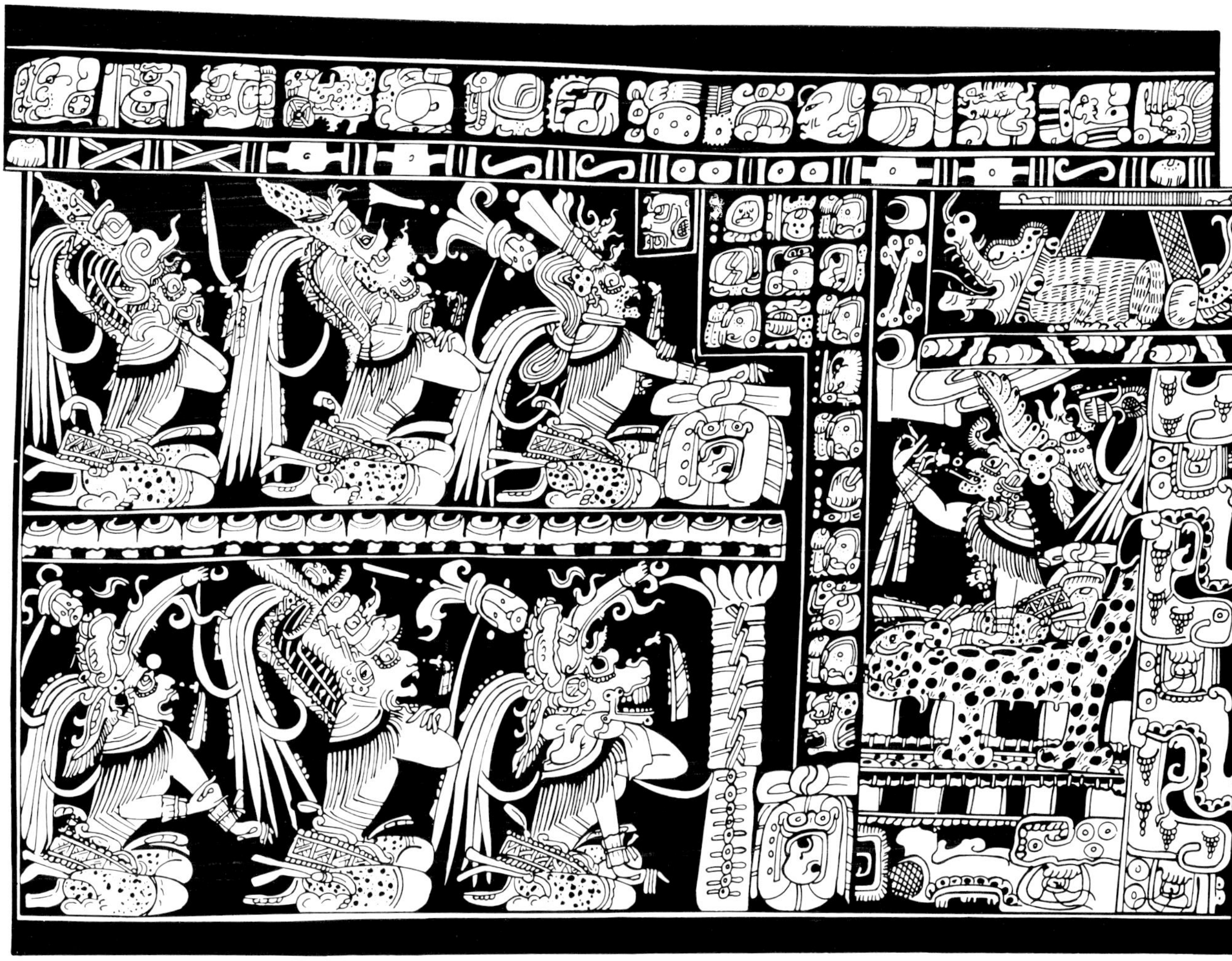

Figure 161. Vase of the Seven Gods (B-1) (after Michael D. Coe)

mouth and shell ear ornaments, is identified as one of the gods of the Palenque Triad.[4] Figure 7 has no distinctive sign. There are two bundles in the scene, one in front of God 2, the other at the bottom of the vertical glyph column. On both is the compound of Ix Star Earth. In front of God 5 is a columnar torchlike object, similar to the one held by the lord on Lintel 33 at Yaxchilán.[5]

For a discussion of the hieroglyphic text see Michael D. Coe.[6]

Another forceful example of the art of the Maya potter is the Vase of the Jaguar Deer (B-2), a polychrome ceramic vessel, probably made in Petén during the Classic period (now in a private collection). It has the conventional orange base, upper and lower red rims, and an upper band of hieroglyphs. Nine figures appear on the vase, as well as four short glyph columns masterfully worked into the composition. The glyphic text and probably also the action begin with the figure of a large animal with the body of a deer and the head of a jaguar. Two little gnomelike spirits are sitting on this creature, one on his chest with one hand extended as if petting the head of the beast, the other posing cross-legged on its left knee. The two little spirits have peculiarly old faces, bald heads, and small, round, marshmellowlike head ornaments, probably death eyes. Their jaws are not visible; either they are wearing long red beards or, as Coe believes, the nearby God GI has just cut off their jaws, and they are bleeding.

The jaguar-deer, which is lying in a position as if wounded or ready to roll over playfully, is certainly puzzling. It is possible that this unusual beast is Xbalanque, the mythological hero of the *Popol Vuh* in disguise, whose name was translated by Thompson and Lounsbury as Jaguar Sun but by Edmondson as Jaguar-Deer.[7] Turning the vase counterclockwise, we next see the figure of a standing deity painted red with a particularly grotesque mask. He

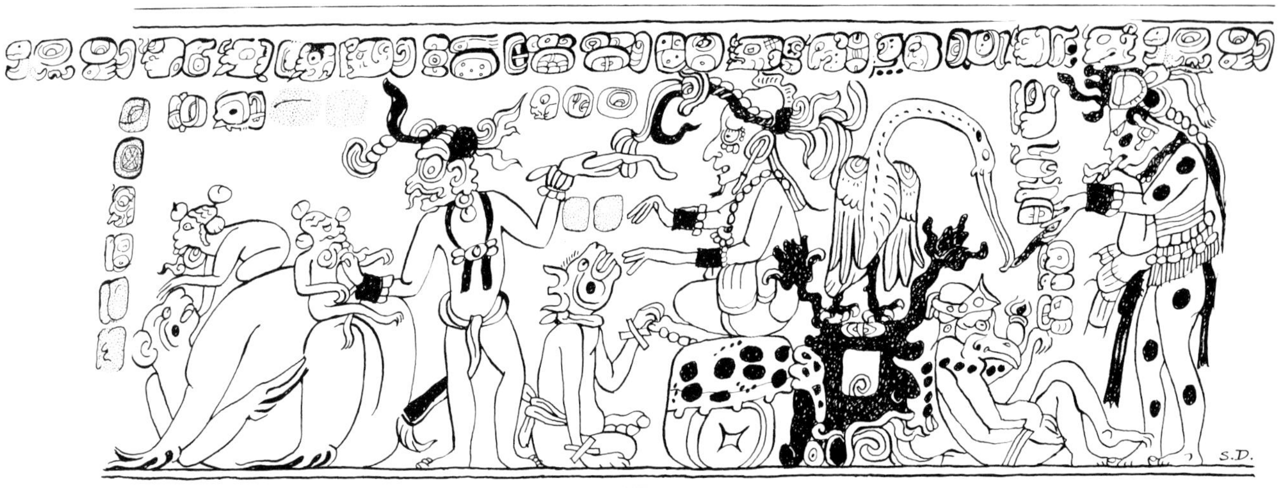

Figure 162. Vase of the Jaguar Deer (B-2)

has an odd short upper lip and eyes as "big as saucers," like the dog in the fairy tale of Hans Christian Andersen. This deity is undoubtedly the gargoylish executioner God GI of the Palenque Triad, appearing on this vase in disguise, who makes similar appearances in the fishing scene of the Tikal bones.[8] He is wearing the mask of a jester god, but his identity is given away by his "trademarks," the shell earplug, the bound hair, the upthrust headdress, and the knotted pectoral. On this vase he is in his characteristic "dancing" position, and he is wielding an ax in his left hand. With his right hand he is touching one of the gnomes. It is tempting to conclude that GI is about to sacrifice the two gnomes and the jaguar-deer. The act may correspond to a story in the *Popol Vuh*, which tells how the Hero Twins, Hunaphú and Xbalanque (Jaguar Deer), triumph over the Lords of Xibalbá by fooling the lords into allowing themselves to be cut apart, after which they bring themselves back to life. In envy the Lords of Xibalbá request that the same thing to done to them, but they are not brought back to life.

At the left of GI, apparently undistrubed by his menacing posture, a red-collared anthropomorphic, canine sits on its haunches. It is wearing a white loincloth and wristlets shaped like crossbands. The dog's association with the Maya underworld is further emphasized by the large death spot on its face.[9] Special attention should be given to the dog's ears, which seem to have large pieces missing, as if something may have taken a bite out of them. This points to an association between the dogs of Maya mythology and Xolotl, the Mexican Dog God, who is usually portrayed with ragged ears. The explanation for this may be the close relationship between Xolotl and Nanautsin, the Mexican god of syphilis. The one can be substituted for the other in the calendar, and they are confused in mythology. Thompson has no doubt that Nanautsin is merely a variant of Xolotl, who is shown with ears that are either eaten away or from which blood is pouring.

The canine is leaning on its right foreleg and appears to be looking at an aged god whom he is touching with his left paw. The old god has large eyes and central pupils and is wearing a cantilevered headdress made of plants and/or feathers, a shell (?) ornament on the crown of his head, a red kilt, a bead necklace, and a pectoral, as well as wide wristlets. He is apparently undisturbed by the possibility of acquiring V.D. from his faithful canine and is seated stoically in tailor fashion on a cube-shaped, jaguar-hide-upholstered cushion that has a *pop* symbol painted on its side. This probably indicates the high position (*ah-po[p]*) of the occupant.[10] Leaning against the back of the cushioned throne but looking away from it as if he is a rear guard, is a large monster head painted green. It has large, angular scroll eyes.[11] A jaguar paw is behind the ear, probably a reference to *chac mol*, "giant paw," one of the many Maya names for the jaguar. From its forehead and the back of its crown rises a peculiar tree-like ornament, similar to the tree forms on the sides of the sarcophagus in the Temple of Inscriptions at Palenque. The jaguar paw, the *tau*-shaped tooth, and the mouth emanations identify the monster as the patron god of the month Pax, whose features recall those of the jaguar but are also reminiscent of the Sun God.[12]

In front of the Pax monster lies a peculiar creature with a human body and head of a bird. The creature probably belongs to the family of Maya vulture gods (Zimmermann's God T5), some of whom are portrayed in the *Dresden Codex* and also on a Classic vase found in southern Campeche.[13] The tubelike element on the god's beak identifies the bird as a *tahol* vulture rather than the king vulture often painted on Yucatec ceramics. The beak element is the T59 locative *ti ta*, homophemous with *ta*, the root for the Maya word for "excrement." The uncomplimentary *tahol* is a common name for the vulture in the Maya area. The identification is further supported by extension substitution of the two graphemes T59 and T747b, "vulture head" in some glyphic texts. Looking down at the fallen vulture is a lord, a powerful, sloe-eyed figure. He is dressed in a wide belt decorated with tassels, a narrow loincloth, and wide wristlets similar to those worn by the Old God. His head is topped with a cloth headdress that has the mask of the Jester God attached to the front.[14] Michael D. Coe has noted that the figures he identifies as the Hero Twins are often shown with such Jester God headdresses.[15] The body and face of the standing lord are painted with large black spots, surely indicating that he is dead. He appears to be holding in his lips either a long cigar or a blowgun.[16] What appears to be the glowing end of the cigar is touched by the long beak of a waterfoul, perhaps a cormorant perched on top of the Pax monster's head.[17]

Glyphic Text

The rim text is a version of the Primary Standard. It begins with a double introductory glyph located between the standing figure and the head of the jaguar-deer. Both of them contain the T617a "mirror" grapheme.

The Secondary Texts are difficult to read because of the dark pigment used as an overlay on some of them. The use of the black overlay appears to be patterned but its meaning cannot be determined.

The first of the two Secondary Texts is located to the left of and above the jaguar-deer. It has a pair of dark compounds, followed by two light ones, then by another pair of dark ones. It is introduced by a side view of the T617a mirror grapheme. The second text above the jaguar-deer has the same introductory glyph consisting of a pair of light and dark compounds. The second compound appears to be a verbal group with *muluc* as the main grapheme.

The third Secondary Text is located above the dog. It is unclear if the groupings of compounds are to be read as a continuous text or as two separate ones. The group of the glyphs to the right of the head of God GI has an apparent *xib* head as the first grapheme. The second grapheme is *caban*, and the third is the T617a mirror in front view. All three may compose another nominal for GI or the Aged God and perhaps read Ahau Caban Nen, "Lord of the Earth Mirror." *U nen cab* is a title group recorded in the *Motul Dictionary* as a designation of rulers. Here we assume that the description refers to a supernatural rather than a historical person.

The last text is located to the left of the standing figure and above the vulture. It most probably refers to the standing lord. The first compound is a verbal group (T1.617a.670) which is used at Quiriguá and elsewhere as an "accession" phrase. The subject of the verb is recorded most probably in the two following compounds. The "accession" compound apparently expresses the activity of the root *nen* which functioned in a more generic sense than other "accession" expressions. *Nem* is recorded in Quiché[18] as "to succeed in office" and, according to the *Motul Dictionary*,[19] it has the meaning in Yucatec expressions as *u nen cab, u nen cah, "el sacerdote, cacique, governador de la tierra o pueblo, que es espejo en que todos se miran"* (the priest, chief or governor of the earth or pueblo who is the mirror in which all see themselves).

The **Vase of the Drinking and Smoking Lords** (B-3) is a short ceramic cylinder of the late Classic epoch, probably from Petén (now in the Carlos H. Nottebohm Collection). The vase is painted red and black on white with red upper and lower rims and an upper glyphic band. It shows three figures; two of them are apparently rulers, and the third is an attendant. The first figure is a stocky lord seated on a low dais, leaning comfortably against a cushion made of spotted jaguar skin. His high rank is evident from his elegant hairdo, wristlets, bead necklace, and quetzal floaters. He is wearing a red kilt with a circular white motif and a narrow matching apron. Separated from the lord by a vertical red column, perhaps a wall, is the seated figure of either the same lord or another noble of apparently equal rank. He is wearing a similar kilt, an ear ornament, a bead necklace, and a quetzal floater, and he is seated cross-legged on a throne of somewhat different shape but equal height. The portrait may support Coe's theory that under certain circumstances the Mayas may have had dual leadership. While the second noble is apparently a nonsmoker, he certainly does not reject the joys of alcohol, because he is holding in his hands a tall beaker from which yellow dotted semicircles with clearly outlined central points rise. This indicates both pictographically and phonetically that the good lord is going to have a hearty drink of a foaming, fermented beverage, probably *pulque* or *balché*, the native Maya mead flavored with the bark of the *Lonchotarpus* tree.[20] In front of him kneels a light-skinned attendant wearing a costume similar to his lord's but an unusual headdress, probably made of lengths of stiff cloth folded like "fancy starched napkins at a dinner party." Michael D. Coe believes that such headdresses may be associated with specific deities.[21]

Glyphic Texts

The rim text is Primary Standard. The introductory glyph is above and slightly to the front of the smoking lord. It is the GI (T1011) variant.

The Secondary Text is composed of two compounds and is located above and in front of the smoking lord. It is badly eroded but appears to be nominal. The first compound has the water group prefix and the T168 *ahpo*

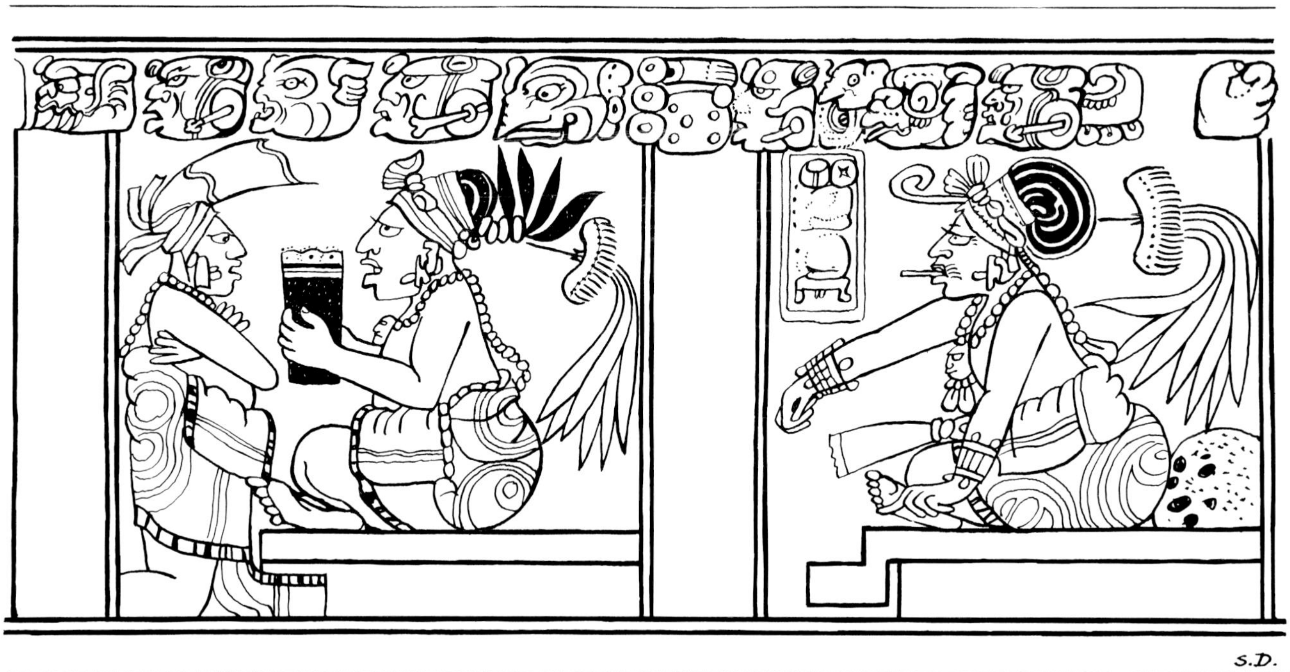

Figure 163. Vase of the Drinking and Smoking Lords (B-3)

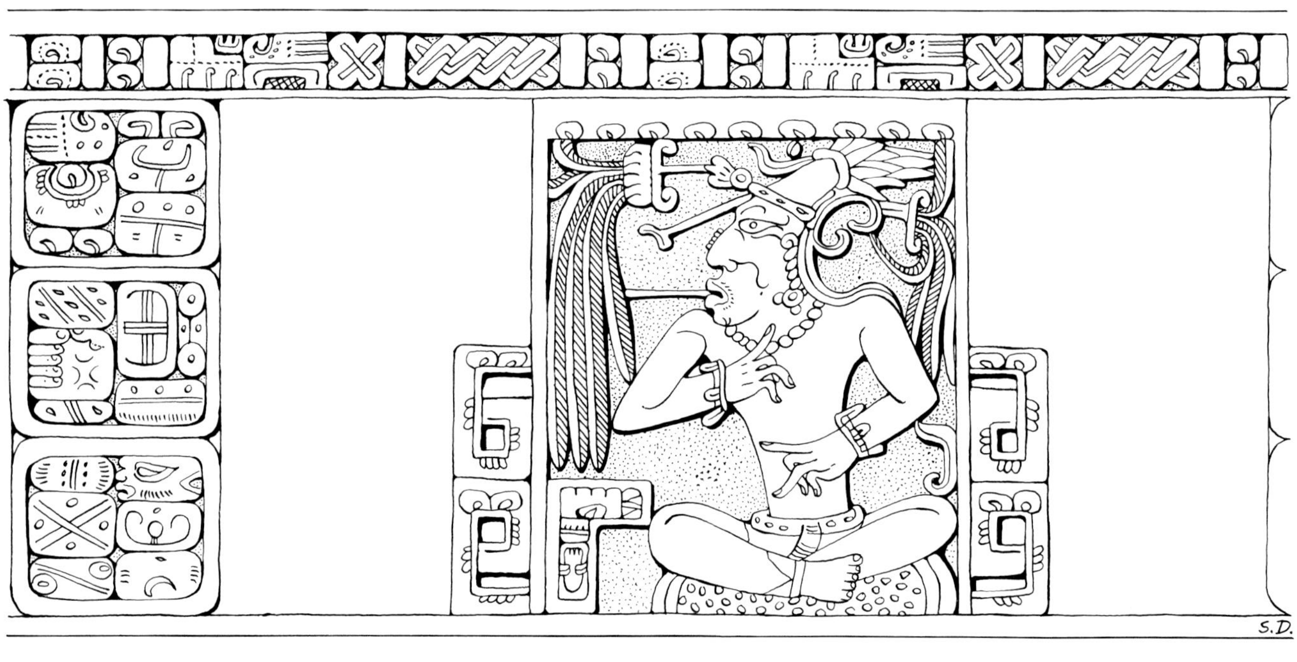

Figure 164. Vase of the Seated Ruler (B-4)

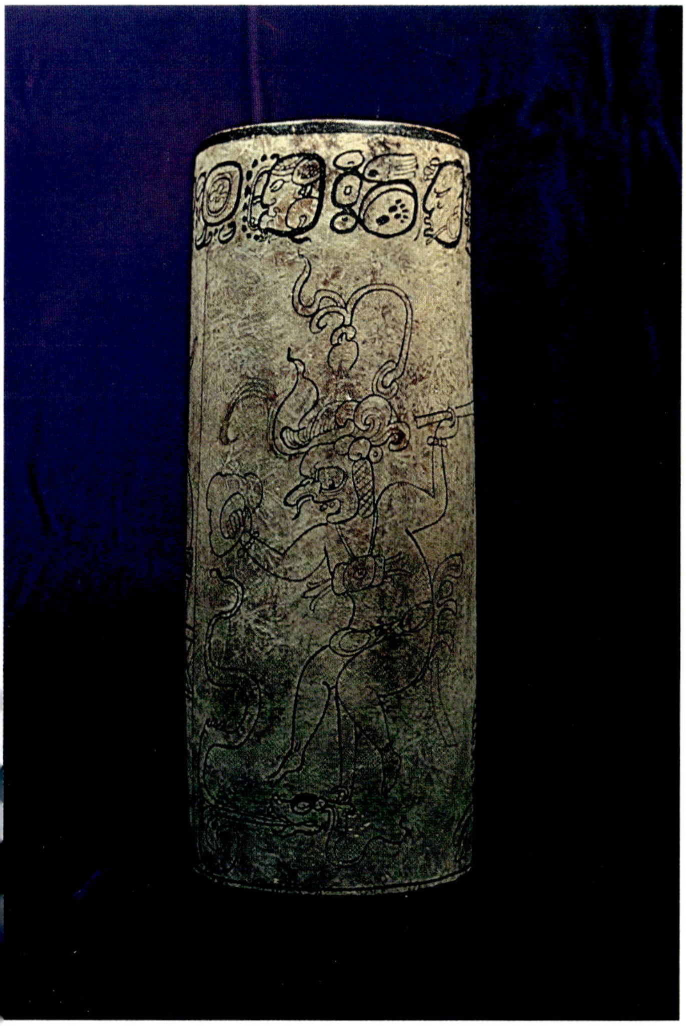

Plate 133

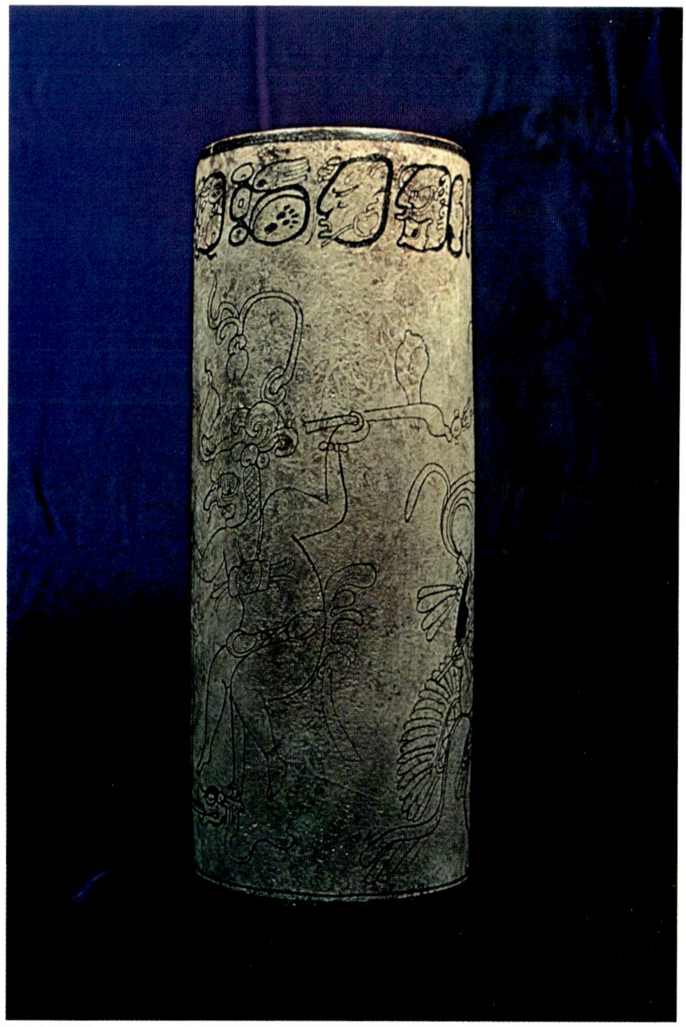

Plate 134

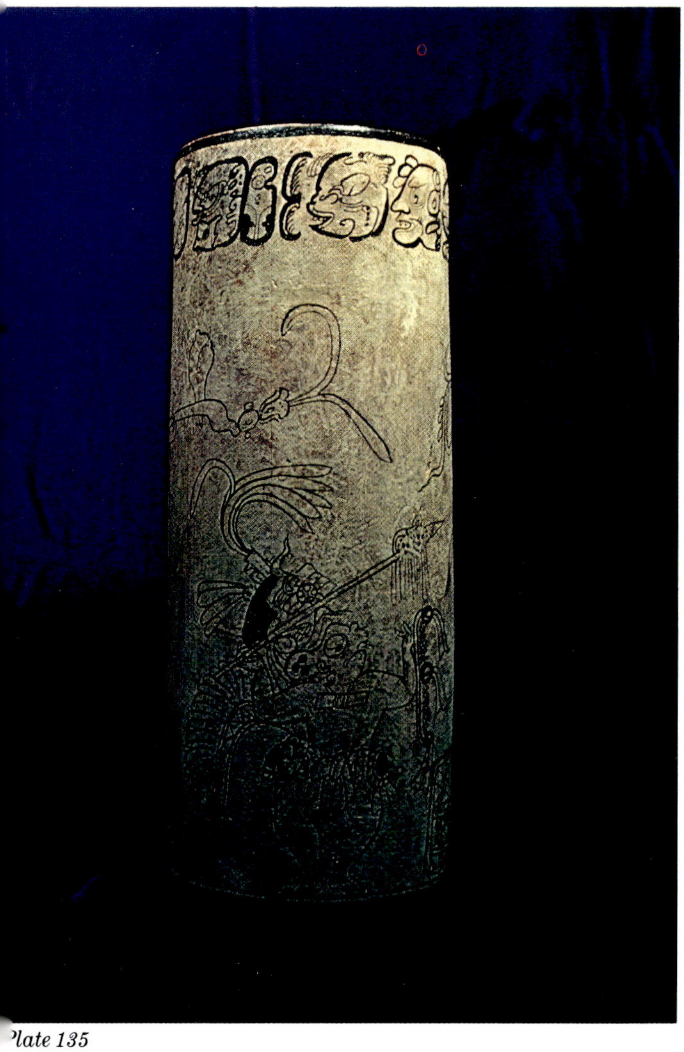

Plate 135

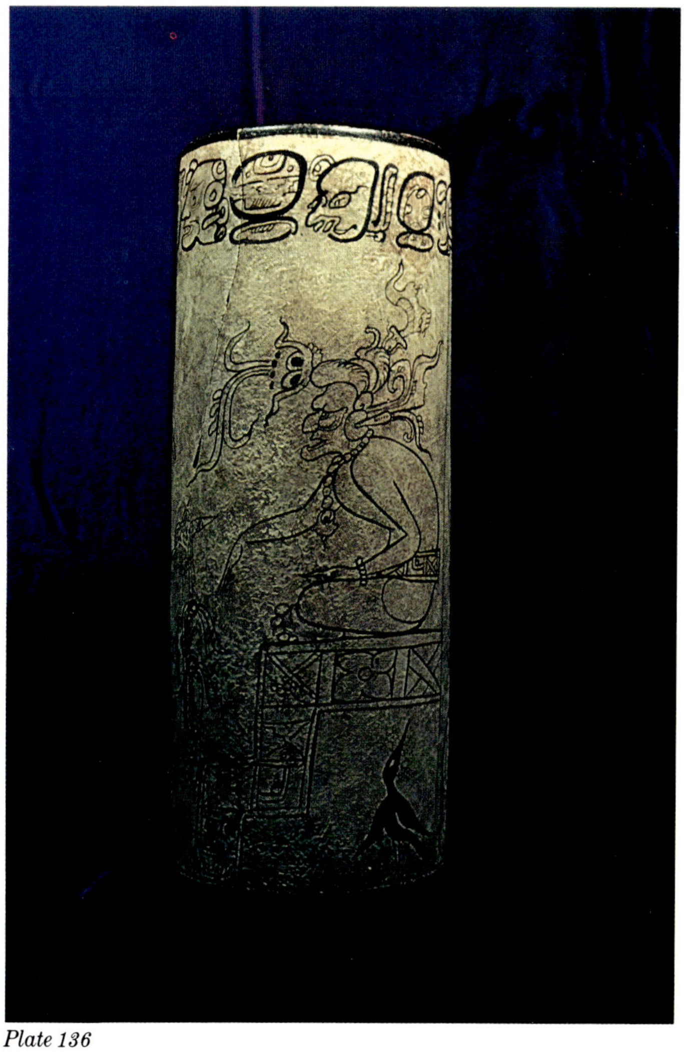

Plate 136

Plates 133-136 Vase of the Black Line (A-18)

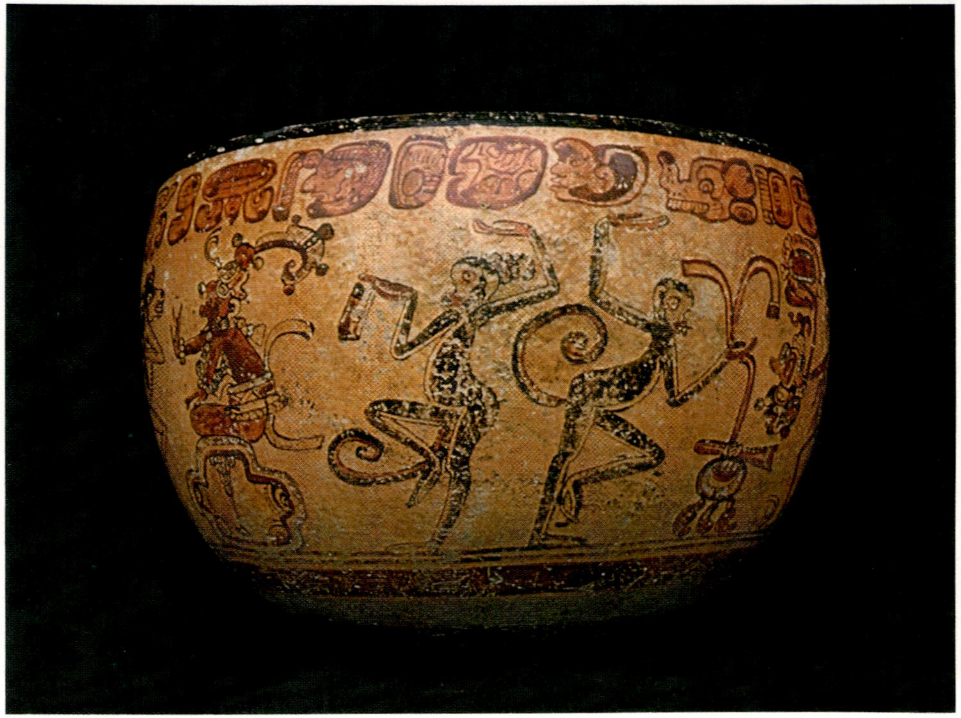
Plate 137

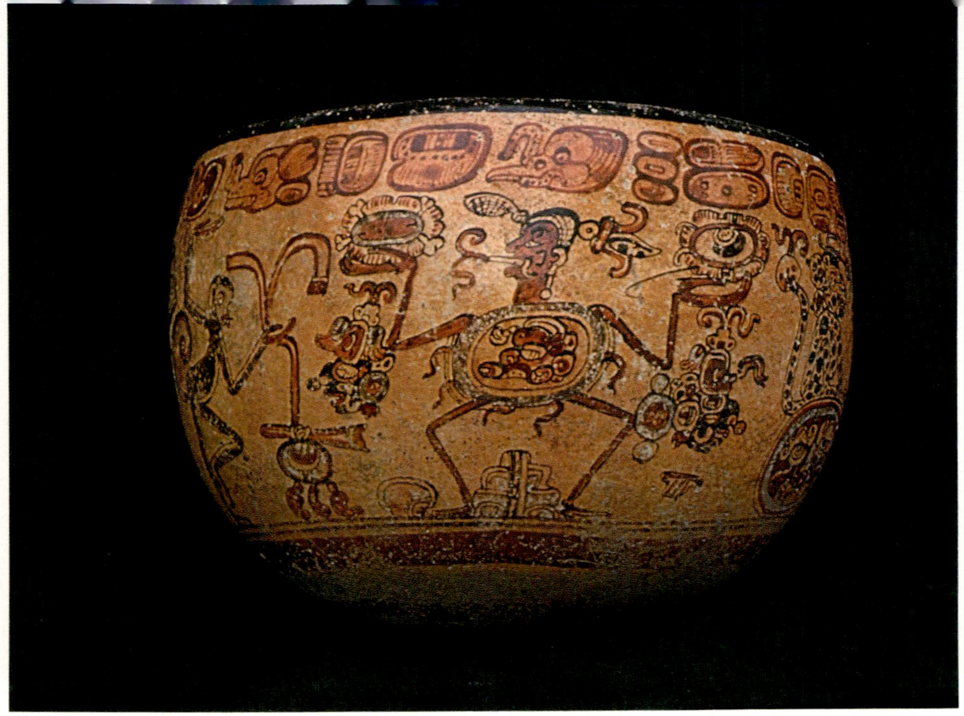
Plate 138

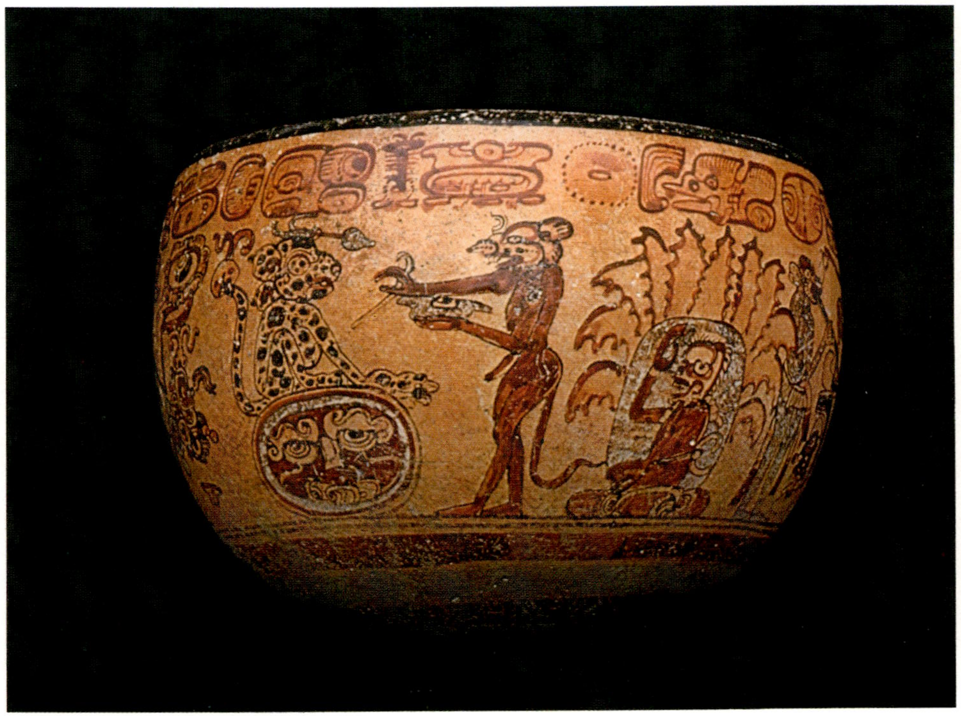
Plate 139

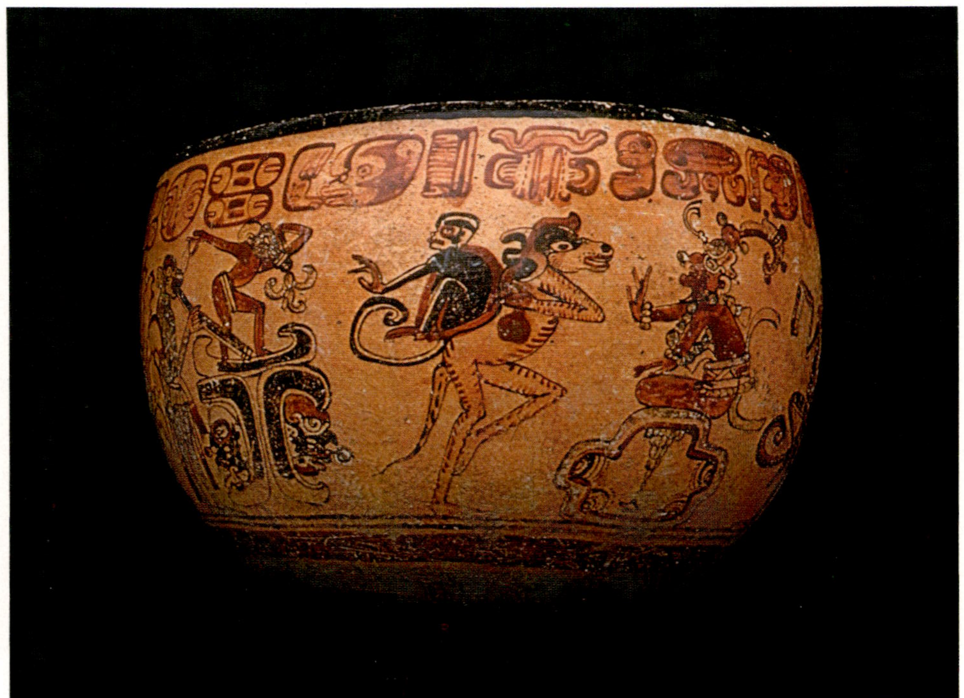
Plate 140

Plates 137-140 Bowl of God N (A-17)

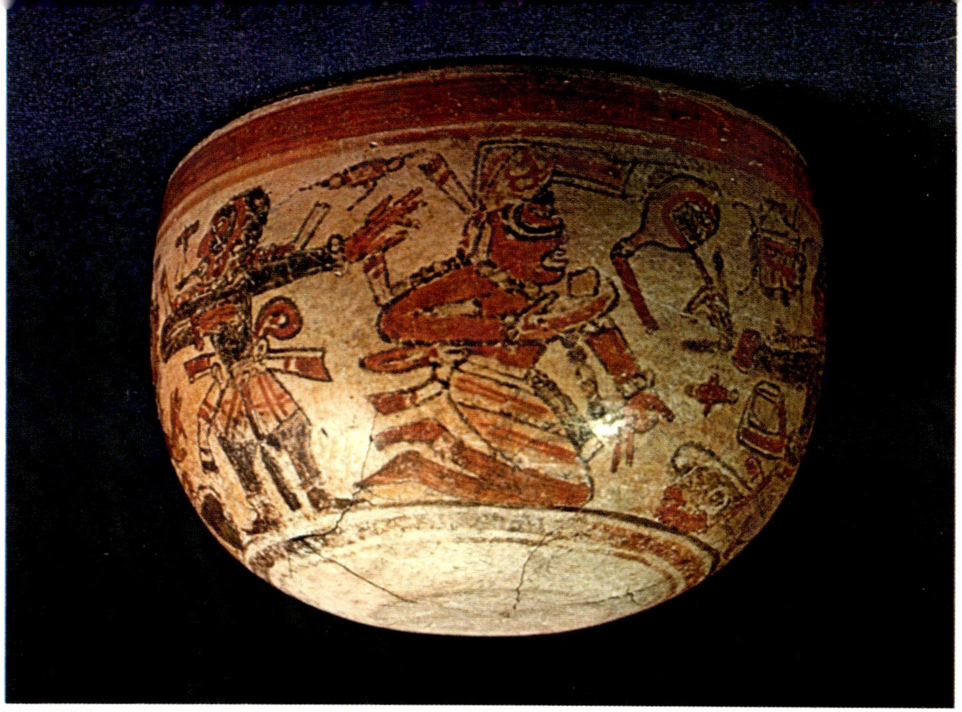

Plate 141

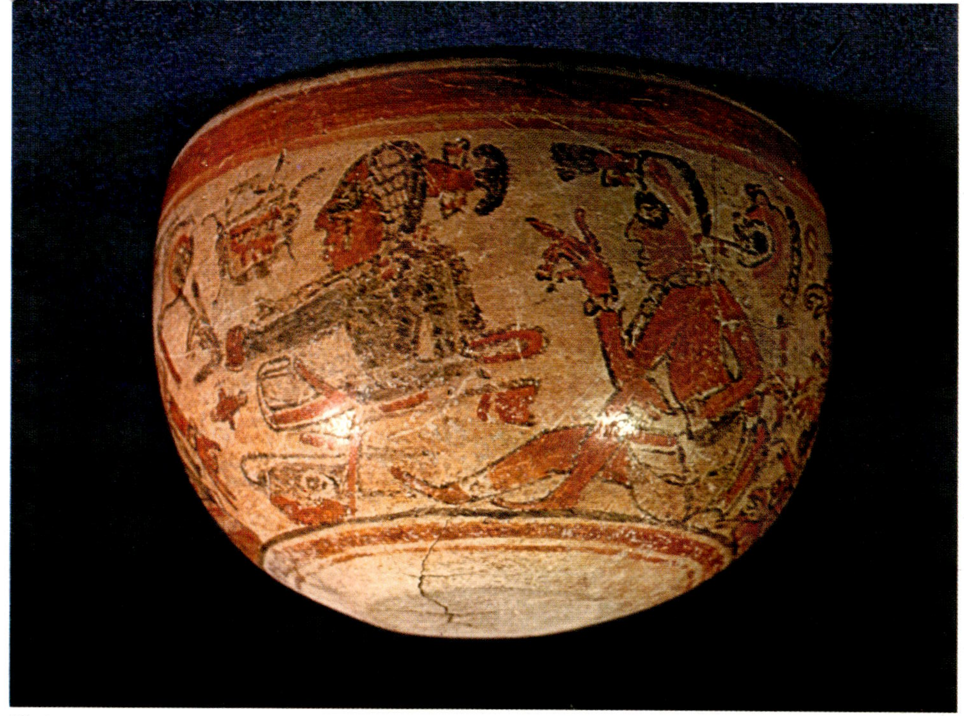

Plate 142

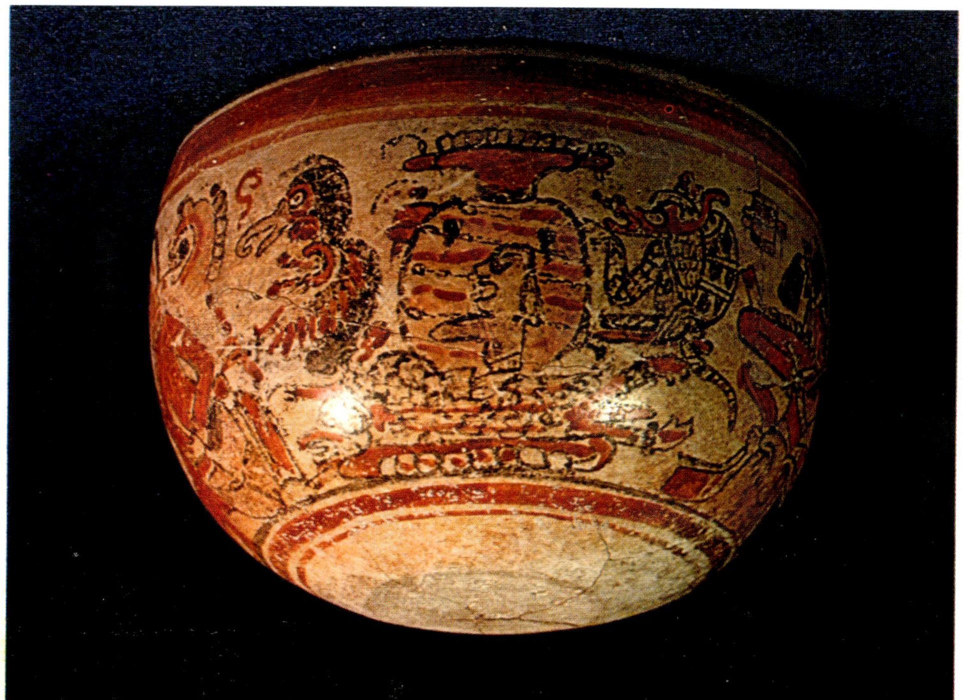

Plate 143

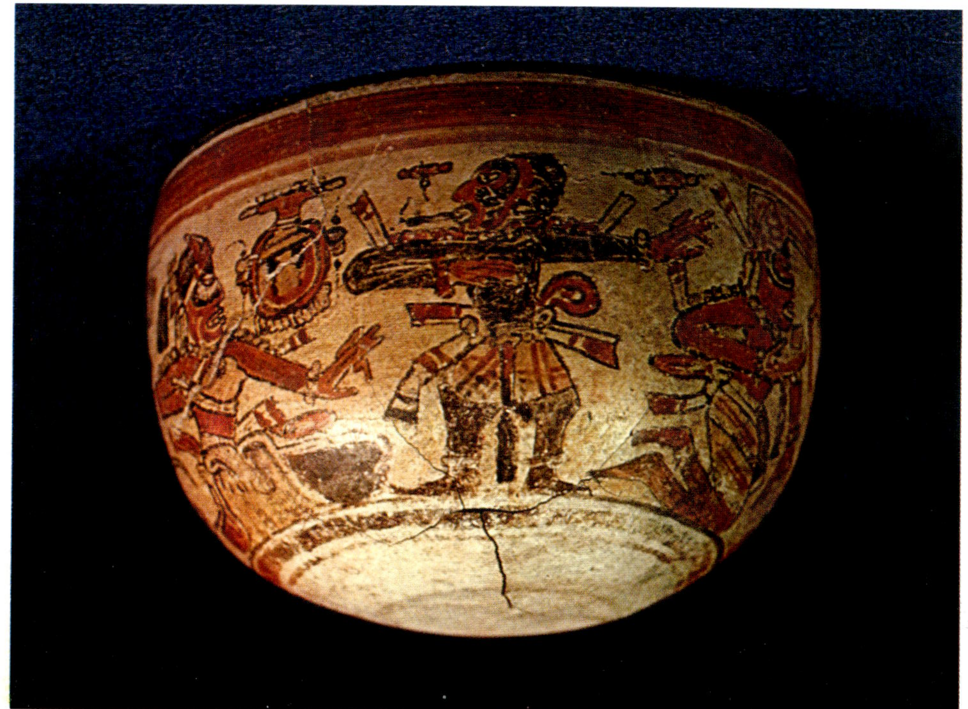

Plate 144

Plates 141-144 Bowl of the Smoking Black Lord (A-19)

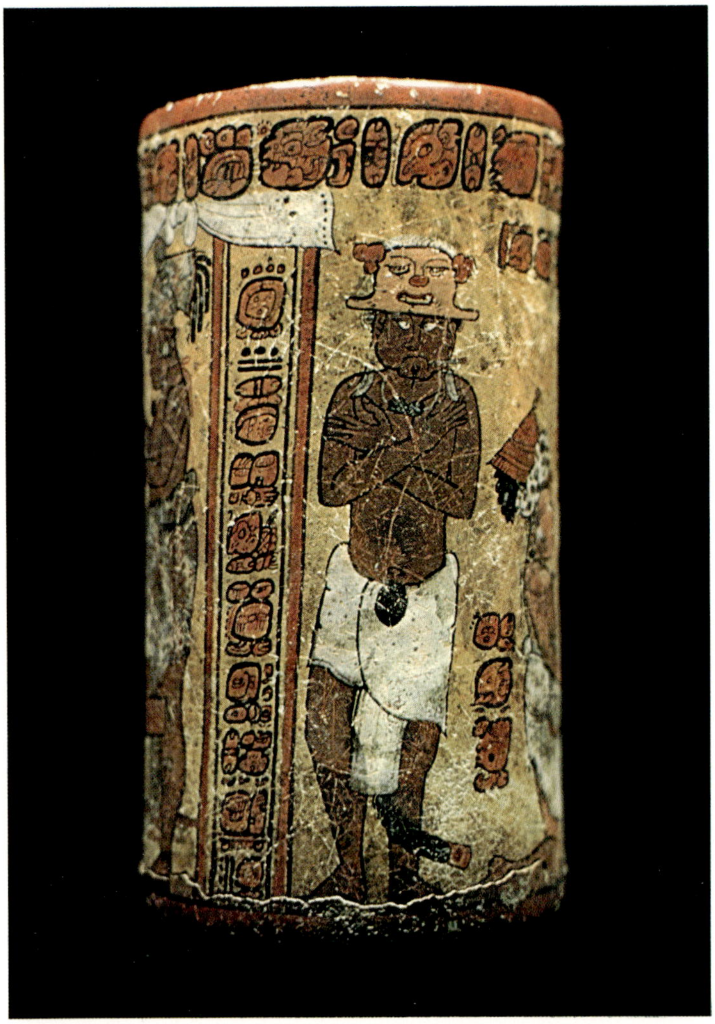

Plate 145

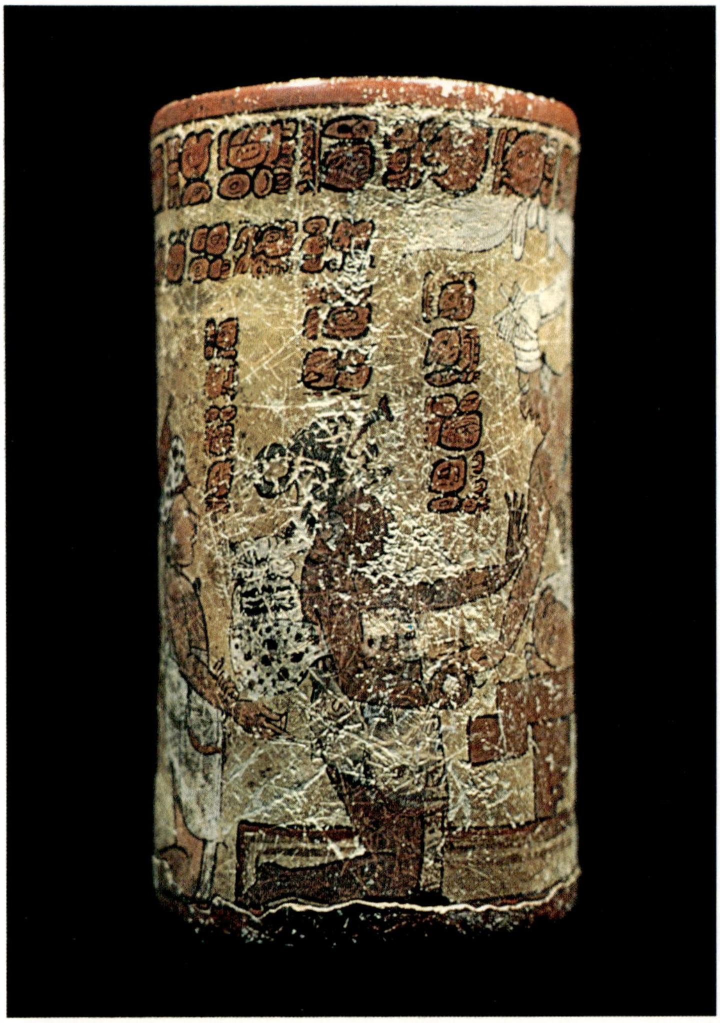

Plate 146

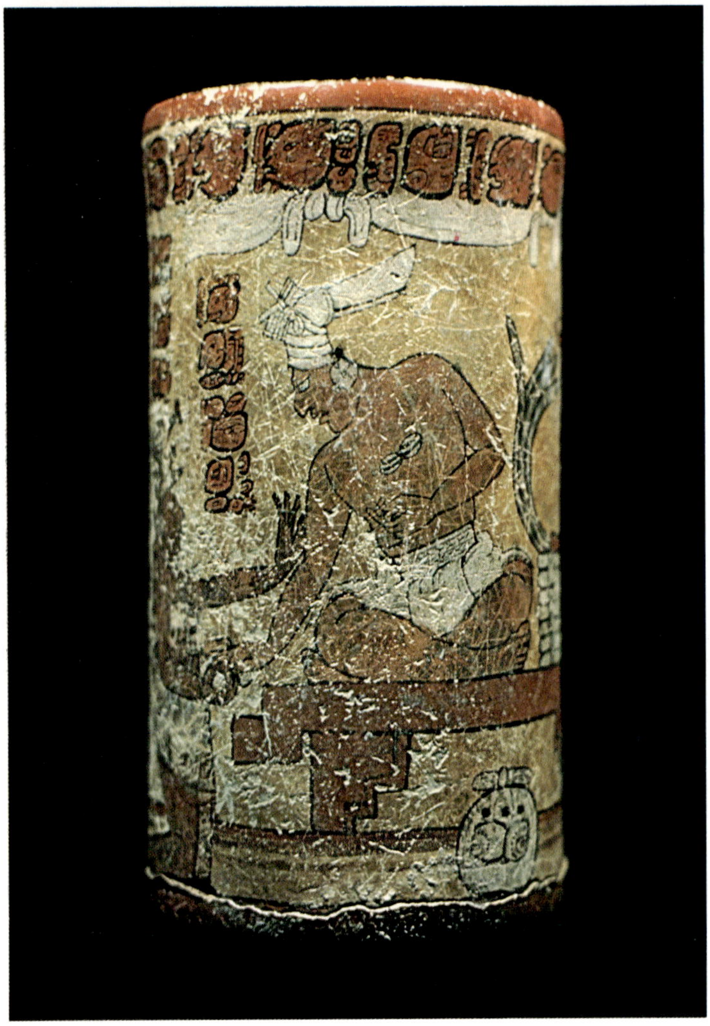

Plate 147

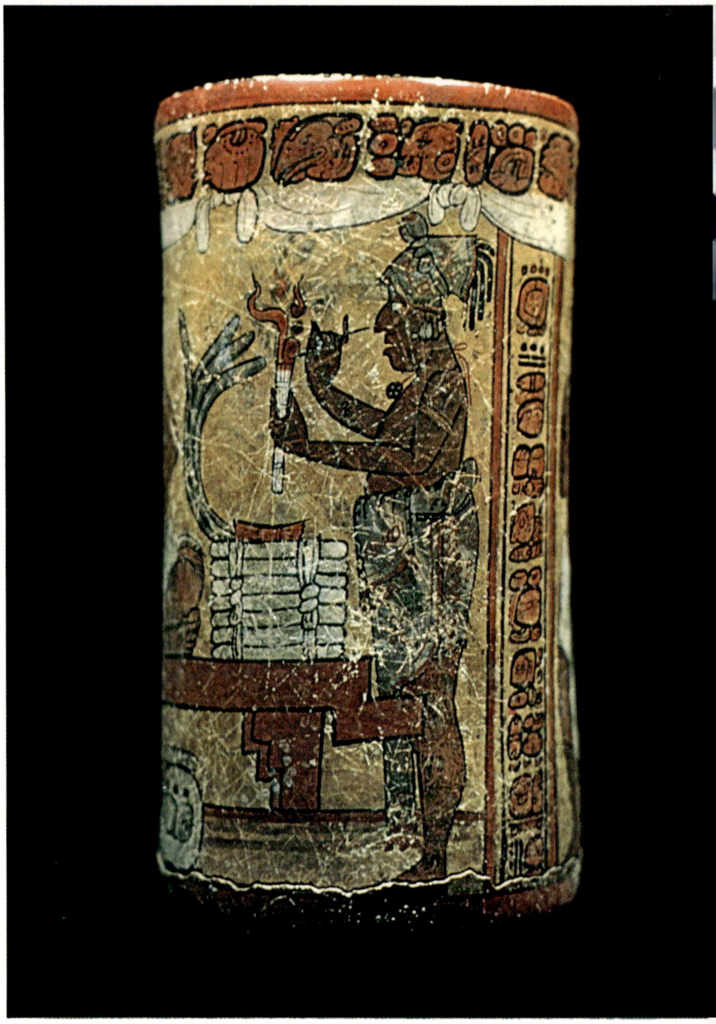

Plate 148

Plates 145-148 Vase of the Palace (A-22)

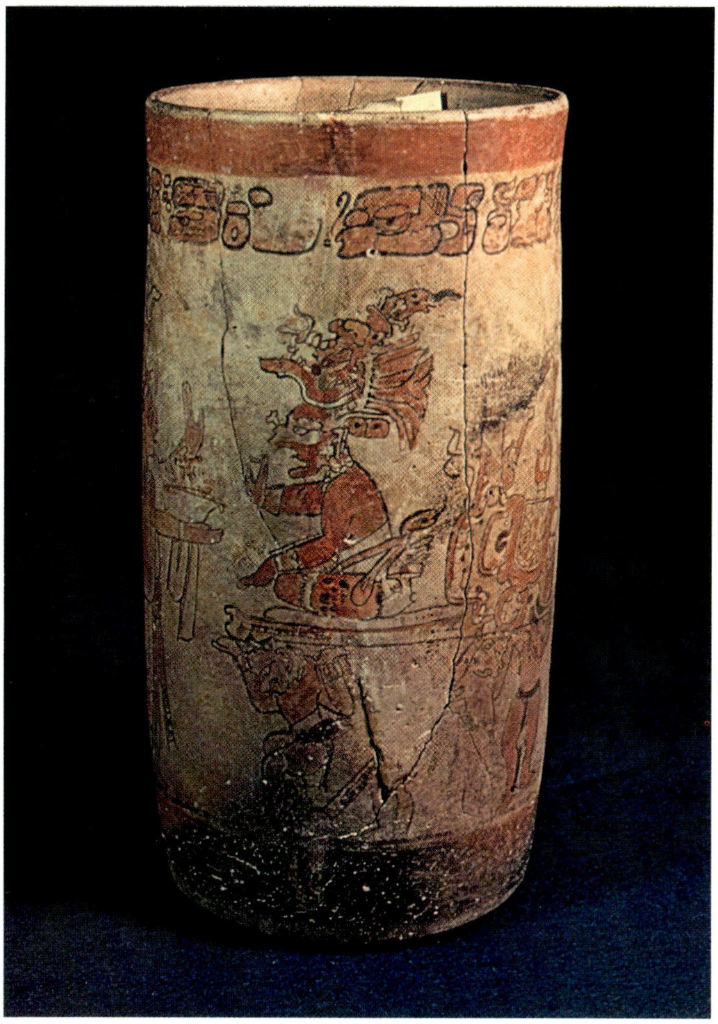

Plate 149

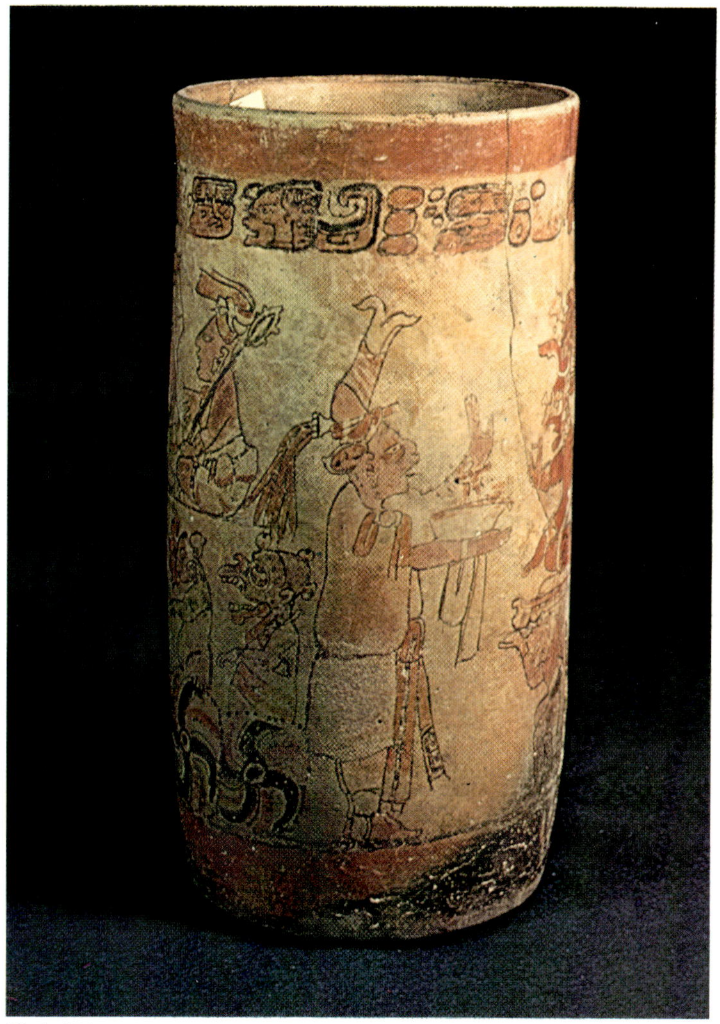

Plate 150

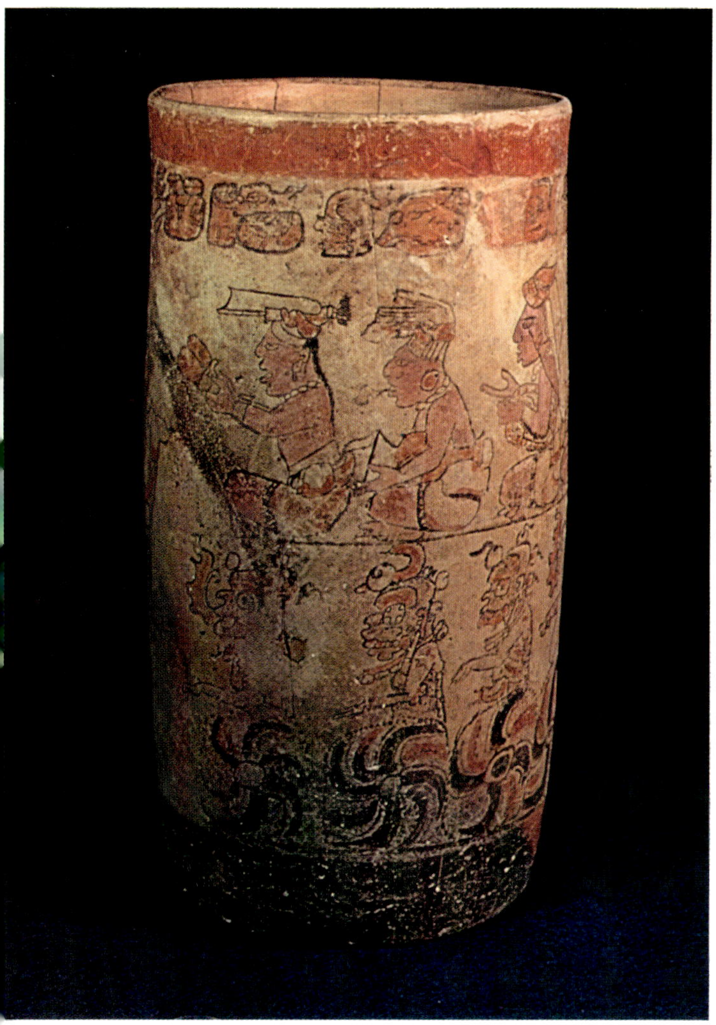

Plate 151

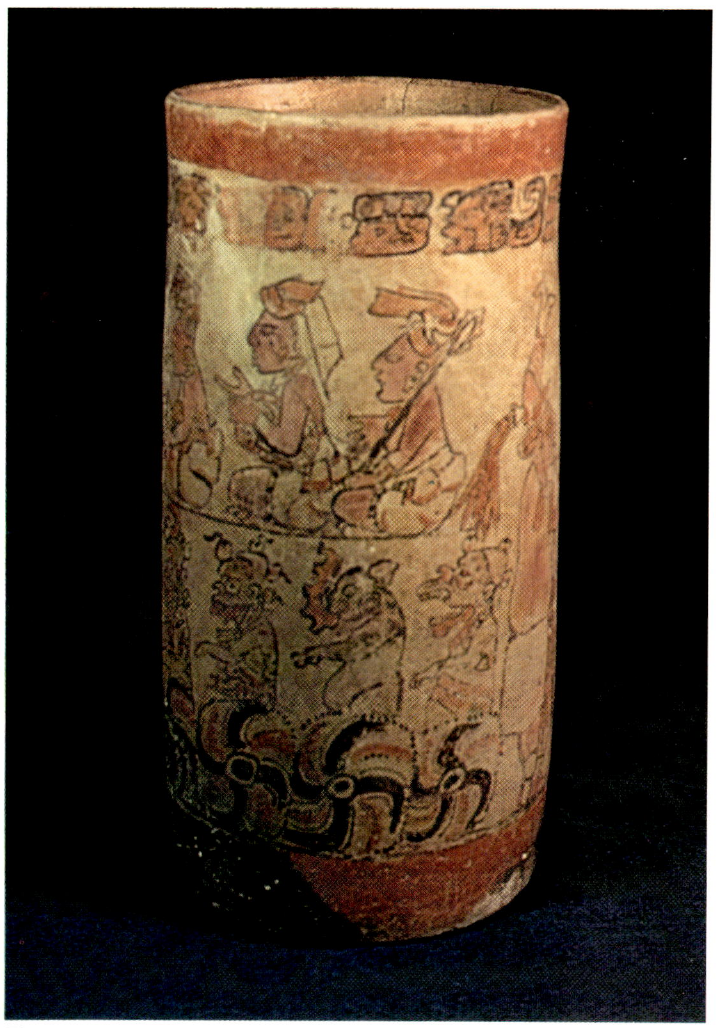

Plate 152

Plates 149-152 *Vase of the Seated Processional (A-20)*

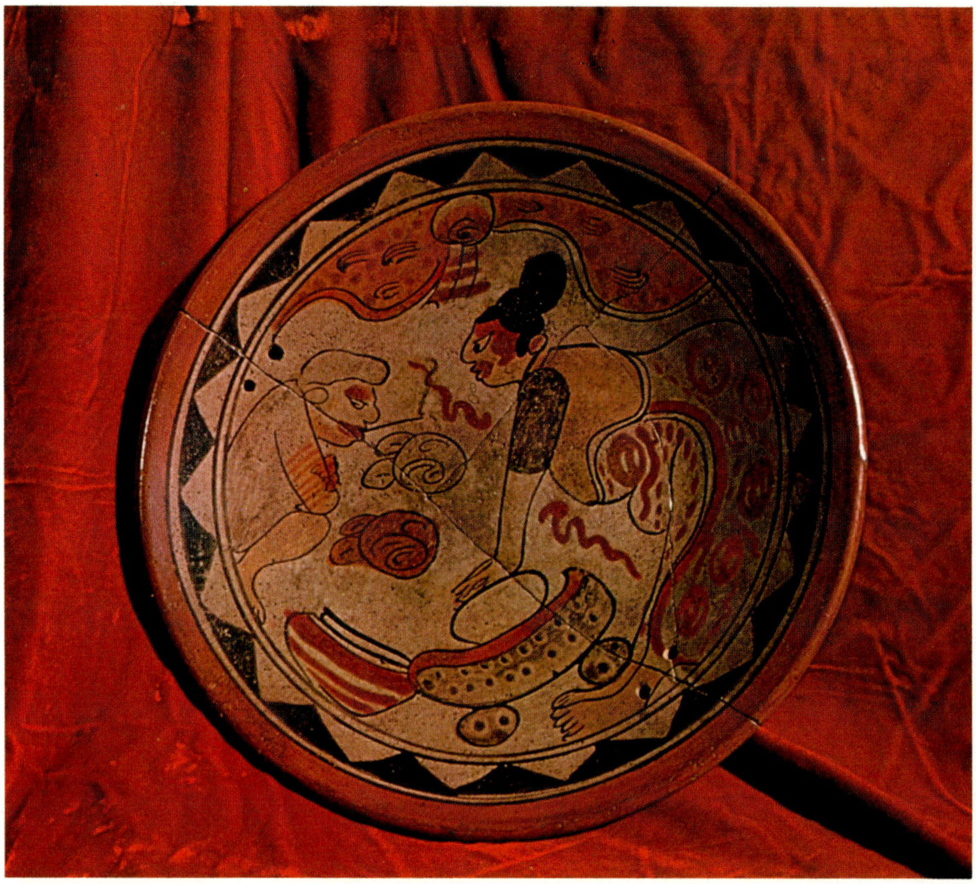

Plate 153 Plate of the Smoking Husband (A-23)

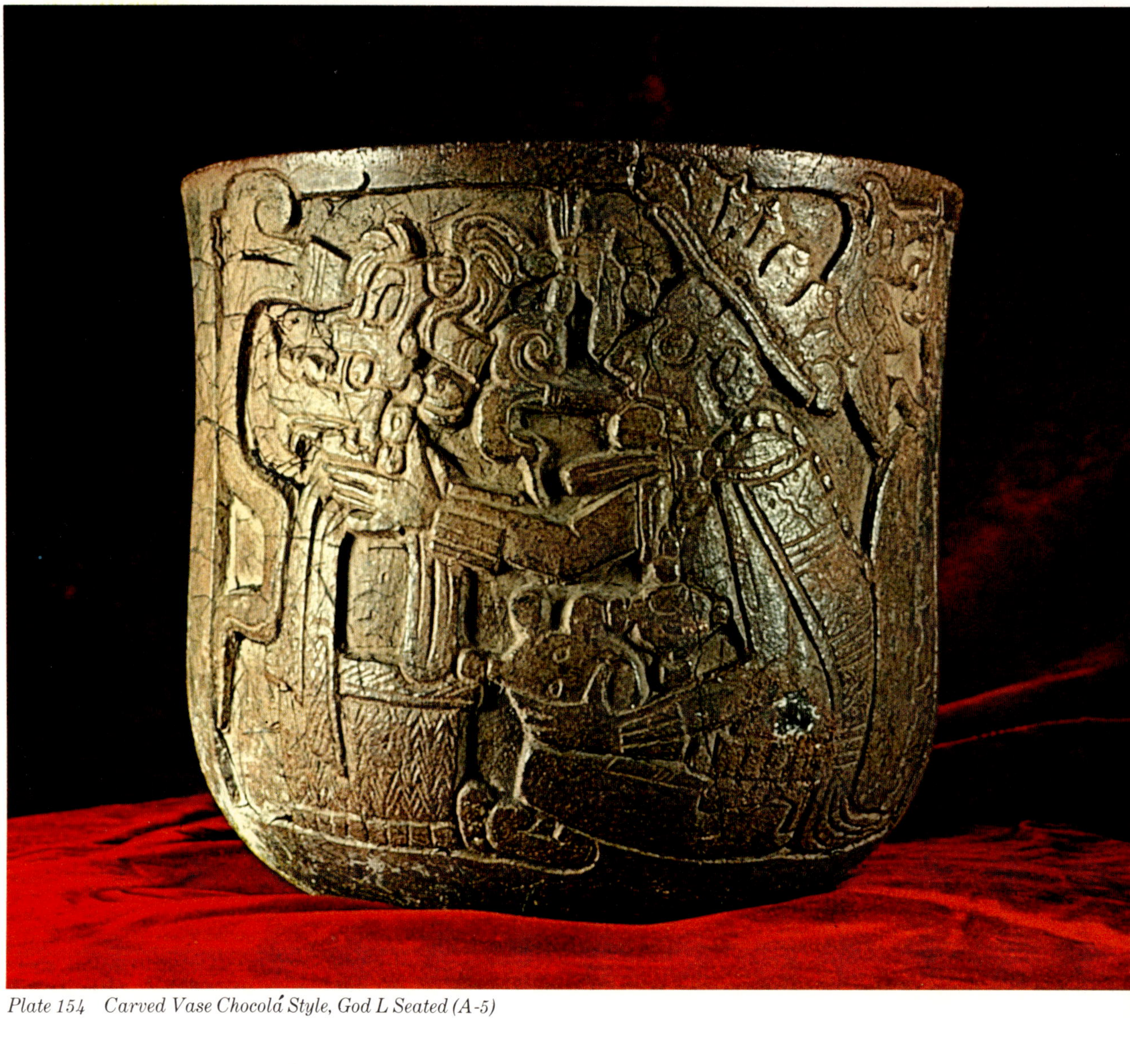

Plate 154 Carved Vase Chocolá Style, God L Seated (A-5)

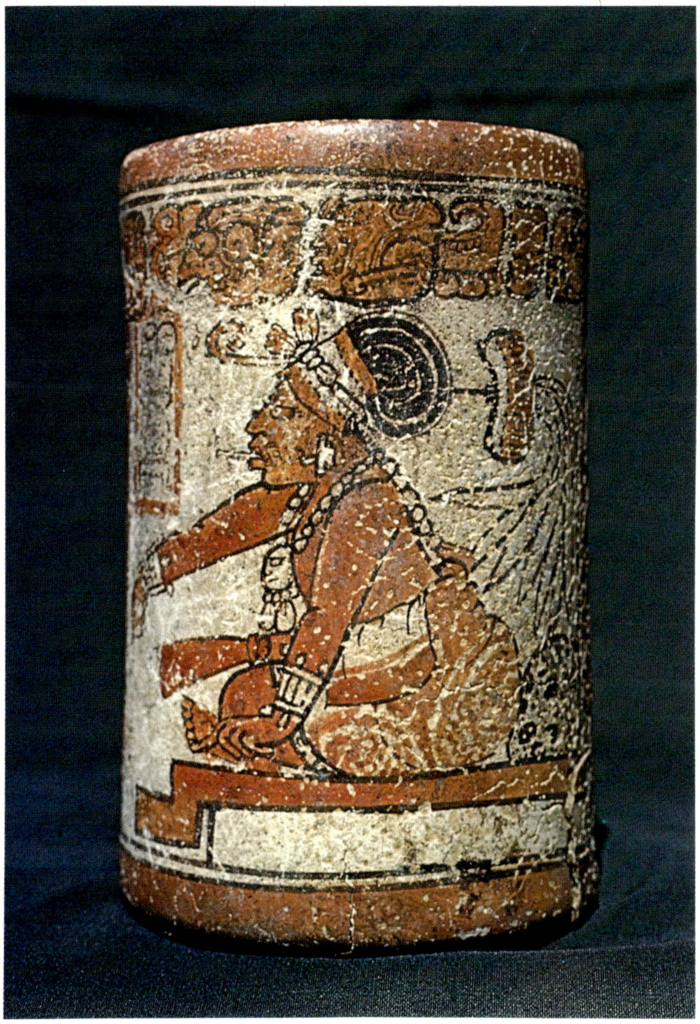

Plate 155

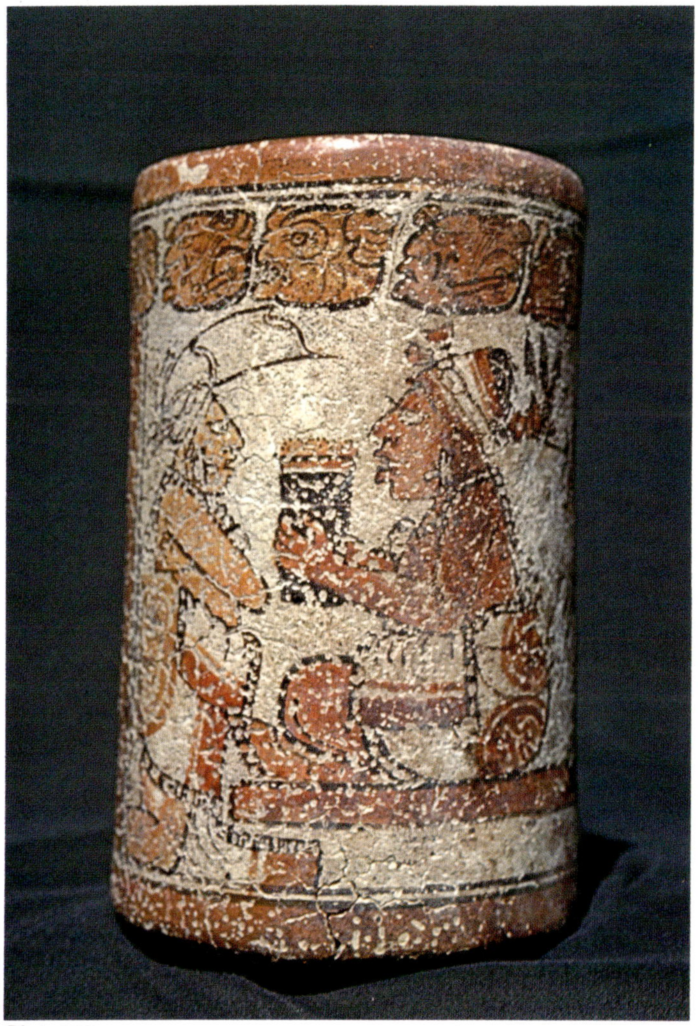

Plate 156

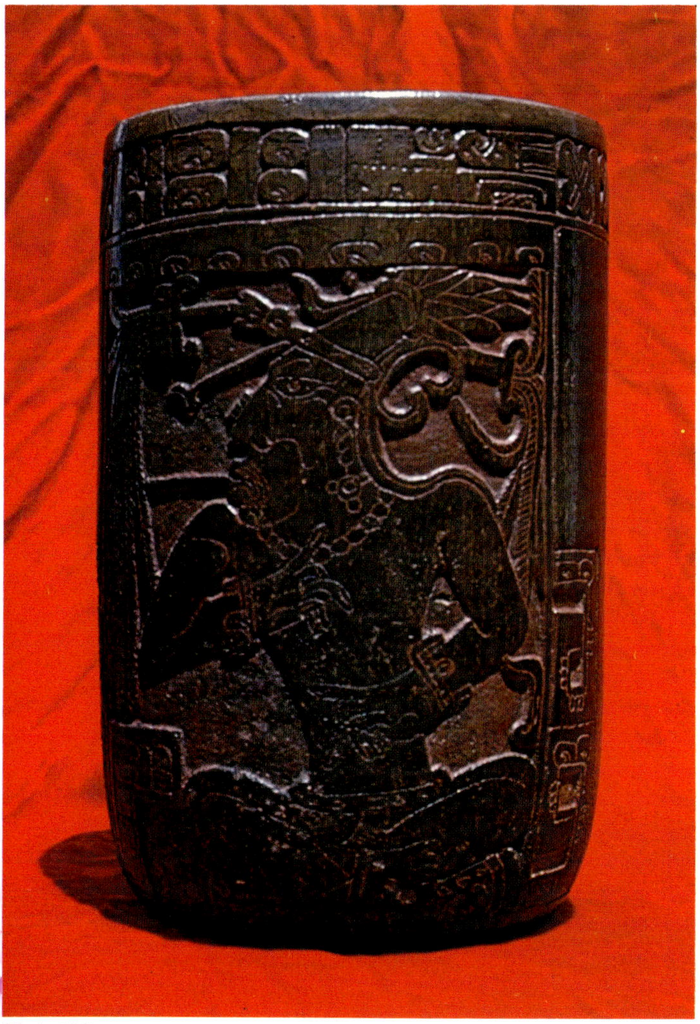

Plate 157

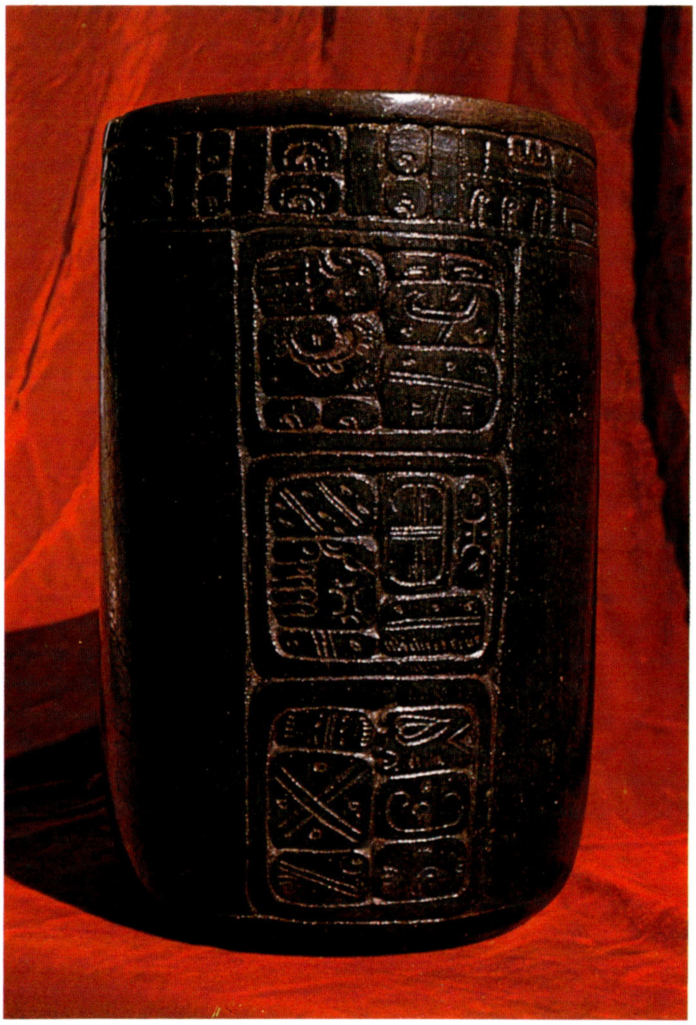

Plate 158

Plates 155-156 Vase of the Drinking and Smoking Lords (B-3)

Plates 157-158 Vase of the Seated Ruler (B-4)

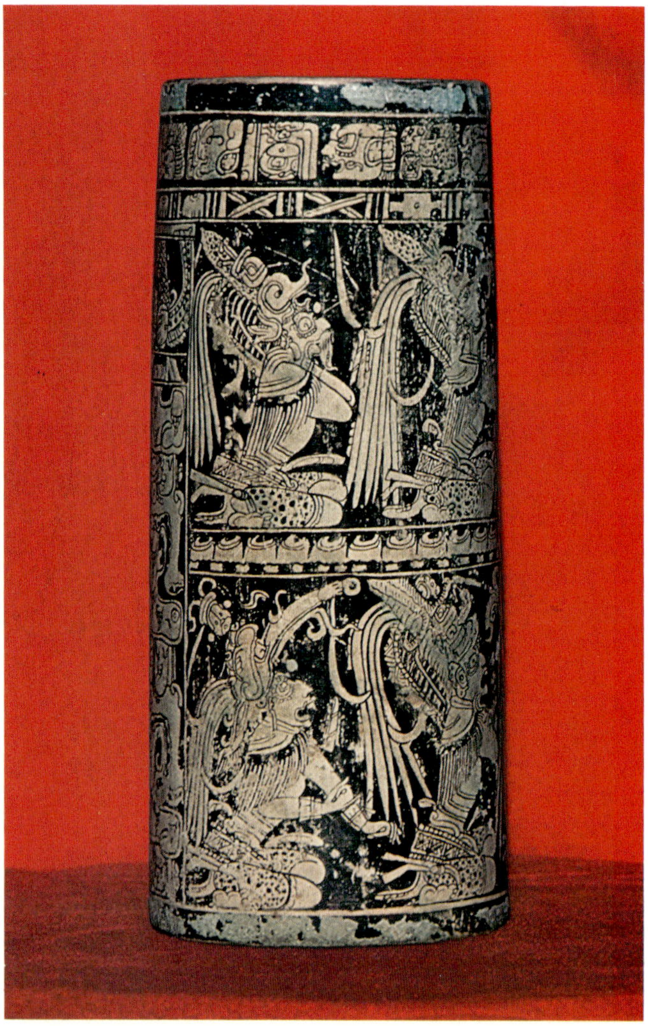
Plate 159

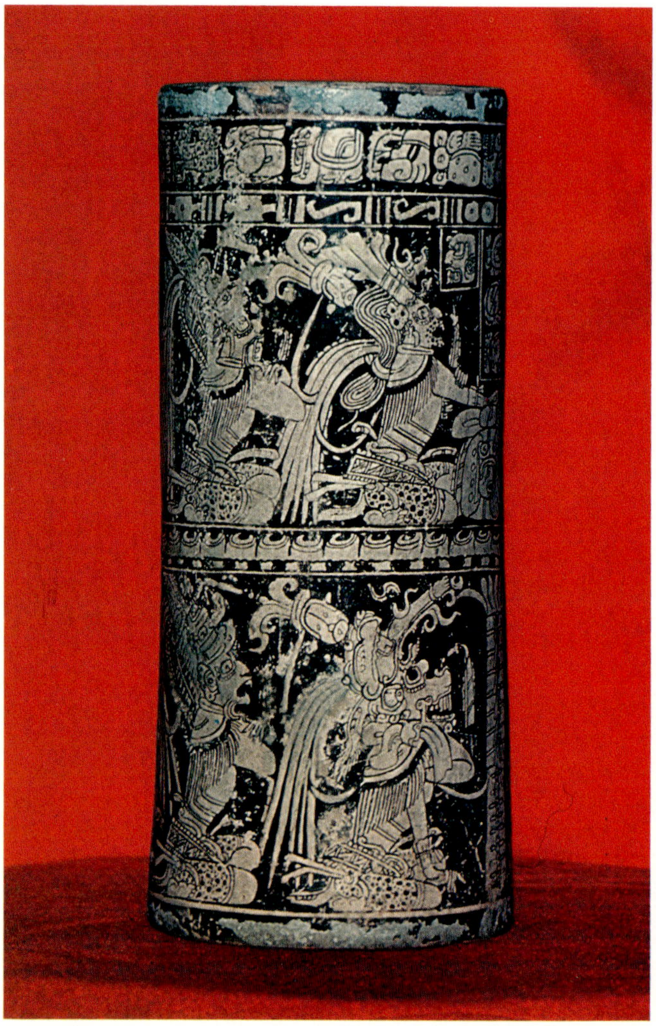
Plate 160

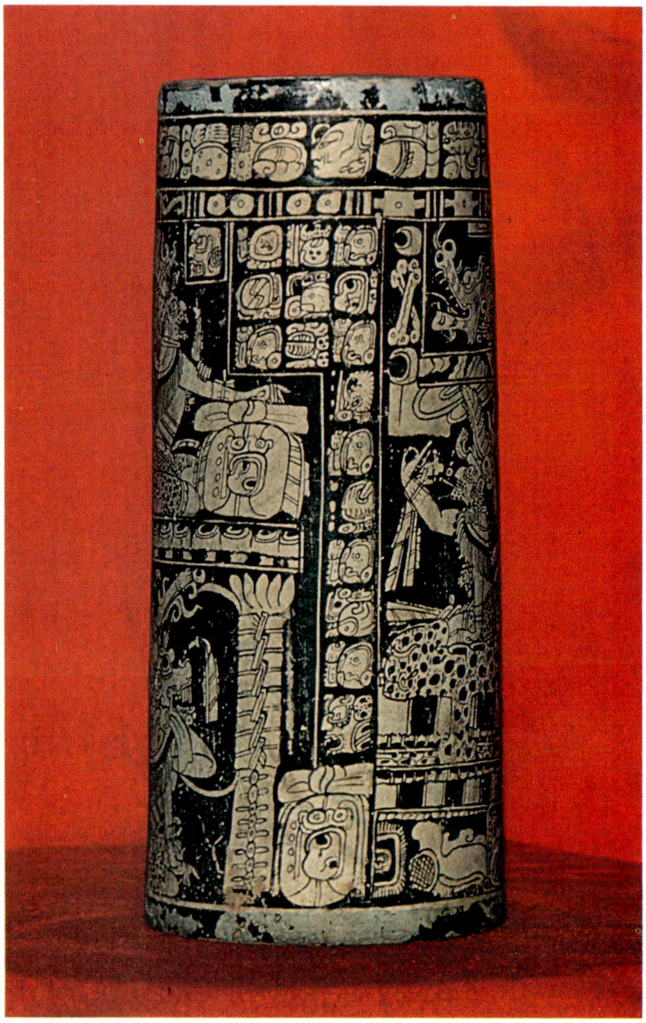
Plate 161

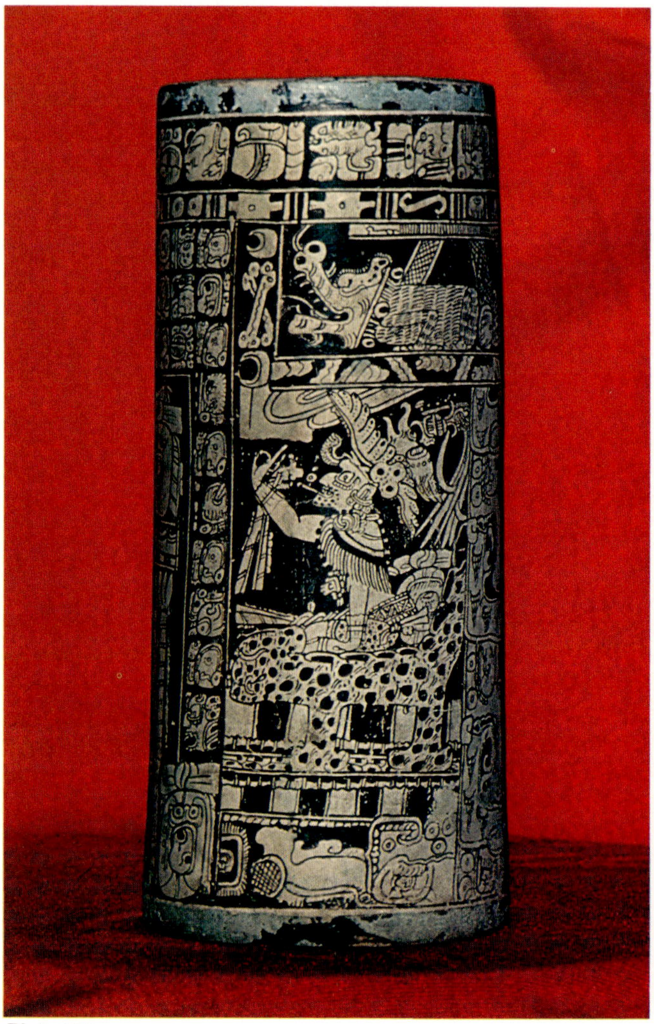
Plate 162

Plates 159-162 Vase of the Seven Gods (B-1)

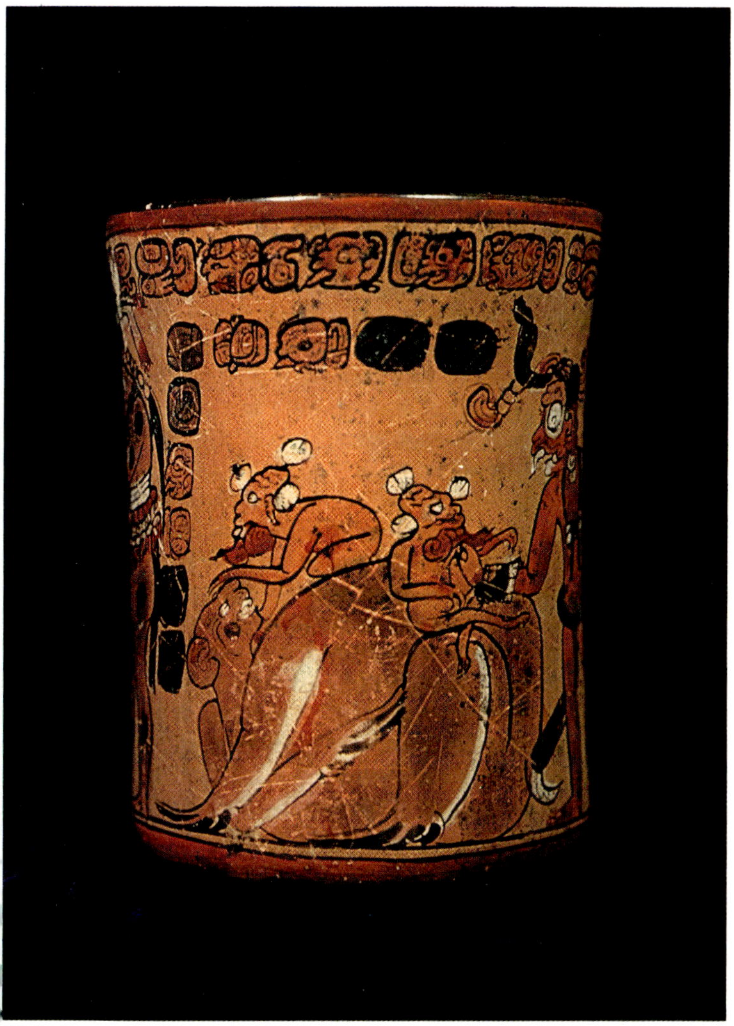

Plate 163

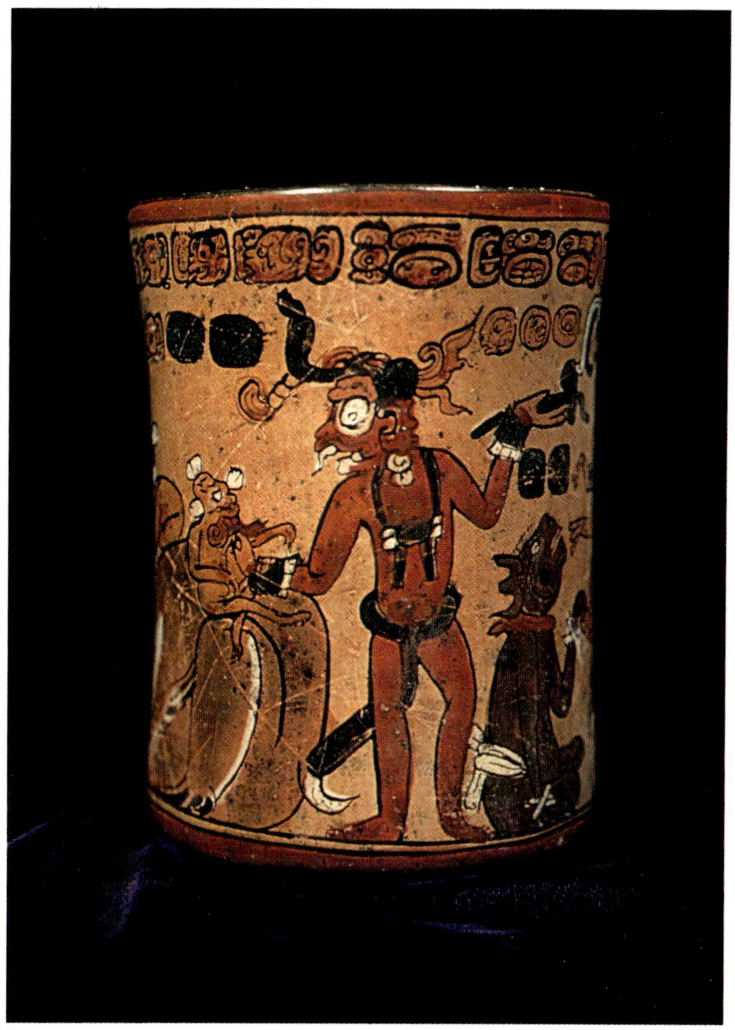

Plate 164

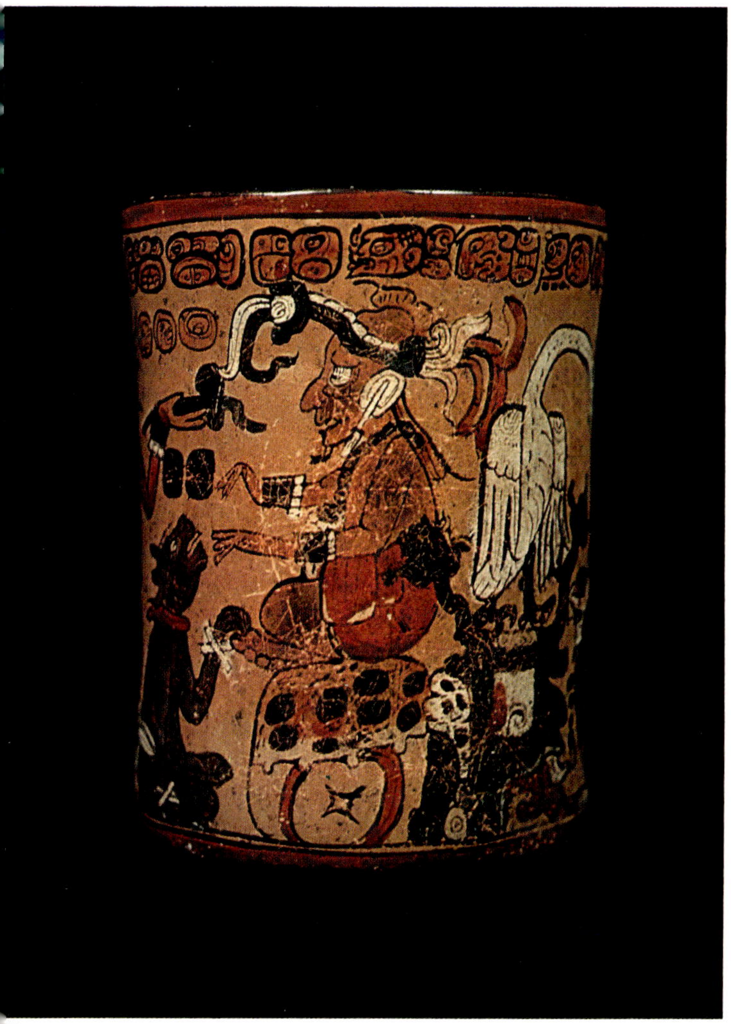

Plate 165

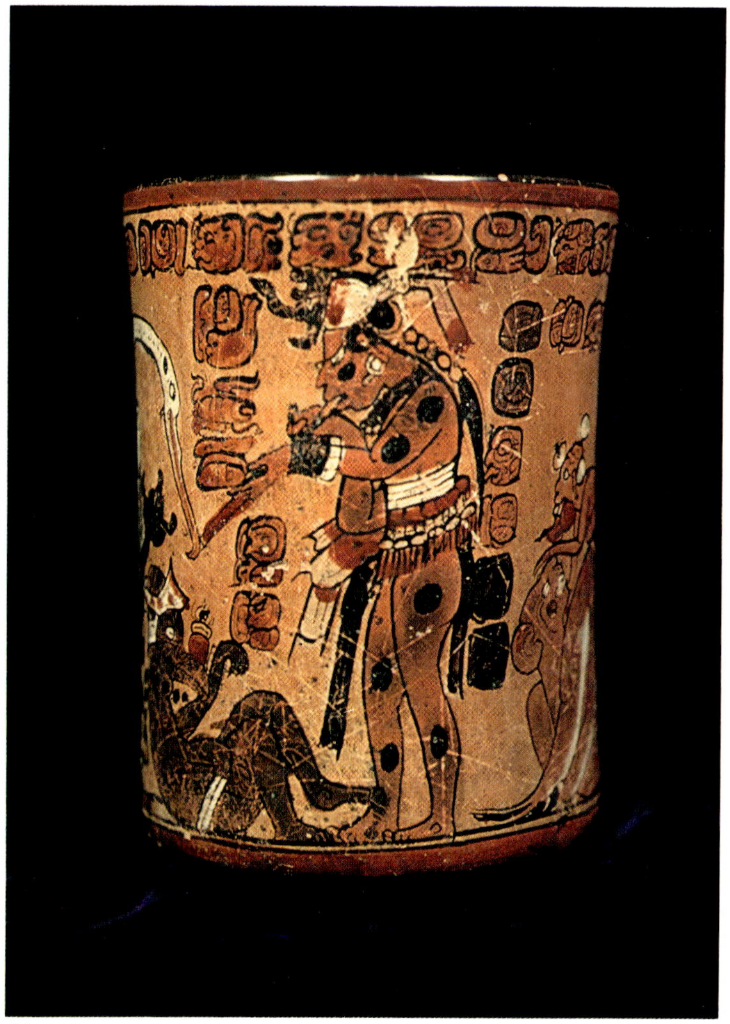

Plate 166

Plates 163-166 *Vase of the Jaguar Deer (B-2)*

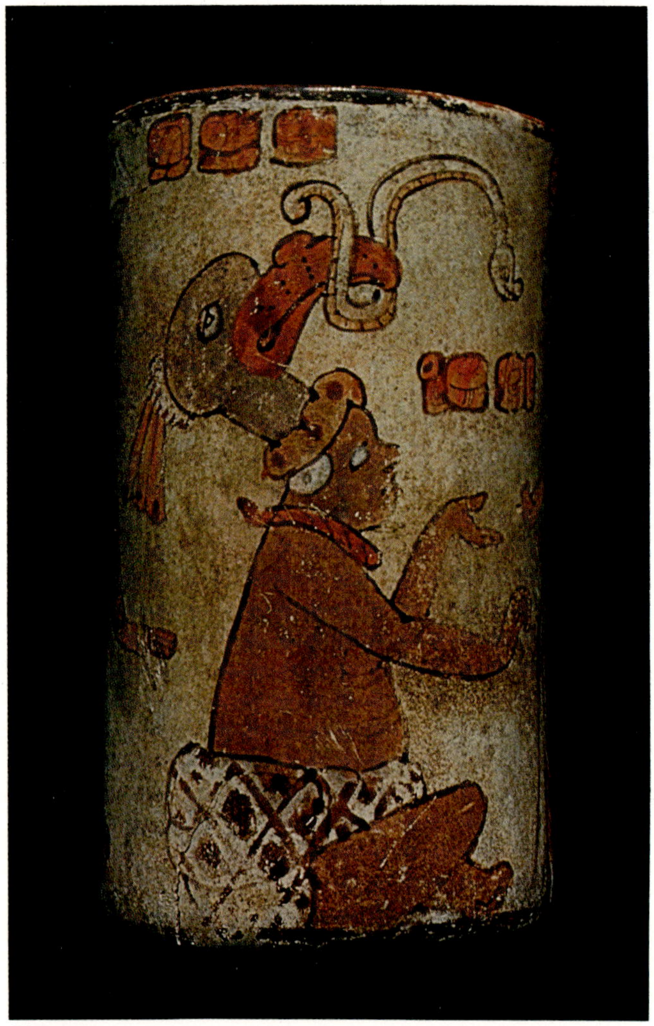

Plate 167

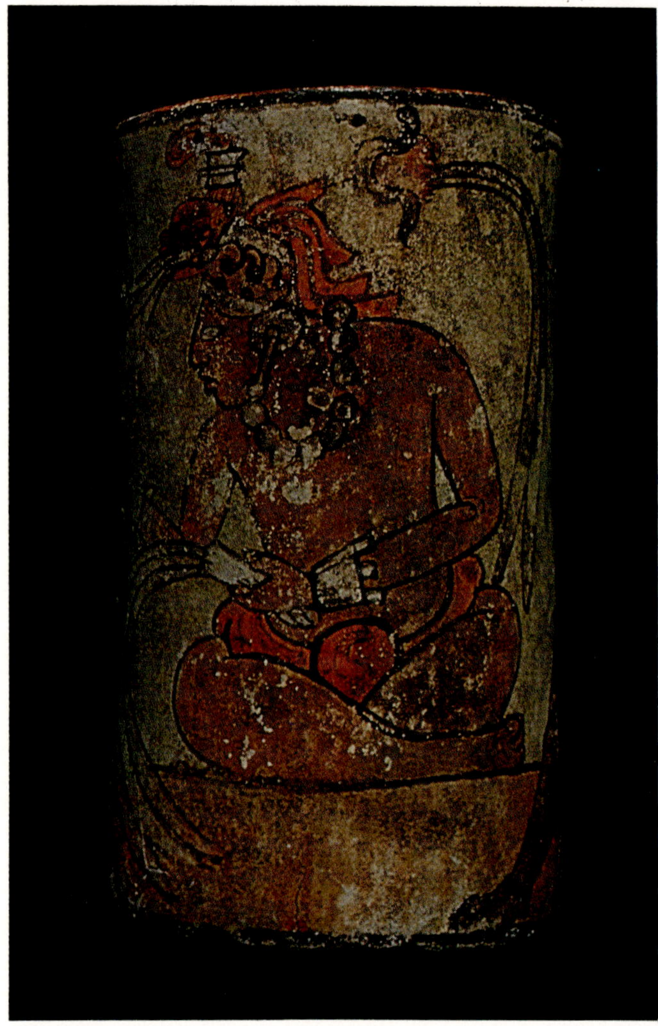

Plate 168

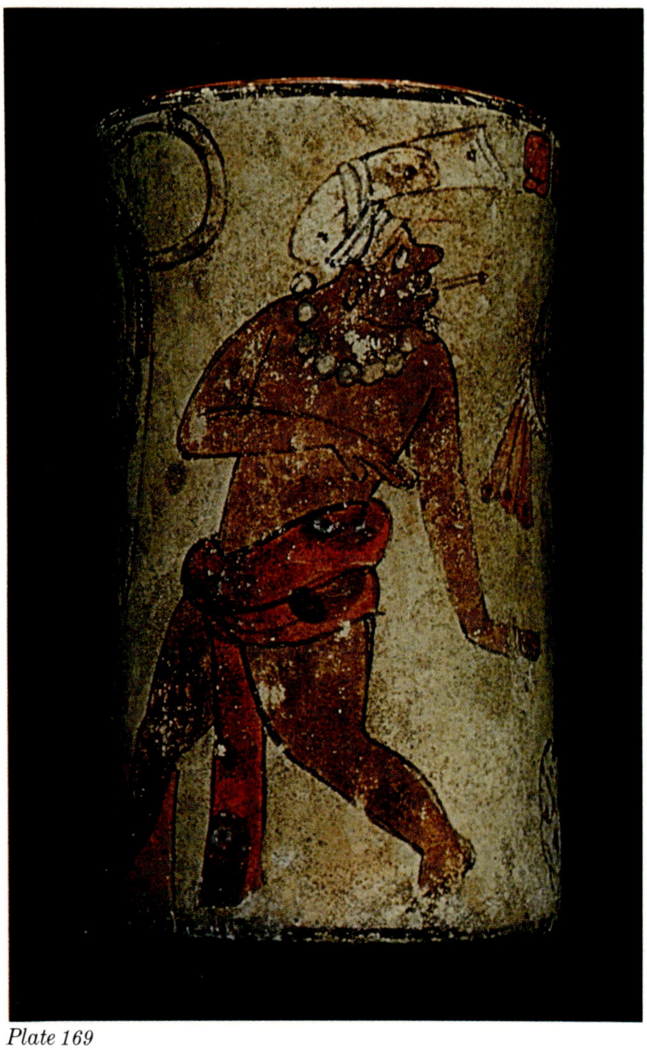

Plate 169

Plates 167-169 *Vase of the Smoking Attendant (B-5)*

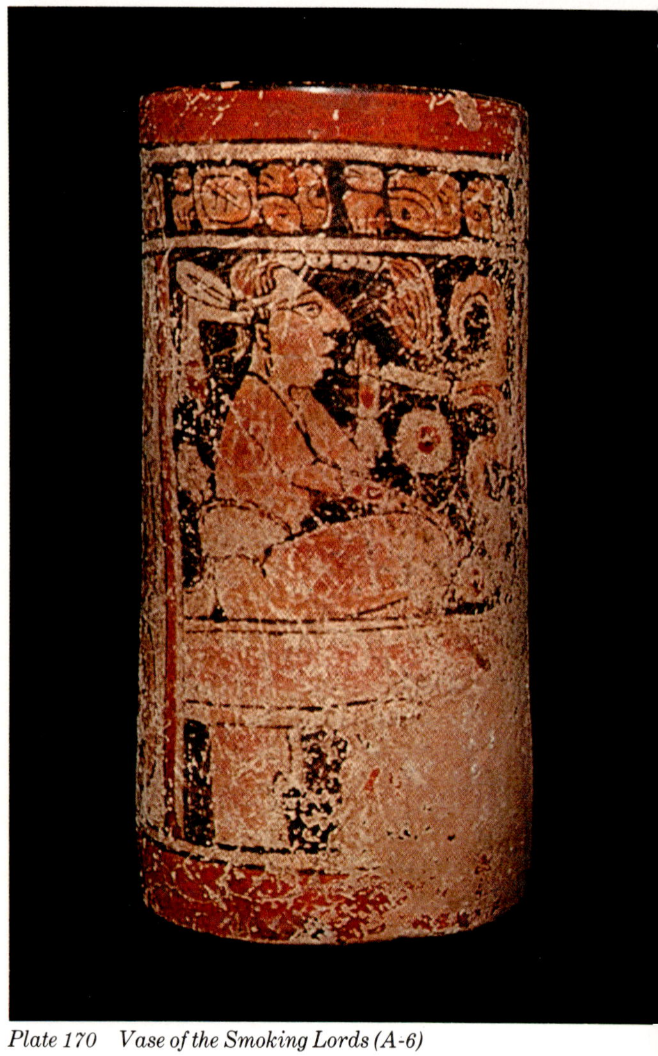

Plate 170 *Vase of the Smoking Lords (A-6)*

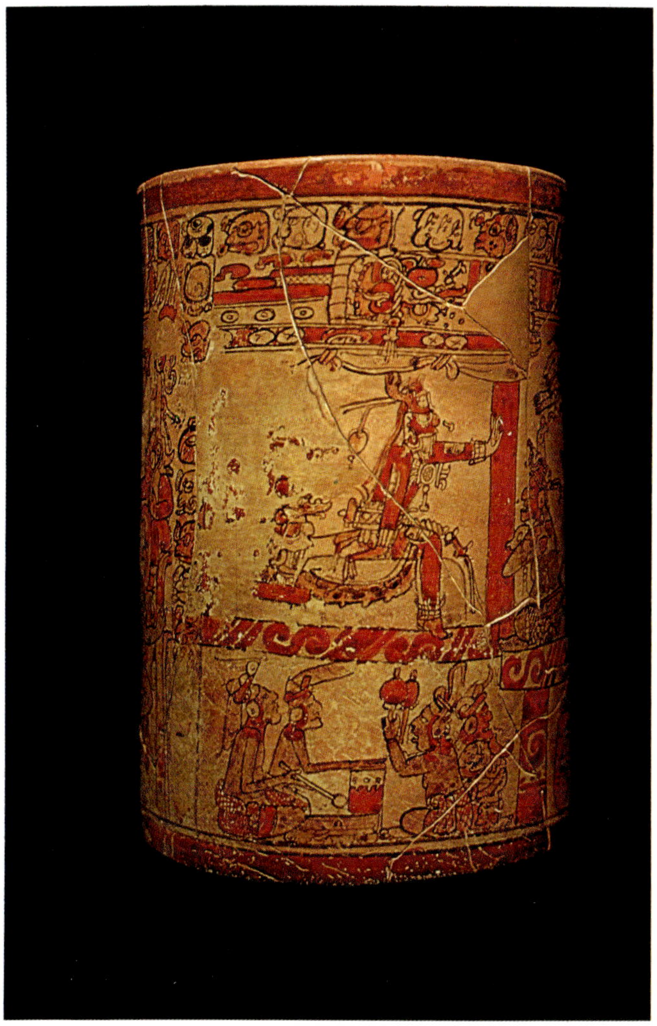

Plate 171

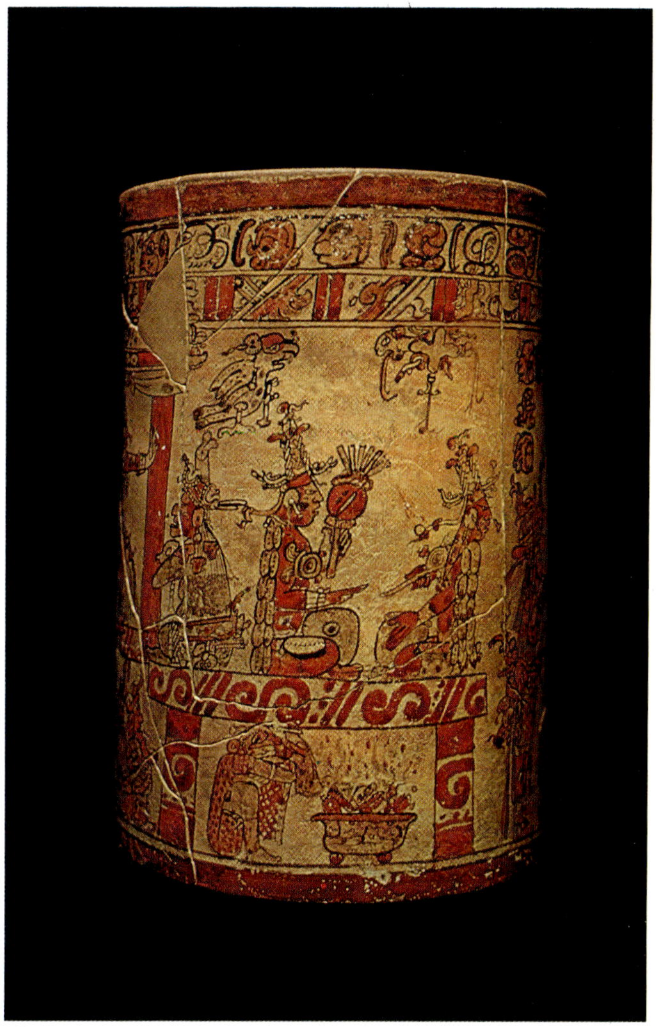

Plate 172

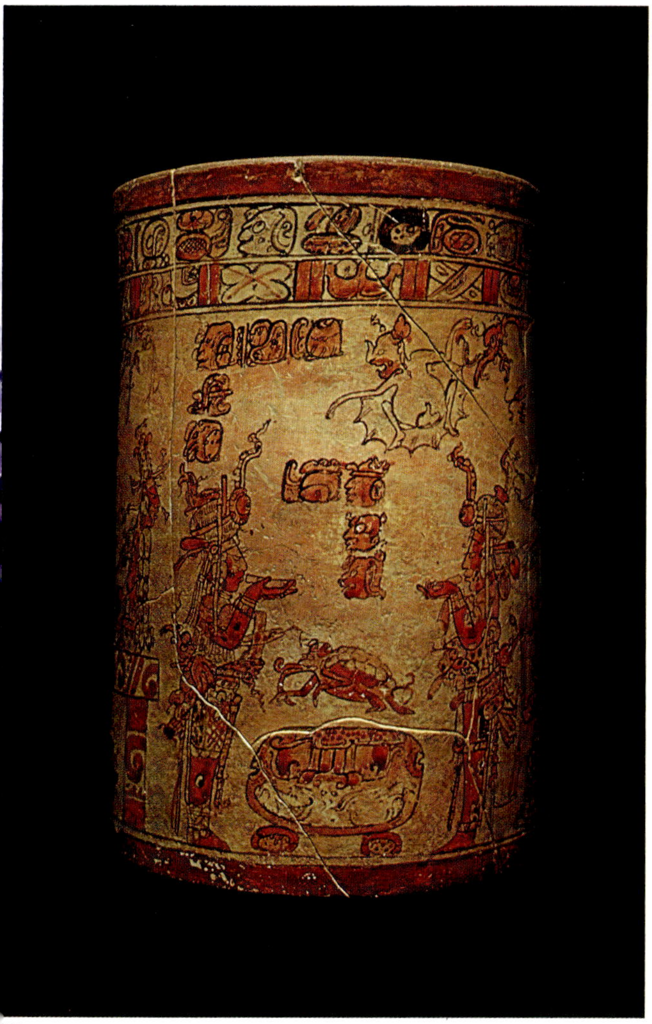

Plate 173

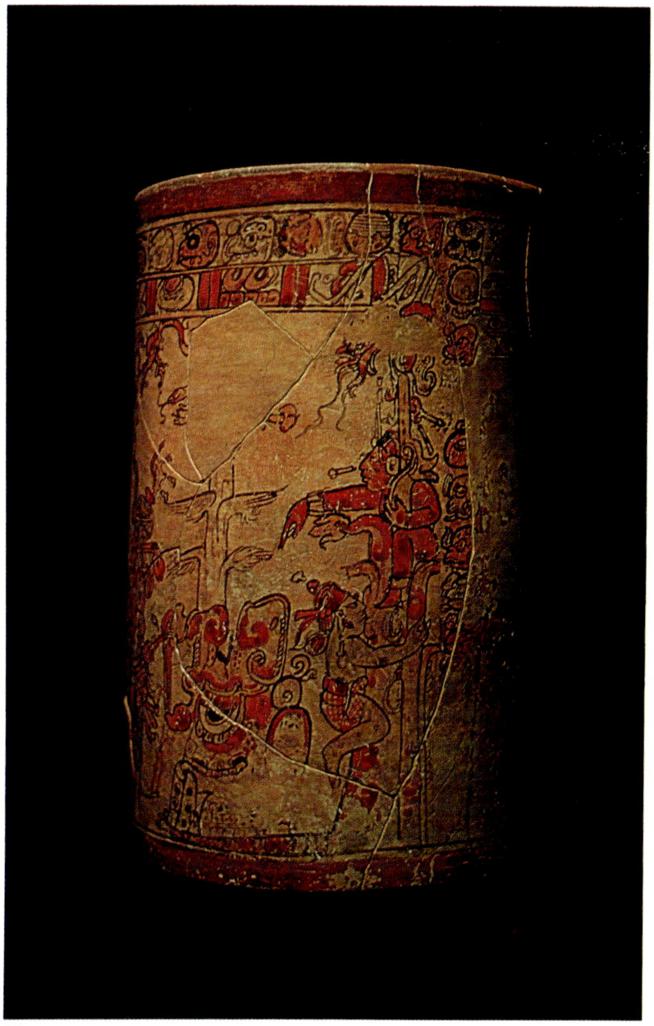

Plate 174

Plates 171-174 The Bloomington Vase (B-8)

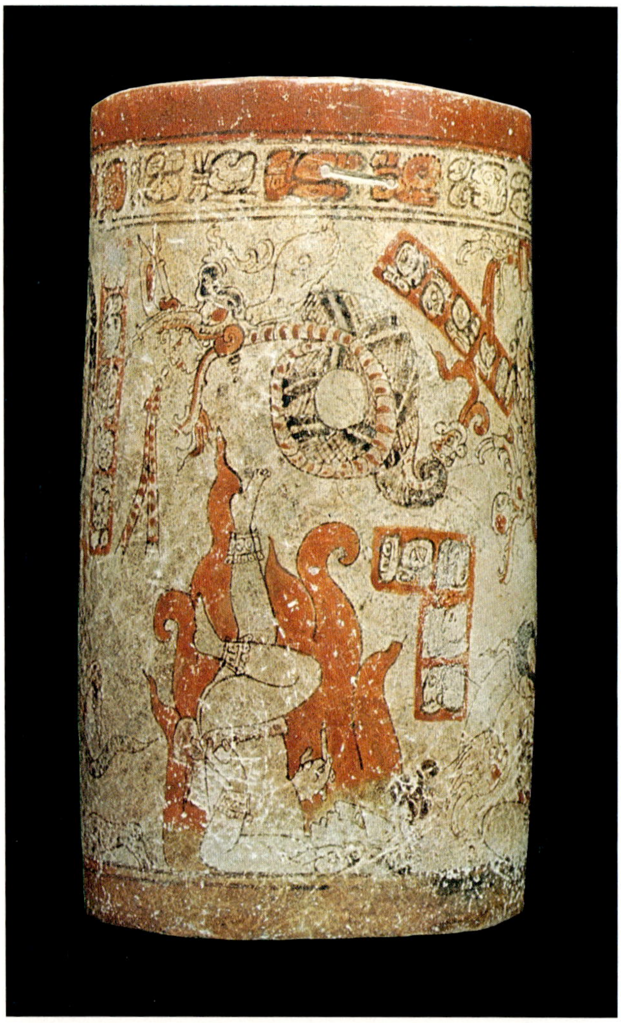

Plate 175

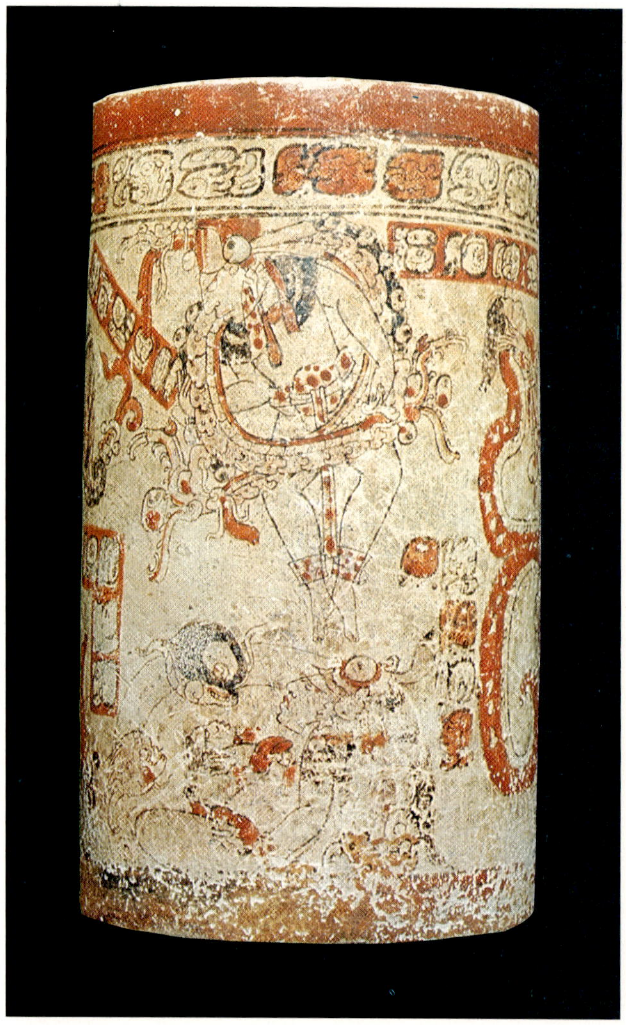

Plate 176

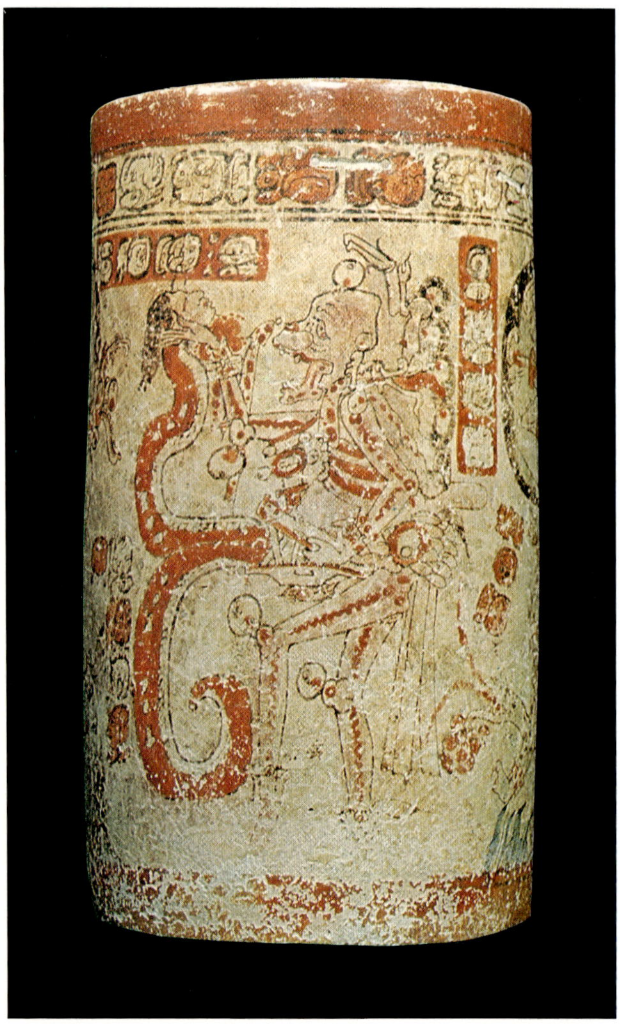

Plate 177

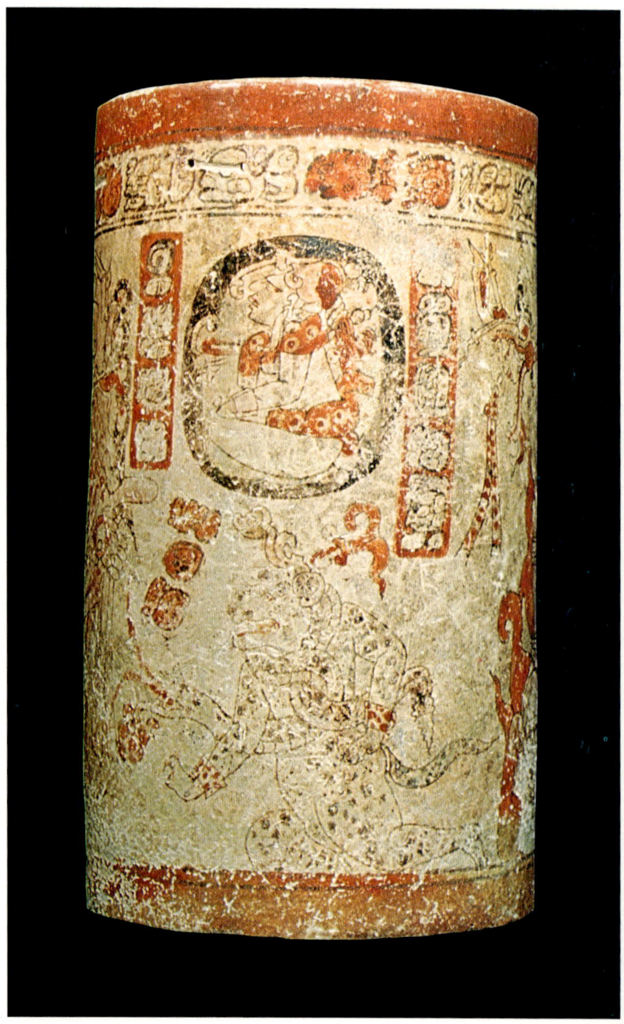

Plate 178

Plates 175-178 Vase of the Falling Lord (B-7)

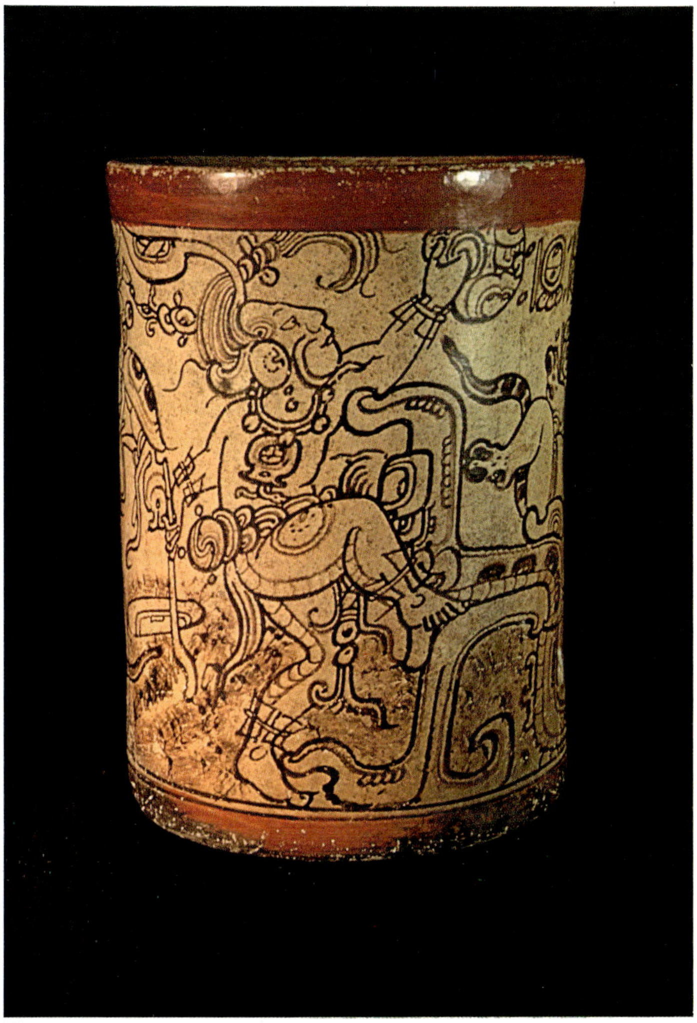

Plate 179

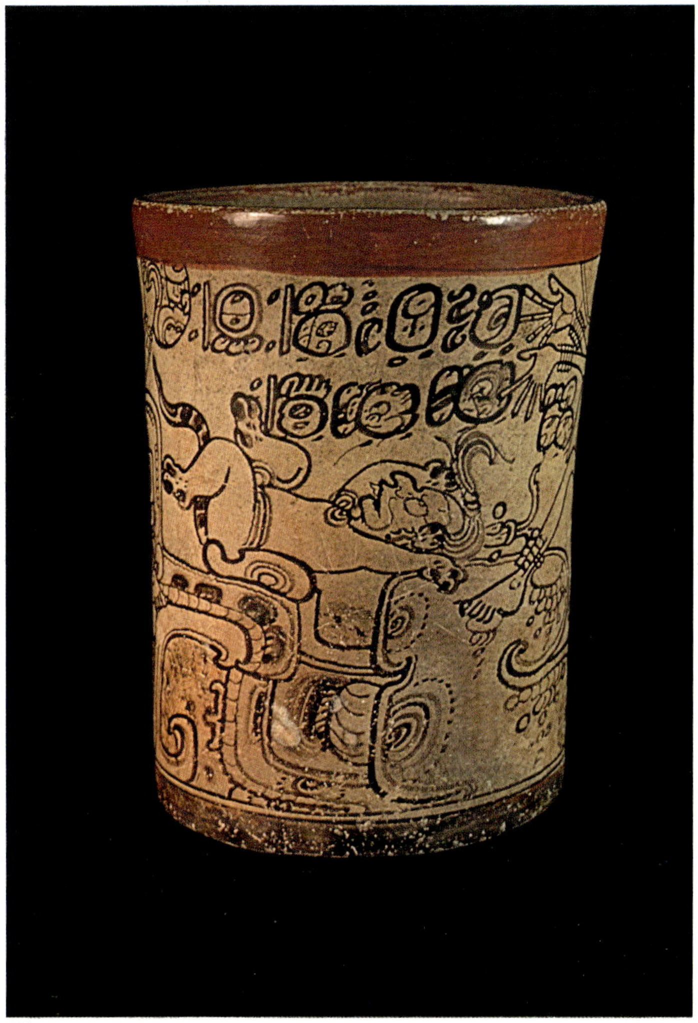

Plate 180

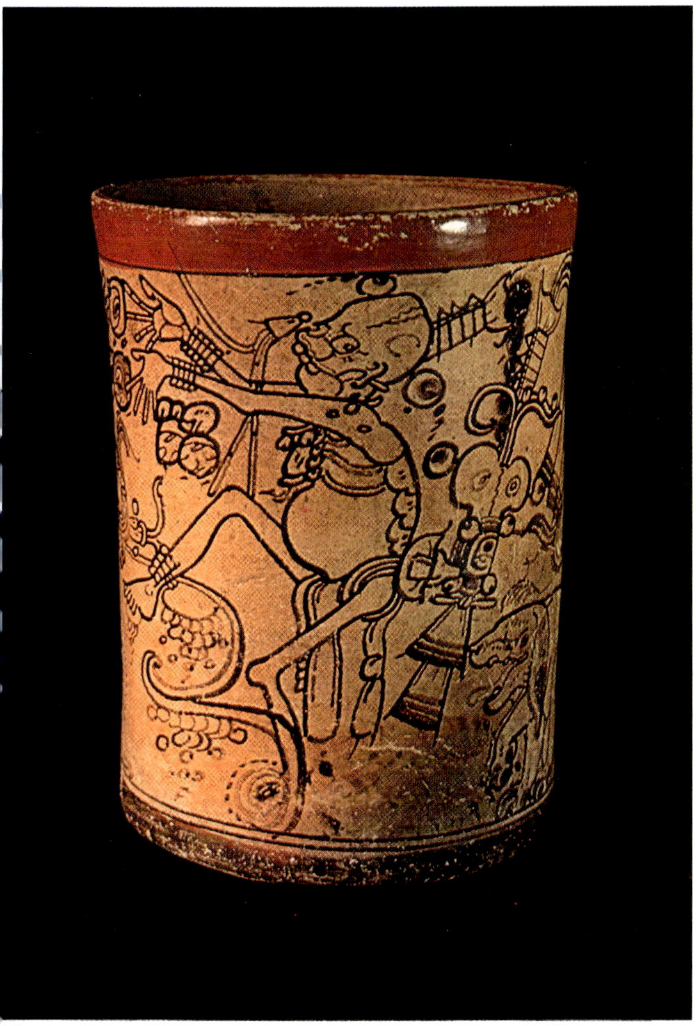

Plate 181

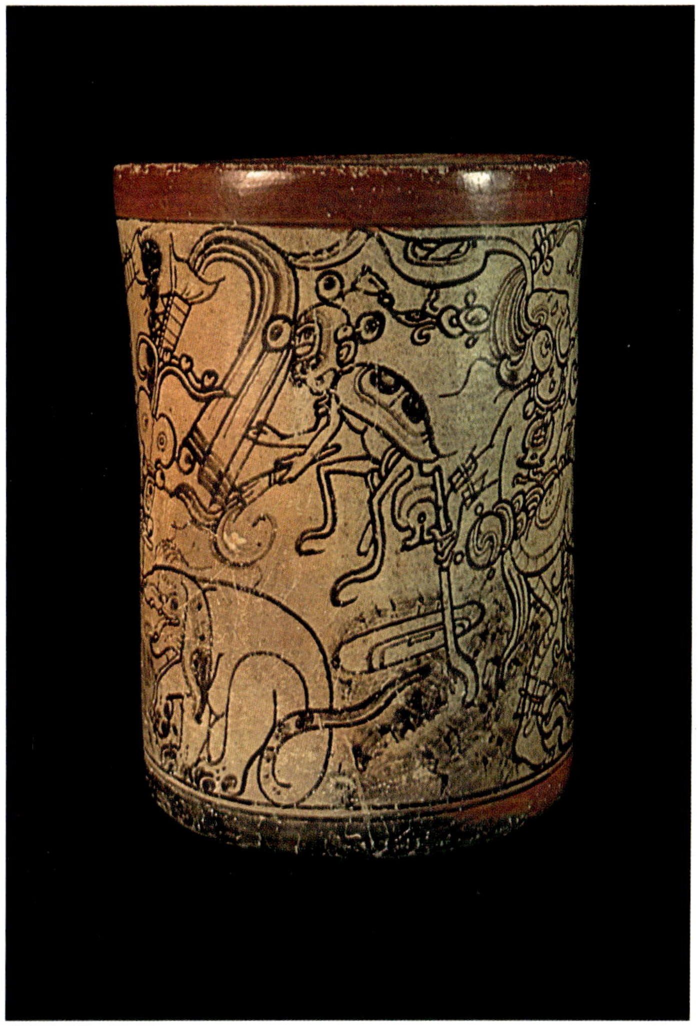

Plate 182

Plates 179-182 Vase of the Underworld Ritual (C-4)

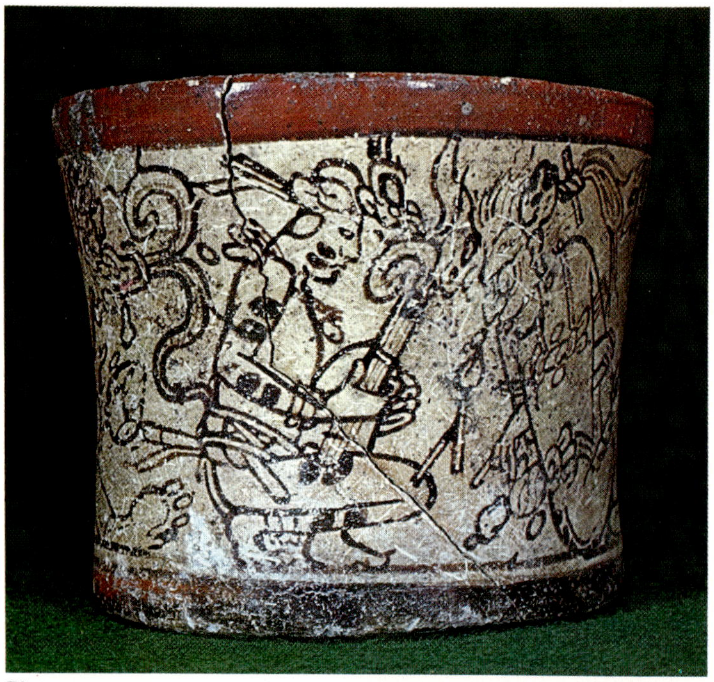

Plate 183

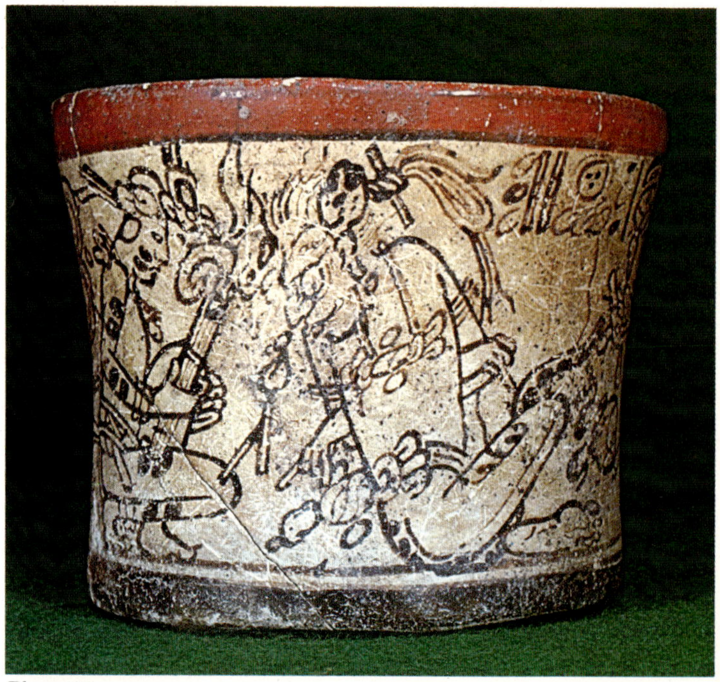

Plate 184

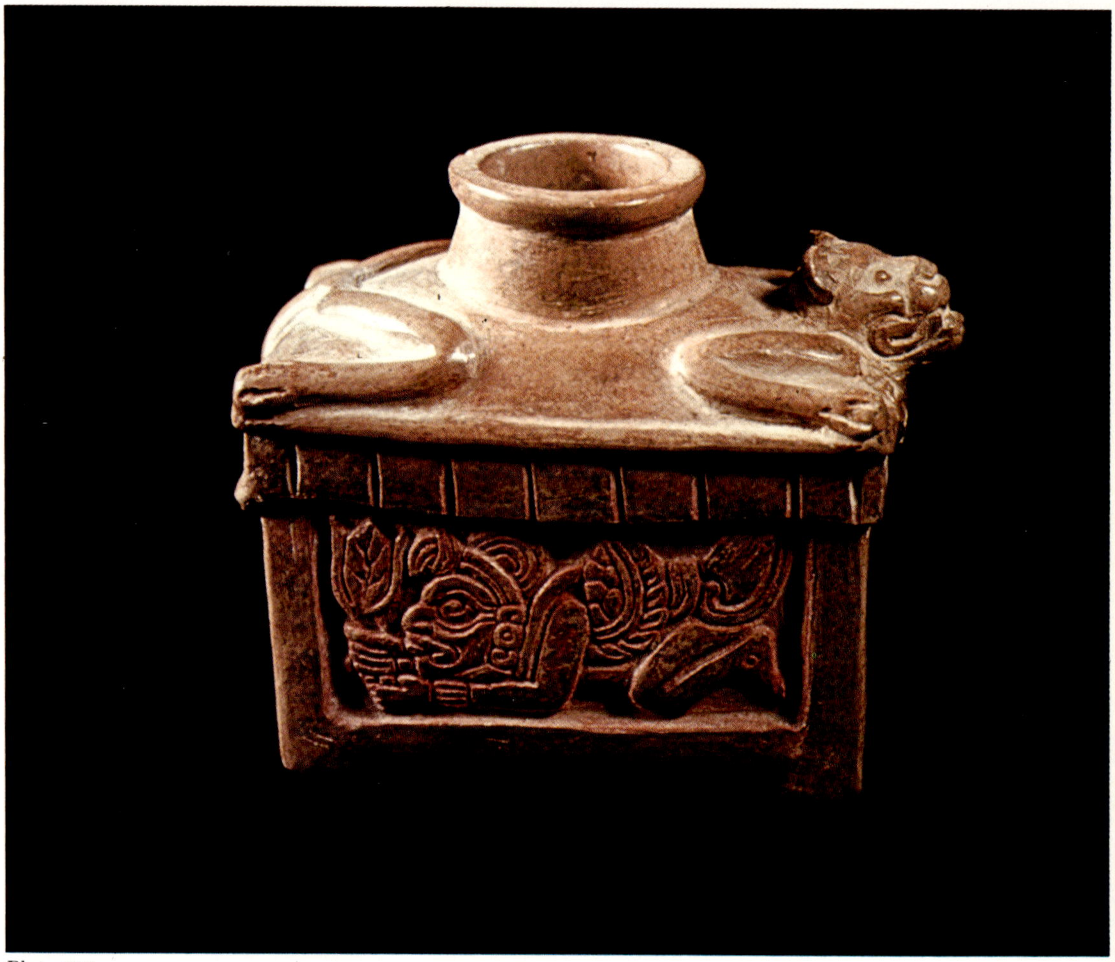

Plate 185

Plates 183-184 Vase of the Jaguar God (C-1) /Photographs courtesy Nicholas M. Hellmuth/

Plate 185 Ceramic container with a God holding tobacco leaf

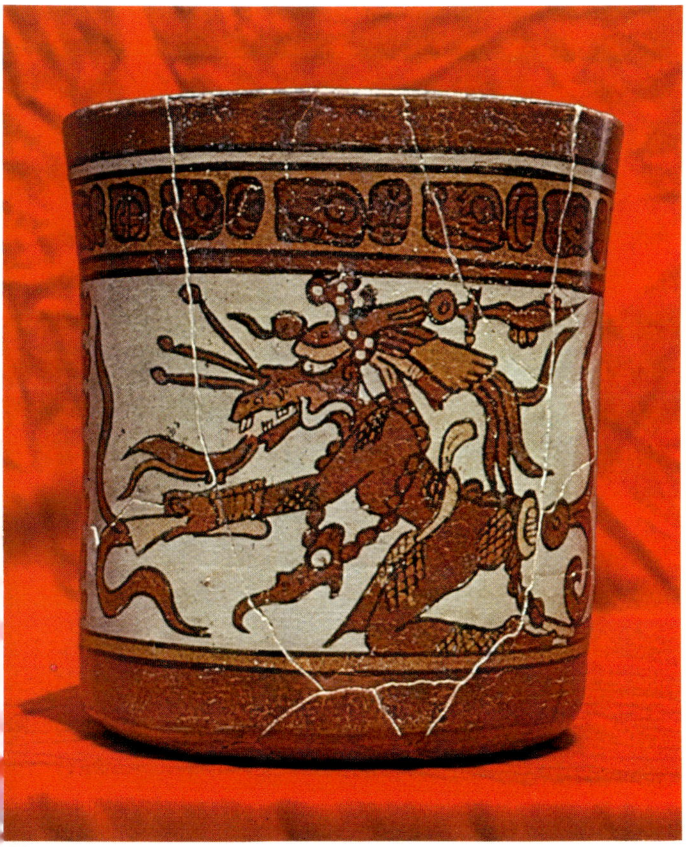
Plate 186

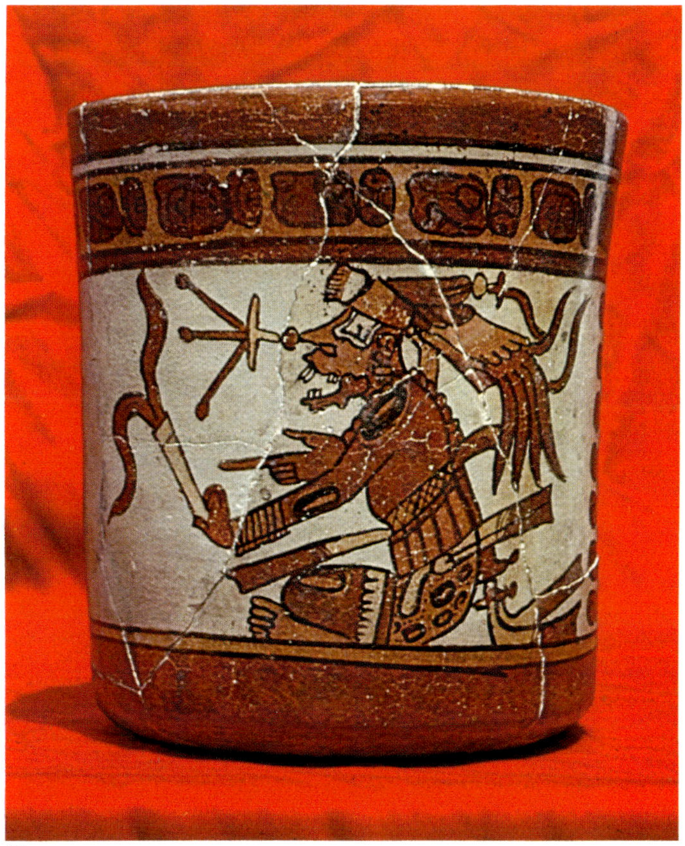
Plate 187

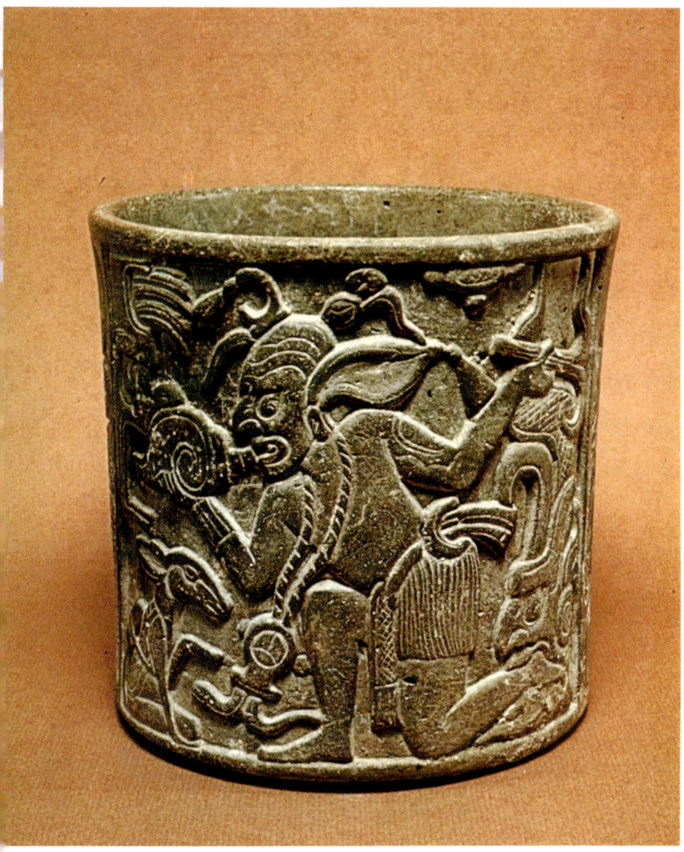
Plate 188

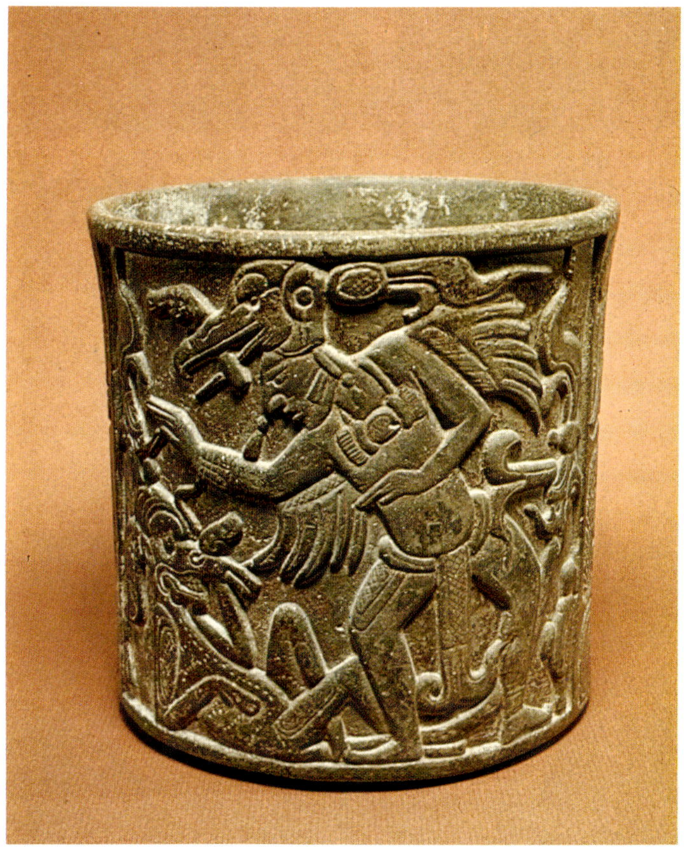
Plate 189

Plates 186-187 Vase of the Fire Breathing Jaguar (C-2)

Plates 188-189 Vase of Chocolá Style. Two Mythological Scenes (C-7) /Photographs courtesy Dumbarton Oaks Collection/

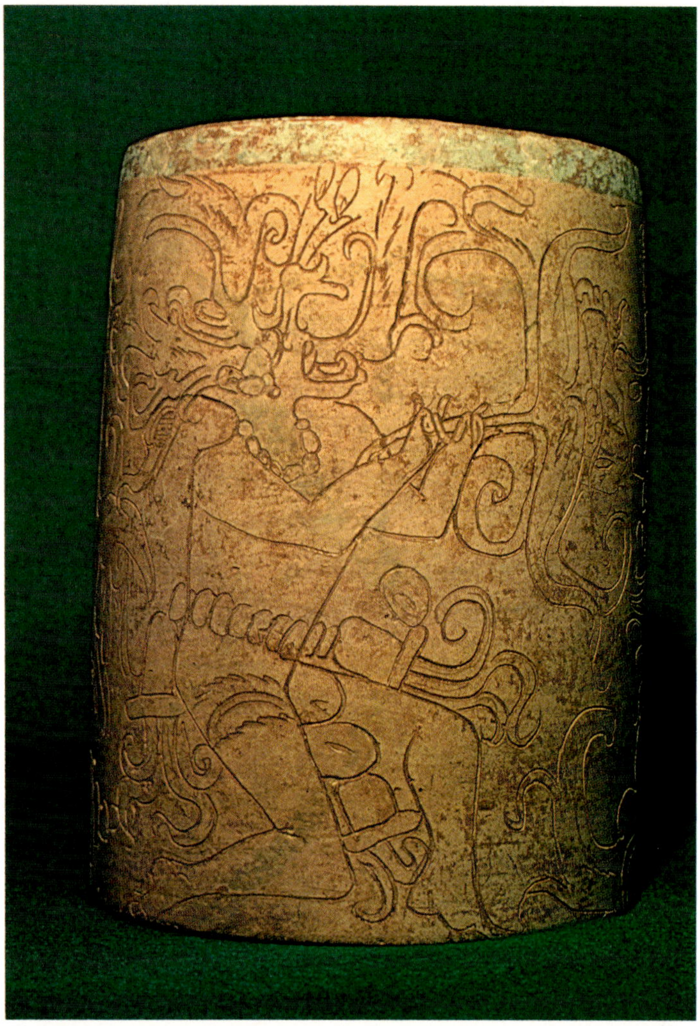

Plate 190

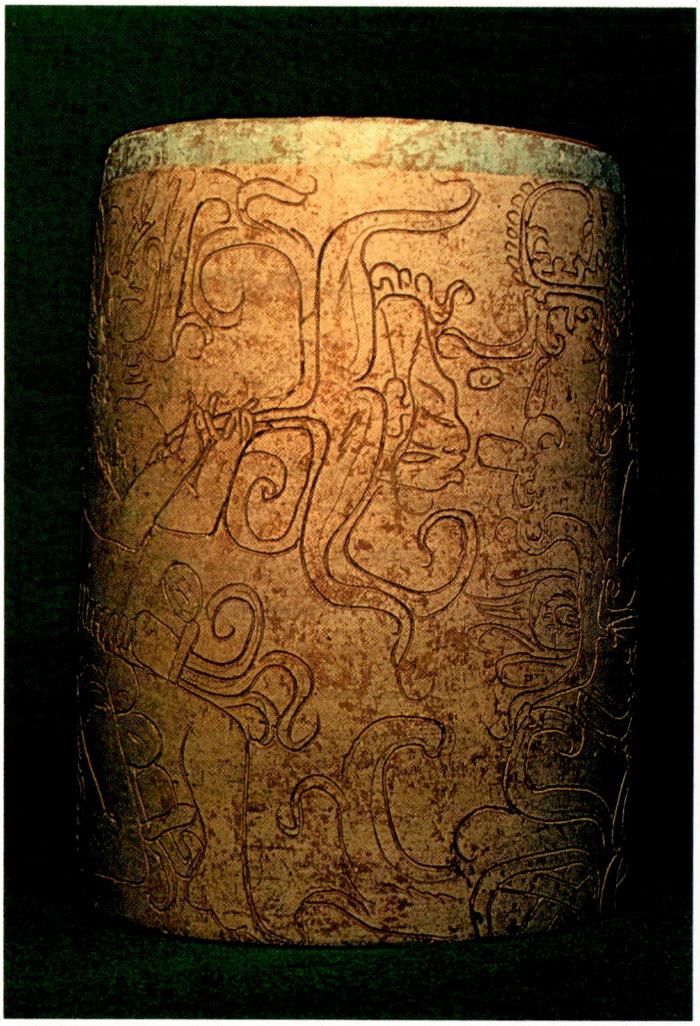

Plate 191

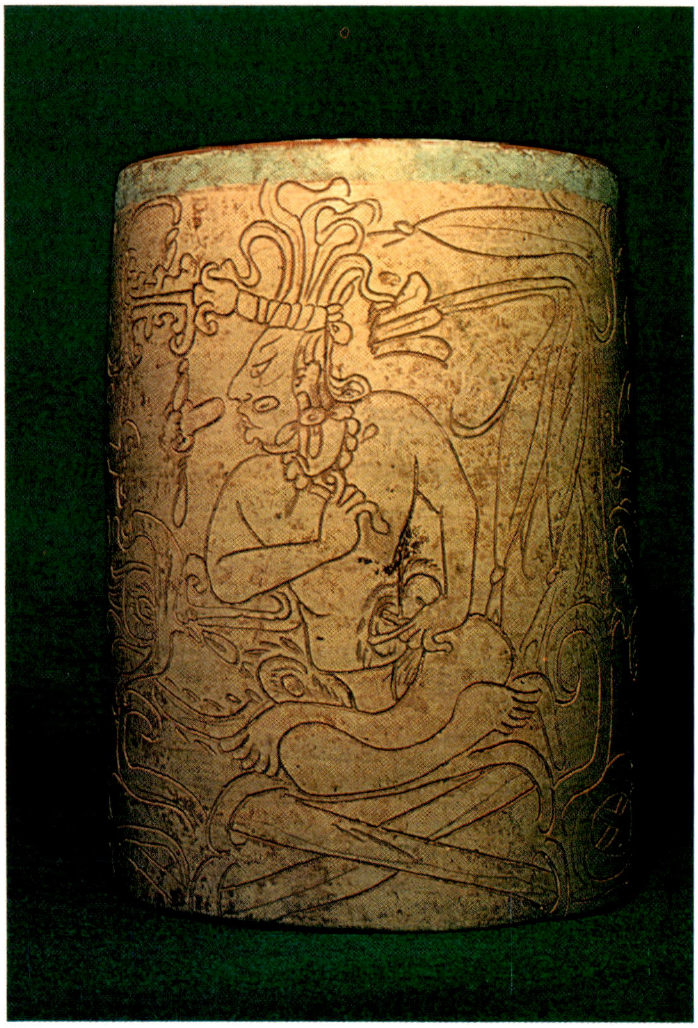

Plate 192

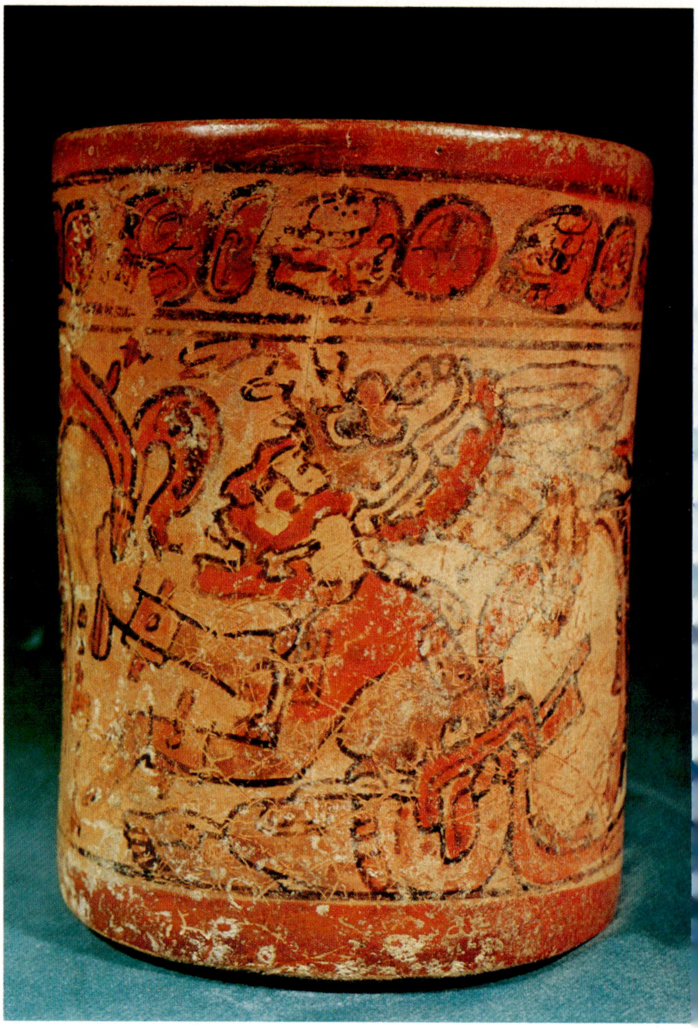

Plate 193 Vase of the Old God Holding a Flare (C-6)

Plates 190-192 Vase of the King's Vision (C-5) /Photographs courtesy Nicholas M. Hellmuth/

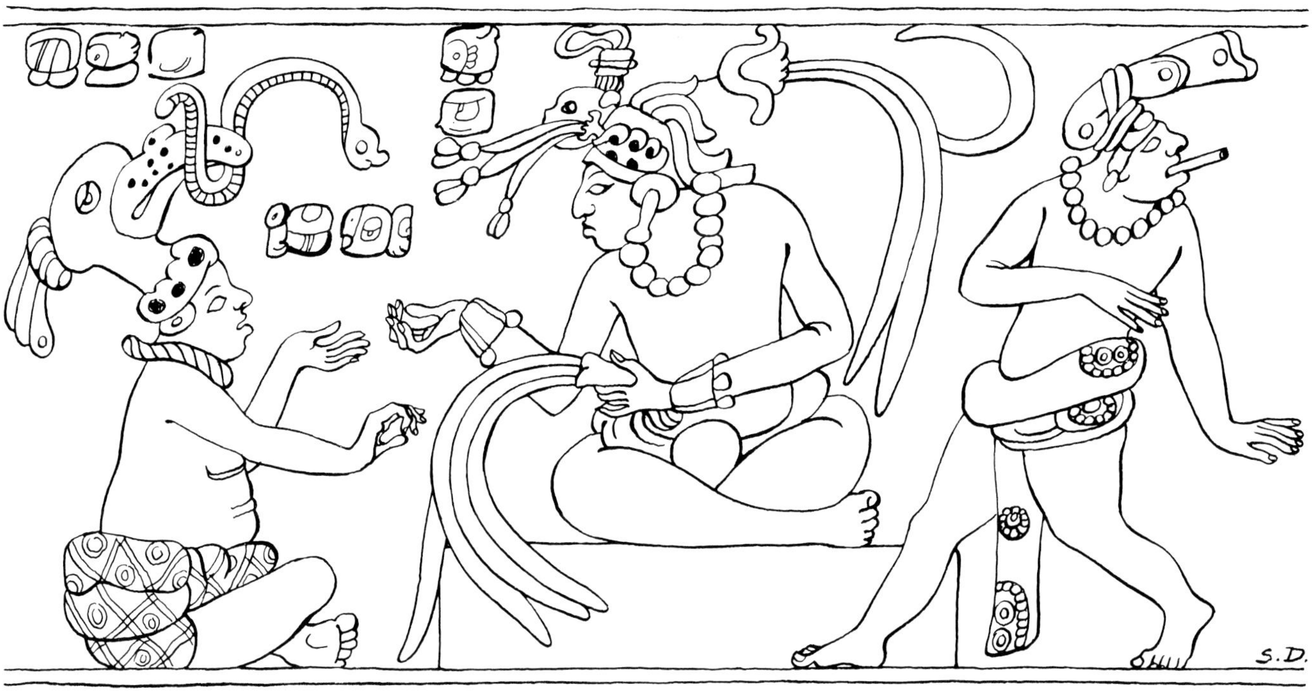

Figure 165. Vase of the Smoking Attendant (B-5)

superfix. It is probably an emblem glyph. The second compound appears to refer to the object upon which the ruler is seated. If its identification as a throne is accurate, then it may function as the *bacab* (T505. . . 5:501) and the *batab* (T10301) do when they appear with an emblem glyph. In other words, the phrase may refer to the seated person as the "seated lord of a particular place."

Another of the carved ceramic objects in this series of artifacts is the **Vase of the Seated Ruler** (B-4), a magnificent black ceramic vessel, probably made by a Tikal artist during the Classic period (now in a private collection). The vase is a greenish-black color with traces of cinnabar. It shows the seated figure of a scantily dressed ruler, a narrow upper glyph band, a wider vertical glyph column, a small angular glyph block, and four large serpent frets. The lord is highborn, as indicated not only by his fancy conical cap, floaters of quetzal feathers, ear pendants, bead necklace, wristlets, anklets, and narrow loincloth but also by the upper glyph band, which contains two large horizontal mat symbols, signs of worldly power and nobility. There are ritual scarifications on his cheek, and he has a long cigarette in his mouth.

Glyphic Text

The inscription is not Primary Standard and it does not relate to any other vase text known to us.

The **Vase of the Smoking Attendant** (B5) is a polychrome vessel found by Hellmuth in Burial 196, Structure 5D-73 at Tikal (now in the Tikal Museum).[22] It is painted on a white base in brown, red, and green. It is slightly flared, is about eight inches tall, and has dark-red upper and lower rims. It is painted with three figures and five interspersed hieroglyphs. The dominant figure is that of a lord seated cross-legged on a dais. He is wearing an elaborate headdress with a water-lily back ornament, the flower par excellence of the Maya underworld;[23] a quetzal feather floater emerging from his coiffure; a jade-bead necklace; a simple red loincloth; wide wristlets; and flared jade earplugs. His right hand is extended in an inviting gesture; in his left hand he is holding a wisklike object similar to those presented by women to the ascending lords on the niched stelae of Piedras Negras. He appears to be conversing with a corpulent seated attendant, who is wearing a breechcloth with a colorful geometric design, a twisted shawl, and a magnificent helmet shaped like the head of a huge bird (a king vulture?) holding the figure of a recoiling snake in its beak, which recalls the flamboyant headgear worn by some of the figures on the Bonampak mural. The most interesting person on the vase is the second attendant; he apparently "steals" the scene. Because the vase is round, he appears to be looking both at and away from the seated lord at the same time, an amusing device sometimes used by Maya vase painters. He has an especially graceful walking pose, as if he were sneaking away. His heavy breechcloth is painted with circular designs, and he also wears a necklace of heavy jade (?) beads and a simple headdress. In his mouth is a white cigarette.[24] Five scattered hieroglyphs complete the composition.

Glyphic Text

The first text is composed of the compounds located over the vulture headdress of the seated attendant. The initial grapheme appears to be the T617a "mirror," but it is very unusual in its delineation. The second compound has the T713a "hand" as its main grapheme. It is possible that the two compounds together were meant to record the expression T617a:713a, a known accession phrase from Palenque, Yaxchilán, and Quiriguá. The third compound is not legible, but it may be the affix cluster to the verbal phrase. The first phrase of three glyphs may then record the activity portrayed in the scene.

The two remaining phrases are composed of two graphemes each. Probably they represent both subject of the verbal compound and nominal phrase for the adjacent figures. The phrase next to the ruler has the T713a hand in the first compound and a head variant in the second. The main grapheme of the first compound of the second phrase is T506 *kan*, "maize." The prefix appears to be T238, one of the graphemes that function as the third-person male particle. The second compound has T743 *ac*, "turtle" as the main grapheme; the postfix T4 does not have an accepted proposed reading. The nominal compound for the seated attendant may have read Ah Kan Ac.

The only ceramic sculpture that to my knowledge shows the act of smoking is the unique figurine **Lord Deer Seated** (B-6), from Jaina (now in the possession of a private collector) (Plate 123). The statuette is about nine inches tall and portrays a young nobleman. He is wearing a bluish-purple kilt, wristlets, a heavy bead necklace, and a tall deer helmet, very similar to that of the lord carved on the Cleveland Shell Gorget, which may indicate that he is a hunter. He has a dreamy expression on his face. In his right hand he is holding a small cigar to his mouth as if preparing to inhale its smoke. The superb artistry makes this figurine one of the most beautiful examples of pre-Columbian miniature sculpture.

A most interesting polychrome vessel from the late Classic epoch is the **Vase of the Falling Lord** (B-7) from the southern Maya lowlands (now in a private collection). The upper and lower rims of the tall cylinder are painted red, and it has a white background. The vase has a circular upper glyph band and several short, angular, vertical, and diagonal glyph columns. The principal scene on the vase shows a young lord wearing only a narrow belt, anklets, wristlets, and ear ornaments. He is lying on his back; one of his legs is upright, the other bent at both hip and knee. His left hand appears to be supporting his hip. This peculiar posture is backed by what appear to be red flames shooting upward. In his right hand he is holding what may be a long

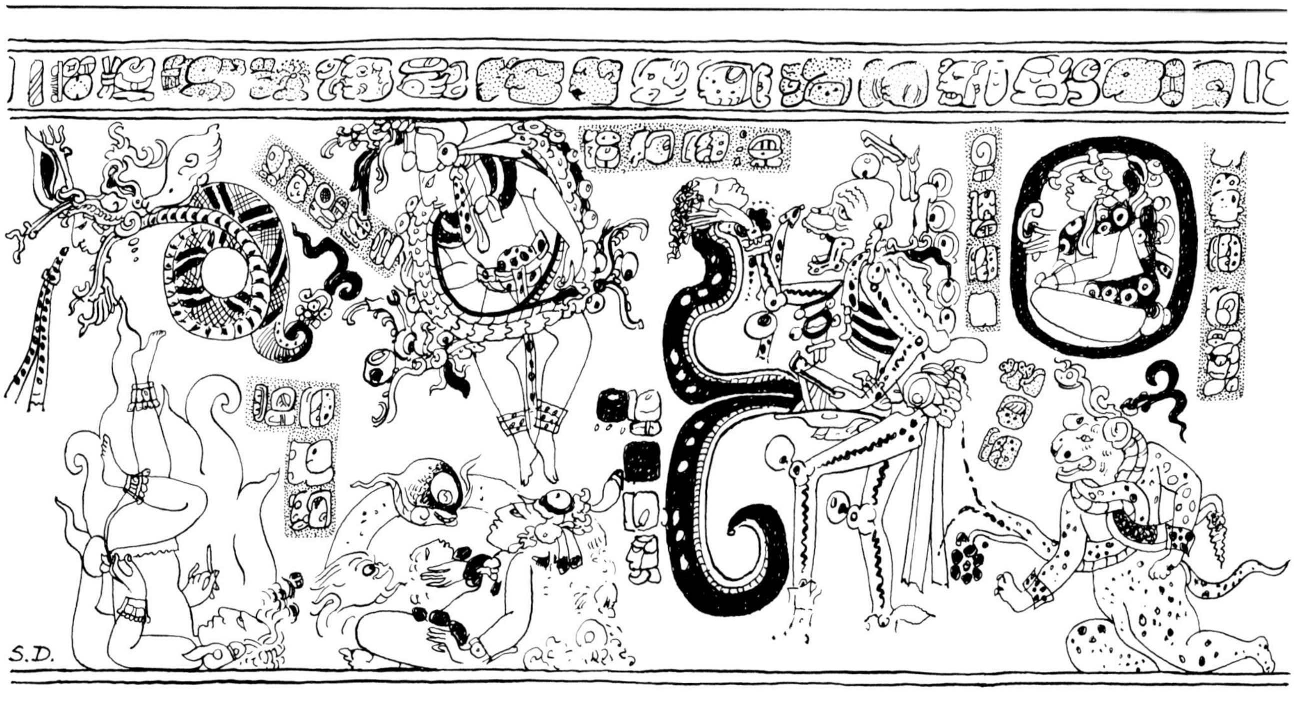

Figure 166. Vase of the Falling Lord (B-7)

cigarette with a glowing tip, a bloody bone perforator, or a stylus used for carving or etching, like the one seen on a vase from Copán and on the Tablet of the Scribe at Palenque.[25] Above his head hovers the figure of a serpent with the ears and horns of a deer and the head of a young man, perhaps a god, protruding from its throat. The scene recalls the vision scenes of the Yaxchilán lintels and may suggest that the object the falling lord is holding is not a cigarette with a glowing tip but an instrument used in blood letting.

As the vase is turned clockwise, the upper body of another young deity appears, flanked by horizontal and oblique glyph blocks. He is floating in the air, enframed by the skeletal figure of a double-headed serpent. This bony snake is bicephalic and represents the snaggle-toothed version of a dragon head. It may refer to the bicephalic serpent bar of Maya rulers, and its skeletal state also suggests underworld identification. The figure holding it is the same as the attendant figures in Altar V and Lintel 2, Temple III, at Tikal, and is often shown on ceramics. The analysis of these combined appearances leaves little doubt that the figure is GIII (the Jaguar God of the underworld), who is many times paired with another figure, who is probably GI. Below this personage is another young deity in a semireclining position in similar head-gear and cape, holding a human head, identical with that held by the Death God, looking at two floating fish. One may speculate that he is intended to be shown underwater.

Next to this young lord stands the skeletal figure of the Death God. His right leg is raised as if he is dancing. He is naked but for his back ornament and the several death eyes decorating his body and neck pendant. In his left hand he is holding a plate containing different ceremonial objects. In his right hand he is holding up the head of a young person. From the abdomen of the Death God arise two large scrolls, probably signifying putrefaction and stench. Next appears a most unusual realistic portrait of the Jaguar God. He is unclad except for cufflike wristlets. The god is in a kneeling position with a rattlesnake wrapped around his neck in a seeming death grip. Above the jaguar, wedged between two vertical glyph columns, is a medallion showing a young lord dressed in a short cape and a loincloth. This highly complicated scene probably represents the ritual of human sacrifice by decapitation and the transcendence of the sacrificed lord through the different stages of the Maya underworld.

Glyphic Text

The rim text is a variant of the Primary Standard. The introductory glyphs are white and located above the head of the serpent-enwrapped figure. The first glyph T238.1011 is the name of GI of the Palenque Triad. The second compound, T617a:713a.181, is one of the most frequent introductory glyphs. Both T1011 (GI) and T617a (mirror) occur in count indicator groups at Quiriguá and at Tortugero. Their common function in count groups may explain their presence as introduction to the Primary Standard. The Primary Standard Text is presented in paired compound with alternated pattern of clear red.

The Secondary Texts appear to be nominal and refer to adjacent figures. The key to the interpretation of the Secondary Texts mayaybe the phrase adjacent to the Water-Lily Jaguar, which includes T524 *ix*, "jaguar" in the second position. By extension, the text adjacent to the figure falling in flames and next to the figure in the water perhaps refers in nominal function to both figures.

The diagonal and horizontal bands flanking the serpent-enwrapped figure both appear to refer to him, but it is possible that the right horizontal text refers to the Death God. The texts flanking the serpent-enwrapped figure and the figure in water include the jaguar-spotted *ahau* (T1.535), which we believe to be a nominal indicator glyph.

The problem of reference determination also applies to the two vertical texts flanking the cartouche above the Water-Lily Jaguar. While the texts may refer to the figure infixed in the cartouche, the left text may refer to the Death God. It is likely that only one of the adjacent texts refers to the Death God.

Two of the Secondary Texts are presented without a framing device. The remaining texts are enframed by the red background itself. The distinction between the two types of texts in this scene and on late Classic monuments is not fully understood.

An act that may represent smoking is also shown on the very important **Bloomington Vase** (B-8), a polychrome vessel made during the late Classic period in nothern Petén (now in the collection of the University Museum of Indiana). The tall yellow cylinder has upper and lower red rims, an upper circular glyphic band, and a plantetary band, as well as twelve human and five animal figures. The scene begins both in text and in action with the enthroned figure of an angry-looking lord seated on the upper level of a palace. His body is painted red, and he is wearing a wide belt and breast, ankle, and wrist ornaments. He is identified by the jaguar paw above his ear, the ornament at his mouth, and his name glyph at the top of the glyph column behind him as the patron god of the month Pax. His head reappears behind the throne on which he is seated. Under Lord Pax, on the lower level of the building, is a quartet of musicians in concert. One is beating the sacred drum, the *tunkul*, with a stick; the second is playing the ceramic drum; the third is shaking rattles; and the fourth is beating a turtle carapace.

In Yucatec, *Pax* means "musical instrument," "drum," or "to play a (musical) instrument by hand."[26] The scene with the musicians may thus

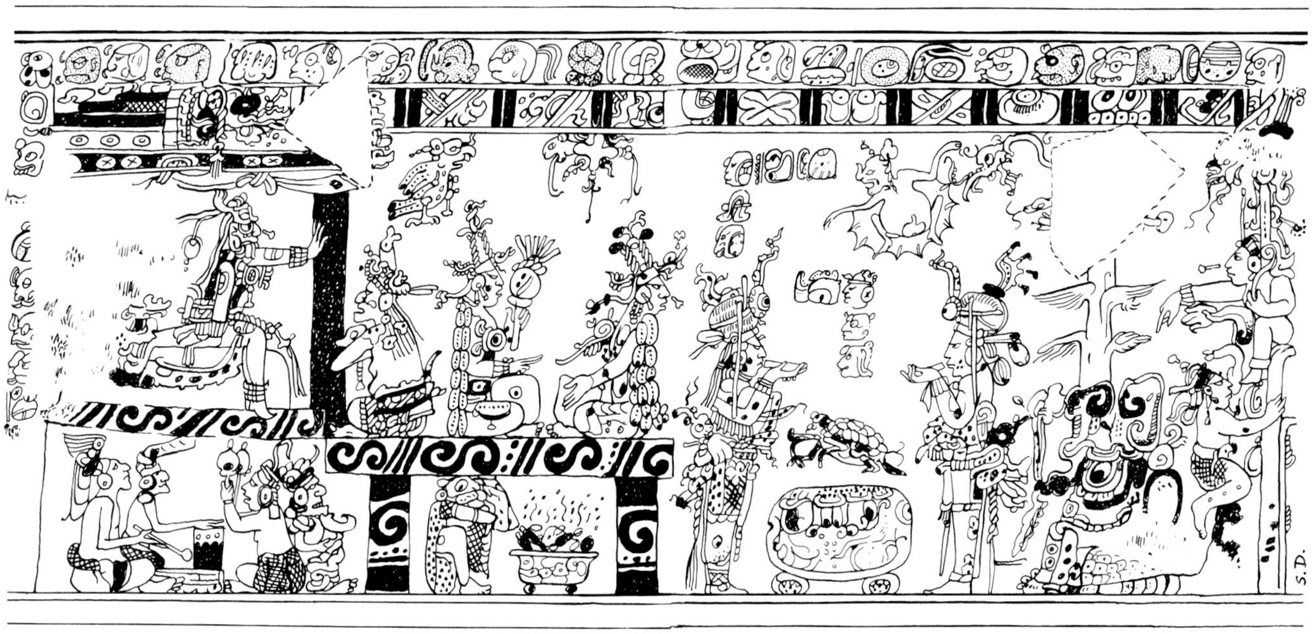

Figure 167. Bloomington Vase (B-8)

represent a normal palace activity, but it may also refer to the patron Pax above. Above the walls of the palace floats a sky band stretched into the shape of the bicephalic celestial monster seen on the niched stelae at Piedras Negras, on House E at Palenque, and in the *Dresden Codex.* The rear head is painted over the head of Lord Pax; the front head reaches to the opposite end of the scene. It has the usual configuration, including the saurian forelegs, but the nose merges with the tree motif that marks the boundary of the scene.

At the right of the composition are two trees. The first, which has the same leaf motifs as the "ancestor" trees on the sarcophagus sides at Palenque, is being climbed by two youthful, chubby figures, one of them painted red, wearing simple caps and kilts. This "Jack and the Beanstalk" composition reminds one of the scene carved on the sides of Stela H at Copán. The upper figure is pointing at the head of a (*cauac*?) monster from which the second tree emerges. Instead of leaves, this tree appears to have bird wings. In the crown of the winged tree perches a long-necked skeletal monster, probably a bird. The part of the vase that showed the body of the bird is missing, but it appears that the bird may be holding a snake in its claws. Such trees or treelike structures constitute important elements of Maya iconography and are often portrayed in the codices, as well as on ceramics. According to legend a giant ceiba, the sacred tree of the Mayas, stands in the center of the earth. Its roots penetrate the underworld, and its branches reach up into the heavens. The Mayas believed that their ancestors had ascended into the world by these roots and that the dead could climb the trunk and branches to the skies.[27]

The next part of this vivid tableau shows two elegantly dressed young lords flanking another *cauac* water monster, which probably represents a stone zoomorph like the ones at Copán and Quiriguá. The word *cauac* and its variants in all Maya languages are associated with "rain" and "storm." The Aztec parallel to this word is *quiahuitl,* which also means "rain." According to David Joralemon, the *cauac* monster has strong underworld connotations and may represent "standing water."[28] In front of their mouths both lords are holding objects that appear to be cigars. Above the zoomorphic altar floats a peculiar animal that appears to be descended from a turtle mother and an armadillo father, or vice versa. It may represent a constellation. Above the entire scene hovers a male bat, probably Cama Zotz, the Bat God. The two young lords may represent either GI and GIII of the Palenque Triad, but more likely they are the Hero Twins in the House of the Bats in Xibalbá, the Maya underworld. The left figure has serpent marks; the right figure has jaguar spots.

As one rotates the vase counterclockwise, the stage changes again into a two-tiered palace scene. On the upper level is another pair of young lookalike princes, their bodies painted red, wearing pointed hats, shell-tassel back hangers, and flared earplugs. With their left hands they are beating on drums; in their right hands they are holding what may be rattles or ladles. Their serenade may be directed to Lord Pax, who is seated in the adjoining chamber. The head of one of the young lords is turned away, as though he dislikes the music they are playing. Next to these young percussionists sits an old man with arms crossed. He is wearing a heavy belt, an elaborate cap, and a shoulder cape, and he is turned toward the wall or entrance of the palace. Above the heads of these figures hover two animals, a bird (screech owl?) and a flying rabbit. The latter has an infixed Venus (?) sign. At the lower level of the building, below these figures, is a frog seated before a brazier. It should be noted that in Maya glyphic texts the month Pax occurs several times as a croaking frog. Lounsbury believes that the frog may relate to *pax* because of the sound it makes, often characterized in Maya as *tun-tun*.[29] *Tun* is also the highland root for "drum" and the word for "to play music."[30] The drum grapheme is used for both the sign for Pax, the month, as well as, in Long Count Inscriptions, the glyph for *tun*, "year." It is possible that the scene on this vase is an allegory of the "seating" of the Pax God as Lord of the Night.

Glyphic Text

The rim text does not represent any variation of the Primary Standard known to us. We suspect that it describes the scene or scenes below, and it may relate to the *Popol Vuh* stories of the Hero Twins. The text begins with the column of three glyphs to the left of the palace. It does not have any of the known introductory glyphs for the Primary Standard. The event phrase (if there is only one) is directly over the standing "twins," with the event compound over the left figure. It is followed by a deer head with an infixed *caban* sign and a throne glyph. Another throne or "jaguar bundle" glyph occurs in the next-to-last compound.

The Secondary Texts apparently refer to the names of the adjacent figures. The first phrase is below the beginning of the Primary Text and behind the Pax God. The initial glyph is G7 of the "Nine Lords of the Night Series," which appears with or instead of the head of the Pax God as the infix for Pax in the initial series introductory glyph. The last (five) compound appears frequently as "accession compound" throughout the Classic period. It may also refer to the "seating" of the Pax God as the Lord of the Night.

The remaining Secondary Texts are adjacent to the standing pair. The phrases have no obvious verbal compounds and thus most likely are the names of the two figures.

These sagas are the expression of the Maya soul. Wherefore in seeking decipherments we must rove in the green pastures of poetry, which we prosaically call mythology.

Sir J. Eric S. Thompson

Like other masterpieces depicted here, it is the product of an alien mentality that me may never fathom, but which we can surely admire and appreciate.

Michael D. Coe

10. Scenes Showing Objects That May Be Large Cigars

Ceramic Artifacts

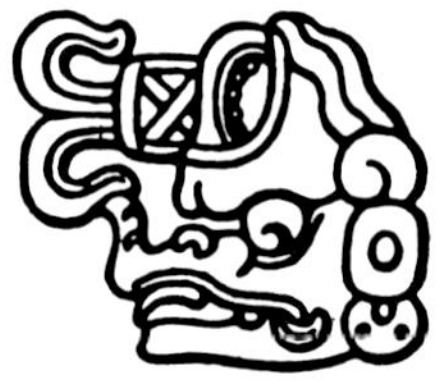

Smoking objects, probably large cigars, are depicted on the **Bowl of the Jaguar God** (C-1), a flared vessel probably made in Petén during the late Classic epoch (now in a private collection). The vessel is decorated in a manner that Michael D. Coe calls "codex style." It has a white base, and the figures are painted with bold, black brush strokes. Probably the most important person in the scene is a reclining near-nude figure, whose divine nature is revealed by his body markings, large eyes, central pupils, aquiline nose, and jaguar tail and ears. His large mane is in a knot, he wears a breast pendant, and his arms are tied behind his back. The jaguar ears, the *tau* booth, the bound long hair, the square eyes, the knotted pectoral and belt, and the jaguar tail identify him as GIII of the Palenque Triad, or the Jaguar God of the underworld.[1] His tail ends in the iconic version of T535. His role as a very important deity was recognized by Sir J. Eric S. Thompson in 1950: "Jaguar is the god of the Underworld, and appears to be merged with the night sun. . . . He is the Maya equivalent of Tepeyollotl."[2] His relationship to war and death has also been clearly demonstrated.[3] Above and in front of him are a few scattered glyphs.

At the right of the captive Jaguar God is the squatting figure of a young man wearing a narrow belt with cross-band motifs and a turban topped by a jawless head. On his visible arm and leg are death spots. The bound paper wristlets and gaiters identify him as a supernatural being.[4] This youthful figure, who is in a position to guard the Jaguar God, is holding in his hand another object, which could be either a torch or a large cigar. This object may relate to two other important occurrences. On the Tablet of the Sun at Palenque this object appears as the superfix to the name of God GIII, the other figure portrayed on this vessel. The full name compound is composed of the Mah K'ina title (T74.184), the cigar-and-torch grapheme, and the jaguar-spotted Ahau (T539). The second occurrence is on Stela 35 at Naranjo, where the object is in such large scale in proportion to the person portrayed that it suggests a torch rather than a cigar.

The scene of the captive god and his youthful guard is repeated on the bowl with only minor differences, such as the position of the guard who is bending slightly forward in the second scene.

Glyphic Text

The scene of two figures is repeated on both sides of the pot. The text for each scene is composed of the same three glyphs positioned in an inverted L around the figure of GIII.

In both cases A and B record the Calendar Round 11 Ix 7 Cumhu. It is not possible to ascertain in the Long Count position of the Calendar Round, but most probably it occurred in prehistoric times.

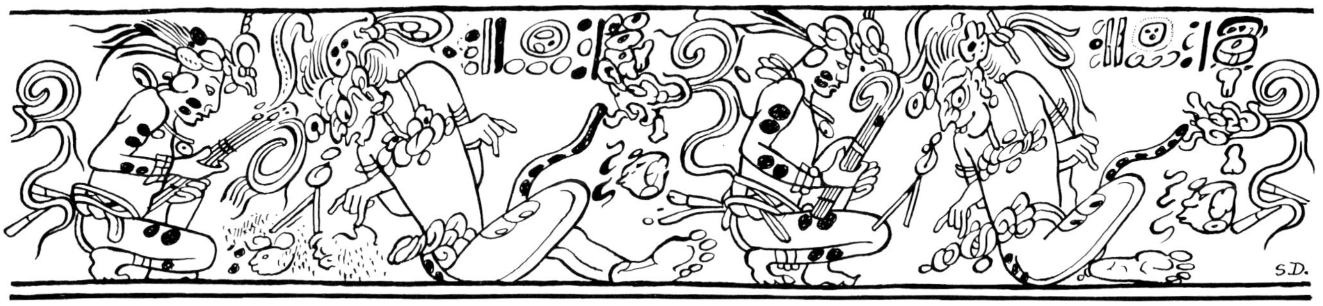

Figure 168. Bowl of the Jaguar God (C-1)

The third compound occurs in both texts by the knee of the reclining figure. The compound (T122.1010) may be a verbal or nominal compound. The main sign is the *kin* sun variant. The prefix is the variant of T122 that appears in the "marriage" compound of the *Dresden Codex* (T122.522:23). This particular configuration of the T122 affix may have carried the value *a* so that the compound may have read *Ah Kin*, "He of the Sun," a title known from ethnohistoric sources and appropriate to GIII.

Another probable example of the smoking cigar is seen on the **Vase of the Fire-breathing Jaguar** (C-2), a low polychrome vessel found in the Guatemalan highlands (now in a private collection). Apparently it is from the late Classic period, and it is painted on a white base in brown, red, and black. The vessel has the usual upper and lower red marginal bands and a circular upper row of hieroglyphs. The rest of the painted surface is divided into two sections by two double rows of repetitive vertical columns of sky-variant glyphs. Both sections show creatures with human bodies, bizarre heads, and fleshless jaws, which identify them as Death Gods. The first god, seated Turkish fashion, is wearing a wide belt, a jaguar kilt, a narrow apron, and a conical cap with a floater of quetzal feathers. The figure is similar to that seen on Lintel 2 at Piedras Negras. His eyes are a quadrangular shape, which Quirarte calls "squint eyes," and his teeth are those of a carnivore.[5] His jaw is bleached, and in his left hand he is holding an object that may be either a flaming torch or a huge smoking cigar. The second figure is in a kneeling position and is pointing his cigar (or torch) forward. He is wearing a rear ornament, a narrow loincloth, a cap with plumage floaters, and a pendant, made of a jawless head. His face has the characteristics of a jaguar, and he seems to be breathing fire. A symbol

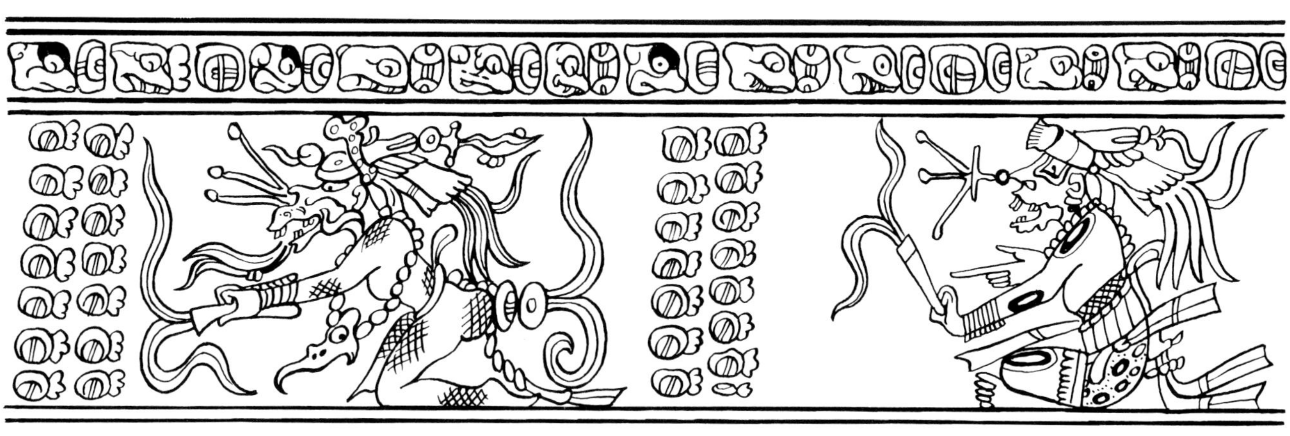

Figure 169. Vase of the Fire-breathing Jaguar (C-2)

resembling three converging drumsticks touches the noses of both deities. In some respects the first creature resembles, though does not seem to be identical with, Coe's Zip monster. The divine character of both creatures is emphasized by their body markings.

Thomas Barthel has recently compared the iconography and the glyphic text of this vessel with the "Burners Almanac" of page 31b of the *Dresden Codex* (Ah Toc) and identified the kneeling god on this vase as Tohil, the mythological figure of the Quichés, and the seated monster-headed deity as Balam Quitzé, the Jaguar God of the *Popol Vuh*. In Barthel's opinion the scene pictured on this vase not only shows two deities holding cigars or torches but also presents the allegory of Tohil, the Fire Giver, regent of the time period of the waxing moon, transmitting fire (light) to Balam Quitzé, the Receiver of Fire, regent of the waning moon. He regards the exposed teeth and bared gums (*quich*) of the monster gods as a picture rebus for the word Quiché and the triple drumstick-shaped nose ornament as a reference to the tripartite nature of that nation.[6]

Glyphic Text

The rim text repeats a few compounds with variation in the suffixes and is written in mirror image and emphasizes the repetition of zoomorphic heads that feature characteristics of the T758 or the *chiccnan* animals. T518 appears in two cases as the main sign of a compound. It is not possible to determine the meaning of the text, but the repetition suggests the structure of a chant.

The Secondary Text, like the Primary, repeats one compound. The main sign is apparently T617a, which has been identified as the grapheme of a mirror, probably read as *nen*.

See also Barthel's interpretation.

The **Vase of the Thirty-One Gods** (C-3) is a ceramic bowl of the late Classic period from southern Campeche or northern Petén (now in a private collection). The artist, who in Michael D. Coe's opinion was a professional codex painter, illustrated the vase with thirty-one deities arranged in four horizontal tiers, using the "four-color-negative" technique. The gods represent a wide variety of known and unidentified members of the Maya pantheon with both underworld and astral associations, among them Old Gods with and without jaguar characteristics, anthropomorphic insects, alligators, and jaguars, as well as young deities probably painted in the image of the Hero Twins. To describe this cavalcade of figures in detail would require a book of its own (see Coe's *The Maya Scribe and His World).*[7] For the purpose of this book, God 12 on the vase is of special interest. The god is a seated figure of a frog, an animal that was revered by the Maya's because of its association with water and rain. Frogs are the patron deities of the month Zac and are also associated with the months Pax and Uo in the Maya calendar. This particular frog is

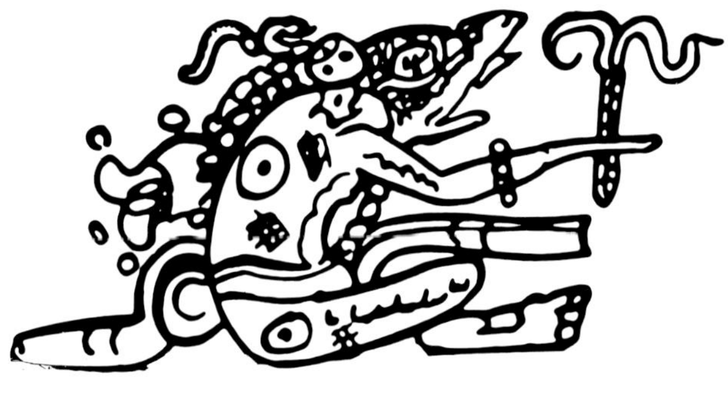

Figure 171. Detail, Vase of the Thirty-one Gods (C-3) (after Michael D. Coe)

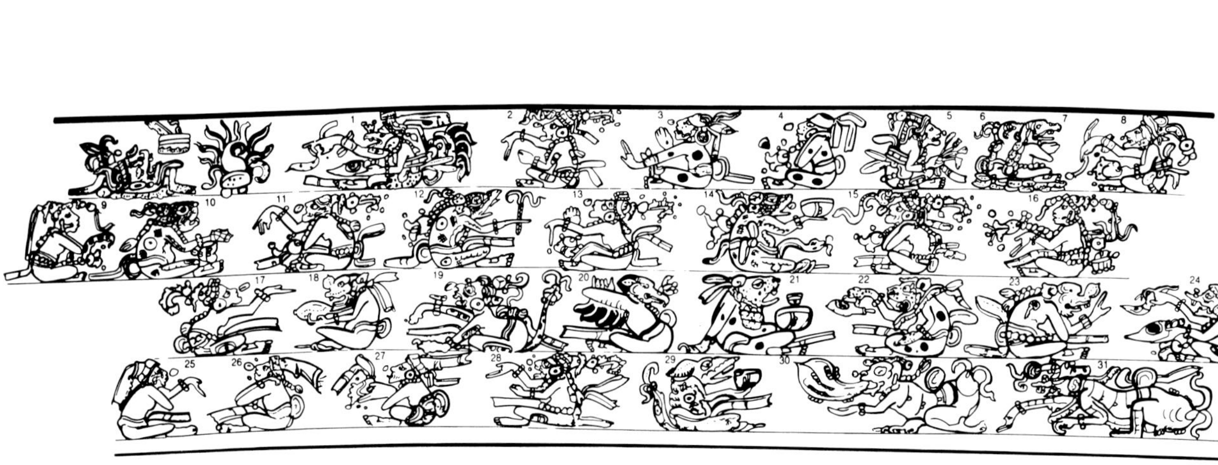

Figure 170. Vase of the Thirty-one Gods (C-3) (after Michael D. Coe)

distinguished by a bead necklace, wristlets, a loincloth, a water-lily back pendant and a king-sized smoking cigar that he is holding in his right hand. The vase has no glyphic text on the outside, but there is a repetitive band of hieroglyphs on the inner rim. Coe sorrowfully remarks: "This is a shame, since it would have been a windfall to have name glyphs attached to thirty-one gods!"[8] It certainly would have constituted a real *Who's Who in Xibalbá*.

The **Vase of the Underworld Ritual** (C-4) is a ceramic container from the late Classic period, probably from northern Petén or southern Campeche (now in the Primitive Art Collection of the Metropolitan Museum of Art).[9] It is painted on a cream background in brownish black. The composition on the vase includes two short rows of hieroglyphs and a tightly packed scene centering around the huge head of a water monster identified by *cauac* water symbols and large serpent frets. On top of his head, as on an altar, lies a small were-jaguar with feline legs and tail but the face of the Sun God. He is probably a sacrificial victim. On the two sides of the monster head (altar?) are two dancers, a young man who is GI of the Palenque Triad, and a skeletal figure. The divine nature of the young man is indicated by his headdress, reptilian skin marking and fish-barbel mustache, similar to the figure of GI on Stela B at Copán.[10] GI is a fish-derived image and is associated in some manner not yet understood with fish (*xoc*). He wears a narrow belt, a death collar on which four death eyes and a pouch with an *akbal* glyph are suspended, identical with that seen on the Vase of the Smoking Monkey.[11] In his left hand he is holding a round object, probably an eccentric flint, similar to that on Stela 18 at Etzná.[12] The purpose of such flints is not yet understood. They may have served as ceremonial weapons or a sacrificial paraphernalia. In his right hand he is brandishing a club or a narrow-blade ax, undoubtedly for use on the were-jaguar. He is probably identical to the ax-wielding GI, the young dancing god of the Dumbarton Oaks Relief Panel.[13]

At the right of the water monster prances the skeletal figure of the Death God (God A of the codices), "as if waiting to receive the sacrifice."[14] He is wearing a collar with death eyes, wide wristlets, and a most peculiar back ornament, described by Michael D. Coe as the "head of a long bone with death eyes and smoke motif," suggesting the bad odors associated with one of the Death God's names, Cizin. The Death God is represented with extraordinary frequency on painted ceramics, as well as in the ancient Maya manuscripts. Not only does the figure of the god himself occur but his attributes are found even in many places where his portrait is missing. Death evidently played an important role in the mythology of the Mayas and was intimately bound up with human sacrifice.[15]

At the far right of the scene sits a "jaguar-dog," an unpleasant-looking canine with drooping mouth, jaguar paws, and a spotted tail. That the dog belongs to the Death God is proved beyond doubt by its regular occurrence in connection with this deity. Coe believes that the animal might be the spotted dog who, like Charon of Greek mythology, was supposed to carry the souls of the dead across the river on their way to the underworld. Above the head of the jaguar dog hovers a flying insect about half the size of the dog. It has a helmetlike head,

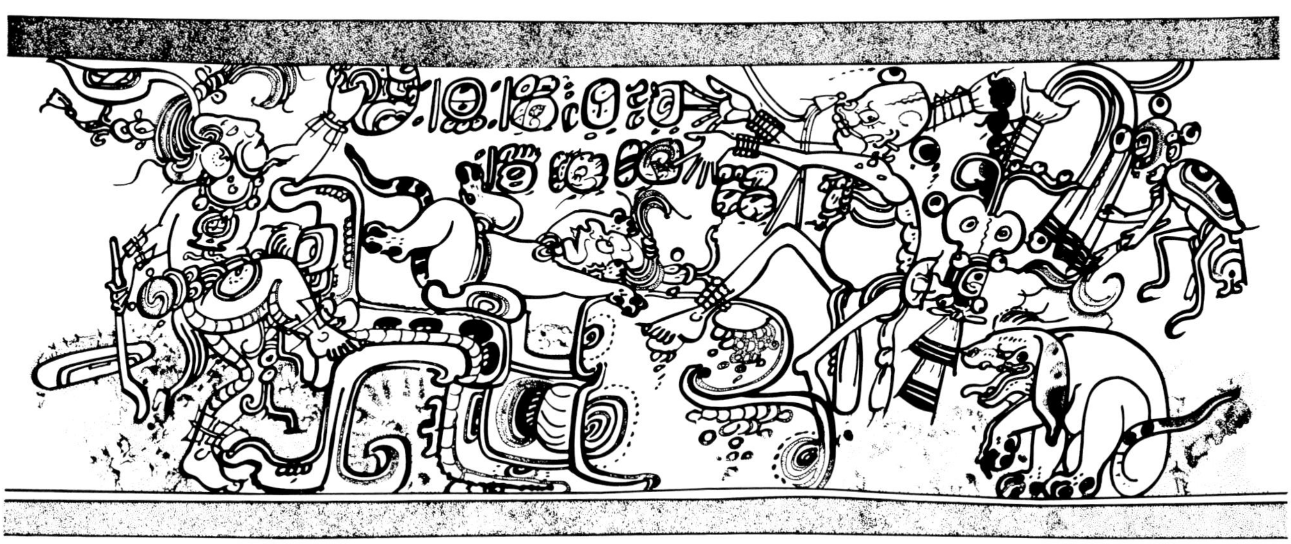

Figure 172. Vase of the Underworld Ritual (C-4) (after Michael D. Coe)

death eyes, and human hands. In one hand it is holding a large cigar (or torch) from which large coils of smoke are rising. Coe raises the question whether this insect, which seems to have escaped from a Vincent Price movie, may be the legendary firefly of the *Popol Vuh.* The cigar in his hand may refer to the incident in which the Hero Twins fasten fireflies to the ends of their unlighted cigars to mislead the guards in the House of Darkness.[16] He believes that the artist made a pun on the cigar incident: the firefly carries the cigar, instead of the other way around.[17]

The hieroglyphic text on the vase includes the Calendar Round date 7 Muluc 7 Kayab, which is one day after the heliacal rising of the Venus of the Morning Star, a day thought to be most malignant in both Aztec and Maya astrology.[18] In the *Codex de Cuauhtitlán* the day is associated with the sacrifice of various deities, including the God of Maize and warriors of the Jaguar Order.[19] On pages 46 to 50 of the *Dresden Codex* spearings of the Old God, a jaguar, the God of Maize, a frog deity, and another young god are depicted in connection with this ominous day. The only other deciphered part of the glyphic text is the hieroglyph of God K, although, bewilderingly, God K does not appear on this vase.

For a discussion of the glyphic text see Michael D. Coe.[20]

The **Vase of the King's Vision** (C-5) is a yellow incised vessel from the late Classic period, probably from Yucatán or Campeche (now in the Dallas Museum of Fine Arts, gift of Mr. and Mrs. Eugene McDermott, McDermott Franklin, and Mr. and Mrs. Alger H. Meadows and the Meadows Foundation, Inc.). It depicts some events that are probably associated with smoking. The center of the scene is occupied by a figure of a ruler sitting on a low throne in the traditional cross-legged pose. His hair is held together with a narrow flowered band and a bouquet of long quetzal feathers. He is looking intently at a creature who has a human body and a monster head and who is naked except for a narrow belt and a bead necklace. The creature has a large god eye and fangs, and the scrolls emerging from his mouth may indicate that he is breathing fire. In his hands he is holding tightly a tubular object (cigar?) from which also scrolls (smoke?) emanate. From the smoke emerges a youthful head facing the seated ruler. Three more pouchlike objects with scrolls decorate the monster's belt and apron. Behind the ruler is a dividing wall decorated with circular designs and surrounded by huge flowers.

One is inclined to believe that the scene portrays the visions of the seated ruler, produced by smoking hallucinogenic herbs. This interpretation can be supported by comparing it to the blood-letting scenes on lintels 13, 14, 15, and 24 at Yaxchilán. On all four monuments men and women of royal rank, are shown in the aftermath of a blood-letting act. Lintels 13 and 14 show the participants holding the blood-letting deity. A young head is emerging from the mouth of the serpent, probably representing a vision that is in form, impact, detail, and graphic approach almost identical to the head emerging from the smoke pictured on this vase. This figure, however, lacks the characteristics that would identify it with absolute certainty. Nonetheless

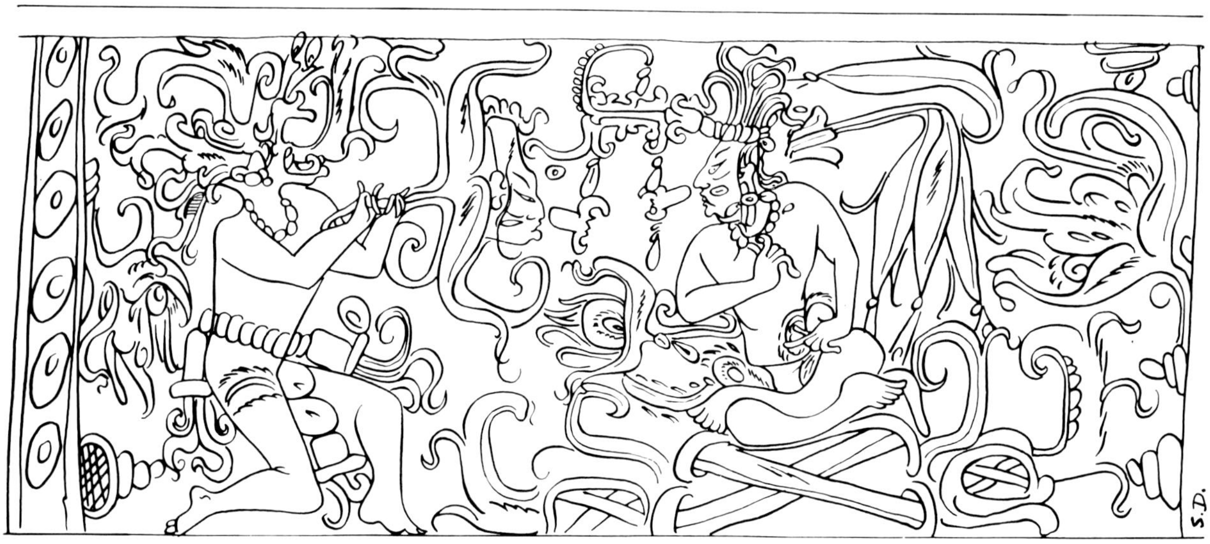

Figure 173. Vase of the King's Vision (C-5)

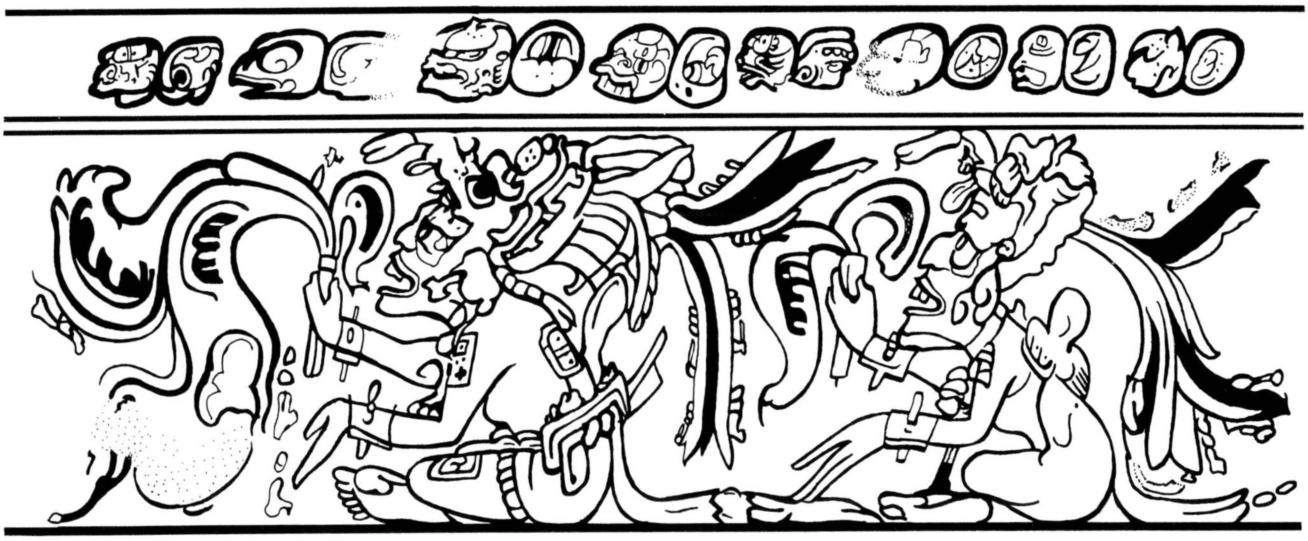

Figure 174. Vase of the Old God Holding a Flare (C-6)

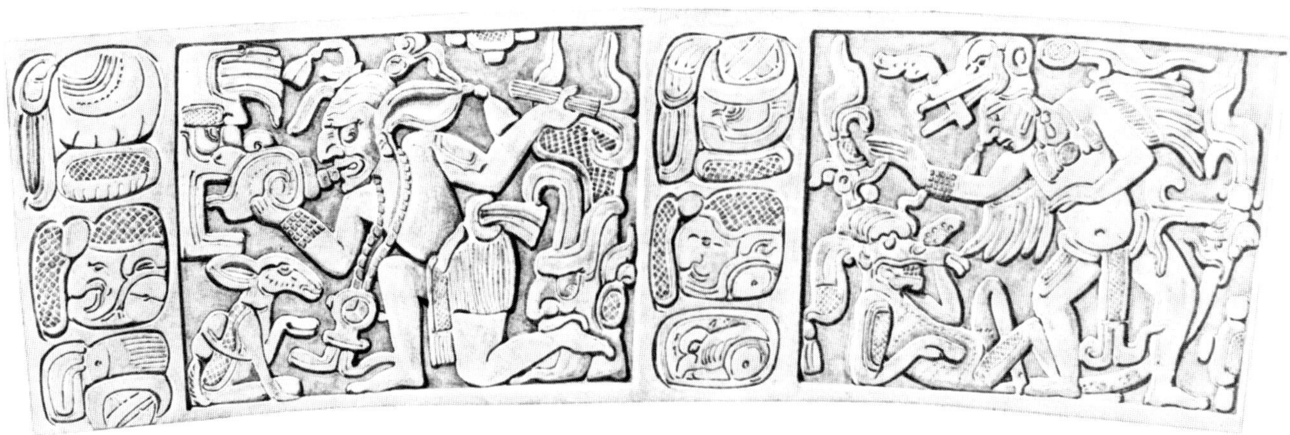

Figure 175. Vase in Chocolá Style, Two Mythological Scenes (C-7) (after Michael D. Coe)

there is a close resemblance between the head of the creature and that of God K as he appears in the codices.

Deities holding large cigars or torches are also portrayed on the late **Classic Vase of the Old God Holding a Flare** (C-6), from Petén (now in a private collection). The vase shows two aged seated gods with large eyes, filed incisors, and aquiline noses. Their faces and upper chests are painted red, and they are wearing monster helmets, kilts, elaborate belts, and wristlets. Their right hands are extended in an inviting gesture. In their left hands they are holding objects that are probably large smoking cigars. The vase has the usual upper and lower red rims and an upper glyphic band.

Glyphic Text

The Primary Text is composed of a pair of compounds with the first grapheme in each case a zoomorphic head. The text may well have been a series of names or a chant, but it is not related in any known descriptive way with the scenes.

A beautiful carved ceramic that Michael D. Coe calls **Vase in Chocolá Style, Two Mythological Scenes** (C-7) also depicts a deity bearing a cigar (or torch).[21] The vessel, which was made during the late Classic period in the Chocolá or Maxcanú area, is now in the Dumbarton Oaks Collection. Coe's condensed description of the vase is as follows:

[The vase is carved with] two highly narrative scenes

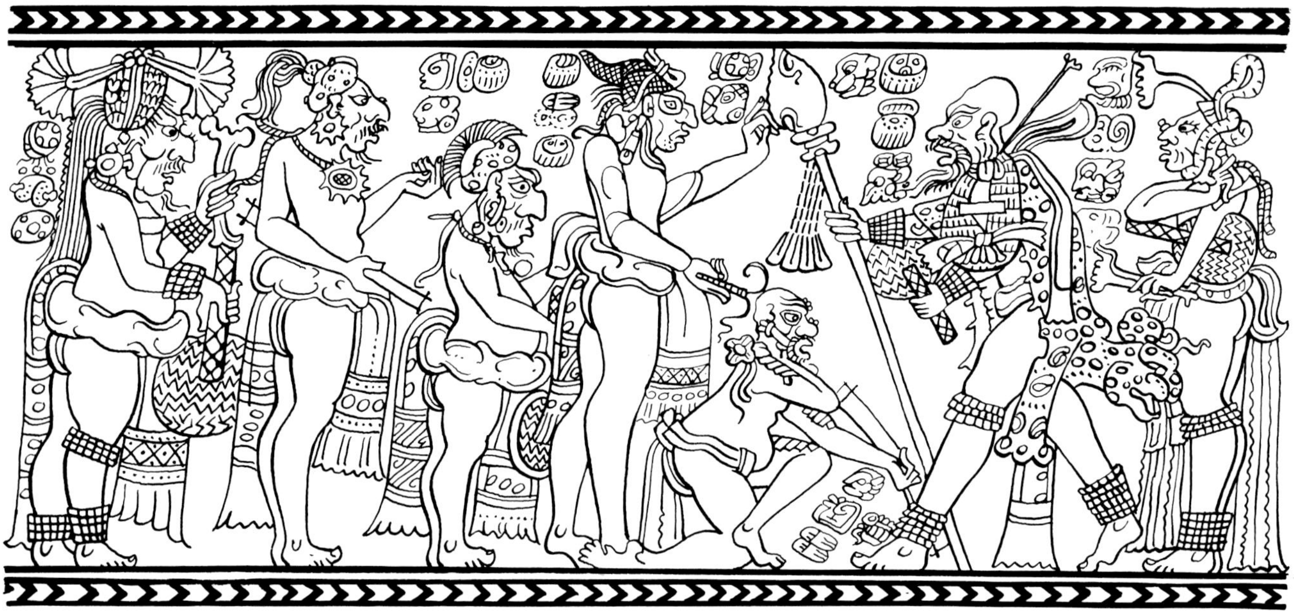

Figure 176. Vase of the Black Warrior (C-8) (after Gordon)

without precedent or reasonable explanation, connected in some way with deer. In the tableau on the left, a wrinkled old god with "god-eye," Roman nose, and deer's ears blows a conch-shell trumpet into the open mouth of a gigantic head resembling God C. In his left hand he holds a cigar or smoking tube. Suspended from a necklace around his neck is an inverted jar, probably with a broken-down *akbal* glyph, from which curls a snake; this device is an attribute of infernal gods, especially GI. His foot is being bitten by another large deity's head; this head is jawless, and there is a quill-like device in the mouth, so we are perhaps dealing with the God of Pax (see Grolier 49, Figure 3). Before the main figure is a little deer, either a doe or a deantlered buck, probably to be identified with the deer on Grolier 66, which also has a pack or container strapped to the back.

To the right we apparently have a scene of ritual combat between a triumphant, youthful divinity with wings over a recumbent, anthropomorphic, aged, and even semi-skeletalized deer-monster. The winged deity wears a long-beaked bird headdress which grasps in its beak the severed left leg of the vanquished enemy. Attached to the scalp and extended tail of the deer-monster are heads of the Jester God; interestingly, the one on the right has the tube with smoke protruding from the forehead, a trait of God K, with whom he has a close iconographic connection.[22]

For a discussion of the glyphic text see Michael D. Coe.[23]

The **Vase of the Black Warrior** (C-8) is an old favorite of Mayanists.[24] It is mentioned here because one of the main characters pictured on it is carrying a smoking cigar in his hand (Plate 119). The vase was made by a Chamá artist during the late Classic period (now in the University Museum of the University of Pennsylvania). The vase has an orange base and seven figures in black and red. They are attired in elegant loincloths, ornamented caps, and body jewelry and are holding various ceremonial objects, including large round fans. The latter may indicate that they are merchants and that the painting depicts a gathering of the Chamá "Chamber of Commerce."[25] Because the personal characteristics of these figures are so faithfully rendered, Morley concluded that they are portraits of historic persons.[26] Michael D. Coe, on the other hand, believes that the scene is mythological and that the first of the two black figures is God L or a combination of God L and God N and the second is God M, known in Yucatán as Ek Chual, the black god of traveling merchants and of war.[27] I find no real conflict in the two views. Just as the merchant princes of Renaissance Italy were often portrayed in the company of the Holy Family and associated saints, the merchant lords of Chamá may have also prided themselves on being shown in the company of supernatural beings.

Recently this important artifact received special attention from Kornelia Kurbjuhn of the Institut der Völkerkunde at Tübingen. Because of the significance of her comments they are quoted at some length:

> The scene on this vase depicts a meeting of seven persons. Only two of them, the ones with the black body paint seem to be communicating with each other, their

mutual concern appears to be the fifth individual at whom they are pointing implements, and who in turn touches both of them.

Inspecting the attire of the seven persons it becomes apparent that only the ear ornaments are individual, all the other motifs are repetitious. The first four persons share more with each other than with the remaining three individuals. The fifth figure has only four shared motifs, a further indication of his special status. The overall costume, objects worn and the glyphs unite rather than separate the participants of the scene.

Facial hair growth indicates that seclusion in preparation of a ritual has taken place. The ceremony in question involves holding fans, large bones and the as yet unidentified planks or tubes. Fans and the presence of a lance characterized by a double T171 suggest sacrificial death by beheading. The victim, figure 5, accordingly has a T544:116 on his neck: *canil* means "agony of death" in Yucatec. His extraordinary state is further emphasized by 'single statements': there are markings in the places of the so-called 'God Markings', but no deity has *pop* painted on arms and legs. There is the mat motif, but the frontal loincloth hanging signifying political-dynastic responsibility of rulers is missing. The posture is actually one of sitting, but there is no seat. Hand-on-shoulder gestures performed this way (hand out of sight) usually have the other hand showing under the armpit or out of the bend of the elbow.

Both figures 4 and 6 seem to be responsible for this sacrifice: both touch the instrument of killing, both are painted black thus belonging to 'darkness' and both are connected with T122 *kak* = fire, power, anger. The presence of the large bones in the scene may indicate war as a likely reason for black body paint: Landa (Tozzer 1941:120) mentions that victorious warriors carry such bones. Black warriors, bones and fans occur together in Bonampak in Structure 1 Room 2, the resulting beheading takes place in Room 3.

The glyphs all seem to be descriptions of status and/or 'profession' but not all of them can be read yet. They do not describe the event.

I suggest therefore that this vase is an historical record of the sacrificial death of a Lord captured in war. His demise, however, was not of interest to the group as a production of a deified ancestor to a ruling lineage but seems to have trophy character.[28]

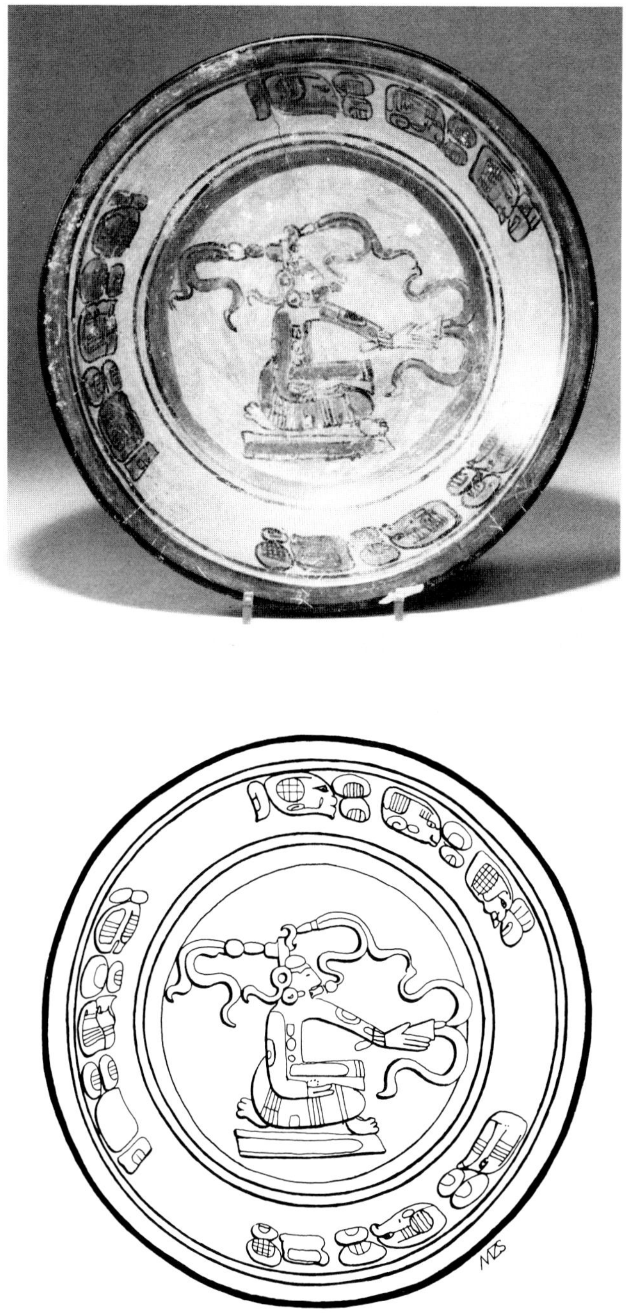

Figure 177. Plate of the Young God with the Flare (C-9)

A young deity holding a torch or a large smoking cigar in his left hand is depicted on the **Plate of the Young God with the Flare** (C-9), a ceramic probably of the late Classic period of Petén (now in a private collection). The plate is painted in red and black on an orange base, with two concentric bands enclosing six repetitious glyphs. The center of the dish is occupied by the figure of a young god seated on a low cushion. His divine nature is revealed by the "god spots" on his arms. He wears a short kilt with linear motifs. The unclothed parts of his body are partly painted in red. He wears a large earplug and a small headdress with rear and front floaters.

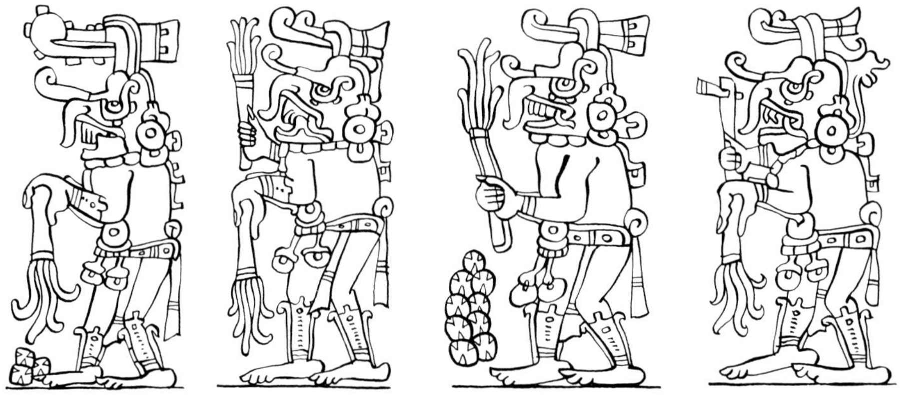

Figure 178. Torch-bearing gods in the Madrid Codex

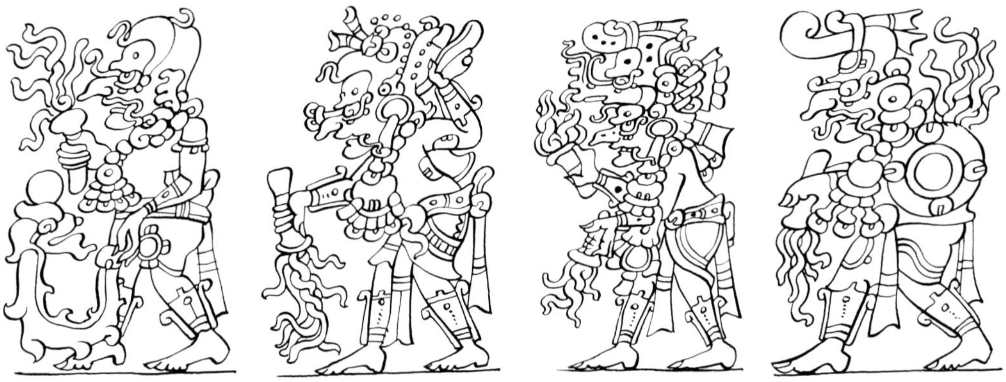

Figure 179. Torch-bearing gods in the Dresden Codex

Codices

In this chapter it is appropriate to include the so-called torchbearer gods of the Maya codices. These deities, most of them manifestations of God B, characteristically carry large, uniform flares in their hands.[29] The flares are slightly conical, and the tongues of flame or perhaps smoke emerging from them are represented as short coils.

On pages 3 to 5 of the *Madrid Codex* torchbearing God B is portrayed with the Sacred Serpent in a water scene.[30] He is pointing his flare upward in the first picture, downward in the second, and two flares in both directions in the third. It is noteworthy to observe one of his "trademarks," the peculiar curl ornament rising from the forehead. This may indicate that God B may be related in some way to God K, a deity with a similar though not identical forehead ornament.

That God B is portrayed on page 34 of the *Dresden Codex* holding a torch and wearing the head of God K as a headdress strengthens this hypothesis.[31] Forehead curls also decorate some of the Aztec deities, some of whom are definitely related to their Maya "colleagues." Examples of this can be seen in the *Codex Fejervary-Mayer* and the *Codex Laud.*[32] These and other similarities led Michael D. Coe to suggest that God K, the Flare God of the Mayas, and Tezcatlipoca (Smoking Mirror), one of the principal deities of the Aztecs, are the same, just as Zeus of the Greeks is identical to Jupiter of the Romans.[33]

God B, holding a flare, reappears on page 32 of the *Dresden Codex.* With his left hand he is wielding an ax, and he certainly appears to be in a destructive mood. He is also pictured holding flares on pages 34, 36, and 37 of the codex. These scenes are described by Thompson as "Chac with probably a torch in his hand," "Chac with torches in both hands indicating

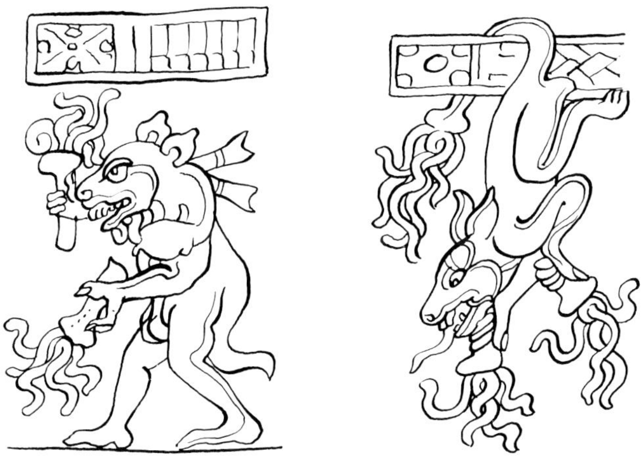

Figure 180. Torch-bearing animals in the Dresden Codex

drought," and "Chac holding a burning torch." Apparently God B (Chac?) did not have an absolute monopoly on torches, because the *Dresden Codex* is also illustrated with two scenes in which animals, possibly dogs, are holding flares. The one on page 36 is described by Thompson as follows:

> A dog with tail hooked around a sky-band dives earthward. His front paws and tail hold flaming torches. This is a pictorial *rebus*: *pek* in Yucatec is dog, but it has a homonym *pek*, bad weather which seemingly produces little rain. The dog's out-thrust tongue probably refers to lightning, for dog's tongue is a term for lightning in Kekchi.

A second dog walking with a torch in each paw beneath a celestial dragon strip is portrayed on page 39.[34]

The question naturally arises: Do these torch-bearing scenes in the codices relate to smoking? Could the flares represent large cigars? Most interpreters of these manuscripts think that the flares are torches, an opinion with which I concur.[35] There are four reasons for this conclusion:

1. The codex flares are shown with small red curls, similar to the image of fire usually portrayed in these books. The flares on ceramics are usually seen with large coils similar to those associated with readily identifiable acts of smoking.
2. The flares in the codices are usually portrayed in context with weather scenes, and thus they may represent lightning or drought. On ceramics the flares usually occur in esoteric scenes, palace rituals, and so on.
3. The theory that the flares in the codices represent torches is also supported by the accompanying glyphic text. On page 40b of the *Dresden Codex* a parrot is portrayed holding a pair of such flares. The pertinent glyphic text reads, "The fire in heaven of Four X Macaw drought."[36] The glyph for "fire" also appears in the passage in which God B and dogs are shown with flares.
4. In the codices the hand-held flares are distinctly different from the recognizable smoking scenes in which deities are shown with cigars in their mouths.

The conclusion seems clear that flares held by various individuals painted and carved on ceramics probably represent cigars while flares held by individuals in the codices probably represent torches.

Beasts with upturned snouts are a genus of many species in Maya iconography. All of whatever form are clearly celestial. Maya mythological beings are composite products of features abstracted from various species of diverse orders. Identification of their component features is sometimes fairly obvious, more often difficult, and sometimes impossible.

Floyd G. Lounsbury

Visual images of Maya art are subject to strict control much like verbal language.

Linda Schele

The enigmatic God K, the serpent-footed deity of the "manikin scepters," is not very rare, but so little is known about him iconographically and functionally that he cannot be yet explained.

Michael D. Coe

11. God K on Ceramic Artifacts and Frescoes

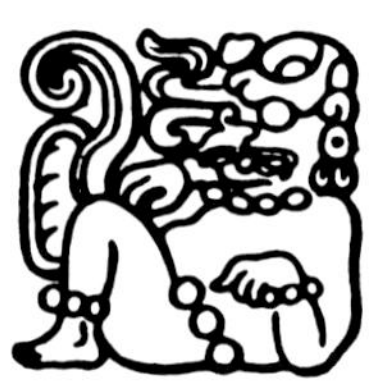

To describe all the monumental sculptures and to analyze in detail all the codex pages on which God K is represented would double the size of this book and, because of the many published studies on this subject, would be an unnecessary duplication. God K iconography on painted and carved ceramics has received only occasional mention, however, and I believe that it deserves detailed presentation.

Ceramic vases and bowls related to God K iconography can be divided into two groups: true portrayals and scenes possibly related to God K iconography.

Portrayals of God K

A most important example of early Classic Maya ceramic art is the **Bowl of God N and God GI with God K Scepter** (D-1). It was first photographed by Hellmuth (now in a private collection). The style of the bowl clearly points to the Tikal-Uaxactún region. The bowl's composition and high-relief sculpture and bas-relief etching (the latter evidently done after firing) make it one of the masterpieces of Maya ceramic art.

The engraved design provides some of the most important iconographic information known from ceramic vessels. The style is very much like that of the early monuments of Tikal, including Stelae 29, 18, 31, 1, 2, and 28. The monuments span the period 8.12.0.0.0. to 9.1.10.0.0. This vessel also dates from that period.

On the front of the bowl is the figure of God N, who is shown in his characteristic pose, emerging from a conch shell. The face of the deity is painted red with cinnabar. He is wearing large round earplugs, a peculiar cylindrical headdress, and a bead necklace. He is resting his chin comfortably in his right palm.

Another engraved figure is seated in profile and holds a God K scepter. Details of the figure, including the shell earpiece, the hair curl in front of the forehead, and the knotted pectoral, identify him as GI of the Palenque Triad. This is the earliest known appearance of GI in which the full iconographic program of the late Classic version is complete. The headgear includes a *kan*-cross-and-leaf motif and a cantilevered motif. The motif above the shell earpiece is a dragon's head; the same kind of detail is found on Stela 31. The head and leg of the figure have the serpent-scale markings typical of GI. The top half of an *akbal* sign is infixed in front of the ear, and "mirror" or "reflective" god markings appear on his upper arm, leg, and back. His necklace, wristlets, and belt are made of beads, and his rear loincloth has a zoomorphic motif very similar to the early version of the Tikal emblem glyph as it appears on Stela 4 and on the Leiden Plate. He

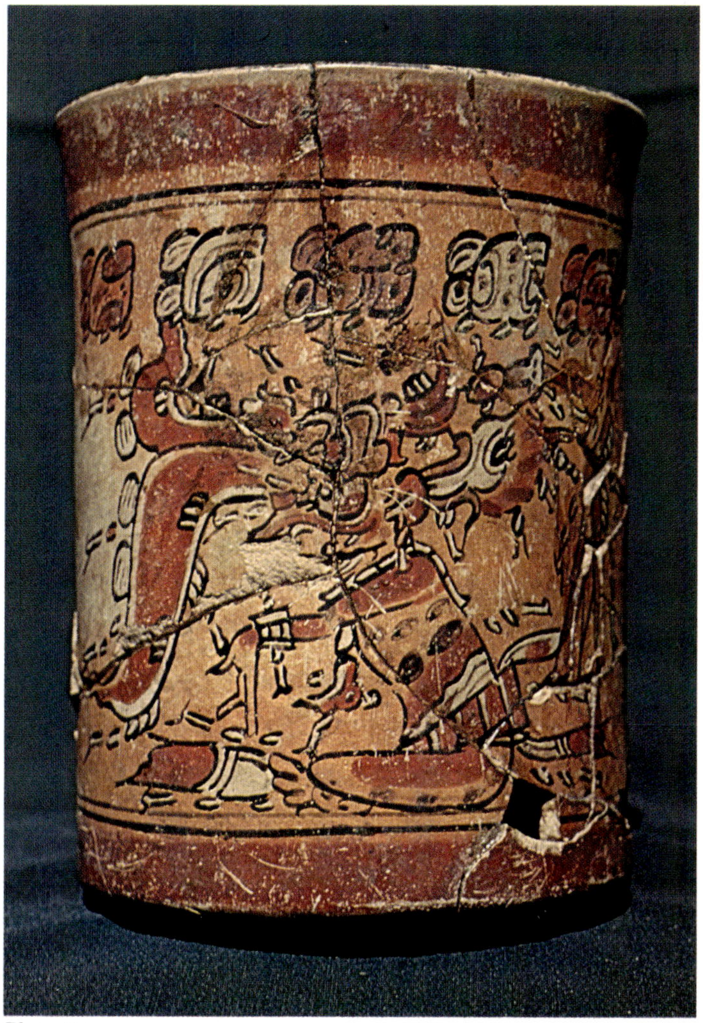

Plate 194

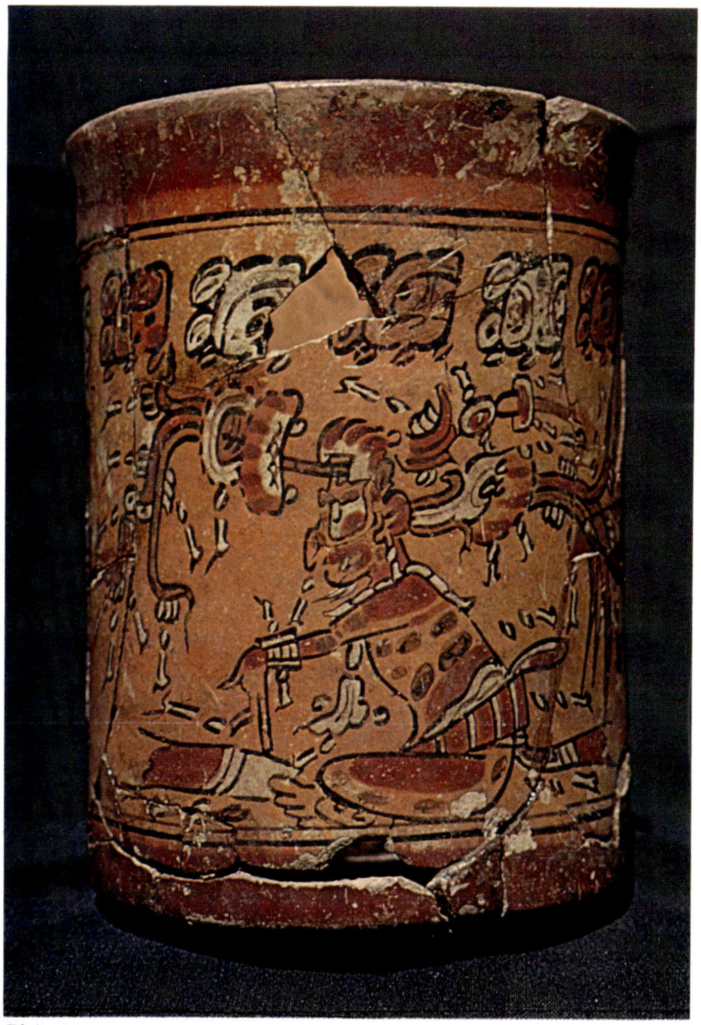

Plate 195

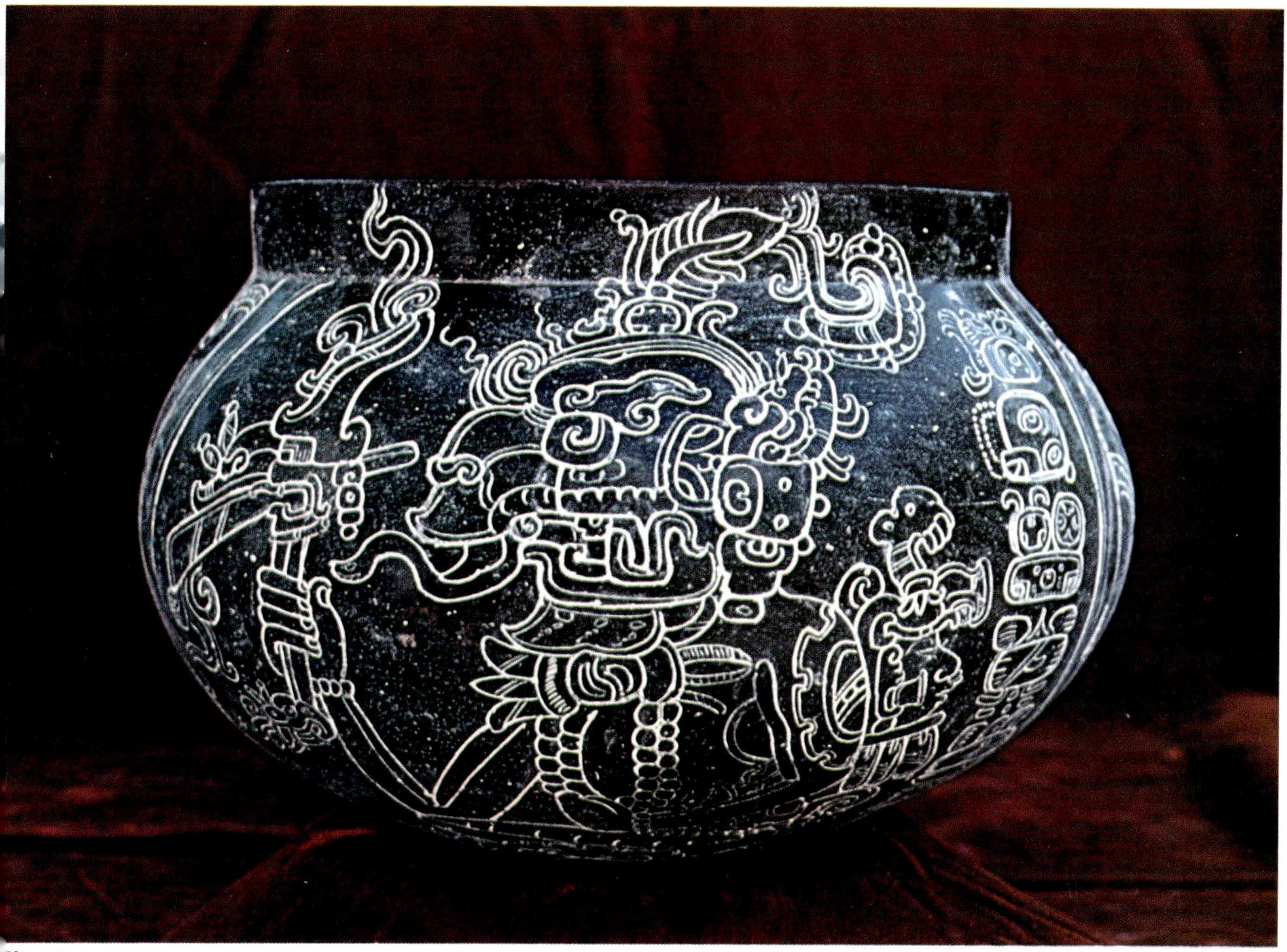

'late 196

'lates 194-195   Vase of God K and the Old God (D-6)

'late 196   Bowl of God N and God G I with God K Scepter (D-1)

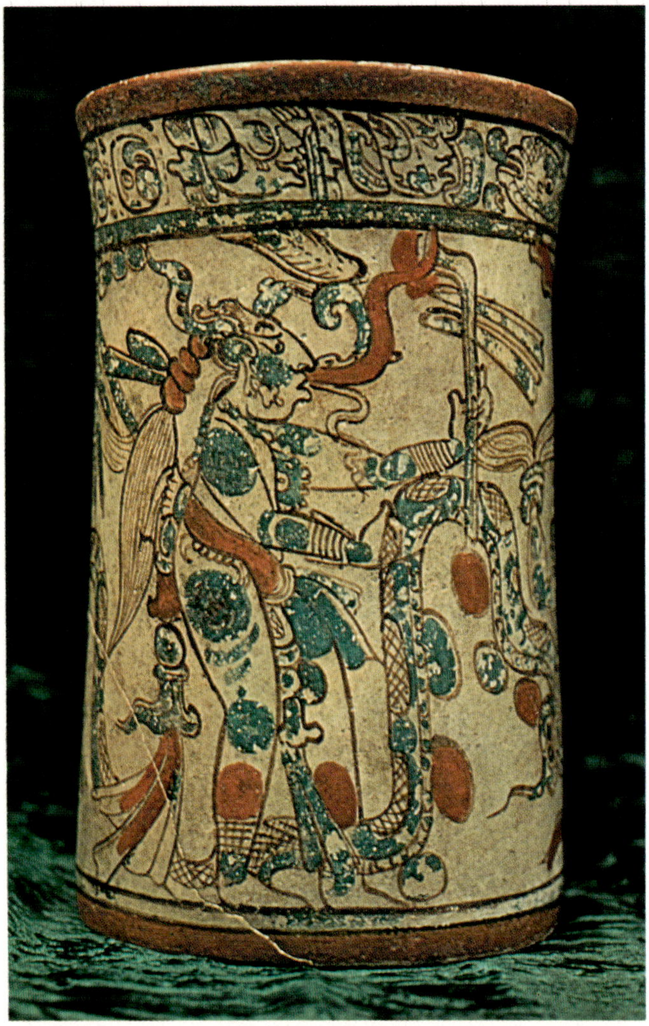

Plate 197

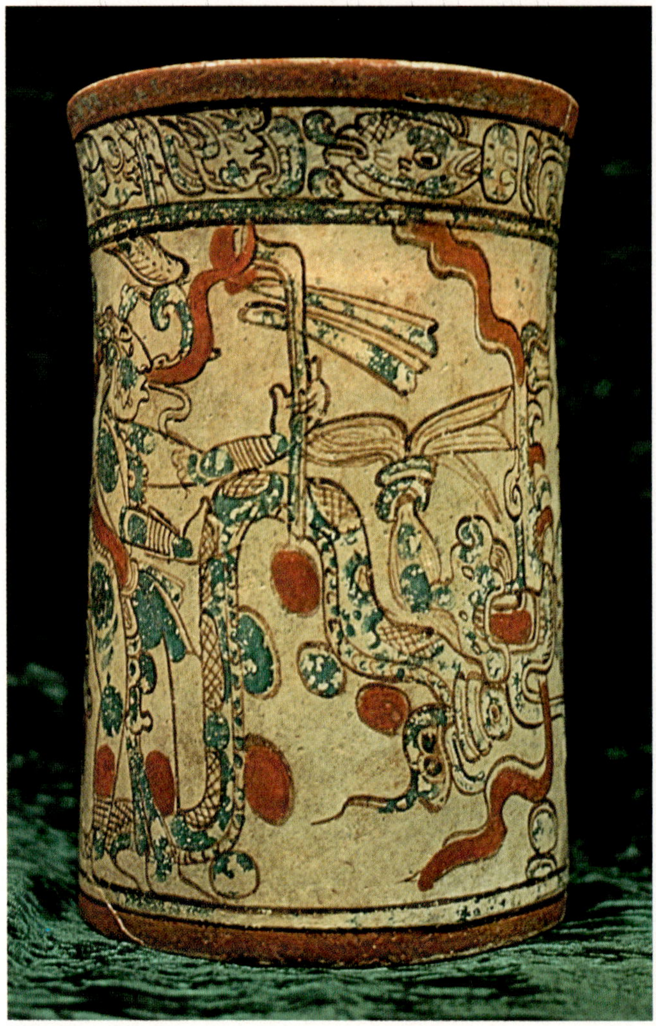

Plate 198

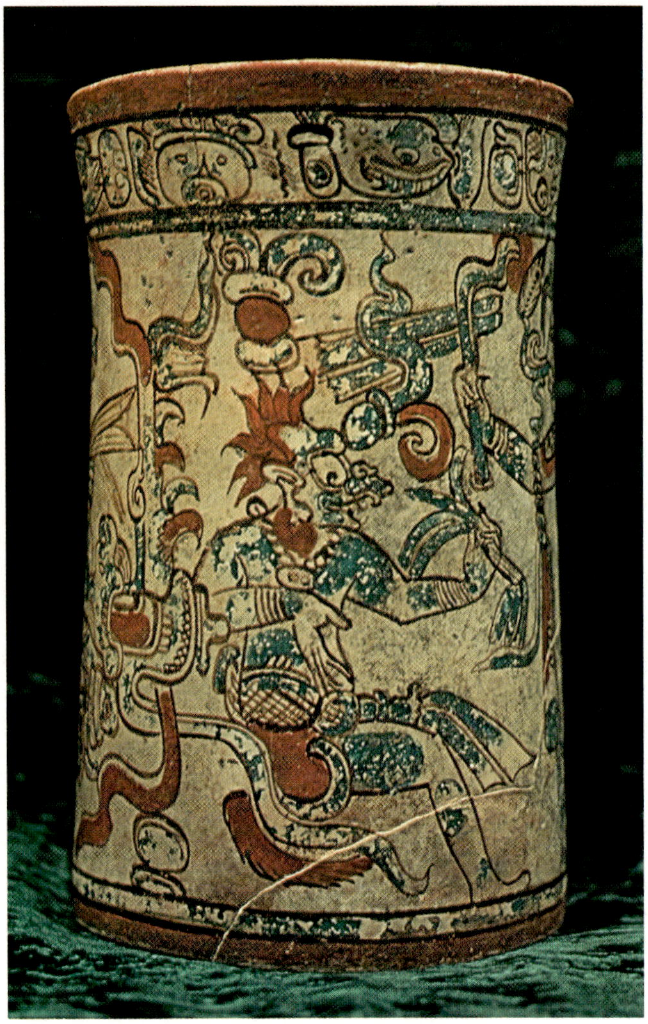

Plate 199

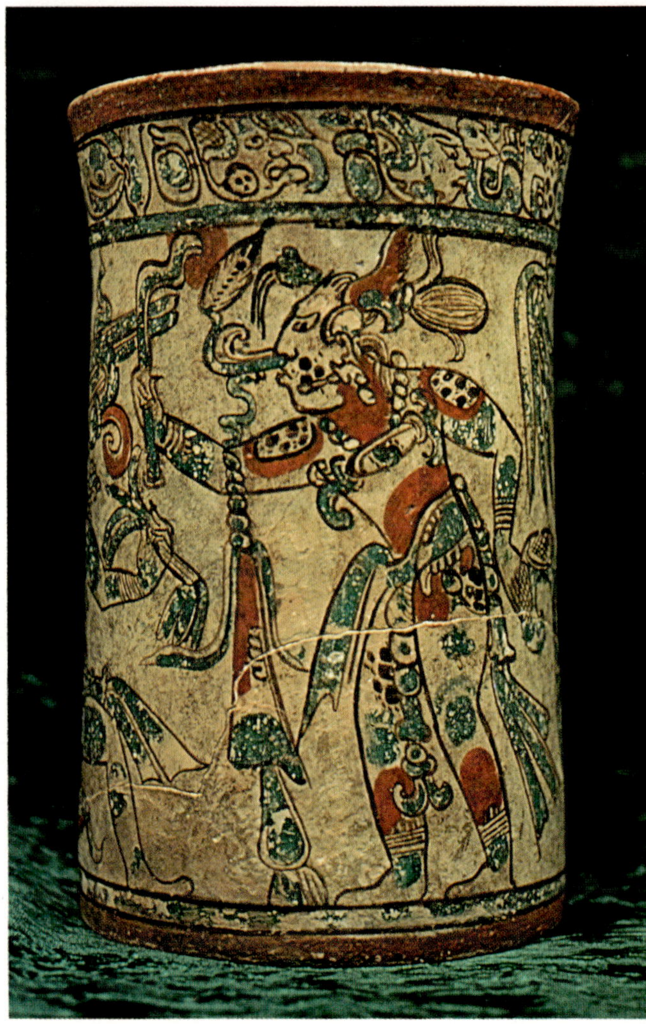

Plate 200

Plates 197-200 Vase of God K and the Serpent Dragon (D-2)

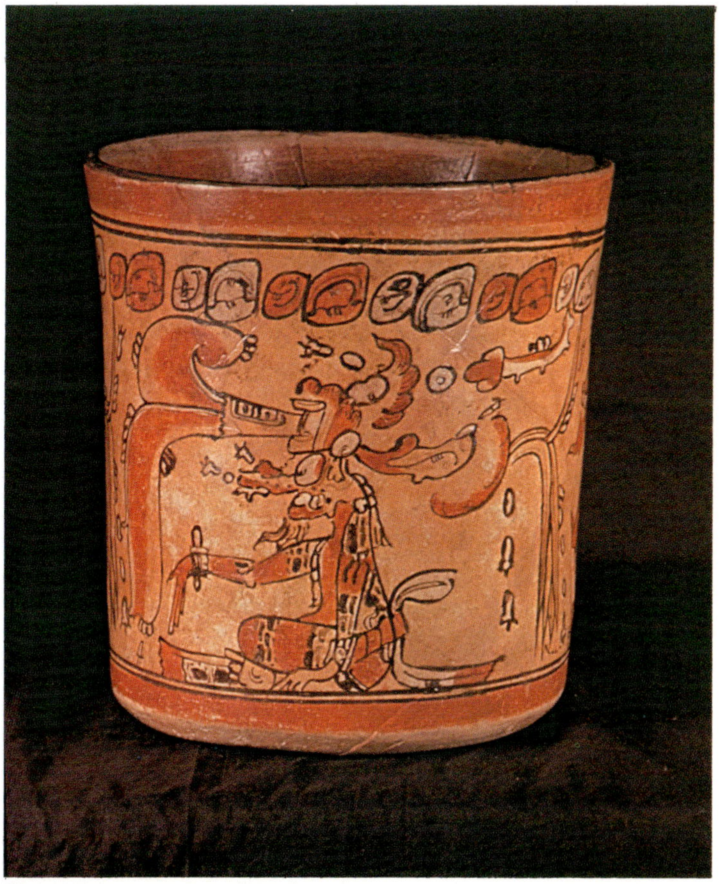

Plate 201

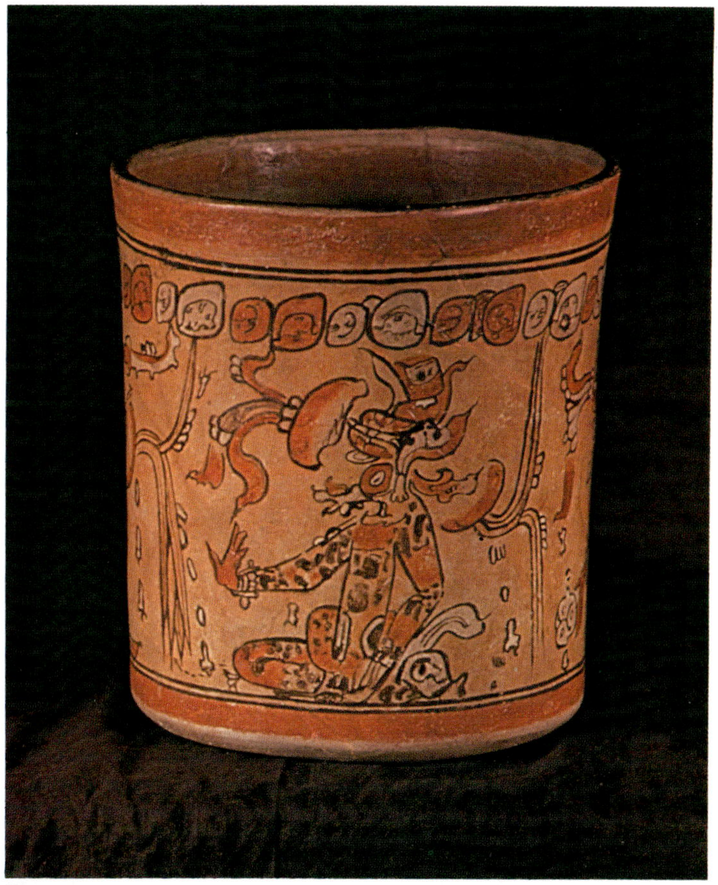

Plate 202

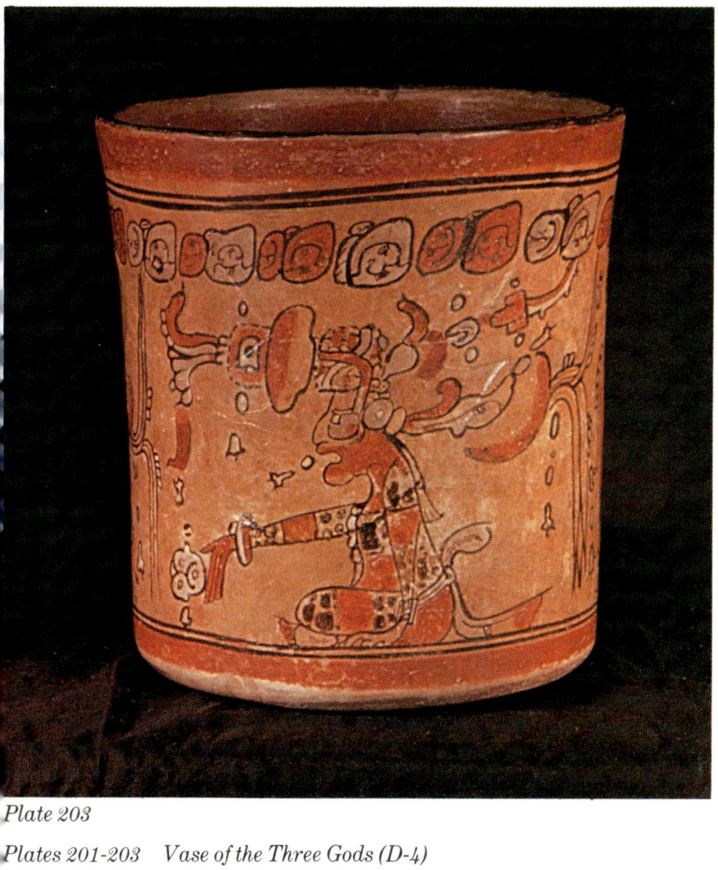

Plate 203

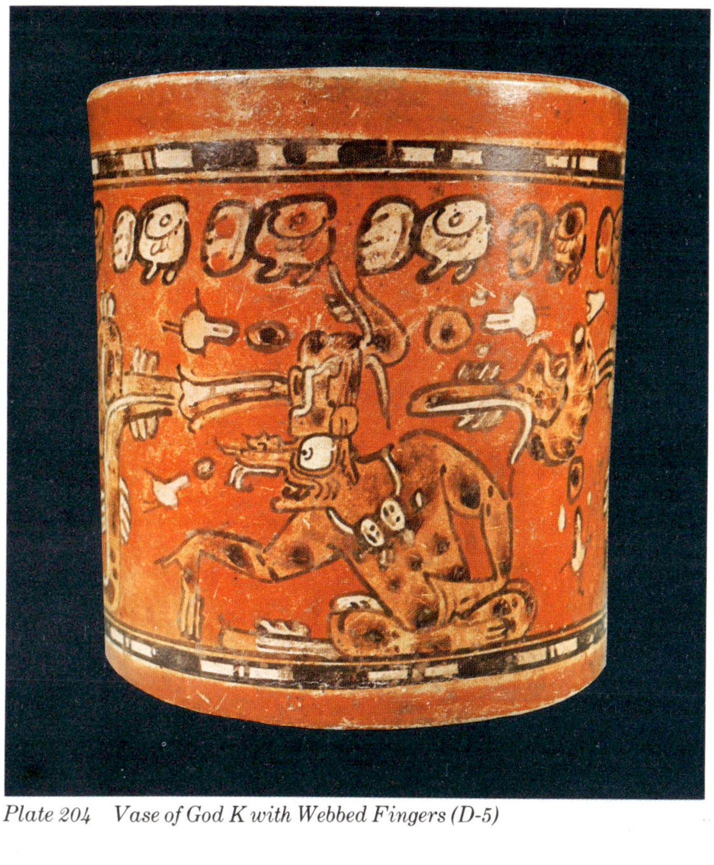

Plate 204 *Vase of God K with Webbed Fingers (D-5)*

Plates 201-203 Vase of the Three Gods (D-4)

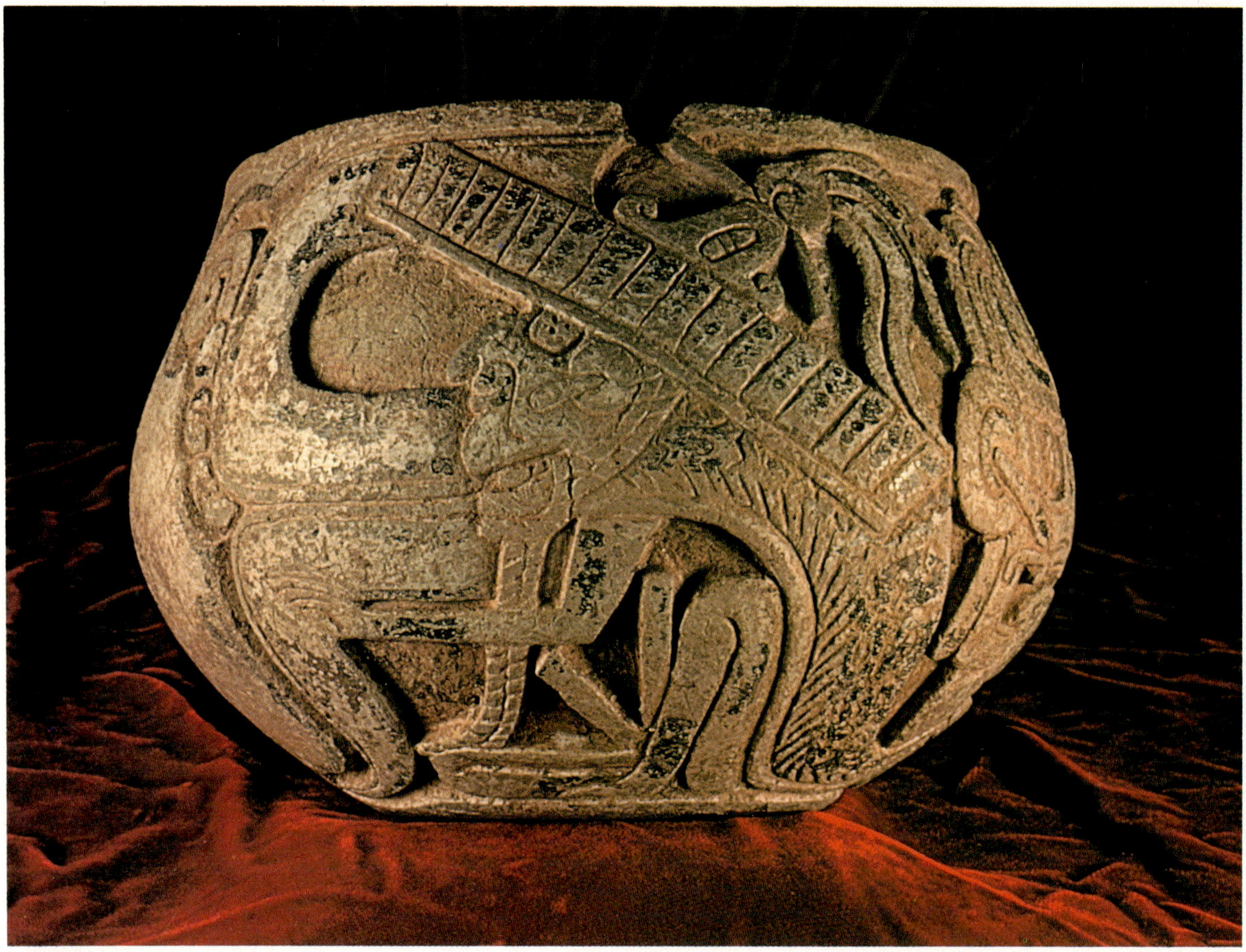

Plate 205

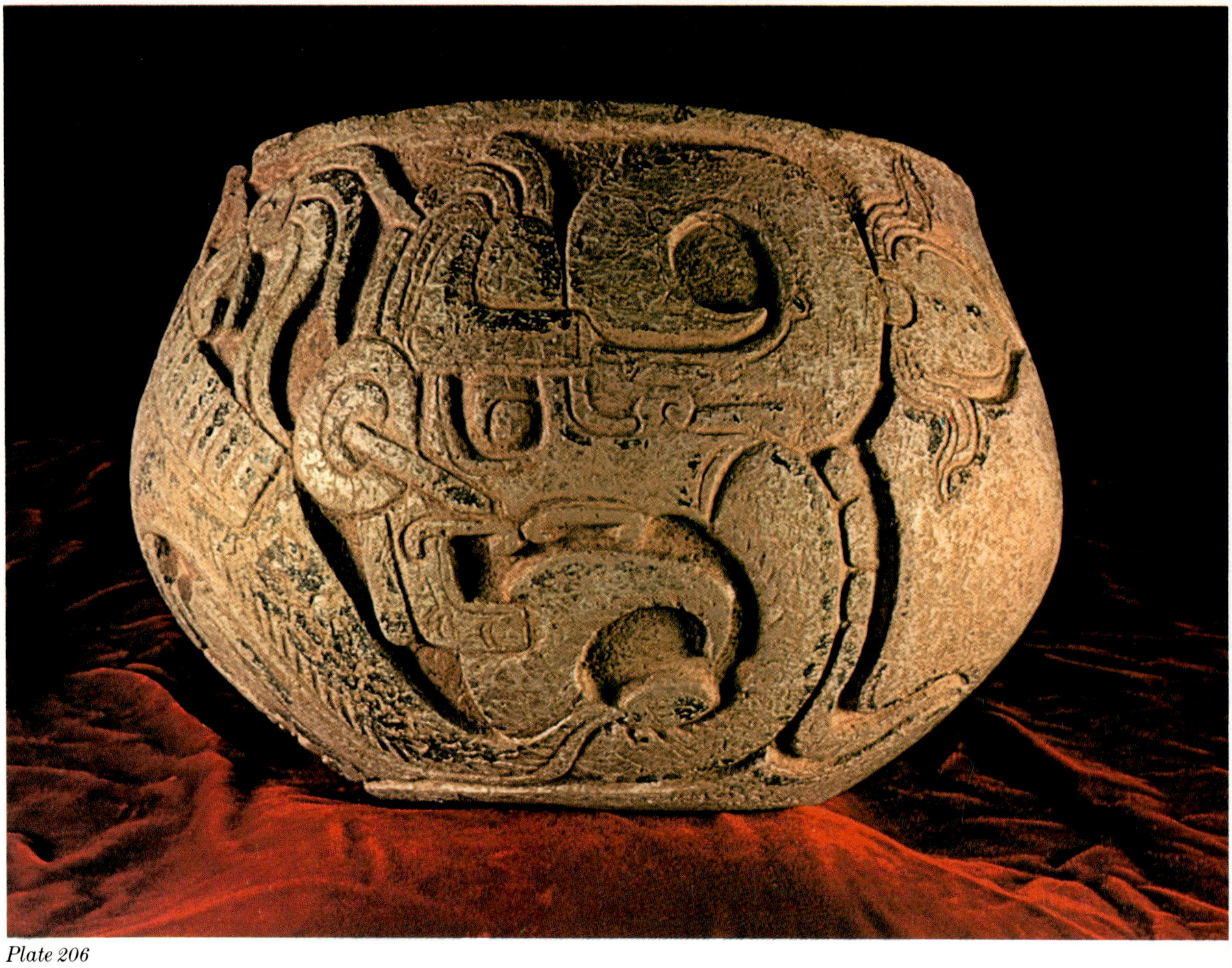

Plate 206

Plates 205-206   Bowl of God K and God L I (D-8)

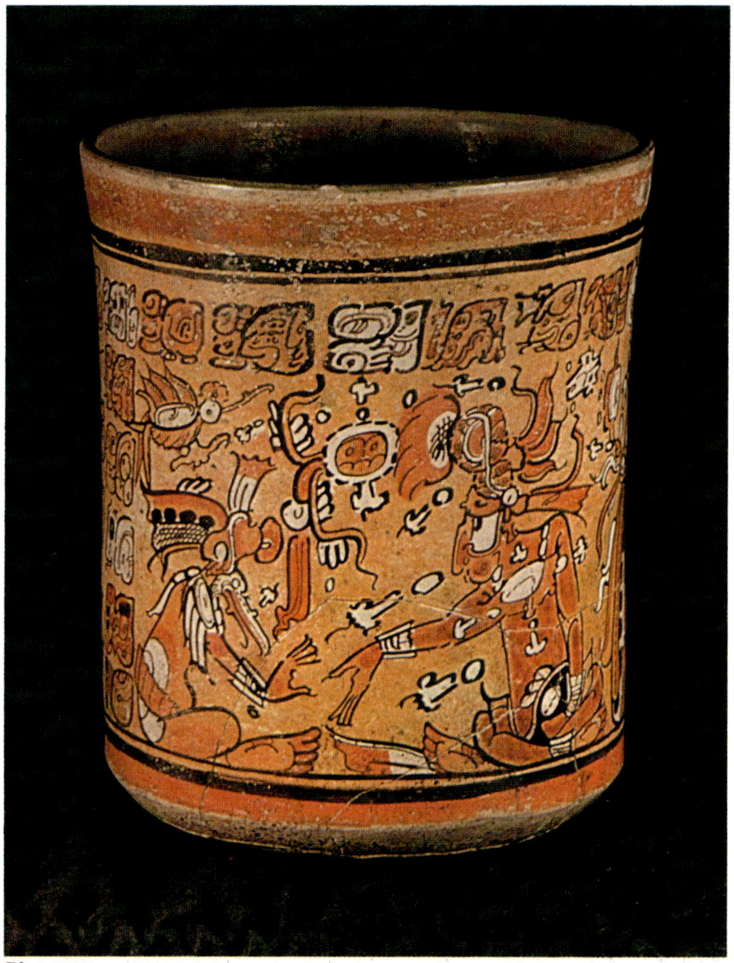

Plate 207

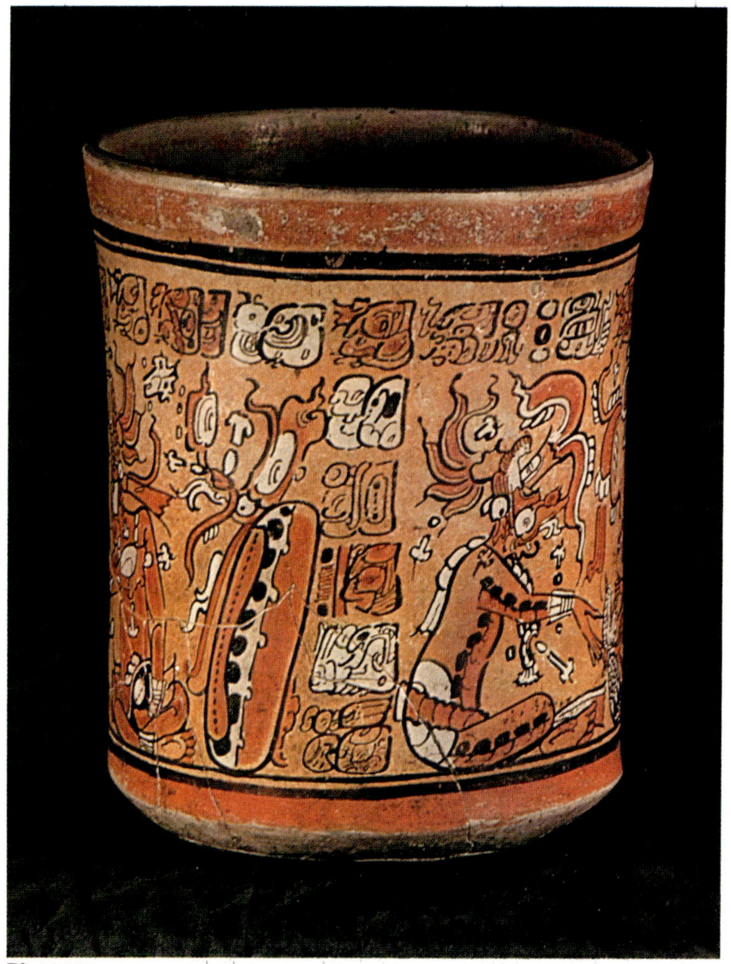

Plate 208

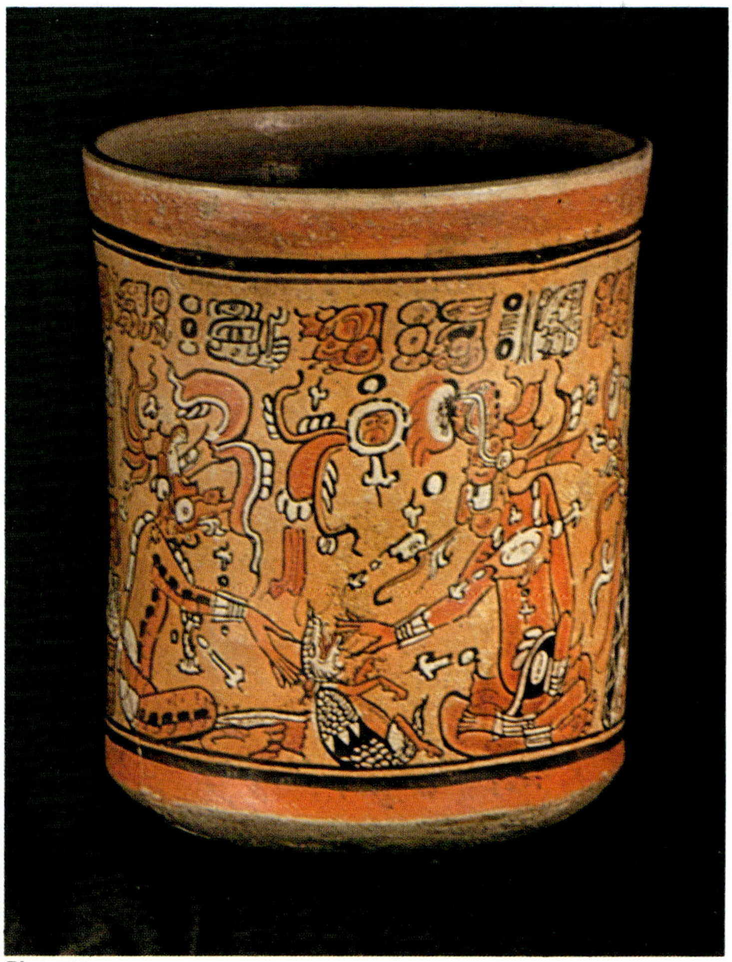

Plate 209

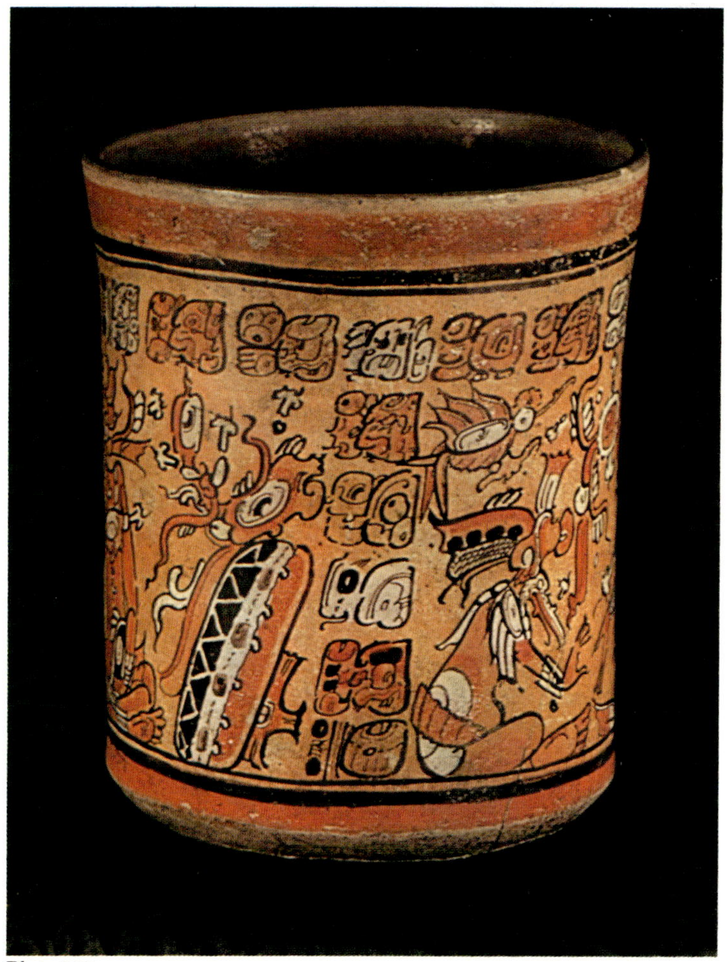

Plate 210

Plates 207-210 Vase of the Four Gods (D-7)

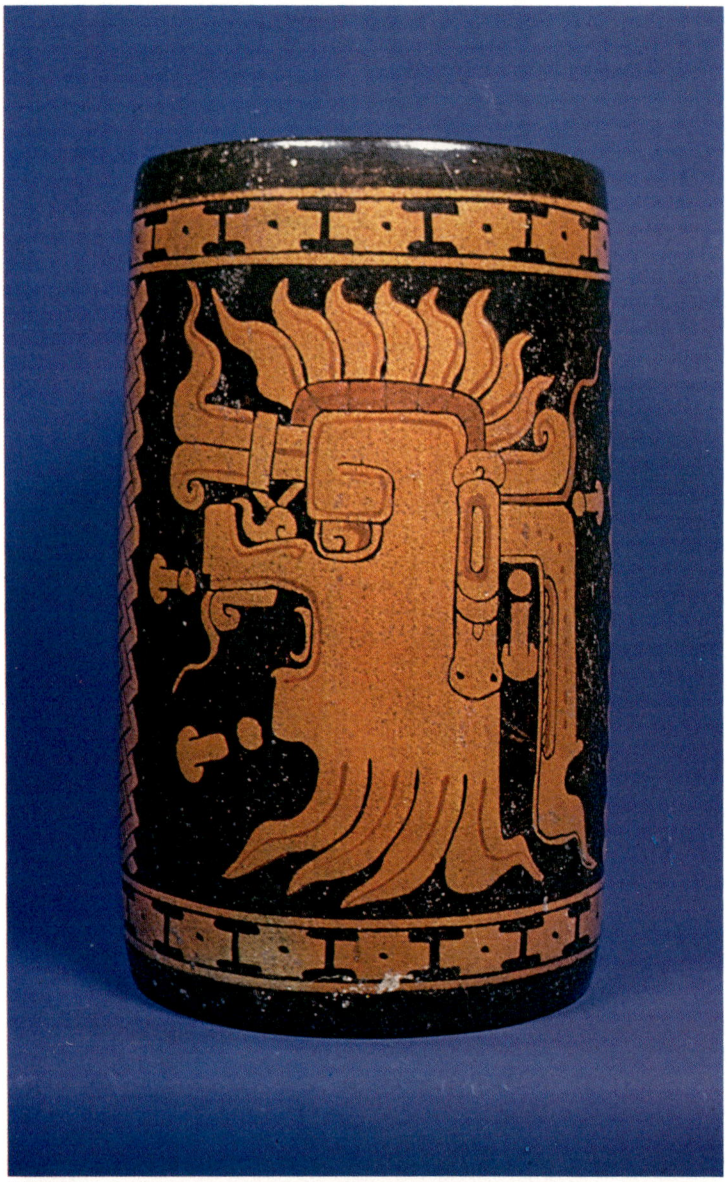

Plate 211

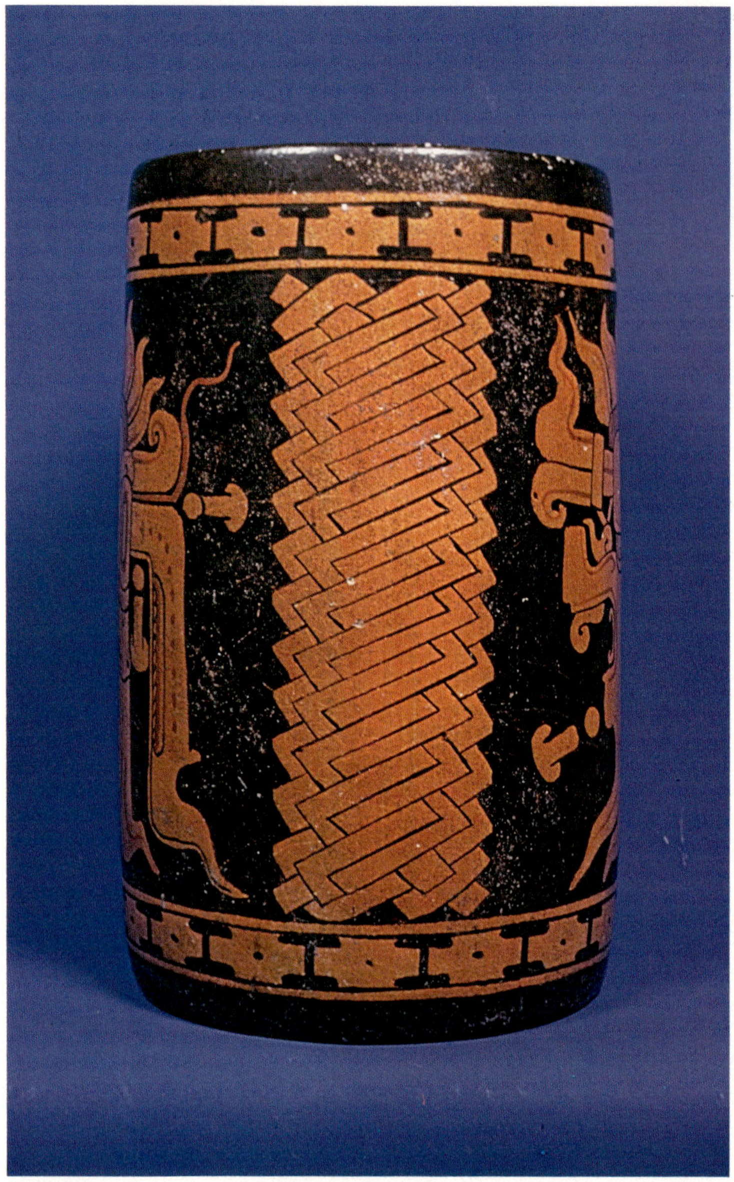

Plate 212

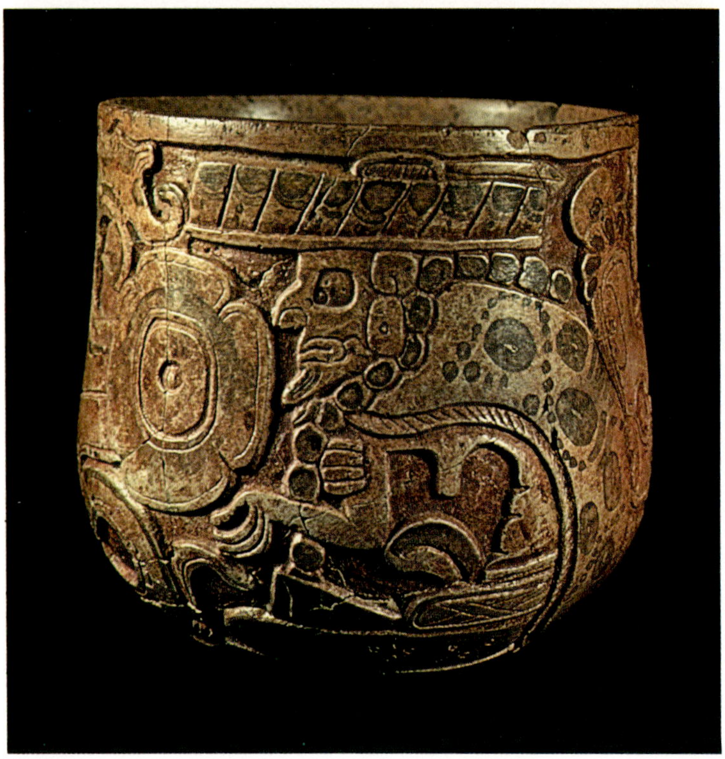

Plate 215

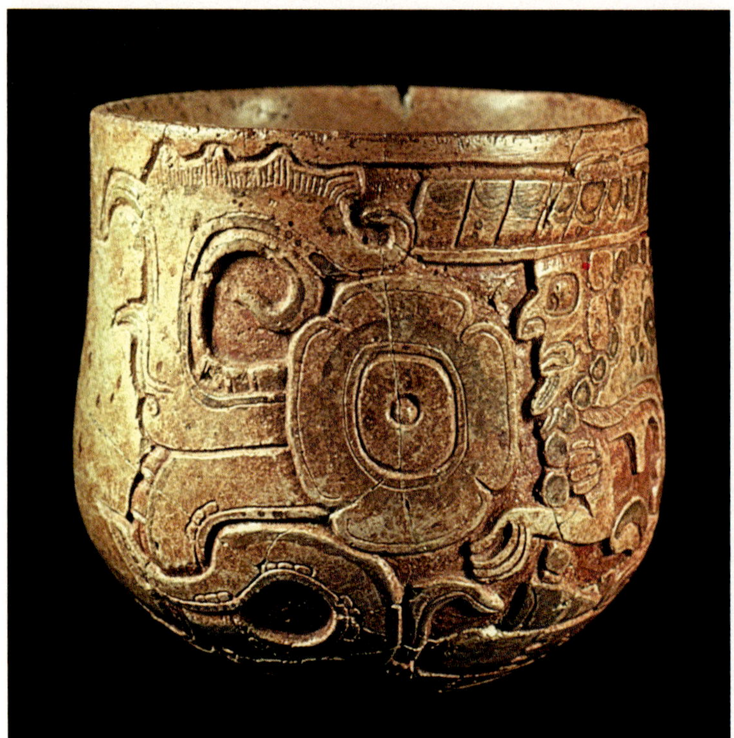

Plate 216

Plates 211-212 Vase of God K with the Flaming Hair (D-24)

Plates 215-217 Bowl of God K and God L II (D-9)

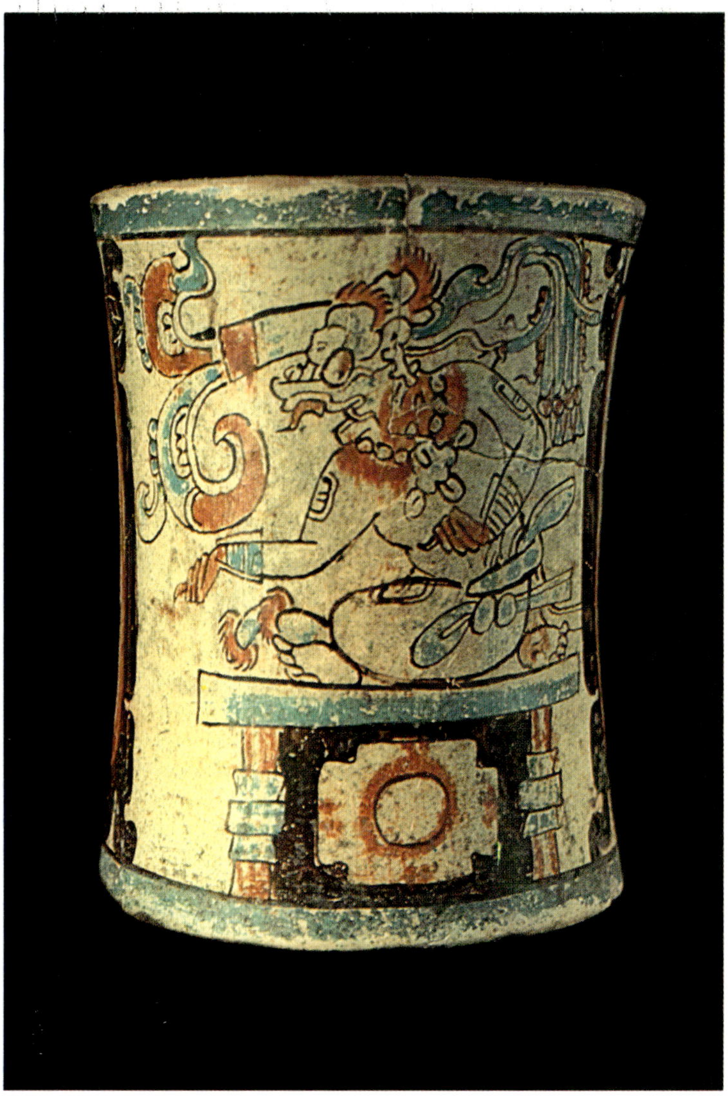

Plate 213

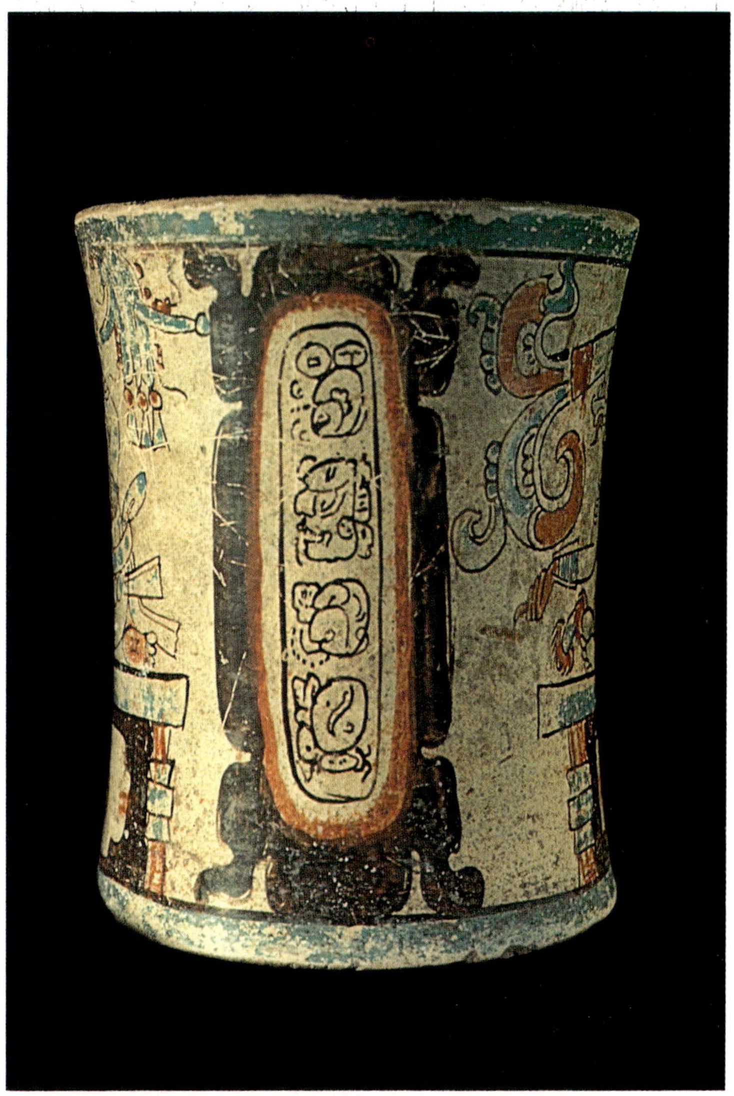

Plate 214

Plates 213-214 Vase of God K Enthroned (D-24)

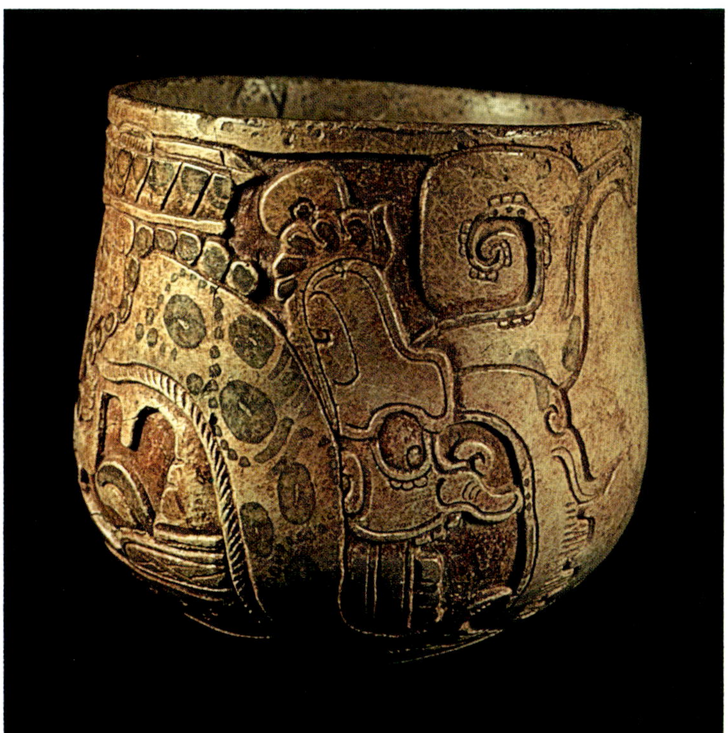

Plate 217

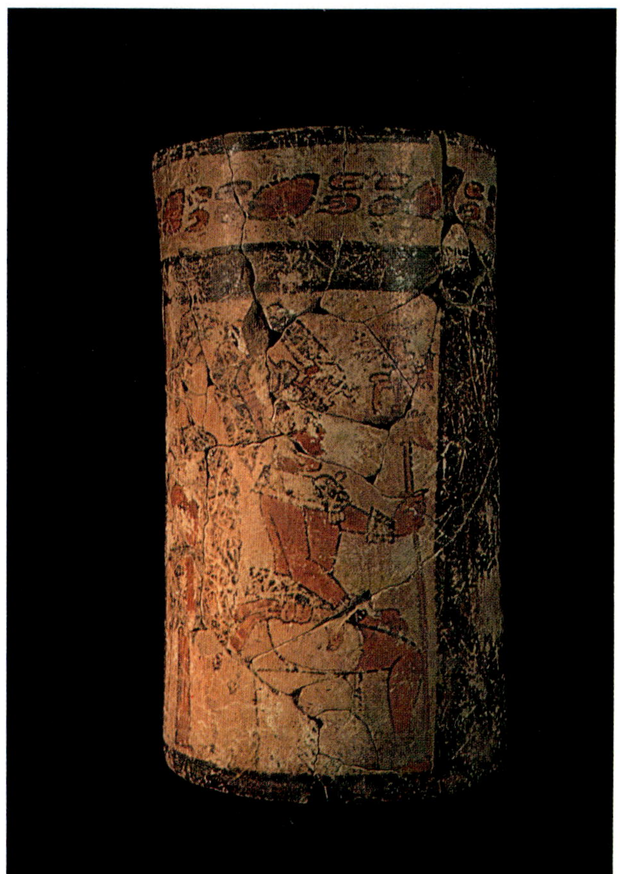

Plate 218 Vase of the Lord Seated with Manikin Staff (D-19)

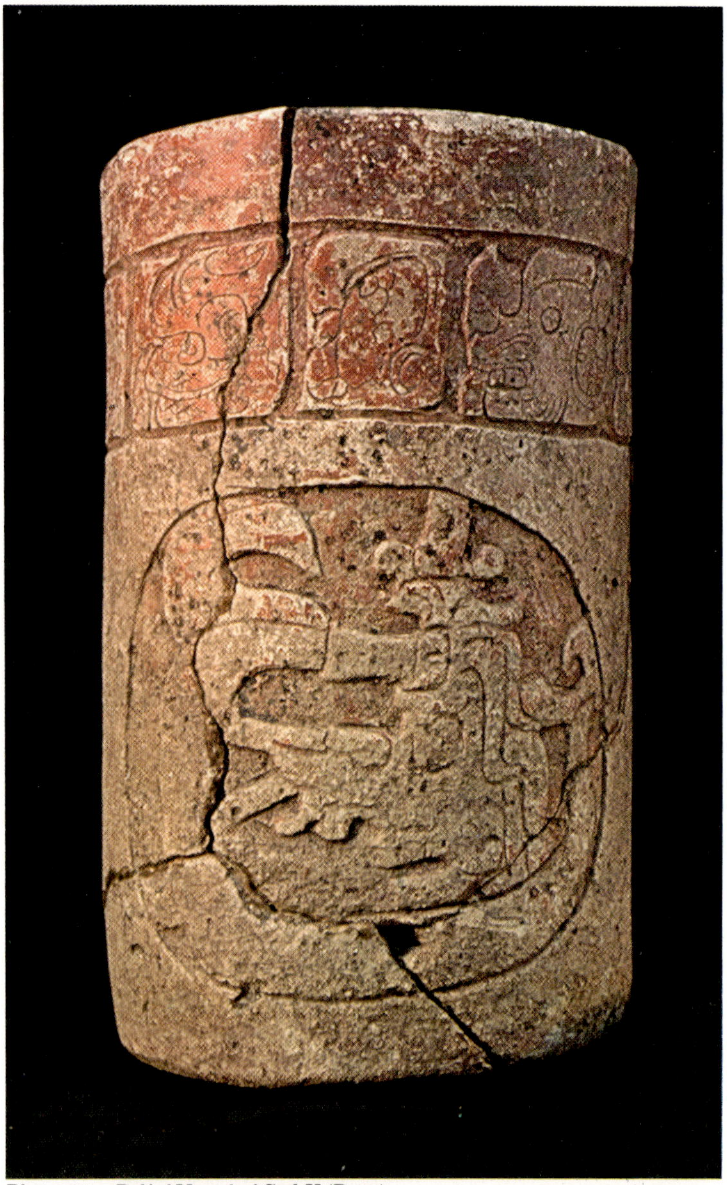

Plate 219 Relief Vessel of God K (D-20)

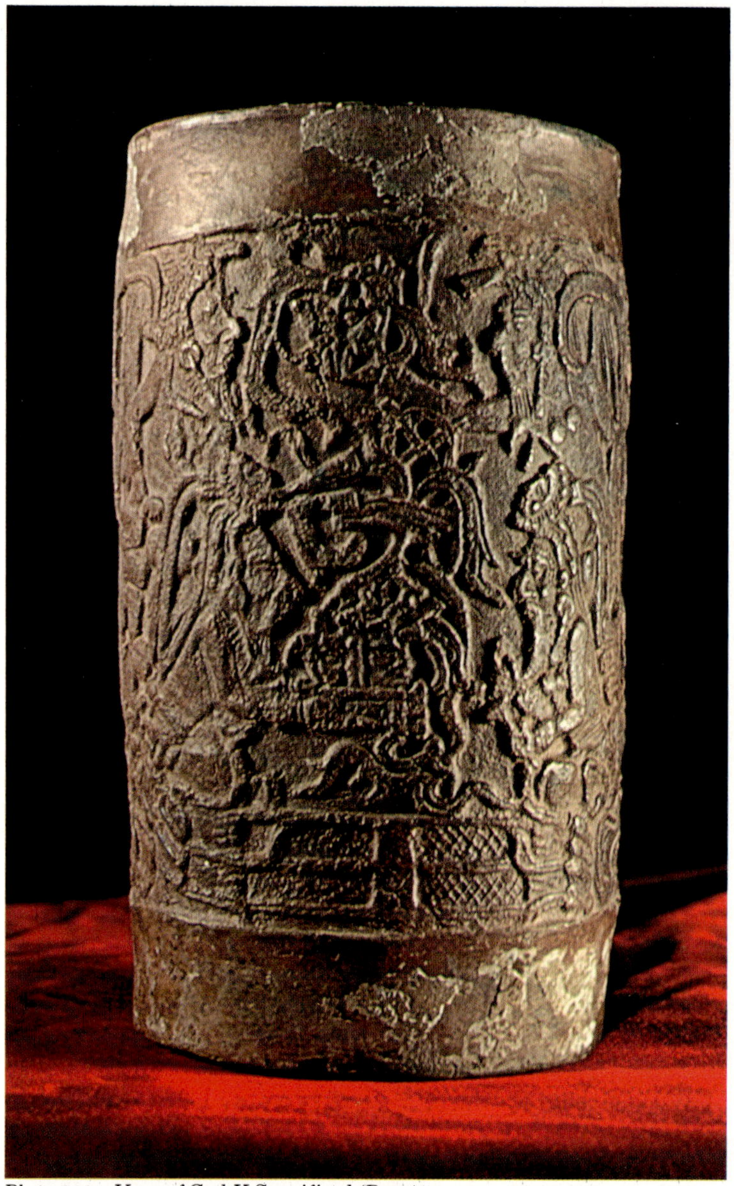

Plate 220 Vase of God K Sacrificed (D-22)

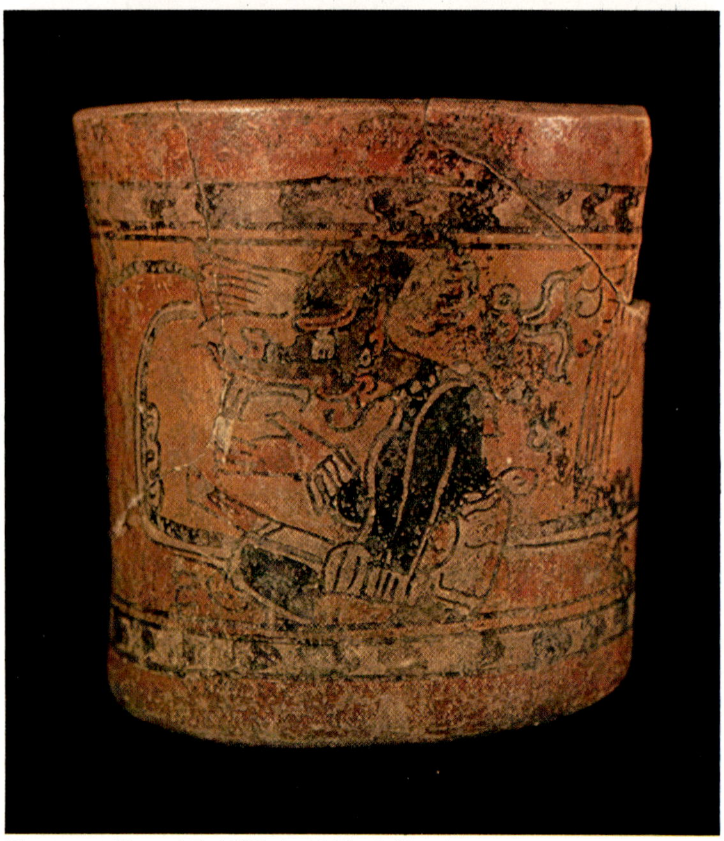

Plate 221 Vase of God K Painted Black (D-12)

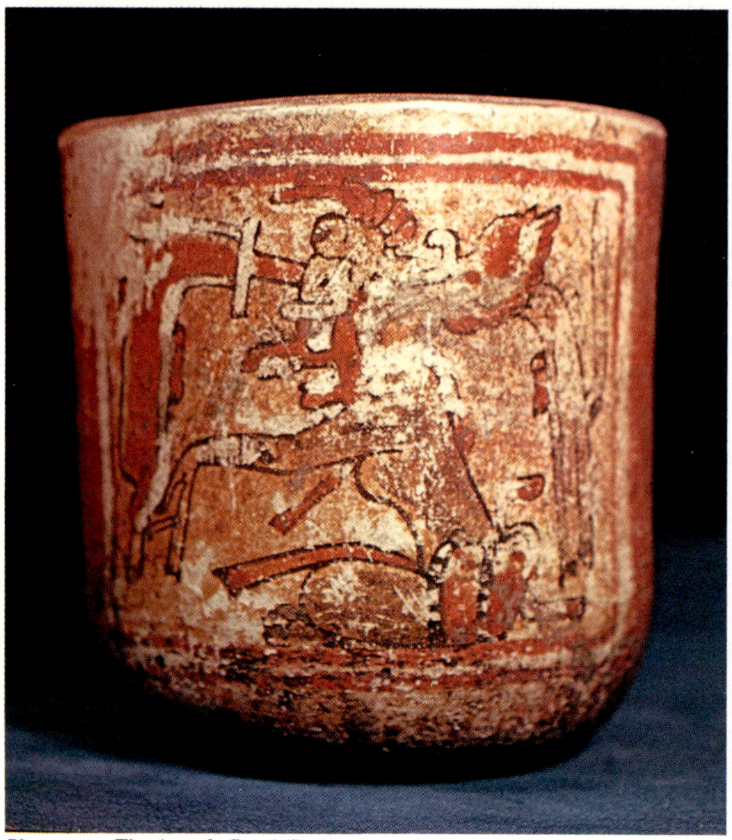

Plate 222 The Amado Bowl (D-13)

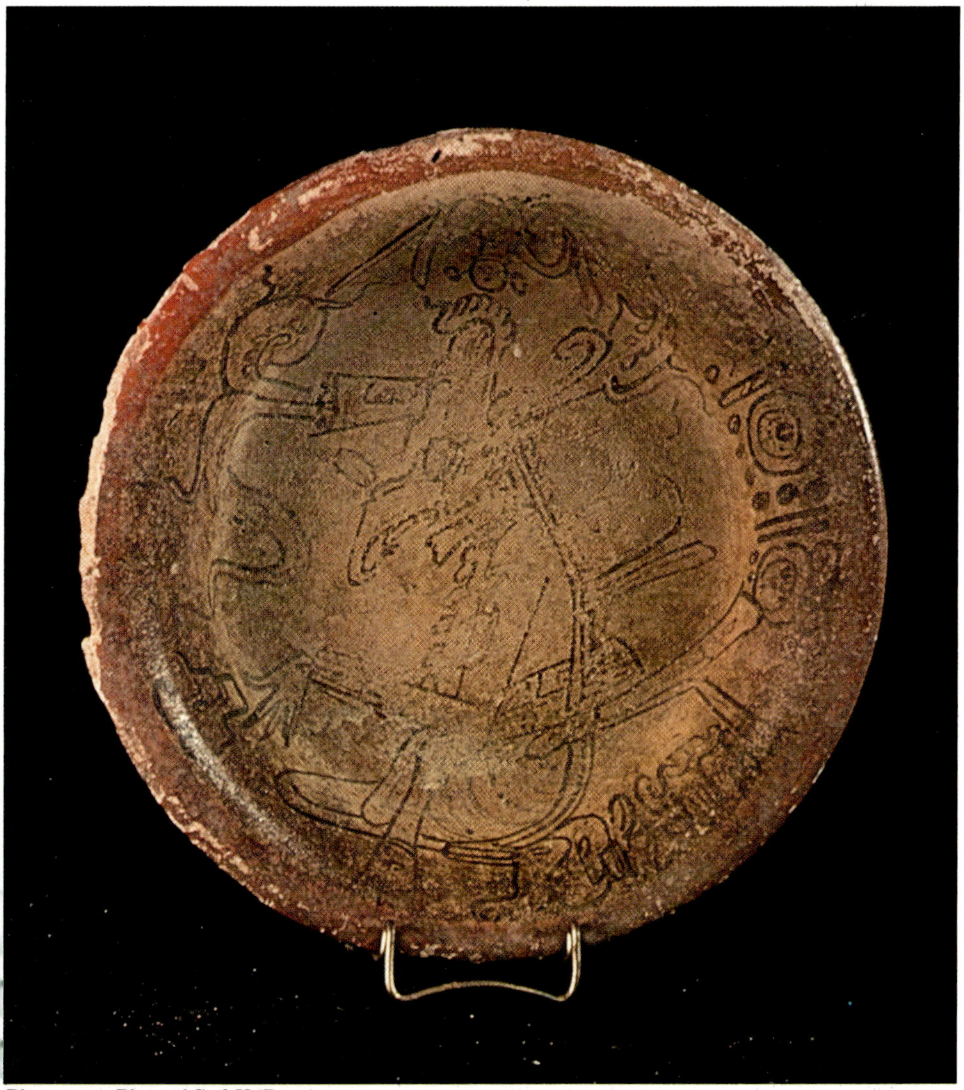

Plate 223 Plate of God K (D-17)

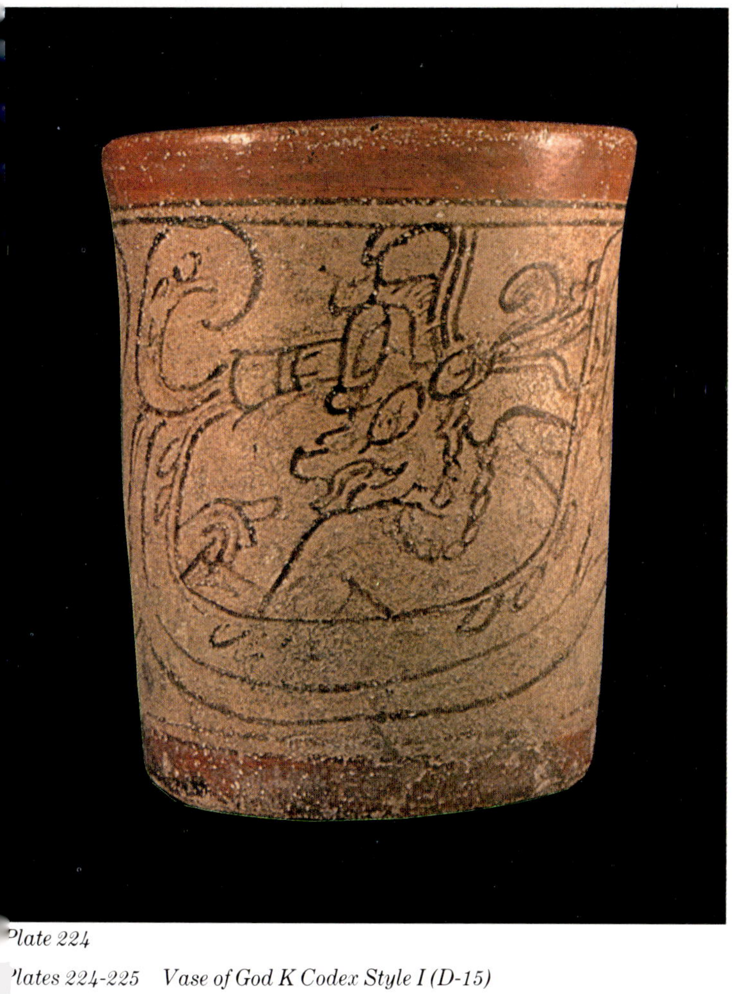

Plate 224

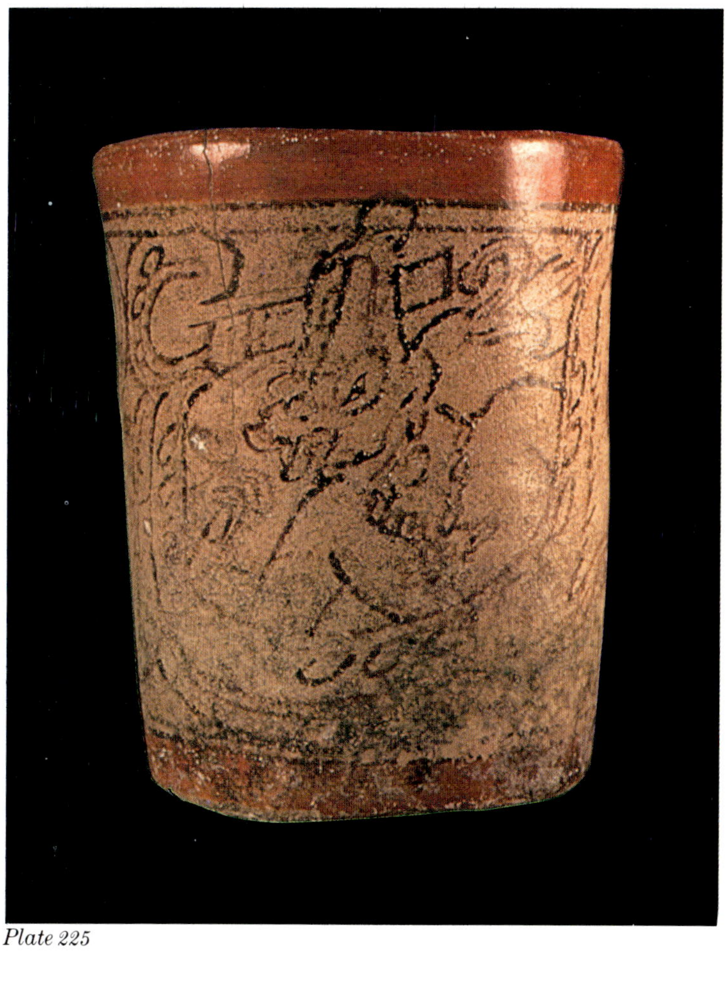

Plate 225

Plates 224-225 Vase of God K Codex Style I (D-15)

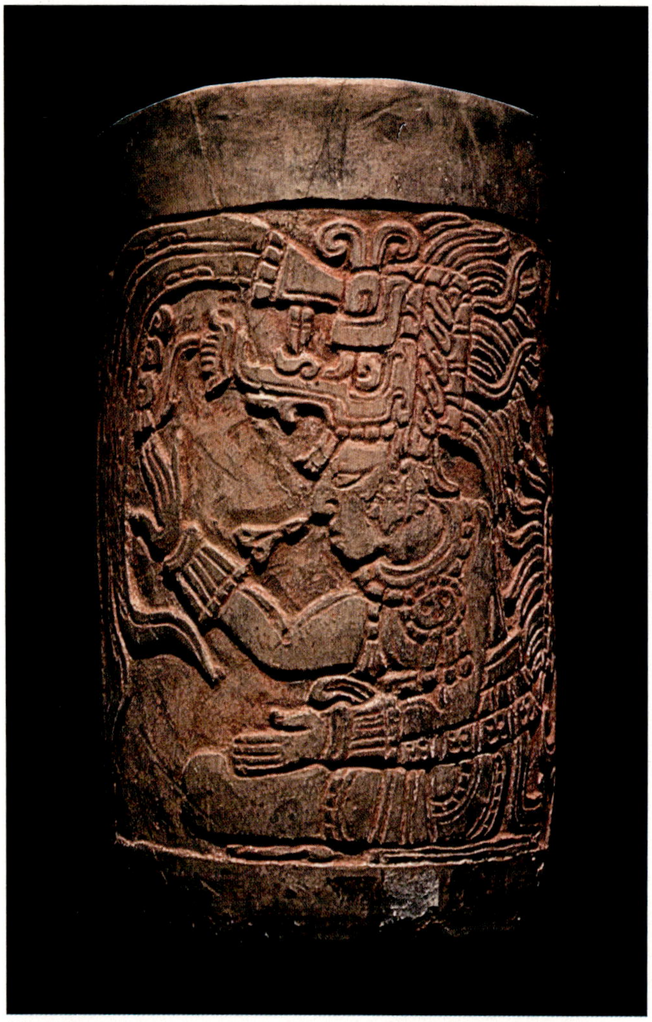

Plate 226

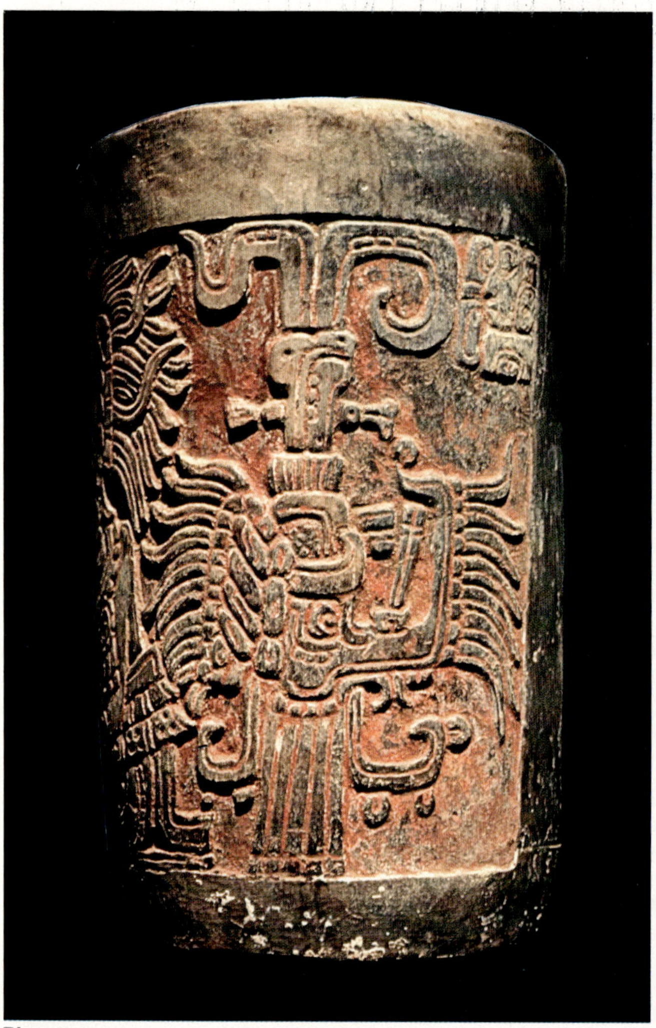

Plate 227

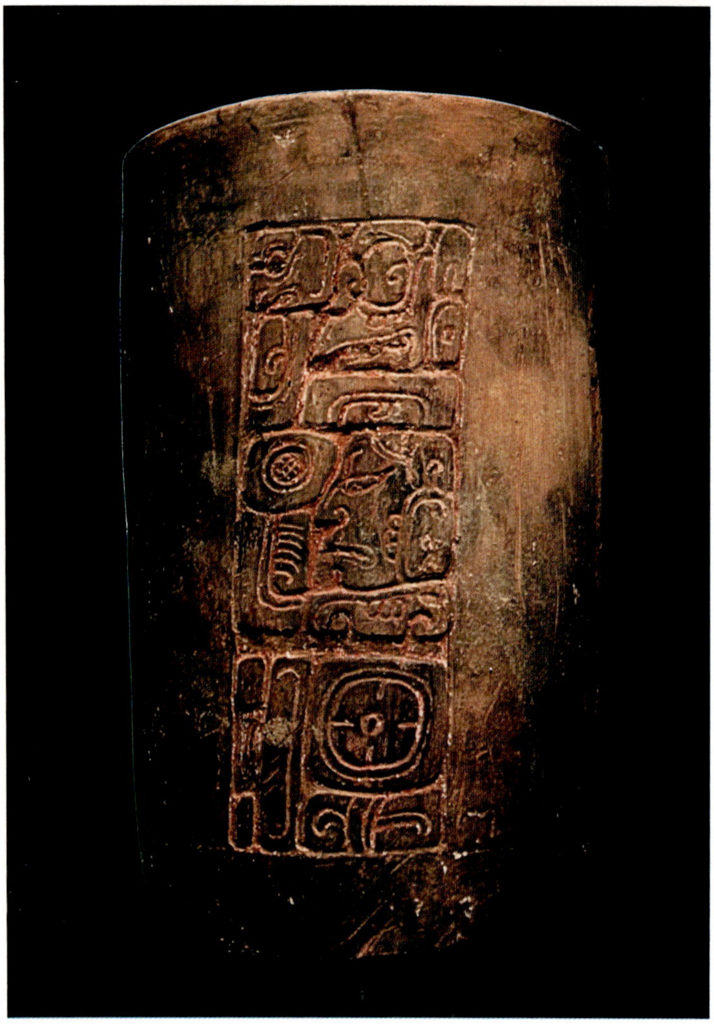

Plate 228

Plates 226-228 Vase of the Ruler wearing God K Helmet (D-35)

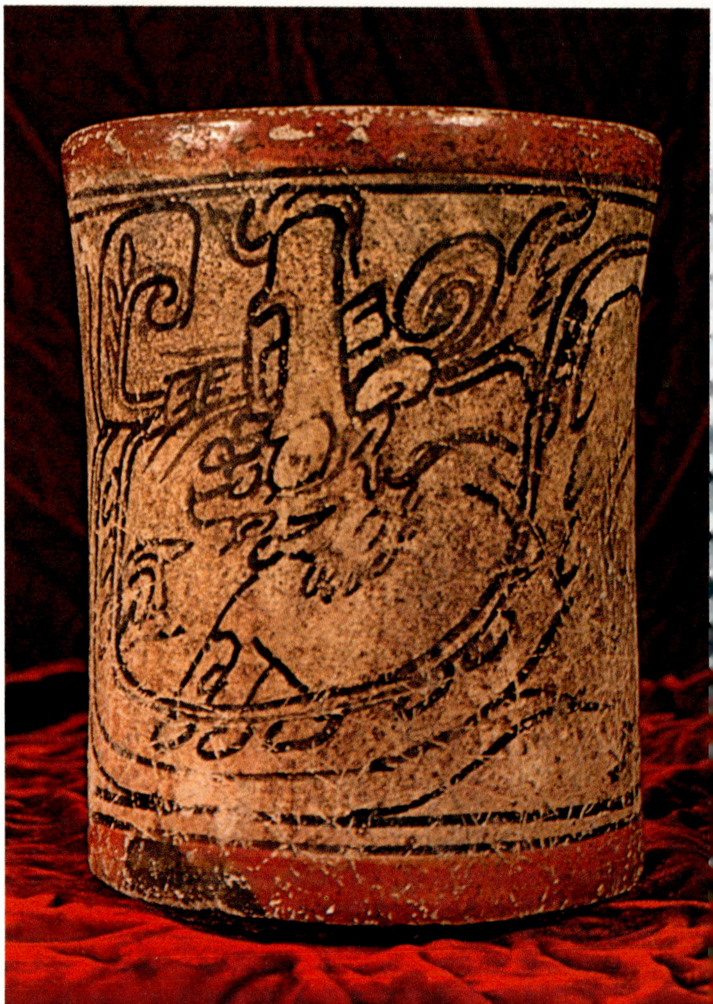

Plate 229 Vase of God K Codex Style II (D-16)

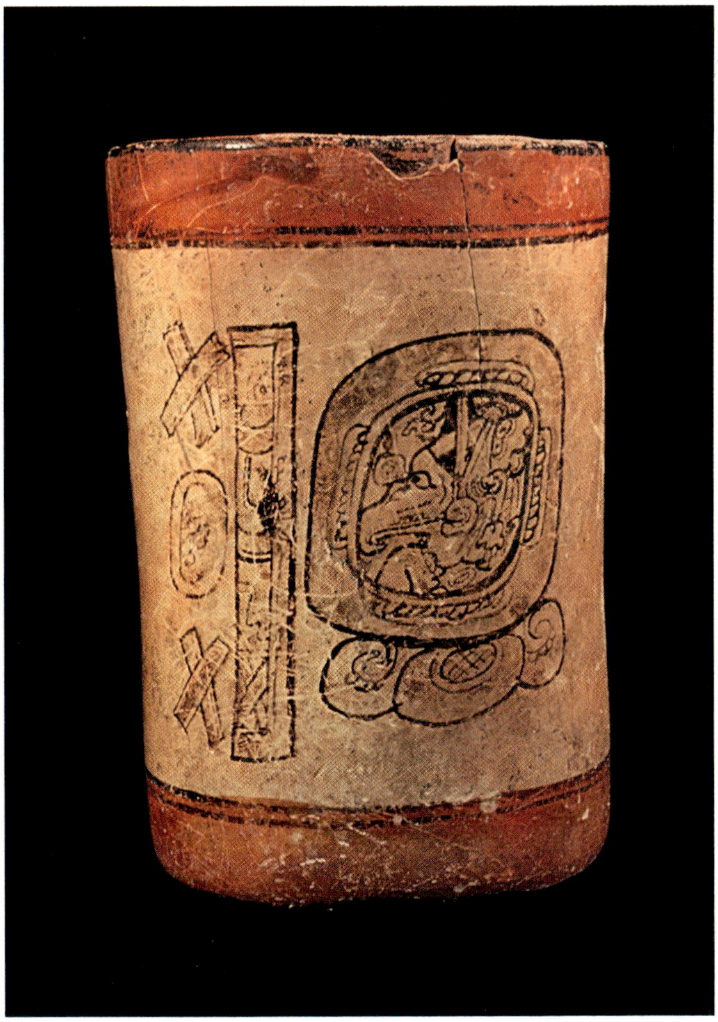

Plate 230

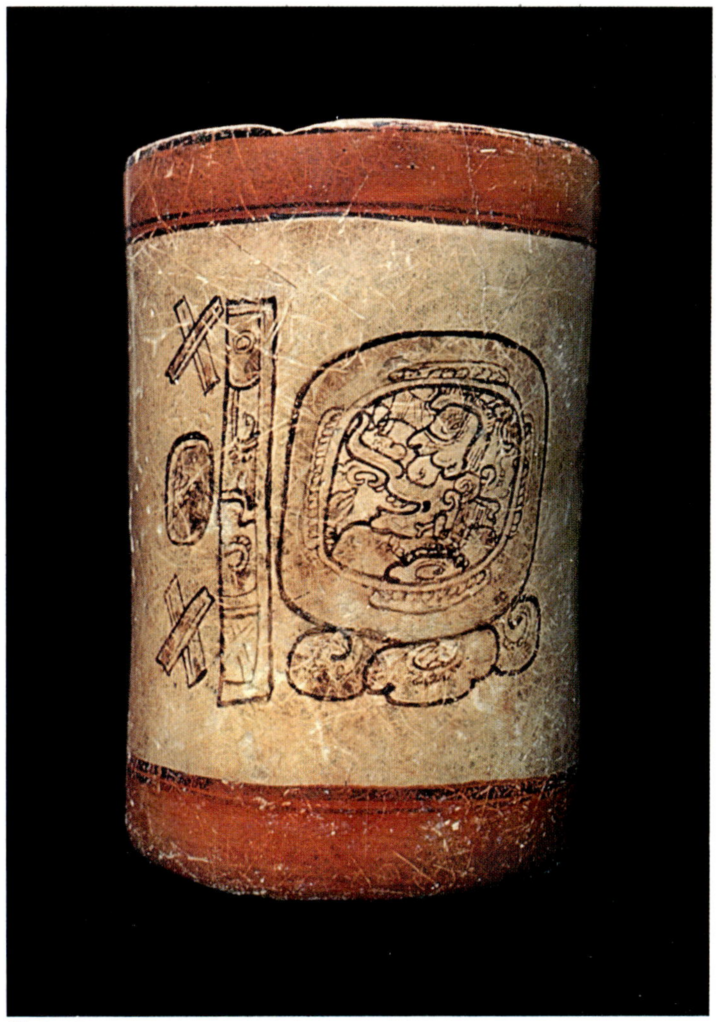

Plate 231

Plates 230-231 Vase of the Two Day Glyphs (D-14)

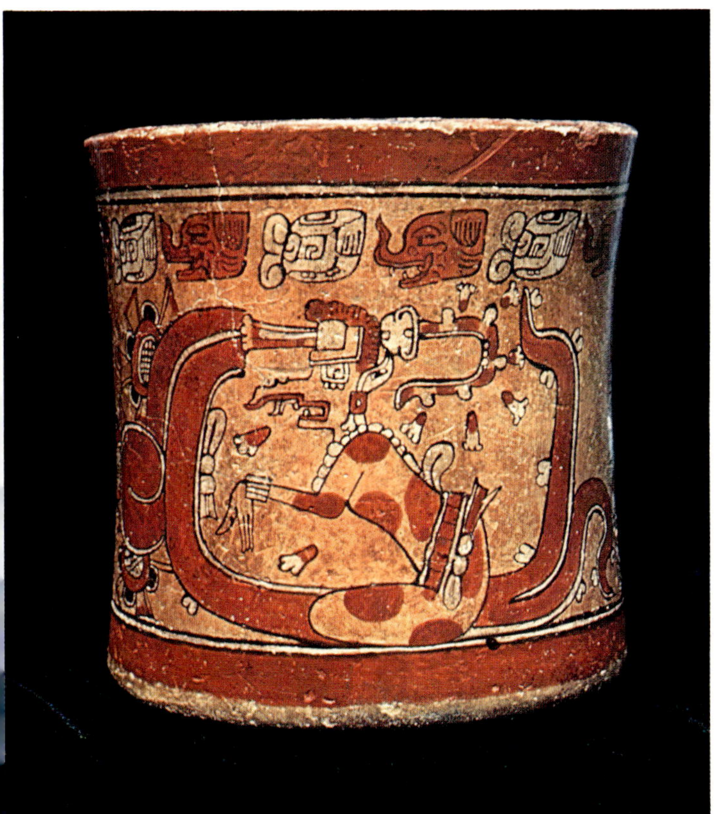

Plate 232 Vase of God K Seated in a Smoke Curl (D-21)

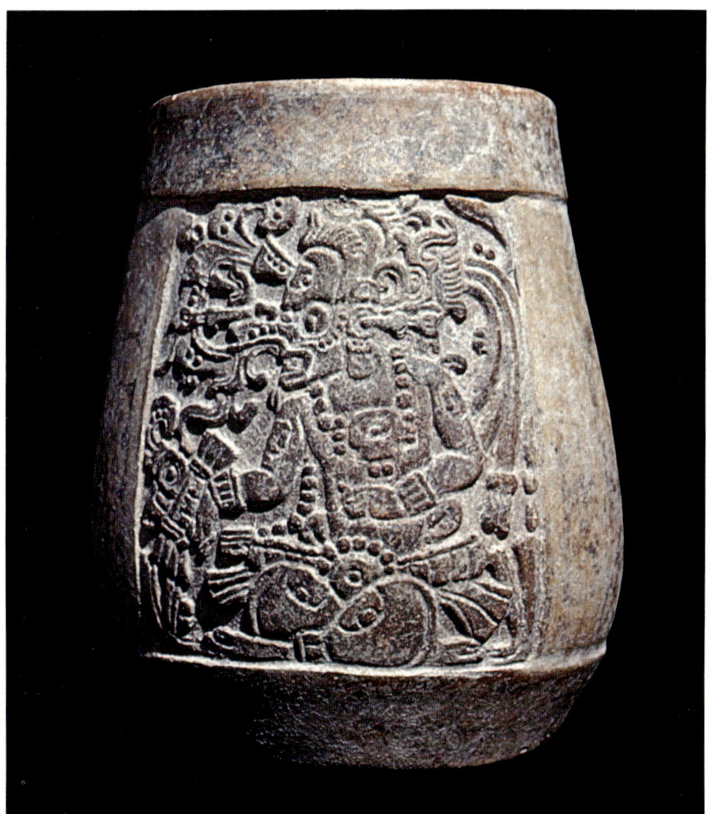

Plate 233 Vase of God K from San Agustin Acasaguastlan (D-18)

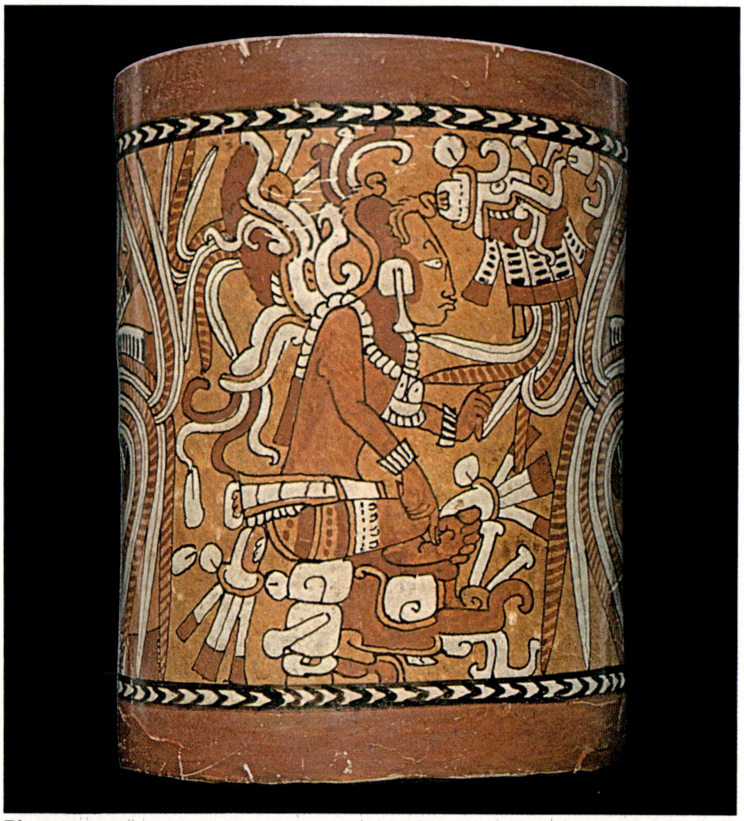

Plate 234

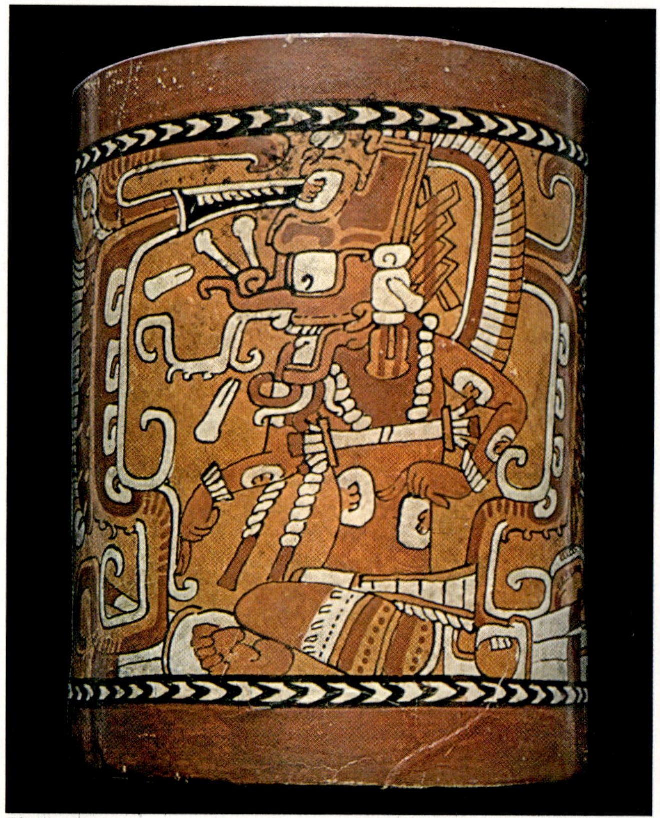

Plate 235

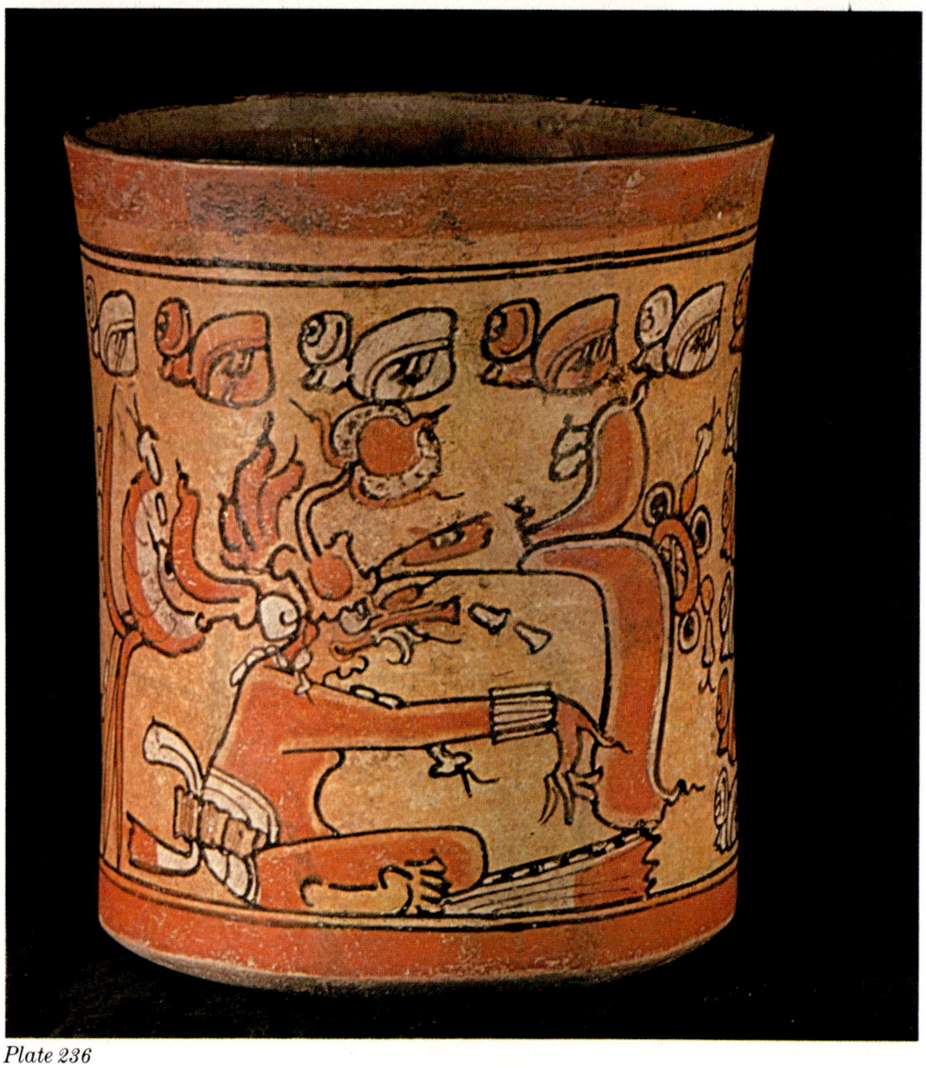

Plate 236

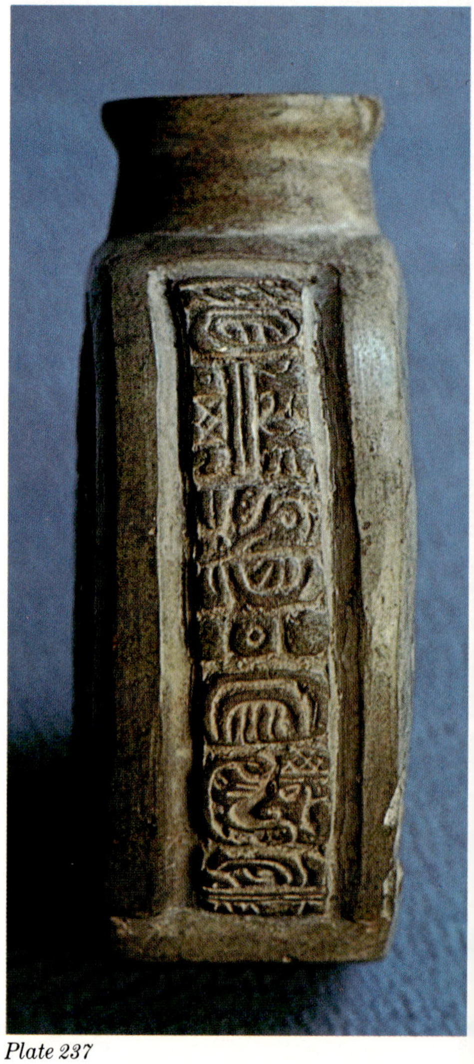

Plate 237

Plates 234-235 Vase of God K and the Young God (D-26)

Plate 236 Vase of the Two Flare Gods (D-3)

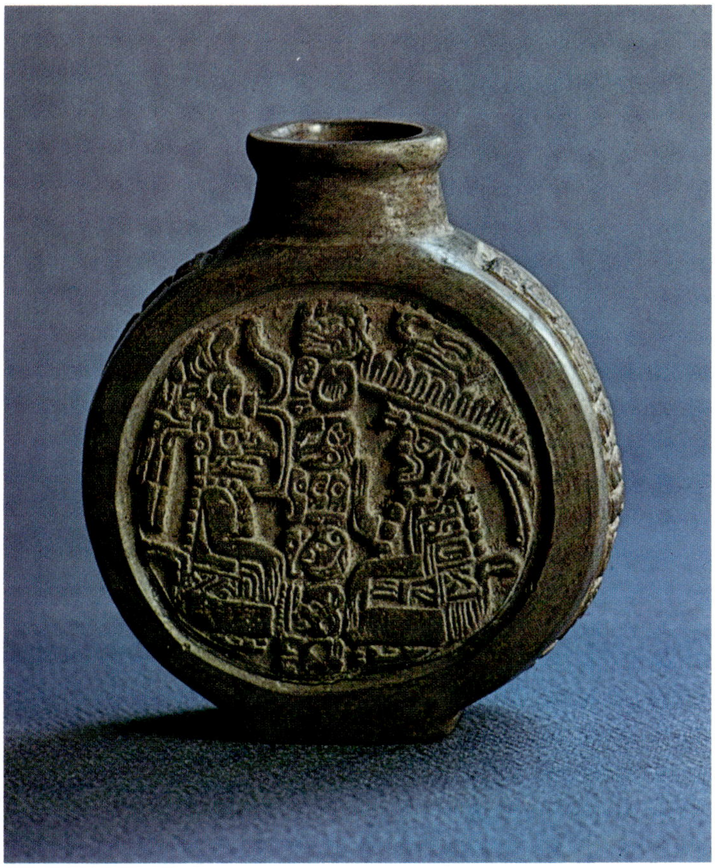

Plate 238

Plates 237-238 Poison Flask of God K and God L (D-30) /Photographs courtesy Middle American Research Institute/

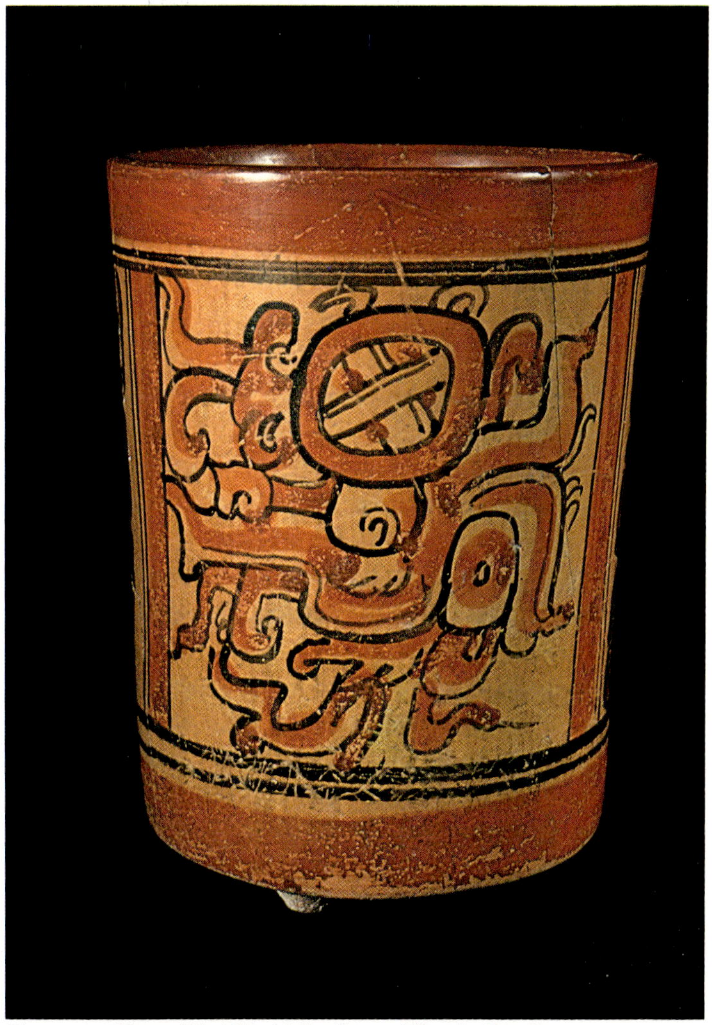

Plate 239 Vase of God K with the Crossed Bands (D-10)

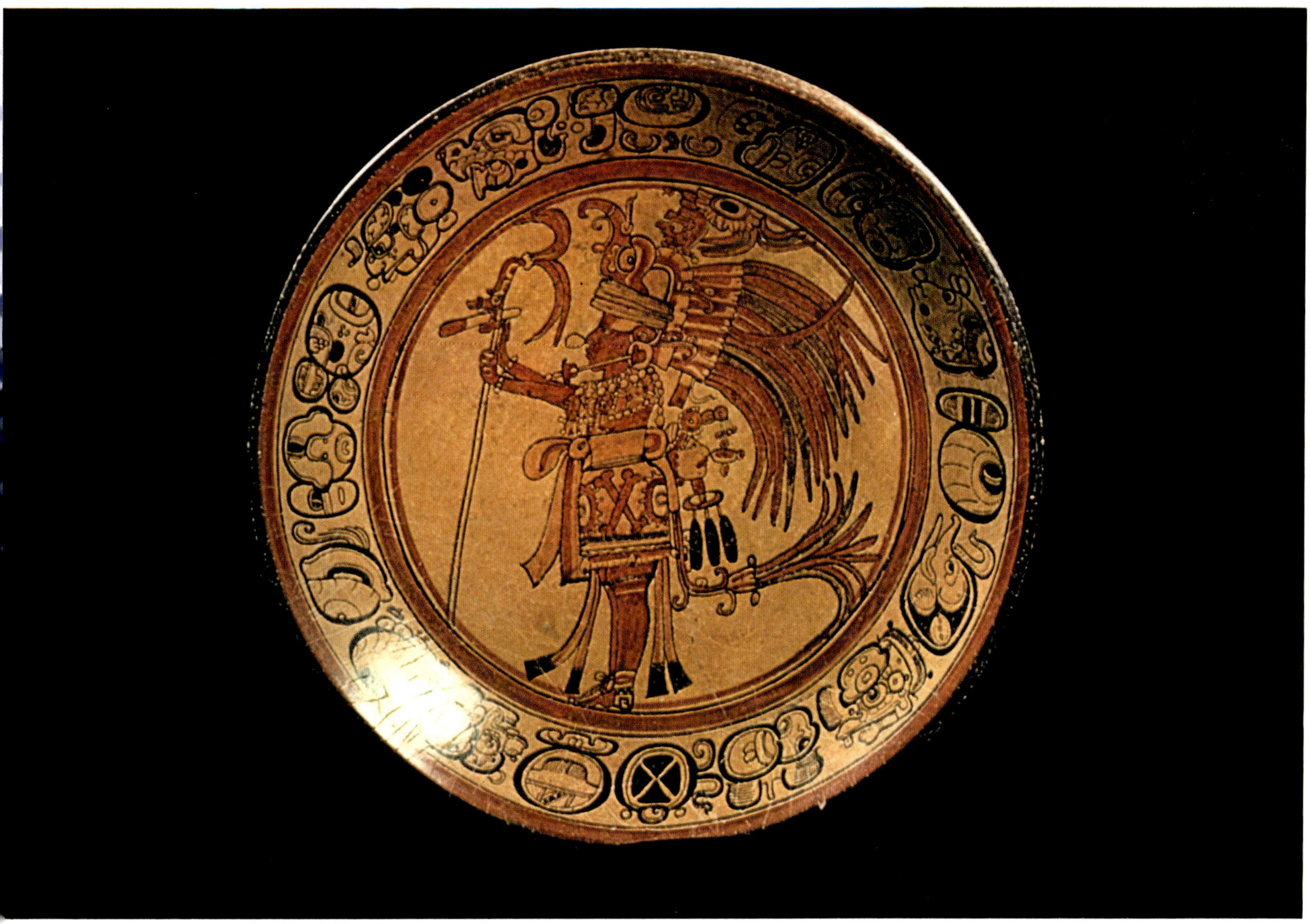

Plate 240 Plate of the Standing Lord (D-23)

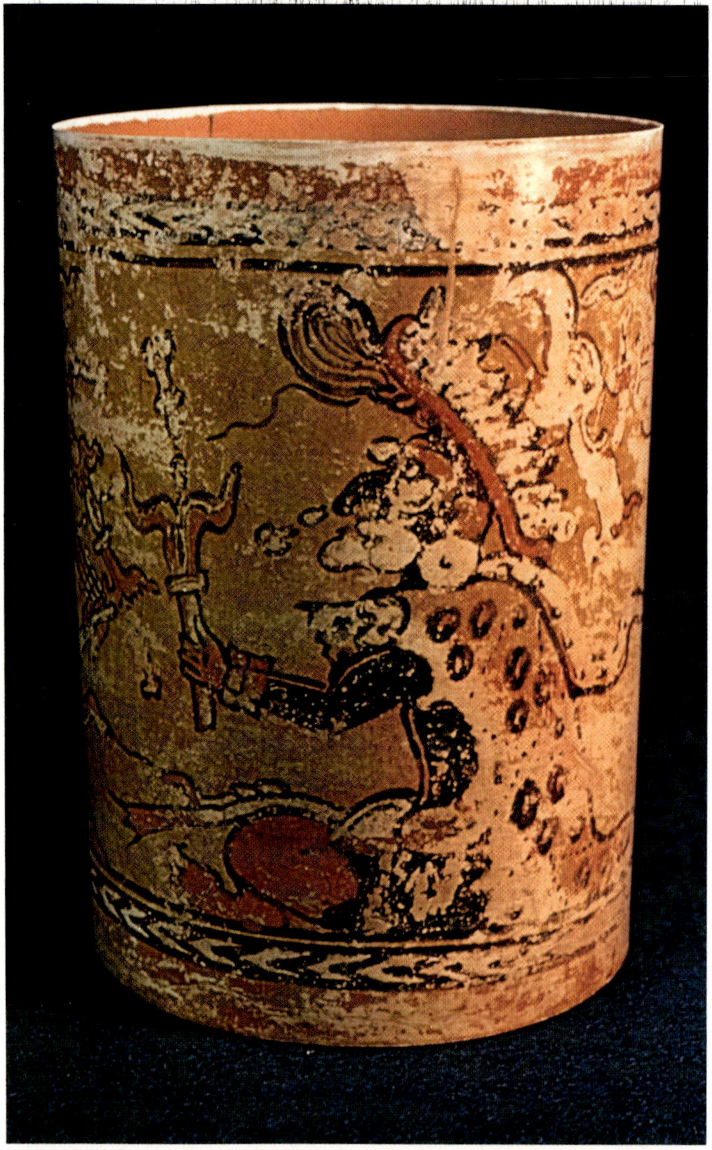

Plate 241

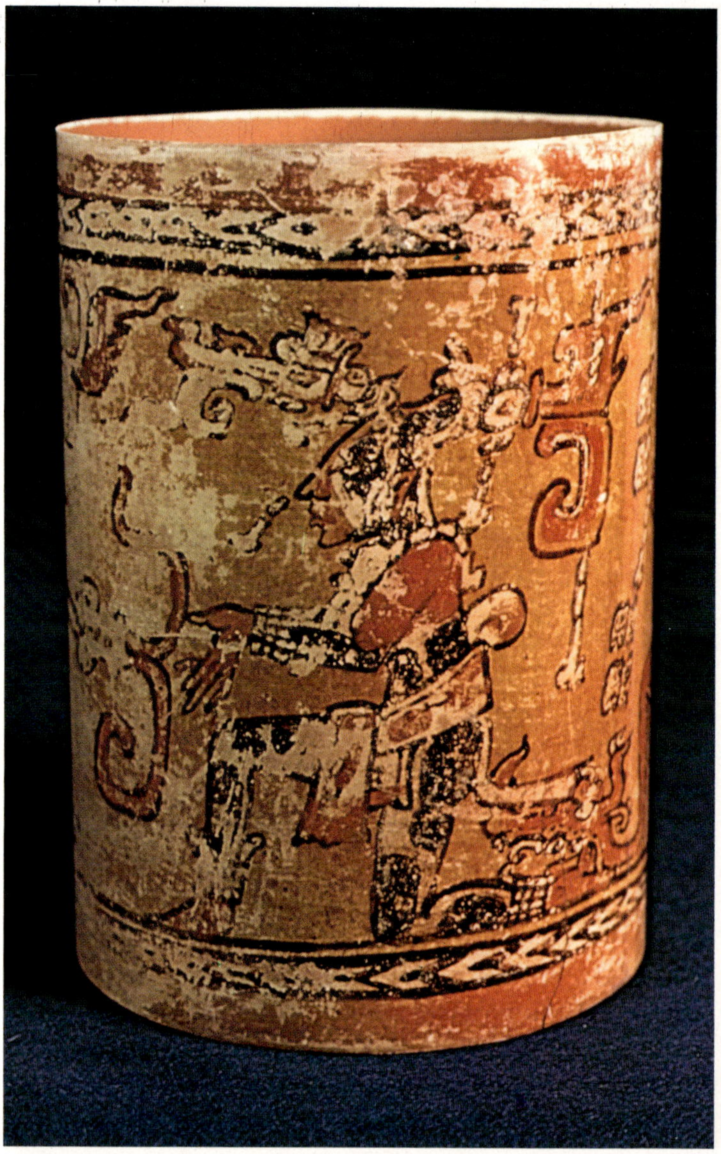

Plate 242

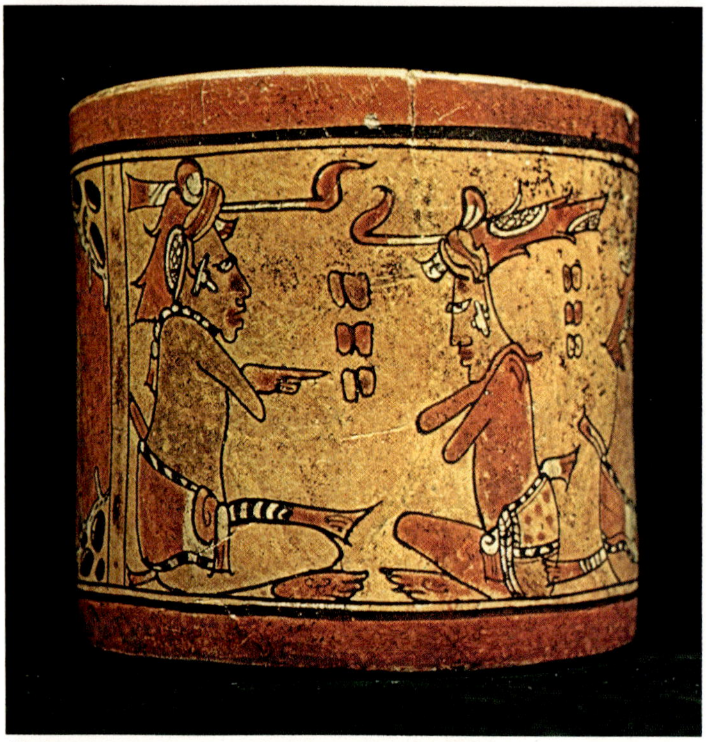

Plate 245

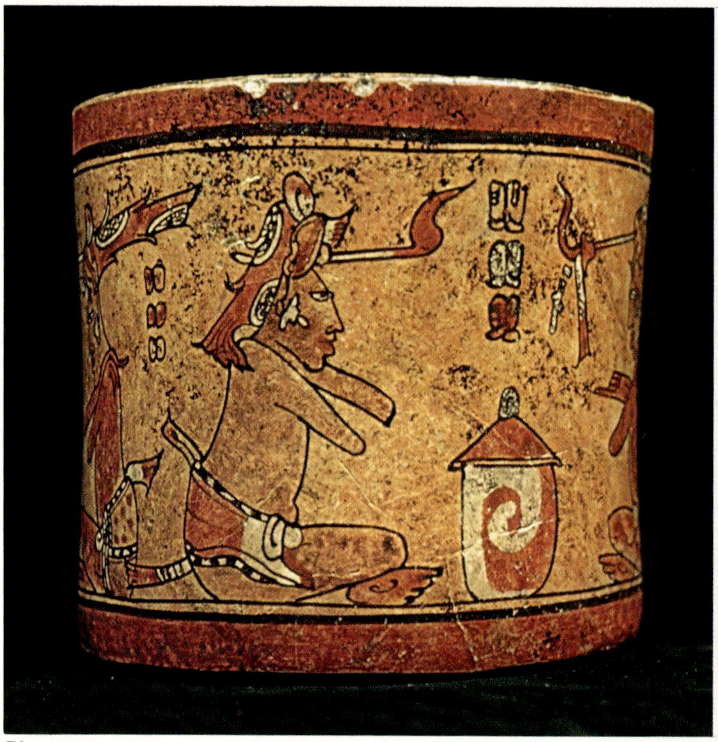

Plate 246

Plates 241-243 Vase of God K, the Young God and God L (D-29)

Plates 245-248 Vase of the Four Lords (E-4)

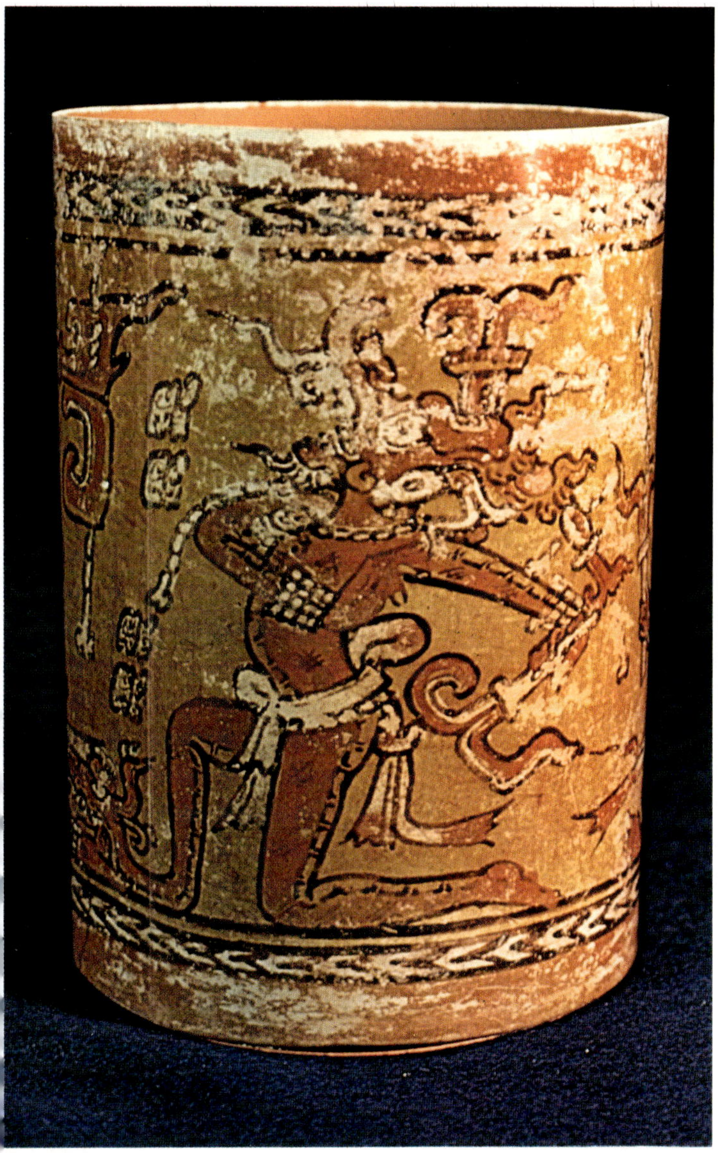

Plate 243

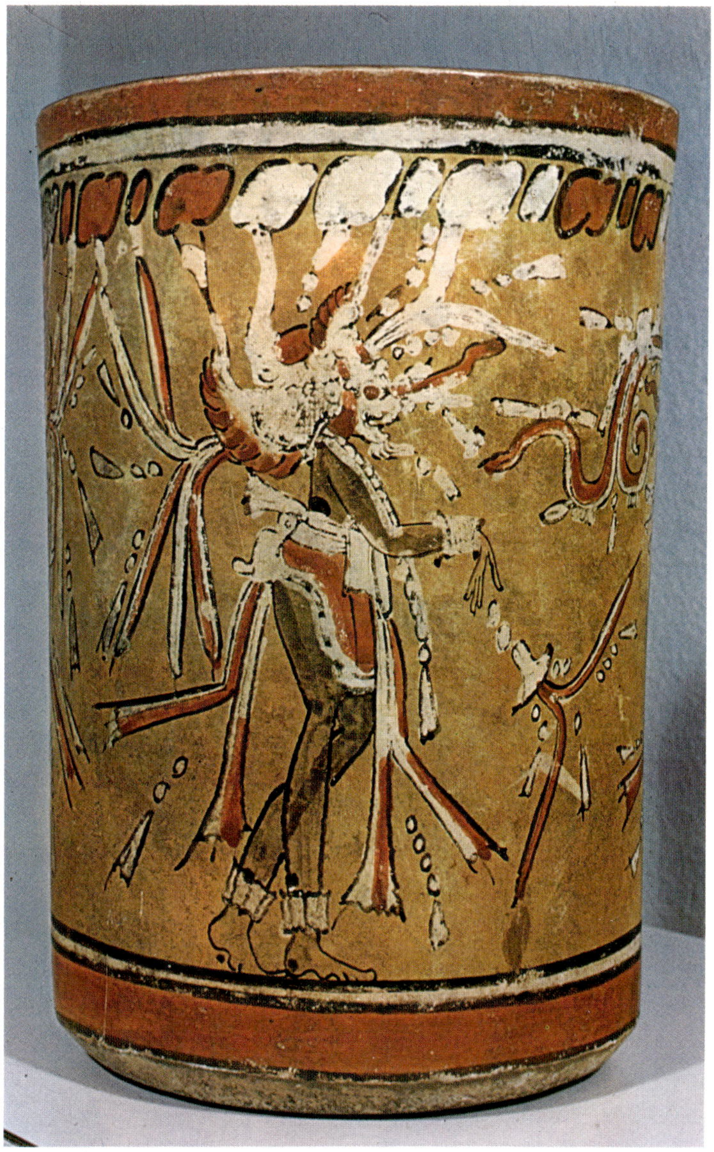

Plate 244   Vase of the Four Standing Monster Gods (D-34)

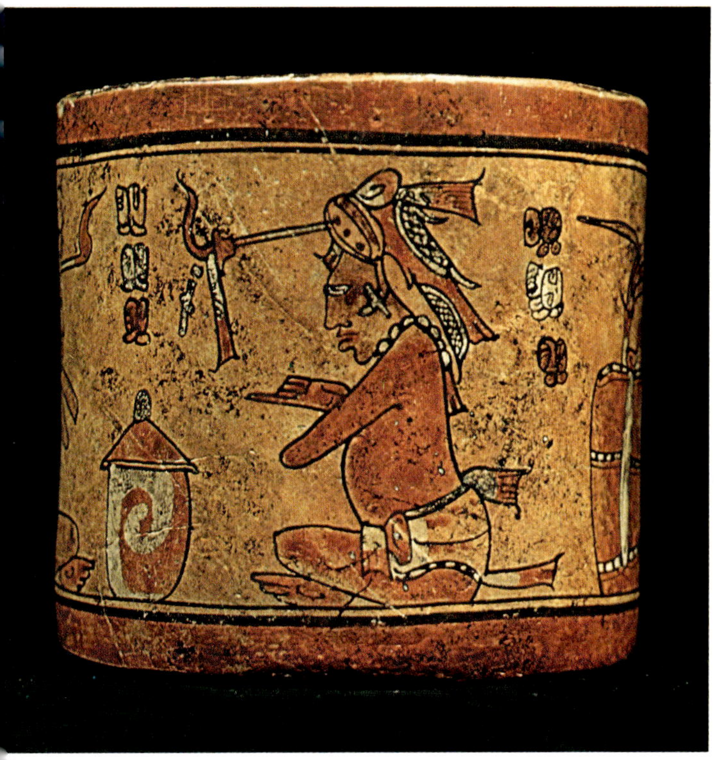

Plate 247

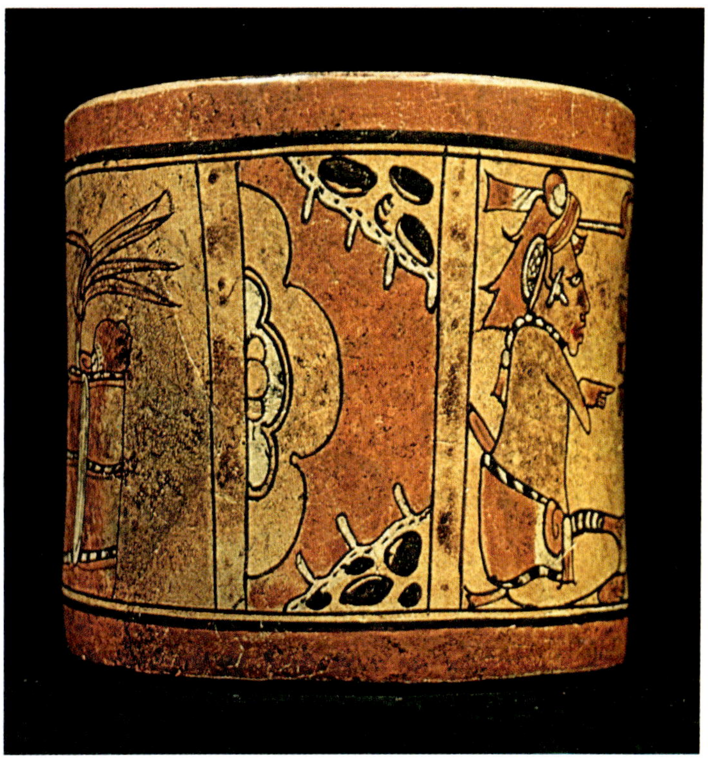

Plate 248

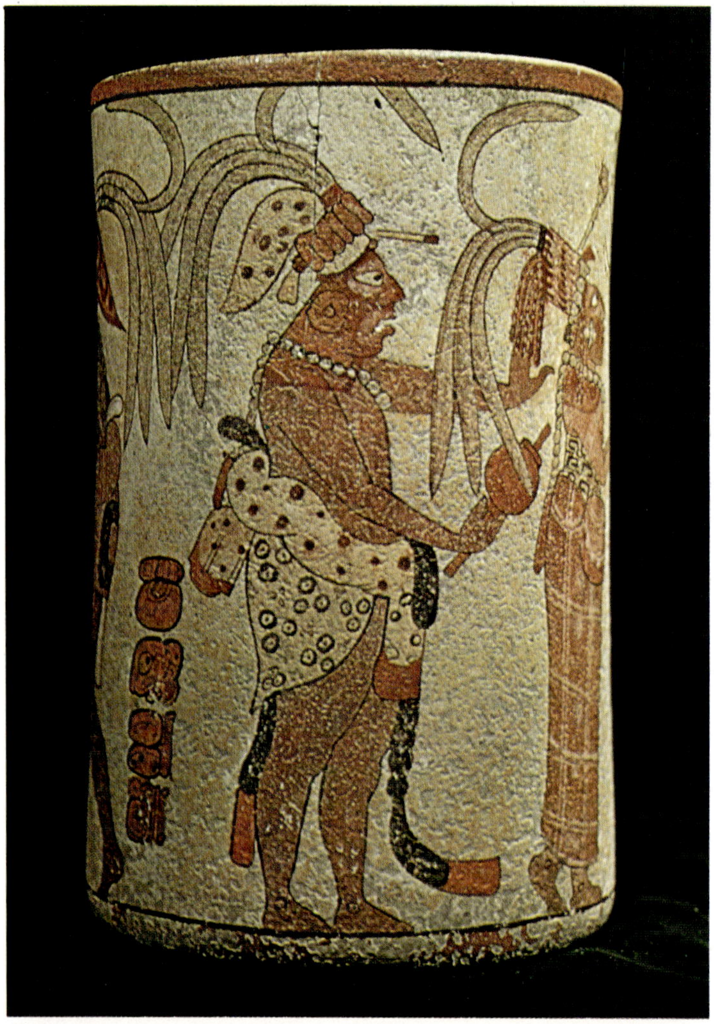

Plate 249

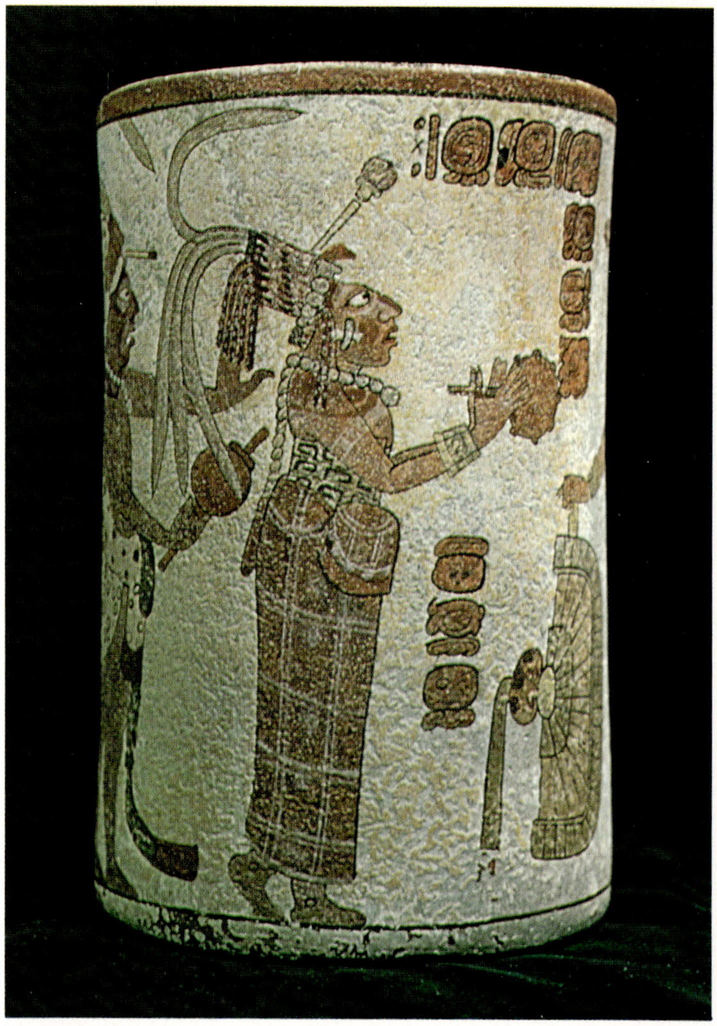

Plate 250

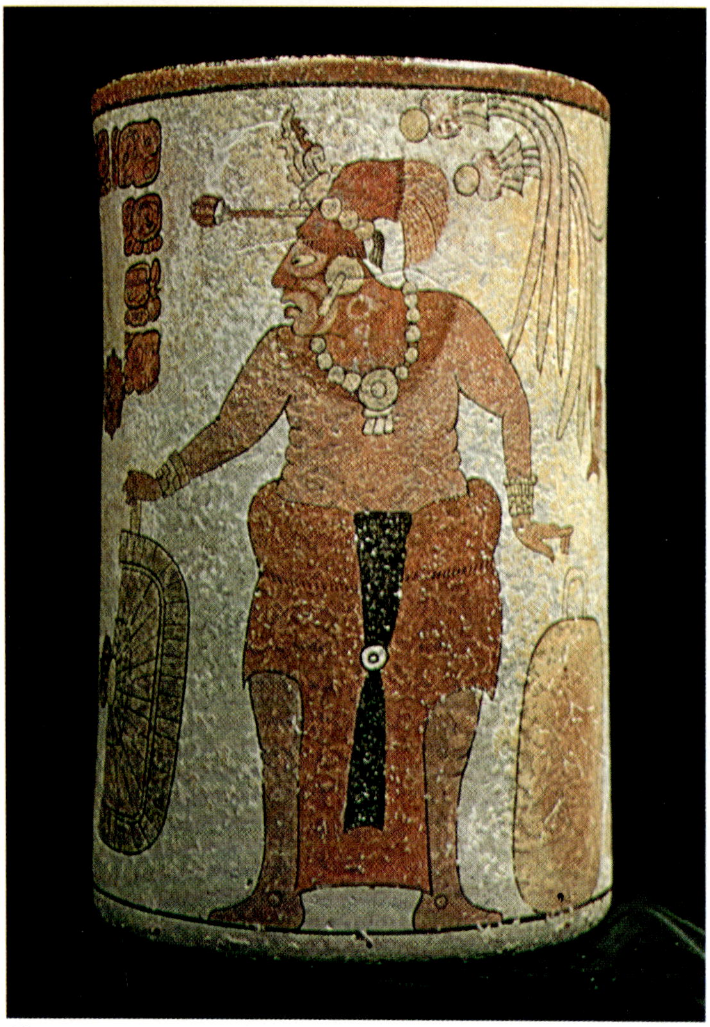

Plate 251

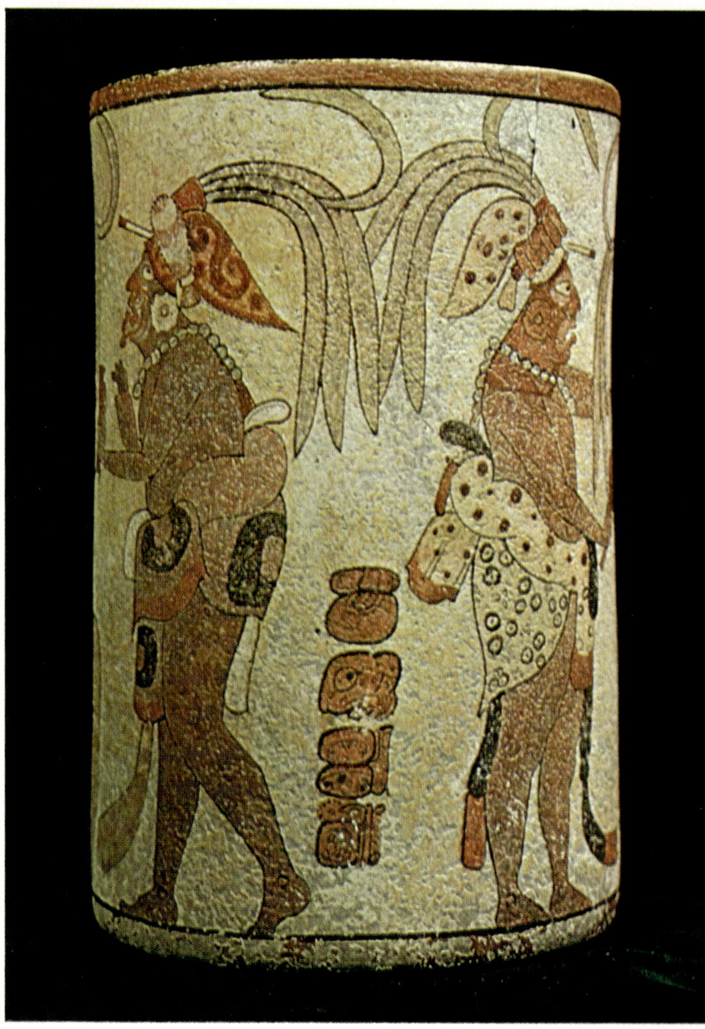

Plate 252

Plates 249-252 Vase of the Court Attendants (E-1)

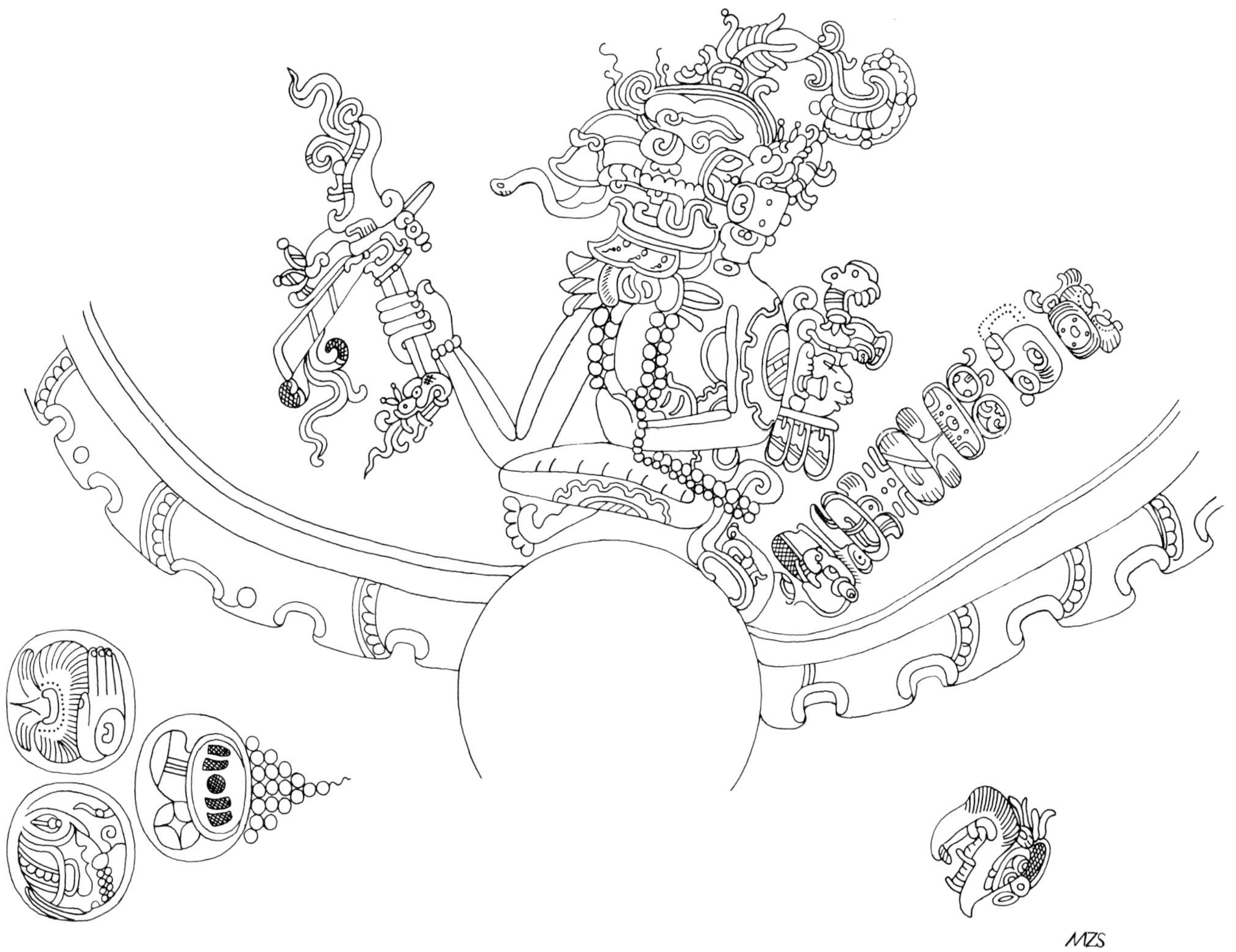

Figure 181. Bowl of God N and God GI with God K Scepter (D-1)

wears a back-rack head with three plaques hung below. The plaques bear the same markings as those on Stela 31. The headdress of the back-rack head is composed of a blunt-nosed animal with a skeletal lower jaw emerging from behind the eye. The head is very similar to the name glyph of the ruler named Kan Boar, who reigned at Tikal around 9.2.0.0.0. This vessel, however, is stylistically too early for the time of that ruler. The jawbone glyph differentiates this head from the name glyph of the later ruler.

The God K scepter is also very important in terms of iconography. It is probably the earliest known example in which the serpent appendage is clearly shown. It consists of only the head of God K with the serpent head attached to the handle of the scepter. This arrangement resembles the God K staffs of a later date in Palace House A and in the Temple of Inscriptions Tomb at Palenque. God K has a long, curving forehead with a mirror pattern inside. Smoke swirls emerge from the mirror. The smoking tube normally shown thrust into the forehead here emerges from the mouth. Smoke also filters from the side of the tube.

The entire figure is surrounded by a band that has the same scalloped motif as the god markings on the body.

Glyphic Text

The text is difficult to interpret because it is from the early Classic. The sample of early Classic texts, especially from pottery, is very small. The interior text appears to name the person for whom the pot was made. This cannot be absolutely confirmed at this time, but many of the same glyphs appear on other texts in the same style from the same period.

The group of three glyphic cartouches appears to be a part of an early Classic sequence of glyphs found on these engraved, unslipped vessels. Their present meaning and function are unknown. The other isolated glyph is a large-toothed, pointed-nosed zoomorph. Its meaning is unknown, but its position is apparently important. It is located in the center of the spiral of God N's shell.

A flare god, probably God K, is portrayed on another Classic vase (D-2), probably from the Maya Lowlands (now in a private collection). I call it Vase of God K and the Serpent Dragon. The tall vase is painted bone yellow with an upper and a lower red rim and an upper circular glyphic band. The most distinguished characteristic of the vase is the beautiful, well-preserved "Maya-blue" color with which the figures are richly endowed. The composition is centered around the undulating figure of the Serpent Dragon. In his huge gaping mouth sits a wide-eyed deity with a long nose and an upturned lip, wearing a headdress of feathers or leaves. From his forehead emerges a large flare with two coils of billowing smoke. This flare does not have the long, cylindrical shape usually seen penetrating God K's forehead. It has instead a disk shape that resembles the *zac* grapheme. The posture of the deity is also atypical; therefore, identification of this figure as God K is not absolutely certain, though the frequent association of God K with the bicephalic serpent bar (a motif apparently associated with the adjacent Serpent Dragon) supports the identification. The body and narrow belt of the god are also painted blue. In his right hand he is holding a small ceremonial wisk, possibly of feathers. His left hand is extended in an inviting gesture. In front of him stands the erect figure of another god, his body painted red and blue with round spots and god markings. He wears a narrow belt and a loincloth. In his left hand he brandishes an object that appears to be a ceremonial weapon, probably an ax.

Both figures have characteristics, including their hair styles and clothing, that suggest that they are associated with the Palenque Triad. The left figure does not have the expected feline ear of the Jaguar God of the underworld (GIII of the Palenque Triad). The jaguar paw may identify him as the patron of Pax, or it may be flame, which would suggest that he is one of the lesser-known deities of the complex. The jaguar spots on the face recall the head variant of figure 9 rather than figure 7, which is represented by the Jaguar God of the underworld. There is a near-identical figure on the left, paired and opposed to the Jaguar God. I believe that the two are also shown in the Palenque GI–GIII iconography and in the canoe-paddling scenes at Tikal. In such a case, the three deities on the bowl may be variants of or members of the Palenque Triad complex. The axes, the bound hair, and the jaguar motif support that hypothesis.

Glyphic Text

Some of the graphemes are part of the Primary Standard Text, but the rest are not. The first two glyphs may be nominal compounds. None of these compounds relate in any known or suspected manner to smoking, tobacco, or the known names of the deities portrayed.

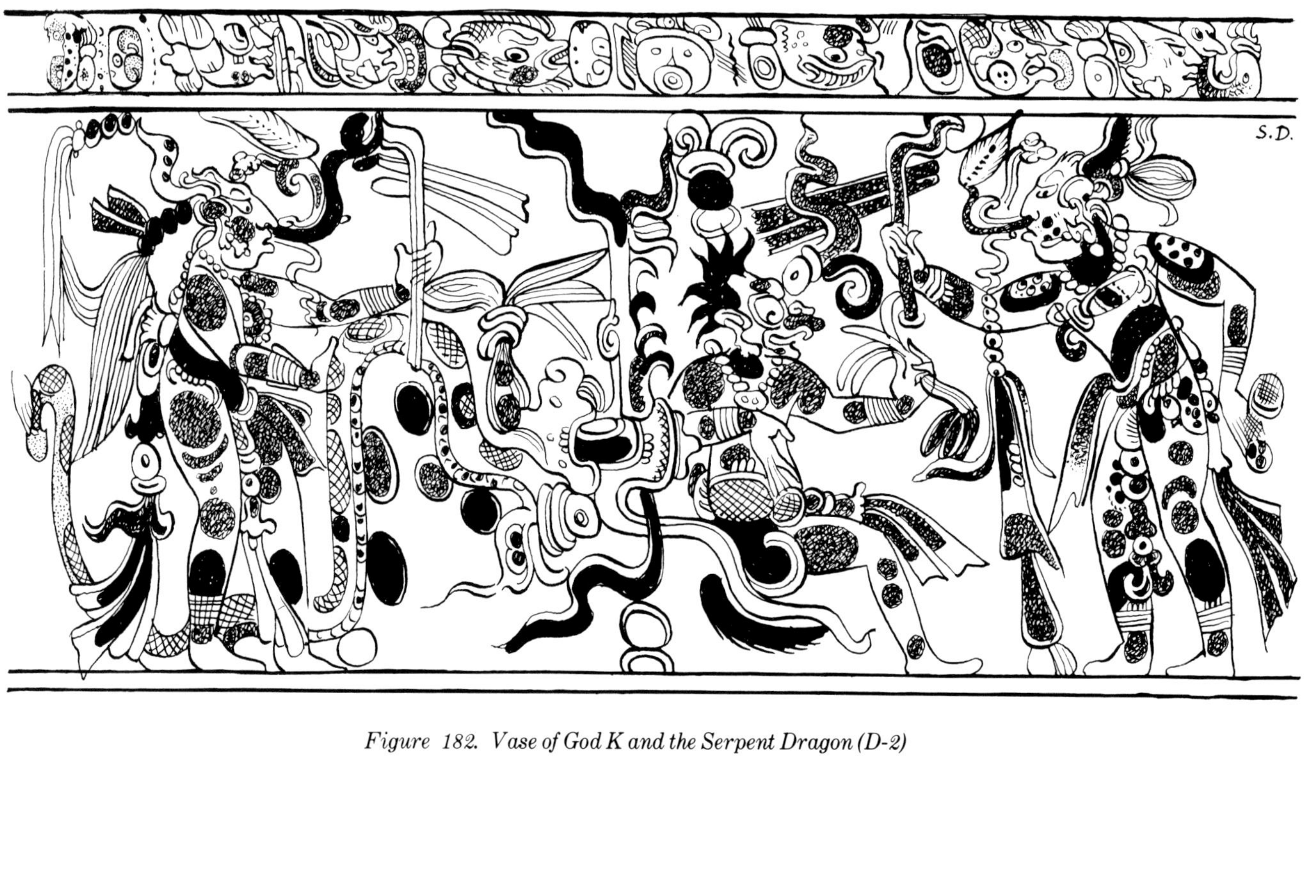

Figure 182. Vase of God K and the Serpent Dragon (D-2)

Two figures of God K with huge forehead flares are portrayed on the **Vase of the Two Flare Gods** (D-3), from the late Classic period, probably made in the Maya highlands (now in a private collection). The vase is painted red and black on an orange base with an upper and a lower red band and one circular and two vertical rows of repetitious glyphs, which may represent a name or chant. The deities seated cross-legged, have long-nosed monster heads, large god eyes, and humwn bodies. They are wearing flamboyant headdresses of flowers and feathers. The two versions are slightly different in detail and appear to be in paired opposition. Each figure has the earplug foliage and serpent-wing-headdress element suspended behind the head. A water lily (not normally associated with God K) is cantilevered from the top of the head. Both figures have mirrors infixed in their foreheads. One mirror is painted red; the other is white with reflection lines infixed. Both foreheads are penetrated by the large cigar motifs with smoke elements. Completion (?) signs are attached to the smoke coils. The heads have the long-muzzled, large-eyed conventions for God K.

The bodies, like the mirrors, are somewhat different. The red-mirrored God K is painted red across the center of his body, while the edges of his body contour are clay color. His lower back is painted with a large white circle with interior lines. A belt of rectangular motifs surrounds his waist, and the end of the loincloth emerges in front of his legs. The opposing God K (he of the white mirror) has red paint on his shoulders, lower arms, lower back, and thighs. He has the same belt and loincloth motifs, but the body markings on his lower back are black rather than white.

Glyphic Text

Both the rim text and the Secondary Texts are composed of one repeated compound, T85.528?. The main sign appears to be a version of the *cauac* grapheme, but the meaning of the compound and its association with God K is not understood.

A very similar representation of God K appears on the **Vase of the Three Gods** (D-4), a vase of a period and provenance identical to those of D-3 (now in a private collection). In fact, the vessel was probably painted by the same artist. Its color is orange with the conventional upper and lower red bands and a row of repetitive hieroglyphs. God K appears as one in a sequence of three figures. Since all the figures face toward the left, it is doubtful that the sequence was designed as an integrated scene. Rather, it appears to be a seated processional. God K is seated Turkish fashion. His body is painted red, and he is wearing a composite belt, a short kilt, and a cossack helmet. Streaks of jaguar spots are painted on his arm and thighs. He has a snout resembling a crocodile's and large god eyes. A double coil of smoke as large as his body emanates from his forehead tube, which is probably a cigar.

The figure at the right of God K is a creature of uncertain species. The muzzle suggests a carnivorous animal or dragon, but the ringed eye is unusual, and he has a deer ear attached above his ear. He has a spangled turban headdress with an elaborate cantilevered motif.

The third figure is an aged god. He has a spangled headdress with the characteristic cantilever in front. The shell element extends from the rear of the head, but it should be pointed out that God K also has this noncharacteristic shell element. The aged god holds an Ahau sign in his hand. The side elements on the object identify it as the capped Ahau (T535) rather than the usual *ahau*, which means "lord." The capped Ahau often appears as the tail element of jaguars, and as a relationship glyph it signals a male parent.

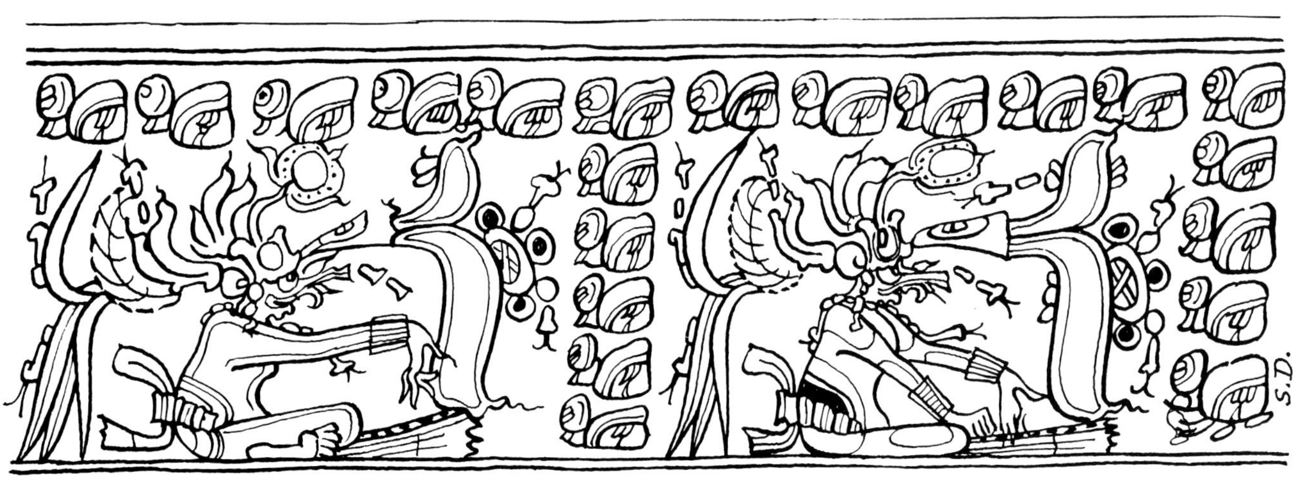

Figure 183. Vase of the Two Flare Gods (D-3)

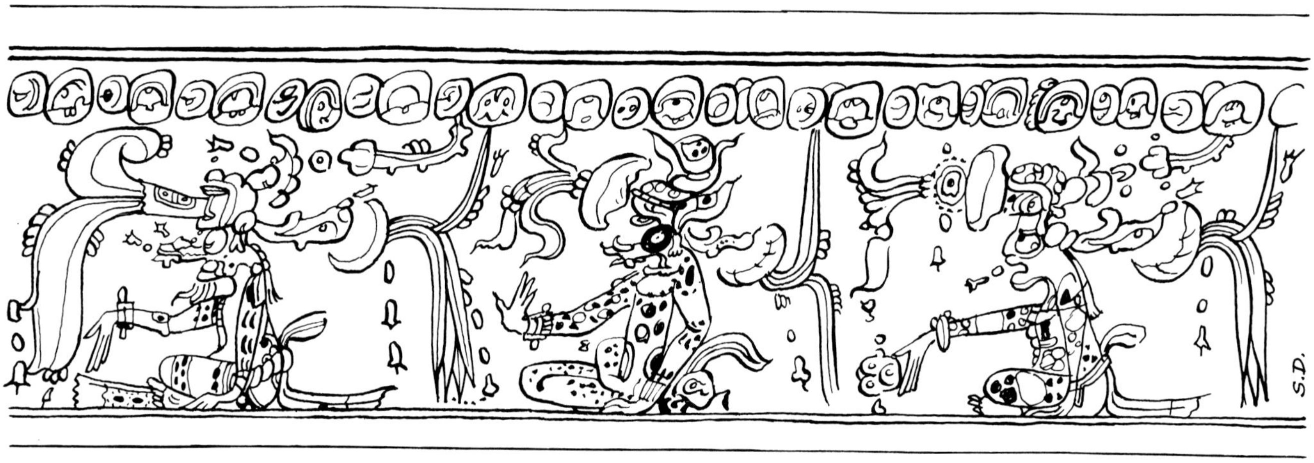

Figure 184. Vase of the Three Gods (D-4)

All three figures have elaborate foliage motifs extending from their earplugs. This elaboration may be a characteristic of this painter or school of painters. The head configuration, the cigar, and the clothing are typical of God K, but the alternation of red and spotted streaks is unusual. The pattern appears on all three of the figures. The spotted streaks may refer to the jaguar or signify body painting.

Glyphic Text

The rim and Secondary Texts are composed of one repeated compound. Both the affix and the main grapheme were not included in Thompson's catalog, but the combination appears in other similar repetitive pottery texts.

A twin portrait of God K decorates the **Vase of God K with Webbed Fingers** (D-5), found in the Maya highlands (now in a private collection). The vessel is from the late Classic period. It is painted reddish orange on a light-orange base and has a peculiar black-and-white upper-and lower-band design and a circular band of repetitious glyphs. The deity sits in his usual cross-legged pose, but his body is in front view—an unusual position in Maya art. His body, his element, even his forehead tube and the smoke coils erupting from it are painted with jaguar spots. Behind him is a free-floating head ornament with large feathers and tassels, the same ornaments that appear on his smoke coils. The god is wearing a loincloth and a simple pectoral. The typical large eyes, long forehead, upturned nose, short lower jaw, and long snout are all present in this portrait, which is repeated on the back of the vase.

Glyphic Text

The text is composed of a single, repeated compound written in mirror image.

The same artist who painted the **Vase of the Two Flare Gods** (D-3) created the **Vase of God K and the Old God** (D-6), a late Classic vessel from northern Petén (now in the Carlos H. Nottebohm Collection). The vase is painted on an orange base in red and black with an upper and lower red rim, upper glyphic band. Two vertical hieroglyphic columns divide the composition into two sections, one occupied by God K himself, the other by an aged deity. The portrait of God K is fairly conventional. He is seated cross-legged, wearing only a wide belt, an ornamental headdress, and simple body jewelry. His shoulders, arms, and waist and part of his hip are painted red; the rest of his body is decorated with jaguar spots. He has the usual long snout, extended forehead, and two huge coils of smoke. The smoke tube itself cannot be seen.

God K's companion is the Old God identified elsewhere in this book. He has the diagnostic characteristics, the cantilevered-head motif, the shell pectoral, and the aged features. He has the same headdress elements, body painting, and clothing as those of God K, and his hand makes the same down-pointing gesture as that of God K.

Glyphic Text

The texts on this pot are not very clear as to function or meaning. They appear to be divided into two sections, but no clear verbs are discernible. In each case there is a slightly wider space between the vertical column behind each figure and the first horizontal glyph thereafter. The wider space is presumably a reading cue to indicate that each half is to be read in the order of an inverted L. The text begins with the horizontal section above each figure and continues down the vertical column. Each text may name and/or describe the figure.

God K is also a participant in a complex scene pictured on the beautiful **Vase of the Four Gods** (D-7), a polychrome vessel probably painted in the

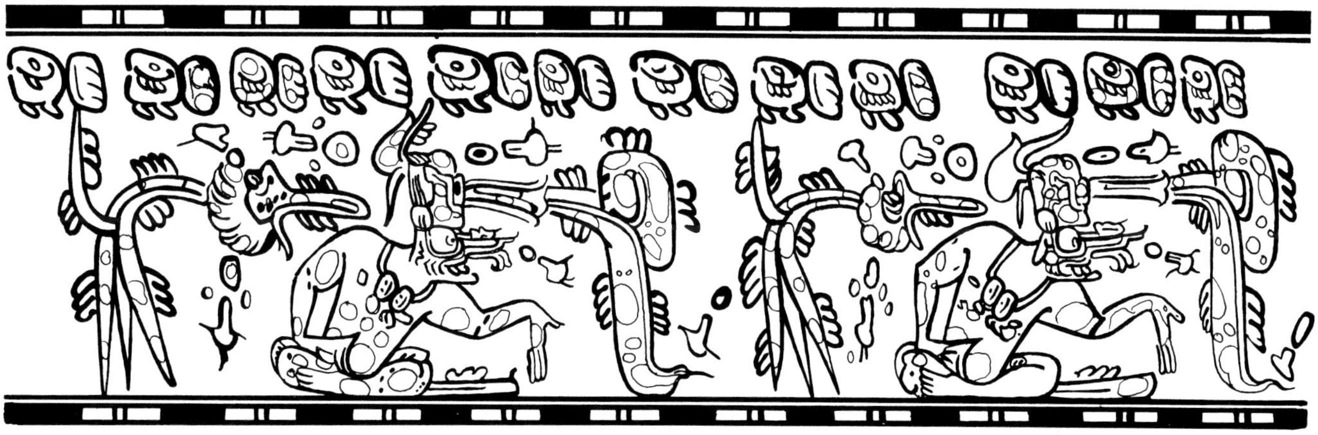

Figure 185. Vase of God K with Webbed Fingers (D-5)

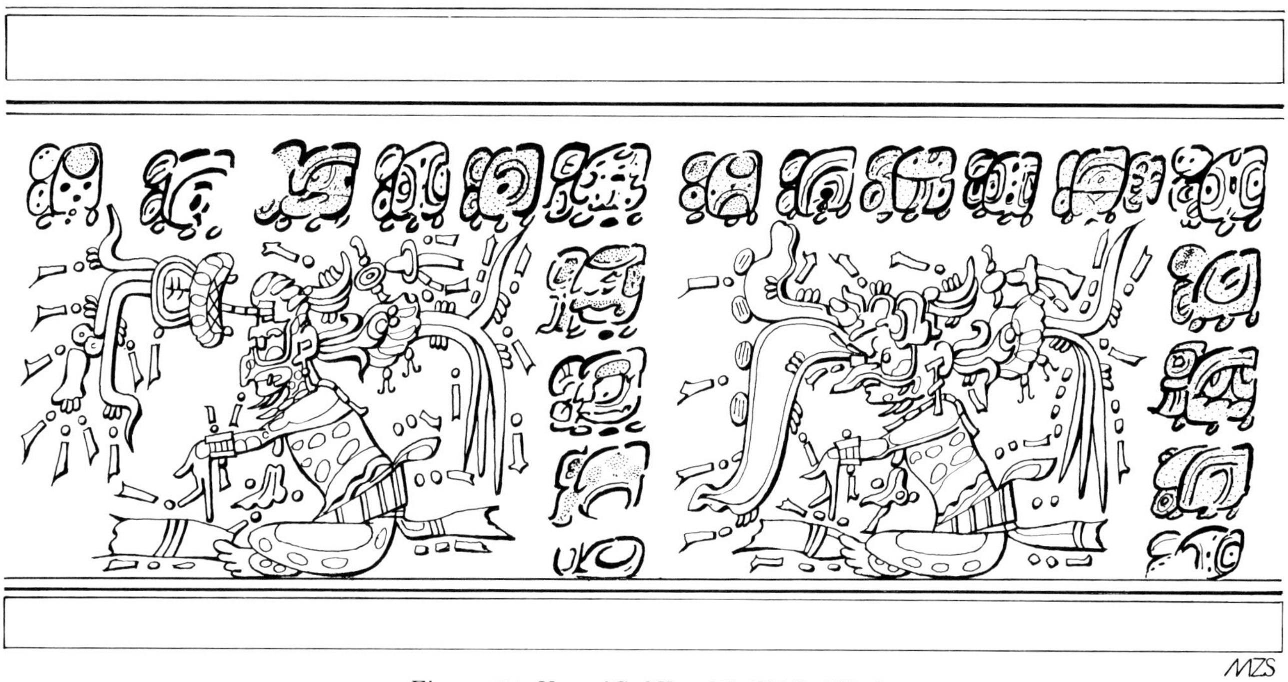

Figure 186. Vase of God K and the Old God (D-6)

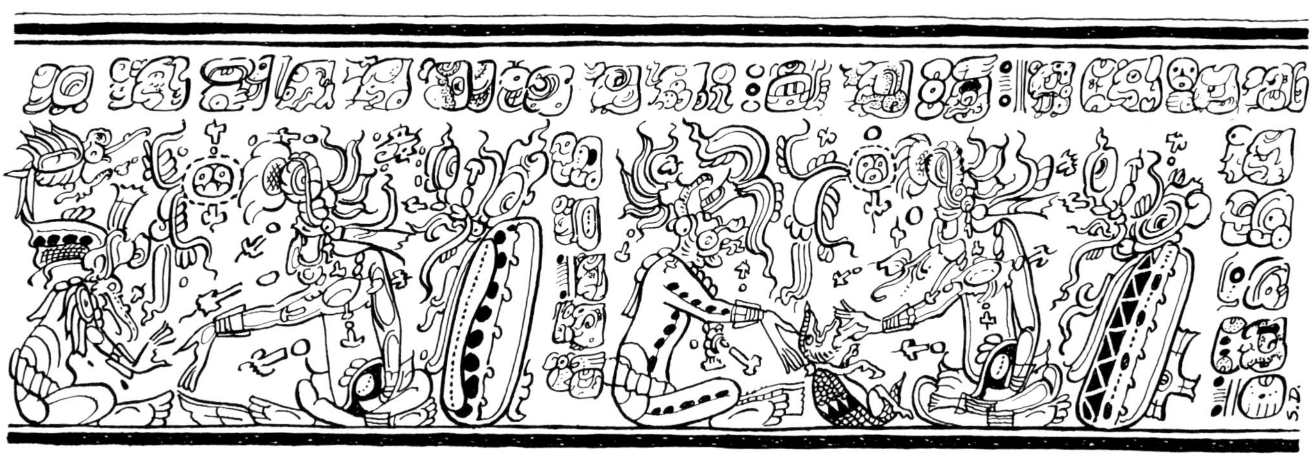

Figure 187. Vase of the Four Gods (D-7)

Maya highlands during the Classic period (now in a private collection). Four deities are represented on the orange vase, which has red bands on its upper and lower rims with circular and vertical hieroglyphic texts. The surface is divided into two scenes, each showing paired figures. In each scene the left figure is the Old God identified on several ceramics previously described. His distinguishing characteristics are the cantilevered headdress with medallion and shell pectoral. In one scene the headdress medallion has an *akbal* infix, and in the other an *ahau*. In both scenes the Old God sits front view in the cross-legged position and gestures to his right toward the accompanying figure or to a central motif.

In the first scene (determined by the rim text) the accompanying figure is an anthropomorphic waterfowl, a bird deity with a human body. It wears an elaborate headdress, a belt, wristlets, and shoulder ornaments resembling epaulets. Another bird of unidentified species is shown above the seated avimorph.

The accompanying figure in the opposite scene is God K. Like the bird god, he also has a large white spot on his lower back and is wearing a simple belt composed of rectangular elements. His body is marked on the abdomen, legs, and arms with the scales and the side pattern of the serpent. The serpent markings echo the serpent leg often shown on God K on Classic monuments. His head is a good example of the Classic version of God K. The forehead mirror is painted white and is penetrated by a cigar. He lacks the characteristic smoking *ahau* topknot, but he has the backswept hair style usually seen on his portraits. His eye is large and square, his face is fleshed, and he has an extended, fanged snout. He has a receding lower jaw, suggesting serious problems with underbite.

In both scenes there is a cushion in a diagonal position behind the Old God. In the bird scene the pillow has jaguar markings, and in the God K scene what appear to be serpent symbols. In each scene is also the head of a deity attached to the cushion. The deity is fleshed and open-jawed, with the lower jaw extending along the front of the pillow. It is an unusual variant of the Jester God. He has a white mirror infixed in the forehead and a small head of typical configuration. The muzzle has an enlarged shape normally associated with the Serpent Dragon but very unusual for the Jester God. A fish barbel extends from the nostril region. This is also unexpected for the Jester God; however, the version of the Jester God portrayed on the Tablet of the Foliated Cross at Palenque sets a precedent for such fish motifs. The mouth and teeth configurations show the expected characteristics.

An armadillo in belly-up position rests between the Old God and God K. The hands of both figures appear to be touching the head of the animal.

Glyphic Text

The rim text appears to be a variant of the Primary Standard. The introductory glyphs are found over the bird god and consist of the T228:617a "mirror," the T1011/GI, and the T617:713a "mirror-hand" compounds.

The Secondary Texts are located between each pair of figures and presumably follow normal reading order; i.e., the column relates to the scene on the right. They do not have clearly verbal or nominal compounds. The first compound of each column is probably an introductory glyph, but if it is so, they are not the ones previously so identified. Each text includes one compound with a T12 affix, which is thought to be the third personal pronominal article, *ah* or "he." The presence of affix T12 suggests that the compounds are nominals.

A most unusual combination of images is seen on the **Bowl of God K and God L I** (D-8), probably made in northern Petén during the middle Classic period (now in a private collection). The bowl portrays two deities, God K and an aged god with a serpentlike body, probably representing God L.

The Old God is facing left. He has a snakelike body, but his head and extremities are human. He

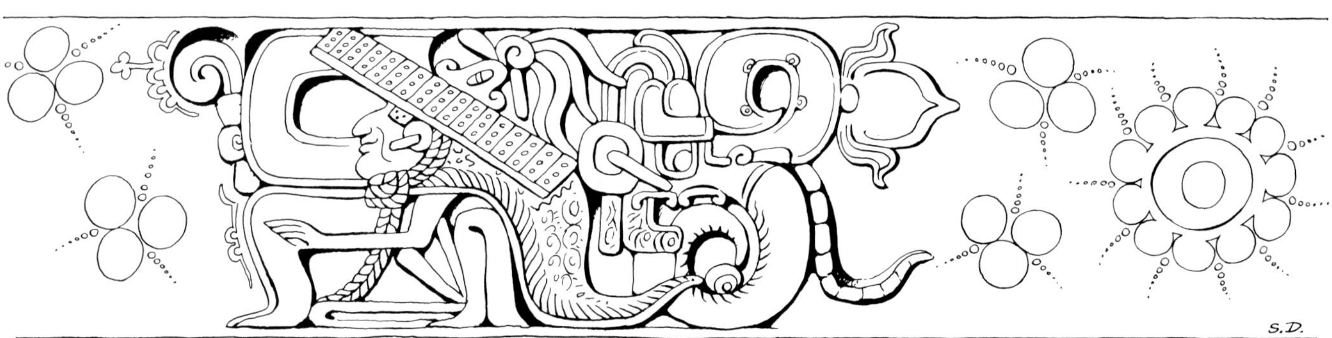

Figure 188. Bowl of God K and God LI (D-8)

wears a magnificent wide-brimmed hat, similar to that seen on Stela 9 at Chinkultik with the cord tied under his chin like a sombrero. The brim of the hat has a row of vertical decorations, probably representing the feathers of the Moan bird, a motif characteristic of God L. This "snake with human arms and legs" is most unusual in Maya iconography. An alternative interpretation is that what appears to be a serpent body may be a long cape somewhat similar to that worn by God L on the tablets in the Temple of the Sun and the Temple of the Cross at Palenque and on a ceramic box from Petén (A-1). From his body emanate two swirling coils of smoke to which a T60 knot motif is attached.

Behind the Old God rests the bodiless head of God K. A large smoke motif emerges directly from his forehead without the conventional cigar tube. Such "cigarless smoke" is also portrayed with God K in the 819-day count and in the Palenque Triad expressions in Classic texts. Except for the absence of the cigar tube, God K has all his usual characteristics: the forehead mirror, the square-cut, backswept hair, the long muzzle, and the receding jaw. Attached to his smoke coils is a water lily similar to that seen on Vase D-4.

If we accept Michael D. Coe's theory that the Maya God K and the Aztec deity Tezcatlipoca are two aspects of the same god, this vase certainly generates interesting thoughts. Tezcatlipoca is a known adversary of Quetzalcoatl, the Sacred Serpent.

The same artist who made the previous vessel or a potter of the same school undoubtedly created the **Bowl of God K and God L II** (D-9), which is also probably from northern Petén (now in the Norweb Collection, Cleveland, Ohio, Museum of Art). The participants in the scene, God L and God K, are identical, and there is also great similarity both in the technique and in the composition. The two vases differ in only minor details, such as the posture of the figures and the shape of God L's hat. It is also noteworthy that on the second vase the aged deity is wearing a beaded necklace, and his cloak has jaguar markings like those on his image in the Temple of the Cross and Sun at Palenque. In his headdress are feathers of the Moan bird, and in front of him rises a huge smoke scroll with a symbol that closely resembles a completion sign. This bowl is the third in the series of carved ceramics on which the full figure of God L is depicted with the head of God K. This phenomenon undoubtedly has some as yet undeciphered mythological significance.

Two near-identical heads of God K also decorate the **Vase of God K with the Crossed Bands** (D-10), made during the late Classic period, probably in Yucatán (now in a private collection). The vessel is painted on an orange base in red and black with a conventional upper and lower red rim and parallel black lines. Two pairs of vertical lines divide the composition into four panels. Two narrow panels enclose repetitious glyph blocks, and the two larger panels frame the heads of the god. The heads are resting on small pedestals held by T129 affixes. They are highly stylized with no lower jaws. Coils emanate from the foreheads and the napes, and the craniums are replaced with large T552 (crossed bands) signs. The crossbands make the identification of the deity as God K somewhat uncertain.

The **Vase of the Triple Portrait of God K** (D-11) was made in the very late Classic period in the Nebaj area (now in the Museo de Popol Vuh, Francisco Marroquín University, Jorge and Ela Castillo Collection). The composition is arranged around three symmetrically placed windowlike panels through which near-identical heads of God K gaze at the observer. The heads are enframed in oblique bars painted in the colors of jaguar hide. There are

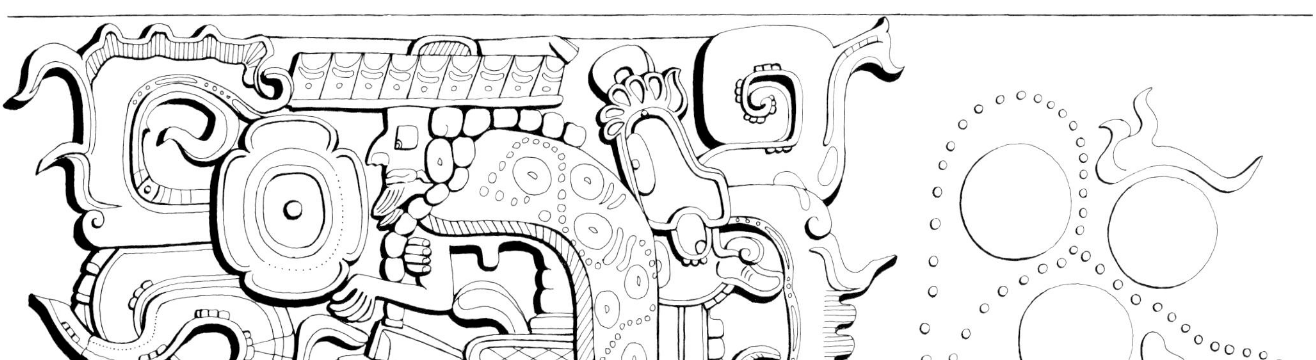

Figure 189. Bowl of God K and God LII (D-9)

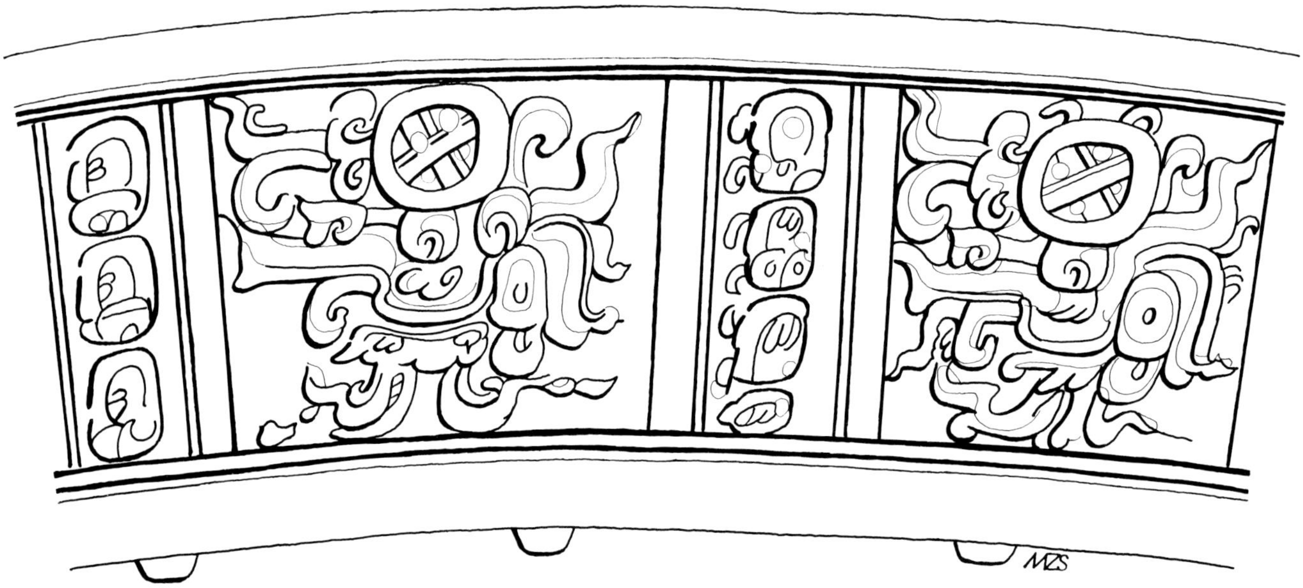

Figure 190. Vase of God K with the Crossed Bands (D-10)

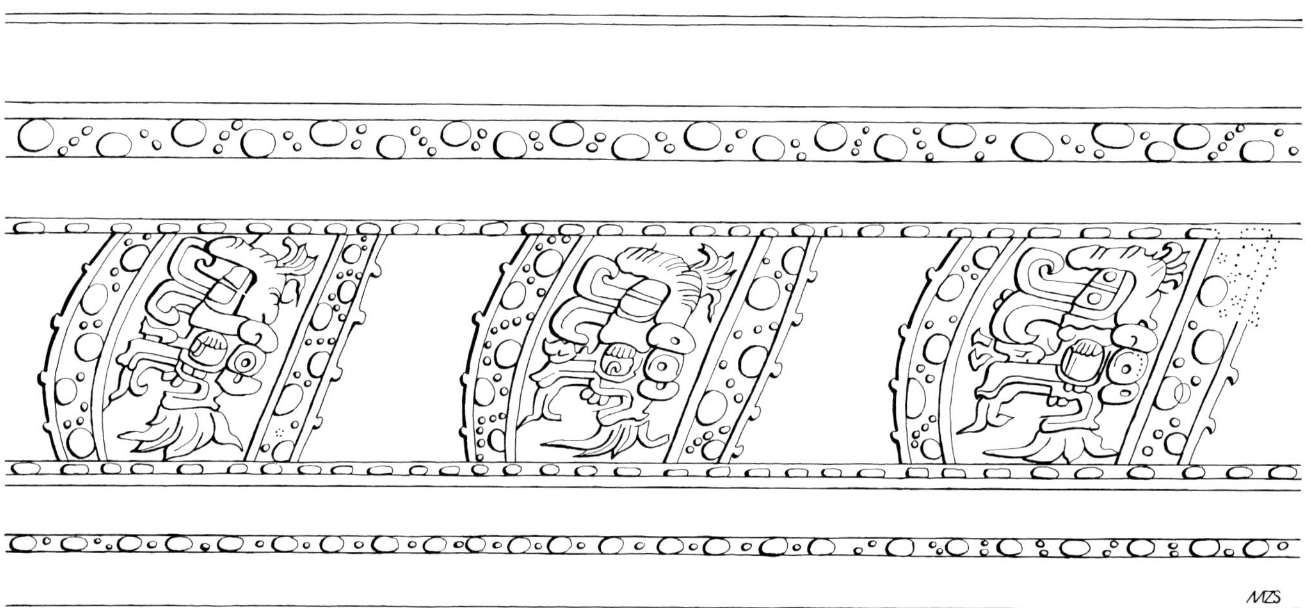

Figure 191. Vase of Triple Portrait of God K (D-11)

also three circular bands of the same design and color, as well as an upper and a lower row of glyphs painted on a cherry-red base. The heads are conventional except that the forehead tube is missing and the twin smoke coils emerge directly from the forehead. It is also interesting to note that, as on vase D-15, the heads rest on flame or feather appendages.

An unusual full-figure representation of God K is seen on the **Vase of God K Painted Black** (D-12), a low ceramic vessel made by a Chamá artist during the last Classic epoch (now in the Museo de Popol Vuh, Francisco Marroquín University, Jorge and Ela Castillo Collection). The vase is also unusual in that God K is painted black, one of the two dark representations of this deity I know of (the other appears in the *Dresden Codex,* where he is shown painted deep blue). The vase is orange with upper and lower red and black-and-white chevron bands. The deity is shown in his usual seated position, leaning on his left hand and gesticulating with the other. His short kilt is painted red and white, and he is wearing a bead necklace and wristlets. He has a flower nape ornament, and from his white forehead

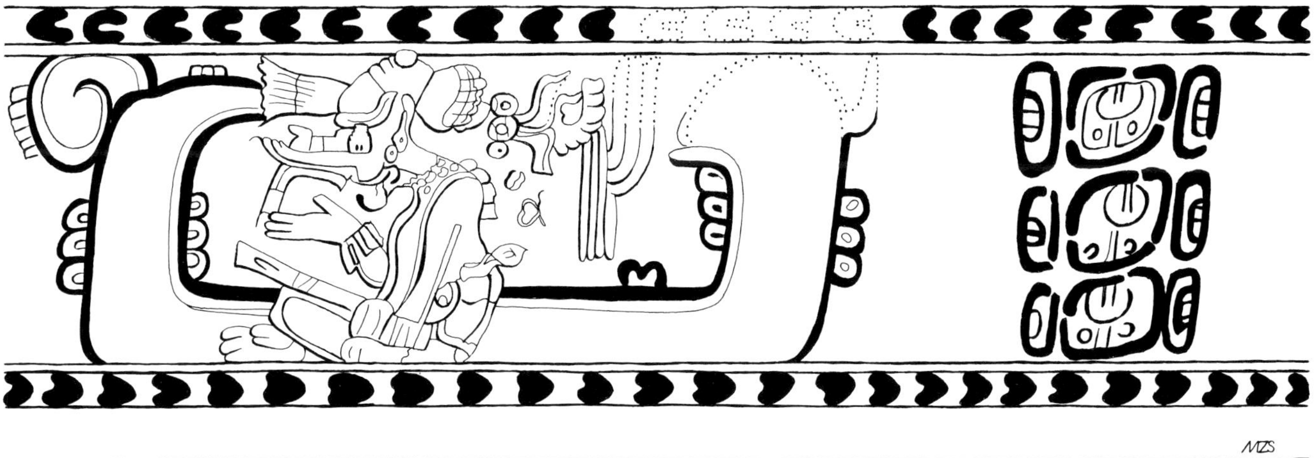

Figure 192. Vase of God K Painted Black (D-12)

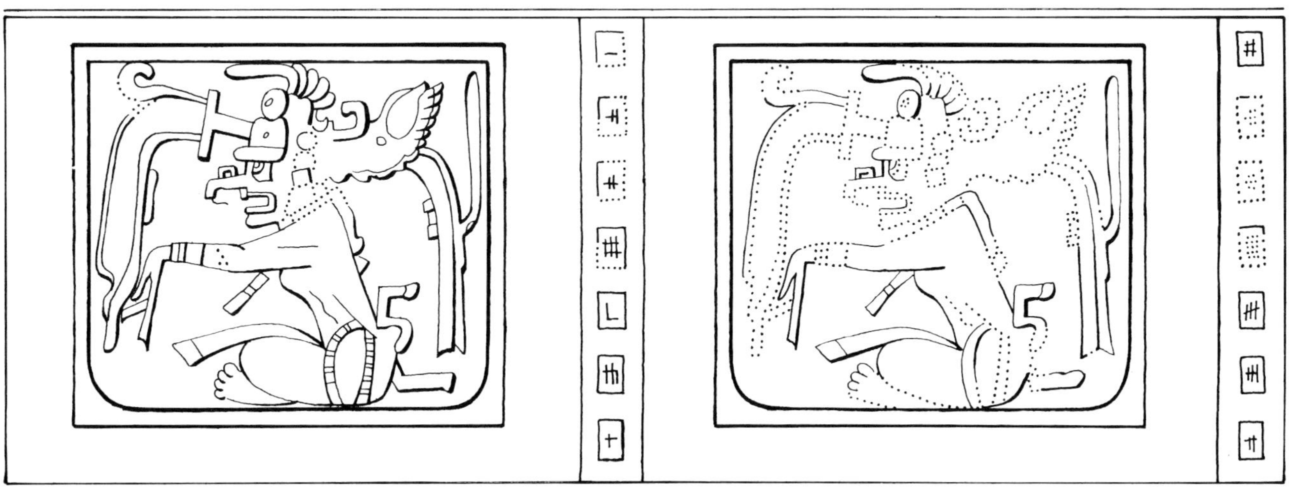

Figure 193. Amado Bowl (D-13)

tube two huge red smoke coils emerge, the lower one framing his body.

Glyphic Text

The glyphic text consists of a single vertical row of repeated glyphic compound T.33 (inverted *ahau*), which has the phonetic value of *la*. A very similar compound T533 (*ahau*) frequently occurs in the *Dresden Codex* as a deity attribute. It is associated especially with Itzamná and other benevolent deities. The inversion of the *ahau* compound might make the statement different from codical texts, but the function of both compounds may be related.

Still another ceramic vessel showing God K is a faded polychrome vessel, the **Amado Bowl** (D-13), found in the Motagua Valley, probably made during the late Classic period (now in the collection of Dr. Guillermo Mata Amado). The composition is divided by two vertical bars and vertical rows of glyphoids into two framed tableaus, both inhabited by the rather primitively painted seated figure of God K. The deity is scantily dressed, wearing only a loincloth and a headdress. The second one has a back floater of feathers and an ornament that could be the wing of a large bird. From his forehead a T-shaped tube emerges. From the vertical portion of the T two large coils of smoke emanate.

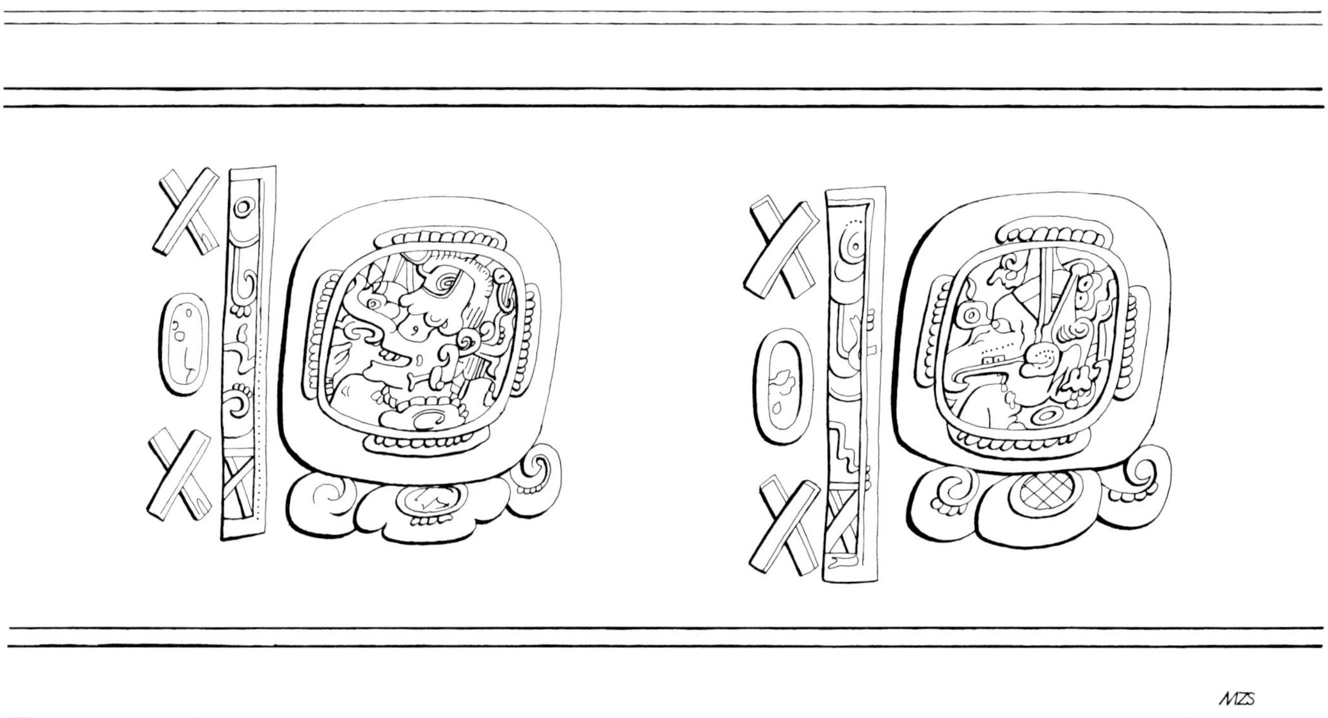

Figure 194. Vase of the Two Day Glyphs (D-14)

The **Vase of the Two Day Glyphs** (D-14) has some very beautiful and interesting details. The polychrome vase was probably made during the late Classic period in Petén (now in a private collection). The surface has a white slip with an upper and lower red band. The two glyphs are painted black and red. Both glyphs have the numerical prefix "six." One of the God C variants is drawn inside the bar of the "six." This God C variant is a version of the Sun God who appears on loincloths throughout the late Classic period. The dot of the "six" appears to have been detailed with a skeletal head. The shapes of the eye and nostrils can be clearly seen.

The day signs are apparently the same. The bird-like head is the vulture variant of Ahau. The diagnostic features of this vulture are the scarf tied around the forehead and the Jester God attached to the head. Such an Ahau vulture appears on the Palace Tablet at Palenque (D-17), which also has the Jester God as a semantic determinant. In most other instances the semantic determinant is an *ahau* glyph. However, the Jester God appears throughout Maya sculpture in the headdress gear or paraphernalia of rulers. He was apparently one of the gods of rulership. It appears here as a determinant to designate this vulture as an ahau "ruler." The vulture also appears to have T59 *ti* attached to his beak. The presence of T59 in this *ahau* glyph is unique because elsewhere T59 is the phonetic complement that ensures that the vulture head is read as the locative *ti* rather than as *ahau*. There is little doubt of the *ahau* reading here, because the cartouche around the head is specialized for day glyphs. *Ti* as a reading for the glyph has no relationship to any known day glyph.

The opposite day glyph has God K inside the cartouche. His identity is confirmed by the presence of the smoking tube in front of the forehead mirror. His presence here is apparently explained by his association with the rites and regalia of rulership, He is paired with the Jester God, and he appears in a large number of ruler portraits painted throughout Maya history.

Both glyphs record the day 6 Ahau. When Ahau day signs occur in isolation and unmarked in position in the Long Count or Calendar Round, they most often funtion as the name glyphs of particular *katúns*. Only three *katúns* in the entire history of the Maya were named 6 Ahau: 9.1.0.0.0, 9.14.0.0.0, and 10.7.0.0.0. The first and the last are much too early or late. 9.14.0.0.0 fits the style of painting. The vase apparently celebrates Katún 6 Ahau, the end of the thirteenth Katún of the tenth Baktún.

A second masterpiece by the same artist or school of artists is the **Vase of God K Codex Style I** (D-15), also probably from Petén (now in a private collection). It is painted black on a whitewashed base with a conventional upper and lower red rim but with no glyphic text, in a style that Lin Crocker describes as "a technique utilizing monochrome painting on a cream background accompanied by

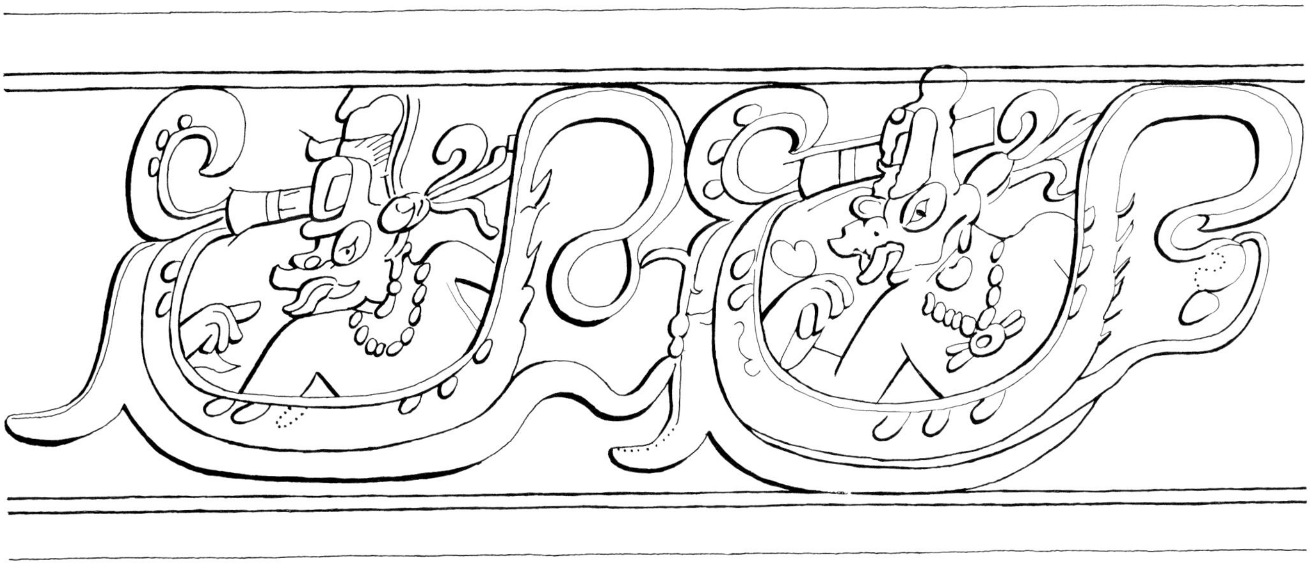

Figure 195. Vase of God K Codex Style I (D-15)

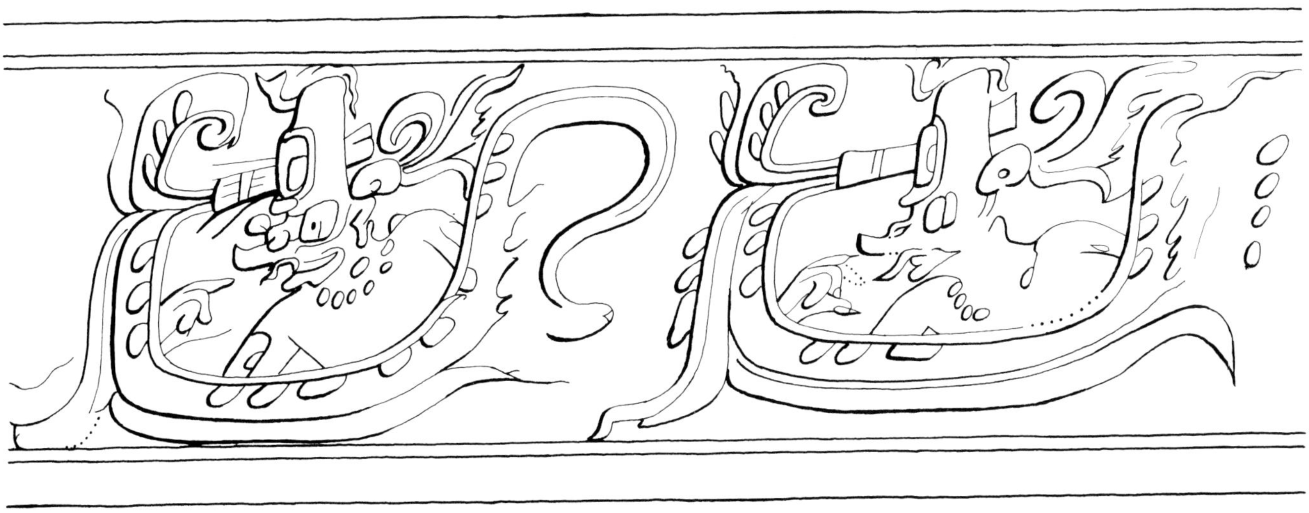

Figure 196. Vase of God K Codex Style II (D-16)

extreme variations in line tone."[1] As mentioned earlier, Michael D. Coe calls it "codex style" because in his opinion the artists who created these vases were early codex painters, who inscribed the folding screen books that the Mayas made of bark paper coated with stucco.[2] Each of the twin portraits of God K shows only the upper body, bending slightly forward, virtually enframed by the lower large coil of smoke emanating from his forehead tube. The tube, similar to the ones shown on the ceramics described above, has a thin binding element. The lines drawn around the bundle resemble wrinkle marks in soft material and suggests the compression of bound tobacco leaves. There are, as usual, some differences between the two representations: the shape and design of the foreheads differ somewhat, and on the second portrait the forehead tube completely perforates the head and emerges at the nape. Interestingly, these portraits show God K sloe-eyed rather than with the usual large scroll eyes. A large black circle is painted around the eye, and one wonders whether the large god eye with which the Maya deities are usually shown may have developed from this "circle-around-the-eye"—or, on the contrary, whether some of the lords were painted thus to impersonate deities.

The "sister vase" of the D-15 is the **Vase of God K Codex Style II** (D-16), a ceramic of identical style, workmanship, color, and provenance (now in a

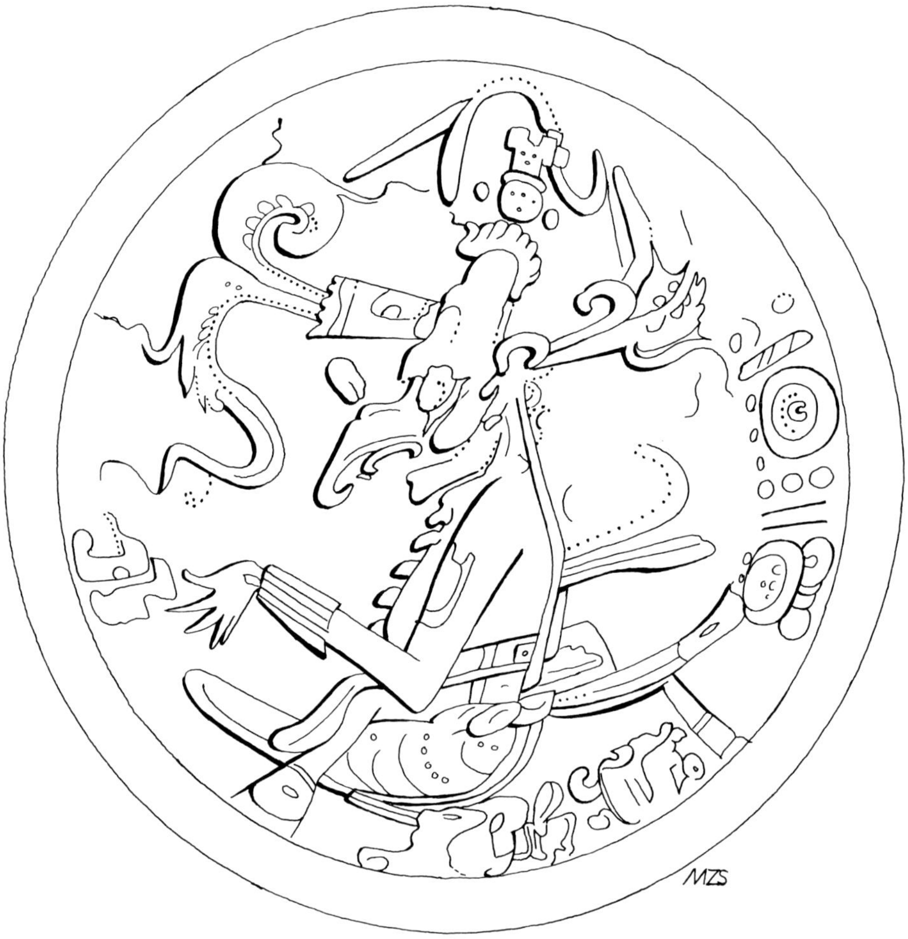

Figure 197. Plate of God K Seated (D-17)

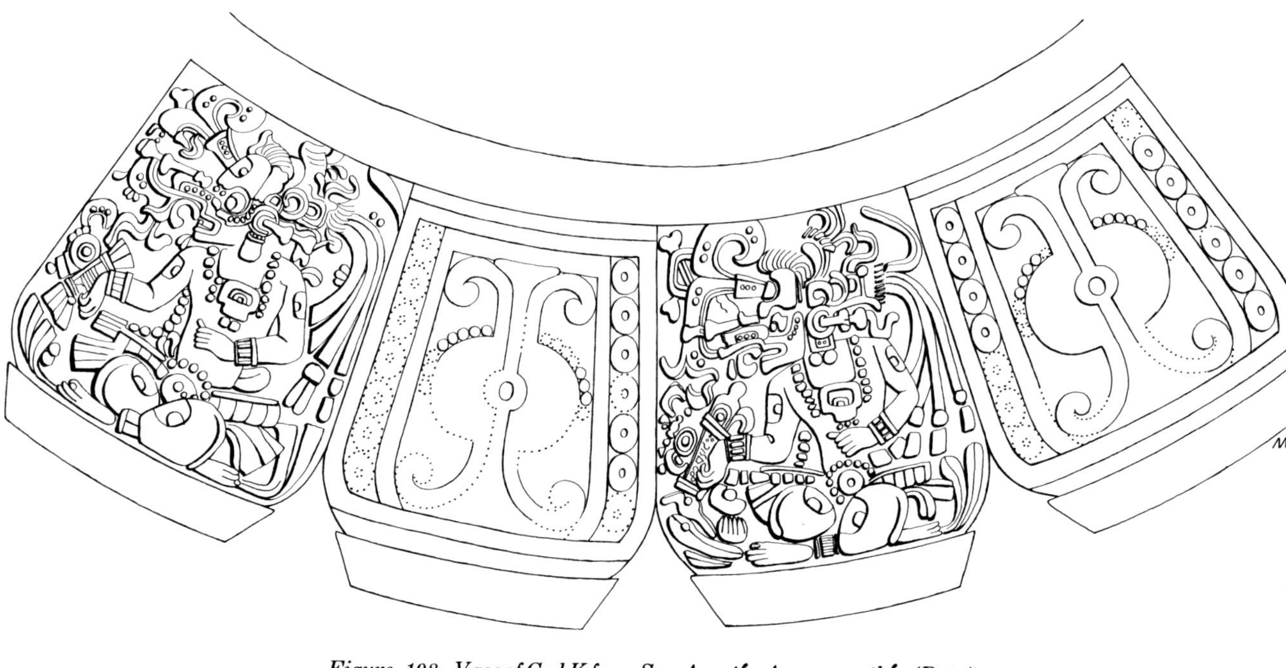

Figure 198. Vase of God K from San Agustín Acasaguastlán (D-18)

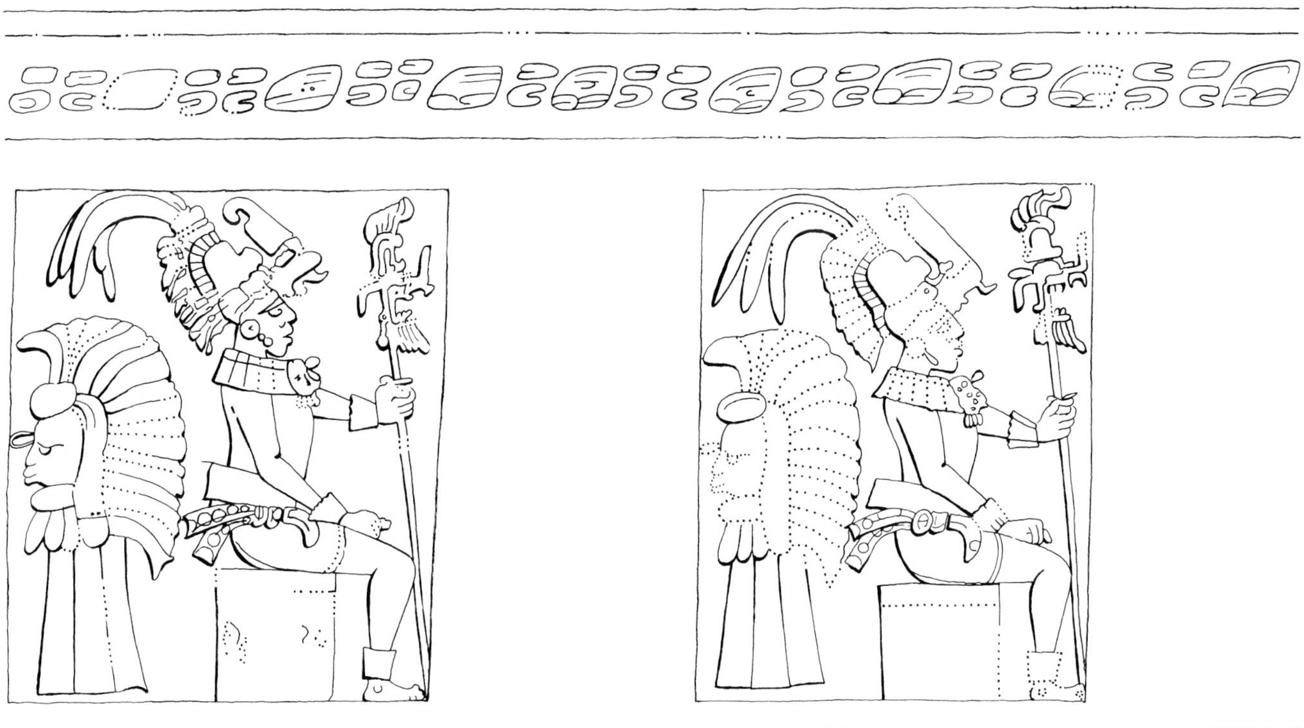

MZS

Figure 199. Vase of the Lord Seated with Manikin Staff (D-19)

private collection). As is usually the case, there are minor subtle differences: the eyes of the gods are somewhat larger and more angular, and the handles of the smoking cigars emerge from the napes of both representations.

The artist who painted D-14, D-15, and D-16 must have been deeply interested in God K, because he used the same style and technique to create the **Plate of God K Seated** (D-17), to my knowledge the only flat ceramic object with a full-figure portrait of this multifaceted god. The plate was made in the late Classic period in Petén (now in a private collection).

The plate is painted on a whitewashed background in reddish brown with a red rim and a semicircular glyphic text. The image of God K is the conventional one, showing him seated cross-legged, dressed in kilt, belt, and beaded necklace. From his high forehead emerges a large tube, probably a cigar, from which come two magnificent curls of smoke.

God K is also shown in full figure on the **Vase of God K from San Agustín Acasaguastlán** (D-18). The vessel is from the late Classic period (now in the collection of the Museo Nacional de Arqueología y Etnología de Guatemala).[3] Each of the figures of the deity faces left. He is seated cross-legged and is wearing an elaborate belt, wide wristlets, counterweighted flexible ear flares, a round pectoral, a belt buckle, and a necklace. The headdress is magnificently ornamented with quetzal feathers. A tube with two coils of smoke emerges from the forehead. In his hand he holds an object that appears to be a miniature mat throne like the one held by the left figure on the Tablet of the Temple of the Sun at Palenque. The figure on the right is distinguished by a glyphic sign (T553a?) emerging from the smoke coils. The smooth space between the two figures was once decorated with a negative painting, of which only traces remain. Some circular and perhaps floral designs are still identifiable. I have seen a photograph of a vase made of the same mold (now in a private collection).

The **Vase of the Lord Seated with Manikin Staff** (D-19) was made during the late Classic period probably in the Usumacinta River valley (now in a private collection). The vase is painted on an orange base in subdued shades of red, blue, and black. The upper rim is painted with a band of repetitious glyphoids. The composition is undoubtedly a palace scene, divided by two black vertical walls into two chambers, both of them occupied by near-identical portraits of the same lord seated on an altar block. He is dressed in short kilt, a cloth collar, wide wristlets, and gaiters. He wears a

God K helmet without the forehead flare but with a characteristic upturned nose and with blue plumage. In his hand he is holding a long ceremonial staff topped with the head of God K. Behind him, on a pedestal, is a head (perhaps a trophy), painted blue, also decorated with blue feathers.

The head of God K appears twice on the **Relief Vessel of God K** (D-20), made during the Classic period, probably in northern Petén, (now in the collection of the Mint Museum of Art). The vase is partly covered with stucco, with traces of red cinnabar and "Maya blue." The upper rim is carved with a row of hieroglyphs, under which are two conventionally represented heads of God K. The coils of smoke emanating from the forehead tube (cigar?) are quite large, especially the lower one, which partly encircles the head.

Glyphic Text

The text begins with the second compound in the drawing. It appears to be the T1011/GI introductory glyph to the Primary Standard Text, but the rest of the glyphs do not appear to be either in sequence or in configuration to be identified as part of Primary Standard. The second compound consists of a T670 "hand" holding double *cauac* (T528) glyphs. The third glyph in the text represents a long-eared animal, probably a deer. The fourth grapheme is one of the "smoking" heads, probably one of the God K variants found in titles and the 819-day-count texts. The fifth grapheme is an unusual head variant of the *cauac* glyph, and the sixth is the *kin* bat. The seventh compound includes an unidentifiable animal head. The text closes with a glyph which appears to be the head of a "jog" animal. This glyph is drawn as the first one of the line rendering of the vase.

The text does not have a readily identifiable verbal compound, unless we accept the "hand-*cauac*" compound as such. The script may be one of the texts which records a chant or names of various deities.

Twin portraits of God K are presented on the **Vase of God K Seated in a Smoke Curl** (D-21), provenance probably Petén, late Classic period (now in a private collection). The vessel is painted red and black on an orange base with an upper horizontal and two vertical repetitive glyph bands. In the space between the vertical bands are two seated images of the deity. The figures are practically identical, the only significant difference being the jaguar-hide pillow (throne motif) behind one of the portraits. The deity is shown wearing a loincloth, which leaves his large jaguar spots uncovered. He has large, angular god eyes, and his hand is extended palm forward. The forehead of God K resembles a *nen*, mirror, grapheme, and from it emerges a large (cigar?) tube with two tremendous smoke coils. Completion signs are attached to the front side of the smoke and a T60 knot to the rear. The huge curls pass behind the deity.

Glyphic Text

The horizontal rim text is composed of a bat glyph and a head that we cannot identify. The bat head is painted red, while the other head is white. A very unusual feature of the text is that the first glyphs are placed in a red cartouche. The vertical text is made of the repetition of the white head compound only.

The **Vase of God K Sacrificed** (D-22) is a small molded gray cylinder, probably from northern Petén, made during the late Classic period (now in a private collection). The vessel is decorated with an

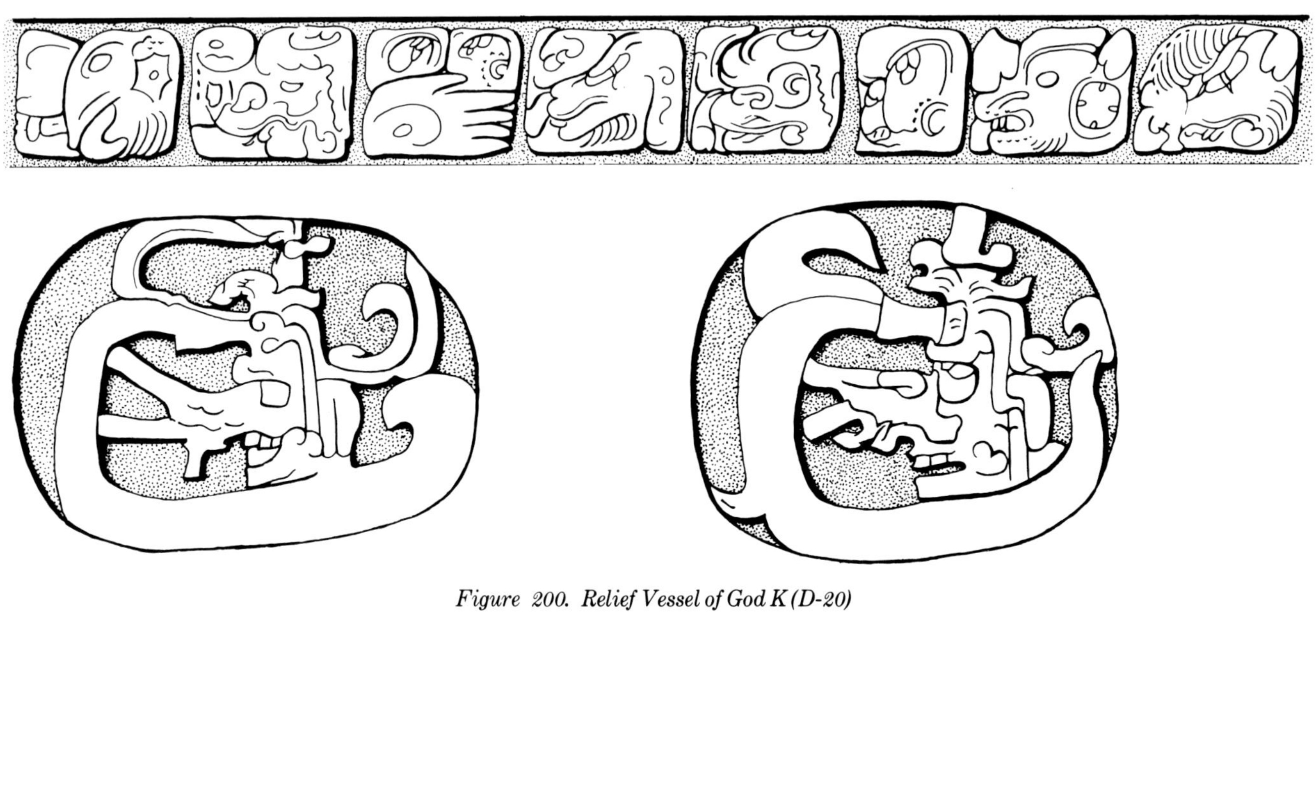

Figure 200. Relief Vessel of God K (D-20)

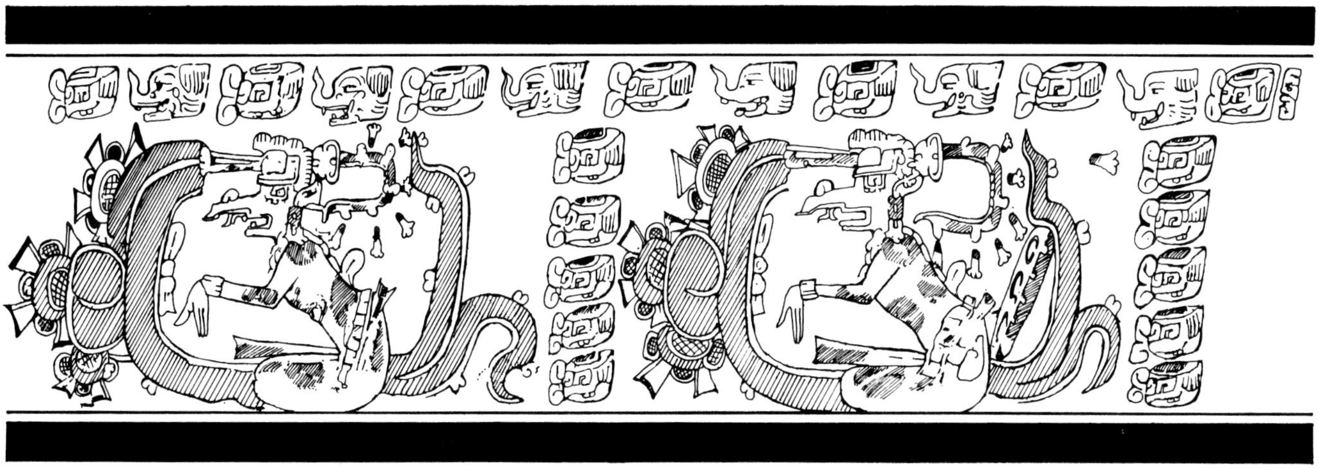

Figure 201. Vase of God K Seated in a Smoke Curl (D-21)

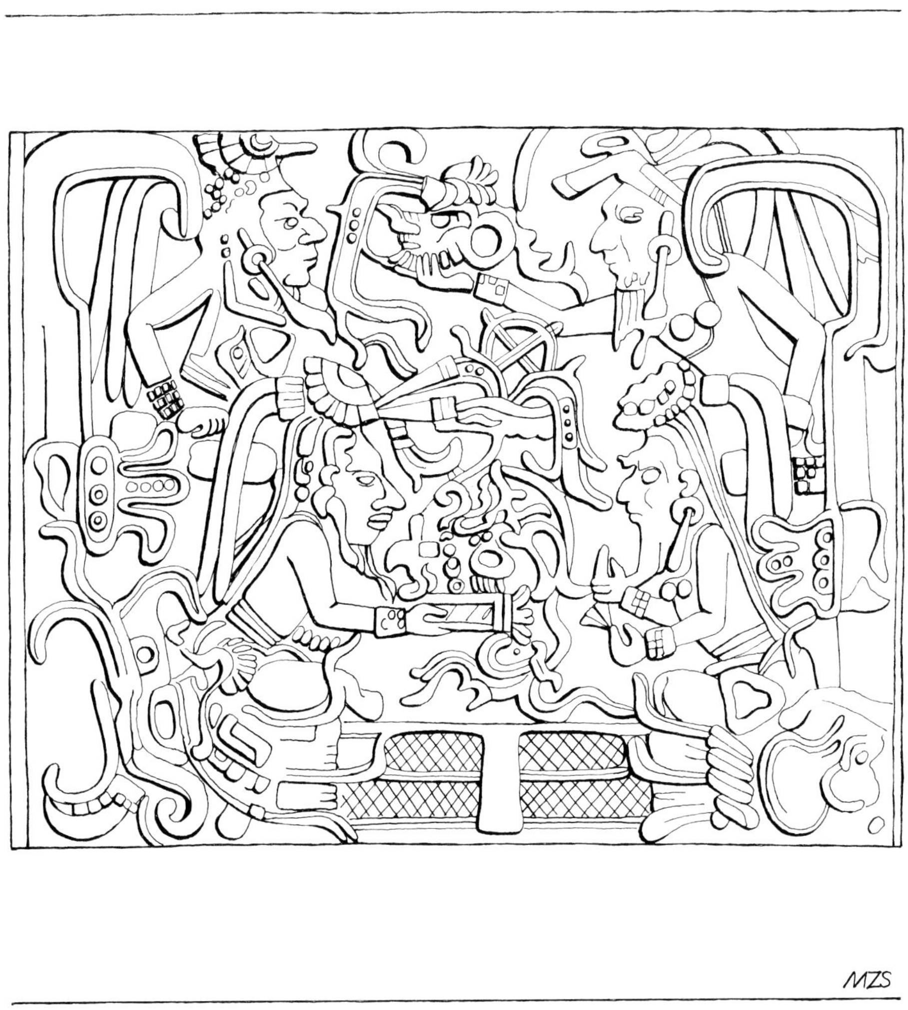

Figure 202. Vase of God K Sacrificed (D-22)

esoteric scene that is reproduced on the back. The somewhat eroded but still identifiable composition shows four male personages dressed only in loincloths, wearing headdresses with floaters of quetzal feathers, flexible ear ornaments, and wide wristlets. They are arranged on two levels: two of them are seated cross-legged and facing each other on a low dais on the lower level; the other two seem to be emerging above and from behind them, also face to face. The space around the human figures and the objects they are holding is filled with scrolls, mat symbols, flowers, and leaves.

The personage sitting on the left is a younger individual with long, gathered hair, who is distinguished by a round composite headgear. He is holding a "presentation" plate similar to those seen on the Palenque monuments, holding the head of an aged deity. He appears to be offering the head to a bearded old man sitting across from him whose hand is held as if to receive the plate. His headdress and his hairdo match those of his counterpart, but he wears a "double pompom" (totemic?) breast pendant, which is seen not infrequently on Classic vases and on highland and Jaina figurines. I believe that it represents tribal association.

Above the old man's head, his upper body leaning forward, is the figure of a person holding the head of another deity in his outstretched hand. He is also wearing the peculiar round breast pendants and therefore probably belongs to the same tribe or caste. The head, which by the long snout, forehead tube, and smoke coils can be identified as that of God K, is being offered to the fourth individual sitting across from him in an intense, expectant posture.

For an interpretation of the tableau there are two possibilities: either it is a mythological scene in which the heads of sacrificed deities are being presented by their executioners, or it represents a historical event in which tribal symbols were exchanged.

The **Plate of the Standing Lord** (D-23) is a beautiful example of late Classic ceramic art. It was probably made in northern Petén (now in a private collection). The plate is illustrated with the magnificent figure of a standing noble framed by two red circles and a circular band of well-executed hieroglyphs. The lord is richly ornamented with a nose plug, a flexible ear decoration, a heavy necklace, and bead wristlets. He is wearing a complicated helmet with a large floater of quetzal feathers. The small head of a Jester God appears to be perched on the top of the headdress. The bead-and-rod necklace also partly covers the shoulders. He wears a wide sash folded around his waist with a feather back ornament and a trophy-head back pendant. His lower body is covered with an elegant kilt decorated with crossbones and water-lily motifs, and he also wears garters and sandals. The crossbones may indicate underworld connections. In his hands the lord holds a long staff with the head of God K on the top. The staff strongly resembles some the manikin staffs of Palenque, with one notable exception: the tube, instead of emerging from the forehead, seems to perforate the head of the god at a lower level and protrude from the mouth. The appearance of the lord is most impressive. It was probably representations like this one that led Miguel Covarrubias to exclaim that "the spectacle of these Maya lords dressed in garments of golden jaguar skins lined with blood-red cloth and covered with green jade and iridescent feathers must have been magnificent!"[4]

Glyphic Text

The text on this plate is almost identical to MT217, a plate found in Burial 195 at Tikal. MT217 names a ruler, who it has been assumed is the occupant of the tomb. The full nominal phrase of MT217 includes a statement of the parentage of the protagonist, who is named at D–J. The mother is named at K–O and the father at P–C. The protagonist of MT217 and the father reappear in the text of this plate.

This text begins with the Primary Standard Sequence. The introductory glyph is located immediately above the head of the figure. The first six glyphs are part of the Primary Standard Sequence.

The father of the MT217 protagonist is named on this plate in the seventh through the twelfth glyphs. The seventh glyph, an inverted *ahau* head variant, and the ninth glyph, a crossbands compound, both occur on MT217. The remainder of the glyphs are recognized titles that appear at Tikal and elsewhere. The last two glyphs of the nominal phrase include a sky glyph and a zoomorphic figure holding the *cauac* sign and compose a title especially prevalent at Quiriguá. The deity also appears at Palenque (Temple of the Cross, Jamb Panel) and at Tikal hung from the belts of rulers.

The function of the *imix* compound that follows the deity head is not known. It may be part of either nominal phrase or a relationship glyph between the two persons. The nominal phrase of the MT217 protagonist begins immediately after the *imix* compound and completes the text. All of the glyphs with slight variation appear on MT217. The nominal phrase begins with the "eye" glyph which is followed by black *ahau*, sky over earth God C variant, Mah K'ina, Lizard. The last glyph is the personal name. The *ahau* title appears on MT217 but where it is prefixed by "black" on this plate, it is prefixed by *yax*, "blue," on MT217.

Two rather primitively painted heads of God K decorate the **Vase of God K with the Flaming Hair** (D-24), late Classic period, probably made in northern Petén (now in a private collection). The vessel is painted red and yellow on a black base. It has an upper and a lower band composed of

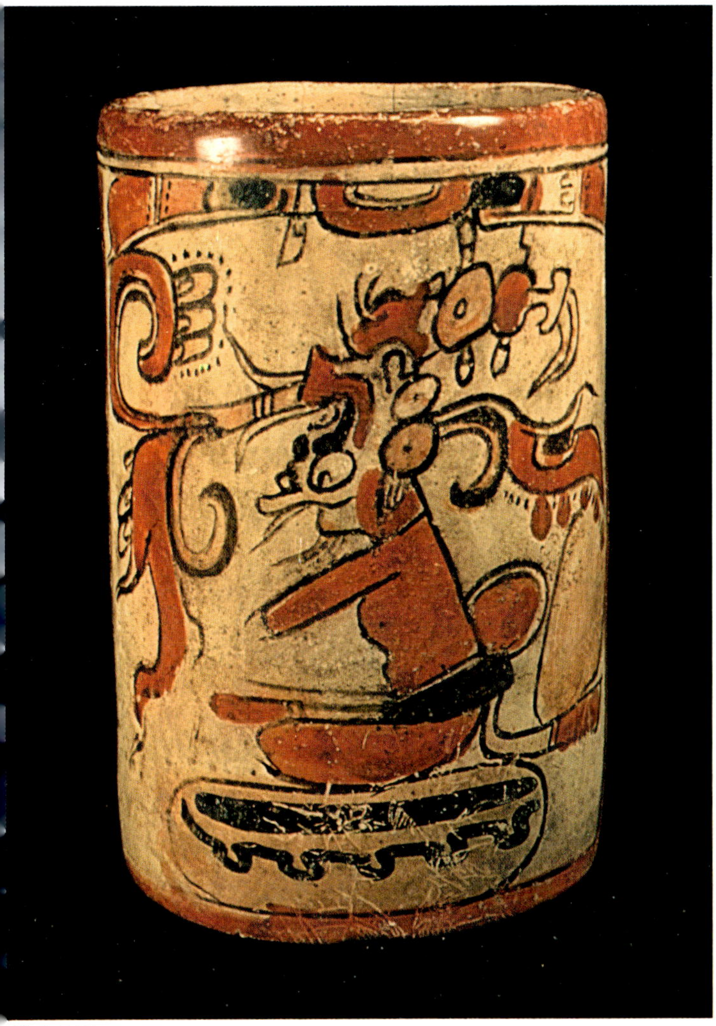

Plate 253

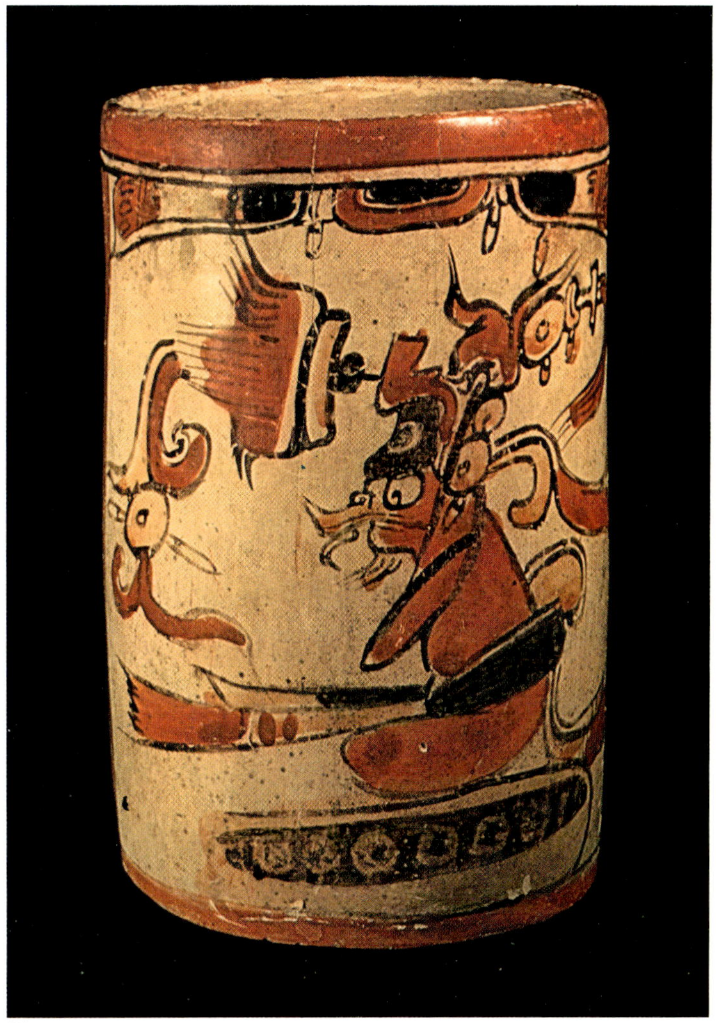

Plate 254

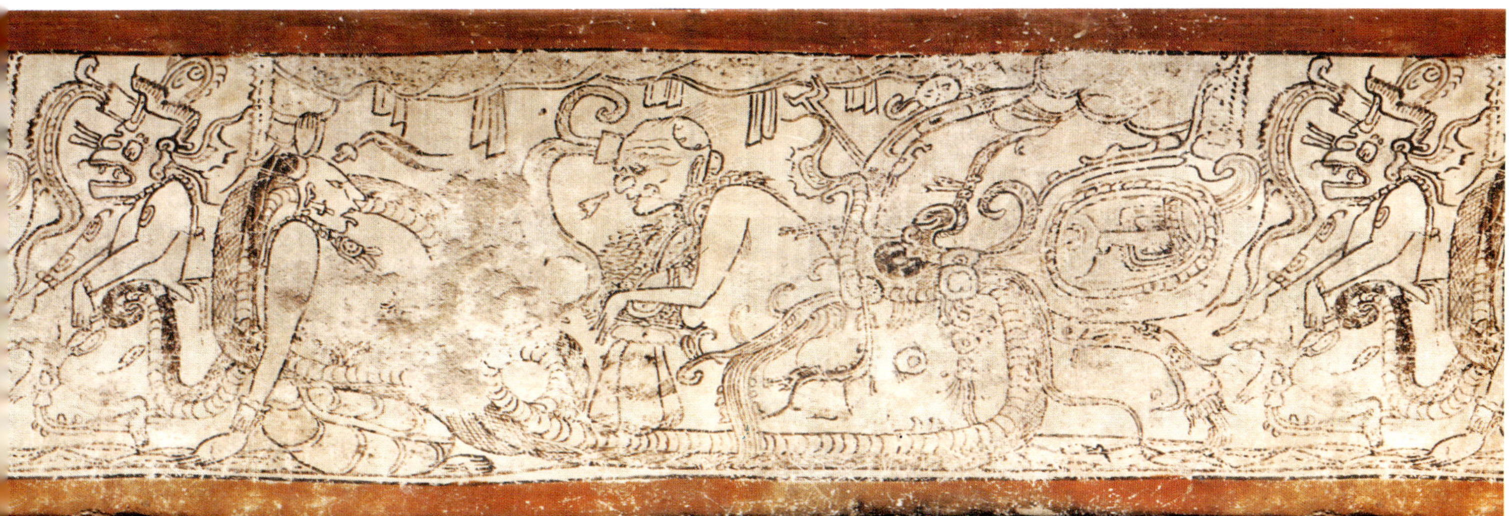

Plate 255

Plates 253-254   Vase of God K with a Beard (D-28)

Plate 255   Vase of the Gods with Forehead Flares (E-6) /Photograph copyright Justin Kerr, 1978/

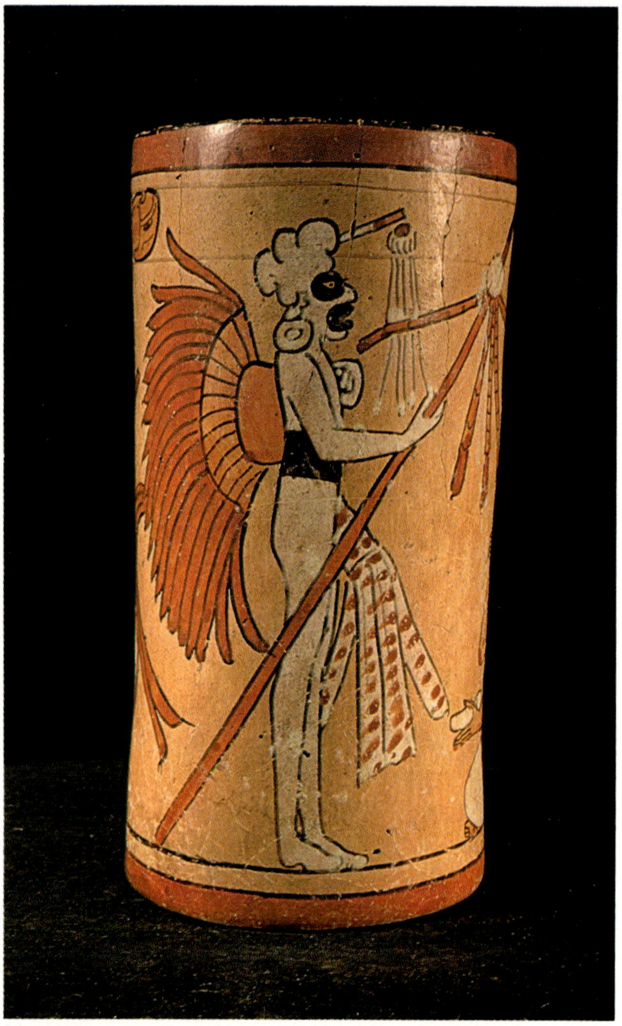

Plate 256

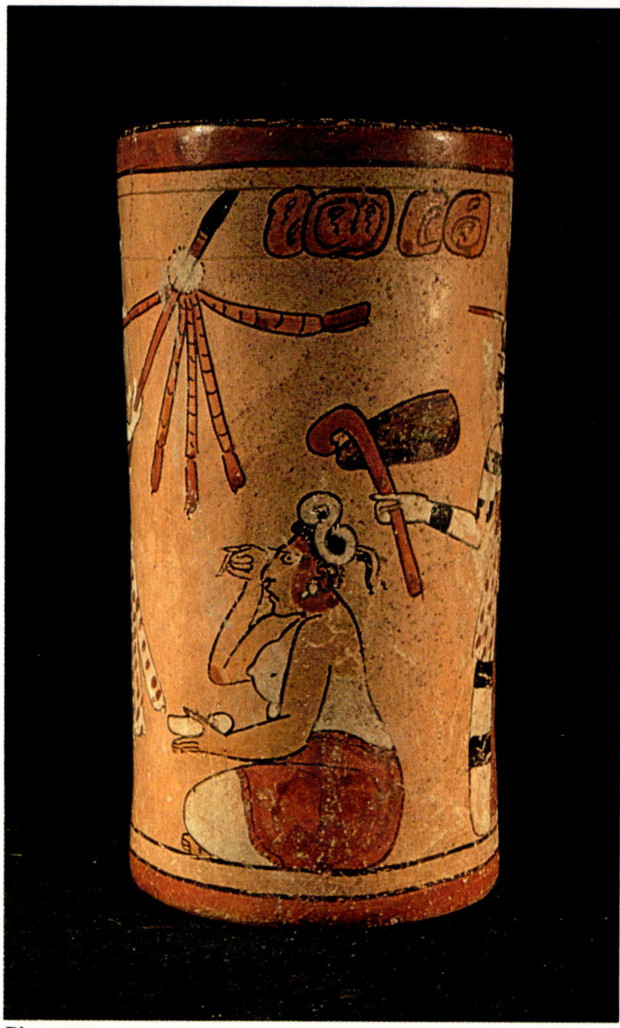

Plate 257

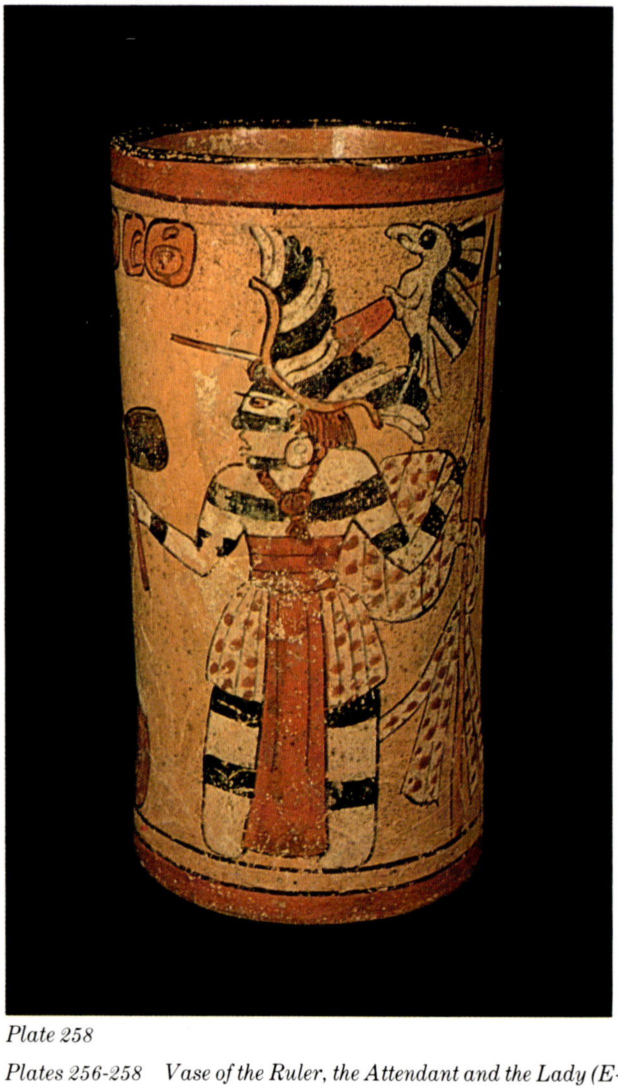

Plate 258

Plates 256-258 Vase of the Ruler, the Attendant and the Lady (E-2)

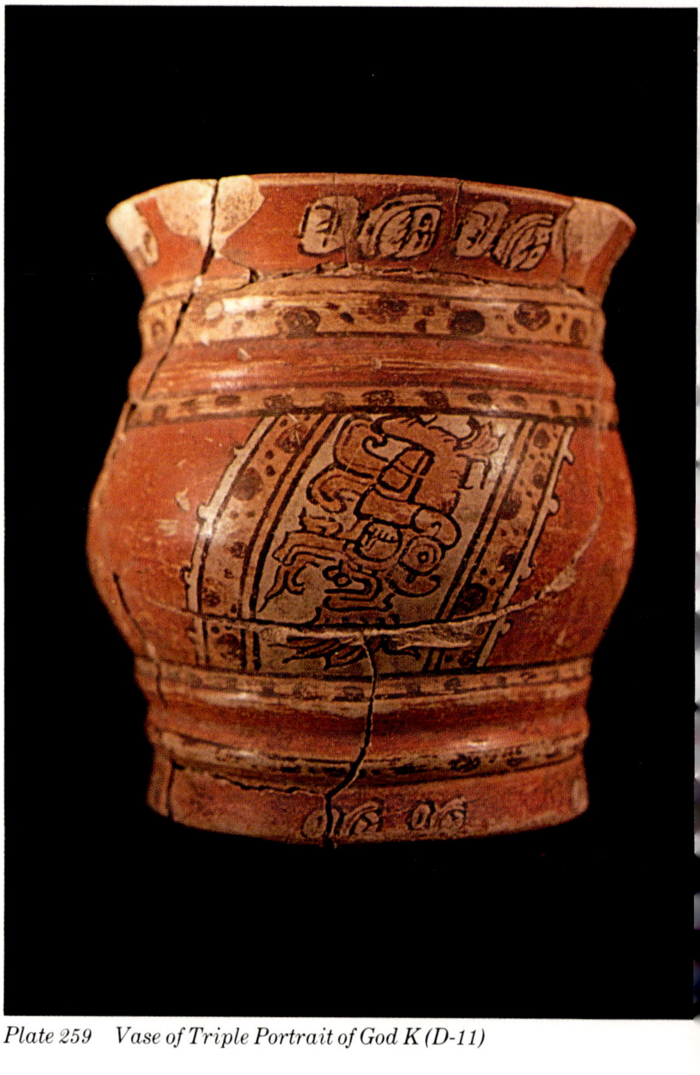

Plate 259 Vase of Triple Portrait of God K (D-11)

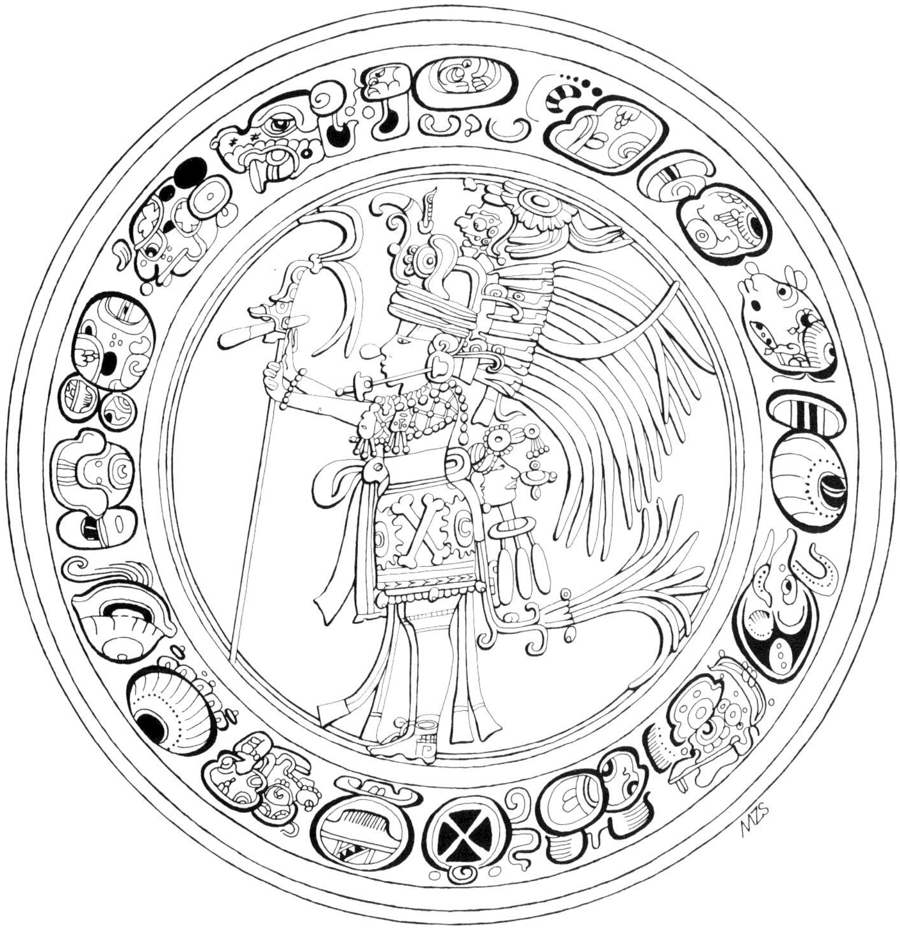

Figure 203. Plate of the Standing Lord (D-23)

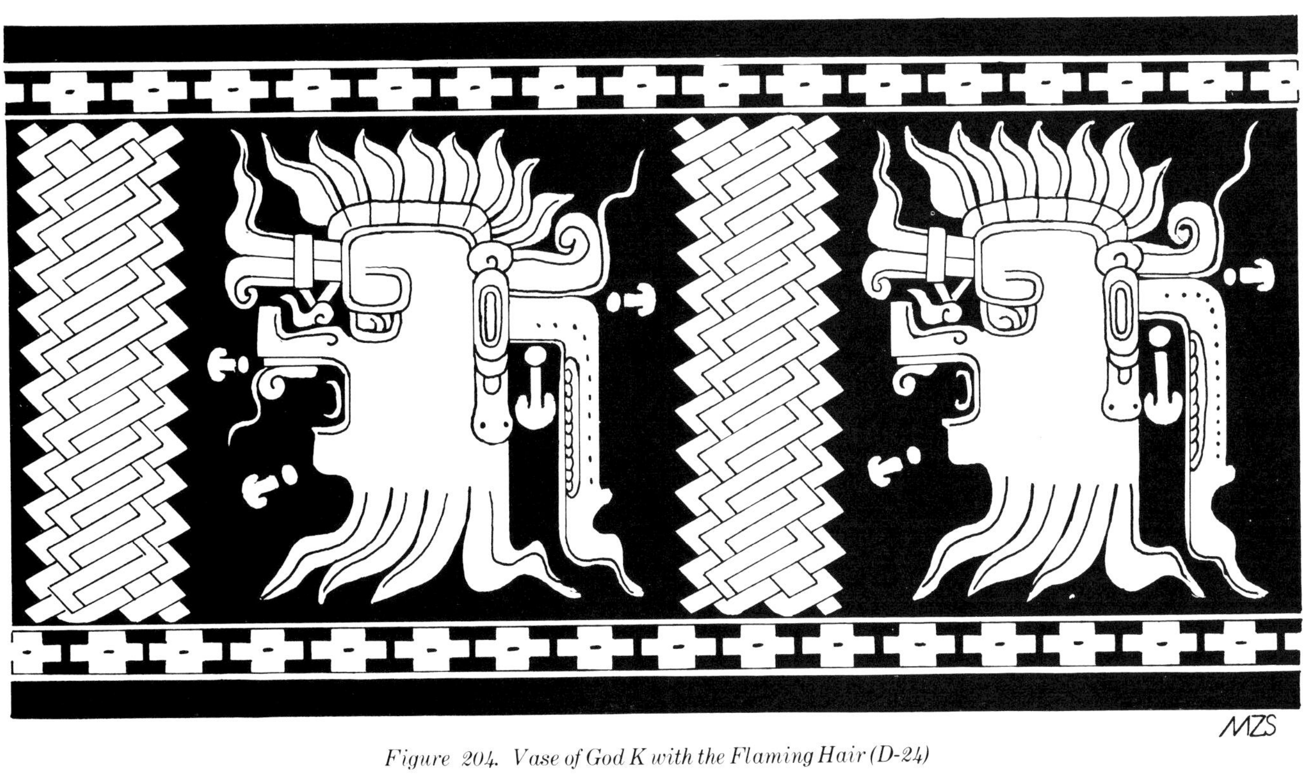

Figure 204. Vase of God K with the Flaming Hair (D-24)

repetitious *kan*-cross symbols. The flamelike motif on top of his head represents hair, characteristic of some Classic monumental representations of God K and also on vase D-4. A similar motif is seen at the base of the neck. The two heads are divided by two large, elaborate mat designs, symbols of power and overlordship.

As usual, there are subtle differences between the two portraits. The first head has two tadpolelike appendages, one in front of the nose, the other in front of the chin, as well as a barb-shaped whisker. The second head has only one tadpole motif in the front of the chin and has no whisker.

The **Vase of God K Enthroned** (D-25) is a polychrome vessel of the middle Classic period, provenance unknown (now in the collection of the American Museum of Natural History). The deity is portrayed seated Turkish fashion on a high throne decorated with a crosslike symbol with a circle in its center, similar to some of the architectural drum motifs found in Puuc centers. God K is wearing quetzal-feather floaters, and from his forehead the usual cigar tube and twin coils of smoke emerge. He has a shell pectoral, wide wristlets, and a narrow belt. His hands are painted red, and he has god markings on his arms and thigh. An identical

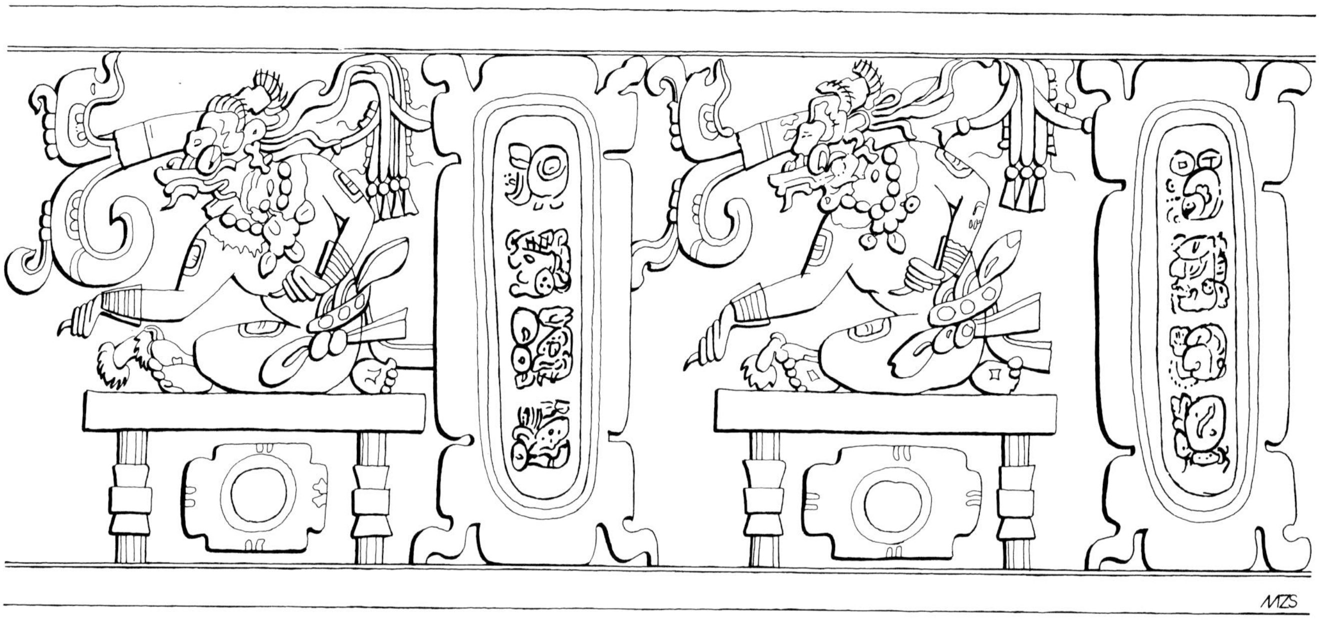

Figure 205. Vase of God K Enthroned (D-25)

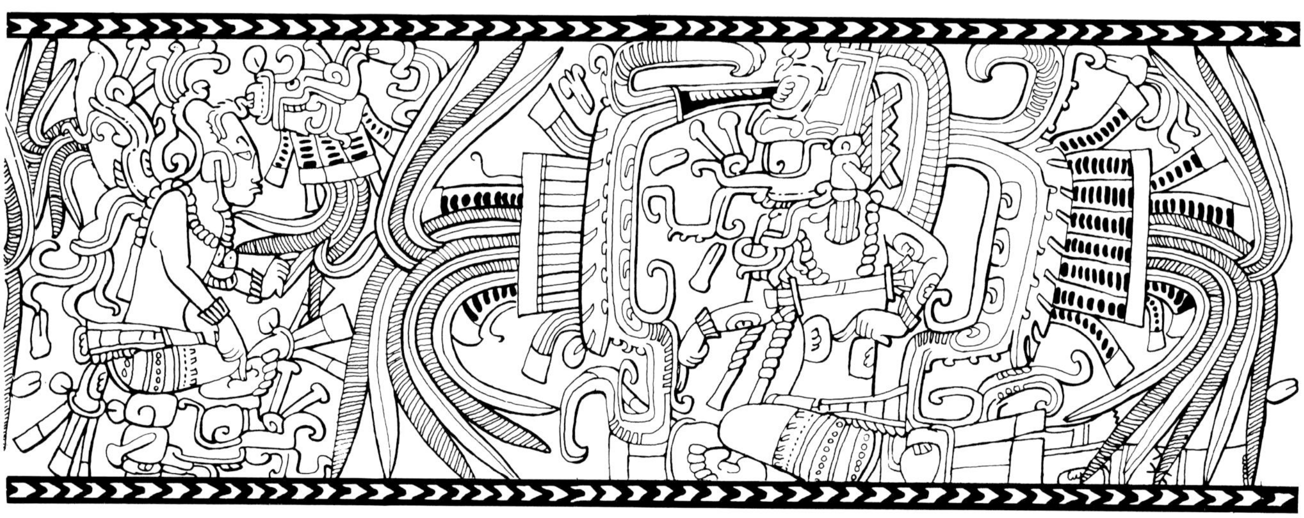

Figure 206. Vase of God K and the Young God (D-26)

portrait of the deity is painted on the opposite side of the vase. The two portraits are divided by vertical glyph blocks framed by elegant cartouches. The vase is painted on a whitewashed base in dark red and beautiful Maya blue.

Glyphic Text

The text of this pot is split into two sections of four glyphs each, but it was meant to be read continuously. The text begins with the Primary Standard Sequence. The introductory glyph is in the first position. The God N and "winged-*quincunx*" groups follow it at positions A2 and A3. The text from the opposite side names a historical (?) person. The second glyph is the Mah K'ina (T1010.184) title. The following two glyphs are the personal names. Presumably the second part of the text names the person who commissioned the pot. The glyphs appear to have been added to the pot after the figural drawing was completed. The quality of the calligraphy is certainly much worse in the glyphs than in the figures and probably was done by a different person.

A most unusual artistic style is expressed on the **Vase of God K and the Young God** (D-26), a vessel from the late Classic period, found in the Guatemalan highlands (now in a private collection). Its upper and lower chevron bands suggest that it was painted by an artist from the Chamá region. The kind of brush stroke used, however, and the "restlessness" of the composition, which comes through strongly, even though the figures are seated, is not seen on other Chamá vases.

The vessel is black and white on a base of Indian red. It portrays God K and a young deity whom I cannot identify. God K is seated with his shoulders twisted to face the viewer, but the rest of his body turned sideways. His "trademarks"—the huge double flare emanating from a forehead tube, the "mirror" grapheme from which the cigar tube rises, the large god eye, the upturned nose, the receding lower jaw, and the oval body markings—are all present. He is wearing a short kilt, a bead necklace, a horizontal ceremonial bar, and a most complicated headdress, which is as large as the god himself. The headdress is composed of huge coils and mat and flower motifs, as well as feathers of the quetzal bird. The hands of God K are in a peculiar gesturing position, with the index fingers pointing downward.

The second personage on the vase is a young man with classic Maya features, seated cross-legged on a monster head. He is scantily dressed in a kilt and a wide belt. Like God K, he wears a bead necklace and short wristlets and is holding his hands as though pointing at something or someone. Attached to his shoulderblades and forehead are complicated ornaments. His back ornament is winglike, with tassels, quetzal feathers, and short bars. The forehead ornament is composed of a small grotesque animal head, feathers, and floral motifs.

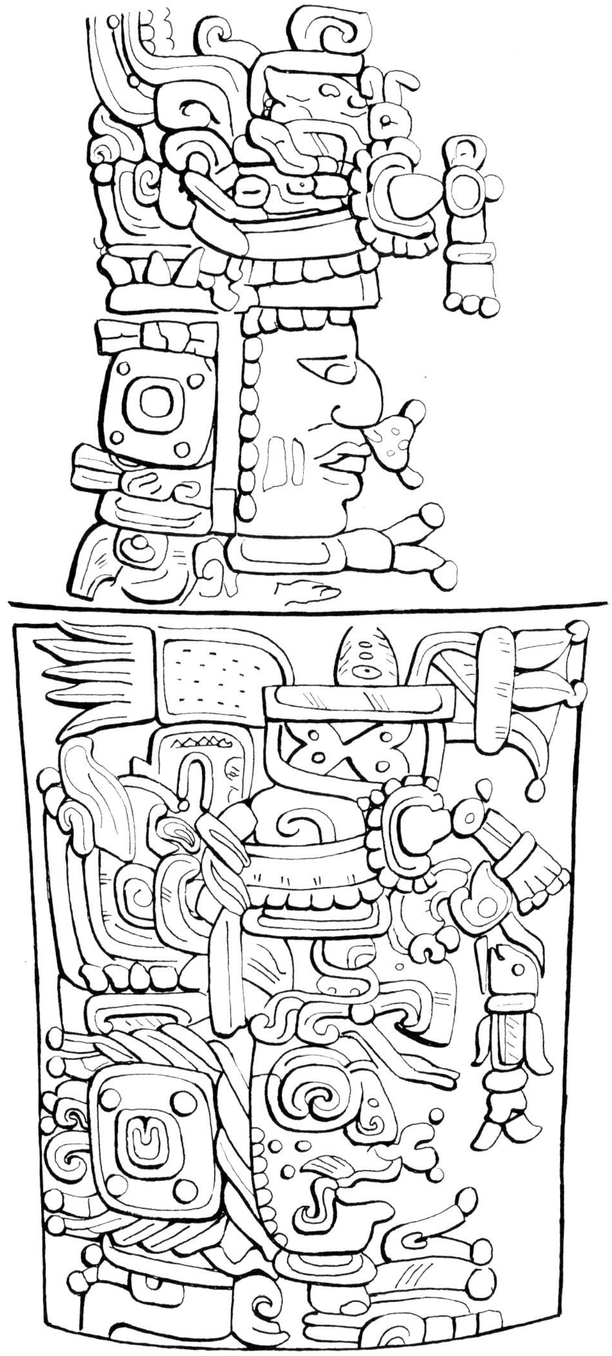

Figure 207. Ceramic Container of the Lord with God K Helmet (D-27)

God K is hidden in the headdress of a dignitary portrayed on the **Ceramic Container of the Lord with God K Helmet** (D-27) probably made in the early Classic period of northern Petén (now in a private collection). The top of the container is light reddish brown and is carved with the head of the noble. The container is one of a number of similar vessels from the early Classic period, part of an iconographic tradition that appears in the late Classic period on the stelae of Copán and the *incensarios* of Palenque. The lid of the cylindrical container shows an apparently human lord. He wears a flanged headdress (see Stelae 31, 1, and 2 at Tikal), which includes a chinstrap, marked with a zoomorphic head, a large disk earpiece with a knot (or "knot god") above and below, and a zoomorphic head. The upper part of the headdress includes a headband with a decorated object in front and an upended dragon head behind. The dragon head is attached to the headdress flange (see Stela 31 at Tikal). A God K head appears just above the headband with its hand raised behind his head. A lovely jaguar head rests above the God K head, and a trifoil motif extends behind it. The God K jaguar heads may name the young lord.

The lower half of the container has the head of a deity, which is marked by the fish barbel on the mouth, the aquiline nose, and the scroll eye as GI of the Palenque Triad. He also wears a flanged headdress, which includes zoomorphic heads at the chin and the same knotted flange and the upended dragon head. The zoomorph in the headdress wears the headband that is worn by the human above, but the tripartite (shell, sting-ray, and crossed-bands) symbol and the *kin* helmet appear above the headband and mark the zoomorph as the quadripartite monster associated with GI. A water bird perches on the muzzle of the monster and feeds on a fish. This entire set of iconography is related to the *incensario* complex of Palenque and is very closely associated with the iconography of the Temple of the Cross.

The colorful **Vase of God K with Beard** (D-28) is a ceramic masterpiece of a late Classic painter of Petén (now in a private collection). It shows two figures seated under a canopy. The figures are vivid red on a white base. The vase has several unusual features besides the beards. While the first figure has most of the characteristics of God K, such as the long snout and the extended forehead, it has neither a forehead tube nor smoke coils. Instead, a large flower ornament emerges from his forehead. The second figure is almost identical, but instead of the forehead flower it sports a forehead cigar. In both representations the god is seated on cushions. The left cushion is decorated with circular designs, while the design on the right cushion has the scalloped line representing the stretcher hole used in the curing process. There is a large bundle behind the right figure. The two figures are divided by wide red vertical lines. The vase raises some disturbing questions: Does it show a Jester God in two different aspects (one of them as God K)? Or is God K able to shed his principal "trademark," the forehead cigar, and display another emblem, a flower ornament, in its place?

An iconographically unique masterpiece is the **Vase of God K, the Young God, and God L** (D-29), by a Classic artist of Chamá (now in the November Collection). The vase has an orange base, and is painted in brilliant red, black, and white. It has upper and lower red rims, as well as black-and-white chevron bands.

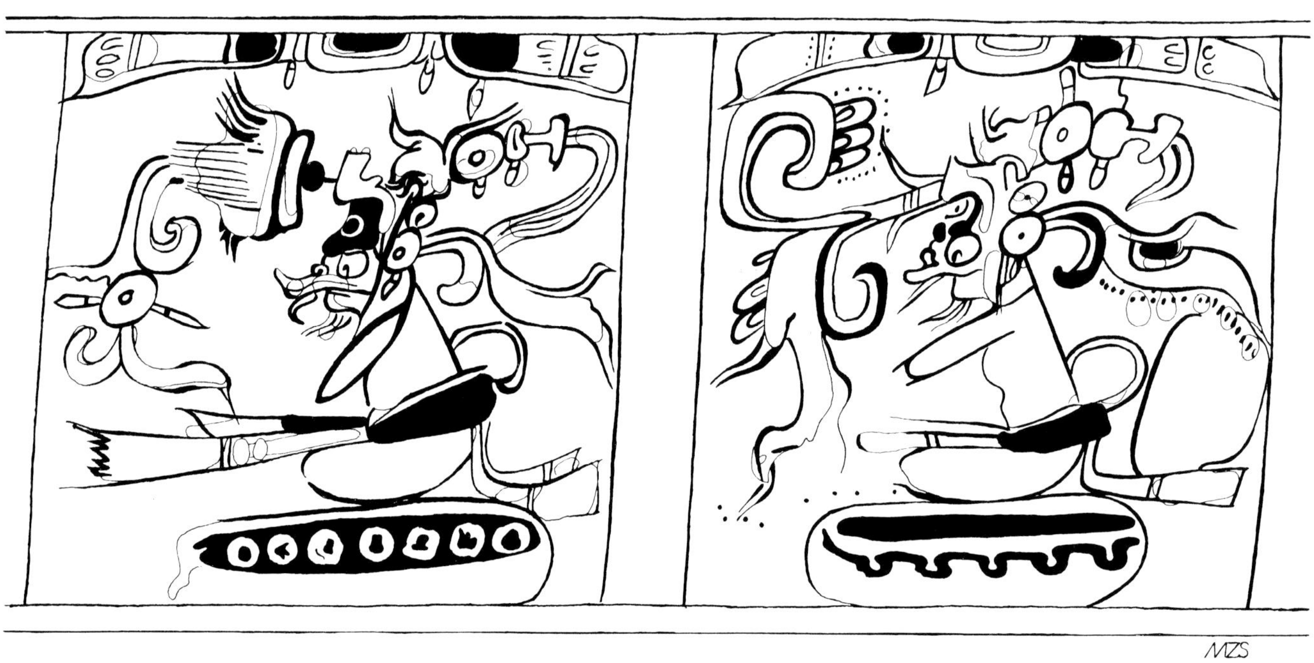

Figure 208. Vase of God K with Beard (D-28)

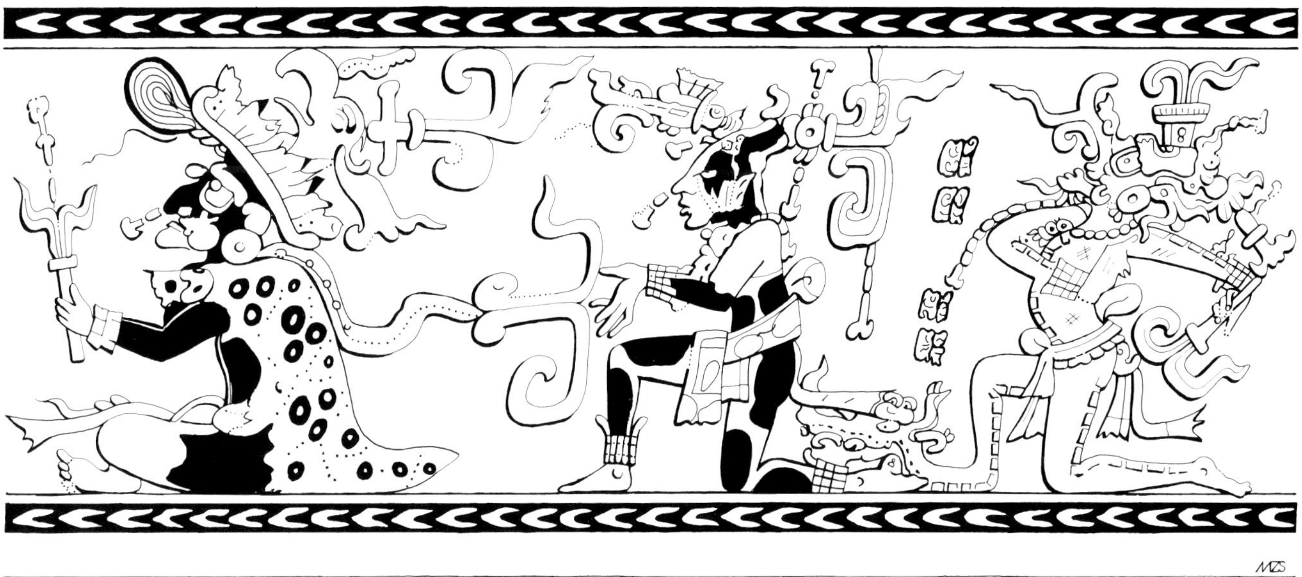

Figure 209. Vase of God K, the Young God, and God L (D-29)

The composition consists of the figures of three deities and four repetitious glyphs. The first figure is that of God K, who is in a peculiar half-kneeling posture. His head and right shoulder are thrown back, and his chest is pushed forward. His right arm is extended, and his left hand is touching his right shoulder, as if he is rehearsing a love scene in a play. His facial features are the characteristic ones. He has a short beard, saurian markings run downward along the contours of his body, and in place of his left foot is a large serpent head. He is wearing a narrow belt with front and back hangers, a bead necklace, wide wristlets, an earplug, and a flower headdress. His tall forehead is pierced by a well-painted smoking tube. In his right hand he is holding a second such "cigar," probably a spare.

In front of God K is the figure of a youthful male, probably one of the Young Lords, also in a half-kneeling posture, but with his upper body bent slightly forward. His attire and jewelry are identical with that of God K, but his body is covered with large black spots, almost certainly indicating that he is dead. He wears his hair gathered with a double-scroll back and a front floater. The foot of the Young God is apparently caught in the jaws of God K's serpent leg, an iconographic feature I have not seen elsewhere.

The last figure in the composition is the seated figure of God L. His body, legs, chin, and forehead are painted black. He wears his usual Moan-bird-and-feathers sombrero and long jaguar-hide cloak. Attached to the tail of the hide is a scrolled glyph symbol. In his hand he is holding a smoking cigar tube identical to those worn and held by God K.

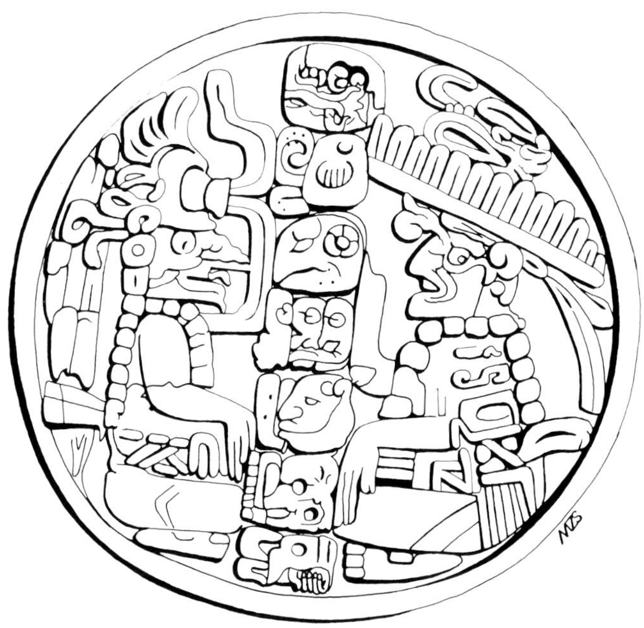

Figure 210. "Poison" Flask of God K and God L I (D-30)

The small **"Poison" Flask of God K and God L I** (D-30) is said to be from El Progreso, Honduras (now in the Middle American Research Institute, Tulane University).[5] The beautifully executed molded *venenero* has a hieroglyphic band running across the narrow shoulder from the spout to the bottom. An additional column of glyphs also divides the composition on the front, where Gods K and L are shown seated facing each other. Both deities serenely pose on a low dais wearing belts and long bead necklaces. The high forehead of God K is

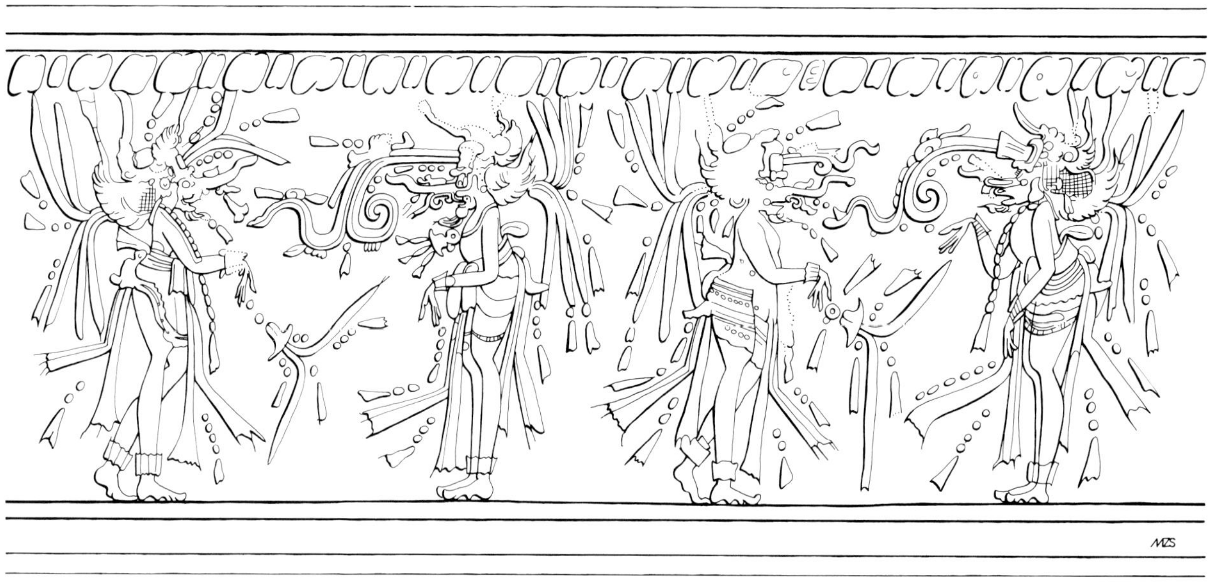

Figure 211. Vase of the Four Standing Monster Gods (D-33)

pierced by a smoking cigar. God L's head is covered with his usual wide-brimmed hat, on top of which the Moan bird perches. I know of two other such flasks of the same mold, D-31 and D-32 (both in private collections). The text on these little flasks does not appear to name either God K or God L. The fifth glyph includes T12 (the male article) and a bat head; it is apparently a name, but without a verbal glyph.

The **Vase of the Four Standing Monster Gods** (D-33) is a masterpiece of late Classic polychrome pottery. The vessel (now in a private collection) was probably made by an artist of Petén. It is painted on a brilliant orange base in red, black, and white, with upper and lower conventional rim bands and an upper circular row of glyphoids.

The four figures are arranged in pairs of two, standing face to face. Their reptilian heads, god eyes, recessed lower jaws, high foreheads, and proboscis-shaped noses place them in the God K category. The noses of these monsters are especially long and curled and are best described with the words of Edmond Rostand about Cyrano de Bergerac:

> *Insolent:* Sir when you smoke, the neighbors must suppose Your chimney is on fire. *Cautious:* Take care—a weight like that might make you topheavy. *Thoughtful:* Somebody fetch my parasol—those delicate colors fade so in the sun! *Pedantic:* Does not Aristophanes mention a mythological monster called Hippocampelephantocamelos? Surely we have here the original![6]

The faces of all four figures are painted red, but the third and fourth have white areas around their eyes and on their chins. The red paint extends downward to the chests of the second and third figures. All of them wear red headdresses composed of matted elements, birds' wings, and feather floaters. From the forehead mirrors of figures 1, 2, and 3 double smoke coils emerge, but only the first monster has the usual forehead tube. The forehead mirror of the fourth creature spurts feathers instead of smoke coils. The bodies of all four figures are covered with black paint, leaving only their feet and hands unpainted. Monsters 1 and 3 are standing still; the other two creatures are apparently walking. All four figures wear kilts and belts of red and white, back and front hangers, wristlets, and gaiters, as well as bead and shell necklaces.

This vase, like vase D-28, seems to indicate the existence of more than one God K, each with different regalia and possibly a different function.

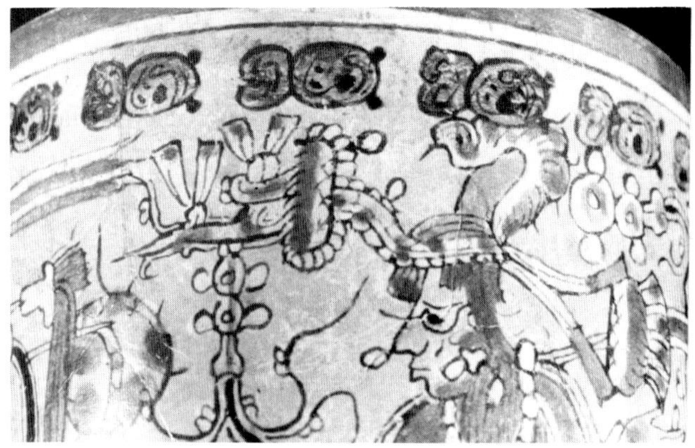

Figure 212. Detail, polychrome vase, Classic Period, showing a dignitary with monster headdress

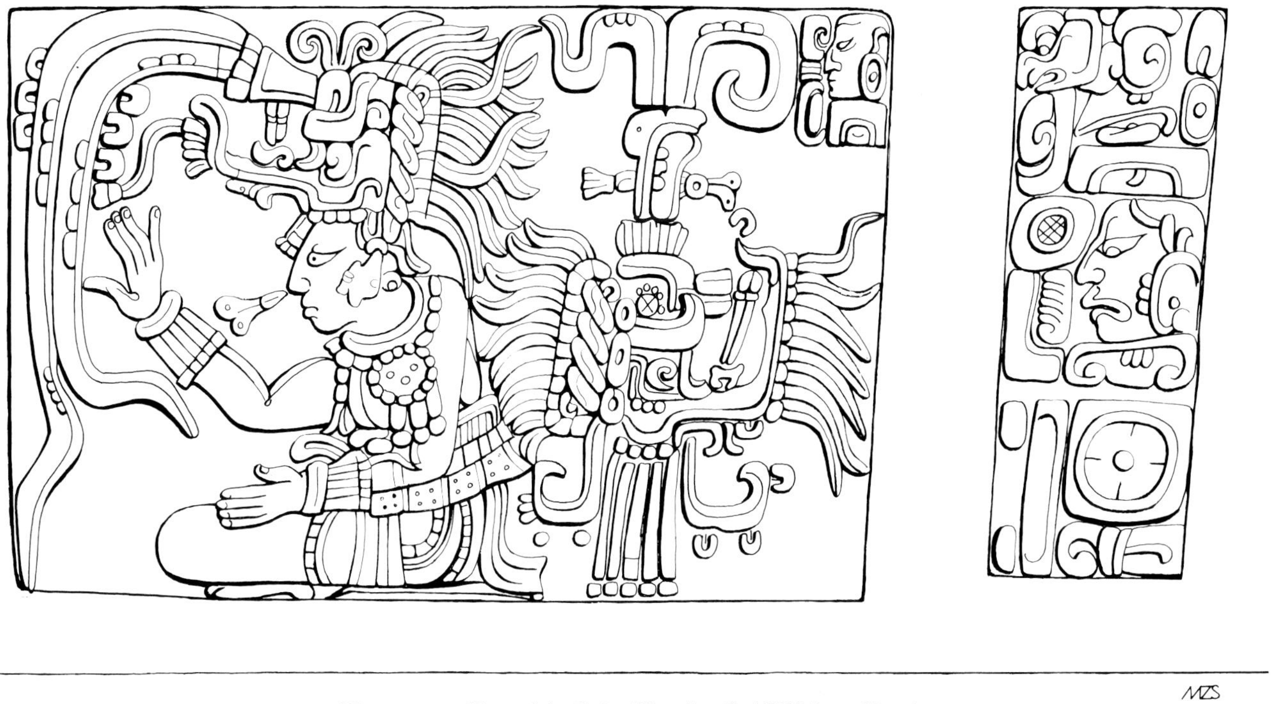

Figure 213. Vase of the Ruler Wearing God K Helmet (D-34)

The **Vase of the Ruler Wearing God K Helmet** (D-34) was first exhibited in 1969 by the Instituto Italo-Latino Americano in Rome.[7] It was probably created in Petén during the late Classic period (now in the Museo de Popol Vuh, Francisco Marroquín University, Ela and Jorge Castillo Collection). The vase is carved with the image of a seated dignitary wearing a broad necklace, a breastplate, wide wristlets, and a kilt. His right hand is extended as if in greeting. He is distinguished by the magnificent head of God K that he wears as a helmet. From the forehead of the helmet rises a huge smoking cigar, and there are additional smoke coils emanating from the top of the headdress, which is also richly decorated with quetzal feathers.

Closely behind but not attached to the lord's attire is a large, long-nosed ophidian monster head. The monster has no lower jaw but a long, quill-shaped appendage emerges from its mouth. It is decorated with a helmet composed of a mat symbol, a net element, and quetzal (?) feathers and is topped with a smoking-Ahau-variant glyph complex. A tube emerges from the ridge of the nose, and feathers also flare from the upturn of the fretted nose. There is a small glyph block in the upper right corner of the composition and three additional large glyphs on the back of the vase, probably naming the seated ruler.

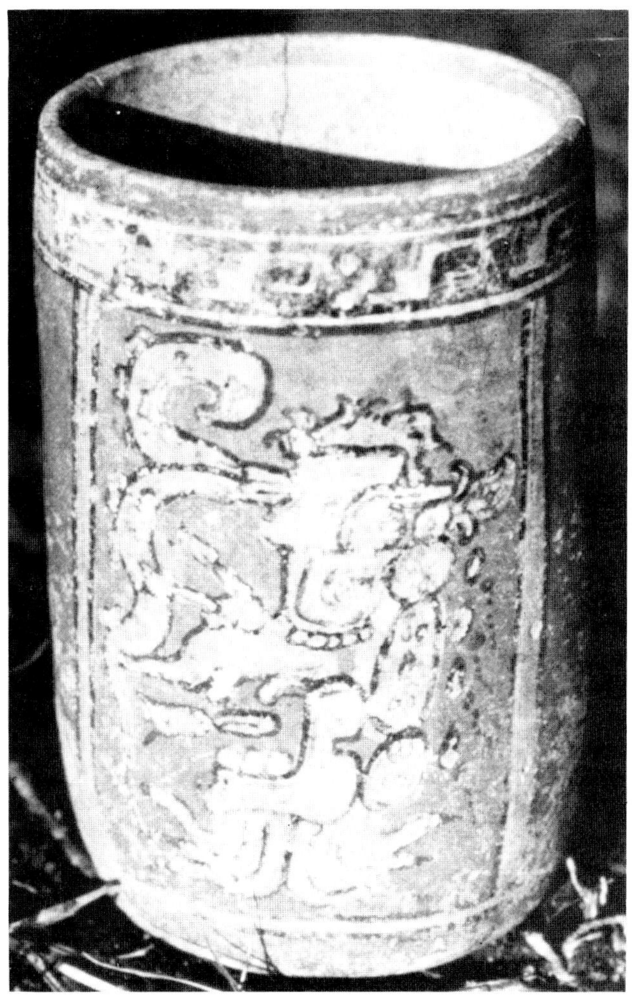

Figure 214. Red Vase of God K (D-35)

God K's head decorates the two sides of the **Red Vase of God K** (D-35). The vessel was found in northern Petén (now in a private collection) and was

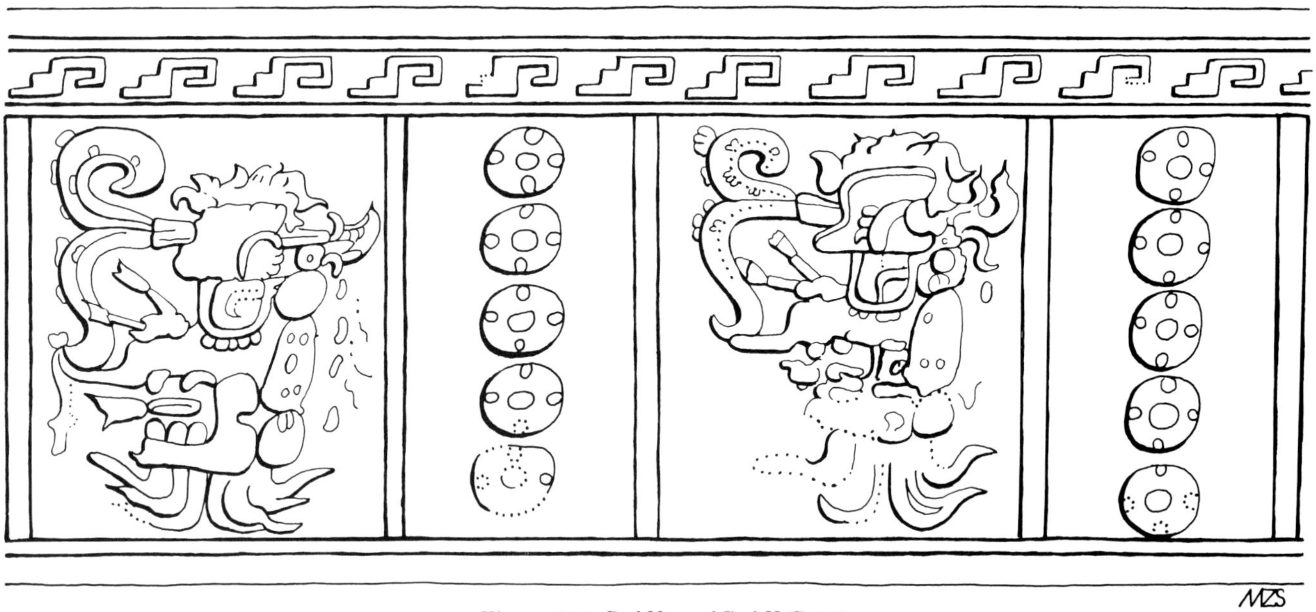

Figure 215. Red Vase of God K (D-35)

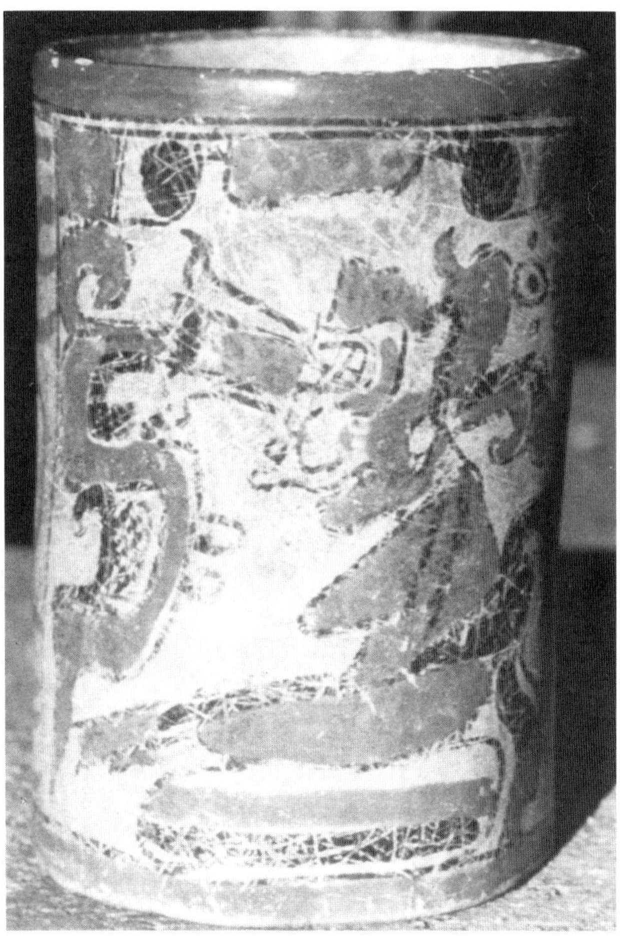

Figure 216. Vase of God K with the Nape Scroll (D-36)

probably made during the late Classic period. It is painted with an upper band with a repetitious motif that strongly resembles the "monkey tail" symbol of the Nicoyas of Costa Rica. The area below the band is divided into four bright-red panels, two of them painted with rather conventional heads of God K and two with vertical columns of five *kin* glyphs. The heads have flamelike hair, rather fleshy noses from which drumsticklike appendages emerge, and bleached jaws. Both heads rest on scrolls.

The **Vase of God K with the Nape Scroll** (D-36) is a product of a late Classic Petén potter (now in a private collection). In style and color the vase strongly resembles the Vase of God K with Beard (D-28). The vase features two identical seated portraits of the deity, each shown on a low dais or pillow. The body of the god is painted red with black shoulder markings. He wears a narrow belt with back and front hangers. His facial features and his large forehead cigars are typical. Special features of the portraits are the large scroll rising from the nape and an unusual tadpolelike ornament above the winged crown of each head. The red-and-black design suspended on the upper red rim band may indicate cushions and thus that the scene takes place inside a building. The two portraits are divided by semicircular symbols in a paneled arrangement.

From the studio of one of Petén's "codex-style" ceramic artists came the **Vase of the Head of God K Painted Codex Style** (D-37). The vase (now in a

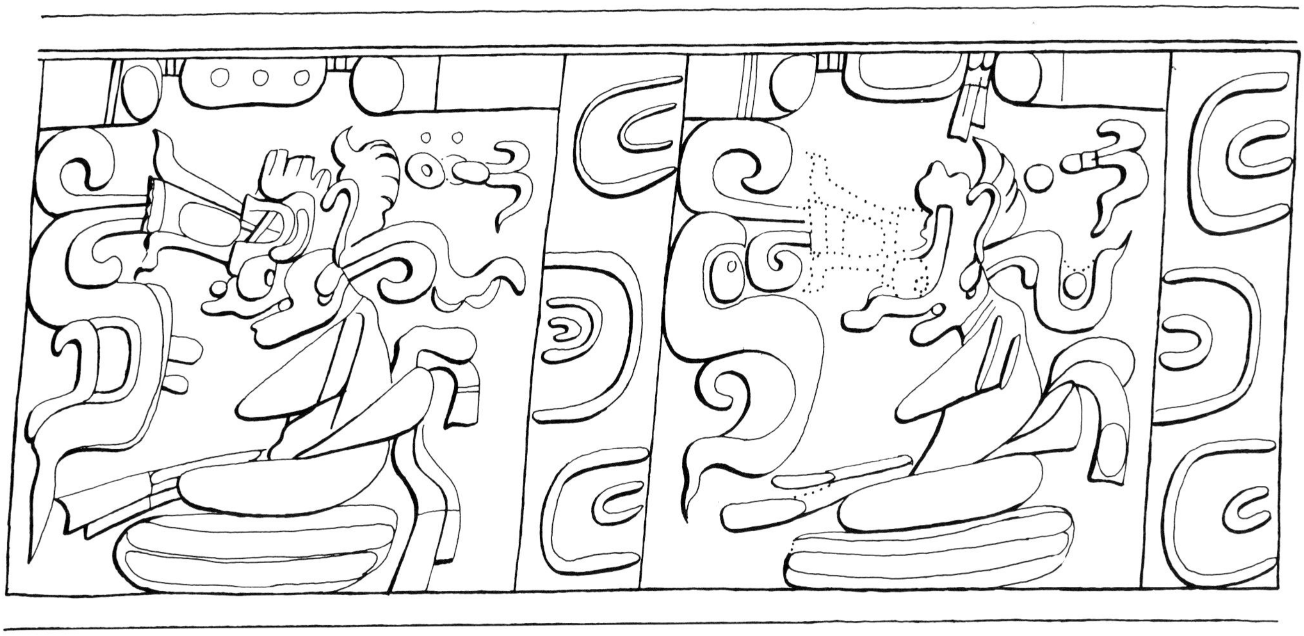

Figure 217. Vase of God K with the Nape Scroll (D-36)

private collection) was made during the late Classic period. It is painted with a fine black line on a white base with a red rim band. The composition consists of two presentations of the head of God K. The god is shown without a lower jaw, with unusually large smoking forehead tubes, a manelike hairdo, and large *kin* crosses instead of "mirror" graphemes infixed in his forehead. There are also round "pompoms" on the forehead tube, a decoration I have never seen before on similar representations.

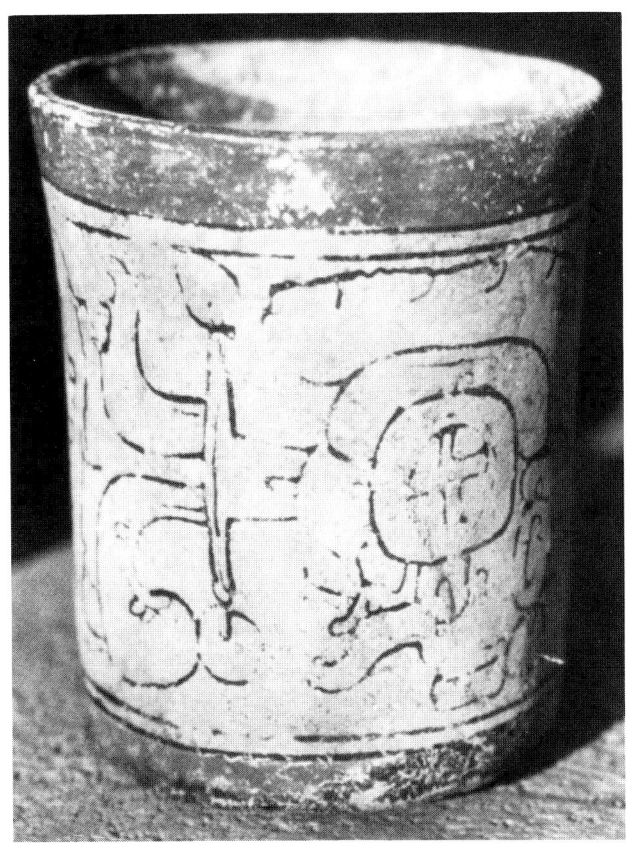

Figure 218. Vase of the Head of God K Painted Codex Style (D-37)

What may be one of the most beautiful works of the ancient Maya potters is the heavily eroded **Vase of God K Painted on a White Base** (D-38), made during the middle Classic period of northern Petén (now in a private collection). The vessel is painted on a white base with masterful black brush strokes. The composition originally included two portraits of the God with the Forehead Cigar seated in semicircular cartouches. The figures of the deity, the forehead mirrors, the flamelike hair, the smoking cigar tubes are clearly recognizable, but most of the bodies and all of the faces have eroded away. The upper and lower black rims and the beautiful glyphic band are virtually intact.

A simple drawing of another vase is shown in Smith and Kidder, where a double portrait of God K's head appears in a paneled arrangement.[8]

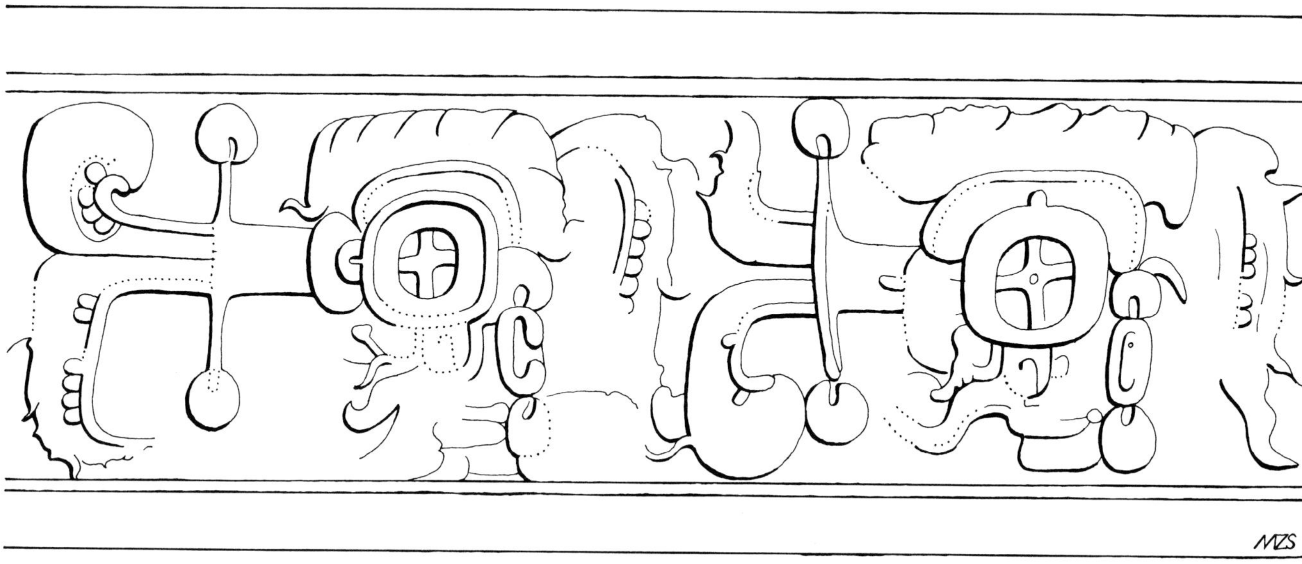

Figure 219. Vase of the Head of God K Painted Codex Style (D-37)

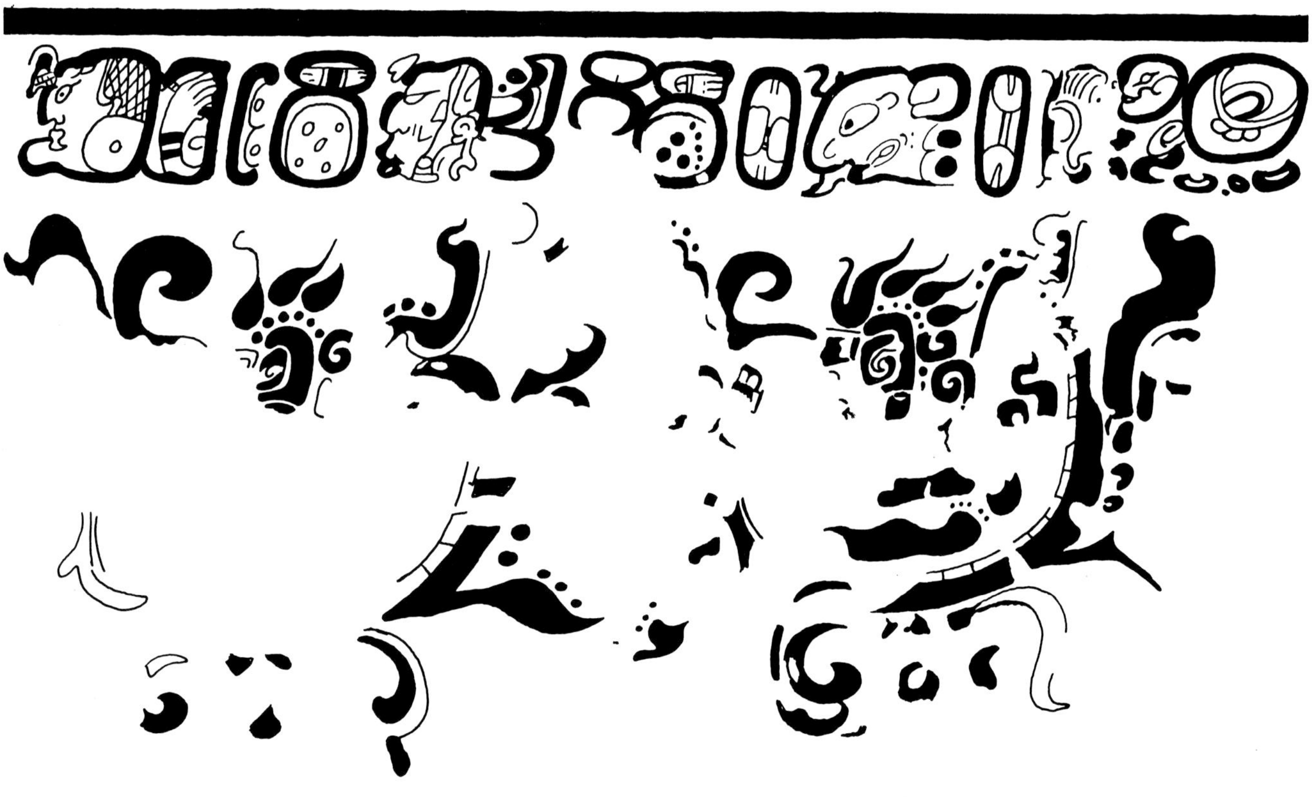

Figure 220. Vase of God K Painted on a White Base (D-38)

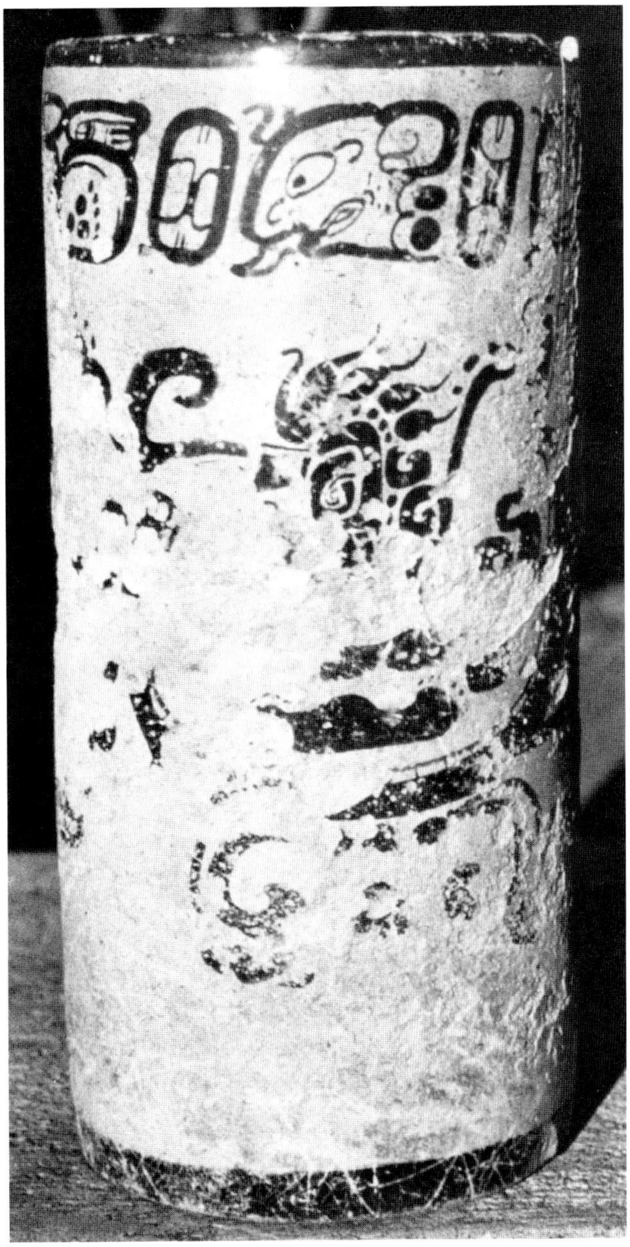

Figure 221. Vase of God K Painted on a White Base (D-38)

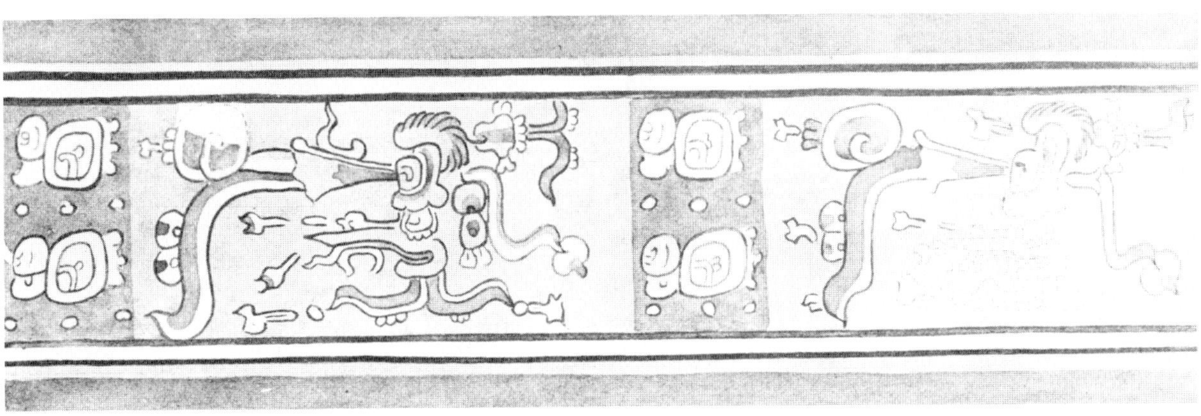

Figure 222. Drawing of vase D-38 (after Smith and Kidder)

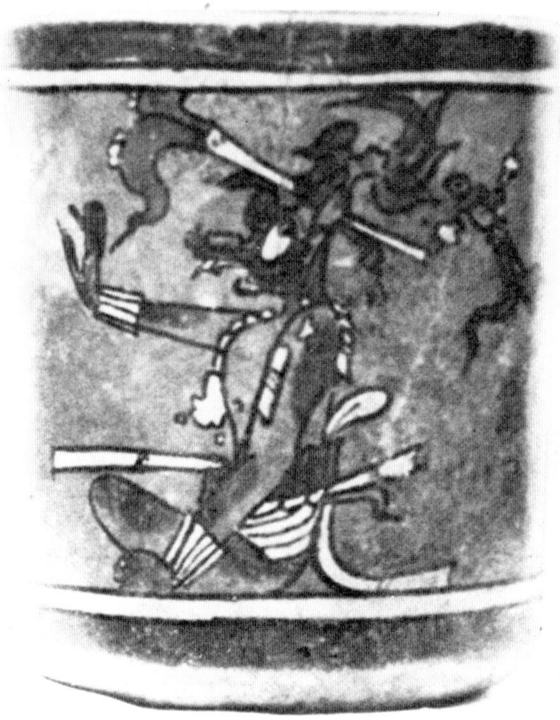

Figure 223. Bowl of God K with Long Cigar (D-39)

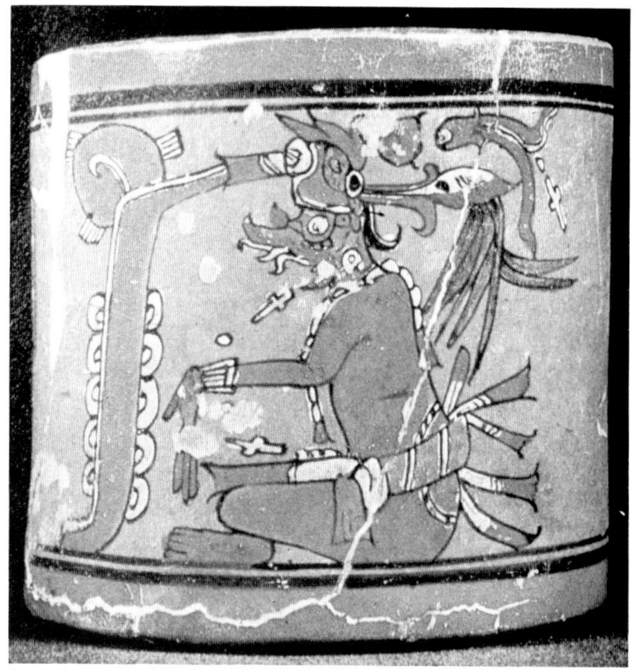

Figure 224. Vase of God K Painted Red (D-40)

To complete the God K ceramic series, listed below are three additional ceramic artifacts to which I had no personal access and about which I have limited documentation.

The **Bowl of God K with Long Cigar** (D-39) is a polychrome vessel from the late Classic period of Nebaj (now in the Morant Collection, Paris). The aspect of the vase presented in the catalogue of a 1963 exhibition shows a rather primitively painted seated figure of God K wearing a white kilt, wristlets, a bead necklace, and a flowered headdress back floater. An unusual feature of the portrait is the exceptionally long, slender white "cigar" that perforates the God's cranium and emerges at the nape.[9]

The **Vase of God K Painted Red** (D-40) is probably from Petén, made during the late Classic period (present whereabouts unknown). It shows the seated figure of God K in brilliant red on an orange base. The god's nape is decorated with stylized flowers and quetzal feathers. He wears a narrow belt and a bead necklace, and from his forehead emerges a large tube with a double coil, the upper one foreshortened.[10]

God K is also seen on the **Potsherd from Santa Cruz Quiche** (D-41), published in Spinden's *A Study of Maya Art.*[11] The drawing shows the snake-footed deity in a reclining posture.

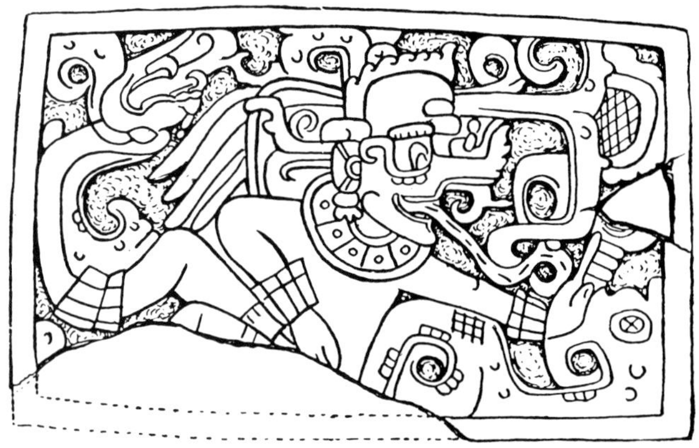

Figure 225. Potsherd from Santa Cruz Quiché (D-41) (after Spinden)

Frescoes Showing God K

One of the two frescoes bearing the image of God K is the **Painted Capstone of Xnucbec.** The small, beautifully executed wall painting was found in 1973 by Evaristo Pacheco Azul at the archaeological site at Xnucbec, south of Bolonchén, Campeche, and was first described by Eric Von Euw.[12] The frescoe (now in the Centro de Restauración, Mérida) has also been studied by Pavon Abreu and by Cynthia Gutiérrez-Gonzales.[13] It shows the standing figure of God K with a large forehead tube and smoke coils. He is holding a smoking flare (cigar?) in his right hand. He is wearing a loincloth, a bead necklace (perhaps jade), gaiters, and feather floaters. He appears to be standing in front of a large container.

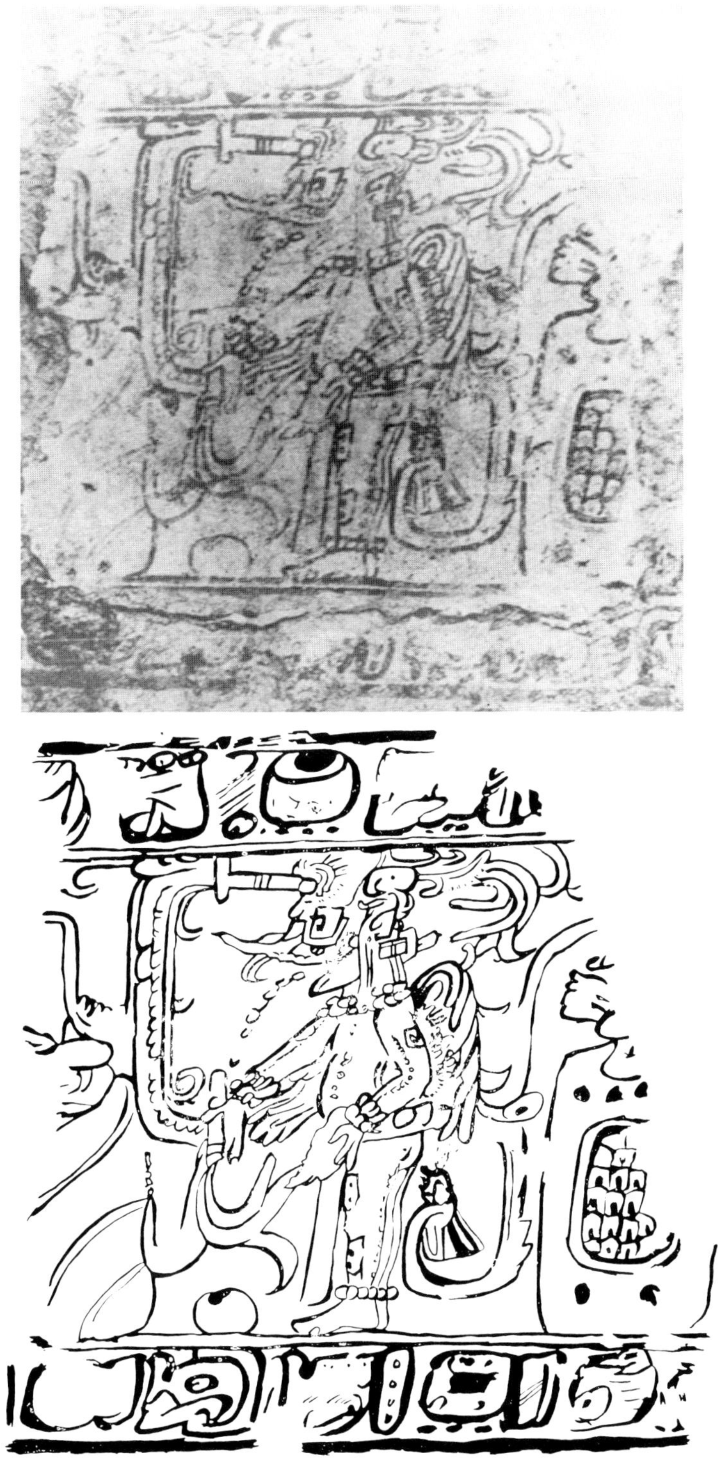

Figure 226. Painted capstone of Xnucbec (photograph after von Euw, drawing after Gutierrez-Gonzales)

An upper and a lower row of glyphs partly covered with stucco complete the composition.

The only other known possible representation of God K on a **wall painting is in the right middle register of the fresco in the Temple of the Frescoes (Structure 16) at Túlum.** This detail of the surprisingly well preserved mural shows an elaborately dressed personage standing in front of a ceremonial object, probably an *incensario.* His facial features and his attire indicate strong Toltec influence. There is a small long-nosed deity on the lord's back, and he is holding a second small god in his hands. The latter also has a long nose, and, as well as can be determined from the painting, which is somewhat eroded in that area, a flare is emerging from his forehead.

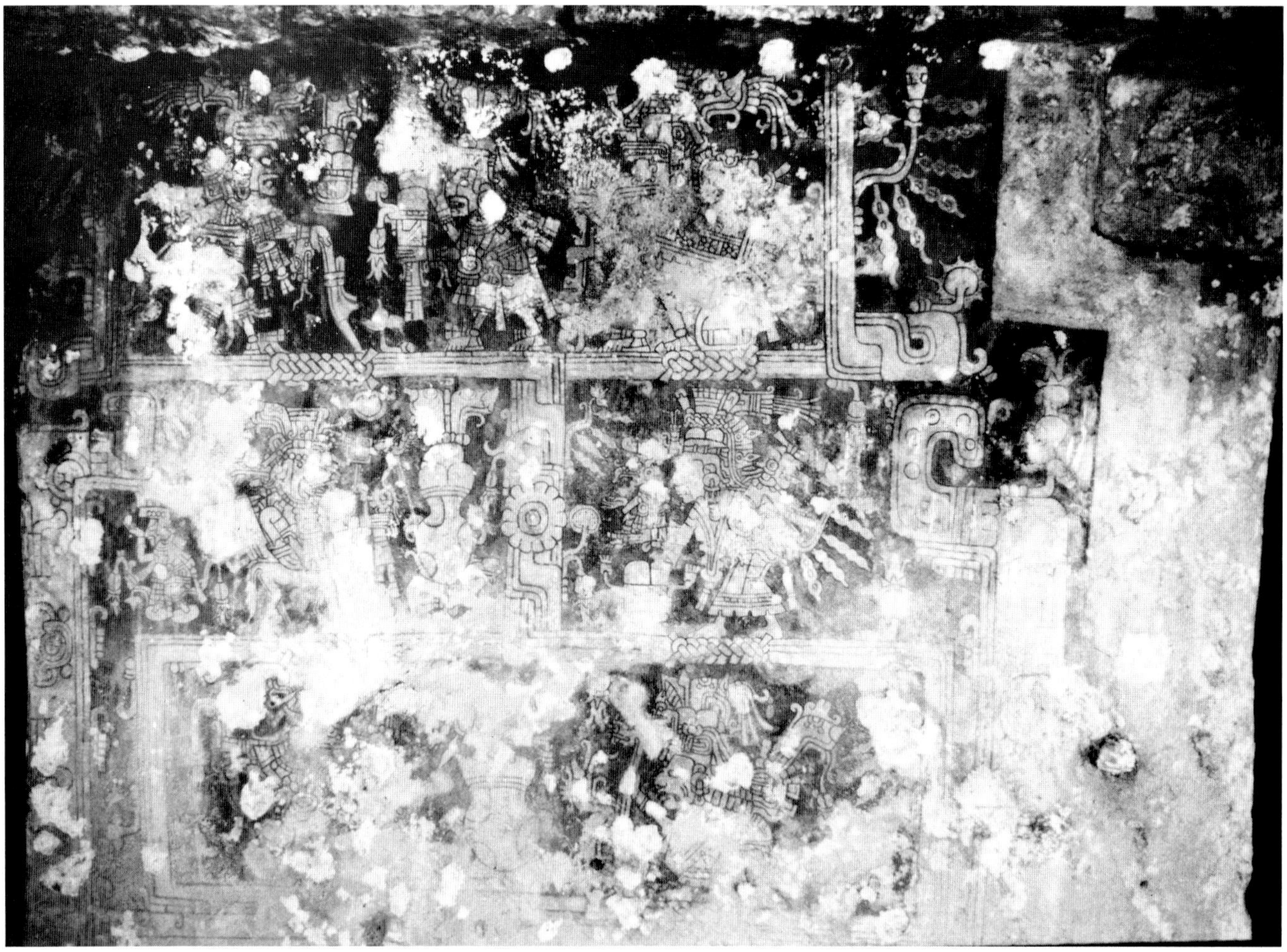

Figure 227. Wall painting, Temple of the Frescoes, Structure 16, Tulúm (photograph courtesy William A. Garrett, drawing after Dávilos)

Scenes Related to God K Iconography

The **Vase of the Court Attendants** (E-1) is a tall, slightly flared container from the middle Classic period, probably from southern Campeche (now in a private collection). It is painted in subdued red and purple on a white base. It depicts a scene that may be related to the "flare-god" concept. Four figures appear in the scene. The most important person is apparently a hefty warlord. Parts of his face and neck are painted red. He is bare to the waist, wearing a red kilt and an apron, a heavy jade-bead necklace, ear flares, and a cap with floaters of bird plumage. His right hand rests majestically on a battle shield. He is looking at the figures of two musicians, a woman and a man, who are shaking rattles. They wear headdresses with rich floaters of quetzal feathers, bead necklaces, and ear ornaments. Parts of their faces are painted red. The man is wearing a short jaguar kilt; the woman is dressed in a long, elegant skirt with a quadrangular pattern. Behind the lord is a large cube-shaped bundle flanking the figure of another attendant dressed in a purple kilt with a circular pattern and a helmet topped with quetzal feathers. In his left hand is a small unidentifiable fish-shaped object. The scene is broken by one angular and two vertical glyph columns. One detail of the headdresses deserves special attention: on every one of these helmetlike head ornaments s a slim attachment that rises at a forward angle from the portion covering the forehead. It is a narrow white rod with a distinct cigarette appearance. On the figures of the warlord and the woman musician the rods end in small, round, red ornament (glow?), but not on the male musician and the attendant. This object may or may

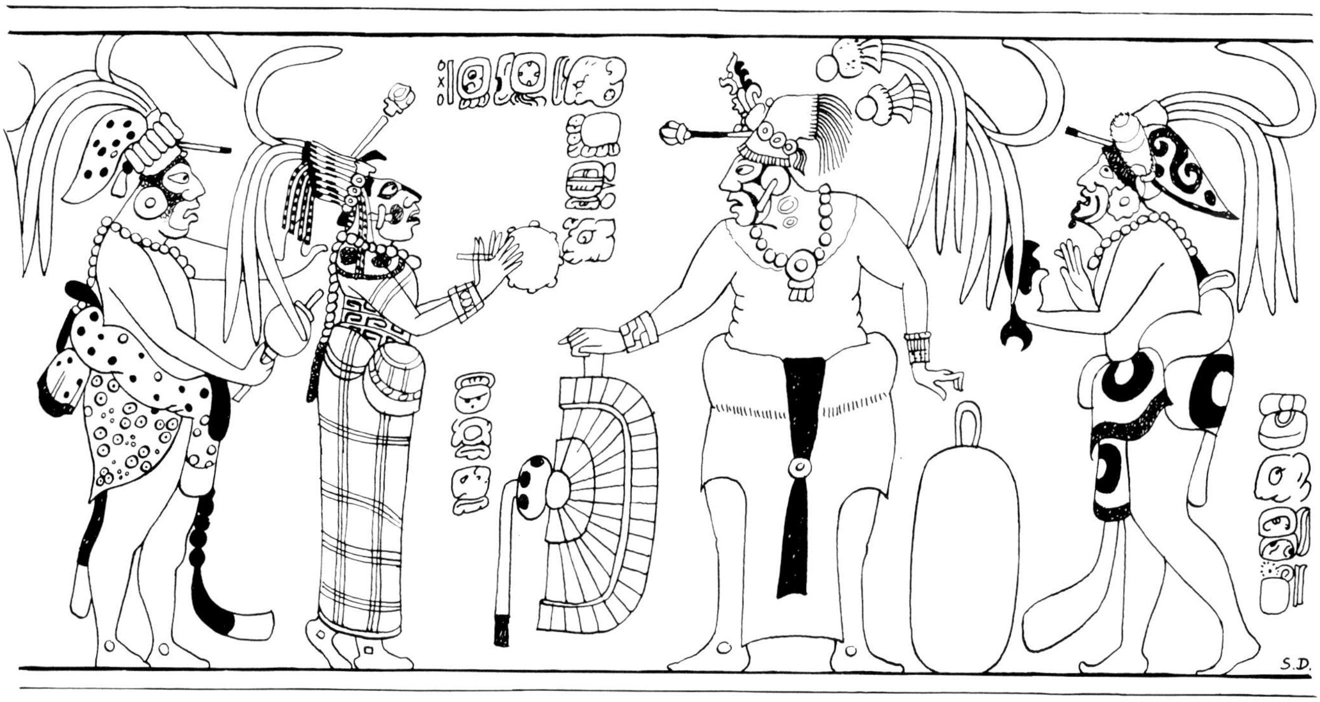

Figure 228. Vase of the Court Attendants (E-1)

not be related to the "forehead-cigar" concept. Similar drumsticklike forehead ornaments frequently occur on *copador* vessels, but they seem to have undergone an evolution into the shape of a flower rather than into a smoking object.

Glyphic Text

The first text is L-shaped and located between the main figure and the left attendant. The first compound appears to be 7 Cimi, but a *haab* position is not present to complete the Calendar Round. It must be assumed, therefore, that in this case the scene occurred in the 260-day *tzolkin* and the Calendar Round notation was not important.

The second compound appears to be T204.544:116. T204, "flattened fish," is known to function as one of the third-person possessive pronouns, but on Stela J at Quiriguá it also substitutes for T205, which is the *xoc*, "count," indicator. The main grapheme is the *kin* sign, and the affix is T116, accepted by Lounsbury and us as the phonetic determinative *n*. This compound may read either "his day" or "count of the days."

The third compound is T1.757, which is regarded by Proskouriakoff as a phrase introductory expression. [Schele has] proposed a reading as *umulih* or *t'ulah*, which in both readings means "to follow behind," and in this context may have read "it follows . . . ." The fourth compound is T238 (?). T238 is Landa's first and often serves as the male article. The main grapheme cannot be seen well enough for proper identification. The fifth compound is T561:23.74, *caan: (na).mah*, which may be read as "sky" with the superlative adjective "great." The final grapheme is a zoomorphic head. The verbal compound of the passage is located in the third position,

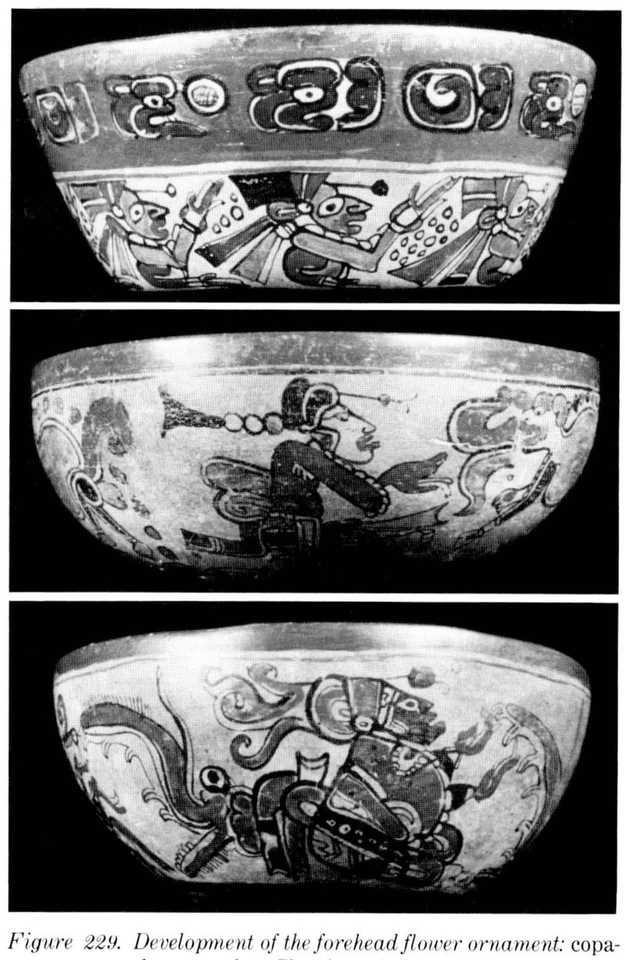

Figure 229. Development of the forehead flower ornament: copador *vases, late Classic period*

and the subject of the verb is recorded in the final three compounds of the passage. It is possible that the subject of the verb also identifies the main figure.

The second text is composed of three compounds located in a column to the right of the opposing attendant. The passage may name the attendant, but the grapheme again cannot be seen well enough to read. The third text is located on the opposite side of the vase between the opposite facing attendants. Again the details of the graphemes are unidentifiable. It is unlikely that it names either of the attendants since both face away from the text. The second grapheme looks very much like the T758 "rodent head." The third compound (T521:526.117?) is composed of *uinal* and *caban* graphemes.

A cigarettelike object also emerges from the forehead of a lord on the **Vase of God L, the Attendant, and the Lady** (E-2), a late Classic ceramic vessel, probably from Petén (now in a private collection). In the scene painted on the vase are three persons: an axwielding lord, a seated woman, and a pale, eerie individual holding a lance. The vase has an orange background, an upper and a lower red rim, and a short row of hieroglyphs. The principal figure is that of the lord, who is portrayed in profile with his body turned toward the viewer and his head looking sideways. Either his feet are omitted or he is kneeling. On his short, well-groomed coiffure he wears a pointed hat with a feather crown on which a bird perches. He probably represents God L. The feathers, as well as the entire body of the lord, are painted in alternate black-and-white stripes in a manner often seen in Aztec iconography but seldom encountered in Maya art. He is dressed in a short, "polka-dotted" skirt with a large, unconventional back hanger. Around his waist is a wide, red sash that falls forward as a narrow apron. In his left hand is a javelin; in his right he holds a battleax. From his forehead emerges a narrow white rod with a red tip, which may represent a cigarette with a glowing end. He wears an elegant necklace and a round ear ornament.

The lord is looking at an individual standing opposite him who is also holding a lance or javelin in his hand. The tip of his ornamented weapon is tilted toward the lord as if he is saluting him. The appearance of this figure is unparalleled on Maya ceramics. He is painted entirely white except for a black circle around his eye and mouth, and he is naked except for his black belt and magnificient red back ornament of feathers. He wears a round pectoral and a (shell?) earplug. His hair is indicated by a wavy outline, but is not colored. From his forehead, as from the ruler's, rises a narrow, red-tipped rod on which an ornament representing either an unusual flower or a death eye is suspended.

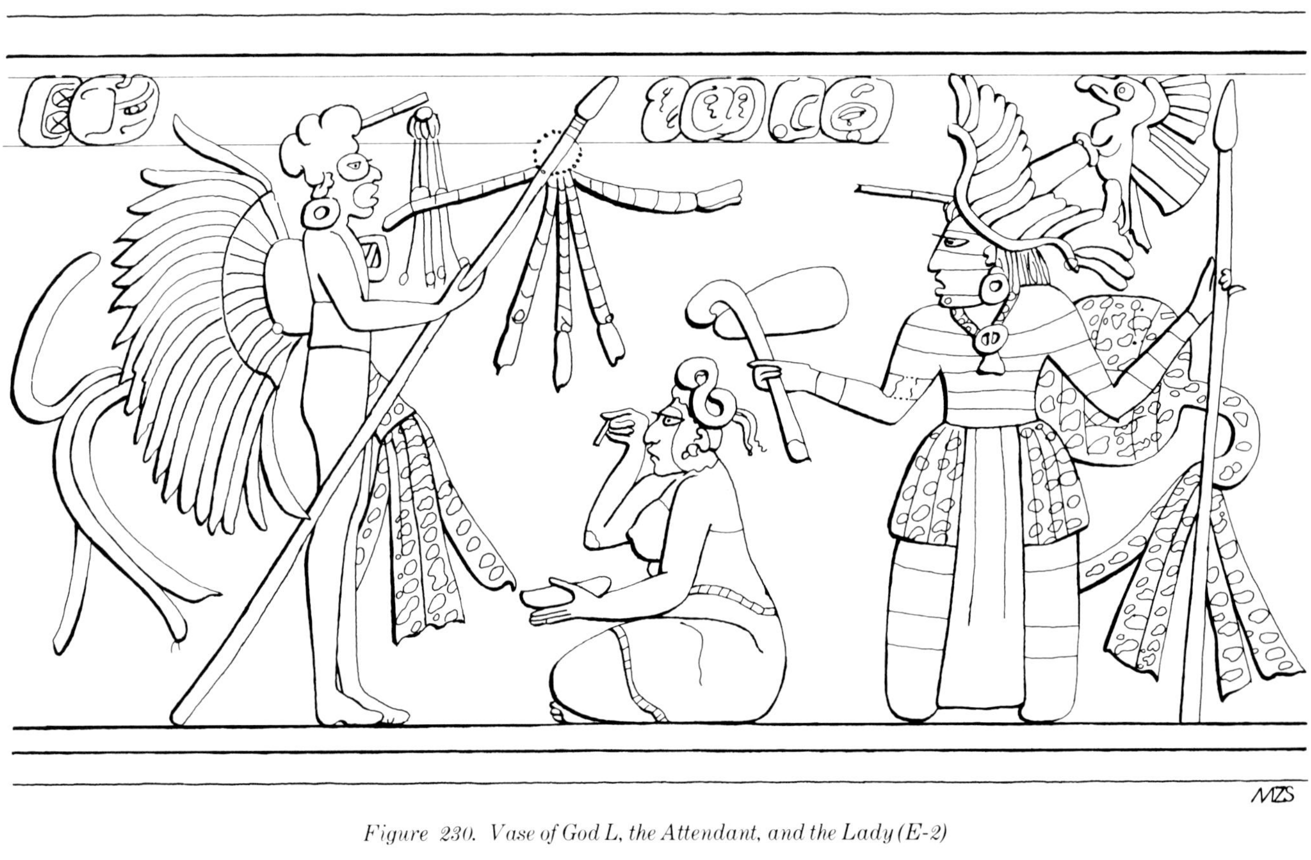

Figure 230. Vase of God L, the Attendant, and the Lady (E-2)

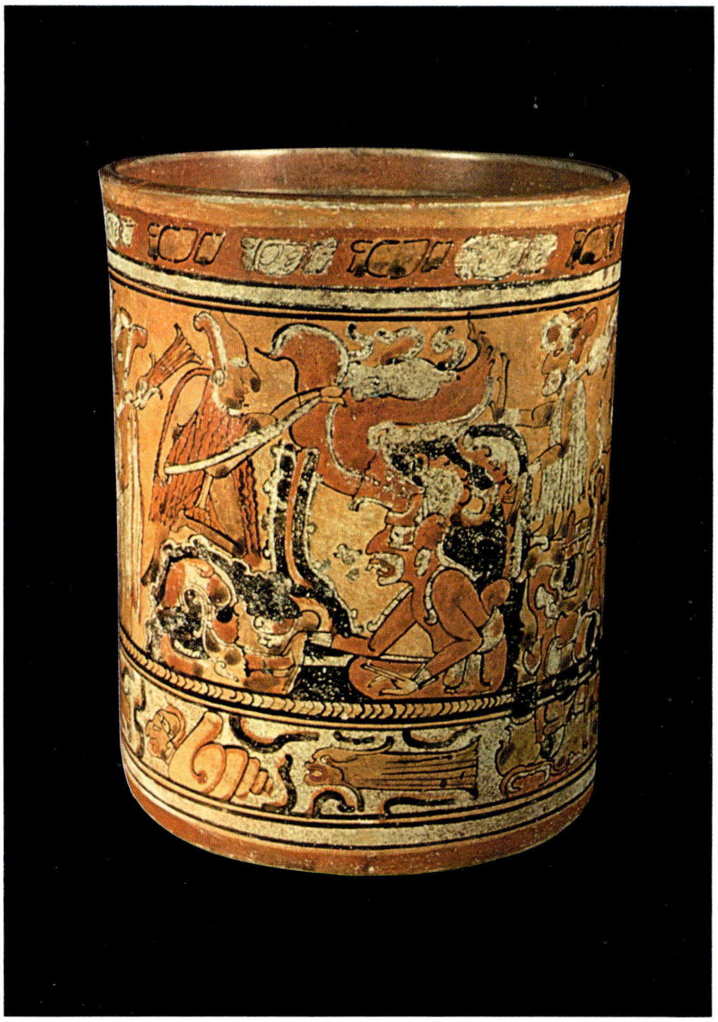

Plate 260

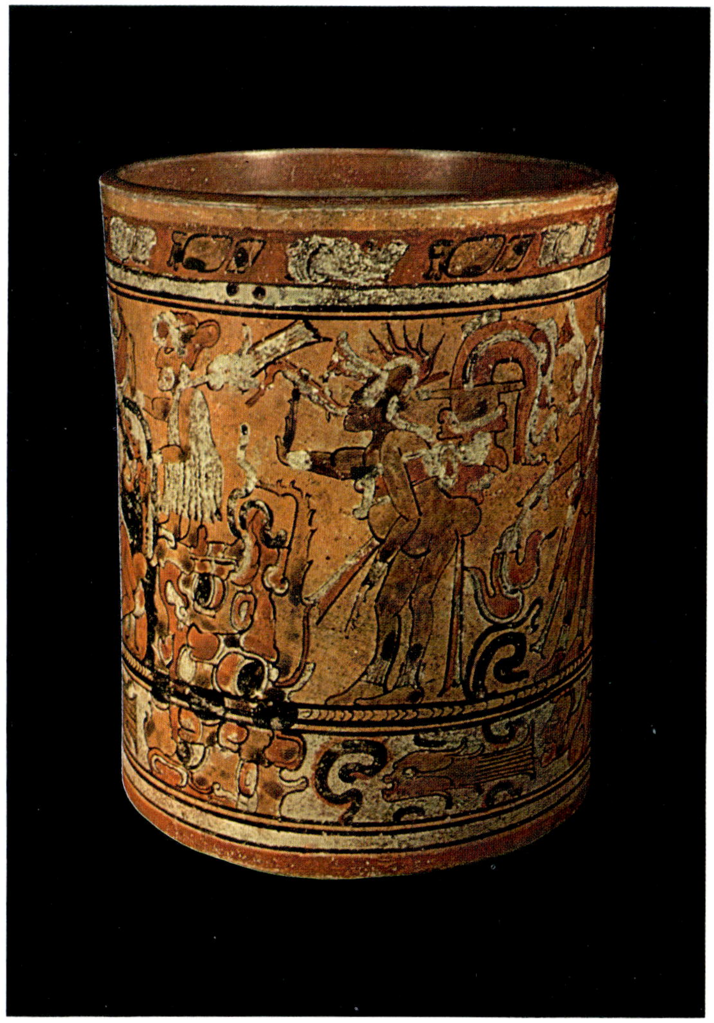

Plate 261

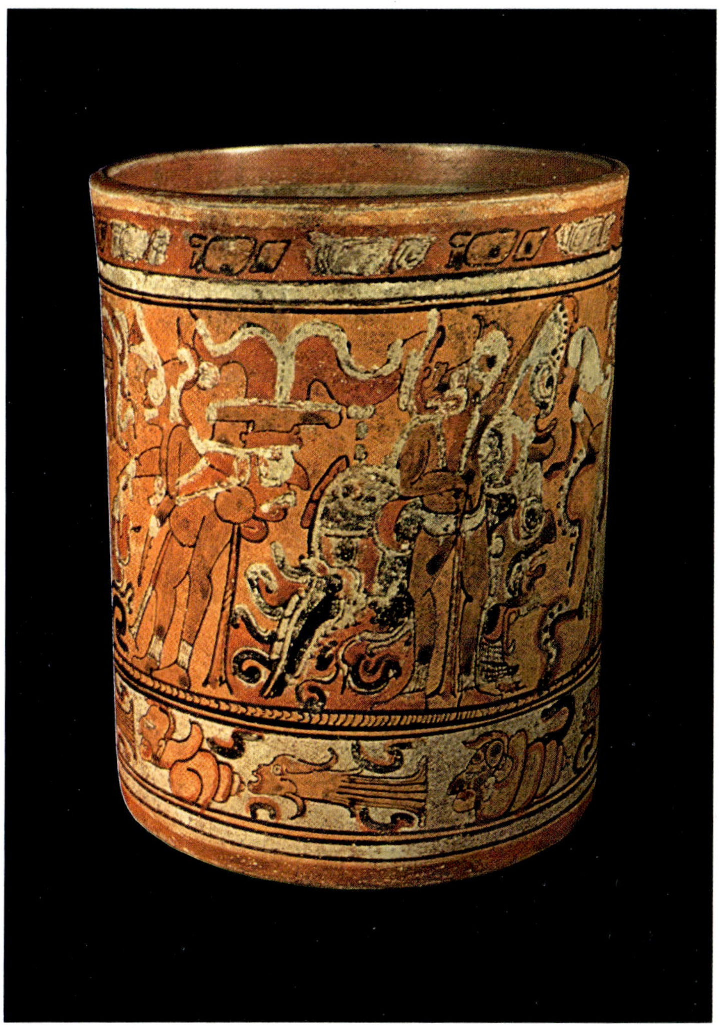

Plate 262

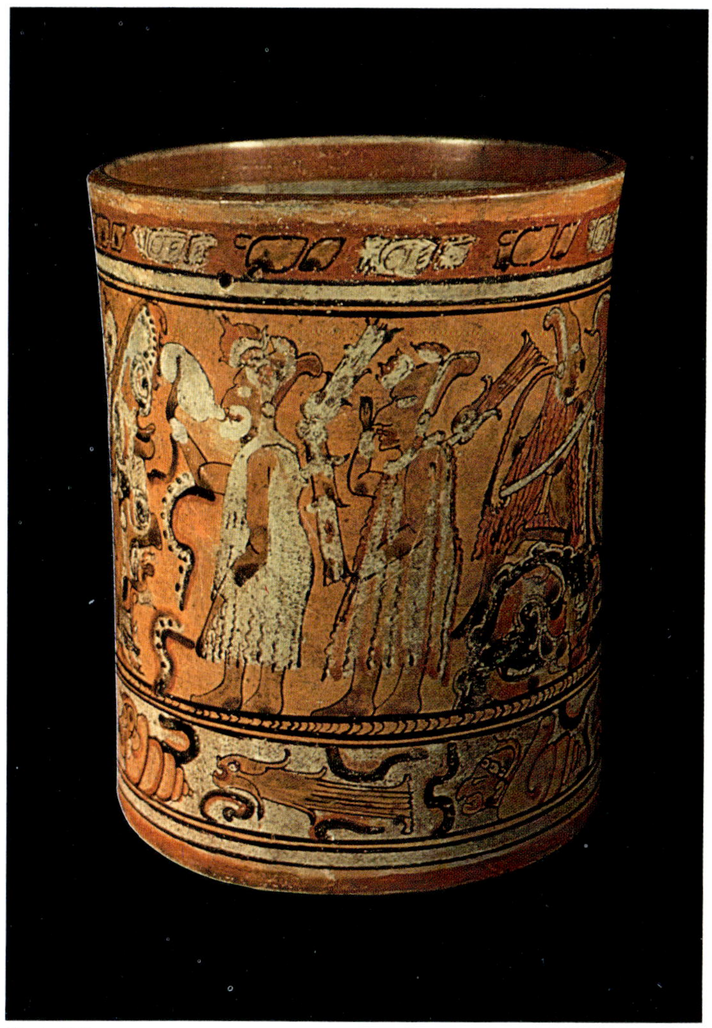

Plate 263

Plates 260-263 Processional Vase of the Old God (E-15)

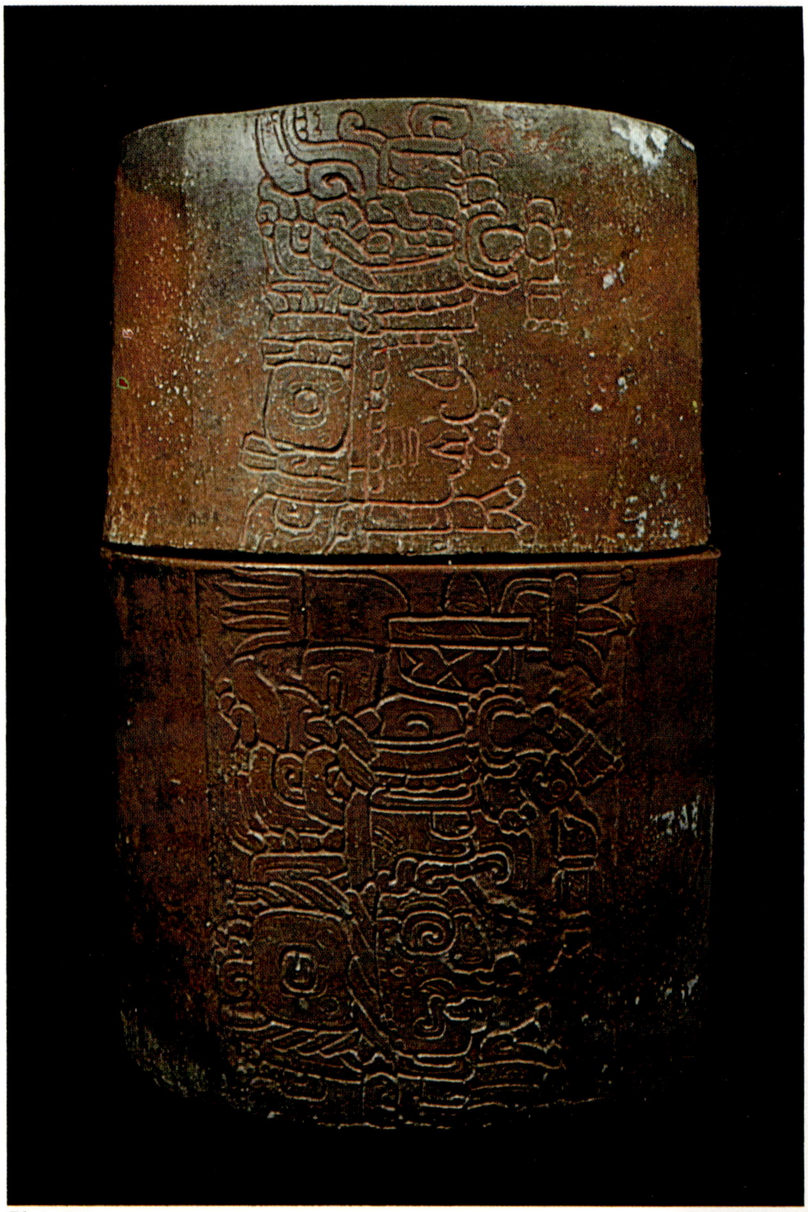

Plate 264

Plate 264 *Ceramic container of the Lord with God K Helmet (D-27) /Photograph courtesy Donald M. Hales/*

Plate 265 *Effigy Vessel of the Old God with Forehead Spout (E-3)*

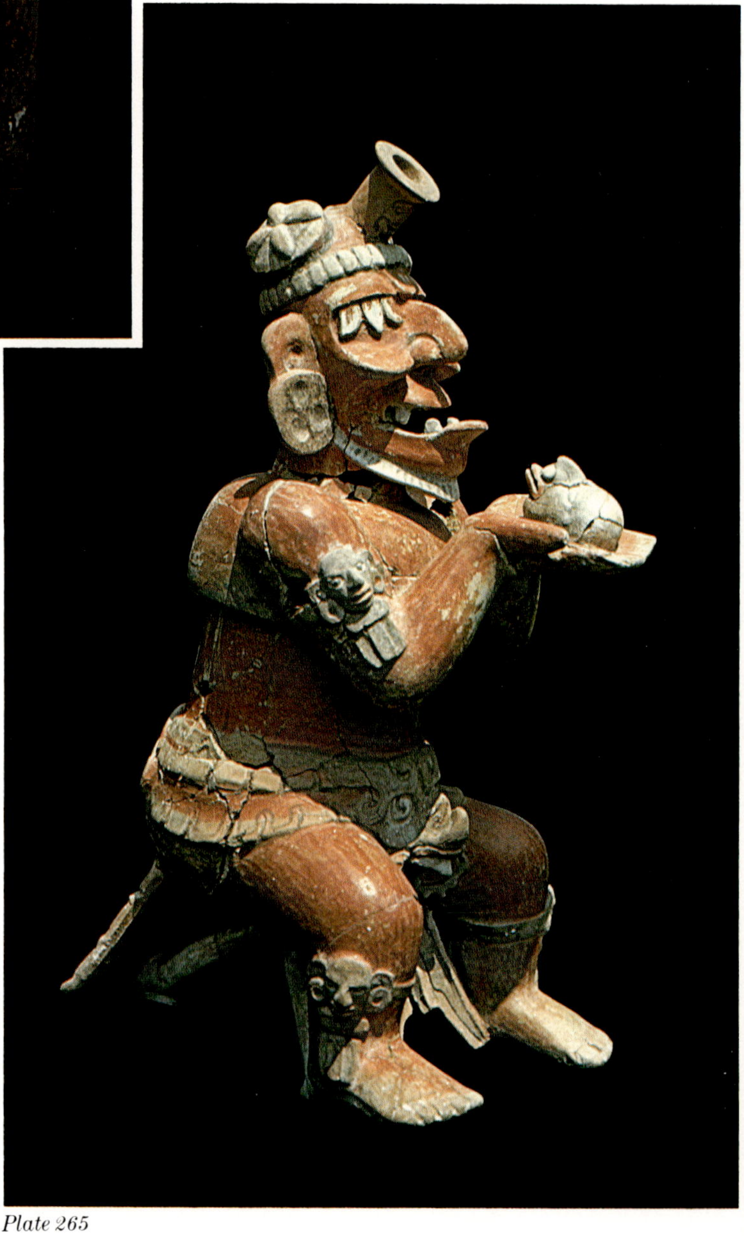

Plate 265

Between these figures the artist painted the seated figure of a woman. Her entire body from the shoulders down is painted white. Her forehead and part of her face are painted red. She wears a red skirt and a simple turban made in the shape of a ∞. Except for a flexible earplug she wears no jewelry. In her left hand she is holding an unidentifiable object, perhaps a codex, in her right hand a stylus or paintbrush.

Glyphic Text

The glyphs in the text are too badly drawn to be read. The guidelines for drawing the glyphs still survive.

The **Effigy Vessel of the Old God with the Forehead Spout** (E-3), a two-piece ceramic vessel, may also be related to God K iconography. The hollow statuette was found by William R. Coe at Tikal.[14] It was probably used as an incenser with the smoke emanating through the forehead tube à la God K. The vessel is of early Classic origin. It depicts the Old God, painted red, holding a human head (or mask) on a plate. Moser believes that the statue represents the god Itzamná receiving a severed head as an offering.[15] The deity is in a seated position and is naked except for a narrow belt and body jewelry consisting of an elaborate belt buckle, armlets, gaiters, a headband, and earplugs.

The **Vase of the Four Lords** (E-4) is a polychrome vessel probably from northern Petén (now in a private collection). It is painted red and black on an orange base, with an upper and a lower red rim and two short vertical columns of glyphs or glyphoids. The vase shows the figures of four dignitaries, all seated Turkish fashion and all with slim white rods (cigarettes?) with flamelike appendages (similar to that seen on the Polychrome Vase, Three Gods in the Dumbarton Oaks Collection) emerging from their foreheads. They all wear loincloths and elegant headdresses. The first two personages appear to be looking at a lidded vase, the other two at each other. The composition also includes a halved "shamrock" design and a container with ceremonial paraphernalia.

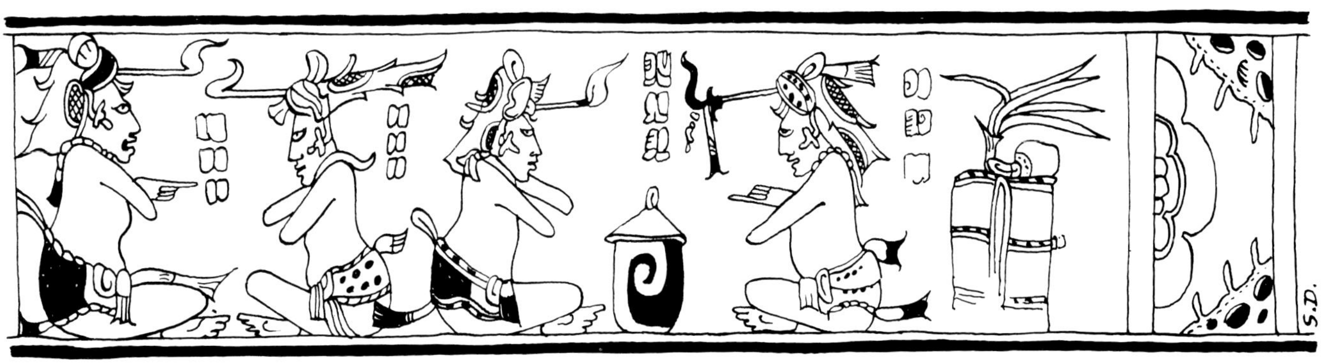

Figure 231. Vase of the Four Lords (E-4)

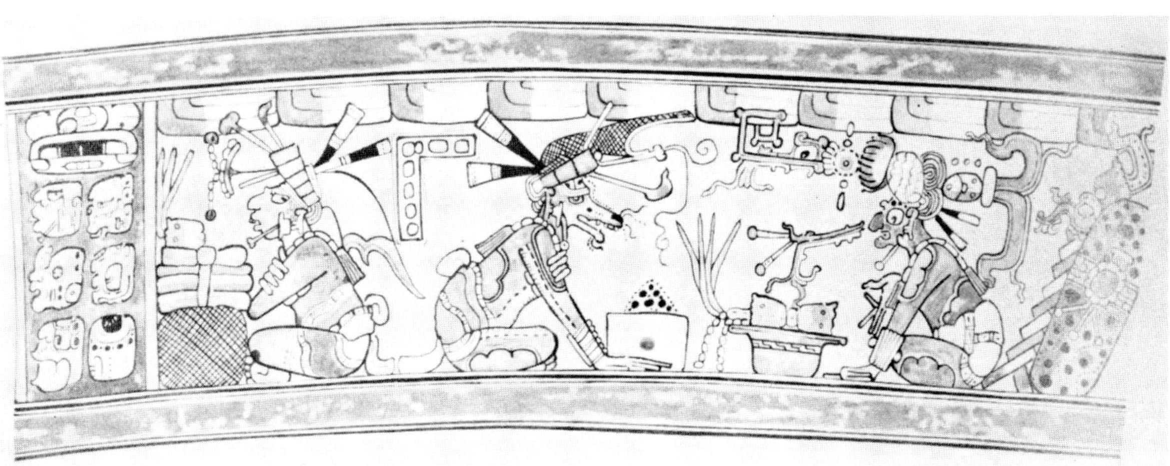

Figure 232. Polychrome Vase, Three Gods (Dumbarton Oaks Collection), (drawing by Diane Triffith Peck, after Benson)

The Processional Vase of the Old God (E-5) is an orange cylinder painted red, white, and black by a Petén artist during the late Classic period (now in a private collection). The vessel has the conventional upper and lower red margins, an upper circular glyph text, and a fine chevron band, probably representing the mat (earth), which subdivide the composition in an upper, wider and a lower, narrower panel. The upper panel shows eight individuals gathered in three groups. Four of them have divine characteristics; the rest are probably human attendants.

The principal person of the first group is a seated deity with filed incisors whose attire consists of a wide belt, bracelets, a bead necklace, and an elaborate headdress. From his forehead rises a funnel-shaped tube emitting two giant smoke coils, identifying him with great probability, though not with certainty, as God K. He is gesticulating with his hands as though pushing away an iconographic complex painted in the shape of a stylized serpent throat. Kneeling on the serpent head is the figure of an attendant dressed in a pleated white skirt. Over the head of the seated god he holds a long-stemmed flower much as black slaves used to hold palm-leaf fans over the heads of rulers of desert kingdoms. Behind the seated god is a monster head resembling some of the "altars" of Copán and Quiriguá, on which stands a second skirt-clad attendant. The monster head rests on three affixlike strollers, which penetrate the chevron band and are connected to a similar but smaller monster head. The three figures and the two heads are set in a beautifully arranged composition resembling a medieval coat of arms.

This *menagerie de trois* is followed by a strange pair of saucer-eyed deities, who, like Japanese *sumo* wrestlers, have pendulous bellies and large rumps and are dressed only in wide belts. The first one is empty-handed; the second has a large flower in his hand. As the walk, they both carry large god heads strapped to their backs. Large coils emanate from the god heads.

The last group is composed of three standing figures. The first stands surrounded by iconographic symbols of god and dragon heads, scrolls, and flowers. He is dressed only in a wide belt and gaiters and either has a zoomorphic head or is wearing an animal mask. Next to him are two attendants, one of them blowing a conch shell, the other holding an unidentified ceremonial object. Their identical garments and flower nape ornaments strongly suggest that they may be the same attendants as those described in the first subgroup.

The panel under the chevron band probably represents water. It is filled with alternating images of God N emerging from a conch shell and of floating fish.

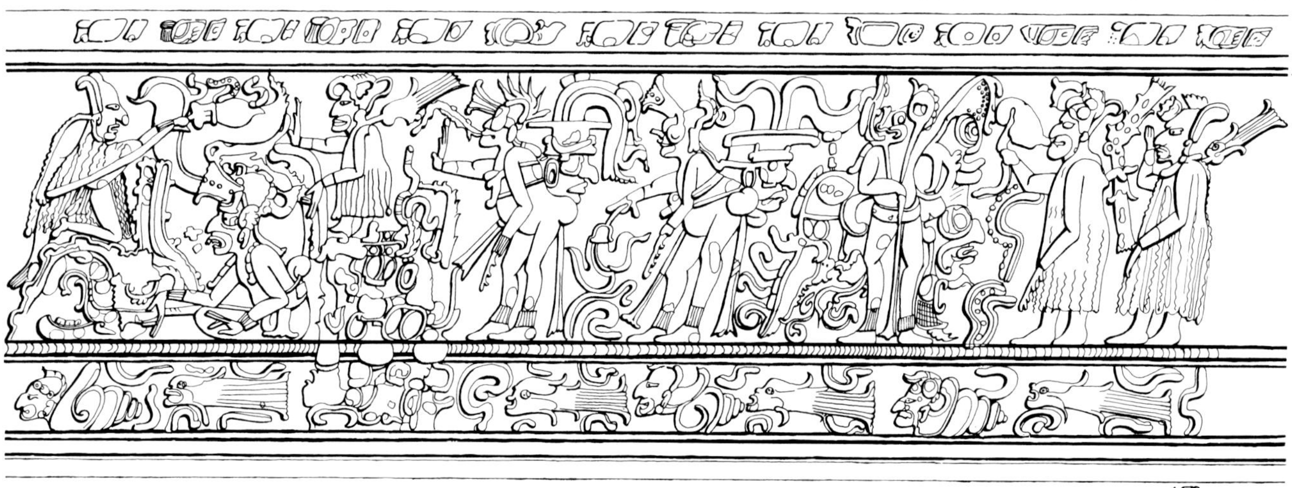

Figure 233. Processional Vase of the Old God (E-5)

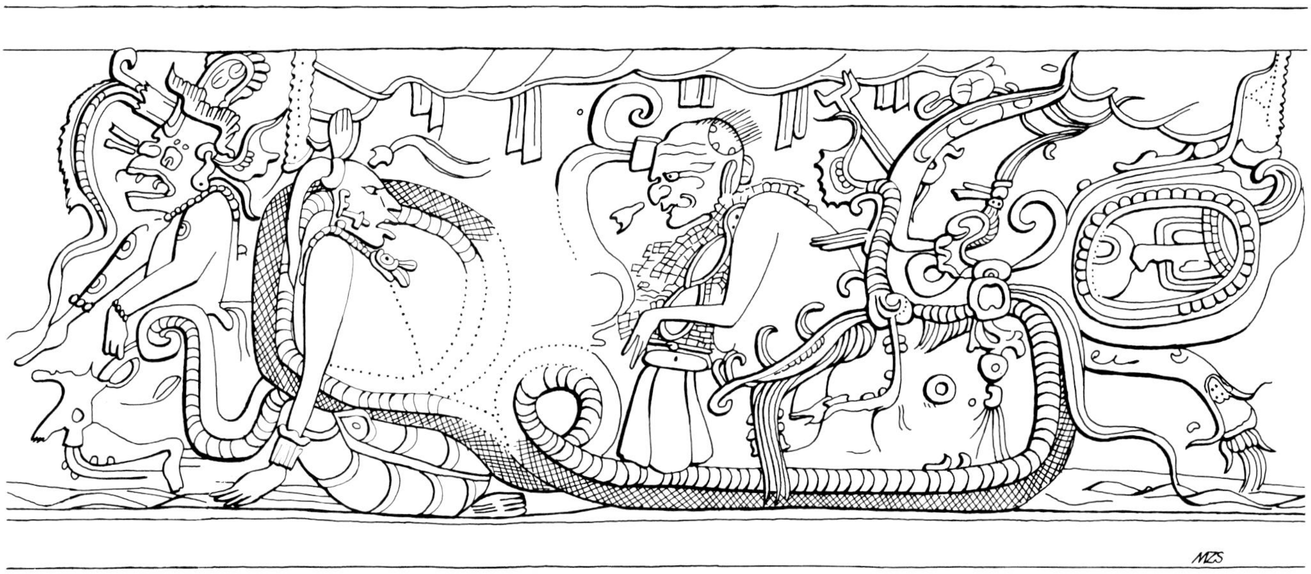

Figure 234. Vase of the Gods with Forehead Flares (E-6)

The **Vase of the Gods with Forehead Flares** (E-6) is the last of the "codex-style" vases to be discussed in this book. This vessel (now in a private collection), like vases D-14, D-15, and D-16 and plate D-17, was undoubtedly created in the studio of the same artist or artists. Their high-quality work is characterized by extremely fine black-line work on a white base and upper and lower brownish-red rims. The vase was probably made during the late Classic period of northern Petén.

Through the entire composition of the vase coils the serpentine body of the bearded dragon, a favorite iconograph of both the ancient Mayas and Michael D. Coe. Representations of this beast can be seen on other codex-style vases, such as on Princeton vases 3 and 5.[16] The dragon on the vase described here has an exceptionally long beard, bound hair, an ear pendant, and a large ear-extension plate with a masklike motif, comparable in size to that seen on Princeton vase 3.

There are three other participants in the scene. The first is an aged god, bald-headed, with a sunken face and a wrinkled forehead. He wears a small skullcap with a *kin* (?) sun sign and a heavy neck pendant, which seems to be made in part of some kind of woven material. From the forehead of the god emerges a short smoking tube with two large smoke coils. The next figure, that of a young, seated personage, is enframed in the large coil of the serpent. His hair is bound, and he wears a shell (?) necklace and a long kilt.

At the tail of the dragon kneels the third god, a monster-headed deity with a human body. Except for his modest body jewelry he is unclad. Despite the absence of the proboscis-shaped nose, his high forehead, forehead cigar, crewcut hairdo, and oval body markings identify him with great probability as God K.

The well-drawn drapery extending above the entire scene indicates that the events portrayed are taking place inside a building.

We had the opportunity to see but not to photograph another vase undoubtedly by the same artist showing a similar scene with the same participants, undoubtedly a "followup" to the story presented on this artifact.

I am a brother to dragons and a companion to owls.
Job 30:29

Conclusion

The worldwide diffusion of tobacco is one of the most interesting phenomena in the cultural history of mankind, one that furnishes food for many reflections. In less than a century after the Spanish sailors first saw Indians smoking on the shores of Hispaniola, tobacco established itself as a universal necessity without which man is now very reluctant to live.

Besides smoking tobacco in pipes, cigars, and cigarettes, the American Indian also consumed the herb by chewing, eating, and drinking it. Among the North American Indians pipe smoking prevailed. The Aztecs smoked both pipes and cigars. Pipes were unknown to the Mayas who consumed tobacco as cigars, cigarettes, or chew.

According to ethnological and historical data, tobacco smoking played an important role in ancient Maya religion, folklore, and healing. In describing the ceremonies, several contemporary observers attributed narcotic and hallucinogenic effects to tobacco, a claim that is not easy to explain. Why did *Nicotiana tabacum*, a herb consumed without such effects by millions, produce those effects in some regions? The question could be answered in one of the following ways: (1) the Mayas smoked a different, more potent tobacco; (2) they smoked *tabacum*, but in very large quantities; (3) they potentiated the effects of tobacco by nonpharmacological means; or (4) on some occasions they consumed herbs with hallucinogenic properties simultaneously with or instead of tobacco.

In scenes pictured in the codices, carved on stone monuments, and portrayed on ceramics, smoking is seen more often than had been previously thought. I believe that many of these scenes do indeed represent tobacco smoking. In the foregoing pages these artifacts have been grouped into categories according to the degree of certainty that they portray individuals smoking. Scenes that almost certainly portray individuals smoking objects resembling cigars or cigarettes occur in the *Madrid Codex*, on the Palenque tablets, on twenty ceramic vases, on two plates, and on one shell carving. Scenes that probably portray smoking but do not show the actual smoke are shown on seven vases and one ceramic sculpture. Scenes that show objects that could be large cigars either held in the hands of figures or interspersed in the composition are shown on eight vases. The Flare Gods—a subject that is presently a matter of speculation and debate—are shown on thirty-seven vases, two plates, and a large number of other artifacts.

The objects the Mayas appear to have smoked can be divided by size into cigars and cigarettes. We also see large, flarelike objects held in the hands, which may be very large cigars. There are also distinct categories in the style in which the smoke is portrayed. In some cases (where the color is determinable) the cigarettes are painted white, giving the impression that the Mayas either wrapped their tobacco in some other substance, like cornhusks, in much the same manner that cigarettes are wrapped nowadays, or applied some coating, such as lime, to the tobacco.

In the examples I found, the impression of the act of smoking is very strong. Performed by persons of apparently high rank, smoking in the times of the ancient Mayas was probably an activity of considerable importance rather than just a pleasurable act. The examples of Maya art and other archaeological artifacts examined in this book prove beyond doubt that the herb tobacco and the act of smoking not only had a deep religious meaning in the life of the ancient Mayas but also deeply penetrated art and had important significance in iconography.

The possibility that smoking for pleasure was also a widespread custom among the ancient Maya population can be neither proved nor disproved. The lives of ordinary people were simply not considered important enough to record and preserve for posterity.

Notes

Chapter 1

[1]Arents, 1936, Vol. 1, p. 10.
[2]Setchell, 1921, pp. 397–414.
[3]Hall, 1973; Lowie, 1956, p. 274.
[4]Mangelsdorf et al., 1971, Vol. 1, p. 443.
[5]Webb et al., 1957.
[6]Killebrew et al., 1906, p. 27.
[7]Arents, 1936, Vol. 1, p. 5.
[8]Dixon, 1921, pp. 19–49.
[9]Navarette, 1890, p. 41.
[10]Columbus, 1871.
[11]Las Casas, 1937, Chap. 66, p. 8.
[12]Mason, 1937, p. 32; Heiman, 1960, p. 6.
[13]Oviedo y Valdés, 1932, Bk. 62, Chap. 11, p. 96.
[14]Díaz de Castillo, 1973, p. 227.
[15]Clavijero, 1887, p. 87.
[16]Torquemada, 1723, Vol. 1, p. 168.
[17]Sahagún, 1950–69, Bk. 9, Chap. 7.
[18]Hall, 1976, pp. 16–17.
[19]Las Casas, 1877, p. 18.
[20]Acosta, 1591.
[21]Benzoni, 1568, Bk. 4, p. 116.
[22]Arents, 1936, Vol. 1, p. 22.
[23]Chita de la Calle, in Porter, 1948, p. 139.
[24]Dodoens, 1553.
[25]Brongers, 1964, p. 20.
[26]Moeteren, 1635; Brongers, 1964, p. 17.
[27]Arents, 1936, Vol. 1, p. 46.
[28]Hariot, 1584.
[29]James I, 1604.
[30]Rych, 1617.
[31]Molière, 1665, Act 1, Scene 1.
[32]Conti, 1932, p. 20.
[33]Linton, 1924, p. 14.
[34]Porter, 1948, pp. 231–32.
[35]Bancroft, 1875, p. 287; Stahl, 1932, p. 105.
[36]Porter, 1948.
[37]Ibid., p. 233.
[38]Lothrop, 1926, p. 282; Porter, 1948, p. 233; Furst, 1974, p. 11.
[39]Porter, 1948, pp. 231–32.
[40]Stahl, 1937, p. 1.
[41]Linton, 1924, pp. 9–10.
[42]Wafer, 1699, p. 102.

Chapter 2

[1]J. E. S. Thompson, 1970a, p. 108.
[2]Díaz, 1958.
[3]Oviedo y Valdés, 1936.
[4]Wiener, 1921, pp. 86–87; Arents, 1936, Ref. 1535.
[5]Díaz de Castillo, 1973, p. 227.
[6]Ibid., p. 33.
[7]Ibid., p. 38.
[8]Ibid., p. 38.
[9]J. E. S. Thompson, 1970a, p. 108.
[10]Tozzer, 1966, p. 106.
[11]J. E. S. Thompson, 1970a, p. 108.
[12]Tozzer, 1966, p. 104.
[13]Villagutierre de Soto-Mayor, 1933, p. 5.
[14]García de Peláez, 1968, p. 73.
[15]Hellmuth, 1972, p. 23.
[16]Valenzuela, 1695.
[17]Edmonson, 1971, p. 113.
[18]Girard, 1969, pp. 219–20; J. E. S. Thompson, 1970a, p. 108.
[19]Roys, 1967, p. 97.
[20]Ibid., p. 152.
[21]Ibid.
[22]Linton, 1924; Hind, 1960; Lowie, 1919; Setchell, 1921; Wilson, 1920.
[23]Mangelsdorf et al., 1971, pp. 431; 440.
[24]Wisdom, 1940, p. 51.
[25]Radin, 1970, p. 68.
[26]Blom et al., 1955, p. 148; Redfield et al., 1934, p. 46.
[27]Wisdom, 1940, p. 185.
[28]Cardos, 1959, p. 40; J. E. S. Thompson, 1970, p. 154; Duby et al., 1964, p. 281.
[29]Sahagún, 1950–69, Bk. 10, Chap. 26.
[30]Roys, 1939, pp. 394–97.
[31]Cockburn, 1735.
[32]Boggs, 1944, p. 68.
[33]Douglass, 1889, pp. 348–53.
[34]J. E. S. Thompson, 1946, p. 5.
[35]Morris et al., 1931, Pl. 21.
[36]Brainerd, 1958, p. 76.
[37]*Relación de Michoacán*, 1903; J. E. S. Thompson, 1962, p. 31.
[38]Fuentes y Guzmán, 1882, Bk. 12, Chap. 3.
[39]Valenzuela, undated.
[40]Anonymous, 1855, pp. 349–50; J. E. S. Thompson, 1970, p. 109.
[41]Wisdom, 1940, pp. 185–86.
[42]Heiman, 1960, p. 7.
[43]Wisdom, 1940, p. 186.
[44]Torquemada, 1723, Vol. 1, pp. 168, 340; Clavijero, 1887, Vol. 1, pp. 158, 294; Sahagún, 1938, Bk. 2, Chap. 349; Tezacomoc, 1944, p. 234; *Relación de Asuchitlán*, 1946, p. 43.
[45]Linton, 1924, p. 9.
[46]Arents, 1936, Vol. 1, p. 15.
[47]Wisdom, 1940, pp. 117–18.
[48]J. E. S. Thompson, 1946, p. 5.
[49]La Farge and Beyers, 1931, pp. 123–24.
[50]Mason, 1924, p. 39.
[51]Oviedo y Valdés, 1851–55, Bk. 6, Chap. 20.
[52]Arents, 1936, Vol. 1, p. 5.
[53]Vespucci, 1507.
[54]Arents, 1936, Vol. 1, p. 19.
[55]Mason, 1924, p. 39; Starr, 1900–1901, Vol. 2, p. 71; J. E. S. Thompson, 1970, p. 110; La Farge et al., 1931, p. 123; Anguiano, 1908, Vol. 8, p. 389.
[56]Starr, 1902, Vol. 9, p. 71.
[57]*Descripción de San Bartolomé*, 1965, pp. 262–76.
[58]Brongers, 1964.
[59]Furst, 1976, pp. 28–29; Nordenskjold, 1930, pp. 189–95; Wassen, 1972, p. 64.
[60]Karsten, 1935, pp. 508–509.
[61]Furst, 1976, pp. 27–28.
[62]Pomar, 1941, p. 59; Cruz and Badianus, 1939, p. 259; Coe, and Furst, forthcoming.
[63]El Conquistador Anonymo, 1941, p. 31.
[64]Bartholin, 1661.
[65]Buc'hoz, quoted in Brongers, 1964, p. 30.
[66]Kelley, 1968.
[67]Seler, 1902–1903, Vol. 1, p. 397.
[68]Beyer, 1937, p. 82.
[69]J. E. S. Thompson, 1968, pp. 142–43.
[70]Seler, 1900, Vol. 1, p. 551.
[71]Kelley, 1968, pp. 142–43.
[72]Ibid., p. 158.
[73]J. E. S. Thompson, 1939, p. 12.
[74]Stuart, personal communication.
[75]Von Euw, personal communication.

Chapter 3

[1]Arents, 1936, Vol. 1, p. 23.
[2]Hall, 1975, pp. 7–8.
[3]Torquemada, 1615.
[4]J. E. S. Thompson, 1970, p. 121.
[5]Ibid.
[6]Alarçón, 1928, p. 131.
[7]Durán, 1971, p. 262.
[8]Seler, 1904, pp. 146–47.
[9]Nicholson, 1969, p. 437.
[10]Durán, 1971, pp. 105–106.
[11]Ibid., p. 232.
[12]Caso, 1971, p. 917.
[13]Durán, 1971, p. 296.
[14]Ibid., p. 297.
[15]Weitlaner et al., 1969, p. 520.
[16]Muñoz Camargo, 1892, p. 62.
[17]*Relación de Mechuacán*, 1903.
[18]Villa Rojas, 1955, p. 114.
[19]Sahagún, 1938, Bk. 2, Chap. 33.
[20]Weitlaner et al., 1969, p. 520.
[21]Cervantes y Salazár, 1914, Vol. 1, p. 18.
[22]Harvey et al., 1969, p. 675.
[23]Sahagún, 1950–69, Bk. 11, Chap. 5.
[24]La Barre, 1964, p. 30.
[25]Bennett, 1935, p. 45.
[26]Janiger, 1973.
[27]Koskowski, 1957, pp. 66, 69; Fenton, 1940, p. 426.
[28]Koskowski, 1957, p. 69; Metraux, 1949, p. 377.
[29]Cooper, 1949, pp. 534.
[30]McGuire, 1898, p. 56.
[31]Whitman, 1940, p. 402; McGuire, 1898, p. 568.
[32]Laski, 1958, p. 140.
[33]Steward, 1949, p. 21; Brooks, 1938, pp. 249, 289; Cooper, 1949, p. 535; Kirschoff, 1949, p. 48; Levi-Strauss, 1961, pp. 339, 365–384, 516, 578.
[34]Seler, 1961, Vol. 5, p. 388.
[35]Tozzer, 1907, p. 158.
[36]Brinton, quoted in Seler, 1960, Vol. 1, p. 550.
[37]Blom et al., 1955, pp. 147–51.
[38]*Codex Vindobonensis Mexicanus*, 1967, p. 18a; Jill Furst, personal communication.
[39]Duby, personal communication.
[40]Tozzer, 1907, p. 158; Tozzer, 1913, pp. 497–509.
[41]Tozzer, 1907, pp. 142–43.
[42]J. E. S. Thompson, 1970, p. 112; Redfield et al., 1939; Loughlin, 1969, p. 185.
[43]Holland, 1963, p. 97.
[44]Gutiéras-Holms, 1961, pp. 209, 217, 235, 264.
[45]Holland, 1963.
[46]Wisdom, 1940, p. 349.
[47]Fought, 1972, p. 281.
[48]B. Berlin, 1974, p. 445.
[49]Harner, 1973, p. 17.
[50]Wisdom, 1940, p. 337.
[51]Fought, 1972, p. 264.
[52]Wisdom, 1940, p. 337.
[53]Arents, 1940, pp. 347–48.
[54]Wisdom, 1940, pp. 347–48.
[55]Schellhas, 1904, p. 13; J. E. S. Thompson, 1970, p. 300; M. D. Coe, 1973a, p. 12.
[56]Tozzer, 1941, p. 132.
[57]Kelley, 1965; A. Miller, 1974, p. 45; J. E. S. Thompson, 1970, p. 300; M. D. Coe, 1973a, p. 12.
[58]Giblin, 1951, p. 116.
[59]Wisdom, 1940, p. 349.

Chapter 4

[1]Stahl, 1931, p. 577; Arents, 1936, Vol. 1, p. 19; Pané, 1867, quoted by Arents, 1936, Vol. 1, pp. 19–20.
[2]Pomar, 1941, p. 59.
[3]Sahagún, 1938, Bk. 3, Chap. 92.
[4]Sahagún, 1938, Bk. 3, Chap. 95.
[5]Sahagún, 1938, Bk. 3, Chaps. 99, 204.
[6]*Relación de Alahuistlán*, 1905, Vol. 6, p. 111.
[7]Cruz and Badianus, 1940, p. 259.
[8]Durán, 1971, pp. 117–18.
[9]Fairholt, 1859, p. 23.
[10]Vogel, 1969, p. 37.
[11]Vogel, 1969, p. 247.
[12]Cooper, 1949, p. 535.
[13]McGuire, 1898, p. 403.
[14]Oviedo y Valdés, quoted by Cortí, 1932, p. 41.
[15]Monardes, 1565.
[16]Nicot, 1573, p. 478.
[17]Brongers, 1964, p. 25.
[18]McGuire, 1899, p. 362.
[19]Jonson, 1601.
[20]Quoted by Arents, 1936, Vol. 1, p. 36.
[21]Anguiano, 1908, p. 395.
[22]Cervantes de Salazar, 1914, Vol. 1, p. 18.
[23]Tozzer, 1966, pp. 153–54.
[24]Tozzer, 1966, p. 94.
[25]Roys, 1972, p. 94.
[26]*Diccionario de Motul.*
[27]Roys, 1972, p. 94.
[28]Roys, 1972, p. 94.
[29]Roys, 1931, p. xix; Seler, 1902, Vol. 3, p. 114.
[30]Roys, 1965a.
[31]Roys, 1931, p. 259.
[32]Roys, 1965, p. 57.
[33]Roys, 1965, p. 59.
[34]Roys, 1965, p. 45.
[35]Roys, 1965, p. 37.
[36]Osado, *Libro del Judeo.*
[37]Roys, 1931, pp. 1, 2, 33, 41, 80–81, 86–87, 100–101, 137, 168, 259; Sotuta Manuscript, undated, pp. 5, 11, 28, 34, 37, 41, 43; Osado, *Yerbas y Hechicerías del Yucatán*, pp. 41, 53, 82, 85; Mena Manuscript, 1921, p. 287; *Book of Chilam Balam de Nah*, p. 101.
[38]*Descripción de San Bartolomé*, 1965, pp. 262–76.
[39]Valladores, 1957, p. 120.
[40]Cangas y Quinones, 1938, p. 179.
[41]Furst, 1974, p. 155.
[42]Berlin, 1974, p. 445.
[43]Redford, personal communication to Spinden, 1950, p. 51.
[44]Martínez, 1933, p. 569.

Chapter 5

[1]Fuentes y Guzmán, 1932–33, Bk. 12, Chap. 3.
[2]Mendieta, 1870, Vol. 2, Chap. 19.
[3]Benzoni, 1568, Bk. 4, p. 116.
[4]Oviedo y Valdés, 1526.
[5]Sahagún, 1938, Bk. 3, Chap. 78.
[6]Arents, 1936, Vol. 1, p. 81.
[7]*Relación de Nuchistlán*, 1947, p. 66.
[8]Porter, 1948, p. 168.
[9]Gilbert, 1972, p. 55.
[10]Janiger, 1973, p. 2; Dobkin de Ríos, 1974, p. 161.
[11]Marin, p. 262.
[12]Hall, 1975, p. 8.
[13]Janiger, 1973, p. 149; Poindexter, 1962; Testor, 1965; Curin, 1967.

[14]Janiger, 1973, p. 149; Lewin, 1928; Rennes, 1957.
[15]Janiger, 1973, p. 149.
[16]Furst, 1972, p. xvi.
[17]Schultes, 1972, pp. 53–54.
[18]Wilbert, 1972, p. 56.
[19]Hernández, 1651, p. 403.
[20]Crumrine, 1974, p. 153.
[21]Roys, 1965, Chap. 80.
[22]*Descripción de San Bartolomé*, 1965, p. 273.
[23]Siegel, 1976, p. 475.
[24]Weiss, 1973, p. 43.
[25]Stohl, 1925, p. 316; Wilbert, 1972, pp. 56–57.
[26]L'Obel, 1570, Fig. 3; Thévét, 1558.
[27]Harner, 1973, p. xiii.
[28]J. E. S. Thompson, 1970, pp. 172–76; Landa, 1941, pp. 113–14; Joralemon, 1974, pp. 59–75.
[29]Benson, 1974, p. 109.
[30]Furst, 1976b, pp. 181–93.
[31]Paso y Troncoso, 1971, p. 970.
[32]Spores, 1969, p. 985.
[33]Dobkin de Ríos, 1974, p. 147.
[34]Ibid., pp. 147–48; Ludwig, 1969, pp. 15–16.
[35]Gumilla, 1741.
[36]McGuire, 1899, p. 369.
[37]Clavijero, 1887, Vol. 1, p. 87; Linton, 1924, p. 8.
[38]Hall, 1975, p. 4.
[39]Beverly, 1702, p. 24.
[40]Pomar, 1941, p. 59.
[41]Sahagún, 1938, Bk. 1, Chap. 146.
[42]*Relación de Ostomán*, 1905, p. 111.
[43]Siegel, 1976, p. 475.
[44]Ibid.
[45]Vespucci, 1534.
[46]Arents, 1936; Loven, 1921, p. 361; Reis Altschul, 1964, p. 3.
[47]Wassen, 1965, p. 15; La Barre, 1974, p. 157.
[48]Durán, 1971, pp. 115–16.
[49]Wilbert, 1972, p. 78; Dobkin de Ríos, 1968, p. 193.
[50]La Barre, 1964, pp. 66, 78–79.
[51]Seler, 1904, p. 146.
[52]Escheverría y Veytia, 1836.
[53]Clavijero, 1787.
[54]Spores, 1960, p. 972.
[55]Gage, 1958, p. 225.
[56]J. E. S. Thompson, 1970, p. 120.
[57]Furst, 1972b; Dobkin de Ríos, 1974; Tozzer, 1910; Gordon, 1898; Schellhas, 1904, p. 39; Wassen, 1934, pp. 617–23; Shook, 1952, p. 57; Sharer, 1968.
[58]J. E. S. Thompson, 1954, p. 230.
[59]Fabing, 1957; Ewarts, 1956.
[60]Felton, 1976, p. 197.
[61]Sedat, personal communication.
[62]M. D. Coe, 1971.
[63]Furst, 1974a, p. 154; Schultes, 1972, p. 28; Carmack, 1972, p. 28.
[64]Roth, 1915; Carneiro, 1970, pp. 331–41; Knab, 1974.
[65]Furst, 1974a, p. 154.
[66]Alger, 1974, pp. 152–53; R. N. Crumrine, 1974, p. 153; Hellmuth, 1974, pp. 155–56; Lowy, 1974, p. 158; Proskouriakoff, 1974, p. 159; J. E. S. Thompson, 1974, p. 160.
[67]J. E. S. Thompson, 1974, p. 160.
[68]Heim, 1958, p. 175.
[69]Dobkin de Ríos, 1974, p. 148.
[70]Garibay, 1961, p. 91.
[71]Meyer, 1977, p. 33.
[72]Ibid.; Wassen, 1957; Heim, 1959.
[73]Borhegyi, 1961, 1963, 1965; Meyer, 1977.
[74]Kohler, 1976; Rose, personal communication.
[75]Lowy, 1971, pp. 983–93.
[76]Lowy, 1972, p. 819.
[77]*Codex Tro-Cortesianus*, 1969, p. 49; Villacorta, 1930, p. 230.
[78]Förstemann, 1902, p. 78.
[79]Lowy, 1971, p. 983.
[80]Borhegyi, 1965, p. 281.
[81]Greene Robertson, 1972, p. 8.
[82]Furst, 1974, p. 155.
[83]Greene Robertson, 1973, personal communication to Rubel.
[84]Furst, 1976a, p. 191.
[85]Meyer, 1977, p. 3.
[86]Proskouriakoff, 1963, pp. 149–201.
[87]Greene Robertson, 1972, p. 8.
[88]J. E. S. Thompson, 1970, p. 117.
[89]Heimann, 1960, p. 7.
[90]Pané, 1867, p. 192.
[91]Oviedo y Valdés, 1936.
[92]Ibid.
[93]Arents, 1936, Vol. 1, p. 393.
[94]Las Casas, 1937, Chap. 66, p. 8.
[95]Friederici, 1941, p. 190.
[96]Ibid.
[97]Stahl, 1937, p. 11.
[98]Brasseur de Bourbourg, 1861, p. 84.
[99]Recinos, 1953, p. 112.
[100]Seler, quoted by Stahl, 1937, p. 2.
[101]Stall, 1888, p. 188.
[102]Dixon, 1921, p. 47.

Chapter 6

[1]Maudslay, 1896–1902; Spinden, 1913, pp. 50–54.
[2]W. R. Coe, 1967, p. 92; H. Spinden, 1957; Greene Robertson, 1972, p. 4.
[3]Spinden, 1913, p. 52; Maudslay, 1889–1902, Pl. 76, 81, 88; Lothrop, 1936, Fig. 78c; Hellmuth, personal communication; Girard, 1969, Fig. 204.
[4]Schele, 1973, pp. 48–49.
[5]Schellhas, 1904, p. 32; Anders, 1963, p. 274.
[6]Schellhas, 1904, p. 34; Spinden, 1913, pp. 50–52; M. D. Coe, 1966, p. 116.
[7]Spence, 1930, p. 34; J. E. S. Thompson, 1970, p. 226.
[8]Coe and Benson, 1966, p. 19.
[9]M. D. Coe, 1973a, p. 16; Kelley, 1976, p. 133; M. D. Coe, personal communication.
[10]Schele, personal communication.
[11]Kelley, 1976, p. 133.
[12]W. R. Coe, 1970.
[13]Easby, 1963, p. 97; Lothrop, 1936, pp. 53–60; Smith and Kidder, 1951, pp. 32–35; Woodbury and Trik, 1953, Fig, 281g.
[14]Sedat, personal communication.
[15]Hellmuth, personal communication.
[16]Morris et al., 1931, Pl. 119, 126.
[17]Kelley, personal communication.
[18]Schellhas, 1904, p. 34; Lounsbury, personal communication.
[19]Förstemann, 1901, pp. 215–16.
[20]Barthel, 1952, pp. 73–112.
[21]Seler, 1961, Vol. 1, p. 361.
[22]Kelley, 1976, p. 65.
[23]J. E. S. Thompson, 1943, pp. 137–51.
[24]J. E. S. Thompson, 1950, pp. 212–15.
[25]Berlin and Kelley, 1961, p. 11.
[26]Kelley, 1976, Fig. 17.
[27]Kelley, personal communication.
[28]Seler, 1902–1903, pp. 367–89; Seler, 1961, Vol. 1, p. 377; J. E. S. Thompson, 1970, p. 220; M. D. Coe, 1973a, p. 116; Roys, 1967, pp. 104–105; J. E. S. Thompson, 1970, p. 227.
[29]Seler, 1961, Vol. 1, p. 378; Kelley, 1976, p. 65.
[30]Spence, 1930, p. 350; J. E. S. Thompson, 1973, p. 65.
[31]J. E. S. Thompson, 1934, pp. 224–27.

[32]Kelley, 1965, p. 112.
[33]Kelley, 1976; Berlin, 1963.
[34]Kelley, 1976.
[35]Seler, 1902, Vol. 2, pp. 17-19; Berlin, 1963; Kelley, 1965, pp. 93-95.
[36]Kelley, 1965, pp. 93-95.
[37]J. E. S. Thompson, 1934, pp. 226-27; Kelley, 1965, p. 112; Seler, 1961, Vol. 1, p. 372; Schele, 1976, p. 12.
[38]J. E. S. Thompson, 1973, p. 65.
[39]Kubler, 1974, p. 33.
[40]Morley, 1968, p. 22.
[41]Maudslay, 1889-1902, Pls. 55-56; Griffin, 1974, p. 12.
[42]Spinden, 1913, p. 52.
[43]Greene Robertson, 1974.
[44]Schele, 1976, p. 15.
[45]Greene Robertson, 1976, p. 82.
[46]Ibid.
[47]Kelley, personal communication.
[48]Brongers, 1964.
[49]Lounsbury, personal communication.
[50]Kelley, personal communication.
[51]Moran and Kelley, 1969, p. 161.
[52]Benson, 1975, p. 118.
[53]Kelley, 1965, p. 106.
[54]Ibid.
[55]Seler, 1904, p. 33.
[56]Ibid., p. 34.
[57]M. D. Coe, personal communication.
[58]M. D. Coe, 1973a, p. 116; Durán, 1973, p. 98.
[59]Burland, 1975, p. 56.
[60]Caso, 1970, p. 39.
[61]Burland, 1975, p. 55.
[62]Roys, 1967, p. 100; Barthel, personal communication.
[63]M. D. Coe, 1973a, pp. 8-9.
[64]L. Schele and J. Miller, forthcoming.
[65]Ekholm.
[66]Carlson, 1978.
[67]Kelley, 1965.
[68]Kelley, personal communication.
[69]Barthel, 1978.
[70]Lounsbury, personal communication.
[71]Ríos Meneses, 1976, pp. 16-17.

Chapter 7

[1]Sharer, personal communication.
[2]Carlyle, 1976, p. 72.
[3]M. D. Coe, 1976, personal communication.

Chapter 8

[1]Baynes, n.d.
[2]*Codex Tro-Cortesianus*, 1967, p. 79b; J. E. S. Thompson, 1970, p. 106.
[3]*Codex Tro-Cortesianus*, 1967, pp. 86b-87b; Villacorta et al., 1930, p. 397.
[4]J. E. S. Thompson, 1970, p. 107.
[5]*Codex Tro-Cortesianus*, 1967, p. 88.
[6]Kelley, 1976, p. 158.
[7]Ruz, 1952, Fig. 12.
[8]De la Fuente, 1965, Plate 50.
[9]Spinden, 1950, p. 8.
[10]Benson, 1976, p. 13.
[11]Robicsek, 1975.
[12]Benson, 1973, p. 116; De la Fuente, 1965, p. 138.
[13]Stahl, 1937, Fig. 1.
[14]L'Obel, 1576, Fig. 3; Muñoz, 1783, Book 3, p. 96.
[15]Spinden, 1950, p. 3.
[16]J. E. S. Thompson, 1970, p. 106.
[17]Bunzel, 1932; A. Miller, 1974, p. 47.
[18]Schele, 1976, p. 18.
[19]J. L. Stephens, 1969, p. 72.
[20]M. D. Coe, 1978.
[21]M. D. Coe, 1978.
[22]J. E. S. Thompson, 1973, pp. 52-73; M. D. Coe, 1978, p. 28.
[23]M. D. Coe, 1978, p. 45.
[24]J. E. S. Thompson, 1962, p. 354; Proskouriakoff, 1968.
[25]J. E. S. Thompson, 1970a, p. 46.
[26]Robicsek, 1975, p. 169.
[27]M. D. Coe, 1973a, p. 16.
[28]Hellmuth, 1976, personal communication.
[29]M. D. Coe, 1973a, pp. 13-14.
[30]Ibid., p. 101.
[31]Ibid., p. 89.
[32]M. D. Coe, 1978, p. 15.
[33]Matthews, P., personal communication, in ibid.
[34]Ibid., p. 21.
[35]M. D. Coe, 1973a, p. 89.
[36]Quirarte, 1976.
[37]Groth-Kimball, 1964, Fig. 1.
[38]Barthel, personal communication.
[39]Kelley, 1965; Kelley, personal communication.
[40]Schele, 1976, p. 10.
[41]J. E. S. Thompson, 1969, p. 80.
[42]Matthews and Schele, 1974.
[43]M. D. Coe, 1978, p. 28.
[44]J. E. S. Thompson, 1972, p. 8.
[45]Hellmuth, forthcoming.
[46]Benson, 1976, p. 53.
[47]Hellmuth, 1976, personal communication.
[48]*Codex Vindobonensis*, 1967.
[49]M. D. Coe, 1973a, p. 99.
[50]J. E. S. Thompson, 1970, Plate 14a.
[51]Ibid., p. 302.
[52]Spinden, 1950, p. 11.
[53]J. E. S. Thompson, 1970, p. 302; M. D. Coe, 1973a, p. 15.
[54]R. L. Holland, 1963, p. 129.
[55]Roys, 1972, p. 75.
[56]Quirarte, 1976, pp. 20-26.
[57]Kelley, 1976, personal communication.
[58]Pío Pérez, 1866-77, p. 18.
[59]*Diccionario Motul*, p. 136.
[60]Smith, 1955, Fig. 2g.
[61]J. E. S. Thompson,1960, p. 143.
[62]Greene Robertson, 1973, p. 80.
[63]Cohodas, 1973, p. 102.
[64]J. E. S. Thompson, 1930, pp. 119-40; J. E. S. Thompson, 1960, p. 11.
[65]M. D. Coe, 1973a, p. 14.
[66]Ibid., pp. 85-86.
[67]Ibid.
[68]Ibid., p. 14.
[69]Edmonson, 1971, p. 93.
[70]Ibid.
[71]M. D. Coe, 1978, p. 15.
[72]M. D. Coe, 1973a, p. 14.
[73]Ibid.
[74]Ibid., p. 91.
[75]Kubler, 1977, p. 23; Willey, 1972, p. 207.
[76]Folley, 1932, p. 388; Helfrichy, 1973, pp. 139-41.
[77]Kubler, 1977, p. 23.
[78]M. D. Coe, 1973a, p. 91.
[79]J. E. S. Thompson, 1960, p. 144; Greene Robertson, 1967.
[80]M. D. Coe, 1978, p. 46.
[81]Ibid.
[82]M. D. Coe, 1973a.

[83]J. E. S. Thompson, 1950, pp. 170–71.
[84]A. Miller, 1974, p. 47.
[85]A. Miller, 1974, p. 47.
[86]Bardawil, 1976.
[87]Gann, 1900.

Chapter 9

[1]M. D. Coe, 1973, pp. 107–109.
[2]Robicsek, 1975, p. 35.
[3]M. D. Coe, 1973, p. 107.
[4]Ibid., p. 108.
[5]Robicsek, 1975, p. 234.
[6]M. D. Coe, 1973, p. 108.
[7]J. E. S. Thompson, 1950, p. 234; Lounsbury, personal communication to Schele; Edmonson, 1971, p. 37.
[8]Schele, personal communication.
[9]Schellhas, 1903; J. E. S. Thompson, 1969, p. 86; M. D. Coe, 1973b, p. 93.
[10]Lounsbury, personal communication; Robicsek, 1975.
[11]J. E. S. Thompson, 1960, p. 79.
[12]Ibid.
[13]*Dresden Codex;* M. D. Coe, 1973, pp. 91–93.
[14]Schele, personal communication.
[15]M. D. Coe, personal communication.
[16]Hellmuth, 1976, personal communication.
[17]Edmonson, 1965.
[18]M. D. Coe, 1978, pp. 58–60.
[19]*Diccionario Motul*, p. 671.
[20]Lounsbury, 1974, pp. 104–105.
[21]M. D. Coe, 1973a, p. 71.
[22]Hellmuth, personal communication; W. R. Coe, 1970, p. 52.
[23]M. D. Coe, 1973a, p. 43.
[24]W. R. Coe, 1970, p. 52.
[25]Robicsek, 1970, Fig. 118.
[26]Pío Pérez, 1877, p. 271.
[27]J. E. S. Thompson, 1970, p. 195.
[28]Joralemon, personal communication to Schele, 1976, p. 22.
[29]Lounsbury, 1976, personal communication.
[30]Edmonson, 1965, p. 130.

Chapter 10

[1]Lounsbury, 1974.
[2]J. E. S. Thompson, 1950, p. 134.
[3]Kelley, 1965.
[4]Schele, 1976, p. 14.
[5]Quirarte, 1976.
[6]Barthel, 1978.
[7]M. D. Coe, 1973a, pp. 81–83.
[8]J. E. S. Thompson, 1950, p. 144; M. D. Coe, 1973a, p. 83.
[9]J. E. S. Thompson, 1970, Plate 14; M. D. Coe, 1973a, p. 98; M. D. Coe, 1978.
[10]Robicsek, 1970, Plate 86.
[11]Schellhas, 1904, p. 10.
[12]Proskouriakoff, 1950, p. 95.
[13]M. D. Coe and E. P. Benson, 1966, Fig. 6.
[14]M. D. Coe, 1978, p. 34.
[15]Schellhas, 1903, p. 43.
[16]M. D. Coe, 1973a, p. 99.
[17]Ibid.
[18]Sahagún, 1938, Bk. 7, Chap. 3; J. E. S. Thompson, 1960, pp. 217–18.
[19]Seler, 1909, pp. 384–87.
[20]M. D. Coe, 1973a, p. 99.
[21]M. D. Coe, 1975, p. 15.
[22]Ibid.
[23]Ibid.
[24]Gordon, 1925, Pl. 1.
[25]J. E. S. Thompson, 1969, p. 23.
[26]Morley, 1937–38, p. 399.
[27]M. D. Coe, 1977, personal communication; M. D. Coe, 1978, p. 62.
[28]Kurbjuhn, 1977, personal communication.
[29]Schellhas, 1904, p. 17.
[30]*Codex Tro-Cortesianus*, 1969, p. 34.
[31]J. E. S. Thompson, 1972, *Dresden Codex*, p. 34; ibid., p. 99.
[32]Kingsborough, 1831, Vol. 4, p. 187; ibid., Vol. 3, p. 317.
[33]M. D. Coe, personal communication.
[34]J. E. S. Thompson, 1972, pp. 99–100; ibid., *Dresden Codex*, pp. 32, 34, 36, 37, 39.
[35]Villacorta and Villacorta, 1930; Förstemann, 1902.
[36]Kelley, 1968, p. 142.

Chapter 11

[1]M. D. Coe, 1978, p. 15.
[2]Ibid.
[3]Shimbunsa, n.d., p. 12.
[4]Covarrubias, 1966, p. 228.
[5]Kelemen, 1956, Pl. 140a.
[6]Rostand, 1951, p. 31.
[7]Instituto Italo-Americano, 1963.
[8]Smith and Kidder, 1943.
[9]Instituto Italo-Americano, 1968.
[10]Westheim, 1969, p. 211.
[11]Spinden, 1913, Fig. 48.
[12]Von Euw, 1974, pp. 19–26.
[13]Ibid.; Gutiérrez-Gonzales, 1976, pp. 41–51.
[14]W. R. Coe, 1970, p. 60.
[15]Moser, 1973.
[16]M. D. Coe, 1978, pp. 29, 45.

References

Acosta, J. de
1396 *Concilium limense.* Madrid, 1591. Reproduced in G. Arents. *Tobacco* New York.

Adams, R. N.
1952 *Un Análisis de las creencias y prácticas médicas en un pueblo indígena de Guatemala.* Guatemala City: Instituto Indigenista Nacional Publicaciones especiales, no. 17.

1955 A nutritional research program in Guatemala. In B. Paul, ed. *Health, culture, and community.* New York: Russell Sage Foundation.

Aguirre Beltrán, G.
1963 *Medicina y magia: el proceso de aculturación en la estructura colonial.* Mexico City: Instituto Nacional Indigenista.

1967 Función social de la medicina precortesiana. *Etnohistoria* 1: 22–25.

Alarçón, R. de
1928 Tratado de las superticiones de naturales de esta Nueva España. Escrito en 1629. Anales del Museo Nacional. Mexico City.

Alger, N.
1974 In Marlene Dobkin de Ríos: The influence of psychotropic flora and fauna on Maya religion. *Current Anthropology* 15: 153–54.

Altschul, S. von Reis
1964 *A taxonomic study of the genus* Anadenonthera. Contributions from the Gray Herbarium of Harvard University, no. 193.

1973 *Drugs and food from little-known plants.* Cambridge, Mass.: Harvard University Press.

Anders, F.
1963 *Das Pantheon der Maya.* Graz. Akademisch Druck und Ferlaganstalt.

Anderson, A.
1961 Medical practices of the Aztecs. *Palacio* 68: 113–18.

Anguiano, R. de
1908 *Un Nuevo proyecto para reducir a la de a los indios xicakes. Relaciones históricos y geográficos de América central.* Madrid.

Anonymous
1850 *Paris fumeur.* Paris: Maison Crapelet.

Anonymous
1855 Cocom. *Registro Yucateco.* Mérida.

Anton, F.
1968 *Kunst der Maya.* Leipzig.

Arents, G.
1936 *Tobacco.* New York.

Armitage, A. K.
1965 Effects of nicotine and tobacco smoke on blood pressure and release of catecholamines from the adrenal glands. *British Journal of Pharmacology* 25: 515–26.

Arnold, L.
1960 Altered states of consciousness. In C. Tart, ed. *Altered states of consciousness.* New York.

Bancroft, A.
1875 *Native races of the Pacific States.* New York.

Bandelier, F. R.
1976 *Ancient Mexico.* New York: Fisk University Press.

Bardawil, L. W.
1976 The principal bird deity in Maya art: an iconographic study of form and meaning. *Proceedings, Segunda Mesa Redonda de Palenque.* Vol. 3. Robert Louis Stevenson School. Ed. M. Greene Robertson. Pebble Beach, Calif.

Barrera-Vásquez, A.
1963 Fuentes para el estudio de la medicina nativa de Yucatán. Rev. ed. *Médica del Hospital O'Horan.* Mérida.

Barthel, T.
1963 Die Stele 13 von Tikal. *Tribus* 12:159–241. Stuttgart.

1964 Comentarios a las inscripciones clásicas tardías de Chichén Itzá. *Estudios de Cultura Maya* 4: 223–77. Mexico City.

1968 *El Complejo emblema.* Primera Seminario Internacional para el Estudio de la Escritura Maya. Mexico City.

1974 Neue Lesungen für Mayaschrift. *Tribus* 23: 175–211. Stuttgart.

1978 Tohil des Feuerbringer. *Festschrift Gerdt Kutscher.* Berlin.

Bartholin, T.
1936 *Historiarum anatomicarum et medicorum rariorum.* Copenhagen, 1961. In G. Arents, *Tobacco.*

Baynes, A.
N.d. *The Mayas had tobacco.* MS. Arents Collection, New York City Public Library.

N.d. *Tobacco comes to the ancient Mexicans.* MS. Arents Collection, New York City Public Library.

Beard, J. S.
1944 Climax vegetation in tropical America. *Ecology* 25: 128–58.

Becerra, M.
1925 Exploración biológica en el estado de Chiapas. *Boletín División Estudios Biológicos* 3: 27-29. Mexico City.

Behar, M.
1968 Food and nutrition of the Maya before the Conquest and at the present time. In *Biomedical challenges presented by the American Indian.* Pan

American Health Organization, Scientific Publication no. 165, pp. 114–19. Washington, D.C.

Benítez, F.
1975 *The magic land of peyote.* Austin: University of Texas Press.

Bennett, W. C., and Lingy, R.
1935 *The Tarahumara: an Indian tribe of northern Mexico.* Chicago.

Benson, E. P.
1970 *The Maya world.* New York: Crowell.

1975 Gestures and offerings. *Proceedings, Primera mesa Redonda de Palenque.* Vol. 1. Robert Louis Stevenson School. Ed. M. Greene Robertson. Pebble Beach, Calif.

1976 Ritual cloth and Palenque kings. *Proceedings, Segunda Mesa Redonda de Palenque.* Robert Louis Stevenson School. Ed. M. Greene Robertson, Pebble Beach, Calif.

Benzoni, G.
1568 *Historia del Mondo Nuovo.* Venice.

Bequert, J. C.
1933 Botanical notes from Yucatán. In G. G. Shattuck. *The peninsula of Yucatán.* Carnegie Institution of Washington Pub. 431, pp. 505–25. Washington, D.C.

Berlin, B.
1972 Speculations on the growth of ethnobotanical nomenclature. *Journal of Language in Society* 1: 63–98.

———, Breedlove, D. E., and Raven, P. H.
1973 *Principles of Tzeltal plant classification: an introduction to the botanical ethnography of Mayan speaking people of highland Chiapas.* New York: Academic Press.

Berlin, H.
1963 The Palenque Triad. *Journal de la Societé des Americanistes* 52: 91–99. Paris.

1965 The inscription of the Temple of the Cross at Palenque. *American Antiquity* 30: 330-42.

1968 The Tablet of the 96 Glyphs at Palenque, Chiapas, Mexico. *Middle American Research Institute, Tulane University, Pub. 26.* New Orleans.

———, and Kelley, D. H.
1961 The 819-day count and color direction symbolism among the Classic Maya. *Middle American Research Institute, Tulane University, Pub. 26* New Orleans.

Beverly, R.
1702 *History and present state of Virginia.* Bk. 2, p. 24. London.

Beverwijk, J. van
1649 *Schat der Ghesondhijt.* Amsterdam.

Beyer, H.
1937 Studies on the inscriptions of Chichén Itzá. Carnegie Institution of Washington Pub. 21. 4:29–175. Washington, D.C.

Blom, F., and Duby, G.
1955 *La Selva Lacandona.* Mexico City.

Boggs, G.
1944 In Longyear, J. N. Archaeological investigations in El Salvador. *Memoirs, Peabody Museum, Harvard University.* Cambridge, Mass.

Book of Chilam Balam of Kaua MS. Copy in the Latin-American Library, Tulane University, New Orleans.

Bordeson, W. E.
1972 Tobacco from a Tihuanacoid culture period. *Etnografiska Studier* 32: 177–89. Göteborg.

Borhegyi, S. A. de
1961 Miniature mushroom stones from Guatemala. *American Antiquity* 26: 498–504.

1963 Pre-Columbian pottery mushrooms from Mesoamerica. *American Antiquity* 28: 328–38.

1965 Archaeological synthesis of the Guatemalan highlands. In *Handbook of Middle American Indians.* Vol. 2. Ed. Wauchope and G. R. Willey. Austin: University of Texas Press.

Brainerd, G. W.
1958 The archeological ceramics of Yucatán. University of California anthropological records. Los Angeles.

Brasseur de Bourbourg, C. E.
1861 *Popol Vuh: Le livre sacre et les mythes de l'antiquité americaine avec les herioques de Quichés.* Paris: A. Bertrand.

Breedlove, D. E.
1971 *History of botanical exploration in Chiapas, Mexico.* Modroño.

———, and Hopkins, N. A.
1970 A Study of Chuj (Mayan) plant names with notes on uses. *Wasman Journal of Biology.* Vol. 28, pt. 1, no. 2, pp. 275–79.

Brinton, D. G.
1898 Mushroom-shaped Images. *Science,* n.s. 8: 127.

Brongers, G. A.
1964 *Nicotiana tabacum.* Amsterdam: H. J. Bechts.

Bronzon, J. E., and Hehaka, Sapa
1953 *Les Rites secrets des indiens Sioux.* Paris.

Bruhn, J., and Holmstedt, B.
1974 Early Peyote research: an interdisciplinary study. *Economic Botany* 28: 353–90.

Bunzel, R.
1932 Introduction to Kuni ceremonialism. Bureau of American Ethnology Annual Report, no. 47. Washington, D.C.

1952 *Chichicastenango.* Seattle: University of Washington Press.

Burland, C. A., and Foreman, W.
1975 *Feathered Serpent and Smoking Mirror.* New York: G. P. Putnam's Sons.

Cabrera, L. G.
1943 *Plantas curativas de Mexico.* Mexico City: Ediciones Cicerón.

Cangas y Quinones, S. de
1928 Descripción de la villa de Espíritu Santo. In *Biblioteca de la revista mexicana de estudios históricos.*

Caravajao, P. A.
1961 *Plantas que curan, plantas que matan.* Mexico City: Ediciones del Caribe.

Cárdenas, J. de
1913 *Primera parte de las problemas y secretos maravillosos de las Indias.* Mexico City: Pedro Ocharte Ediciones.

Cardos, A. M.
1959 Comercio de las Mayas antiquos. *Acta Antropología.* Mexico City.

Carlson, J. B.
1978 *Olmec Concave Iron Ore Mirrors Technology and Purpose.* MS. under publication.

Carlyle, T.
1976 Quoted by C. Stephenson in Preserving everyday history. *Scanorama.*

Carmack, R. M.
1972 Personal communication in R. E. Schultes. An Overview of hallucinogens in the Western Hemisphere. In *Flesh of the gods: The ritual use of hallucinogens.* Ed. P. T. Furst, New York: Praeger.

1973 *Quichean civilization.* Berkeley: University of California Press.

Carnegie Institution of Washington
1963 Botany of the Maya area. Carnegie Institution of Washington Miscellaneous Papers, no. 461, pp. 1–13.

Carneiro, R. L.
1970 Hunting and hunting magic among the Amahuaca. *Ethnology* 10: 331–41.

Caso, A.
1970 *The Aztecs: people of the sun.* Norman: University of Oklahoma Press.

1971 Lapidary work, goldwork, and copperwork from Oaxaca. In *Handbook of Middle American Indians.* Vol. 1. Ed. R. Wauchope and G. R. Willey. Austin: University of Texas Press.

———, and Gutiérrez, T.
1965 El Deporte prehispánico. *Artes de México,* No. 75–76. Mexico City.

Castiglioni, L.
1790 *Viaggio negli Stati Unite dell'America.* Milan: Stamperia di Guiseppe Marelli.

Catlin, G.
1973 *Letters and notes on the manners, customs and conditions of North American Indians.* New York.

Cervantes y Salazár, F.
1914 *Papeles de Nueva España.* Tercera serie historia. Ed. F. de Paso y Troncoso. Madrid.

Chilam Balam de Kaua, Book of
MS. Gates reproduction in Peabody Museum, Harvard University, Cambridge, Mass.

Chilam Balam de Nah, Book of
Photostat. Middle American Research Institute, New Orleans.

Ciba Zeitschrift
1949 *Ciba Zeitschrift.* Vol. 116. Basel.

Clavijero, F. S.
1887 *The history of Mexico.* Philadelphia.

Cockburn, J.
1735 Quoted in G. Stahl. *Die Zigarre. Eine Kulturhistorische Betrachtung. Der Tabak.*

Codex Borbonicus
1940 *A sacred almanac of the Aztecs (Tonalamatl of the Codex Borbonicus).* Ed. G. C. Vaillant. New York: American Museum of Natural History.

———,
1969 Reproduction of original MS. Graz: Akademische Druck und Verlaganstalt.

Codex Borgia
1963 *Codice Borgia.* Ed. Eduard Seler. 3 vols. Mexico City: Fondo de Cultura Económica.

Codex Fejervary-Mayer
1901 MS. Mexique pre-Columbien des Free Public Museums de Liverpool (M120143n Paris: Duc de Loubat.

Codex Laud
1961 *Códice Laud.* Ed. C. Martínez Marín. Serie investigaciones, no. 5. Mexico City.

Codex Mendoza
1938 Mexican MS. in Bodleian Library. Ed. and trans. J. C. Clark. London: Waterloo and Sons.

Codex Nuttall
1902 *Codex Nuttall: facsimile of an ancient Mexican codex belonging to Lord Zouche of Harynworth, England.* Intro. Zelia Nuttall. Cambridge, Mass.: Peabody Museum, Harvard University.

Codex Tro-Cortesianus (Codex Madrid)
1967 Reproduction of original MS. Museo de America, Madrid. Graz: Akademische Druck und Verlaganstalt.

Codex Vaticanus B
1939 *Códice Vaticano.* MS. Pictorico mexicano no. 3773 de la Biblioteca Vaticana. Mexico City: Librería Anticuaria.

Codex Vindobonensis
1963 *Codex Vindobonensis mexicanus 1.* Osterreichische Nationalbibliothek, Vienna. Ed. O. Adelhofer. Graz: Akademische Druck und Verlaganstalt.

Coe, M. D.
1962 *Mexico.* New York: Praeger.

1966 *The Maya.* New York: Praeger.

1970 Olmec jaguars and Olmec kings. In *The cult of the*

feline. Ed. E. P. Benson. Dumbarton Oaks: Trustees of Harvard University.

1971 The shadow of the Olmecs. *Horizon* 13: 67–76.

1973 Personal communication in R. E. Schultes. An overview of hallucinogens in the Western Hemisphere. In *Flesh of the gods: The ritual use of hallucinogens.* Ed. P. T. Furst, p. 28. New York: Praeger.

1973a *The Maya scribe and his world.* New York: Grolier Club.

1973b Death and the ancient Maya. In *Death and the afterlife in Pre-Columbian America.* Ed. E. P. Benson, pp. 87–103. Washington, D.C.

1975 *Classic Maya pottery at Dumbarton Oaks.* Washington, D.C.

1978 *Underworld masterpieces of Classic Maya ceramics.* Princeton: Princeton University Press.

———, and Benson, E. P.
1966 *Three Maya relief panels at Dumbarton Oaks.* Studies in Pre-Columbian Art and Archeology, no. 2. Dumbarton Oaks: Trustees of Harvard University.

Coe, W. R.
1965 Tikal: ten years of study of a Maya ruin in the lowlands of Guatemala. *Expedition* 8:5–56. Philadelphia: University Museum, University of Pennsylvania.

1970 *Tikal: a handbook of the ancient Maya ruins.* University of Pennsylvania Museum Publication, Philadelphia.

1970a Two incised obsidians from Cache 161 at Tikal. MS. University of Pennsylvania, Philadelphia.

———, Shook, E. D., and Satterthwaite, L.
1961 *The carved wooden lintels of Tikal.* Museum Monographs, Tikal Reports, no. 6. Philadelphia: University of Pennsylvania Museum.

Coggins, C. C.
1976 Painting and drawing styles at Tikal: an historical and iconographic reconstruction. PhD. dissertation, Harvard University. Ann Arbor: University Microfilms.

Cohodas, M.
1973 The iconography of the Panels of Sun, Cross, and Foliated Cross at Palenque; Part 2. *Proceedings, Primera Mesa Redonda de Palenque.* Vol. 1. Robert Louis Stevenson School. Ed. M. Greene Robertson. Pebble Beach, Calif.

Colmeiro, M. de
1871 *Diccionario de los diversos nombres vulgares de muchas plantas usuales o notables del antiguo y Nuevo Mundo.* Madrid.

Columbus, C.
1871 *Logbook, 1492–1493.* Amsterdam: Wereldbibliotheek.

Comas, J.
1968 La medicina aborigen mexicana en obra de Fr. Agustín de Vetancourt (1698). *Anales de Antropología* 5: 129–62.

El Conquistador Anonymo
1941 *Relación de algunas cosas de la Nueva España y de la gran Ciudad de Temestitlán, México. Escrito por un compañero de Hernán Cortés.* Mexico City.

Conti, E. C.
1930 *Die Trockene Trunkenheit.* Berlin.

1932 *A history of smoking.* New York.

Contreras, R. C.
1936 *50 plantas medicinales indígenas.* 2d ed. Mexico City.

Coon, N.
1963 *Using plants for healing.* New York: Hearthside Press.

Cooper, J. M.
1949 Stimulants and narcotics. In *Handbook of South American Indians.* Vol. 5. Ed. J. Steward. Washington, D.C.: Smithsonian Institution.

Cosminsky, S.
1975 Changing food and medical beliefs and practices in a Guatemalan community. *Ecology of Food and Nutrition* 4: 183–91.

Coury, C.
1969 *Médicine précolombienne.* Paris.

Covarrubias, M.
1966 *Indian art of Mexico and Central America.* New York: Alfred A. Knopf.

Crumrine, R. N.
1974 In Marlene Dobkin de Ríos. The influence of psychotrophic flora and fauna on Maya religion. Current anthropology 15: 153–54.

Cruz, M., and Badianus, J.
1940 *An Aztec herbal of 1552.* Trans. E. W. Emmart. Baltimore: Johns Hopkins Press.

Cudell, R.
1927 *Das Buch vom Tabak.* Cologne.

1930 *Die Sammlung Haus Neuburg. Von Rauchen und Rauchgerate.* Cologne.

Cuevas, B.
1913 *Plantas medicinales de Yucatán y guia médica práctica doméstica.* Mérida.

Cuevas, M.
1928 *Historia de la iglesia en México.* El Paso.

Cuzin, J.
1967 *Chemical abstracts.* Vol. 66, p. 92491.

Davis, E. A., et al.
1969 Effects of harmine on the cat's visual system. American Association of Anatomists abstract. *Anatomical Record* 163: 2–175.

Delay, J.
1967 Psychopharmacology and psychiatry: towards a classification of psychotropic drugs. *Bulletin on Narcotics* 19: 1–5.

Delgado, M. C., Urzua, A., and Cassels, B.
1972 Tribal and urban healing with hallucinogenic beverages in Amazonian Peru. *American Journal of Pharmacy* 144: 187-90.

Del Pozo, E.
1965 La Botánica medicinal indígena de México. *Estudios de Cultura Nahuatl* 5: 57-74.

1967 Empiricism and magic in Aztec pharmacology. In D. Efron, ed. *The ethnopharmacological search for psychoactive drugs*, pp. 59-76. Public Health Service Publication no. 1645. Washington, D.C.: U.S. Government Printing Office.

———, et al.
1964 Symposium sobre el códice de medicina azteca de Martín de la Cruz y Juan Badiano. *Gaceta Médica de Mexico* 94: 1159-1201.

Descripcion de San Bartolomé
1965 *Anales de Sociedad de Geografiá y Historia de Guatemala* 38: 262-276.

Díaz, J.
1958 *La Primera noticia del descubrimiento de la Nueva España en Europa. El Itinerario de la espedición del Capitán Juan de Grijalva (año 1518).* Mérida.

Díaz de Castillo, B.
1973 *The conquest of New Spain.* Middlesex.

Dibble, C.
1966 La Basa científica para el estudio de las yerbas medicinales de los Aztecas. *Actas y memorial del 36 Congreso Internacional de Americanistas* 2: 63-67.

Dickson, S. A.
1954 *Panacea or precious bane: history of tobacco in the sixteenth century.* New York: New York City Public Library.

Dieseldorff, E. P.
1922 Welchen Gott Stellen die Steinidolen der Mayavölker dar? In *Festschrift Edward Seler.* Stuttgart.

Dixon, R. B.
1921 Words for tobacco in American Indian languages. *American Anthropologist*, n.s. 23: 19-49.

Dobkin de Ríos, M.
1968 *Trichocereus Pachanoi:* a mescaline cactus used in folk healing in Peru. *Economic Botany* 22: 191-94.

1970 A note on the use of Ayahuasca among urban mestizo populations in the Peruvian Amazon. *American Anthropologist* 72: 1419-22.

1973 The non-Western use of hallucinatory agents. In *Drug use in America: problem in perspective.* Vol. 1, *Patterns and Consequences of Drug Use*, pp. 1179-1226. Appendix, 2d Report of the National Committee on Marihuana and Drug Abuse. Washington, D.C.: U.S. Government Printing Office.

1974 The influence of psychotropic flora and fauna on Maya religion. *Current Anthropology* 15: 147-64.

1975 Man, culture, and hallucinogens: an overview. In V. Rubin. *Cannabis and Culture.* The Hague.

1977 The Maya and the water lily: some Musings. *New scholar* (in press).

———, and Smith, D. E.
1976 Using or abusing? An anthropological approach to the study of psychoactive drugs. *Journal of Psychoactive Drugs* 8: 263-66.

Dodoens, L.
1554 *Cruijdeboeck.* The Hague.

1936 *Trium priorum stirpium historia.* Antwerp. Reproduced in G. Arents. *Tobacco.* New York.

Dorsifang-Smets, A.
1976 *Art mesoamerique.* Bruxelles: Societe generale de baroque.

Douglas, W.
1969 Illness and curing in Santiago Atitlán. Ph.D. dissertation, Stanford University.

Douglass, A. E.
1889 A portrait pipe from Central America. *American Antiquarian* 2: 348-53.

Duby, G., and Blom, F.
1969 The Lacandón. In *Handbook of Middle American Indians.* Vol. 7. Ed. R. Wauchope and G. R. Willey. Austin: University of Texas Press.

Dulante Gutiérrez, E.
1970 La medicina primitiva en México. *Artes de México* 17.

Dunhill, A.
1924 *The pipebook.* London.

Durán, Fray Diego
1951 *Historia de las Indias de Nueva España y islas de tierra firme.* Ed. J. F. Ramirez. 2 vols., atlas. Mexico City: Editorial Nacional.

1964 *The Aztecs: the history of the Indies of New Spain.* Trans. and ed. D. Heyden and F. Horcasitas. New York: Orion Press.

1971 *Book of the Gods and Rites and the Ancient Calendar.* Trans. and ed. F. Horcasitas and D. Heyden. Norman: University of Oklahoma Press.

Duran, C. M.
1963 Surgery of the Mayas. *Abbot Tempo*, vol. 1, no. 4. London: Interspan.

Dütting, P.
1965 Algunas consideraciones sobre el trabajo de H. Berlin, "The Palenque Triad." *Estudios de Cultura Maya* 5: 135-44.

Easby, E. K.
1963 Un "Díos Hacha" de las tierras altas Mayas. *Estudios de Cultura Maya* 3: 97-106.

———, and Scott, J. F.
1970 *Before Cortés: sculpture of Middle America.* New York: Metropolitan Museum of Art.

Edmonson, M. D.
1965 Quiché-English dictionary. *Middle American Research Institute, Tulane University, Pub. 5, 30.* New Orleans.

1971 The Book of the Council. The Popol Vuh of the Quiché Maya of Guatemala. *Middle American Research Institute,* Tulane University, *Pub.* 35. New Orleans.

Edwards, E.
1966 *Painted walls of Mexico: from prehistoric times until today.* Austin: University of Texas Press.

Efron, D.
1967 *The ethnopharmacological search for psychoactive drugs.* Public Health Service Publication no. 1645. Washington, D.C.; U.S. Government Printing Office.

Ekholm, G. F.
1970 *Ancient Mexico.* New York: American Museum of Natural History.

Emmart, E.
1937 Herb medicine of the Aztecs. *Journal of the American Pharmaceutical Association* 26: 42-45.

1940 *The Badianus Manuscript: an Aztec herbal of 1552.* Baltimore: Johns Hopkins Press.

Encyclopedie du Tabac
1975 *Encyclopedie du tabac et des fumeurs.* Paris: Le Tempo.

Ernst, A.
1899 On the etymology of the word tobacco. *American Anthropologist* 2: 169.

Escalante, Solis, E.
1941 *Botánica.* Mérida: Publicaciones de la Universidad de Yucatán.

Euw, E. von
1974 Las ruinas de Itzimté. *Boletín de Instituto Nacional de Antropologiá y Historia* 10: 19-26. Mexico City.

Evarts, E. V.
1956 Some Effects of bufotenine and LSD on the monkey. *A.M.A. Archives of Neurology and Psychiatry* 75: 49-53.

Ewers, J. C.
1957 Blackfoot Indian pipes and pipe-making. *Smithsonian Institution Bulletin* 186. Washington, D.C.

Fabing, H. D.
1957 Toads, mushrooms, and schizophrenia. *Harper's* 214: 1284-90.

———, and Hawkins, J.
1956 Intravenous injections of Bufotenine in man. *Science* 123: 886.

Fabrega, H.
1971 Some features of Zinacantan medical knowledge. *Ethnology* 10:25-43.

———, Metzger, D., and Williams, G.
1970 Psychiatric implications of health and illness in a Maya Indian group. *Social Science and Medicine* 3: 609-26.

———, and Silver, D.
1973 *Illness and shamanistic curing in Zinacantan: an ethnomedical analysis.* Stanford: Stanford University Press.

Fairholt, F. W.
1876 *Tobacco: its history and association.* London.

Farías Galindo, J.
1956 *El simbolísmo médico indígena.* Mexico City: Vargas Rea.

Feinhals, J.
1936 *Vom Tabak Ein Spaziergang durch das Raucherparadies.* Cologne.

Felton, B., and Fowler, M.
1975 *Best, worst, and most unusual.* Greenwich, Conn.: Information House Books.

Fernández del Castillo, F.
1958 Historia de la medicina. La cronología y la medicina Nahuatl. *Medico:* 7: 54-59.

Fields, F. H.
1969 Rivea corymbosa: notes on some Zapotec customs *Economic Botany* 23: 206-209.

Folgairolle, E.
1897 *Jean Nicot.* Paris.

Follett, Prescott
1932 War and weapons of the Maya. *Middle American Research Institute, Tulane University,* Pub. 4. New Orleans.

Foncerrado de Molina, M.
1976 Reflecciones en torno a Palenque como necropolis. Proceedings, Primera Mesa Redonda de Palenque. Vol. 3. Robert Louis Stevenson School. Ed. M. Greene Robertson. Pebble Beach, Calif.

Förstemann, E.
1901 Drei Mayahieroglyphen. *Zeitschrift für Ethnologie,* pp. 215-21.

1902 *Commentar zur Madrides Maya Handschrift.* Danzig.

Foster, G.
1953 Relationships between Spanish and Spanish-American folk medicine. *Journal of American Folklore* 66: 201-17.

Fought, J. G.
1972 *Chorti Maya texts.* Philadelphia: University of Pennsylvania Press.

Friederici, G.
1947 *Amerikanistiches Wörterbuch.* Hamburg.

Fuente, B. de la
1965 *La escultura de Palenque.* Mexico City: Universidad Nacional Autónoma de Mexico.

———, and Schaverzon, D.
1976 Algunas noticias poco conocidas que sobre Palenque se publicaron en el siglo 19. *Proceedings,*

Segunda Mesa redonda de Palenque. Vol. 3. Robert Louis Stevenson School. Ed. M. Greene Robertson. Pebble Beach, Calif.

Fuentes y Guzman, F. A.
1882 *Historia de Guatemala y recordación Florida*. Reprint, 1932. Guatemala City: Sociedad de Geografía y Historia Guatemala.

Furst, P. T.
1972 [Editor] *Flesh of the gods: the ritual use of hallucinogens*. New York: Praeger.

1972a The Olmec were-jaguar motif in the light of ethnographic reality. In M. P. Leone: *Contemporary Archaeology*. pp. 444–64. London.

1972b Symbolism and psychopharmacology: the toad as earthmother in Indian America. *Religión en Mesoamérica. XII Mesa Redonda*, pp. 37–45. Mexico City.

1974 Archaeological evidence of snuffing in Prehispanic Mexico. *Botanical Museum Leaflets*, Harvard University, vol. 24, no. 1, pp. 1–28. Cambridge, Mass.

1974a In Marlene Dobkin de Ríos. The influence of psychotropic flora and fauna on Maya religion. *Current Anthropology* 15: 154–55.

1974b *Les Chairs des dieux*. Paris.

1976 Fertility, vision quest, and auto-sacrifice: some thoughts on ritual blood-letting among the Maya. *Proceedings, Segunda Mesa Redonda de Palenque*. Vol. 3. Robert Louis Stevenson School. Ed. M. Greene Robertson. Pebble Beach, Calif.

1976a *Hallucinogens and culture*. San Francisco: Chandler and Sharp.

———, and Coe, M.
1971 Maya ecstatic intoxication. *Natural History*, March.

———, and Myerhoff, B.
1966 Myth as history: the jimson weed cycle of the Huichols of Mexico. *Antropológica* 17: 3–39.

Gage, T.
1958 *Thomas Gage's travels in the New World*. Norman: University of Oklahoma Press.

Gann, T.
1900 Mounds in northern Honduras. In *Nineteenth Annual Report of the Bureau of Ethnology to the Secretary of the Smithsonian Institution, 1897–98*. Pt. 2, pp. 655–92. Washington, D.C.: U.S. Government Printing Office.

García de Peláez, F. de P.
1968 *Memorias para la historia de antiguo reino de Guatemala*. Guatemala City: Biblioteca Goathemala de la Sociedad de Geografía y Historia.

Garibay, K. A. M.
1961 *Vida económica de Tenochtitlán*. Mexico City.

Gerste, A.
1909 *Notes sur la médicine et la botanique des anciens Mexicains*. Rome: Vatican.

Giblin, J.
1936 Bahama River Caribs, British Guiana. *Papers, Peabody Museum, Harvard University*. Cambridge, Mass.

1951 The culture and security in San Carlos: a study of Indians and Ladinos. *Middle American Research Institute, Tulane University*, Pub. 16. New Orleans.

Gilbert, J.
1972 Tobacco and shamanic ecstasy among the Warao Indians of Venezuela. In P. T. Furst, *Flesh of the gods: the ritual use of hallucinogens*. New York: Praeger.

Gillin, J.
1956 The making of a witch doctor. *Psychiatry* 19: 131–36.

Girard, R.
1969 *Die Ewigen Mayas*. Zurich: Rhein Verlag.

Gleason, H.
1961 *Introduction to descriptive linguistics*. New York: Holt.

Gómez-Pompa, A.
1965 La vegetación de México. *Boletín de la Sociedad Botanica de México* 29: 76–120.

González Ulloa, M.
1959 *La Medicina en México: síntesis histórica*. Mexico City: Cyanamid de México.

Gonzatti, C.
1910 *Flora sinóptica mexicana*. Mexico City: Secretaria de Fomento.

Goodspeed, T. D.
1954 *The genus Nicotiana*. Waltham.

Gordon, G. B.
1925 *Examples of Maya pottery in the museum and other collections*. Philadelphia: University of Pennsylvania Museum.

Gorter, R. A.
1953 *De Tabaksrookklisteer als Reanimator My. tot Redding van Drenkelingen*. Amsterdam.

1957 De Tabaksrookklisteer-Machine. *Het Reddingwezen*, vol. 1957, no. 5.

Graham, I.
1967 Archaeological explorations in El Petén, Guatemala. *Middle American Research Institute, Tulane University, Pub.* 33. New Orleans.

1970 The ruins of La Florida, Petén, Guatemala. In *Monographs and Papers in Maya Archaeology*. Ed. W. R. Bullard, Jr. *Papers, Peabody Museum, Harvard University*, vol. 61, pp. 425–55. Cambridge, Mass.

1971 *The art of Maya hieroglyphic writing*. Cambridge, Mass.: Harvard University Printing Office.

———,and Von Euw, E.
1975–77 *Corpus of Maya hieroglyphic inscriptions.* Cambridge, Mass.: Harvard University Printing Office.

Greene, E. L.
1909 *Landmarks of botanical history: a study of certain epochs in the development of the science of botany.* Smithsonian Miscellaneous Collections, vol. 54. Washington, D.C.: Smithsonian Institution.

Greene Robertson, M.
1967 *Ancient Maya relief sculpture.* New York: Museum of Primitive Art.

1972 *The ritual bundles of Yaxchilán.* Tulane University Symposia on the Art of Latin America, Apr.15. New Orleans.

1973 The quadripartite badge: badge of rulership. *Proceedings, Primera Mesa Redonda de Palenque.* Vol. 1. Robert Louis Stevenson School. Ed. M. Greene Robertson, Pebble Beach, Calif.

1974 Stucco techniques employed by ancient sculptors on the Palenque Piers. *Proceedings of the 41st International Congress of Americanists.* Mexico City.

1977 *A correlation between the Codex Tro-Cortesianus and Palenque's East Subterraneo.* 42d Annual Meeting, Society for American Archaeology, Apr. 28–30. New Orleans.

———, Scandizzo, M. D., and Scandizzo, J. R.
1976 Physical deformities in the ruling lineage of Palenque and the dynastic implications. *Proceedings, Primera Mesa Redonda de Palenque.* Vol. 3. Robert Louis Stevenson School. Ed. M. Greene Robertson, Pebble Beach, Calif.

Griffin, G. G.
1975 Early travelers to Palenque. *Proceedings, Primera Mesa Redonda de Palenque.* Vol. 1. Robert Louis Stevenson School. Ed. M. Greene Robertson, Pebble, Beach, Calif.

1976 Portraiture in Palenque. *Proceedings, Segunda Mesa Redonda de Palenque.* Vol. 3. Robert Louis Stevenson School. Ed. M. Greene Robertson, Pebble Beach, Calif.

Groth-Kimball, I., and Aveleyra Arroyo de Anda, L.
1964 *Obras selectas del arte prehispánico.* Consejo para la planeación y instalación del Museo Nacional de Antropologiá, ser. 1964. Mexico City.

1971 Aspectos de la medicina popular en el area rural de Guatemala. *Guatemala Indígena,* vol. 6, no. 1.

Guerra, F.
1950 *Bibliografía de la materia médica mexicana.* Mexico City: La Prensa Médica Mexicana.

1964 Maya medicine. *Medical history.* vol. 8, pp. 31–43.

1966 Aztec medicine. *Medical history,* vol. 10, pp. 315–38.

1966a Mexican phantastica: a study of the early ethnobotanical sources on hallucinogenic drugs. *British Journal of Addiction,* vol. 62, pp. 171–87.

Gumilla
1741 *El Orinoco illustrado.* Quoted in R. E. Schultes. An overview of hallucinogens in the Western Hemisphere. In *Flesh of the gods: ritual use of hallucinogens.* Ed. P. T. Furst. New York: Praeger, 1972.

Gutiéras-Holms, C.
1961 La Magia en la crisis del embarazo y parto en los actuales grupos mayances de Chiapas. *Estudios de Cultura Maya,* vol. 1, pp. 159–66.

1961a *Perils of the soul: the world view of the Tzotsil Indian.* New York: Free Press of Glencoe.

Gutiérrez de González, C.
1976 Identificación de un personaje de Xnucbec, Campeche. *Pictografías Mayas,* no. 1, pp. 41–50. Mérida.

Hall, R. L.
1975 Ideology and adaptation in eastern North America: some thoughts on kinnikinnick, mound burial, and the Sun Dance as metaphor and ritual. MS. University of Illinois at Chicago Circle, Chicago.

1976 An anthropocentric perspective for eastern United States archaeology. MS. University of Illinois at Chicago Circle, Chicago.

Harmon, R.
1969 La Medicina preventiva en una communidad india de los altos de Chiapas. *Anuario Indigenista,* vol. 29, pp. 277–83.

Harner, M. J.
1973 *Hallucinogens and shamanism.* New York: Oxford University Press.

Harper, Goodspeed, T.
1950 *Plant hunters in the Andes.* New York.

Harrington, H. D., and Durrell, L. W.
1957 *How to identify plants.* Denver: Sage Books.

Harvey, H. R., and Kelly, I.
1969 The Totonac. In *Handbook of the Middle American Indians.* Vol. 8. Ed. R. Wauchope and G. R. Willey. Austin: University of Texas Press.

Hay, C. L., et al., eds.
1940 *The Maya and their neighbors.* New York: Appleton-Century-Croft.

Hedrick, V. P.
1964 *Sturdevant's notes on edible plants.* State of New York, Department of Agriculture, 27th Annual Report, vol. 2, pt. 2. New York.

Heim, R.
1967 Nouvelles investigations sur les champignons hallucinogéns. *Archives du Musée National d' Histoire Naturelle,* vol. 7, no. 9.

———, and Wasson, R. G.
1958 Les champignons hallucinogéns de Mexique. *Archives du Musée National d' Histoire Naturelle,* ser. 7, vol. 6.

Heimann, R.
1960 *Tobacco and Americans.* New York.

Heiser, C. B., Jr.
1969 *Nightshades: the paradoxical plants.* San Francisco.

Helfrich, Klaus
1969 Menschenopfer und Tötungstrituale im Kult der Maya. *Monumenta Americana,* no. 9. Berlin: Ibero-Amerikanisches-Institut Preussicher Kulturbesitz, Gebr. Mann Verlag.

Hellmuth, N.
1972 An ethnological study of the southern Maya lowlands in the 16th and 17th centuries. MS. Library, Foundation for Latin American Anthropological Research. Guatemala City.

1974 In M. Dobkin de Ríos. The influence of psychotropic flora and fauna on Maya religion. *Current Anthropology* 15: 155–156.

1978 *Classic Maya art showing dances after decapitation sacrifice.* In press.

Hernández, F.
1615 *Quatro libros de la naturaleza y virtudes medicinales las plantas y animales de la Nueva España.* Ed. F. Ximénez. Mexico City.

1651 *Rerum medicarum novae hispaniae thesaurus, seu plantarum, animalum, mineralium mexiconorum historia.* Rome: Mascardi.

Herodotus
1855 *Herodoti historiae.* New York: Henry Clay.

Hill, A. F.
1945 Ethnobotany in Latin America. In F.Verdoorn, ed. *Plants and Plant Science in Latin America,* pp. 176–81. New York: Ronald Press.

Hind, G.
1960 *The Canadian Red River.* London.

Hoffman, A.
1966 The active principles of the seeds of *Rivea Corymbosa* and *Ipomoea Tricolor.* In *Summa antropologíca en homenaje a Robert Weitlander,* pp. 349–57. Mexico City: Instituto Nacional de Antropología y Historia.

Holland, R. L.
1963 *Medicina Maya en los altos de Chiapas.* Mexico City: Instituto Nacional Indigenista.

Holland, W., and Tharp, R.
1964 Highland Maya psychotherapy. *American Anthropologist* 66:41–52.

Holmstedt, B., and Lindgren, J. F.
1967 Chemical constituents and pharmacology of South American snuffs. In D. Efron, ed. *The ethnopharmacology search for psychoactive Drugs.* U.S. Public Health Service Publication no. 1645. Washington, D.C.: U.S. Government Printing Office.

Instituto Italo-Latino Americano
Arte Maya del Guatemala. Exhibition catalogue.

James I, King of England
1924 Quoted by B. Laufer. Introduction of tobacco into Europe. Field Museum of Natural History Anthropology Leaflet no. 19. Chicago.

1936 Quoted by G. Arents: *Tobacco.* New York.

Janiger, O., and Dobkin de Ríos, M.
1973 Suggestive hallucinogenic properties of tobacco. *Medical Anthropology Newsletter* 4:6–11.

1976 Nicotiana an hallucinogen? *Economic botany.*

Jones, C.
1977 Inauguration dates of three late Classic rulers of Tikal, Guatemala. *American Antiquity* 42:28–60.

Jonson, B.
1601 *Every man in his humour.* London.

Joralemon, D.
1974 Ritual blood sacrifice among the ancient Maya; Part 1. *Proceedings, Primera Mesa Redonda de Palenque.* Vol. 2. Robert Louis Stevenson School. Ed. M. Greene Robertson, pp. 59–76. Pebble Beach, Calif.

Karsten, R.
1935 Headhunters of the western Amazonas: life and culture of Jibaro Indians of east Ecuador and Peru. *Societas Scientiarum Fennica.* Vol. 7. Helsinfors.

Kelemen, P.
1943 *Medieval American art: a survey in two volumes.* New York: Macmillan.

Kelley, D. H.
1965 The birth of the gods of Palenque. *Estudios de Cultura Maya* 5:93–134.

1968 Maya fire glyphs. *Estudios de Cultura Maya* 7:141–57.

1976 *Deciphering the Maya script.* Austin: University of Texas Press.

Kensinger, K.
1973 Banisteriopsis usage among the Peruvian Cashinahua. In M. Harner, ed. *Hallucinogens and Shamanism,* New York: Oxford University Press.

Kidder, A. V., and Ekholm, G. F.
1951 Some archaeological specimens from Pomona, British Honduras. Carnegie Institution of Washington, Notes on Middle American Archeology, Publication no. 102. Washington, D.C.

Killebrew, G., Myrick, J.
1906 *Tobacco leaf.* New York.

Kingsborough, Lord
1831 *Antiquities of Mexico.* 4 vols. London.

Kinross-Wright, V. J.
1959 Research on ololiuqui: the Aztec drug. In P. Bradley, ed. *Neuropsychopharmacology,* pp. 453–56. Amsterdam: Elsevier.

Kirchhoff, P.
1943 Mesoamérica: Sus limites geográficos, composición étnica y caracteras culturales. *Acta americana* 1:92–107. Sociedad Interamericana de Antropología y Geografía, Washington, D.C.

1949 Tribes north of the Orinoco River. In *Handbook of South American Indians.* Vol. 4. Ed. J. Steward. Washington, D.C.

Knab, T.
1974 Personal communication to P. T. Furst in M. Dobkin de Ríos. The influence of psychotropic flora and fauna on Maya religion. *Current Anthropology* 15:154.

Knauth, L.
1961 El juego de pelota y rito de la decapitación. *Estudios de Cultura Maya* 1:183-98.

Knorozov, Y.
1967 *Selected chapters from the writings of the Maya Indians.* Cambridge, Mass.: Peabody Museum, Harvard University.

Koenig, P., ed.
1940 *Die Entdeckung des Reinen Nicotins.* Reprint of 1828 ed. Bremen: Artur Geist Verlag.

Kool, J. A.
1854 *Geschidkunolige Beschouwing van de My tol Redding van Drenkeligen te Amsterdam.* Amsterdam.

Koskowski, W.
1933 The habit of smoking. In G. Arents. *History of Tobacco.* London.

Krieg, M.
1964 *Green medicine: the search for plants that heal.* New York: Rand McNally.

Kubler, G.
1960 Studies in Classic Maya iconography. Memoirs of the New Haven Academy of Arts and Sciences no. 18.

1974 Climate and iconography in Palenque sculpture. In *Art and environment in native America.* Ed. M. E. King and I. R. Traylor, Jr. Special Publications, Museum, Texas Tech University, no. 7, pp. 103-13. Lubbock: Texas Tech Press.

1974a Mythological ancestries in Classic Maya inscriptions. Proceedings, *Primera Mesa Redonda de Palenque.* Vol. 2. Robert Louis Stevenson School. Ed. M. Greene Robertson, Pebble Beach, Calif.

1976 The double-portrait lintels of Tikal. In *Actas del 23 Congreso Internacional de Historia del Arte: España entre el Mediterraneo y el Atlántico* (Granada, 1974) 1:165-75. Granada: Universidade de Granada, Departamento de Historia del Arte.

1977 *Aspects of Classic Maya rulership on two inscribed vessels.* Dumbarton Oaks: Trustees of Harvard University.

La Barre, W.
1960 Twenty years of peyote studies. *Current Anthropology* 1:45-60.

1964 *The peyote cult.* New York: Schocken Books.

1970 Old and New World narcotics: statistical question and an ethnological reply. *Economic Botany* 24:73-80.

1974 In M. Dobkin de Ríos. The influence of psychotropic flora and fauna on Maya religion. *Current Anthropology* 15:157.

La Farge, O., and Beyers, D.
1931 The Year Bearer's people. *Middle American Research Institute, Tulane University,* Pub. 3. New Orleans.

Landa, D. de
1928 *Relación de las cosas de Yucatán.* Mexico City.

Langman, I. K.
1947 *Bibliography on plant sciences in Mexico.* Master's thesis, University of Pennsylvania.

Las Casas, B.
1877 *Historia de las Indias.* Mexico City.

Laski, V.
1958 *Seeking life.* Philadelphia: American Folklore Society.

Laughlin, R. M.
1969 The Tzotsil. In *Handbook of Middle American Indians.* Vol. 7. Ed. Robert Wauchope and G. R. Willey. Austin: University of Texas Press.

N.d. *Tzotsil dictionary.* Washington, D.C.; Smithsonian Institution.

Lavadores, V. G.
1969 *Las 119 plantas medicinales más conocidas en Yucatán.* Mérida.

León-Portilla, M.
1973 *Time and reality in the thought of the Maya.* Boston: Beacon Press.

Leopold, A. S.
1950 Vegetation zones of Mexico. *Ecology* 31:507-18.

Levi-Strauss, C.
1950 The use of wild plants in tropical South America. In J. Steward, ed. *Handbook of the South American Indians.* Washington, D.C.: Smithsonian Institution.

1961 *Le Miel aux cendres.* Paris.

1964 *Le Crue et le cult.* Paris.

Lewin, L.
1928 Untesuchungen über Banisteria Caapi Spruce. *Archiv. für Experimentelle Pathologie und Pharmokologie* 129: 133-49.

Libermann, G., and, Libermann, C.
N.d. *Les civilisations du tabac.* Paris.

Linton, R.
1924 *Use of tobacco among North American Indians.* Field Museum of Natural History Anthropology Leaflet, no. 15. Chicago.

Logan, M.
1973 Digestive disorders and plant medicinals in highland Guatemala. *Anthropos* 68: pp. 537-47.

1974 Humoral medicine: a study of change in a traditional medical system of highland Guatemala. Ph.D. dissertation, Pennsylvania State University.

Long, J. K.
1974 In M. Dobkin de Ríos: The influence of psychotropic flora and fauna on Maya religion. *Current Anthropology* 15:158.

López, A.
1971 De las plantas medicinales y de otras cosas medicinales. *Estudio de Cultura Nahautl* 9:125-230.

Lothrop, S. K.
1924 *Tulúm: an archaeological study of the east coast of Yucatán.* Carnegie Institution of Washington Publications, no. 335. Washington, D.C.

1926 *Pottery of Costa Rica and Nicaragua.* Contributions of the Museum of the American Indian, Heye Foundation. Vol. 8. New York.

1936 *Zacualpa: a study of ancient Quiché artifacts.* Carnegie Institution of Washington Pub. 472. Washington, D.C.

1940 South America as seen from Middle America. In *The Maya and their neighbors.* Ed. C. L. Hay et al., pp. 417-29. New York: D. Appleton-Century Co. Reprint. Salt Lake City: University of Utah Press, 1962.

Lounsbury, F. G.
1974 *On the derivation and reading of the Ben Ich Prefix.* Dumbarton Oaks Conference on Mesoamerican Writing Systems. Washington, D.C.: Trustees of Harvard University.

1974a Explanation of a frontispiece. *Proceedings, Primera Mesa Redonda de Palenque.* Vol. 1. Robert Louis Stevenson School. Ed. M. Greene Robertson. Pebble Beach, Calif.

1976 A rationale for the initial date of the Temple of the Cross at Palenque. *Proceedings, Segunda Mesa Redonda de Palenque.* Vol. 3. Robert Louis Stevenson School. Ed. M. Greene Robertson, pp. 211-30. Pebble Beach, Calif.

Loven, S.
1921 *Über die Weitzeln der Tainischen Kultur.* Göteborg.

Lowie, R. H.
1919 The tobacco society of the Crow Indians. *American Museum of Natural History, Anthropological Papers*, vol. 21. New York.

1956 *The Crow Indians.* New York.

Lowy, B.
1971 New records of mushroom stones from Guatemala. *Mycologia* 63:983-93.

1972 Mushroom symbolism in Maya codices. *Mycologia* 64:816-21.

1974 *Amanita Muscarina* and the thunderbolt legend in Guatemala and Mexico. *Mycologia* 66:188-91.

Luján Muñoz, L.
1970 *Apreciación de la cultura Maya.* Guatemala City: Editorial Universitaria Centroamericano.

Lundell, C. L.
1939 The vegetation of Petén. Carnegie Institution of Washington Pub. 478, pp. 1-243. Washington, D.C.

1942 Flora of eastern Tabasco and adjacent Mexican areas. Contributions of the University of Michigan Herbarium 8:1-74.

McGuire, J. D.
1896-1897 Pipe and smoking customs of the aborigines. *U.S. National Museum annual report*, Publication no. 1899, pp. 351-641. Washington, D.C.

1899 *Pipes and smoking customs of the American aborigines, based on the material in the U.S. National Museum.* Washington, D.C.: Smithsonian Institution Publications.

Maler, T.
1901 Researches in the central portion of the Usumacinta Valley. *Memoirs, Peabody Museum, Harvard University*, vol. 2, no. 1. Cambridge, Mass.

1908 Explorations of the upper Usumatsintla and adjacent region: Altar de Sacrificios; Seibal; Itsimté-Sácluk; Cankuen. *Memoirs, Peabody Museum, Harvard University*, vol. 4, no. 1. Cambridge, Mass.

Mangelsdorf, P. D., MacNeish, R. S., and Willey, G. R.
1971 Origins of Agriculture in Middle America. In *Handbook of Middle American Indians.* Vol. 1. Ed. R. Wauchope and G. R. Willey. Austin: University of Texas Press.

Marin, A. B., Vázquez, A. G., and Franco, R. M. L.
1976 Nomenclature etnobotánica Maya. Una Interpretación taxonómica. Instituto Nacional de Antropología y Historia, serie etnológica, no. 36. Mexico City.

Martin, R. T.
1970 The Role of coca in the history, religion, and medicine of South American Indians. *Economic Botany* 24: 422-437.

Martínez Hernández, J.
1929 *Diccionario Motul: Maya-Español.* Atribuido a Fray Antonio de Ciudad Real. Mérida: Tipografía Yucateca.

Martínez, M.
1933 *Las Plantas medicinales de Mexico.* Mexico City: Editorial Botas.

Mason, J. A.
1924 Use of tobacco in Mexico and South America. Field Museum of Natural History, Anthropology Leaflet no. 16. Chicago.

Matthews, P., and Schele, L.
1974 Lords of Palenque—the glyphic evidence. *Proceedings, Primera Mesa Redonda de Palenque.* Vol. 1. Robert Louis Stevenson School. Ed. M. Greene Robertson. Pebble Beach, Calif.

Maudslay, A. P.
1889-1902 *Biología Centrali-Americana.* London: R. H. Porter and Dulan & Co.

Mellon, G.
1974 El Uso de las plantas medicinales en Guatemala. *Guatemala Indígena* 9:99-179.

Mena Manuscript
1921 Transcript in A. M. Tozzer. *A Maya grammar with bibliography and appraisement of Words Noted.* Peabody Museum Papers. Cambridge: Harvard University.

Mendieta, G.
1870 *Historia eclesiástica indiana.* Ed. G. García Tcazbalceta. Mexico City.

Meteren, E. van
Historie der Nederlanden. Grevenhage.

Metraux, A.
1949 The Tucuna. In *Handbook of South American Indians.* Vol. 5. Ed. J. Steward. Washington, D.C.: Smithsonian Institution.

Metzer, D., and Williams, G.
1963 Tenejapa medicine: the curer. *Southwestern Journal of Anthropology* 19: 216–34.

Meyer, H. K.
1977 *The mushroom stones of Mesoamerica.* Ramona, Calif.: Acoma.

Miles, S. W.
1965 Sculpture of the Guatemalan Chiapas.

1965a Sculpture of the Guatemalan Chiapas highlands and Pacific slopes, and associated hieroglyphs. In *Handbook of the Middle American Indians.* Vol. 2. Ed. R. Wauchope and G. R. Willey. Austin: University of Texas Press.

Miller, A. G.
1972 The iconography of the painting in the Temple of the Diving God, Tulúm, Quintana Roo: a tentative hypothesis. In *Religión en Mesoamérica, 12 Mesa Redonda,* pp. 329–33. Mexico City: Sociedad Mexicana de Antropología.

1973 Archaeological investigations of the Quintana Roo Mural project: a preliminary report of the 1973 season. In *Studies in ancient Mesoamerica.* Ed. J. Graham. Contributions of the University of California Archaeological Research Facility, no. 18, pp. 137–48. Berkeley.

1973a The mural painting in Structure 12 at Tancah and in Structure 5 at Tulúm, Quintana Roo, Mexico: implications of their style and iconography. *Atti del 40 Congreso Internazionale degli Americanisti, Roma-Genova, 3–10 Settembre 1972.* Vol. 1, pp. 465–71. Genoa: Casa Editrice Tilgher.

1973b *The mural painting of Teotihuacán.* Dumbarton Oaks: Trustees of Harvard University.

1974 The iconography of the painting in the Temple of the Diving God, Tulúm, Quintana Roo: The Twisted Cords. In N. Hammond, ed. *Mesoamerican archaeology: new approaches,* pp. 167–86. Austin: University of Texas Press.

1974a West and East in Maya thought: death and rebirth at Palenque and Tulúm. *Proceedings, Primera Mesa Redonda de Palenque.* Vol. 2. Robert Louis Stevenson School. Ed. M. Greene Robertson. Pebble Beach, Calif.

1977 The Maya and the sea: trade and cult at Tancok and Tulúm, Quintana Roo, Mexico. In *The sea in the pre-Columbian world,* ed. Elizabeth P. Benson. Conference at Dumbarton Oaks, Washington, D.C.

Miller, V. E.
1977 A reinterpretation of Maya gestures of submission. *Proceedings,* 42d Annual Meeting, Society of American Archaeologists, New Orleans.

Millspaugh, C. F.
1895 Contribution to the flora of Yucatán. Field Museum, Botanical Publications, 4th ser., 1:3–56.

Miranda, F.
1951 La Selva del ocote. *Revista Ateneo Chiapas* 3:35–47.

1953 *Vegetación de Chiapas* Vols. 1, 2. Mexico City.

1953a Plantas nuevas o notables de la flora de Chiapas. *Boletín Instituto Biológica de México* 24:69–96.

1953b Un botánico en el borde de la selva lacandona. *Memorias Congreso Científio de México* 6:285–303.

1957 Vegetación de la vertiente del Pacífico de la Sierra Madre de Chiapas (México) y sus relaciones floristicas. *Proceedings of the 8th Pacific Science Congress* 4:438–53.

———, and Sharp, A. J.
1950 Characteristics of the vegetation in certain temperate regions of eastern Mexico. *Ecology* 31:331–33.

———, and Xolocotzi, E. H.
1963 Los tipos de vegetación de México y su classificación. *Boletín de la Sociedad Botánica de México* 28:29–180.

Molière, J.
1665 *Don Juan.* Paris.

Monardes, N.
1565 *Historia medicinal de las cosas de la sirven al uso de medicina.* Seville.

1925 *Joyful newes out of the newe founde worlde.* Ed. S. Gaselle.

Moran, H. A., and Kelley, D.
1953 *The alphabet and the ancient calendar signs.* Palo Alto, Calif.: Daily Press.

Moriarty, F. R., III
1970 *A general syllogy of Maya medicine.* University of Northern Colorado, Museum of Anthropology, Miscellaneous Series, Publication no. 15.

Morley, S. G.
1937–38 *The inscriptions of Petén.* Carnegie Institution of Washington Pub. 437. Washington, D.C.

Morris, E. H., Charlot, J., and Morris, A. A.
1931 The Temple of Warriors at Chichén Itzá. Carnegie Institution of Washington Pub. 406. Washington, D.C.

Morton, J.
1968 The calabash (*Crescentia Cujete*) in folk medicine. *Economic Botany* 22:273–80.

Moser, C. L.
1973 *Human decapitation in ancient Mesoamerica.*

Dumbarton Oaks: Trustees of Harvard University.

Motul, Diccionario
MS., sixteenth century. Original missing; copy Middle American Research Institute Library, New Orleans.

Munn, H.
1973 The mushrooms of language. In M. Harner, ed. *Hallucinogens and shamanism*, pp. 86–122. New York: Oxford University Press.

Muñoz, J. B.
1783 *Historia del Nuevo Mundo.* Madrid.

Muñoz Camargo, D.
1892 *Historia de Tlaxcala.* Ed. A. Chavero. Mexico City.

Nadoillae, Marquis de
1885 *Les Pipes et le tabac.* Paris.

Naranjo, C.
1967 Psychotropic properties of harmala alkaloids. In Efron, D., *Ethnopharmocologic Search for Psychoactive Drugs.* Public Health Service Publication no. 1645, pp. 385–91.

Nash, J.
1967 The logic of curing behavior: curing in Maya Indian town. *Human Organization* 26:132–40.

Nash, M.
1958 *Machine-age Maya: the industrialization of a Guatemalan community.* Chicago: University of Chicago Press.

Navarette, D. F.
1890 *Die Reisen des Cristoph Columbus 1492–1504.* Leipzig.

Nicholson, H. B.
1969 Religion in pre-Hispanic Central Mexico. In *Handbook of Middle American Indians.* Vol. 10. Ed. R. Wauchope and G. R. Willey. Austin: University of Texas Press.

1976 *Origins of religious art and iconography in preclassic Mesoamerica,* Los Angeles: University of California at Los Angeles, Latin America Center.

Nicot, J.
1573 *Dictionnaire Franco-Latin.* Paris.

Nordenskjold, E.
1930 The use of enema tubes and enema syringes among the Indians. *Ethnografiska Studier,* no. 8, pp. 184–86. Göteborg.

Nuttall, Z.
1922 *The gardens of ancient Mexico.* Annual Report, Smithsonian Institution. Washington, D.C.

Obel, M. l'
1576 *Stirpium adversariae nova.* Antwerp.

Ocaranza, F.
1934 *Historia de la medicina en México.* Midy Press.

Osado, R.
N.d. Libro de Judeo. MS. Peabody Museum, Harvard University.

N.d. Yerbas y hechicerías del Yucatán. MS. Library, Middle American Research Institute, New Orleans.

Osmond, H.
1955 Ololiuhqui: the ancient Aztec narcotic. *Journal of mental science* 101: 526–37.

Oviedo y Valdés, G. F. de
1851 *Historia general y natural de las Indias islas, y tierra firme del mar oceano.* Madrid: Ediciones de la Academia Real de Historia.

1932 *Historia general y natural de las Indias.* Quoted by E. C. Conti in *A history of smoking.* New York.

1936 *La Historia general de las Indias.* Seville, 1535. Reproduced in G. Arents. *Tobacco.* New York.

Paso y Troncoso, F. de
1905 *Papeles de Nueva España.* Madrid: Sucesores Rivadenly.

Patten, N. Van
1931 *The medical literature of Mexico.* Chicago.

Patten, W. J.
1977 *A catalog of Maya rubbings.* Guatemala City.

Penn, W. A.
1901 *The soverane herbe.* London.

Pennes, H. H., and Hoch, P.
1957 Psychomimetics: clinical and theoretical considerations: harmine, win-2299, and nalline. *American Journal of Psychiatry* 113: 887–92.

Pennington, C.
1963 Medicinal plants utilized by the Tepehuán of southern Chihuahua. *America Indígena* 23: 31–47.

1963a *The Tarahumara of Mexico: their environment and material culture.* Salt Lake City: University of Utah Press.

1973 Plantas medicinales utilizadas por el Pima montañas de Chihuahua. *America Indígena* 33: 213–32.

Pennington, T. D., and Sarukhan, J.
1968 *Arboles tropicales de México.* Mexico City: Instituto Nacional de Investigaciones Forestales.

Pineda, E.
1845 *Descripción geográfica del departamento de Chiapas y Soconusco.* University of Pennsylvania Museum Publication. Philadelphia.

Pío Pérez, J. P.
1877 *Diccionario de la lengua Maya.* Mérida.

Pitt-Rivers, J.
1970 Spiritual power in Central America: The Naguals of Chiapas. In M. Douglas, ed. *Witchcraft confessions and accusations,* London: Tavistock.

Pliny the Younger
1806 *The letters of Pliny the consul.* London: Bohn Edition.

Poindexter, E. H., and Carpenter, R. D.
1962 Isolation of harman and norharman from tobacco and cigarette smoke. *Phytochemistry* 1: 215–21.

Pomar, J. B.
Relación de Texcoco. Mexico City.

Porter, M. N.
1948 Pipas precortesianas. *Acta Antropológica*, vol. 3, no. 2. Mexico City.

Prescott, W. H.
1843 *History and conquest of Mexico*. Philadelphia: J. B. Lippincott.

Proskouriakoff, T.
1950 A Study of Classic Maya sculpture. Carnegie Institution of Washington Pub. 593. Washington, D.C.

1960 Historical implications of a pattern of dates of Piedras Negras, Guatemala. *American Antiquity*, vol. 25, no. 4. Salt Lake City.

1963 Historical data in the inscriptions of Yaxchilán, Parts 1–2. *Estudios de cultura Maya* 3:149–201. Mexico City.

1974 In M. Dobkin de Ríos. The Influence of Psychotropic Flora and Fauna on Maya Religion. *Current Anthropology* 15:158.

Quirarte, J.
1976 The underworld Jaguar in Maya vase painting: an iconographic study. *New Mexico studies in the fine arts* 1:20–26.

1976a *Methodology in the study of pre-Columbian Art*. Presentation at the 42d International Congress of Americanists, Sept. 2–9, Paris.

Radin, P.
1970 *The Winnebago tribe*. Lincoln, Nebr.: University of Nebraska Press.

Raffour, L.
1900 *La medicine chez les Mexicains Précolombiens*. Paris: Jouvet et Boyer.

Ramírez Aguirre, A.
1959 Medicina precortesiana en Mexico. *Revista de la Sociedad Médica* 1:47–55.

Rands, R. L., and Rands, B. C.
1965 Pottery figurines of the Maya lowlands. In *Handbook of Middle American Indians*. Vol. 2. Ed. R. Wauchope and G. R. Willey. Austin: University of Texas Press.

Raven, P. H., Berlin, B., and Breedlove, D. E.
1971 The origins of taxonomy. *Science* 174:210–13.

Recinos, A.
1947 *Popol Vuh. Antiquas historias de Quiché*. Mexico City.

———, Goetz, D., and Morley, G. S.
1950 *Popol Vuh: The sacred book of the ancient Quiché Maya*. Norman: University Oklahoma Press.

Redfield, R.
1934 Culture changes in Yucatán. *American Anthropologist* 36:57–69 pp. 57–69.

1940 *Disease and its treatment in Dzitas, Yucatán*. Carnegie Institution Pub. 523. Washington, D.C.

1941 *The folk culture of Yucatán*. Chicago: University of Chicago Press.

———, and Villa-Rojas, A.
1934 *Chan Kom: A Maya village*. Carnegie Institution of Washington, Pub. 448, Washington, D.C.

1939 *Notes on the ethnography of Tzeltal communities of Chiapas*. Carnegie Institution of Washington, Pub. 509, Contribution no. 28. Washington, D.C.

Reichel-Dolmatoff, G.
1972 The cultural context of an aboriginal hallucinogen: *Banisteriopsis Caapi*. In P. T. Furst, ed. *Flesh of the gods*, pp. 84–114. New York: Praeger.

Reko Blas, P.
1919 De los nombres botánicos aztecas. *El Médico Antiguo*, 1:113–57.

1938 *De los nombres botánicos aztecas*. Mexico City: H. Beyer.

Relación de Alahuiztlán
1905 *Papeles de la Nueva España*. Ed. F. de Paso y Troncoso. 2d ser. 6:100–105. Madrid.

Relación de Asuchitlán
1946 *Papeles de Nueva España*. Ed. F. de Paso y Troncoso. 2d ser. 7:7–49. Mexico City.

Relación de Mechuacán
1903 *Ceremonias y ritos y Población y gobernación de la provincia de Mechuacán*. Morelia.

Relación de Nuchistlán
1947 *Papeles de Nueva España*. Ed. F. de Paso y Troncoso. 2d ser. 8:59–74. Mexico City.

Relación de Ostomán
1905 *Papeles de Nueva España*. Ed. F. de Paso y Troncoso. 2d ser. 6:105–15. Madrid.

Ríos Meneses, M. B.
1976 *Relación de los dioses Mayas con algunas ceremonias que aun se practican en Yucatán*. Mérida: Yucatán.

Robertson, D.
1970 *The Tulúm murals: the international style of the late post-Classic*. Verhandlungen des 38 Internationalen Amerikanistenkongresses, Stuttgart-Munich, 12–18, Aug. 1968. Vol. 2, pp. 77–88. Munich: Kommissionsverlag Klaus Renner.

Robicsek, F.
1970 *Copán: home of the Mayan gods*. New York: Museum of the American Indian, Heye Foundation.

1975 *A study in Maya art and history: the mat symbol*. New York: Museum of the American Indian, Heye Foundation.

Rodríguez Rouanet, F.
1970 Prácticas médicas tradicionales de los indígenas de Guatemala. *Guatemala Indígena* 4:52–86.

Rose, R. M.
1977 Mushroom stones and the Maya ballgame. Proceedings, 42d Annual Meeting, Society of

American Archaeologists, New Orleans.

Rossell, R.
1964 La Medicina entre los Mayas. Guatemala City: Editorial "Rosales."

Rostand, E.
1951 *Cyrano de Bergerac.* Trans. B. Hoover. New York: Bantam Books.

Roth, H. L.
1929 The aborigines of Hispaniola. *Journal of the Anthropology Institute of Great Britain and Ireland,* vol. 16. London.

Roth, W. A.
1915 An inquiry into the animism and folklore of Guiana Indians. In *Thirteenth Annual Report of the Bureau of American Ethnology,* pp. 1908–10.

Roys, R. L.
1931 *The ethno-botany of the Maya. Middle American Research Institute,* Tulane University, Pub. 2. New Orleans.

1939 *The Titles of Ebtun. Carnegie Institution of Washington,* Publ. 505.

1965 Lowland Maya native society of Spanish contact. In *Handbook of Middle American Indians.* Vol. 3, *Archaeology.* Ed. R. Wauchope and G. R. Willey. Austin: University of Texas Press.

1965a *Ritual of the Bacabs.* Norman: University of Oklahoma Press.

1967 *The Book of Chilam Balam of Chumayel.* Norman: University of Oklahoma Press.

1972 *The Indian background of colonial Yucatán.* Norman: University of Oklahoma Press.

Rubel, A.
1974 In Marlene Dobkin de Ríos. The influence of psychotropic flora and fauna on Maya religion. *Current Anthropology* 15:159.

Rubin, V.
1975 *Cannabis and culture.* The Hague: Mouton & Co.

Ruppert, K., Thompson, J. E. S., and Proskouriakoff, T.
1955 Bonampak, Chiapas, Mexico. *Carnegie Institution of Washington,* Pub. 602. Washington, D.C.

Ruz Lhullier, A.
1952 Exploraciones en Palenque: 1950. *Anales Instituto Nacional de Anthropología y Historia.* Mexico City.

1973 El Templo de las Inscripciones, Palenque. *Colección científica: arqueología,* no. 7. Mexico City: Instituto Nacional de Anthropologia y Historia.

Rych, B.
1614 *The honestie of this age, prooving by good circumstances, the world Was never honest till now.* London.

Rzedowski, J.
1965 Relaciones geográficas y posibles origenes de la flora de México. *Boletín de la Sociedad Botánica de México* 29:121–77.

1966 La vegatación de Nueve Galicia. *Contributions from the University of Michigan Herbarium* 9:1–123.

Saenz de Santa María, C.
1940 *Diccionario Cakchiquel Español.* Guatemala City: Tipografía Nacional.

Sahagún, B. de
1938 *Historia general de las cosas de Nueva España.* Mexico City.

1950–1969 *The Florentine Codex: general history of things in New Spain.* Santa Fe, N. Mex.: School of American Research and the University of Utah.

1976 *A history of ancient Mexico.* Trans. F. R. Bandelier. Glorieta, N. Mex.: Rio Grande Press.

Saler, B.
1964 Nagual, witch, and sorcerer in Quiché village. *Ethnology* 3:305–27.

Schaffer, G.
1955 Aztec herb therapy. *Palacio* 62:357–66.

Schele, L.
1973 Observations on the cross motif at Palenque. *Proceedings, Primera Mesa Redonda de Palenque.* Vol. 1. Robert Louis Stevenson School. Ed. M. Greene Robertson. Pebble Beach, Calif.

1976 Accession iconography of Chan-Bahlum in the group of the cross at Palenque. *Proceedings, Segunda Mesa Redonda de Palenque.* Vol. 3. Robert Louis Stevenson School. Ed. M. Greene Robertson. Pebble Beach, Calif.

1977 *The Bundle in Maya hieroglyphics and icongraphy.* Proceedings, 42d Annual Meeting of the Society of American Archaeologists. New Orleans.

———, and Miller, J.
1977 The "Mirror," the "Rabbit," and the "Bundle": accession phrases in Classic Maya inscriptions. MS.

Schellhas, P.
1897 *Die Göttergestalten der Mayahandschriften: Ein Mythologisches Kulturbilde aus dem alten Amerika.* Dresden.

1904 *Representation of deities in the Maya manuscripts.* Cambridge, Mass.

Schlieffer, H.
1973 *Sacred narcotic plants of the New World Indians: an anthology of texts from the sixteenth century to date.* New York: Hafner Press.

Scholes, F. V., and Roys, R.
1948 The Maya Indian of Acalán-Tixchel: a contribution to the history and ethnography of the Yucatán Peninsula. Carnegie Instituition of Washington Pub. 560. Washington, D.C.

Schotel, G. D. J.
1876 *Letterkundige Bijdrogen tot de Geschiedenis van de Tabak.* The Hague.

Schounberg, P., and Paris, F.
1972 *Guía de las Plantas Medicinales.* Barcelona: Editorial Omega.

Schultes, R. E., and Hoffmann, A.
1973 *The botany and chemistry of hallucinogens.* Springfield, Mass.: Charles C. Thomas.

Schultes, T. W.
1941 *A contribution to our knowledge of* Rivea Corymbosa (Ololiuqui) *of the Aztecs.* Cambridge, Mass.: Botanical Museum, Harvard University.

1951 *Recent advances in American ethnobotany.* Waltham, Mass.

1960 Tapping our Heritage of Ethnobotanical Lore. *Economic Botany* 14: 257–62.

1962 The role of the ethnobotanist in the search for new medicinal plants. *Lloydia* 25: 257–66.

1969 Hallucinogens of plant origins: interdisciplinary studies of plants sacred in primitive cultures yield results of academic and practical interest. *Science* 163: 245–54.

1972 An overview of hallucinogens in the Western Hemisphere. In *Flesh of the gods: the ritual use of hallucinogens.* Ed. P. T. Furst. New York: Praeger.

1972a The future of plants as sources of new biodynamic compounds. In T. Swain, ed. *Plants in the development of modern medicine.* Pp. 103–24. Cambridge, Mass.: Harvard University Press.

Schwarzer, H.
1956 *Die Heilige Pfeife.* Freiburg.

Seaforth, C.
1962 Drugs from the West Indies. *Caribbean Quarterly* 7: 198–202.

Seler, E.
1902-1903 *Gesammelte Abhandlungen zur Amerikanischen Sprach und Altestumkunde.* Berlin.

1902-1903a *Codex Vaticanus No. 3773 (Codex Vaticanus B): An old Mexican pictorial manuscript in the Vatican Library.* Trans. A. H. Keane. Berlin.

1904 Altmexikanischer Schmuck und soziale und militärische Rangabzeichen. In *Gesammelte Abhandlungen zur Amerikanischen Sprach- und Altertumskunde.* Vol. 2, pp. 509–619. Berlin: A. Asher & Co.

1904a The Mexican picture writing of Alexander von Humboldt, Mexican and Central American antiquities, calendar systems, and history. *Smithsonian Institution, Bureau of American Ethnology,* Bulletin no. 28. Washington, D.C.

1909 Venus period in the picture writing of the Borgia Codex group. *Smithsonian Institution, Bureau of American Ethnology, Bulletin* no. 28. Washington, D.C.

1917 Die Ruinen von Uxmal. *Abhandlungen der Konigl. Preuss. Akademic der Wissenschaften, Phil.-Hist. Klasse,* no. 3. Berlin.

1937 Quoted by Stahl, G. Die Zigarre. Eine Kulturhisto-rische Betrachtung. *Tabak,* 1:2. Berlin.

1961 *Gesammelte Abhandlungen zur Amerikanischen Sprach und Altertumkunde.* 2 vols. Graz: Akademische Druck und Verlaganstalt.

Setchell, W. A.
1921 Aboriginal tobaccos. *American Anthropologist,* vol. 23, no. 4.

Sharer, J. R.
1968 *Pre-Classic archaeological investigations at Chalchuapa, El Salvador.* Philadelphia: University of Pennsylvania.

1974 The pre-history of the southeastern Maya periphery. *Current Anthropology* 15: 165–87.

Sharon, D.
1972 The San Pedro cactus in Peruvian folk healing. In P. T. Furst, ed. *Flesh of the gods,* pp. 114–35. New York: Praeger.

Shimbunsha, Y.
N.d. *Tesoras Maya de Guatemala.* Guatemala City.

Shook, E. M., and Kidder, A. V.
1952 *Mount E-III-3, Kaminaljuya, Guatemala.* Carnegie Institution of Washington, Publ. 596. Washington, D.C.

Shuttuck, G. C.
1933 *The Peninsula of Yucatán.* Carnegie Institution of Washington, Medical Biological, Meterological and Sociological Studies, Publ. 431.

Siegel, R. K.
1976 Herbal intoxication: psychoactive effects from herbal cigarettes, tea, and capsules. *Journal of the American Medical Association* 236: 473–75.

Silver, D.
1966 Zinacantan shamanism. Ph. D. dissertation, Harvard University.

Smith, A. L., and Kidder, A. V.
1943 Explorations in the Motagua Valley, Guatemala. In *Contributions to American Anthropology and History.* Carnegie Institution of Washington Publ. 546, pp. 101–82. Washington, D.C.

Smith, J.
1627 *The general history of Virginia.* London.

Smith, R. E.
1955 Ceramic sequence at Uaxactún, Guatemala. *Middle American Research Institute Series,* vol. 21. New Orleans: Tulane University.

Somolinos D' A., G.
1965 Historia de la medicina. I. Medicina precortesiana. *Médica Nacional* 1: 9–14.

1968 La medicina Teotihuacana. *Gaceta médica de México* 98: 350–69.

Sorenson, J.
1957 A bibliography for Yucatán medicinal plant studies by William Gates. *Tlalocán* 3: 334-43.

Sotuta Manuscript
N.d. MS. Photostat in Middle American Research Institute. Tulane University, New Orleans.

Souza-Novelo, N.
1945 *Kan Lol (plants medicinal).* Mérida.: Editorial Tamná.

Spence, L.
1930 *Arcane secrets and occult love of Mexico and Mayan Central America.* London.

Spinden, H.
1913 A study of Maya Art. *Memoirs, Peabody Museum, Harvard University,* vol. 6. Cambridge, Mass.

1950 *Tobacco is American: a story of tobacco before the coming of the white man.* New York: New York Public Library.

1957 *Maya art and civilization.* Indian Hills, Colo.: Falcon Press.

Spores, R.
1960 The Zapotec and Mixtec at Spanish contact. In *Handbook of Mesoamerican Indians.* Vol. 2. Ed. R. Wauchope and G. R. Willey. Austin: University of Texas Press.

Stahl, G.
1925 Tabakrauchen in Sudamerika. *Proceedings of the 21st International Congress of Americanists.* Vol. 1. Göteborg.

1930 *Historisches vom Tabakrauchen.*

1931 Zur Frage de Ursprungs des Tabakrauchens. *Anthropos,* vol. 24. Vienna.

1932 Ursprüngliches von der Zigarre. *Schweizerische Archiv für Volkskunde* 32:105.

1937 Die Zigarre. Eine Kulturhistoriche Betrachtung. *Tabak* 1:11 Berlin.

Standley, P. C.
1920–26 Trees and shrubs of Mexico. *United States National Herbarium.* Vol. 23. Washington, D.C.: U.S. Government Printing Office.

1930 *Flora of Yucatán.* Field Museum of Natural History Botanical Series, vol. 3, no. 3. Chicago.

———, and Steyermark, J. A.
1961 Flora of Guatemala. *Fieldiana botany,* vol. 24.

———, and Williams, L. O.
1961 Flora of Guatemala. *Fieldiana botany,* vol. 24.

Starr, F.
1900–1901 *Notes on the ethnography of southern Mexico.* Proceedings of the Davenport Academy of Natural Sciences, vol. 2. Davenport, Iowa.

Steggerda, M.
1941 Maya Indians of Yucatán. *Carnegie Institution of Washington* Publ. 531. Washington, D.C.

1943 Remedies for diseases as prescribed by the Maya Indian herb-doctors. *Bulletin of the Institute of Medicine* 13: 54–82.

1943a Some ethnological data concerning one hundred Yucatán plants. In Smithsonian Institution, Bureau of American Ethnology Bulletin 136, Anthropological Papers, vol. 29, pp. 193–226.

Stephens, J. L.
1969 *Incidents of travel in Central America, Chiapas, and Yucatán.* New York.

Stoll, O.
1888 *Die Mayasparachen der Pokom-Gruppe.* Vienna.

Sturtevant, W. C.
1964 Studies in Ethnoscience. In *Transcultural Studies in Cognition.* Ed. A. K. Romney and R. G. D'Andrade. *American Anthropologist* 66: 99–131.

Swain, T.
1972 *Plants in the development of modern medicine.* Cambridge, Mass.: Harvard University Press.

Switzer, H. A.
1969 *Tobacco, pipes, and cigarettes of the prehistoric Southwest.* El Paso, Texas.

Taylor, N.
1965 *Plant drugs that changed the world.* London: George Allen and Unwin.

Tenzell, J.
1970 Shamanism and concepts of disease in a Mayan Indian community. *Psychiatry* 33: 322–80.

Tesozomoc, H. A.
N.d. *Crónica Mexicana.* Mexico City: Editorial Lazenda.

Testa, A., and Testa, P.
N.d. Nitrogenous heterocyclic compounds in (cigarette) smoke condensates. In *Annuel directoire des études équipementes,* Sec. 1, pp. 163–91.

Thévét, A.
1558 *Les singularités de la France antarctique, autrement nommée Amérique.* Paris.

Thompson, J. E. S.
1930 *Ethnology of the Mayas of southern and central British Honduras.* Field Musuem of Natural History Anthropology Series, vol. 17, no. 2. Chicago.

1934 *Sky bearers, colors, and directions in Maya and Mexican religion.* Carnegie Institution of Washington Pub. 436. Washington, D.C.

1939 *The moon goddess in Middle America: with notes on related deities.* Carnegie Institution of Washington Publ. 509 Washington, D. C.

1943 *Maya epigraphy: a cycle of 819 days.* Carnegie Institution of Washington. Notes on Middle America Archaeology and Ethnology, no. 22.

1946 *Some uses of tobacco among the Maya.* Carnegie Institution of Washington Notes on Middle American Archaeology and Ethnology. Washington, D.C.

1950 Maya hieroglyphic writing: an introduction. Carnegie Institution of Washington Publ. 589. Washington, D.C.

1954 *The rise and fall of Maya civilization.* Norman: University of Oklahoma Press.

1958 Symbols, glyphs, and divinatory almanacks for diseases in the Maya Dresden and Madrid Codices. *American Antiquity,* vol. 23.

1960 *Maya hieroglyphic writing.* Norman: University of Oklahoma Press.

1962 *A Catalog of Maya Hieroglyphs.* Norman: University of Oklahoma Press.

1962a Trade relations between the Maya highlands and lowlands. In *Proceedings of the 35th International Congress of Americanists.* Mexico City.

1966 Merchant gods of Middle America. In *Summa antropológica en homenaje a Roberto J. Weitlaner,* pp. 159–72. Mexico City: Instituto Nacional de Antropología y Historia.

1970 *Maya History and Religion.* Norman: University of Oklahoma Press.

1972 *A commentary on the Dresden Codex.* Memoirs of the American Philosophical Society, no. 93. Philadelphia.

1973 Maya rulers of the Classic period and the divine right of kings. In *The iconography of Middle American sculpture,* pp. 52–71. New York: Metropolitan Museum of Art.

1974 In M. Dobkin de Ríos: The Influence of psychotropic flora and fauna on Maya religion. *Current Anthropology* 15:160.

Thompson, R. A.
1974 *The winds of tomorrow: social change in a Maya town.* Chicago: University of Chicago Press.

Torquemada, J. de
1943–44 *Los Veintidós libros naturales y monarquía indiana.* Heyhoe, Mexico: Editorial Salvador Chávez.

1956 *De los Monarquías indianas,* Seville, 1615. Quoted in G. Arents. *Tobacco.* New York.

Tozzer, A. M.
1907 *A comparative study of the Mayas and Lacandones.* New York.

1913 A Spanish manuscript letter on the Lacandones in the Archives of the Indies in Seville. In *Proceedings of the 18th International Congress of Americanists,* pp. 497–509. London.

1957 Chichén Itzá and its cenote of sacrifice. *Memoirs, Peabody Museum, Harvard University,* vols. 11–12. Cambridge, Mass.

1966 Landa's Relación de las Cosas de Yucatán. In *Peabody Museum Papers,* vol. 18, p. 106 Cambridge, Mass.

———, and Allen, G.
1910 Animal figures in the Maya codices. *Papers, Peabody Museum, Harvard University,* vol. 3. Cambridge, Mass.

Urgina, M.
1904 Plantas comestibles de los antiguos Mexicanos. *Anales de Museo Nacional,* vol. 3. Mexico City.

Valenzuela, N. de
1695 MS. Archivo General de Indias, Seville, Spain. Transcript by Nicholas Hellmuth, in the files of Foundation for Latin American Anthropological Research, Guatemala City.

Valladores, L. A.
1957 *El hombre y el maíz.* Guatemala City.

Vargas Castelazao, M.
1955 La Patología y la medicina entre los Mexicas. *Revista Méxica de estudios antropológicos* 14:119–43.

Vespucci, A., and Martire, P. d'Anghiera
1534 Die welt. Strasburg, 1534. Reproduced in G. Arents. *Tobacco.* New York.

Villacorta, J. A., and Villacorta, C. A.
1930 *Codices Mayas. Reproducidos y desarrollados.* Guatemala City: Tipografía Nacional.

Villagutierre de Soto-Mayor, J.
1933 Historia de la conquista de la Provincia de el Itzá, reducción y progreso de la de el Lacandón. *Biblioteca Guatemala de la Sociedad de Geografía y Historia,* vol. 9. Guatemala City.

Villa Rojas, A.
1955 Los Mazotecos y el problema indígena de la cuenca del Palaloapán. *Memorias del Instituto Nacional Indigenista* 7:114. Mexico City.

Vogel, V.
1969 *American Indian medicine.* Norman: University of Oklahoma Press.

Vogt, E. A.
1969 *Zinacantán: A Maya community in the highlands of Chiapas.* Cambridge, Mass.: Belknap Press of Harvard University Press.

——— and Vogt, C.
1970 *Levi-Strauss among the Maya.* 5:379–92.

von Winning, H., and Stendahl, A.
1968 *Pre-Columbian art of Mexico and Central America.* New York: Harry N. Abrams, Inc.

Wafer, L.
1699 *A new voyage and description of the isthmus of America.* London.

Wagner, P.
1964 Natural vegetation of Middle America. *Handbook of Middle American Indians.* Vol. 1. Ed. R. Wauchope and G. R. Willey. Austin: University of Texas Press.

Wassen, S. H.
1964 Some general viewpoints in the study of native drugs especially from the West Indies and South America. *Ethnos* 29:97–120.

1965 *The use of some specific kinds of South American Indian snuff and related paraphernalia.* Göteborg.

1972 The anthropological outlook for Amerindian medicinal plants. In T. Swain, ed. *Plants in the development of modern medicine,* pp. 1–65. Cambridge, Mass.: Harvard University Press.

1972a A medicine man's implements and plants in a Tihuanacoid tomb in highland Bolivia. *Etnologiska Studier* 32:8–114. Göteborg.

Wasson, R. G.
1963 Notes on the present status of ololiuhqui and other hallucinogens of Mexico. *Harvard University, Botanical Museum Leaflets* 20:161–93.

1966 Ololiuhqui and the other hallucinogens of Mexico. In *Summa antropológica en homenaje a Robert Weitlander*, pp. 329–48. Mexico City: Instituto Nacional de Antropología y Historia.

1968 *Some divine mushroom of immortality*. New York: Harcourt Brace Jovanovich.

1972 The divine mushroom of immortality. In P. T. Furst, ed. *Flesh of the Gods*, pp. 185–200. New York: Praeger.

———, et al.
1975 *Maria Sabina and Her Mazatec mushroom velada*. New York: Harcourt Brace Jovanovich.

Webb, W. S., and Raby, R. S.
1957 *The Andean people*. Cleveland, Ohio.

Weiss, G.
1973 Shamanism and priesthood in the light of the Campa Ayahuasca ceremony. In Michael J. Harner, ed. *Hallucinogens and shamanism*. New York: Oxford University Press.

Weitlaner, R. J.
1960 The Cuicatec. In *Handbook of Middle American Indians*. Vol. 7. Ed. R. Wauchope and G. R. Willey. Austin: University of Texas Press.

———, and Hoppe, W. A.
1969 The Mazatec. In *Handbook of Middle American Indians*. Vol. 7. Ed. R. Wauchope and G. R. Willey. Austin: University of Texas Press.

West, G. A.
1970 *Tobacco, pipes, and smoking customs of the American Indians*. Westport, Conn.

Westheim, P.
1965 *The art of Ancient Mexico*. Garden City, N. Y.: Anchor Books, Doubleday.

———, et al.
1969 *Cuarenta siglos de plástica mexicana*. Mexico City: Editorial Herrero.

Wiener, L.
1921 Africa and the discovery of America. *American Anthropologist* 23:86–87.

Wilbert, J.
1972 Tobacco and shamanistic ecstacy among the Warao Indians of Venezuela. In *Flesh of the Gods. The Ritual Use of Hallucinogens*. Ed. P. T. Furst. New York: Praeger.

1972 Personal communication to P. T. Furst. In *Flesh of the gods: the ritual use of hallucinogens*. Ed. P. T. Furst. New York: Praeger.

Willey, G. R.
1966 *An introduction to American archaeology*. Vol. 1. Englewood Cliffs, N.J.: Prentice-Hall.

Wilson, G. L.
1920 *Agriculture of the Hidatsa Indians*. University of Minnesota Studies in Social Science, no. 9. Ann Arbor.

Winter, E.
1972 *Mexico's ancient and native remedies*. Mexico City: Editorial Fournier.

Wisdom, C.
1940 *The Chorti Indians of Guatemala*. University of Chicago Publications in Anthropology.

Woodbury, P. B., and Trik, A. S.
1953 *The ruins of Zaculeu, Guatemala*. Richmond, Va.

Woods, C.
1968 Medicine and culture in San Lucas Toliman, a highland Guatemalan community. Ph.D. dissertation, Stanford University.

———, and Graves, T.
1973 *The process of medical change in a highland Guatemalan town*. Los Angeles: Latin American Center, University of California at Los Angeles.

Wyman, L. C., and Harris, S. K.
1941 *Navajo Indian medical ethnobotany*. University of New Mexico Bulletin no. 366, Anthropological Series, no. 3., Albuquerque: University of New Mexico.

Ximénez, F.
1888 *Cuatro libros de la naturaleza y virtudes medicinales de las plantas y animales de la Nueva España*. Ed. León Morelia.

Zimmermann, G.
1956 *Die Hieroglyphen der Maya-Handschriften*. Hamburg.

Index

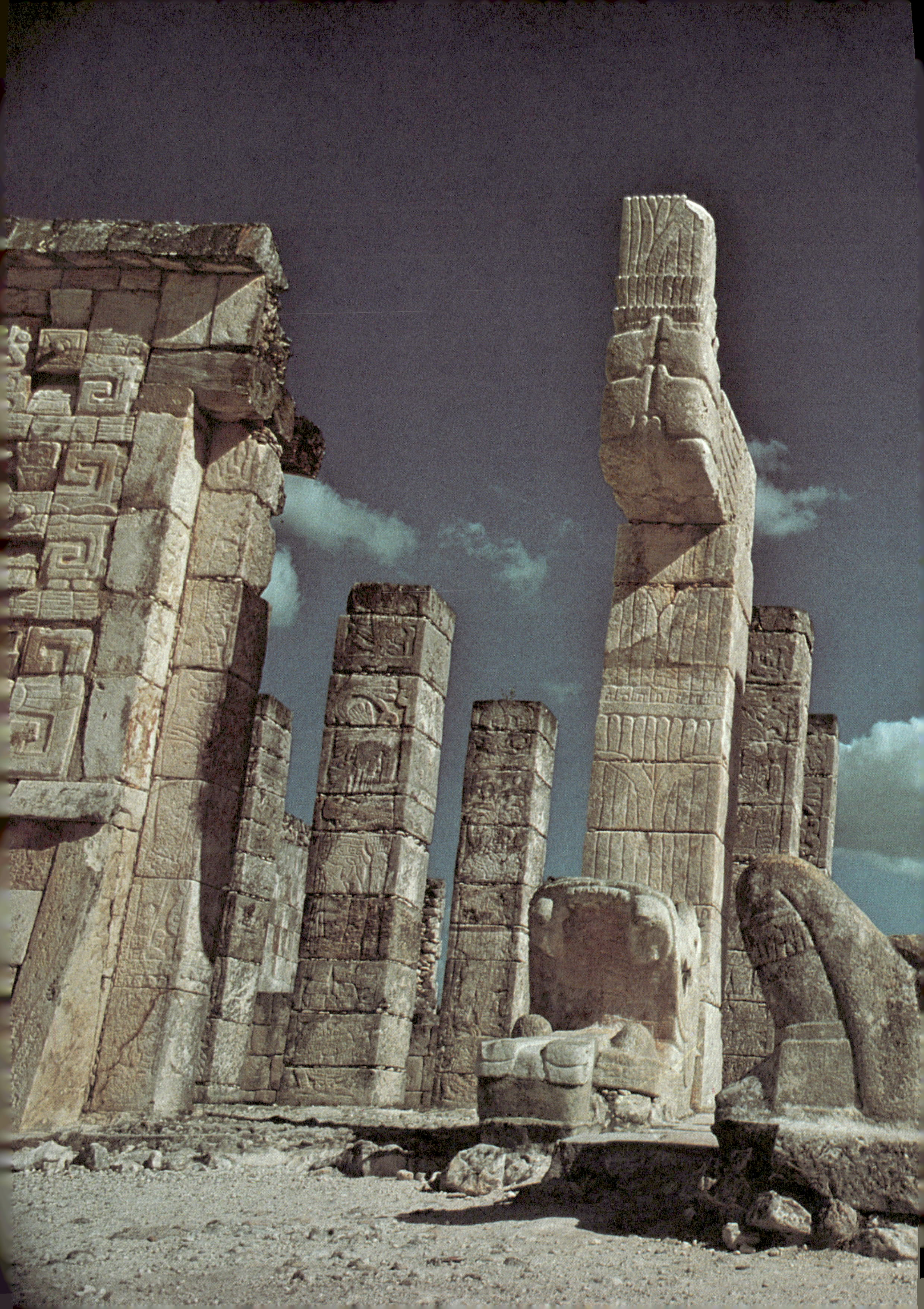